INDUSTRIAL ELECTRONICS

THOMAS E. KISSELL

Terra Community College

PRENTICE HALL
Upper Saddle River, New Jersey   Columbus, Ohio

Library of Congress Cataloging-in-Publication Data

Kissell, Thomas E.
    Industrial electronics/Thomas Kissell.
        p.   cm.
    Includes index.
    ISBN 0-13-121864-6
    1. Industrial electronics.   I. Title.
TK7881.K57   1997
621.3—dc20   96-8644
                                                  CIP

Cover photo: LR Mate 100, A small, five-axis robot demonstrates robotic MIG welding. Courtesy of
    Fanuc Robotics, North America, Inc. All rights reserved.
Editor: Charles E. Stewart, Jr.
Production Editor: Rex Davidson
Production Supervision: Bookworks
Text Designer: Maureen Eide
Cover Designer: Brian Deep
Production Buyer: Laura Messerly
Marketing Manager: Debbie Yarnell
Illustrations: Academy Artworks Inc.

This book was set in Times Roman and Univers by Bi-Comp, Inc. and was printed and bound by
Von Hoffmann Press, Inc. The cover was printed by Von Hoffman Press, Inc.

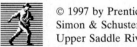 © 1997 by Prentice-Hall, Inc.
Simon & Schuster/A Viacom Company
Upper Saddle River, New Jersey 07458

Printed in the United States of America
10 9 8 7 6 5 4 3 2 1

ISBN 0-13-121864-6

Prentice-Hall International (UK) Limited, *London*
Prentice-Hall of Australia Pty. Limited, *Sydney*
Prentice-Hall Canada Inc., *Toronto*
Prentice-Hall Hispanoamericana, S. A., *Mexico*
Prentice-Hall of India Private Limited, *New Delhi*
Prentice-Hall of Japan, Inc., *Tokyo*
Simon & Schuster Asia Pte. Ltd., *Singapore*
Editora Prentice-Hall do Brasil, Ltda., *Rio de Janeiro*

Students who enter the job market and become electronics technicians and electrical technicians must be prepared to work on industrial electronics in many forms. The job responsibilities for these fields are rapidly changing because electronic devices and circuits have become thoroughly integrated into all aspects of modern industrial control systems during the past ten years. As industries have learned to compete in the global marketplace, the role of technicians has changed to the point where they are expected to work on every aspect of an industrial system from the simplest electrical components, such as fuses and motors, to the most complex electronic components, such as electronic boards, motor drives, and programmable controllers.

This book is designed to be used by students in two-year technical and community colleges and by students in engineering technology programs and engineering programs at four-year universities. It is designed to be used in courses that cover one semester or one quarter, or to be used in a series of courses covering one year. The book also provides sufficient depth to be a useful resource after students finish school and are working on the job.

The book is different from many of the current texts in that it provides detailed industrial applications for each device and circuit discussed in every chapter. The author has over twenty years' experience working on automated industrial systems and has a distinctive position wherein he teaches the theory and operation of these systems to traditional college students several days each week, and spends the remaining time delivering seminars and consulting directly to technicians and engineers currently working on these systems in industry. The author has taken the course material that he uses in teaching technicians and engineers at Ford Motor Company, General Motors, Campbell Soup, Union Carbide, Kelsey Hayes, H. J. Heinz Company, and many other manufacturing and process industries and incorporated it into this text. Each chapter includes examples of the numerous control circuits and systems that technicians and engineers will be expected to understand and troubleshoot when they finish school and arrive on the job.

The chapters in the book are presented in an order where individual components and smaller circuits are introduced in the early chapters, and larger circuits and complete systems are added along the way and specifically in later chapters. Each chapter is also written to stand alone. This allows teachers to change the order of the sequence of chapters if their course syllabus is slightly different from the order of the chapters. Many times the chapters in books are linked so that they only make sense when they are read in the order they are presented. When this text is used, the teacher can feel free to change the order of the chapters and the student will still understand everything in each chapter.

The first two chapters introduce solid-state devices and logic devices used in industrial electronic circuits. Chapter 3 provides a comprehensive study of programmable controllers. This chapter explains the basic parts of all programmable controllers and then provides numerous applications that show how instructions are used.

It also provides much more information and applications than merely reviewing the instruction set from a typical programmable controller. The Allen-Bradley SLC100 family, SLC500 family, PLC2, and PLC5 are used as examples.

The fourth chapter provides an in-depth study of power devices such as thyristors and power transistors and the types of circuit applications where they are commonly used in industrial electronic circuits. The fifth chapter introduces the triggering devices used to control the power electronic devices and the sixth chapter provides in-depth coverage of optoelectronic devices and industrial lasers. A complete set of manufacturer's data sheets for each major device used in these three chapters is provided in the appendix to supplement the information in these chapters.

Chapter 7 provides an in-depth study of power supplies. This information includes inverters and converters that are found in all types of industrial electronic circuits such as AC and DC motor drives, stepper motor and servomotor drives, and other power control circuits. It also goes into great detail to explain switch-mode power supplies used to supply power to electronic boards. The eighth chapter provides detailed information about operational amplifiers and the ninth chapter provides information about open-loop and closed-loop systems found in industrial applications.

Chapter 10 provides a comprehensive study of all types of input sensors and instruments. This chapter provides details about the theory of operation and troubleshooting information for all of the major sensors and instruments used in the manufacturing and process industry today. Chapter 11 covers output devices such as solenoids, valves, relays, AC and DC motor drives, stepper motors, and servomotors. This chapter has perhaps the most information provided in any text today on variable-frequency AC motor drives and industrial DC motor drives including stepper and servo amplifiers. Chapter 12 provides information about traditional industrial DC motors and single-phase and three-phase AC motors.

In-depth case studies of four industrial systems are provided in Chapter 13. This chapter shows how all of the devices and components studied to this point operate together in typical industrial systems.

Chapter 14 is devoted to robots and other motion control systems such as CNC machines. This chapter helps students understand how servo and stepper motors must be integrated into complete multiple-axis systems. Chapter 15 provides an in-depth study of traditional motor control devices and circuits.

The final chapter in the book provides a study of data communications used in modern industrial applications such as linking programmable controllers, robots, and other automated systems to computer networks to pass data. Today students are expected to have experience with networks and understand how different electronic boards and systems, such as programmable controllers, robots, servo and stepper systems, are all able to be interconnected through networks. This chapter helps students to begin to understand all the parts of data communications systems and how they are integrated with modern industrial systems. Examples of the most widely used industrial networks are provided along with a complete explanation of network standards and protocols.

Each chapter begins with a job assignment that each student is expected to solve after reading the chapter. The assignments are actual jobs that technicians are performing today. Each chapter also has a comprehensive list of objectives to provide an outline of what the student is expected to learn from the chapter. At the end of each chapter a variety of true or false, multiple choice, open-ended questions, and problems are provided for the student to complete. The solution to the job assignment is given at the end of each chapter and the answers to the questions and problems are provided in a separate instructor's manual.

This text uses only applied mathematics and each formula is accompanied with example problems worked out. Additional problems are provided at the point in the chapter where the topic is discussed, and solutions to these problems are worked out on a separate page. This allows the student to practice working a problem and immediately receive feedback about the solution. Problems are also provide at the end of each chapter that the teacher can assign to test each student's comprehension. The answers to these problems are available in the instructor's manual.

ACKNOWLEDGMENTS

The author personally thanks the following people and companies for providing pictures, diagrams, and research material.

Scott Lynch, ABB Instrumentation; Ann Smith, ABB Flexible Automation; Robert Greenfield, Acromag Inc.; Dan Shingold, Analog Devices; Richard Stern, Automatic Switch Co.; John Hitch, Belden Wire & Cable Company; Saul Estreicher, Bridgeport Machines Inc.; Robert Roderique, Bussmann; Steven Rosenthal, Danaher Controls; David Darrah, Darrah Electric Company; Louis Ted Keys, Delta Controls Corporation; Joseph Portada, Dynamics Research Corporation; John Reiley, Eaton Cuttler-Hammer; Al Anthony, EG&G VACTEC, EG&G Optoelectronics Group; Jon Bauer, EIT; Jennifer Loffer, Electro-Craft, A Rockwell Automation Business; Beverly Odegaard, Electro-Craft, A Rockwell Automation Business; Carl Fields, Electrical Apparatus Service Association; Jennifer Howe, Fairbanks Scales; Jennifer Dondero, Fanuc Robotics; Emmily Bopp, Fanus Robotics; Gene Kimmel, Fostoria Industries; Daniel Gilmore, The Foxboro Company; Patricia Pruis, GE Motors & Industrial Systems, Fort Wayne, IN; Thomas Hausman, Harris Semiconductors; Robert Rice, Hyde Park Electronics Inc.; Armstrong Wong, Hewlett Packard Company; Richard Dute, Instrumentation & Controls Systems; Donyel Smith, Johnson Controls Inc.; Lori Cahil, Kistler-Morse; Manfred Kersten, Klockner Desma; Lee McBride, Liebert Inc.; Sally Parker, Masoneilan, Dresser Valve and Controls Division; Kathleen Bee, Micro Motion; Marlyn Smit, Honeywell Micro Switch Division; Sally Fairchild, Motoman; Linda Capcara, Motorola; Cy Pfeifer, National Controls Corporation; Dennis Berry, National Fire Protection Association; Eric Heise, MTS Systems Corporation; William Reilly, Novatec; Maria Psychopeas, Omega Engineering; Jim Krill, Omron Electronics; Gary George, Opto 22; Thomas Barnett Jr., Parker Compumotor Division; Robert Mitchel, Parker, Hannifin Corp, Daedal Division; James Lally, PCB Piezotronics; Lawrence Fogel, Philips Semiconductors; Claudia Tapper, POWEREX Inc.; Robert Donaghy, Princo Instrumentation, Inc.; Aaron Weida, Rexroth Corporation; Linda Sorenson, Rockwell Automation's Allen Bradley Business; David Herkkiuen, Sankyo Robotics; Brent Hart, Sensotec Inc.; Mark Thompson, SMC Pneumatics, Inc.; Shoun Kerbaugh, SMS Engineering; Jennifer Toffler, SQUARE D COMPANY/GOUPE SCHNEIDER; John Murray, Superior Electric, Warner Electric; John O'Meara Jr., Tapeswitch Corporation; Gary Honeycutt, Texas Instruments; Brenda Dodson, Pacific Scientific; Gary E. Gramke, Cincinnati Milacron, Plastic Machinery Group; Paul Handle, Eaton Corporation Cutler-Hammer Products

Additionally, the author thanks the following for reviewing the manuscript and providing helpful comments:

Robert Allen, Columbus Technical Institute; A. R. Beets, Iowa Central Community College; Fred Bradford, Ranken Technical College; Leonard Bundra, Lincoln Technical Institute; J. B. Cornwall, NEC, Fort Worth, TX; James R. Davis, Muskingum Area Technical College; Michael C. Hall, Indiana Vocational Technical College; Edward Herman, Corning Community College; Tom Hersh, Orange Coast College; Sam Ho, ITT, Van Nuys, CA; Mary Kryjewski, ITT, LaMesa, CA; Erik Liimatta, Anne Arundel Community College; Dan A. Lookadoo, New River Community College; Gregory S. Romine, Indiana University/Purdue University, Indianapolis

The author has written other books on electric motor controls and programmable controllers and it is his hope that this book provides the information students will use to become better prepared to install, calibrate, troubleshoot, and repair industrial electronic components, circuits, and systems.

CONTENTS

4 SOLID-STATE DEVICES USED TO CONTROL POWER: SCRs, TRIACs, AND POWER TRANSISTORS 129

5 SOLID-STATE DEVICES USED FOR FIRING CIRCUITS 165

7 INDUSTRIAL POWER SUPPLIES, INVERTERS, AND CONVERTERS 236

8 OPERATIONAL AMPLIFIERS 275

9 OPEN-LOOP AND CLOSED-LOOP FEEDBACK SYSTEMS 299

13 CASE STUDIES OF FOUR INDUSTRIAL APPLICATIONS 637

14 ROBOTS AND OTHER MOTION CONTROL SYSTEMS 664

15 MOTOR CONTROL DEVICES AND CIRCUITS 712

16 DATA COMMUNICATIONS FOR INDUSTRIAL ELECTRONICS 752

To Kathy and her three little angels:
Amber, Cassidy, and Shelby

1

SOLID-STATE DEVICES USED IN INDUSTRIAL LOGIC CIRCUITS

OBJECTIVES

When you have completed this chapter, you will be able to:

1. Identify the three parts of a control circuit and explain the function of each.
2. Explain the operation of a relay including the function of the coil and contacts.
3. Explain the terms *poles* and *throws* in regard to relay contacts.
4. Explain how the contacts and coil of a relay can be used to provide logic and make decisions.
5. Explain the function of normally open (NO) and normally closed (NC) relay contacts.

6. Describe the operation of the circuits provided in this chapter and explain how the logic controls the circuit operation.
7. Explain why signal conditioning may be necessary for inputs to logic circuits.
8. Explain why amplifiers may be required for the output stage of logic circuits.
9. Explain the operation of AND, OR, NOT, NAND, and NOR gates.
10. Compare the features of solid-state logic devices and relays.

You are assigned to convert a machine that operates with relay logic so that it will operate on solid-state gates. Your supervisor will provide the original ladder logic diagram of the machine for you. (See Fig. 1-20.) When you have completed converting the machine to solid-state gates, you should turn on the power to the system and test each input switch to determine that the output for the logic is correct. You should also be able to explain the parts of the circuit that may remain in control of relay logic where the relay provides an advantage over the solid-state gate.

1.1 OVERVIEW OF LOGIC USED IN INDUSTRIAL LOGIC CIRCUITS

Machines in industrial applications require control systems that range from basic to complex in order to operate efficiently and to provide safety. Even though these control circuits may seem vastly different, they are in fact very similar. Each electrical control circuit in an industrial system can be broken into three basic parts: *input*, *decision*, and *output*. When you understand the function each part of the system plays, it is easier to troubleshoot complex control systems.

The first part is the *input* section. Its function is to provide a link between the control circuit and the world around the machine. For instance, input switches such as limit switches provide information about the position of various parts of the machine. Pressure switches provide information about the line pressure of hydraulic and pneumatic parts of the machine. Push-button switches provide an interface between the machine and human operators who activate the push buttons to start or stop the machine. A touch screen on a modern machine can be used in place of switches as an interface between the operator and the machine. All of these examples are part of the input section of the control system. Fig. 1-1 shows an example of the three parts of a circuit.

The second part of the system makes *decisions* and is called the *logic* section. The decision-making part of the circuit is sometimes difficult to define because it may use the same input switch contacts that were described as the input part of the circuit. For example, if an industrial oven has two doors that both must be

FIGURE 1-1   Example of the three parts of a control circuit: input, logic, and the output.

closed before the electric heat can be energized, the two limit switches that indicate that the doors are open or closed are part of the input section of the circuit because they tell the condition of each door. Since both door limit switches are wired in series, switch 1 and switch 2 both must be closed to indicate it is permissible to start the oven. Thus the same two switches become part of the decision-making (logic) part of the circuit as well as being the input part of the circuit.

The *output* part of the circuit causes action to occur. In the case of this example, the output will be a large relay called a magnetic contactor. The contactor is similar to a relay except its contacts are larger in order to handle the currents necessary to operate the electric heating element of the oven. Other examples of outputs are motor starters, which are special relays that incorporate overcurrent protection for the motor, solenoid valves, which are magnetic coils that are specifically designed to move a valve open or closed, and indicator lamps. The contactors, motor starters, and solenoids can also be considered amplifiers because they generally are powered with smaller voltage and current than are used by the device they energize and de-energize. For example, the coil of the contactor may be energized by 110 volts ac and use 300 mA to energize, while the electric heating element that it controls may use 480 volts ac and 50 A.

Traditionally, the inputs for the circuit will be found on the left side of the circuit, and the output will be found on the far right side of the circuit. The decision part of the circuit is determined by the way the input switches are connected in series or parallel to enable the output. This means the switches are the inputs for the circuit, and how they are connected is the decision part of the circuit.

1.2 RELAY LOGIC

Relay logic has been the main type of industrial control for electric circuits ever since the relay was first designed. The relay is versatile since it can have more than one set of contacts, which can be any combination of normally open or normally closed contacts.

The relay is an electromagnetic device that has a coil made of wire. The wire is normally coated (laminated) with plastic or a similar substance to insulate the coil wires from each other and other parts of the relay so they don't create an electrical short circuit. When current is passed through the coil, it becomes a strong magnet. Mounted on a plastic carrier, the contacts of the relay are electrically isolated from each other. The plastic carrier is connected to a piece of metal called an armature, which is drawn into the coil's center where it becomes an electromagnet. The movement of the armature causes a normally open set of contacts to move to the closed position, or a set of normally closed contacts to move to the open position. Fig. 1-2 shows examples of four types of armature assemblies found in different types of relays.

The horizontal-action-type armature has a set of stationary contacts shown on the far left, and a movable aramature. The armature in this device moves to

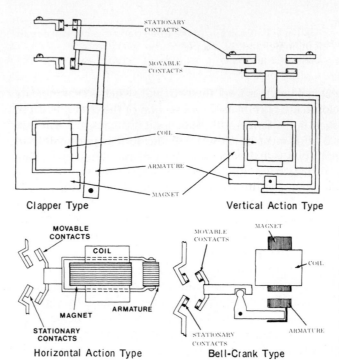

FIGURE 1-2 Types of armature assemblies found on various types of relays (Courtesy of SQUARE D COMPANY/ GROUPE SCHNEIDER. "Square D Company/ Groupe Schneider assumes no liability for accuracy of information.")

the left when the coil is energized, which causes the contacts to close. When the current is turned off and the coil is no longer magnetized, a spring will move the armature to the right.

The bell-crank-type armature has an L-shaped arm connected to a pivot point. The armature in this type of relay pulls one end of the arm up when the coil is energized, which shifts the movable contacts to the left so that they press against the stationary contacts. When current is turned off to the coil, the weight of the armature will cause the end of the arm to drop, which will pull the movable contacts back to the right.

The clapper-type armature has a long arm connected at one end to a pivot point. When the coil is magnetized, the middle of the arm is pulled to the coil, which will cause the movable contacts to move to the left until they touch the stationary contacts. When power is removed from the coil, a spring will pull the contacts back to their open position.

The vertical-action-type armature is shaped like a backward letter C. The contacts are connected to the top part of the armature, and the bottom part is pulled to the coil when it becomes magnetized. When the armature is pulled up, the movable contacts will touch the stationary contacts. When power is turned off to the coil, the weight of the armature will cause it to move down to its original position where the contacts are open.

1.2.1 Terms for Relays

Before you can begin to understand relay logic concepts, you must become familiar with several terms used to describe the action of a relay. Contacts will be described as *normally open* (NO) or *normally closed* (NC). The term *normal* for relays means the position you find the contacts in when no power is applied to the coil. The word *normal* is sometimes referred to as the condition of the contacts when the relay is pulled out of the box that it is shipped in. Relay contacts are always shown in a diagram in their *normal condition,* which means they are drawn in the condition you would find them in when no power is applied to the coil.

When the relay coil is energized, the contacts will move to the opposite condition of their normal condition. For this reason, it is always important to verify whether the coil is energized or de-energized. When the coil is energized, it will be referred to as being *picked up,* and if it is de-energized, it will be referred to as being *dropped out.* These terms may also be used to describe the condition of the contacts.

1.2.2 Types of Relays

The relay can have more than one set of contacts, which can be any combination of normally open and normally closed contacts. For example, in Fig. 1-3 you can see that the relay can have all of its contacts normally open (Fig. 1-3a), all normally closed (Fig. 1-3b), or a combination of two or more sets of normally open and two or more sets of normally closed contacts (Fig. 1-3c). The fourth combination (Fig. 1-3d) is called the double-pole, double-throw (DPDT) switch. The word *pole* refers

FIGURE 1-3 Examples of various combinations of relay contacts.

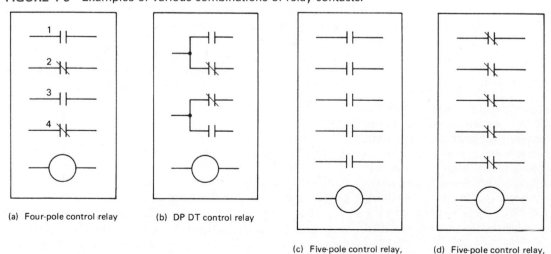

(a) Four-pole control relay

(b) DP DT control relay

(c) Five-pole control relay, all NO contacts

(d) Five-pole control relay, all NC contacts

to the number of sets of contacts and the word *throw* refers to the number of positions or combinations (open or closed) the contacts can have. The main feature of the DPDT contacts is that a set of normally open and a set of normally closed contacts are used together. They have a terminal marked *common* (C), which is electrically common to both sets of contacts.

1.2.3 Voltage Ratings for Coils and Contacts

When you are wiring a relay coil into a circuit, you must be sure that the voltage rating of the coil matches the circuit voltage within $+/-$ 10%. Some industrial-grade relays have color-coded coils so that you can determine the voltage rating for the coil by its color as well as from printed specifications on the relay. A red coil is rated for 110 volts ac and a black coil is rated for 220 volts ac. On the other hand, the contacts can be rated at any voltage equal to or above the voltage of the circuit in which they are used. This means that you can use contacts rated for 208 volts ac in a circuit that has a power supply of 110 volts ac. The current rating of the contacts should never be exceeded. The contacts will be rated for a specific amount of current or power. If the rating is for power, it may be expressed in VA (voltamperes) rather than wattage and you will need to do a calculation

$$\left(\text{Current} = \frac{\text{voltamperes}}{\text{voltage}} \right)$$

to determine the maximum current rating. Some relays will also provide a chart to help you determine their ratings.

1.2.4 Troubleshooting Relay Coils and Contacts

It is important to understand that you can determine the condition of the relay coil and its contacts with an ohmmeter when the relay is out of the circuit or if it does not have voltage applied. Since the relay coil is made from a long piece of wire, the ohmmeter should indicate that the coil has *some amount of resistance.* The precise amount of resistance will vary from coil to coil, since the size and amount of wire used in the coil will vary. Generally, the amount of resistance will be between 20 Ω and 1500 Ω. You need to verify that the coil does not have a short circuit or an open circuit. If the coil is shorted, its continuity (resistance) test will indicate

▼ IMPORTANT SAFETY NOTICE...

When you have a relay disassembled and the coil is removed from its armature, you should never apply power to it because it will burn out from overcurrent. When the coil is mounted correctly in the relay, its current is normally limited by the metal in the armature. If you have the coil out on a bench by itself, there will not be anything to limit the current in the coil and it will act like a simple piece of wire instead of a coil. The low resistance of the wire will allow the coil current to become exceedingly high, which will burn it out.

0 Ω or near zero resistance, and if it has an open circuit, its continuity test will indicate an infinite reading ($\infty\,\Omega$). If the coil shows any amount of resistance other than infinite and zero, you can move to the second part of the troubleshooting test and apply power to the coil. When the proper amount of voltage is applied to the coil, the contacts should move. If you find that the coil is open when you perform the continuity test, or if it will not pull in the contacts when power is applied, you must replace it.

The contacts can also be tested with an ohmmeter if no power is applied to them. Normally open (NO) contacts should show a reading of infinity, and normally closed (NC) contacts should show a reading of zero or near zero ohms. You can manually move the contact carrier to move the contacts from their open to their closed position. You should always test the open and the closed position of the contacts because sometimes the contacts will weld together in the NC position. If the test of the contacts indicates high resistance when they are closed, it means they are damaged or pitted and they must be replaced.

1.3 TYPICAL LOGIC CIRCUITS

Fig. 1-4 shows a typical logic circuit that indicates the condition of doors on a heating oven for a painting process. The oven has two doors that both must be closed during the heating process. A fan must also be running before the electric heating coils can be energized. A limit switch is mounted near each door to indicate if the door is open or closed. A flow switch is mounted near the fan to indicate if the fan is moving air. If both doors are closed, and the start switch is turned on, the coil of control relay CR1 will become energized. When the NO contacts of CR1 close, the fan motor will be energized, causing the fan to move enough air to close the switch. When the flow switch closes, the electric heating element will begin to draw current and heat the oven.

1.3.1 Identifying Inputs and Logic in the Circuit

The inputs in the heating oven circuit are the two door limit switches and the flow switch for the fan. The door limit switches tell the circuit whether the position of each door is open or closed. The flow switch will indicate if the fan is moving air. The condition (open or closed) of each switch is important to the operation of the circuit.

These same switches are also used as part of the logic part of the circuit. This point may be confusing in the beginning, but this example should help make it easier to understand the role of logic. The logic in this circuit can be broken into three parts: the conditions required to allow the coil of CR1 to energize, the conditions that turn on the fan, and the conditions that bring on the electric heat.

The condition that allows the coil of CR1 to energize is that both LS1 on the first door AND LS2 on the second door must both be closed AND the start switch must be turned to the on position. The word *AND* means the logic condition AND. If either door is opened prior to starting the oven, or if they are opened while the oven is running, the coil to CR1 will no longer be energized. Turning the start

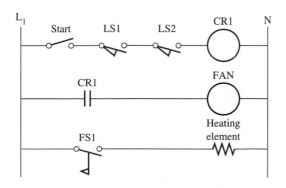

FIGURE 1-4 Relay logic circuit of a heating oven for a painting process. Notice that the limit switch for both doors must be closed to energize the fan, which must be running to energize the heating coil.

button to the off position will also de-energize the coil of relay CR1. The condition of the coil of CR1 (energized or de-energized) will also be part of the logic for the fan and the electric heating coil.

The NO contacts of CR1 must be picked (closed) to pass power to the fan motor. Since the contacts of CR1 can only be closed if the coil of CR1 is energized, you can see that the condition of the start switch, LS1, and LS2 will also affect the fan motor. If the start switch is in the off position, or if either door is opened, the contacts of CR1 will drop (open) and the fan will turn off.

The electric heating element is controlled by the logic of the flow switch. If the fan is operating and air is blowing against the flow switch, it will close and pass power to the heating element. If the start switch is in the off position or if either of the doors are opened, LS1 or LS2 will open and de-energize the coil to CR1. The contacts of CR1 will then drop and turn off the fan. This will cause the air flow from the fan to stop and the flow switch will open and turn off power to the heating element.

This example circuit shows how switches such as LS1 and LS2 can be inputs to the circuit and also be an integral part of the logic that operates the system. In most circuits, the logic is determined by safety and operational conditions. For example, someone may get burned if he or she opened the door while the oven is operating, so the door limit switches will de-energize both the fan and the heating element to prevent this from happening. Sometimes logic is used in the circuit to protect the equipment. For example, the heating element of this oven will burn out if sufficient air is not passing over it when it is energized. The flow switch protects the heating element by interrupting current to it if the fan is not moving air.

1.3.2 Relay Ladder Logic

The logic used in Fig. 1-4 is called relay logic because it uses a relay and its contacts to provide logic functions. The term *ladder* is used because the overall outline of the diagram looks like a wooden ladder. Each line of the circuit looks like a rung of the ladder. Relay ladder logic (RLL) is used quite extensively in industrial electronic circuits. The diagram is drawn to show a sequence of operation. When you are reading simple diagrams, it is usually easier to read them from top to bottom and from left to right. For example, in the electric heat oven circuit, the first line of the logic shows everything that must be energized before the second and third lines of the circuit can become energized. In more complex diagrams that contain many lines, it becomes harder to read the diagram in strict sequence from top to bottom because a set of contacts from the relay coil in the first line may show up in more than one line of the diagram and these lines may not be consecutive.

Another way to read the more complex diagrams is to start with the output you are trying to energize and determine the status of everything in its line of logic. This means that you actually will start from the output point in the diagram and work your way back up the diagram to each occurrence of logic that interacts with this output. For example, if the logic for a circuit is used to extend an air cylinder, and it is not extending, you could start at the line in the logic circuit where the solenoid coil for the cylinder is. Then test to see if current is passing through all of the sets of contacts to energize it. The set of contacts that is stopping the current flow represents the condition that is responsible for keeping the cylinder from extending.

1.4 RELAY LOGIC USED TO CONTROL A PNEUMATIC CYLINDER

Relay logic can be used to control the extension and retraction of a pneumatic (air) cylinder, as discussed in the previous paragraph. Fig. 1-5 shows a drill press that uses an air cylinder to move a drill up and down through its drilling cycle. The cycle begins with the drill at the top of its stroke. The air cylinder is retracted and the drill motor is turned off. When the start button is depressed momentarily, the extend solenoid is energized and the retract solenoid is de-energized so that air is moved to the air cylinder to make it extend, which causes the drill to move down. The drill motor is turned on by the drill motor starter during the down stroke. When the drill is extended to its full down stroke, it will strike a limit switch that will open and cause the retract solenoid to be energized, and the cylinder retracts moving the drill back to the up position.

Fig. 1-6 shows the relay diagram with a timing diagram for the drill press. From the first line of the diagram you can see that CR1 will be energized when the start button is momentarily depressed. The NO CR1 contacts connected in parallel with the start button will close when the coil is energized and seal in the start button. A second set of NO CR1 contacts is also used in line 2 of the diagram to energize the CR2 coil if the drill press is at the top position, which closes LS1. It is important to understand that the limit switch symbol that is used for LS1 indicates this limit switch is *normally open, held closed,* and the symbol for LS2 is a *normally closed limit switch.* This means that any time the drill press is in the up

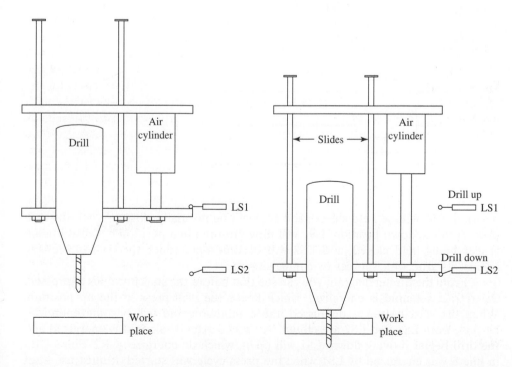

FIGURE 1-5 Diagram of air cylinder used to move a drill up and down through its drilling cycle.

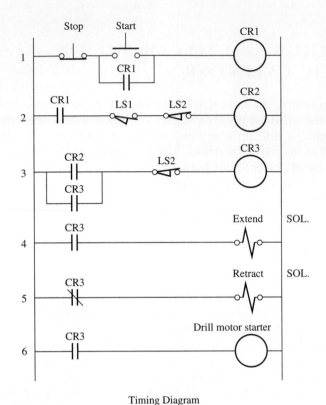

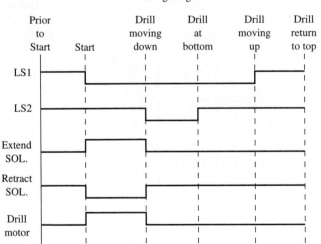

FIGURE 1-6 Example of relay logic that is used to control the extension and retraction of an air cylinder to move a drill through its stroke.

position, LS1 will be held closed, and LS2 will be normally closed. When the drill press is in the down position, LS1 will move to the open position because it is no longer being held closed, and LS2 will become open since the drill is pressing against its arm, which forces its contacts open.

From the timing diagram you can see that before the start button is depressed, the retract solenoid is energized, which keeps the drill press in the up position. When the start button is depressed, the extend solenoid will become energized because both LS1 and LS2 are closed. After the extend solenoid is energized and the drill begins moving down, LS1 will open, which de-energizes CR2. Since CR3 in line 3 was energized by LS1 when the press cycle was started, it must use a set of its own NO contacts to seal in LS1, so that the coil of CR3 will continue to be energized after LS1 opens. The NO contacts of CR3 will keep the extend solenoid energized until the drill moves to the bottom of its stroke and depresses LS2 and causes it to open. When LS2 opens, the coil of CR3 is de-energized, which de-energizes the extend solenoid. The NO CR3 contacts in line 5 will energize the retract solenoid and cause the cylinder to retract the drill press to the up position. Since the drill motor starter coil is energized by CR3 contacts, it will also be de-energized when the cylinder is being retracted. Since the drill depends on the air cylinder to provide speed for the drill, the air pressure on the cylinder must be adjusted to meet the cutting rate of the drill bit.

1.5 USING RELAY LOGIC TO DETERMINE ROBOT PROGRAM FOR INSERTING STUDS INTO TAILLIGHTS FOR AUTOMOBILES

Modern automobile companies make several models of cars that share many common parts. For example, Ford and Lincoln each have similar models such as the Thunderbird and Cougar that are the same size. This means the taillight assembly for a Thunderbird will have identical dimensions as the taillight of a Cougar because these two cars are built on the same body frame. The same is true of the Sable and Taurus models. In fact, some major companies make a foreign model that is based on a domestic body frame. This manufacturing concept makes it easier to manufacture and assemble parts.

If you look closely at the parts of the cars that have a companion model, you will see that they can be somewhat different and still maintain the same size and dimensions for mounting hardware. For example, the Thunderbird taillights are made as two separate assemblies, one for the left and one for the right. The Cougar taillight has the same dimensions, but it has a piece of plastic connecting the left and right assemblies. This means that the left and right assemblies have the exact dimensions for mounting studs for both models, and the Cougar has extra mounting studs in the piece of plastic that connects the two assemblies. These parts are made at a parts plant in large plastic injection molding machines. A small amount of assembly work must be accomplished on each taillight before shipment to the assembly plant where it will be put on the car.

Each taillight must have several studs inserted in it so it can be mounted to the automobile frame. The taillights are placed on an assembly carrier that moves them through a small flexible manufacturing assembly cell. This manufacturing cell consists of a robot that moves from point to point and inserts studs into each taillight at specific points. Since the assembly cell is very expensive to build, it is designed to insert studs into taillights for the Thunderbird, Cougar, Sable, and Taurus models, since all of these cars have similar-size taillight assemblies.

Each carrier on the assembly cell can carry any one of the four taillight assemblies. The assembly cell uses proximity switches to indicate what part is in the carrier so that the robot will execute the proper program. Each carrier has a set of three metal pins called *fingers* that are used to activate the three proximity

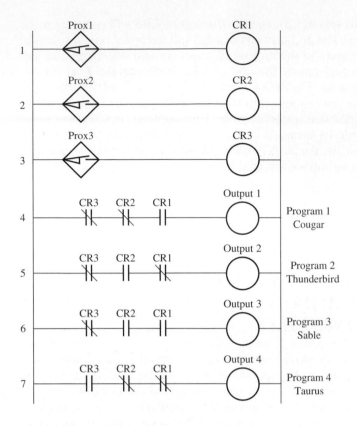

Truth Table				
CR3	CR2	CR1	Binary Count	Program #
0	0	1	001_2	1_{10} (Cougar)
0	1	0	010_2	2_{10} (Thunderbird)
0	1	1	011_2	3_{10} (Sable)
1	0	0	100_2	4_{10} (Taurus)

FIGURE 1-7 Example of the relay logic used to determine the correct taillight assembly. The output of the logic tells the robot which one of four programs it should execute.

switches to indicate which of the four parts are in the carrier. The three proximity switches are connected to three relays. The relay logic is based on the *binary numbering concept* that indicates which of the four programs the robot should execute. Fig. 1-7 shows a diagram and the relay logic for this process of detecting the correct part.

1.5.1 Understanding the Relay Logic

In Fig. 1-7 you can see the operation of the relay logic. Each of the three proximity switches is connected to a relay coil. Each line of the logic has a set of contacts from each of the relays connected in series. The output of each of the four lines of logic is connected to the robot as a *program request*. A combination of NO or NC contacts are used to ensure that the correct output is energized to tell the robot which program to run.

When the Cougar taillight is placed on the carrier, the finger to activate proximity switch 1 is lowered, and fingers 2 and 3 remain up. When this carrier passes the proximity switches, proximity switch 1 is activated and the coil to CR1 that is connected to it is energized. All of the CR1 contacts in the logic part of the program are activated, and all of the contacts from CR2 and CR3 remain in their normal position. Since the normally open CR1 contacts become closed when proximity switch 1 energizes CR1 coil, the first line of logic will become true and the robot will receive the output requesting it to run the program to insert studs into a Cougar taillight.

Lines 5 and 7 of the logic program will be de-energized because they contain NC contacts from CR1, which will move to their open position. Line 6 of the logic will also be de-energized because it has an NO set of contacts from CR2. Since the finger to activate the second proximity switch is not lowered for the Cougar taillight, the NO contact will not be closed and the output in line 6 will remain de-energized.

When a Thunderbird taillight is placed on the carrier, the technician lowers the second finger on the carrier to trip proximity switch 2, and the first and third fingers remain in the up position. When the carrier passes over the proximity switches, proximity switch 2 is activated and it energizes CR2. You can see from the four lines of logic, only the second line of logic will become true, and its output will request the robot to execute the Thunderbird program.

The fingers for the first and second proximity switches are lowered when a Sable taillight is loaded on the carrier. When the carrier passes over these proximity switches, the coils of CR1 and CR2 will be energized, which close their contacts and energize line 6 to request the Sable robot program. When the output for line 6 of the logic is energized, it sends a signal to the robot to request the program for the Sable. Since CR2 coil is energized, it will open the CR2 contacts in line 4 and line 7, causing these outputs to be de-energized. The energized CR1 coil will cause the NC CR1 contacts in the fourth line to open and de-energize output 4.

When the Taurus taillight is placed on the carrier, the third finger is lowered, and fingers 1 and 2 remain up. When the carrier passes over the proximity switches, finger 3 activates CR3. Line 7 of the logic will become activated because it has NO CR3 contacts and NC CR1 and CR2 contacts. Lines 4, 5, and 6 of the logic will be de-energized because they have sets of NO CR1 or CR2 contacts. When output 4 becomes energized, the robot is requested to run the program for the Taurus.

If you look at the output signals in the logic program closely, you will notice that they are activated when the contacts of the control relays provide the correct code. The code in this case is a binary code. If you assign CR1 the *one's position,* CR2 the *two's position*, and CR3 the *four's position,* you would notice that the proximity switches for the Cougar provide the code 001, which equals program 1. The proximity switches for the Thunderbird would provide the code 010, which equals program 2; the switches for the Sable provide code 011, which equals program 3; and the switches for the Taurus provide code 100, which equals program 4. You could get up to eight different combinations (0–7) with the same three proximity switches. The equivalent solid-state logic circuit using logic gates for this application is shown later in this chapter.

1.6 SOLID-STATE DEVICES USED FOR LOGIC

All of the logic in the first examples of this chapter has been accomplished with relay contacts and other input switches. When sets of contacts are wired in series with each other as in Fig. 1-8, their logic function is called an AND function because

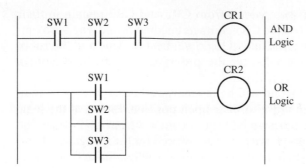

FIGURE 1-8 AND and OR logic functions executed by input switches and a relay coil.

switch 1 AND switch 2 AND switch 3 must all be closed to pass power to CR1, the relay coil for the circuit. If any of the switches is opened, the series circuit is broken, and no power will be sent to the coil.

The second circuit in Fig. 1-8 shows three sets of contacts that are wired in parallel with each other. This type of circuit is called an OR circuit because either switch 1 OR switch 2 OR switch 3 can close and energize the relay coil CR2.

1.6.1 Transistor Equivalent to AND and OR Circuits

Solid-state devices such as transistors can be arranged in circuits to provide AND and OR functions like the relay circuits. (See in Fig. 1-9.) From this example, you can see that inputs of the AND and the OR transistor circuits are identified as input 1, input 2, and input 3. The output of the AND circuit is identified as output 1, and the output of the OR circuit is identified as output 2.

Each part of the transistor logic circuit has an equivalent function in the relay logic. In the transistor logic circuits, applying +5 volts is the equivalent of the closed switch in the relay logic circuit. The +5 volts signal is also called a high, *HI,* or *true* signal. If the input to the circuit is grounded or pulled to 0 volts, it is equivalent to having open contacts in the relay circuit. The grounded input or 0 volts is called a low, *LO,* or *false* signal. When the coil in the relay logic circuit is energized, it becomes magnetic and pulls in its contacts. The transistor logic circuit provides

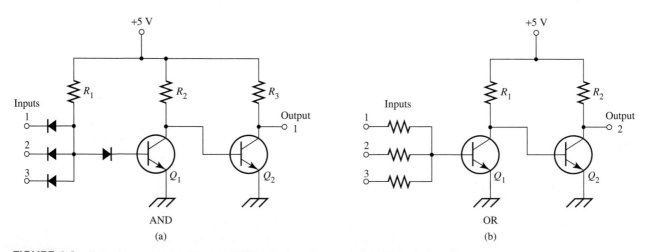

FIGURE 1-9 Transistor circuits for the AND logic function and the OR logic function.

a *logic HI* signal (+5 volts) at the output line of the circuit when the circuit becomes true.

You should notice in the transistor AND logic circuit in Fig. 1-9 that three inputs are available just as in the AND relay logic circuit. To fully understand the circuit operation when a HI signal is present at each input, you must first understand when a LO signal is present.

Notice that a diode is connected in the input part of each circuit. If the input is grounded, the diode will be forward biased and current will flow from the circuit supply through the limiting resistor through the diode to ground. You should remember that when a diode is forward biased, its anode voltage will be approximately 0.6 volts. This will be considered a LO signal. When this occurs at any one of the three inputs, only 0.6 volts will be available at the anode of the diode (D_2) in the base of transistor Q_1, which will not be sufficient to bias it. This means that V_{CE} will be maximum and the transistor is effectively an open circuit. Since transistor Q_1 is open, sufficient voltage will be supplied to the base of Q_2 through resistor R_2. The base voltage at Q_2 will be sufficient to cause it to saturate so its collector will be effectively the same potential as ground, which means output 1 terminal will be LO. If any one of the inputs to Q_1 are supplied with a LO signal, the output of the circuit at Q_2 will also be LO.

When all of the inputs to Q_1 are supplied a HI signal, sufficient base current will be supplied to Q_1 and it will saturate. This will cause the base of Q_2 to become grounded, which will cause its collector-emitter circuit to go to cutoff. This means that voltage from the +5 volts supply will be present at the output, which indicates the circuit has a HI signal. This means that input 1, AND input 2, AND input 3 must all be HI to have a HI output for the logic circuit.

The transistor circuit for the OR logic is similar to the AND logic circuit in that three inputs are provided and two transistors are used. When any one of the input circuits go HI, transistor Q_1 will become saturated, which causes the collector to become grounded. When the collector of Q_1 becomes grounded, the base of Q_2 is also grounded. This causes Q_2 to go to cutoff, which means that the +5 volts supply voltage will be available at the output terminal. The only time the output of the OR circuit is low is when none of the inputs are HI.

1.7 SOLID-STATE LOGIC BLOCKS

There are several variations of the transistor AND circuit and the transistor OR circuit. Once you have understood the operation of all the components within each circuit, the circuit can be replaced with a logic *symbol*. The logic symbol is easier to understand than trying to follow all of the voltages and grounds in a traditional transistor circuit. All you will care about in the logic symbol is the type of symbol that is represented and the conditions of each input. After you have determined the condition of each input, you can use a *truth table* to determine whether the output will be HI or LO.

The basic logic conditions can be expressed with five logic symbols. These are AND, OR, NOT, NAND, and NOR. When the logic symbols are used in a circuit, they are often called *gates*. Fig. 1-10 shows an example of the five basic logic gates and a truth table for each. From this figure you can see that the truth table shows the number of inputs each gate has and the output. A 1 or 0 is showed

AND gate logic matrix

Input 1	Input 2	Output
1	0	0
0	1	0
1	1	1
0	0	0

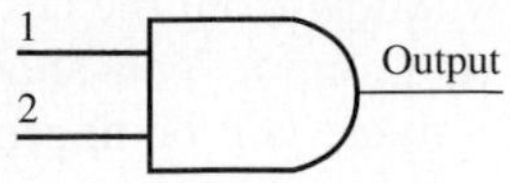

OR gate logic matrix

Input 1	Input 2	Output
1	0	1
0	1	1
1	1	1
0	0	0

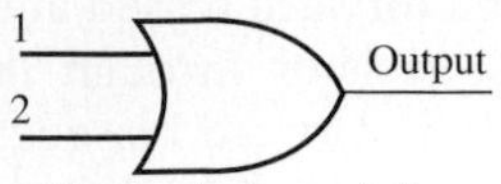

AND gate logic symbol

OR gate logic symbol

NAND Gate

Input 1	Input 2	Output
1	0	1
0	1	1
1	1	0
0	0	1

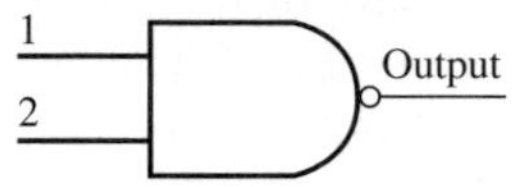

NOR Gate

Input 1	Input 2	Output
1	0	0
0	1	0
1	1	0
0	0	1

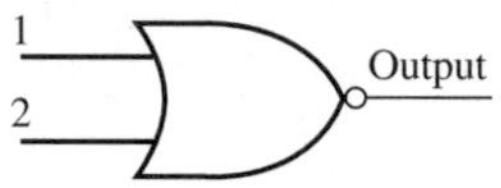

NOT gate logic matrix

Input 1	Input 2
0	1
1	0

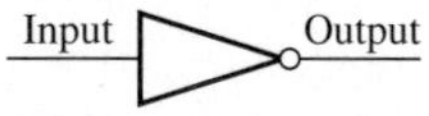

NOT gate logic symbol

FIGURE 1-10 AND, OR, NOT, NAND, and NOR gates with truth tables.

to indicate all of the possible conditions for the input. The 1 indicates the input is HI, and the 0 indicates the input is LO. The condition of the output for each set of inputs is also shown as a 1 or 0 to indicate if it is HI or LO.

The operation of the AND and OR gates is identical to their counterpart in relay logic shown in Fig. 1-8. The NOT gate is similar to a set of normally closed contacts in relay logic. When no power signal is applied to the coil of the relay, power will flow through its set of NC contacts, and when a signal is sent to the coil, no signal can get through the NC contacts because the coil has pulled them open.

The NOT gate is also called an *inverter* because the output is always the *inverse of its input.* You should also notice in this figure that the NOT gate can be combined with the AND gate to make a NAND gate (NOT + AND), and the NOT gate can be combined with the OR gate to make a NOR gate (NOT + OR). The small circle at the output is used to indicate the inverter or NOT condition has been added to the gate.

1.7.1 Early Examples of Logic Gates

Logic gates became widely used in the late 1950s and early 1960s soon after the transistors were mass-produced on integrated circuits (ICs). Since the ICs were made on a single chip of silicon, they also became known as chips. The outline of the device is called *dual in-line package* (DIP) because it has two rows of pins that allow it to maintain a low profile when it is mounted on a circuit board. The integrated circuits provide a variety of logic devices in multiple input AND and OR gates as well as NAND and NOR gates. Since these devices all used similar voltages of approximately 5 volts, they became known as the TTL family (transistor-transistor logic). TTL chips also provide more than one logic gate on an IC. One common IC has four AND gates mounted on it (QUAD AND gate).

The TTL family of chips does have some limitations when larger groups of gates are used together. Then the speed of the logic operation is slowed down, so the inverting-type gates (NAND and NOR) may be substituted because they use fewer transistors and other components. The inverter gates also use less power. The main drawback with inverted logic is that it seems more difficult to follow because humans have been taught to think from "front-to-back" instead of "backwards." It is important to understand that the same logic functions can be achieved by either AND and OR gates or with NAND and NOR gates.

TTL gates are still in use today, but a large variety of similar devices has been designed to help overcome some of the problems associated with them. For example, a complete logic family of complementary metal oxide semiconductors (CMOS) devices have been designed that are fast acting and use less power. The CMOS devices also have some inherent problems such as their susceptibility to damage due to static charges. Since the CMOS devices require very little power to operate, they can also be severely damaged by small amounts of stray power. The thin metal oxide insulation layer at the gates of these devices can also be damaged by static discharges from equipment or humans. Because of this problem, you must adhere to strict procedures for grounding yourself when you work on CMOS circuits.

Other logic circuit gates and components are available as low-voltage technology (LVT) that uses BiCMOS. This family of devices is designed to operate at 3.3 volts instead of the higher TTL voltage levels of 5 volts. Another family of logic devices is called low-voltage CMOS (LVC), which operates at voltage levels of 2.7 to 3.6 volts. These devices use less than 24 mA for drive current and can react at speeds of less than 6.5 ns. These logic devices have evolved to provide reliable, low-power, high-speed logic circuits.

1.7.2 Programmable Logic Arrays

Programmable logic arrays (PLAs) are part of a new family of logic devices called programmable logic devices (PLDs). The PLAs are similar to a programmable read-only memory (PROM). The PLA is made up of a programmable AND plane, which is connected directly to a programmable OR plane. Fig. 1-11 shows an example of the PLA. Notice that the AND plane has three inputs, I_1, I_2, and I_3, which are part of a matrix. Each input also has a parallel input called its *complement*. Multiple columns in the matrix are used to connect the AND plane to the OR plane. A fuse and diode are provided at the point in the matrix where the rows and columns intersect. The logic program is entered into the PLA by blowing out the necessary fuse links. When the fuse link is left in place, the logic path is useful,

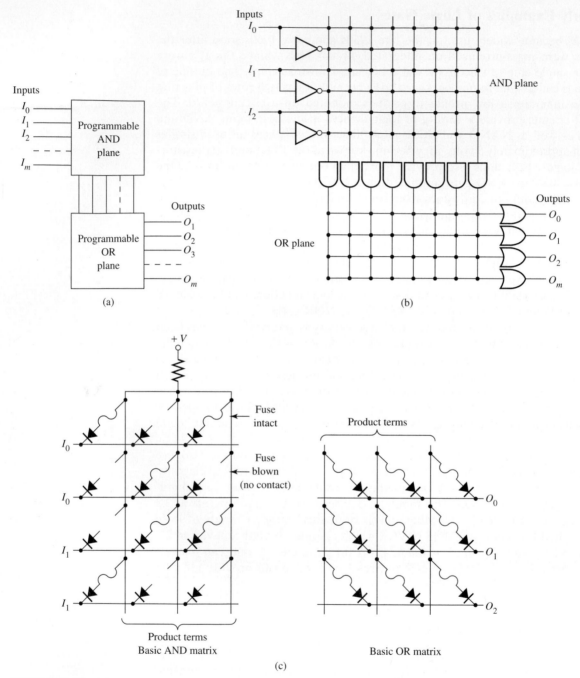

FIGURE 1-11 **(a)** Block diagram of programmable logic array (PLA). **(b)** Diagram that shows the AND plane and OR plane interconnection. **(c)** Internal diagram of basic AND matrix and basic OR matrix with fuses and diodes intact and open. (Reprinted by Permission of Texas Instruments)

and each time the fuse link is blown out the path is interrupted. Together the intact and blown fuse links provide the logic program. Since this technology is similar to the technology used to program a PROM, the PLA can be programmed by a PROM programmer.

The outputs of the AND plane are called the *product terms,* which are connected directly to the OR plane where they act as inputs. The diode and fuse connections are again used to connect each row and column. The fuse is blown at specific intersections between rows and columns to program the logic function of the OR plane. The OR plane has a number of outputs that are numbered O_1, O_2, and O_3, and so on. The exact number of inputs in the AND plane and outputs in the OR plane may be selected when the device is purchased.

Fig. 1-11 shows the connections between each of the logic segments in a PLA. If a program for a system such as the electric heating oven shown in Fig. 1-4 were entered into a PLA, the specific fuses would be blown to provide the logic. The inputs from the door limit switches would be connected to I_1 and I_2. The start switch would be connected to input I_3, and the flow switch would be connected to I_4. The output that indicates it is safe to start the fan would be output O_1, and the output that indicates it is safe to start the electric heating element would be O_2. You can see that all that would remain to have the PLA provide the logic for this circuit would be to open the necessary segments. When the program is printed, each segment that has active logic is shown as a dot on the intersection. This allows the technician to follow the dots when the circuit must be troubleshooted to determine which outputs would be energized.

PLAs are presently used to replace typical quad logic gates or other multiple logic gates that are mounted on a single 14-pin IC. The PLA is slightly larger than the 14-pin IC, yet it is able to provide hundreds of logic functions while using less power. The PLA can also execute the logic at a much faster speed than traditional ICs. PLAs are also available in a variety of logic functions beyond the traditional AND, OR, and NOT logic, such as exclusive OR (XOR).

1.7.3 Programmable Array Logic (PAL)

The programmable array logic (PAL) is very similar to the programmable logic array (PLA) in that it has a programmable AND plane, but it is different because it has a fixed OR plane. This arrangement limits the PAL to a specific number of logic functions. For example, a three-input PAL is limited to four output product terms from the AND logic plane, and other PALs will be needed to provide additional logic functions. This may be a limiting factor, but it provides a specific function that is reliable and low cost.

A similar device called a programmable logic sequencer (PLS) is available to provide sequencer functions. The sequencer is explained in detail in the next chapter. Its function provides excellent logic control for events that have a definite sequence of steps. The numbers of output that are energized can be controlled in a simplified manner as the sequencer turns each output on or off for a given step.

1.7.4 Programmable Logic Controllers

When logic circuits become very large, such as in the control of large industrial machinery, it becomes necessary to change the interaction between logic gates from time to time to change the function of the machine control. In the late 1960s and early 1970s, several companies each designed a programmable logic controller (PLC), which is now known as the programmable controller. These early companies included Modicon, Allen-Bradley, and Texas Instruments, and the controllers they designed were made specifically for General Motors. The programmable controllers represented the logic functions with contacts and coils symbols, much like the original relay logic. These symbols are used instead of logic symbols because the electricians and technicians that use the programmable controllers are used to using relay logic. In fact most of the early PLCs were used as relay replacements because they were very cost-effective in providing complex machine control through logic circuits. The PLC provides several other major advantages over logic chips in that they could be reprogrammed easily, they could be troubleshooted with a video screen that showed which contacts are open or closed, and their logic circuit (program) could be stored and downloaded into the machine control at a later date.

Today PLCs are available as the simple, low-cost replacement for relay logic, or to provide complex functions such as data storage, file manipulation, and process control. Each of these topics will be covered in detail in later chapters of this book.

1.8 SOLID-STATE LOGIC EQUIVALENT TO THE COUGAR TAILLIGHT ASSEMBLY CIRCUIT

Earlier in this chapter we discussed an example of relay logic used to tell a robot which type of taillight was placed on a carrier so the robot could insert mounting studs into it. This same circuit could also be designed to operate using solid-state logic chips. Fig. 1-12 shows the solid-state logic circuit for the taillight detection system. The inputs for this system are proximity switches just as in the relay logic circuit. The outputs are also relays that provide the same signal to the robot as in the previous example. The major difference in the solid-state logic circuit is that

FIGURE 1-12 Solid-state logic gates used to provide the logic circuit for the taillight detection system. The AND gates and inverters are used to provide a binary code signal that is sent to a robot as the proximity switches activate the inputs. The binary code will indicate to the robot which program should be executed.

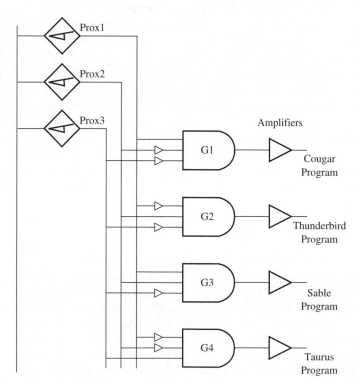

inverters are used with AND gates to provide the logic combinations that allow the circuit logic to produce the binary code, which indicates the program request being sent to the robot. The robot program uses the binary code to determine the insertion program that should be executed.

1.8.1 Operation of the Solid-State Logic Circuit

Previously in this chapter, a system that used relay logic to send a request to robots to execute programs to place studs in taillight assemblies was discussed. Fig. 1-12 shows the equivalent solid-state control circuit that utilizes AND gates and inverters to provide the same function. The hardware for the solid-state taillight detection system uses a set of three fingers that are mounted on each carrier. When they are lowered, the fingers will trigger a proximity switch just like the relay logic circuit. As you remember from the relay logic system, four different types of taillights can be loaded onto the carrier (Cougar, Thunderbird, Sable, or Taurus). The robot must know which type of taillight is in the carrier because it uses a different program for each taillight to ensure that the correct number of studs are mounted and that they are placed in the correct location. When a taillight is placed onto the carrier, the technician lowers the corresponding fingers to indicate which type it is.

In the diagram in Fig. 1-12, the first proximity switch is connected to the first input of each of the four logic gates. The second proximity switch is connected to the second input of each logic gate, and the third proximity switch is connected to the third input of each gate. Inverters are used to provide the correct logic input. Fig. 1-13 shows a truth table for this operation.

When a Cougar taillight is loaded on the carrier, the first finger is lowered and the second and third fingers remain in the up position. When the carrier passes over the proximity switches, the first proximity is triggered and it sends a HI signal to all four logic gates. The second and third proximity switches are not triggered so they both send a LO signal to all four gates. From the truth table in Fig. 1-13 you can see that when this combination of signals reaches each gate, only the G1 gate for the Cougar signal receives the correct combination to provide an output signal. The HI signal that is sent to the first input of gate 2 and gate 4 passes through an inverter, which will cause the signal to go LO so the gate output will be LO, and the LO signal sent to the second input of the third gate (Sable) will cause its output to be LO. When the Cougar logic gate becomes true, it will provide an output to its amplifier and the proper voltage signal is sent to the robot to indicate it should run the Cougar stud insertion program.

When a Thunderbird taillight is placed on the carrier, the second finger is lowered and the first and third fingers remain up. When the carrier passes over the proximity switches, the second switch is activated. You can see in the truth table that this combination of switch 1 (LO), switch 3 (LO), and switch 2 (HI) will only

Prox 1 Condition	Prox 2 Condition	Prox 3 Condition	Binary Value	Program Selected
Off	Off	On	001	#1 Cougar
Off	On	Off	010	#2 Thunderbird
Off	On	On	011	#3 Sable
On	Off	Off	100	#4 Taurus

FIGURE 1-13 Truth table and binary equivalent values for the switches in the circuit in Figure 1-12.

activate the G2 logic gate for the Thunderbird program signal. The first and fourth gates will have a LO logic output because they use an inverter in the line for their second inputs, which negates the HI signal from switch 2. Since the first input is LO for this carrier, the output for the Sable logic gate will also be LO.

When a Sable taillight is placed in a carrier, the first and second fingers are lowered and the third finger remains in the up position. When this combination of fingers passes over the proximity switches, the output of the third logic gate goes HI and the robot is signaled to start the insertion program for the Sable. The Taurus taillight uses the third finger down and the first and second fingers up to trip the proximity switches in the correct sequence.

The sequence of the inputs utilizes a binary code from the proximity switches to provide four possible conditions (outputs) from three input switches. Since the three input signals are interpreted as binary code, up to eight separate signals (0–7) could be generated.

1.9 CONNECTING INPUTS TO SOLID-STATE LOGIC DEVICES

In the logic circuits that have been discussed so far in this chapter, the inputs to the logic chips have been described in their most simplistic form. When a simple switch is connected to a logic circuit as an input, several complex problems occur. First, the switch contacts do not close precisely when the switch is activated. Instead the contacts actually bounce against each other several times before the switch settles into the closed position. The technical term for this is *contact bounce*. If the contacts bounce two or three times, the logic circuit sees this action as though the switch was actuated rapidly open and closed. If the logic circuit is used to count, each bounce of the contacts will be considered one operation of the switch contact closure. This means that the logic circuit would think the switch had actually opened and closed three or four times when the contacts bounce. This should also give you some idea of how fast the logic circuit operates, since the contact bounce occurs in less than several milliseconds.

One method that is used to solve this problem is to provide a capacitive filter between the switch and the logic gate. The filter actually consists of a resistor and capacitor. The resistor is sized to cause the capacitor to take a few milliseconds to fully charge before its voltage is large enough to trigger the input of the logic circuit. The resistor and capacitor act like a traditional time constant circuit. When the contacts first bounce close, the resistor limits the voltage to the capacitor, which makes it charge over a few milliseconds. The time constant is designed to be slightly larger than the time for the contact bounce to occur. Other more elaborate methods of combating the effects of contact bounce may include types of latches that catch the first bounce of the contact closure and ignore all subsequent bounces.

A more serious problem exists for inputs to logic circuits. This is the problem of voltage and current incompatibility. For example, the logic circuit is designed to operate at 3–5 volts dc. A 110 volts ac switch from a photoelectric switch is used as the logic input, which cannot be connected directly to the logic circuit. The primary voltage of control devices used in industrial machinery is 110 volts ac.

The 110 volts ac switch is used because it is immune to signal noise from motors and other high-voltage devices and because it can be transmitted over longer

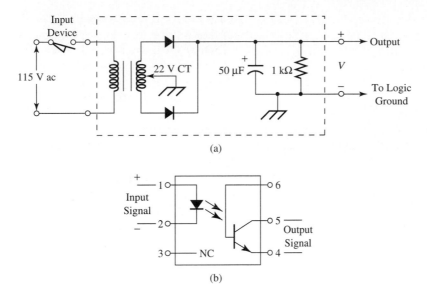

(a)

(b)

FIGURE 1-14 **(a)** Input circuit for a logic gate. This circuit provides signal conversion and isolation for 110 volts ac signal that is used as an input to a logic circuit. **(b)** Circuit of a solid-state optocoupler, which uses a light-emitting diode (LED) and phototransistor to create an isolation circuit for dc signals.

distances. These types of problems can be solved by a signal converter. Fig. 1-14 shows several examples of signal converters. In the first circuit (Fig. 1-14a), you can see that an isolation transformer is used to drop the 110 volts ac to the correct logic voltage level and diodes are used to convert the ac voltage to dc. The second circuit (Fig. 1-14b) provides both the voltage conversion and isolation by using a rectifier and optocoupler. The optocoupler is a device that combines a light-emitting diode (LED) with a phototransistor. These devices are discussed in detail in Chapter 6 of this book.

The resistors in the circuit help to drop the 110 volts ac to the acceptable level and the rectifier changes the ac to dc for the LED. The LED emits a light directly to the phototransistor, since the two devices are encapsulated inside an IC. When the phototransistor sees the light, it will go into conduction and reproduce the action of the input switch. The optocoupler provides the additional feature of isolation in that the LED will not transmit the large transient voltage spikes that may occur on the input to the transistor. The intensity of the LED output will not seriously affect the phototransistor, since it is already in saturation. Both of these circuits will still require a bounce elimination circuit, since the switch contact bounce would be retransmitted by both of these types of circuits.

1.10 OUTPUTS FOR LOGIC CIRCUITS

The signal that comes from the output of the logic circuit is a low-voltage, low-current signal that is too small to drive a solenoid or motor starter. This means the logic output signal must be amplified to be usable. Several methods are available to amplify the output signal. A power transistor can be used as a stand-alone device, or several transistors can be used in an amplifier circuit. Other solid-state devices called *thyristors* can also be used. Some of the thyristors that are commonly used are *silicon controlled rectifiers* (SCRs) and *triacs*. These devices can switch large dc

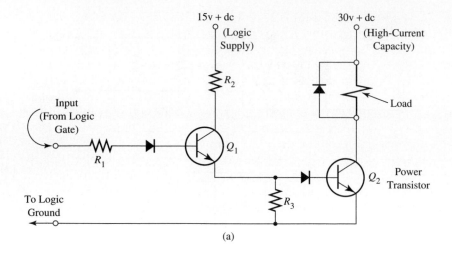

FIGURE 1-15 **(a)** Electronic circuit of two transistors used to amplify the output signal and energize a relay. **(b)** Electronic circuit of a single transistor and two relays used to amplify an output signal so that it can energize and de-energize a 240 volts ac heating element. **(c)** Electronic circuit that shows the output from a logic gate sent to an amplifier, which energizes a relay coil. The amplifier in this circuit is indicated by the amplifier symbol.

and ac voltages and currents with small gate voltages. They are discussed in detail in Chapter 4.

The *operational amplifier* (op amp) is used to drive the solenoid or motor starter. The op amp can be used as the amplifier part of the circuit or it can be incorporated into the logic circuit. A large variety of logic circuits can be designed with op amps. Figs. 1-15a and 1-15b show examples of transistor amplifiers used to amplify the output signal of the logic circuit. A detailed discussion of op amps is provided in Chapter 8.

Two relays can also be used in combination with a transistor to provide amplification and isolation. Fig. 1-15b shows an example of this type of circuit. From the diagram you can see that the output of the logic circuit is connected directly to the base of the transistor. The first relay coil is connected in the emitter-collector circuit of the transistor. A diode is connected in reverse bias across the relay coil (in parallel) to protect the transistor against large transient surges of voltage that occur when the coil is de-energized and releases its inductive energy back into the circuit. The diode will take this spike of voltage and continually reroute it back through the wire of the relay coil until it dissipates. The coil of the second relay is connected in the contact circuit of the first relay. The second relay can be significantly larger because the contacts of the first relay only need to apply power to the coil of the second relay.

The amplification part of this circuit takes place in three steps. First, the transistors increase the small-signal voltage from the logic output sufficiently to energize the low-voltage relay coil. In the second step the contacts of the first relay are used to energize the coil of the second relay. The third step of amplification takes place where the 240 volts ac electric heating element is powered by the contacts of the larger relay called a *contactor*. The small output signal from the logic circuit is able to control the 240 volts ac heating element and turn it on and off as if connected directly to it. Fig. 1-15c shows a diagram of how this circuit would look if the transistors are shown as amplifier symbols. You should also notice that the relay coil is shown as the load to the amplifier. Since the control or logic circuit is shown separately, the heating element that is connected to the relay contacts may not be shown.

 ## 1.11 SOLID-STATE RELAYS

Solid-state relays (SSRs) provide a means of combining electronics with the function of isolation in an electromechanical relay. Since the coil of an electromechanical relay is completely isolated from its contacts, the input signal to the relay coil may be 110 volts ac and the output signal that goes through the contacts can be as small as 3 volts or 5 volts logic level. Another way to use the isolation is through a small coil voltage such as 5 volts dc and allow the contacts to switch 110 volts ac. The solid-state relay is made in a similar design in that its input is isolated from its output. Fig. 1-16 shows a diagram of two different types of solid-state relays.

The original types of SSRs use a silicon controlled rectifier (SCR) if it is rated for dc voltage usage, and it uses a triac if it is rated for ac voltage usage. The SCR and triac are discussed in detail in Chapter 4. These devices required a small gate voltage to operate. When the small gate voltage was applied, they could switch larger voltages. Newer SSRs utilize a photocoupled device that includes a light-emitting diode (LED) mounted so that its light shines directly on a photo device such as a photodiode or phototransistor. The LED and photo device are encapsulated into a single IC so that outside light does not interfere with their operation. When the photo device detects light shining on it, it will go into conduction and pass current and voltage to whatever is connected to it. Some devices use miniature relays as the output device to provide another layer of isolation, while other devices use solid-state components such as metal oxide semiconductor field effect transistors (MOSFETs). In each case the solid-state relay provides an effective means to produce an isolation circuit that can be used for logic functions.

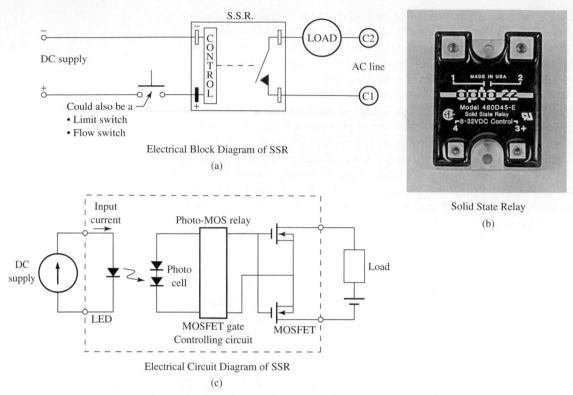

Electrical Block Diagram of SSR

(a)

Solid State Relay

(b)

Electrical Circuit Diagram of SSR

(c)

FIGURE 1-16 Solid-state relays (SSRs) that use silicon controlled rectifiers (SCRs) and triacs for isolation. Newer SSRs use optocoupler devices to provide isolation. These devices mimic the action of a relay coil and its contacts. (Courtesy of Opto 22, Remecula, CA.)

1.11.1 Features of SSRs

The SSRs provide a means of taking a small logic signal and controlling a larger output. The SSRs are also small enough and fast enough to provide some of the same functions as the IC logic gates or full-size relay logic. They also provide an extremely large isolation factor between the input and output and do not cause contact bounce or contact arcing like a relay, which may also add noise to the signal. SSRs are very resilient and can hold up in conditions of vibration. When compared to the logic IC and the relay, the SSR has an equally long life, since it can withstand millions of operations.

1.12 A COMPARISON OF SOLID-STATE LOGIC AND RELAY LOGIC

When solid-state logic was first introduced, the comparison of solid-state logic devices and the traditional relay logic became a fierce competition because it was thought that the newer solid-state logic would replace all relay logic. It is important to understand that both the relay and the solid-state logic devices have unique characteristics that may lead a designer to choose one over the other. Even though the relay is much older and may seem out of date, you will still find a wide variety of relays in industry today. They are no longer used to produce large-scale logic circuits. Instead two or three relays will be used in machine control circuits to provide limited logic functions or to provide a means of connecting smaller control voltages to motor starters that control 50hp motors.

It may be better to relate the comparison of relays to solid-state logic circuits with the comparison of an automobile to a pickup truck. Both the automobile and pickup truck have unique features that make them both good at their particular jobs. Problems arise when you have decided to use one instead of the other and then the conditions change. For example, if you need to haul lumber on Saturday morning, the pickup truck is ideal. Problems start to arise as conditions change slightly because you have to take your two children and large dog with you. The problem becomes more complex when it starts to rain, and the large dog riding in the back wants to be inside the truck with you and the two kids where it is dry. What started out as a good choice of using the pickup truck to make a simple trip now looks like a poor choice because you have to deal with a large dog and two children inside a small truck cab. This is much the same problem that arises when you have selected solid-state logic or relay logic to control a large industrial circuit. By the time you are asked to troubleshoot the circuit, many years have passed and the original conditions of the machine have changed. It is best to understand the strengths and limitations of both the solid-state logic devices and the relays so that you can see potential problems as the control system evolves.

1.12.1 Strengths and Weaknesses of Solid-State Logic Devices

Solid-state logic devices are low cost, reliable, small, and can operate at very high speeds with a high life expectancy. You should understand that in the average control circuit, the inputs may cycle from HI to LO several times a minute. If you compute the number of cycles per day and then per month, you will see that the total would add up to over 1 million cycles per year. This means that if the solid-state logic is specified to last 10–20 million cycles, its life expectancy would be 10–20 years. Fig. 1-17 provides a table that shows the comparison of solid-state logic devices and relays.

One of the reasons that the solid-state logic has such a long life expectancy is that it operates in an enclosed environment, since each device is packaged as an integrated circuit. This is extremely important when the circuit is exposed to the harsh environment of the typical industry where all kinds of dust and dirt may accumulate and cause traditional contacts to wear out or seat incorrectly. Another strength of the solid-state devices is that since they are enclosed in the IC, they do not pose a hazard in explosive atmospheres, such as in control circuits for painting systems or where vapors are used for cleaning.

The fact that solid-state logic devices are smaller means that they need less room in electrical cabinets. Today the floor space in a factory is very expensive and designers try to keep the size of electrical cabinets as small as possible. Some of the critical solid-state ICs in a circuit are now mounted in sockets or the entire circuit is modularized on a single board with an edge connector that plugs into a card cage. These features allow the faulty components or circuits to be removed and replaced more easily during troubleshooting and repair.

Some of the weaknesses of solid-state logic devices are that they are somewhat more difficult to troubleshoot because you need to have a good pin outline of the

Strengths of solid-state devices
   fast
   inexpensive
   small size
   reliable up to 5 million cycles
   encapsulated, good in explosive environment
   some circuits modularized or in socket for easy replacement

Weaknesses of solid-state devices
   difficult to troubleshoot
   some devices soldered
   require power supply
   some CMOS susceptible to static
   do not tolerate heat well
   may need fan or air conditioning
   small-signal voltage picks up noise
   must be mounted near inputs because of low voltage

Strengths of relays
   easy to change out
   durable
   easy to test
   available with plug-in base
   miniature relays available
   immune to noise
   110 volts signals allow relays to be mounted further from inputs
   maintenance personnel already familiar

Weaknesses of relays
   contact wear
   moving parts (springs) limit life expectancy from 1–3 million cycles
   require more power than solid-state devices

FIGURE 1-17 Comparison of solid-state logic devices and relays.

entire circuit and you need a logic probe or scope. If the solid-state IC is soldered into a printed circuit board, replacement is very difficult. Sometimes the circuitry is considered proprietary and the board must be sent back to the original equipment manufacturer. Other manufacturers may encapsulate circuit devices in plastic cases that are not made to be tested or repaired. Instead these circuits are simply swapped with a known good one to determine if they are good. If the circuit is bad, it is discarded or sent back to the factory for repair.

Another problem with solid-state logic circuits is that they tend to add (dissipate) heat into the cabinet where they are mounted, which means the cabinet may require a cooling fan or air conditioning. This is especially critical in industries like the steel industry, glass industry, and some areas of the plastic industry where the controls must be mounted near the high-temperature area of the process. Other problems arise with small-voltage levels of the signals. The small voltages used for inputs are susceptible to induced voltages from large motors or other inductive devices on the factory floor. Since the voltages are usually less than 5 volts, the logic circuitry must be mounted fairly close to the input switches. This may increase the likelihood of picking up interference from the induced voltages.

1.12.2 Strengths and Weaknesses of Relays

The relay has several strengths that make it very useful in certain industrial logic control applications. For example, relays are fairly easy to troubleshoot to determine if the contacts are open or closed and if they are passing voltage signals. Some relays have an indicator lamp mounted where the technician can look quickly at

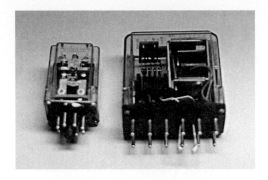

FIGURE 1-18 Examples of relays with indicator lamps and plug-in sockets that make them easy to troubleshoot and replace. (Courtesy of Omron Electronics.)

the cabinet and determine if the relay coil is energized. Also relays have been used in industrial applications for purposes other than logic and their operation is similar to contactors and motor starters, so the technicians are very familiar with them. Some relays used in critical logic circuits are available with plug-in sockets that make them easy to remove and replace for testing and repair. Fig. 1-18 shows examples of relays with indicator lamps and plug-in sockets.

Other strengths of relays are that they are durable and can withstand stray induced voltages without changing state. Relays also hold up well in extremely warm temperatures. Even though relays do require more power than solid-state devices, the power issue becomes clouded if an air-conditioning unit must be added to the electrical panel where solid-state devices are used because the air-conditioning unit will use large amounts of electrical power. Relays are also available in encapsulated packages if they are used in an explosive atmosphere, or conditions such as electroplating processes where the air is corrosive and acid vapors tend to corrode the contacts.

A large variety of miniature relays is available. These smaller relays require less power and take up little space in cabinets. If the circuit is strictly for logic, the smaller relays are very reliable for the job.

Some of the weaknesses of relays include the requirement for larger amounts of power in circuits that use large numbers of relays. Other problems occur when the relay contacts wear prematurely due to excessive arcing. The relay is also a mechanical device that uses springs and other moving components that tend to wear out over time. The life expectancy of a relay may only be 2–3 million operations, which means they could wear out over a period of a few years.

1.13 SOLID-STATE AND RELAY LOGIC CIRCUITS TO CONTROL PART BINS FOR ROBOTIC AUTOMATED WORK CELLS

When robots are used in automated work cells, every aspect of the system must be automated. In some applications, robots must take small parts such as screws or bolts from a parts feeder and insert them into the work pieces as they move by an assembly line. Each parts feeder has to have a bin that holds a 1- to 2-hour supply of the small parts. The bin acts as a storage buffer for the small parts, but

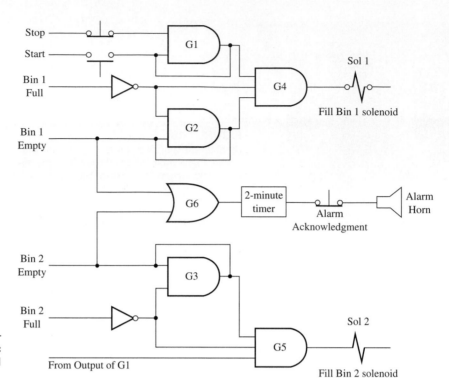

FIGURE 1-19 Solid-state positive true logic for the bin-level control system.

it is important that the level in the bins is automatically controlled to ensure that sufficient parts are available and that the bin does not overfill or become empty.

Fig. 1-19 shows the solid-state logic diagram and Fig. 1-20 shows the relay logic diagram for this system. The system has two bins to feed parts to two robots. Proximity switches are mounted at the top and the bottom of each bin to indicate that the parts bin is full or empty. A gravity-feed system is used to feed screws through a tube so that they can be loaded directly into each bin. A separate solenoid is used to control the amount of parts that drop into each bin. If either bin goes empty and does not begin to fill after 2 minutes, an alarm horn (annunciator) is sounded to warn technicians that the robots are about to run out of parts.

The start and stop switches are connected to the first logic AND gate G1. Since the stop button is normally closed, the gate will provide an output as soon as the start push button is depressed. Since the start push button is a momentary-type switch, the output of AND gate G1 will be used to seal it in. The output of gate G1 is also sent as an input to gate G4. You should notice that gate G5 also uses the output of G1 as one of its inputs. This means the first gate (G1) is used to start and stop the entire circuit.

The fill-bin solenoid (Sol 1) will become energized any time the level in the bin drops below the bin-empty sensor. When the level drops below the bin-empty sensor, a 1 is sent to G2. Since the level is also below the bin-full sensor, G2 will receive its second input from the inverter that is connected to the bin-full sensor. The bin-full proximity switch uses an inverter to make sure the signal that is sent to gate G2 is a 1 when the bin is *not full*. When gate G2 receives both inputs, it will produce an output to gate G4, which is currently receiving an input signal from G1. Notice that the output of G2 is also brought back as an input to gate G2 to seal it in once it is turned on. This is necessary since the bin-empty signal will turn

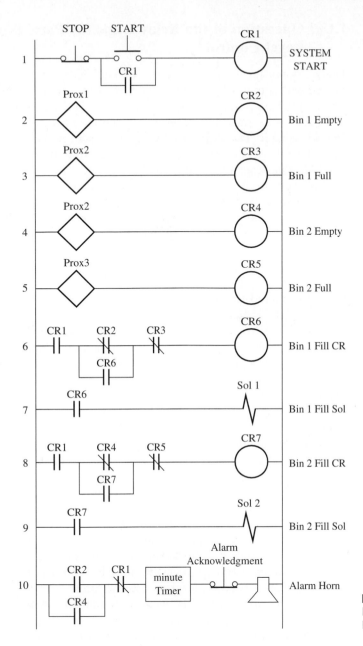

FIGURE 1-20 Relay ladder logic for the bin-level control system.

to a 0 as soon as parts are dropped into the bin. Since G2 is sealed in with its output, it will remain energized until the bin is completely full. When the bin becomes completely full, the bin-full sensor sends a 1 to its inverter, which sends a 0 to G2 and turns it off. When G2 turns off, its output no longer provides a seal-in signal, which keeps the fill solenoid off until the bin becomes empty again. The seal-in signal around G2 ensures that the fill solenoid will remain on any time it is energized until the bin is completely full, and it will remain off any time it is de-energized until the bin is empty again. Gates G2 and G4 provide the logic for the first bin, and gates G3 and G5 operate the second bin.

1.13.1 Operation of the Relay Logic for Parts Bin Level Control

The parts bin level control system is controlled by a control relay CR1, which is started with a manual push button (start/stop station). Notice that the start button is sealed in by a parallel set of CR1 contacts. The seal-in feature allows the control relay coil to be initially energized by depressing the NO start push button once and releasing it. The stop push button acts as a regular stop button and emergency stop, since it will drop out the CR1 coil any time it is depressed.

The *bin-level empty* and *bin-level full* are determined by proximity switches that are activated when parts are in front of them. (A detailed description of the operation of proximity switches is provided in Chapter 10.) This means that the low-level proximity will act like a closed switch as long as parts are near it. If the bin goes empty, this switch will act like an open switch. The bin-full proximity switch will act as an open switch as long as the parts level is not full. When the bin fills up to the top, the parts will be in front of the proximity switch, and it will act like a closed switch. Each proximity switch is treated as an input to the logic circuit and is connected to a control relay. The contacts of these control relays are used to provide the logic for the system.

The logic for the first bin is controlled by the coil of control relay CR6. You should notice that the master start button must be depressed to pull in CR1, since the contacts of CR1 are the first contacts of this circuit. The CR1 contacts are ANDed with NC CR2 contacts (bin empty) and the NC CR3 contacts (bin full). When this line of logic becomes true, the CR6 coil becomes energized and its contacts energize the solenoid valve that opens to allow more parts to drop into the bin.

The important part of this circuit logic is that once the level in the bin is detected as empty, the CR2 contacts will close and pass power through NC contacts CR3 to the CR6 solenoid. After parts begin to fall into the bin, they will fall in front of the low-level proximity switch and turn it off right away. This would normally de-energize the solenoid immediately, which would cause the system to short cycle. When the system is short cycling, it will only allow a few parts to drop into the bin and shut off prematurely. To prevent short cycling from happening, a set of NO CR6 contacts is placed in parallel (OR) with the CR2 bin-empty contacts. When the low bin level is reached and CR2 contacts close for the first time, the CR6 relay coil becomes energized and pulls in its NO contacts that will provide an alternate path around the CR2 bin-empty contacts. When parts start to fill the bin, and the proximity switch indicates the level is no longer too low, the CR2 contacts will open, but the CR6 coil will remain energized until the NC CR3 contacts open to indicate the bin is full. The CR6 contacts that seal in the CR2 contacts provide a type of *dead band*. This means that once the low level is reached and parts start dropping into the bin, they will continue to fill until the bin-full level is reached. The solenoid will not be energized again until the bin level reaches the empty point. The amount of dead band is determined by the height of the bin-full and bin-empty sensors.

The circuit for the second robot's parts bin operates exactly the same as the circuit that was just described. The only difference is that bin 2 uses a bin-empty sensor that activates CR4 and a bin-full sensor that activates CR5. The logic circuit for the second bin uses CR7 to activate the solenoid to allow parts to drop into the bin.

The last part of this automated bin level control is the alarm circuit. The alarm is needed to indicate that one or both of the bins are empty and they have not begun to fill after a 2-minute period. The robot will use up all the parts in the bottom of the bin within 10 minutes of the bin-empty switch activating. The alarm will sound 2 minutes after the bin-empty switch is activated and this will give a technician several minutes to find the problem and fix it.

The logic for the alarm is shown in the last line of this diagram. The NO CR1 contacts ensure that the alarm system is only operational if the system start has been activated and the system is running. You can see that the NC CR2 bin-empty contacts for bin 1 are connected to a 2-minute timer. If the bin goes empty, no parts will be in front of the proximity switch, so it will not energize the coil of CR2 and the CR2 contacts will close and pass power to the timer. If the contacts remain closed for 2 minutes, the timer will *time out* and energize the alarm horn (annunciator). A duplicate circuit is provided for the second bin, and contacts CR4 are connected in parallel as an OR logic condition with the CR2 contacts. This means that the horn will sound if either bin goes empty for more than 2 minutes.

SOLUTION TO JOB ASSIGNMENT

The solution to your job assignment should refer to the relay ladder logic diagram shown in Fig. 1-20 and the solid-state logic circuit diagram shown in Fig. 1-19. This circuit is for the bin-level control system that controls the filling of the bin when it gets empty. Your solution should begin by identifying all the inputs in the relay logic and all of the conditions for the circuit. Each condition will be represented by a logic gate. If the relay logic contacts are in series, an AND gate will be used; if the relay contacts are in parallel, an OR gate should be used.

When you test the circuit, you should test each input with a logic probe to determine that its signal will go HI and LO. Next you should set all inputs to the proper signal to activate the logic. Test the output of the logic gate to determine that it has the proper signal. If you use any amplifiers or inverters, be sure to test them for the proper signal.

Refer to the table in Fig. 1-17 (p. 28) to determine the strengths the relay provides that would better suit it for the job than the solid-state logic device. One reason to keep the relays is where amplification is required to get a small signal from the solid-state circuit to a level where it can energize or de-energize 110 volts ac devices.

QUESTIONS

1. Identify the three parts of the control circuit and explain the function of each.
2. Explain the terms *poles* and *throws* in regard to switches and relay contacts.
3. Identify the basic parts of a relay and explain their operation. Be sure to explain the operation of both normally open and normally closed contacts.
4. Provide an example application that uses normally open relay contacts and normally closed relay contacts.
5. Explain the types of signal conditioning that may be required to interface an input signal to a logic gate.
6. Compare the strengths and weaknesses of relays and solid-state logic gates.

TRUE OR FALSE

1. _____ The start switches LS1 and LS2 in Fig. 1-4 are connected to form OR logic, since any of the three switches can turn on the control relay CR1.
2. _____ A relay with normally closed contacts will provide logic function that is similar to a NOT logic gate.
3. _____ An optoelectronic circuit like the one shown in Fig. 1-14 can provide isolation between two circuits.
4. _____ A solid-state relay can provide both the function of isolation and amplification.
5. _____ One advantage relays have over solid-state logic devices is that relays are faster acting.

MULTIPLE CHOICE

1. A typical relay has _____.
   a. one coil and one or more sets of contacts.
   b. one set of contacts and one or more coils.
   c. multiple coils and multiple sets of contacts.
2. A solid-state relay uses _____ to switch power.
   a. a coil and electromechanical contacts
   b. a light-emitting diode (LED) and phototransistor
   c. a rectifier diode and a relay coil
3. The Bin #1 Fill solenoid in line 7 of Fig. 1-20 will be energized when CR1 is energized and _____.
   a. prox 1 is high and prox 2 is low.
   b. prox 2 is high and prox 1 is low.
   c. prox 1 is high and prox 3 is low.
4. The alarm horn in line 10 of Fig. 1-20 will be activated when _____.
   a. CR2 and CR1 are both energized for more than 2 minutes.
   b. either CR2 or CR4 are energized for more than 2 minutes.
   c. either CR2 or CR4 are de-energized for more than 2 minutes.
5. The Thunderbird program logic in Fig. 1-12 will be selected when _____.
   a. prox 1 is high and prox 2 and prox 3 are low.
   b. prox 1 is low and prox 2 and prox 3 are high.
   c. prox 2 is high and prox 1 and prox 3 are low.

PROBLEMS

1. Draw the symbol for and explain the operation of the basic logic gates: AND, OR, NOT, NAND, and NOR.
2. Draw or sketch the locations where you would place voltmeter leads in order to test the relay logic in Fig. 1-20 if the fill solenoid would not energize.
3. Draw or sketch a circuit that is used for signal conditioning or amplifying an output signal from a logic gate to make the signal interface a 110 volts ac relay coil.
4. Draw or sketch the relay contacts used to create the following gates: AND, OR, NOT, NAND, and NOR.
5. Draw a truth table for the operation of the bin level control circuits shown in Fig. 1-19.

REFERENCES

1. Floyd, Thomas L., *Digital Fundamentals,* 5th ed. New York: Macmillian Publishing Company, 1994.
2. Kissell, Thomas E., *Modern Industrial/Electrical Motor Controls.* Englewood Cliffs, NJ: Prentice-Hall Inc., 1990.
3. *Linear Circuits Data Book.* Dallas, TX: Texas Instruments, 1992.
4. Maloney, Timothy J., *Industrial Solid-State Electronics,* 2nd ed. Englewood Cliffs, NJ: Prentice-Hall Inc., 1986.

2

ADVANCED SOLID-STATE LOGIC:
FLIP-FLOPS, SHIFT REGISTERS, COUNTERS, AND TIMERS

OBJECTIVES

After reading this chapter, you will be able to:

1. Explain the need for solid-state logic devices that can store and shift bits to provide counting and timing functions.
2. Explain the operation of an *RS* latch and describe the condition of each output when each input is applied.
3. Discuss the operation of the edge-triggered *RS* flip-flop, the edge-triggered *JK* flip-flop, and the edge-triggered *D* flip-flop.
4. Provide an example of an industrial application that uses flip-flops and explain the operation of the application.
5. Explain the operation of a bit shift register and describe the condition of the output of each flip-flop as the shift-register clock pulse is applied.
6. Provide an example of an industrial application that uses a bit shift register and explain the operation of the application.
7. Explain the operation of a word shift register.
8. Provide an example of an industrial application that uses a word shift register and explain the operation of the application.
9. Explain the operation of a first-in, first-out (FIFO) register.
10. Provide an example of an industrial application that uses FIFO and explain the operation of the application.
11. Explain the operation of an up counter and a down counter. Provide an industrial application that uses an up/down counter.
12. Explain the operation of a decoder.
13. Describe the operation of a one-shot timer. Provide an industrial application that uses a one-shot timer.
14. Explain the operation of the 555 timer chip. Describe its use in the astable and multistable modes.

You are a technician who has been assigned to design an electronic control system for two spot-welding guns that are operated by two robots. The first robot makes a series of welds on an automobile body on a production line, and the second robot makes a different set of welds. The electrical bus that supplies power to the welders is not large enough to allow both welders to be operated at the same time. The control system that you are designing should use solid-state components described in this chapter and it should provide the following functions: (1) en-sure that both welders do not operate at the same

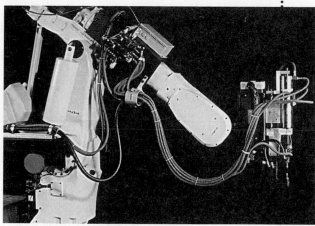

(Courtesy of Fanuc Robotics North America, Inc. All rights reserved.)

time, (2) keep track of good and bad parts, and (3) provide 20 seconds of warm-up time for each welder when it is first energized. For an extra assign-ment, design a means for the system to keep track of present or missing automotive bodies as the production line passes the robots. You can also design a circuit to keep track of faults as they occur. Carefully read the informa-tion provided in this chapter to help you design this control system. The solution is provided at the end of this chapter.

2.1 OVERVIEW OF THE NEED FOR ADVANCED LOGIC

Modern solid-state devices have incorporated advanced logic functions such as latches and flip-flops to store the state (1 or 0) of the logic bits used in the previous chapter. The AND and OR gates provide simple logic functions, but more advanced logic functions are required to store the state of a bit or to count and provide time delay. Counting and timing are basic functions of nearly all industrial applications. This chapter begins with the explanation of a simple latch and several types of flip-flops. It continues through shift registers, counters, and timers. You will see how these solid-state devices are used in various industrial applications.

2.2 LATCHES

In more advanced solid-state logic applications, the state of a logic bit may need to be stored. This can be accomplished by using a latch. A *latch* is a bistable storage device that can reside in either the set or reset state by virtue of feedback from its outputs. Since the latch has two stable states (HI and LO), it is called *bistable* and its output indicates whether a 1 or 0 is stored. Fig. 2-1 shows an *RS* latch, and its equivalent NAND gate logic and truth table.

2.2.1 *RS* Latch

The operation of the *RS* latch shown in Fig. 2-1a can best be explained by the example with two NAND gates shown in Fig. 2-1b and the truth table shown in Fig. 2-1c. The inputs of the latches are identified as *S* for *set* and *R* for *reset,* and the outputs are identified as Q and $\overline{Q}$ (Q NOT). This latch may also be called the SR latch.

When a HI input is applied to the *S* input and the *R* input is held LO, the Q output will be high, and the $\overline{Q}$ will be low. In this state the latch is storing the status or condition of the HI signal that is sent to the *S* input. In most cases this signal is toggled, which means it may only stay HI for an instant. After the input signal goes LO, the latch maintains the HI condition at the Q output. This is called the *latched* condition.

The latch can be reset when a HI signal is sent to the *R* input. The *S* input must be LO at this time for the reset function to be accomplished. When the latch is in the reset state, the Q output will be LO and the Q NOT output will be HI. If a LO signal is sent to both the *S* and *R* inputs, the outputs of the latch become unpredictable and this state is called *invalid.* The *RS* latch is used in many memory applications. Since it has the ability to latch, it is also used as a *bounce eliminator* or in circuits where electromechanical switches are used as inputs to logic circuits. If the contacts bounce when the switch is closed, the logic device will only see the first contact closure.

Another application of the *RS* latch is a control circuit used in sonic-welding or spot-welding systems where multiple welders are used and the amount of current is limited so that only one of the systems may be operated at a time. For example, if the system has two welding heads, a set of latches can ensure that if both systems are energized at the same time, the first one that transitions to the on-state will set its latch and lock out the other's latch. When the first welder's cycle is complete,

FIGURE 2-1 (a) *RS* latch. **(b)** NAND gate logic for *RS* latch. **(c)** Truth table for *RS* latch.

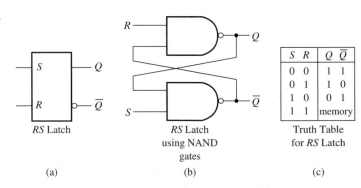

S R	Q $\overline{Q}$
0 0	1 1
0 1	1 0
1 0	0 1
1 1	memory

RS Latch

RS Latch using NAND gates

Truth Table for *RS* Latch

(a)

(b)

(c)

it will enable the latch of the second so its weld can proceed. Since each weld lasts for only 2–3 seconds, the latch circuit can control the systems so that only one fires at a time and allows the second to energize as soon as the first has completed its cycle.

2.2.2 Edge-Triggered Flip-Flops

The *edge-triggered flip-flop* is a synchronous bistable device that only changes state when a specific point on a clock signal occurs. The edge-triggered flip-flop is called *synchronous* because the data available on its inputs are transferred to its outputs when the edge of the triggering clock pulse is present. This makes the edge-triggered flip-flop different from the latch in that it only changes state when the proper clock pulse is received at its clock input, and the latch can change state any time the correct sequence of input signals transitions. Edge-triggered flip-flops include the *R-S* flip-flop, the *D*-type flip-flop and the *JK*-type flip-flop. The *R-S* flip-flop is also called the *S-R* flip-flop.

The edge-triggered flip-flop is available as a positive edge-triggered or a negative edge-triggered device. A positive edge-triggered flip-flop is pulsed by the leading edge of the clock pulse called the positive edge, and a negative edge-triggered flip-flop is pulsed by the trailing edge of the clocked pulse called the negative edge. The positive edge-triggered flip-flop will read its inputs when the clock pulse transitions from LO to HI, and the negative edge-triggered flip-flop will read its inputs when the clock pulse transitions from HI to LO.

Fig. 2-2a shows a positive edge-triggered *RS* flip-flop, the positive edge-triggered *D* flip-flop, and the positive edge-triggered *JK* flip-flop. Fig. 2-2b shows a negative edge-triggered *RS* flip-flop, the negative edge-triggered *D* flip-flop, and the negative edge-triggered *JK* flip-flop. In these figures you should notice the small triangle at the clock input that indicates this is an edge-triggered device. A small circle will be placed in front of the triangle for all of the flip-flops in Fig. 2-2b to indicate the flip-flop is negative edge-triggered.

2.2.3 The Edge-Triggered *RS* Flip-Flop

Fig. 2-3 shows the truth table for the positive edge-triggered *RS* flip-flop. Here you can see that the positive edge-triggered flip-flop will not change the output state when both *R* and *S* inputs are LO when the clock pulse occurs. If the *S* input is LO and the *R* input is HI when the clock pulse occurs, the output will be LO. If the *S* input is HI and the *R* input is LO when the clock pulse occurs, the output

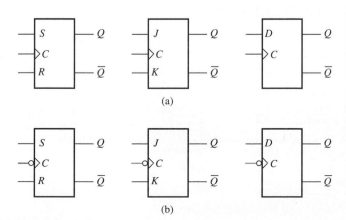

(a)

(b)

FIGURE 2-2 (a) Symbols for the positive edge-triggered *RS* flip-flop, the positive edge-triggered *JK* flip-flop, and the positive edge-triggered *D* flip-flop. **(b)** Symbols for the negative edge-triggered *RS* flip-flop, the negative edge-triggered *JK* flip-flop, and the negative edge-triggered *D* flip-flop.

Inputs			Outputs		
S	R	C	Q	$\overline{Q}$	Comments
0	0	X	Q_0	$\overline{Q_0}$	No Change
0	1	↑	0	1	RESET
1	0	↑	1	0	SET
1	1	↑	?	?	Invalid

FIGURE 2-3 The truth table for the edge-triggered *RS* flip-flop.

will be HI. If both the S and R inputs are HI when the clock pulse occurs, the output state is invalid.

2.2.4 The Edge-Triggered *JK* Flip-Flop

Perhaps the most widely used flip-flop, the *JK* flip-flop is similar to the *RS* flip-flop but has no invalid states. This means that all conditions of the outputs can be predicted for every set of inputs. The letters *J* and *K* do not represent any condition such as the *R* and *S* for reset and set. Fig. 2-4 shows the truth table for a positive edge-triggered *JK* flip-flop where you can see that the flip-flop has no invalid states and its conditions are similar to the *RS* flip-flop. If the *J* and *K* inputs are both LO when the positive clock pulse is received, the Q output will not change state. If the *J* input is LO and the *K* input is HI when the positive clock pulse is received, the Q output will be LO and the $\overline{Q}$ will be HI. If the *J* input is HI and the *K* input is LO when the clock pulse is received, the Q output will be HI and the $\overline{Q}$ output will be LO. If the *J* and *K* inputs are both HI when the clock pulse is received, the output will *toggle,* which means that if the Q output is HI it will go LO and if it is LO it will go HI.

From the diagram in Fig. 2-4 you can see that the output conditions are described as the *reset, set,* and *toggle* modes. The reset and set conditions are exactly like the *RS* edge-triggered flip-flop, except when both inputs are HI the output will toggle when a clock signal is received. For example, if both the *J* and *K* inputs are HI and the Q output is HI, it will transition to the LO state when the clock pulse occurs. If the two inputs remain high, the Q output will transition from the LO state to the HI state when the next clock pulse occurs. This type of flip-flop circuit is useful in applications where the output must be alternately HI and then LO when both inputs are HI.

2.2.5 The Edge-Triggered *D* Flip-Flop

The edge-triggered *D* flip-flop stores the single data bit that is present at the *D* input on its Q output when the clock pulse is received. This means that if a HI signal is present at the *D* input when the clock pulse occurs, the flip-flop will set

Inputs			Outputs		
J	K	C	Q	$\overline{Q}$	Comments
0	0	↑	Q_0	$\overline{Q_0}$	No Change
0	1	↑	0	1	RESET
1	0	↑	1	0	SET
1	1	↑	$\overline{Q_0}$	Q_0	Toggle

FIGURE 2-4 The truth table for the positive edge-triggered *JK* flip-flop.

Inputs		Outputs		
D	C	Q	$\overline{Q}$	Comments
0	↑	1	0	SET
1	↑	0	1	RESET

FIGURE 2-5 The truth table for the positive edge-triggered *D* flip-flop.

and the *Q* output will become HI. If a LO signal is present at the *D* input when the clock pulse arrives, the flip-flop resets and the *Q* output will become LO. The output will remain in its present state until a clock pulse occurs. Fig. 2-5 shows the truth table for the positive edge-triggered *D* flip-flop.

2.2.6 Using Flip-Flops to Cycle Two Air Compressors

Fig. 2-6 shows an example of using flip-flops for industrial control circuits. This circuit shows a positive edge-triggered *JK* flip-flop used to cycle two air compressors. In many industrial applications air supply systems use two small air compressors instead of one large one. This allows one to be the backup if a failure occurs. Since air is used to operate tools, move robots, and open and close clamps and fixtures with air cylinders, it is important that the air source is not interrupted. Since two compressors are used, it is also important that one does not work harder than the other. A flip-flop can be used in the control circuit of the two compressors to ensure that each compressor runs approximately the same amount of time.

When the pressure in the air tank drops below 30 psi, pressure swith PS1 closes and sends a signal to the clock input of the flip-flop. Since both the *J* and *K* inputs are held HI, the flip-flop will toggle the *Q* and the $\overline{Q}$ outputs. The coil (MS1) for the motor starter that controls air compressor 1 is connected to the Q output and the coil (MS2) for the motor starter that controls air compressor 2 is connected to the $\overline{Q}$ output. During the first cycle, the *Q* output will be HI and the coil of motor starter 1 (MS1) will be energized. When the compressor builds up the pressure to 60 psi, pressure switch PS1 will open. When the air pressure drops below 30 psi, pressure switch PS1 will close and provide another clock pulse. This time when the clock pulse occurs, the outputs will toggle and the *Q* output will be LO and the $\overline{Q}$ output will be HI, which will energize the motor starter coil (MS2) for the second air compressor. Each time pressure switch PS1 opens and closes, the outputs of the flip-flop will toggle. This will cause both air compressors to run approximately the same amount of time so that neither of them has to do all of the work. Pressure switches PS2 and PS3 are safety switches that will turn off the air compressors if the air in the system exceeds 90 psi, which is an unsafe condition.

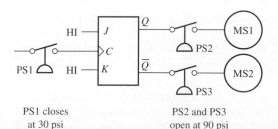

PS1 closes
at 30 psi

PS2 and PS3
open at 90 psi

FIGURE 2-6 A clocked *JK* flip-flop used to control two air compressors to ensure they run approximately the same amount of time.

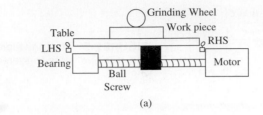

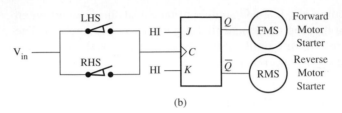

(b)

FIGURE 2-7 **(a)** A diagram of the surface grinder that shows the location of the left-hand switch (LHS) and right-hand switch (RHS). **(b)** Example of a clocked *JK* flip-flop used to control the back-and-forth action of the table of a surface grinder.

2.2.7 Using Flip-Flops to Control an Industrial Surface Grinder

Another application using the edge-triggered *JK* flip-flop is in a circuit to control the operation of a surface grinder. Fig. 2-7 shows a diagram of the surface grinder and the circuit that makes it operate. The surface grinder is used to move a piece of metal back and forth under a grinding wheel to remove small amounts of surface or to polish the surface. The piece of metal that is to be finished is clamped in a fixture that is mounted to a movable bed. If you are standing in front of the grinder table, the back-and-forth movement of the table will look like the table is moving from left to right, and when it reverses, the table will move from right to left. A motor is used to provide this movement as it rotates in the forward and then in the reverse direction.

A second motor is used to control the table's *index motion.* The index motion moves (indexes) the table at the end of each pass. The indexing action is similar to mowing your lawn. If you kept the same path each time you went back and forth across the yard, you would only mow one small swath across the yard. By moving the mower to the next row of grass each time, you ensure that you will cover the entire surface of the lawn.

When the surface of a flat piece of metal 3 inches wide and 15 inches long is polished, the table must move back and forth at least 20 inches so that the entire piece is covered. The amount of distance the table must be indexed at the end of each stroke is determined by the thickness of the grinder wheel. If a 1-inch wheel is used, the grinder table will be indexed approximately 3/4 inch each time to ensure a small amount of overlap occurs. This guarantees the entire surface is machined. The depth of cut can also be controlled by moving the table up and down or by moving the grinding wheel up and down so the grinding wheel cuts deeper into the surface. The depth of cut is usually adjusted after the grinding wheel passes over the entire surface.

For the first part of our example, we will discuss the flip-flop controlling only the back-and-forth motion of the table. A similar circuit would be used for each of the other two axes of motion.

In this diagram you can see the grinder has a limit switch mounted at each end to detect the travel of the table. The limit switches are connected in parallel (OR logic) to the clock input of the flip-flop. The *J* and *K* inputs are connected to a HI signal so that the flip-flop will stay in the toggle mode. You should also notice that the forward motor starter (FMS) is connected to the Q output, and the reverse motor starter (RMS) is connected to the $\overline{Q}$ output. (If you need more information about reversing motors, see Chapter 15.)

When the grinding cycle is started, the FMS is energized and the motor will turn clockwise, which moves the table from left to right. When the table moves the full amount of distance to the right, it will strike the *right-hand limit switch* (RHLS). When the table strikes the limit switch, the switch will transition from open to closed, which will provide the leading edge pulse for the clock signal.

When the flip-flop receives the clock pulse, it toggles and the $\overline{Q}$ output is energized and the Q output is de-energized. The $\overline{Q}$ output energizes the RMS, which makes the motor shaft turn in reverse rotation and causes the table to move back the other direction to the left. As the table moves to the left, it moves away from the RHLS, which allows it to transition to its open position again so that it is ready for the next cycle.

The table moves in the reverse direction (left) until it strikes the *left-hand limit switch* (LHLS). When the table moves against the LHLS, its contacts transition from open to closed. Since it is also connected to the clock input of the flip-flop, the flip-flop will toggle, which will cause the Q output to become energized and the $\overline{Q}$ output to be de-energized again. When the FMS is energized, it causes the table to move back to the right again. As the table moves to the right, it moves off of the LHLS, which allows it to transition to its open position making it ready for the next cycle.

Each time the table makes the travel in both directions, a pulse is provided to a timer that controls the amount of time the index motor for the table is energized. The index motor is usually energized for less than 0.1 second to provide approximately 3/4-inch travel. When the index moves the full amount of travel, it will strike a limit switch, which will provide the clock pulse to its flip-flop and the table index motion is reversed. When the part is indexed in both directions under the grinding wheel, a pulse is sent to a timer that controls the *depth-of-cut* motion. On this machine the depth of cut controls the travel of the grinding wheel down into the surface. When the depth-of-cut limit switch is activated, the machine turns off to indicate the end of the cycle.

2.2.8 Using Flip-Flops for Safety Circuits

Edge-triggered *JK* flip-flops are also useful in safety circuits where the safety switches or devices may be *jumpered out*. In many older industrial control circuits, it was a common practice to place jumper wires around guard safety switches, door switches, palm buttons, and other safety devices to get around their use. This practice created many accidents where operators or technicians were injured. In newer circuits, the edge-triggered *JK* flip-flop is used to ensure the safety device is toggled once each cycle. If the safety device must toggle once during each cycle, it is impossible for the machine to operate if the safety device has been jumpered out.

Flip-flops can also be used on *anti-tie-down* circuits that require the operator to have both hands on palm buttons to start a cycle. These types of safety circuits are specified on presses or other systems where operators may be able to get their hands near the moving parts of a machine when it is started. By requiring operators to have both hands on the palm buttons, it ensures that their hands are in a safe place. The flip-flop is used in toggle mode to ensure that one or both palm buttons are not tied down or jumpered out.

Flip-flops are also used in newer gas-heating systems where exhaust gases are removed with a blower motor. The system will not allow ignition to take place unless the exhaust blower is running and creating the correct amount of pressure drop across a pressure switch. One side of the switch is connected to check the draft (negative pressure from the blower) and the other side of the switch is used to detect the pressure side of the blower. If this switch is jumpered out after ignition is established, the heating system will continue the cycle it is on, but the flip-flop will disable the ignition circuit if the switch does not change positions when the blower stops at the end of the cycle. The flip-flop ensures that the pressure switch sees the higher pressure created by the blower motor at the start of the circuit and then sees the low pressure, which indicates the blower motor has shut off at the end of its cycle. If the switch has been jumpered out, it will not toggle the flip-flop and ignition will not occur. The pressure switch is used in the furnace to detect a faulty exhaust motor, and the flip-flop is designed to prevent anyone from tampering with the pressure switch.

2.3 COUNTERS

Counters are widely used in industrial electronic circuits. The counters are used as up counters, down counters, totalizing counters, reset counters, decade counters, encoders, and decoders. The first part of this section will discuss the chips that are used to provide these functions, and the last part of this section will discuss solid-state devices that are specifically designed for these purposes.

Up counters, down counters, and totalizing counters are used for simple functions such as counting the number of parts a machine has made. These counters can also be used as reset counters that provide an output when the counter reaches a predetermined count. For example, a palletizing machine may need to count the number of boxes as it fills layers on a pallet. Since each layer needs four boxes, the counter is set to 4. When the fourth box is counted, an output is sent from the counter. The output is used to index the machine to the next layer. This feature of the counter is called *reset,* since an output is energized when the stored value (count) reaches the preset value and the stored value is reset to zero.

In their simplest forms the counters may be used for a variety of logic functions: binary counters, flip-flop counters, binary coded decimal counters, decade counters, and cascaded counters. The examples in this text will begin with asynchronous and synchronous counters.

2.3.1 Asynchronous and Synchronous Counters

Counters can be classified as asynchronous or synchronous based on the type of clock pulse the counters use. The flip-flops in an asynchronous counter do not all change state at the same time, whereas all of the flip-flops in the synchronous counter are clocked simultaneously. The asynchronous counter is also called a *ripple counter,* since the signal sent to the input of the first flip-flop is sent through the remaining flip-flops like a wave that ripples outward when a stone is thrown into a pond. The most common types of asynchronous counters are the *four-bit binary counter* and the *decade counter.*

The four-bit binary counter uses four edge-triggered flip-flops to count values from 0–15 in a binary format. Fig. 2-8a shows the four positive edge-triggered *JK*

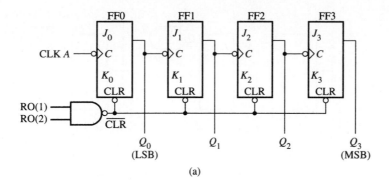

(a)

MSB			LSB	Decimal
Q_3	Q_2	Q_1	Q_0	Value
0	0	0	0	0
0	0	0	1	1
0	0	1	0	2
0	0	1	1	3
0	1	0	0	4
0	1	0	1	5
0	1	1	0	6
0	1	1	1	7
1	0	0	0	8
1	0	0	1	9
1	0	1	0	10
1	0	1	1	11
1	1	0	0	12
1	1	0	1	13
1	1	1	0	14
1	1	1	1	15
0	0	0	0	full reset

(b)

FIGURE 2-8 **(a)** Four flip-flops connected as an asynchronous counter. The output lines can count values 0–15 in binary format. **(b)** The binary bit pattern and equivalent decimal value for the asynchronous counter. (Reprinted by Permission of Texas Instruments.)

flip-flops connected together so that output Q_0 is connected to the clock of FF1. The output of Q1 is connected to the clock of FF2, and output Q_2 is connected to the clock of FF3. Output Q_3 represents the most significant bit (MSB) of the counter, and output Q_0 represents the least significant bit (LSB) of the counter. The table in Fig. 2-8b shows the binary bit pattern and the equivalent decimal value that would be read at the four outputs for each clock pulse of the counter. All of the *JK* inputs are tied HI so that the flip-flops will pulse with each signal sent to the clock input from the previous output. The counter can be reset (cleared) at any time by signals sent to the CLR terminal of each flip-flop.

The four-bit synchronous counter is shown in Fig. 2-9. From this diagram you can see that all four of the flip-flops have their clocks connected to the same line, which is pulsed (strobed) when the counter is incremented. The counter still has the ability to count values from 0–15. The terminals of the counter are identified as Q_0, Q_1, Q_2, Q_3, clock, and clear. Notice that the counter is reset after the value 15 is counted.

2.3.2 Binary Coded Decimal Counters

The binary coded decimal (BCD) numbering system uses a combination of four binary bits to represent decimal values 0–9. As you know, the four binary bits can

FIGURE 2-9 Four-bit synchronous binary counter. Four flip-flops are used to make this counter. All of the flip-flops change state with the common clock pulse (Reprinted by Permission of Texas Instruments.)

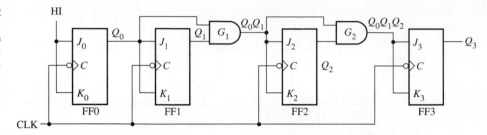

represent values 0–15, but since the decimal numbering system only uses values 0–9, the remaining values (10–15) do not need to be represented. Fig. 2-10a shows an example of the number 467 represented as a BCD number. From this figure you can see that the four bits that represent the hundreds column display the binary value of 0100, which is the decimal value 4. The second set of four binary bits is used to display values in the tens column, and it shows the binary value of 0110, which is the decimal value 6. The third set of four binary bits is used to display values in the ones column and it shows the binary value of 0111, which is the decimal value 7.

The BCD numbering system is used with many electronic devices such as *thumbwheels* that are based on the decimal numbering system (0–9). In the BCD counter the flip-flops for each set of four binary bits are used to count only values 0–9, and then the set of four flip-flops is reset to zero. A second set of four flip-flops is used for the digit (0–9) that represents the ten's column for the decimal number. A third set of four flip-flops is used to represent the digits (0–9) in the hundreds column of the decimal system. Since this type of counter counts ten values (0–9), it is also called a *decade counter*.

It is important to remember that the main feature of the BCD counter is that it changes to the reset value (0000) after it reaches the value for 9 (1001) rather than move to the binary value for 10 (1010). Each additional set of flip-flops in the BCD decade counter counts values that are ten times the value of the previous set. This provides the function of decimal weighted values such as one's, ten's, and hundred's columns for each set of flip-flops.

Figure 2-10b shows an example of a counter that uses three BCD counters to count decimal values 0–999. This type of counter combines three individual BCD

FIGURE 2-10 **(a)** An example of the BCD numbering system displaying the value of 467. **(b)** Three individual BCD counters are combined to make a three-digit BCD counter. The three-digit BCD counter can count decimal values from 000 to 999. (Reprinted by Permission of Texas Instruments.)

HUNDREDS	TENS	ONES	WEIGHTED DECIMAL COLUMNS
8 4 2 1	8 4 2 1	8 4 2 1	WEIGHTED BINARY VALUES
0 1 0 0	0 1 1 0	0 1 1 1	BCD VALUE
4	6	7	DECIMAL VALUE

(a)

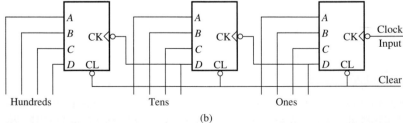

(b)

counters so that one counter counts values for the decimal one's column, a second counter counts values for the decimal ten's column, and a third counter counts values for the decimal hundred's column. This type of counter is called a three-digit BCD counter.

2.3.4 Up/Down Counters

Up/down counters are frequently used in industrial applications where values can be added or subtracted from one register. The up/down counter consists of a load register, a clock line, a reset line, and a max/min output line. The register is incremented (1 is added to its total) or decremented (1 is subtracted from its total) each time the clock line to the counter is transitioned. The initial value can be preset in the register using the four load lines. The counter then counts up or down until it reaches the maximum or the minimum value for the register. The max/min output of the counter goes true when the value in the register reaches either of these conditions. Some counters provide a second output for the full condition. The counter can be cleared by simply loading all 0's from the load inputs. The up/down input determines the direction the counter counts.

 The up/down counter is generally used to keep track of good and bad parts in a production line. Each time a machine makes a part, the up-count input line is transitioned so that the count is incremented. After the parts are inspected, they are sorted and classified as good parts or bad parts. Each time a bad part is detected, the down-count input line is transitioned and the counter will decrement the total. At the end of the shift, the value in the counter register will be the total number of good parts. Another way of thinking about the value in the counter register is to think of it as the total number of parts that were made minus the number of bad parts. If the total number of parts manufactured and the total number of bad parts need to be counted separately, two additional up counters can be used to count these specific parts. It is also possible to cascade one counter to another if larger values are required to be counted.

2.4 SHIFT REGISTERS

Another useful application of flip-flops involves connecting several of them together to form a *shift register*. Shift registers are available as self-contained circuits on a single chip. It is important to understand the circumstances that require the use of a shift register before you can fully understand its operation. A simple example is shown in Fig. 2-11. This application is a painting line for small wooden toys. The toys are moved on a continuous line past the painting nozzles. The spacing of each toy on the line is maintained by an auger system. When the toys move into the painting booth, they pass four painting nozzles. The first nozzle sprays a sealer coat, the second sprays a primer coat, the third sprays a color coat, and the fourth sprays a clear coat as a finish. The line moves the wooden toys continuously past each nozzle and the nozzles are timed so that they are energized to spray paint on each toy as it passes directly in front of a particular nozzle. Since the line runs continuously, the number of toys painted each hour is considerably higher than if the line was jogged and stopped each time a toy was in front of a nozzle. The only problem that arises with this process is when a toy is not present on the line. If a toy has

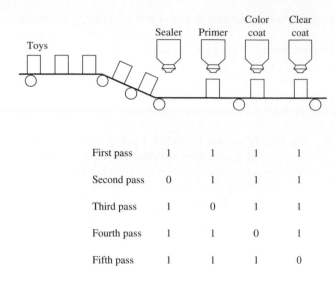

		Sealer	Primer	Color coat	Clear coat
First pass		1	1	1	1
Second pass		0	1	1	1
Third pass		1	0	1	1
Fourth pass		1	1	0	1
Fifth pass		1	1	1	0

FIGURE 2-11 A shift register is used to control four painting nozzles in a toy-painting application. The toys move past the nozzles on a continuous line, and if a toy is missing, the shift register de-energizes each nozzle as the blank space moves past it. If a toy is present on the line, the shift register energizes each nozzle as the toy moves in front of it. A table is provided at the bottom of the figure to show the bit pattern in the shift register to indicate when a toy is missing from the production line (1 indicates toy is present; 0 indicates toy is missing).

fallen off the line prior to entering the spray booth, or if there is a gap in the line of toys, the sprayer nozzle would waste paint trying to paint the space where the toy should have been.

The shift register is used to keep track of whether a toy is present or absent in each space on the line. A sensor is mounted at the point where the toys begin to enter the spraying area. The information from the sensor is stored as a 1 if a part is present and as a 0 if a part is missing, and it is passed to each spray nozzle as the line moves. If a toy is present, a 1 is passed to the solenoid that energizes the nozzle, and if a toy is not present on the line, a 0 is passed to the nozzle solenoid, which will ensure that the nozzle remains de-energized. Since an auger maintains exact spacing as the toys enter the spray booth, it can also be used to provide the clock pulse for the shift register. A photoelectric switch is used as the sensor to determine if a toy is present or absent on the line.

If a condition occurs where the parts must be cleared off the production line in the painting booth, the shift register must be reset. In the reset condition, 0's are placed in each output to ensure all of the spray-painting heads are turned off. The reset condition is accomplished by sending a signal to the reset line of the shift register.

2.4.1 Operation of a Shift Register

The shift register must have a *clock input* and a data *input* to operate correctly. The number of shift locations will depend on the application. In the spray-painting example, there are four nozzles so the shift register has four bit-shift locations. Each bit of the bit shift register energizes one output. The data input is the photoelectric switch that senses whether or not a toy is present on the line. Since the auger that provides the spacing for the toys on the line is chain driven by a sprocket and gear, a proximity switch can sense the tip of each tooth of the gear and determine the amount of movement of the toys on the line. The signal from the proximity switch is used as the clock input for the shift register. It is important for the clock input to be referenced to the spacing of the toys so that the data bits are shifted to the output solenoids to energize them at just the right time when the toy is in front of the spray-painting nozzle.

The example starts with the line running and a toy is present in each position. The table in the bottom of Fig. 2-11 shows the bit pattern in the shift register as the toys pass to be painted. A 0 is used to represent the space where a toy is missing and tracked through the system and a 1 is used to show a toy is present. The first line of the table indicates a toy is present in front of each nozzle. Notice the bit pattern is 1111. Each output that has a 1 activates the spray nozzle connected to it.

The second line in the table shows the bit pattern when the second clock pulse occurs. Since a toy is missing at the entrance of the paint booth when the second clock pulse occurs, the sensor that is connected to the data input senses this and places a 0 in the first position of the shift register. The clock pulse is provided when the auger moves the toys one position. Since a toy is missing in the first position, the bit pattern in the shift register is 0111, so the nozzle for the sealer is de-energized and no sealer is sprayed. Since the other three positions have toys present, the shift register has 1's representing their positions and those nozzles are energized.

In the next step of the cycle, you can see that the line is moved once more, which moves the empty space in front of the second nozzle. and the 0 that represents the missing toy is shifted to the primer nozzle. The bit pattern for this step is 0100. This means that the primer nozzle is de-energized while the other three nozzles are spraying. You should also notice that since the space directly behind the missing toy is now filled with a toy, the shift register will place a 1 in this position to energize the sealer nozzle. The 0 that represents the missing toy will continue to be shifted one space to the right each time the toys are moved by the auger. This ensures that the 0 in the shift register will follow the empty space on the line. This is shown in the table as you follow the remaining clock pulses.

2.4.2 Using Edge-triggered *JK* Flip-Flops to Make a Shift Register

Fig. 2-12 shows a shift register circuit made from edge-triggered *JK* flip-flops. This circuit shows four flip-flops that all have their clocks tied to a common line, which is connected to the sensor in the system that will act as the clock. For example, in the painting circuit this line is connected to the sensor that detects the movement of the auger to ensure exact spacing of the toys on the line. The output for each flip-flop is connected at each *Q* terminal. The *data in* line is connected to the sensor that detects whether or not a toy is present. This line is also connected to the *J* input of the first flip-flop. An inverter also sends an inverted copy of the data (1 or 0) value to the *K* input of the first flip-flop. Each time the clock is transitioned, the contents (1 or 0) of the first flip-flop are sent to the second flip-flop, and the contents (1 or 0) of the second flip-flop are sent to the third, and so on. If the first flip-flop is storing a 1, its *Q* output will send a 1 to the next flip-flop and to the first output. Since each output device, such as the solenoid for the painting nozzle is connected to the *Q* output of each flip-flop, the output device will be energized when the flip-flop output is a 1, and it will be de-energized when the flip-flop output is a 0.

A number of different types of shift registers are available to provide bit shift functions for a variety of applications. Fig 2-13 shows some of the different types of bit shift registers.

FIGURE **2-12** Four
positive edge-triggered
JK flip-flops connected
together to provide a
shift register. The clocks
of each flip-flop are connected together and the
Q output of each flip-flop
is connected to the *J* input of the next. The value
(1 or 0) that is stored in
each flip-flop is shifted
from its output to the input of the next flip-flop
each time the clock is
pulsed. (Reprinted by
Permission of Texas In-
struments.)

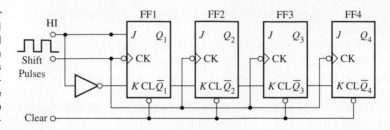

2.4.3 Using Shift Registers to Provide Time Delay

Shift registers are also used in industrial circuits to provide accurate time delay at the microsecond or nanosecond time base. One example of this application is in high-speed aluminum can manufacturing. When aluminum beverage cans are manufactured, they are also inked with the product logo. This process occurs on a continuous line as the cans are made. If the can does not receive the proper amount of ink at the proper location, the can will not be acceptable. A detection system is used to determine if a can is properly inked as it leaves the printing machines. Since the cans move through the production equipment at very high speeds (up to 1500 cans a minute), it becomes difficult to remove the bad can after it has been detected. A *blow-off system* uses high-pressure air to blow the bad can off the line as it passes by. The problem arises in timing the blow-off apparatus so that it removes only the bad can. The blow-off system requires several milliseconds to energize a solenoid valve and get enough air pressure to the nozzle to remove the can.

The shift register is used to provide an accurate amount of time delay to align the position of the bad can with the nozzle at the exact time when the air comes out of the nozzle. This shift register is slightly different from the one used in the spray-painting application in that it only needs one output, which is provided by the *Q* output of the last flip-flop. Since a bit is advanced through one flip-flop each time the clock is transitioned, the number of flip-flops in the shift register will represent the amount of time delay that occurs for the bit to be moved from the input of the first flip-flop to the output of the last flip-flop. By adding or subtracting flip-flops (bit positions) in the shift register, the amount of time delay can be accurately adjusted.

2.4.4 Word Shift Registers

In some applications it is important to shift an 8-bit word or 16-bit word that represents a number rather than a single bit that indicates something should be turned on or off. An example of this type of logic circuit is shown in Fig. 2-14 where a painting system becomes more complex. In this application a set of complex painting robots is used to apply primer, base coats, and clear coats to automobiles on an assembly line. The simple shift register is used to indicate when an automobile is present or not, but a more complex *word shift register* is used to pass the number that represents the paint color to each robot. Some base colors may require a red colored primer coat, while lighter colors require a gray primer coat. The robot that provides the base color can change colors for each automobile as it passes if necessary. New painting technology provides several means to ensure that the nozzle is clean as it begins to spray a new color. The robot must be told which color to apply so that it can energize the proper solenoid.

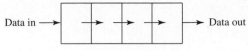

(a) Serial in–shift right–serial out

(b) Serial in–shift left–serial out

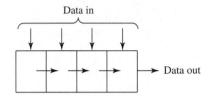

(c) Parallel in–serial out

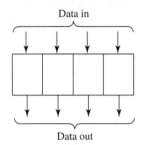

(d) Serial in–parallel out

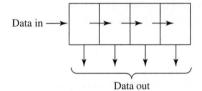

(e) Parallel in–parallel out

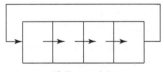

(f) Rotate right

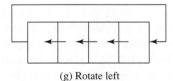

(g) Rotate left

FIGURE 2-13 **(a)** Shift register that shifts bits from left to right in a serial format. **(b)** Shift register that shifts bits from right to left in serial format. **(c)** Shift register that provides parallel (word) data in and serial data out. **(d)** Shift register that provides serial data in and parallel (word) data out. **(e)** Shift register that provides parallel (word) data in and parallel (word) data out. **(f)** Shift register that rotates one bit to the right. Notice the output of the shift register is tied back to the input. **(g)** Shift register that rotates one bit to the left. Notice the output of the shift register is tied back to the input. (Reprinted by permission of Texas Instruments.)

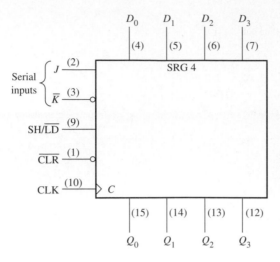

FIGURE 2-14 Word shift register used to transfer values representing paint colors from station to station in a robotic painting line. The numbers are used to indicate the proper paint color a robot should use on the automobile body as it passes on the assembly line. (Reprinted by Permission of Texas Instruments.)

In this application, the color of paint for each automobile is specified when the car is ordered. The number that represents the paint color will accompany the body of the car as it is assembled and moved through the plant. When the car reaches the painting robots, the number for its paint color is entered into the data input line of the shift register, and it is passed to each stage of the painting system so that each robot sprays the correct color of paint.

Each stage of the word shift register has the number of bits to represent the number. You should remember from other courses in digital electronics that you can represent values 0–15 with four bits, and values 0–65565 with sixteen bits. The word shift register can use any number of the 0–15 bits provided to represent the values for the paint numbers.

2.5 FIRST-IN, FIRST-OUT (FIFO) SHIFT REGISTERS

Another example of using logic to store data and retrieve it is called a *first-in, first-out register* (FIFO). FIFO is used to store fault codes. In many complex automated industrial systems, sensors are used to detect a number of faults such as loss of voltage, excessive current, low current, loss of air pressure, loss of motion, and overtemperature. Any one of these conditions may occur and cause the machine to shut down, which will cause other conditions to be sensed. For example, if the sensors detected an overtemperature condition, the system would be shut down and the loss of motion, low current, and loss of voltage sensors may also be affected. When the technician arrives to troubleshoot the system, multiple fault lights would be illuminated, which may make detecting the problem confusing. Fig. 2-15 shows examples of FIFO used in a fault detection system, and Fig. 2-16 shows a diagram of a typical 16-bit FIFO chip. Advancements in technology have provided FIFO chips that have typical stack sizes of 1024 words.

A FIFO register could help this problem by assigning each fault condition a number and storing the fault numbers in a stack (register) as they occur. If the

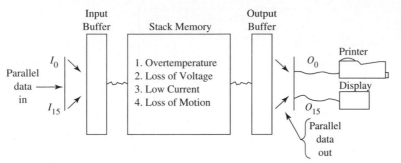

FIGURE 2-15 Example of the first-in, first-out register being used as a fault detection system. Each fault has a fault code number that is placed in the FIFO register in the order that the fault occurs. The printer and display are used to show the order that the faults occurred. This example shows the fault as the name rather than the number that represents the code.

overtemperature condition occurred first, its fault code would be placed in the register first. When the machine begins to shut down in response to the overtemperature and switches are opened and a solenoid de-energizes and shuts off the air, each of these conditions will be recorded in its order of occurrence by placing the fault code in the register and moving the previous code down the stack. When the technician arrives at the machine, the stack can be inspected and each fault code can be noted in the order in which it occurred. In fact most systems use the second part of FIFO to provide the *first-out* capability. The first-out part of FIFO ensures that the fault codes are moved to a printer or display in the order they were recorded.

If the loss of motion occurred prior to voltage being shut off, the technician would know to check for a broken belt or shaft on a motor or a bad gear in a transfer case. If the loss of motion occurred after the loss of voltage, the technician would know that this is a common condition that occurs after voltage has been de-energized by the main relay, and it would not represent a problem. If low current or excessive current is detected, the technician would know that the problem may be in the motor load such as a conveyor system. If excessive current is detected, the system may need lubrication, or the conveyor may be overloaded. If low current is detected, it may be a sign of slippage in belts, gears, or shafts.

2.5.1 The Johnson Counter

The Johnson counter is a shift register with a serial $\overline{Q}$ output that is connected back to its serial input to reset the counter after a predetermined counting sequence. Fig. 2-17a shows a four-bit Johnson counter using four edge-triggered flip-flops that are all pulsed from the same clock signal. Fig. 2-17b shows the bit sequence for this Johnson counter. As you can see, the four-bit Johnson counter can count a total of eight states to provide values 0–7. In the sequence you can see that the shift register is empty when the count begins and all flip-flops are reset to provide the four-bit value of 0000. During the next step when the clock pulse is provided at the D input, the first flip-flop will be set HI. The set of four flip-flops will now show the value of 1000. When the next clock pulse is provided, flip-flop FF0 sends its value of 1 to FF1, so the value of the four flip-flops is 1100.

During the next step when the clock pulse is received, the original two bits are shifted to FF2 and the value of the four flip-flops is 1110. In the next step when the clock pulse is received, the bits are all shifted to the right so the value of the four flip-flops is now 1111.

At this point each flip-flop is set and has the value of 1, so when the next clock pulse is received FF3 sends its $\overline{Q}$ output (0) back to FF0 so that the four flip-

FIGURE 2-16 Diagram
and pin outline of typical
FIFO chip. (Reprinted by
Permission of Texas In-
struments.)

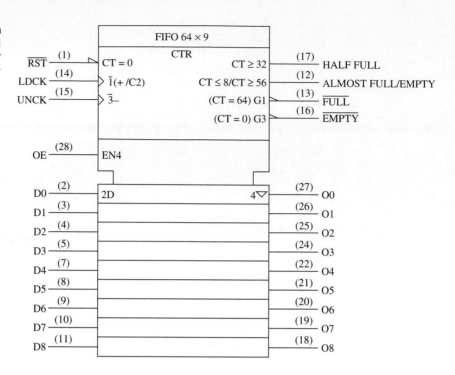

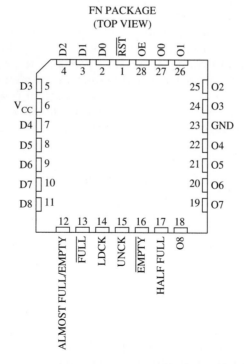

FN PACKAGE
(TOP VIEW)

flops show a value of 0111. During the next steps, each flip-flop is reset to 0 in sequential order until the value of 0001 occurs. At this point the next time the shift register is pulsed, the count starts over again.

From this example you can see that the four-bit Johnson counter counts values 0–7. If a fifth flip-flop is added, the Johnson counter could count values 0–9, and each additional flip-flop in the counter allows the count to move up by a total of two.

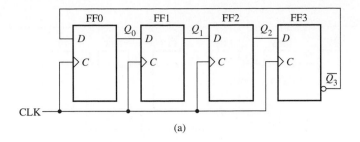

(a)

Clock	Q_0	Q_1	Q_2	Q_3
0	0	0	0	0
1	1	0	0	0
2	1	1	0	0
3	1	1	1	0
4	1	1	1	1
5	0	1	1	1
6	0	0	1	1
7	0	0	0	1

(b)

FIGURE 2-17 (a) Four edge-triggered *D* flip-flops are connected to form a Johnson counter. Notice the *Q* output of the flip-flop is connected to the *D* input of the first flip-flop. **(b)** The sequence table for the four-bit Johnson counter. (Reprinted by Permission of Texas Instruments.)

2.5.2 The Ring Counter

The ring counter is similar to the Johnson counter except the Q output of the last flip-flop rather than the Q NOT is sent back to the first flip-flop. The interstage connections and the clock arrangement are the same as the Johnson counter so that the value of each flip-flop is shifted from left to right through each of the flip-flops. Fig. 2-18a shows a ten-bit ring counter that uses ten flip-flops. Fig. 2-18b shows the sequence table for the ten bits.

From this example you can see that when a 1 is presented to the first flip-flop, it will be shifted through all of the flip-flops in sequence as the clock is pulsed, which allows the counter to count one decimal value for each flip-flop. The ring counter does not require a decoding gate for decimal conversion, since there is an output for each of its each decimal numbers.

2.5.3 The Gray Code Counter

The gray code counter provides values that represent the gray code, which is a nonarithmetic system that does not have weights assigned to each bit. Instead it shows only a single bit change from one code number to the next. This code is used primarily in shaft position encoders in robots and other types of motion control. Encoders are covered in Chapters 10 and 11. The problem with binary code and binary coded decimal systems is that more than one bit may change from a 0 to 1 or from a 1 to a 0 at the same time. This causes a problem in devices such as shaft encoders when the encoder stops at a position that is on a line between two adjacent sectors. Since the encoder stops on the line rather than exactly on a sector, it must judge which sector to select. If the problem occurs between values 1–2 or 4–5, it does not cause a lot of harm, but if the value of the sectors is near to the most significant values, the error is multiplied by the value weight of the two sectors.

The gray code eliminates this problem by changing the state of only one bit at a time as the value increases or decreases. Fig. 2-19 shows an example of the

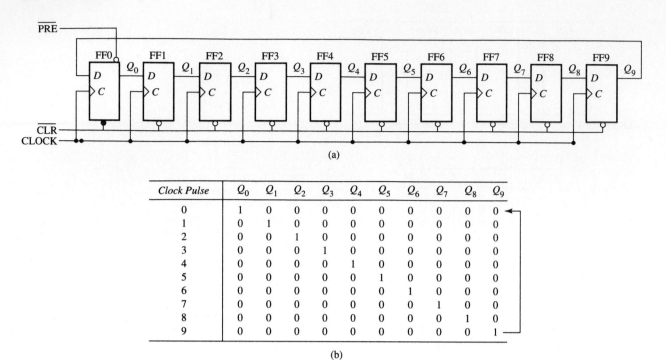

(a)

Clock Pulse	Q_0	Q_1	Q_2	Q_3	Q_4	Q_5	Q_6	Q_7	Q_8	Q_9
0	1	0	0	0	0	0	0	0	0	0
1	0	1	0	0	0	0	0	0	0	0
2	0	0	1	0	0	0	0	0	0	0
3	0	0	0	1	0	0	0	0	0	0
4	0	0	0	0	1	0	0	0	0	0
5	0	0	0	0	0	1	0	0	0	0
6	0	0	0	0	0	0	1	0	0	0
7	0	0	0	0	0	0	0	1	0	0
8	0	0	0	0	0	0	0	0	1	0
9	0	0	0	0	0	0	0	0	0	1

(b)

FIGURE 2-18 **(a)** Ten flip-flops connected to create a ring counter. Notice the Q output of the last flip-flop is fed back to the D input of the first flip-flop. **(b)** The sequence table for a ten-bit ring counter. (Reprinted by Permission of Texas Instruments.)

Decimal Value	Gray Code Value	Binary Code Value
0	0000	0000
1	0001	0001
2	0011	0010
3	0010	0011
4	0110	0100
5	0111	0101
6	0101	0110
7	0100	0111
8	1100	1000
9	1101	1001
10	1111	1010
11	1110	1011
12	1010	1100
13	1011	1101
14	1001	1110
15	1000	1111

FIGURE 2-19 Gray code and binary code equivalents for decimal values 0–15.

four-bit values of the gray code and compares it with the traditional binary code for the corresponding decimal values 0–15. From this figure you can see that the gray code and the binary code for the decimal number 0 are both 0000, and for the decimal number 1 they are both 0001. When number 2 is displayed, you should notice that in the binary system the value in the 1 column changes to a 0 and the value in the 2 column changes to a 1. In the gray code, only the value in the second-bit column changes, while the value 1 in the first-bit column remains the same.

You can see in the binary system when the decimal number changes from a 3 to a 4, the values in the 1, 2, and 4 columns all change. In the gray code, only the value in the third binary column changes. When the binary system changes from the value 7 to 8, the bits in all four columns must change, and in the gray code only the bit in the fourth binary column changes. As the numbers in the binary system increase, more columns must change as the numbers increase or decrease, which increases the possibility of the encoder landing between more sectors. With the gray code, the largest possible error is one bit, and since the values are not weighted, this reduces the total encoder error more significantly.

2.6 TIMERS

Timers are used in a variety of applications in industrial electronic circuit boards and as stand-alone timers. This section will explain the simple one-shot timers and 555 timers, and the next section will explain stand-alone timers. The one-shot timer is a simple timer that can be made by modifying the basic edge-triggered flip-flop.

2.6.1 One-Shot Timers

Fig. 2-20a shows the logic symbol for a one-shot timer, and Fig. 2-20b shows a one-shot timer circuit with an external resistor and capacitor. The one-shot timer is useful because it will take an input pulse of any duration and provide one output (pulse) that lasts for a predefined time. This means a limit switch could provide the input and it could stay closed for 2 seconds, and the one-shot timer could produce an output pulse of 2 milliseconds (0.002 seconds). The amount of time the pulse stays on can be adjusted by changing the values of the resistor and capacitor that are connected to the external terminals of the flip-flop.

The one-shot timer is called a monostable multivibrator, since it has a single stable state. When a trigger pulse is applied to the input of the flip-flop, the capacitor

FIGURE 2-20 **(a)** The logic symbol for a one-shot timer. **(b)** A one-shot timer with an external resistor and capacitor. (Reprinted by Permission of Texas Instruments.)

begins to charge. The rate of charge is controlled by the size of the resistor. When the voltage charge on the capacitor reaches the proper amount, it will cause the flip-flop to transition and produce an output. The formula for setting the output pulse width (t_W) is determined by the values of the external resistor (R_{EXT}) and external capacitor (C_{EXT}). The following formula shows the relationship for the resistor and capacitor.

$$t_W = 0.7\,R_{EXT}\,C_{EXT}$$

$$R_{EXT} = t_W/0.7C_{EXT}$$

or

$$C_{EXT} = t_W/0.7R_{EXT}$$

The one-shot timer is available as a nonretriggerable device or as a retriggerable device. The retriggerable device has a clear input that can be used to retrigger the one-shot timer. The one-shot pulses have a variety of applications for industrial systems. The most usable application is to take the unpredictable operation of the contacts of a field device, such as a limit switch or other switch, and provide a predictable and reliable timed pulse.

2.6.2 The 555 Timer

The 555 timer is widely used in industrial circuits. Fig. 2-21 shows the block diagram of a 555 timer. The inputs include *threshold, trigger, control voltage, discharge,* and *reset.* External capacitors and resistors are used to configure the timer as a *monostable multivibrator,* which can produce a one-shot output, or as an *astable multivibrator,* which will produce a train of pulses as an oscillator. Fig. 2-22a shows a 555 timer as a monostable multivibrator (oscillator) and Fig. 2-22b shows a 555 timer as an astable multivibrator (oscillator).

When the 555 is connected in the monostable (one-shot) mode, it has an external resistor that is connected between V_{cc} and the discharge and threshold terminals. The external capacitor is connected in series with this resistor and ground. This sets up the *RC* time constant for the input. When the 555 is connected as an astable oscillator, the input sees a waveform that is set by the charging and discharging of the C_1 capacitor. Since the capacitor will continue to charge and discharge as long as an input voltage signal is applied, the output will also continue to provide a square wave (on/off) output waveform. The duration of the on and off cycles of the square wave output is determined by the size of the external resistors and

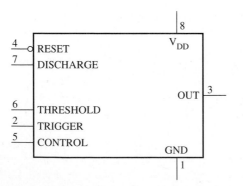

FIGURE 2-21 Block diagram of a 555 timer. (Reprinted by Permission of Texas Instruments.)

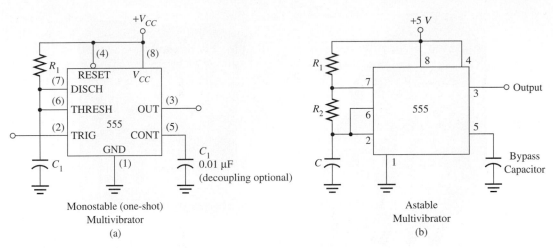

FIGURE 2-22 **(a)** Diagram of the 555 connected as a monostable (one-shot) device. **(b)** Diagram of the 555 connected as an astable oscillator. (Reprinted by Permission of Texas Instruments.)

capacitors. The 555 is used in a variety of industrial timers to provide the clock pulse or time base for the clock because the output is well regulated.

2.7 INDUSTRIAL TIMERS

At first glance it may seem inappropriate to discuss industrial timers in a chapter for advanced solid-state logic. But upon closer examination, you will begin to notice that the same timer circuits that were previously discussed are typically found inside industrial timers. The major difference is that timers that are used in industrial applications must be designed to be easy to operate by someone who has little or no knowledge of solid-state logic. The person who adjusts the time setting for an industrial timer only understands the terms *time delay-on* and *time delay-off*. It is also important for students who are studying to be industrial electronics technicians to understand that a completely different set of symbols is used for industrial timers. Since these symbols predate the electronics age, it may be of help to understand the early pneumatic timer and synchronous motor timers that the solid-state timers are replacing. It will be much easier to understand the solid-state circuits used in industrial timers if you understand the functions of time delay-on and time delay-off in industrial applications. Several industrial applications will be provided that use time delay-on and time delay-off symbols.

The second feature of industrial timers is that they should be able to handle the high temperatures and strong vibrations present on the factory floor. This means additional hardware may be added to the digital circuits you may be used to working with. Industrial timers must also be capable of being interfaced to relay or solid-state logic circuits that require 24 volts ac or 120 volts ac.

This section will explain the four basic types of operators that provide time delay for industrial timers. The time-delay operators can be classified as pneumatic (air chamber), motor and cam, solid-state which may use simple time-constant

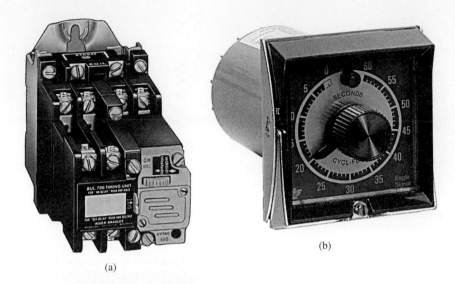

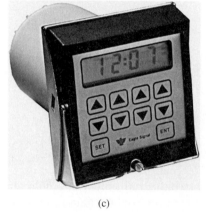

FIGURE 2-23 **(a)** A pneumatic industrial timer. **(b)** A motor-driven industrial timer. **(c)** A solid-state industrial timer. (a. Courtesy of Rockwell Automation's Allen Bradley Business; b. and c. Courtesy of Daraher Controls.)

circuits or more complex circuits like the 555 timer, or programmable timers that use microprocessors. The operation of each of these types of timers will be explained in this section with examples of common applications. Fig. 2-23 shows examples of industrial timers.

The electrical symbol for time-delay devices uses the head or tail of an arrow. The tail of the arrow is used to indicate time delay-on and the head of the arrow pointing down is used to indicate a time delay-off function. Fig. 2-24 shows examples of the time delay-on and the time delay-off symbols.

2.7.1 On-Delay Timers

Industrial timers have contacts and an operator that cause the time delay. On-delay timers provide the function of changing the position of contacts after power is energized to the operator of the timer. Timers are widely used in industrial applica-

TIME DELAY-ON   TIME DELAY-OFF

FIGURE 2-24 Examples of time delay-on and time delay-off symbols.

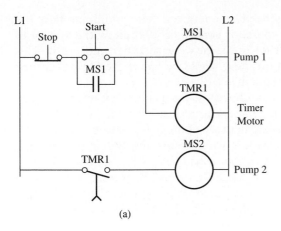

(a)

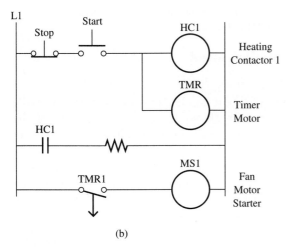

(b)

FIGURE 2-25 **(a)** Time delay-on timer used to delay the time two pump motors start to limit the effects of in-rush current. **(b)** Time delay-off timer used to delay the time a furnace fan stays on after the heating element is turned off. The additional time the fan runs allows extra heat to be removed from the heating chamber.

tions to control the starting time of larger motors so that they do not all try to start at the same time. When a motor starts, it will draw a larger current than when it is running at full speed. This starting current is called *locked rotor amperage* (LRA) or *in-rush current* and it is caused by the armature not turning when power is first applied to the motor. As the motor shaft begins to rotate, the current drops to normal levels as the motor comes up to full speed. If a machine has several large hydraulic pump motors, they will all try to start at the same time when power is turned on. The starting cycle of the motors can be staggered by using on-delay timers so that the effects of in-rush current will be minimized.

The on-delay timer may also be called a *delay-on timer*. When power is applied to the motors, the first motor is allowed to start normally, but power to the second motor starter is controlled through the on-delay timer. If the timer is set for 10 seconds, the timer contacts will close after a 10-second delay. The time-delay period starts when power is applied to the time delay operator. Fig. 2-25 shows an example of this time delay-on circuit.

2.7.2 Off-Delay Timers

In some applications it is necessary to ensure that a motor continues to operate for several minutes after power is turned off to the main system. For example, in large industrial heating systems, the fan may need to run for up to 2 minutes after the heating element or gas valve has been de-energized. The additional time the fan is allowed to run after the heat source is turned off will allow the system to capture all of the heat that has built up in the heating chamber and use it. This provides a degree of efficiency, since this heat would be lost if the fan was turned off at the same time as the heating element.

The time delay also prevents the heating system from overshooting the temperature setpoint on the controller. If the temperature setpoint is set for 150° F, and the controller de-energizes the heating element when the temperature reaches 150°, the heat that remains in the heating chamber would cause the system to overheat. The additional heat that remains in the chamber will act like a flywheel and continue to add heat to the application, which will cause the overtemperature condition.

The off-delay timer allows the heating control to de-energize the heating element several degrees prior to the setpoint, and allows the fan to continue to operate for several additional minutes to dissipate the remaining heat. This type of timer control may also be called *time delay-off timer*. Fig. 2-25b shows an example of the time delay-off timer used to control the furnace fan.

2.7.3 Normally Open and Normally Closed Time-Delay Contacts

The time delay-on and the time delay-off timers can have normally open, normally closed, or both types of contacts. Fig. 2-26 shows the symbol for each of the four possible timers. This figure also shows a timing diagram for each type of timer contact.

The on-delay timer can have normally open (NO) and normally closed (NC) contacts. The symbol for the on-delay timer with NO contacts is called *normally open, timed close* (NOTC). The contacts for this type of timer start out open and after the timing element is energized, the contacts will close after the time delay has elapsed. The symbol for the delay-on timer with NC contacts is called *normally closed, timed open* (NCTO). In Fig. 2-26 you should notice that the NOTC-type timer shows the contacts opening downward, and the tail of the arrow is used to indicate it is an on-delay timer. The NCTO-type timer symbol shows the contact closed.

The timing diagram for each of the four types of timers is also shown in Fig. 2-26. The timing diagram shows the condition of the contacts as either open or closed during the four periods of the timer operation. The first timer period shows the contacts prior to timing. This represents the time *before* the timer operator is energized. The second period represents the time *during timing,* and the third period represents the time *after timing,* but power is still applied to the timer operator. The fourth period represents the time *after timing* when power is turned off to the operator. This condition is called *reset* and it actually occurs twice during the cycle, once at the very beginning of the cycle during the first timing period before timing, and again at the very end of the cycle. This means the very first part of the timing cycle may be called the time before timing, or reset. You should also notice that the timing diagrams show a diagram for the timing operator as well as a diagram for each type of contact used. The diagram that shows power applied to the operator is provided as a reference to indicate when the timing cycle actually begins. The time-delay period for each timer should be counted from the point where power is applied to the time-delay operator.

The off-delay timer can also have NO and NC contacts. If the off-delay timer uses NO contacts, it will be called *normally open, timed open* (NOTO). The first

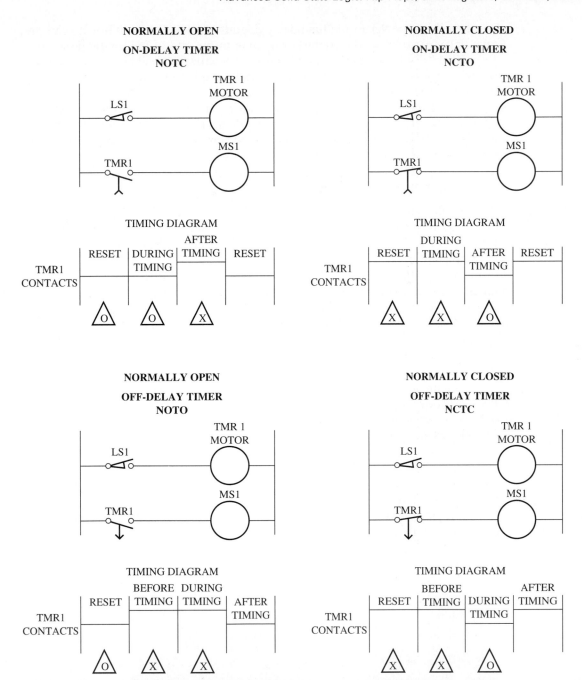

FIGURE 2-26 Symbols for the NOTC, NCTO, NOTO, and NCTC timers as they would be found in typical timing circuits. Timing diagrams are provided for each type of timer.

question that comes to mind about this type of timer is how it can be normally open and then also timed open. The answer to this is explained by the omission of one of the states the off-delay timer goes through. The actual timing cycle for the off-delay timer includes three distinct steps: normally open, *energized close*, and then timed open. The timing diagram for this timer shows the contacts are normally open during the time prior to timing (reset). When power is initially applied to the circuit, the time-delay operator is powered and the NO contacts move to the closed position. When power to the time-delay operator is de-energized, it will begin the time-delay period, and the contacts will open at the end of the timing period.

If the off-delay timer uses a closed set of contacts, it will be named *normally closed, timed closed* (NCTC). The name again seems to be contradictory because the contacts start out closed and end the cycle being timed closed. The part that is missing is the description of the middle part of the cycle where the contacts are opened when power is applied to the time-delay element. The on-delay and off-delay timers can have a single set of NO or NC contacts, or they can have multiple sets of both NO and NC contacts.

2.7.4 Pneumatic Timer Operators

Several different types of operators are used in industrial timers. The simplest type of operator is called a *pneumatic* operator. The pneumatic operator or element uses a bellows that fills with air. Prior to the start of the timing cycle, the bellows is completely filled with air, and when the time delay begins the air is released through a *needle valve*. The needle valve can be adjusted to regulate the time it takes for the air to move out of the bellows. A spring is mounted against the bellows to apply pressure to it to help force the air out. The amount of time it takes for the air to empty out of the bellows will determine the amount of time delay. When sufficient air has emptied out of the bellows, the spring pressure is strong enough to trip the mechanism that ends the timing cycle. The trip mechanism also actuates the timed contacts.

The pneumatic time-delay element is an *add-on* device to a traditional relay. Fig. 2-27 shows a picture of this type of add-on timing device. Since the pneumatic time-delay element is added to a normal control relay, this makes it possible to convert any control relay that is used in a circuit into a time-delay relay. Since the pneumatic element is added to the relay, it is also possible to take advantage of the original relay contacts that will still operate with the relay coil. Since these contacts will open or close instantly when the coil is activated, they are called *instantaneous contacts*.

The pneumatic time-delay element can be converted from time delay-on to time delay-off right in the field. The element is designed so that when it is mounted on the relay in the upright position, it will provide time delay-on functions. If the

FIGURE 2-27 Pneumatic time-delay element. This element is added to a traditional relay to create a time-delay relay. (Courtesy of Rockwell Automation's Allen Bradley Business.)

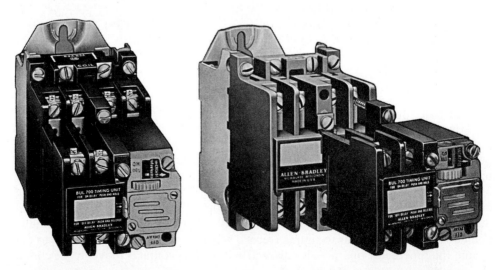

element is turned upside down when it is mounted on the relay, it will provide time delay-off functions. You can determine if the element is mounted for delay-on or delay-off functions by looking closely at the lettering on the element. If the element is mounted for delay-on functions, the words *on-delay* will be visible rightside up, and the words *off-delay* will be visible upside down. When the element is mounted for off-delay functions, the words *off-delay* will be visible rightside up.

The time-delay element generally has two sets of contacts mounted on it. One set is usually normally open, and one set is normally closed. These contacts can be changed from normally open to normally closed or vice versa, by loosening a mounting screw and turning the contact set upside down. This provides the feature of changing the contacts in the field. You can determine if the set of contacts is normally open or normally closed by checking them with an ohmmeter.

2.7.5 Motor-Driven Timers

Motor-driven timers are widely used in industrial applications. These timers are also called *synchronous motor* timers or *electromechanical* timers. The timer uses a motor to turn a shaft on which cams are mounted. The shaft has a clutch that engages or disengages it to the motor. When the shaft is engaged (energized), the motor turns the shaft and the cams actuate several sets of contacts. When the time-delay period has elapsed, the time-delay contacts will change from open to closed or from closed to open, and the timer motor will reset for the next cycle. Fig. 2-27 shows a picture and diagram of a motor-driven timer.

From the diagram in Fig. 2-28, you can see that the timer has a clutch coil and a motor connected in parallel with a common point identified as terminal 2. The timer has four sets of contacts. In the diagram contacts 9-10-C and 6-7-8 are located directly under the clutch coil to indicate they are energized instantaneously by the clutch when power is applied to the timer. Sets 11-12-A and 3-4-5 are located in the diagram directly under the timer motor to indicate they are controlled by the timer motor. These contacts are called the delay contacts.

A second diagram is provided in Fig. 2-28 that shows a terminal diagram for the timer. In a typical application a limit switch is used to control the timer. When the limit switch is closed, power is applied through it to the clutch coil. This same power energizes the motor, but it must go through NC time-delay contacts 11-12. When power reaches the clutch coil, it immediately activates the instantaneous contacts and engages the clutch. Power also reaches the timer motor and it begins to rotate its shaft. Since the clutch is engaged, the motor shaft turns the time-delay shaft and when the correct amount of time delay has passed, the cams located on the shaft will move with the shaft and activate both sets of time-delay contacts. Since the timer motor is connected through the NC set of time-delay contacts 11-12, the timer motor will become de-energized when the contacts open and the timer will be reset and ready for the next cycle. The instantaneous sets of contacts controlled by the clutch coil will remain energized until the limit switch is de-energized.

2.7.5.1 Using Time Delay to Control Pouring Gray Iron ■ An example of the time-delay circuit in an industrial application controls an automatic continuous pouring of cast iron into sand molds moving along a conveyor (see Fig. 2-29). This foundry application is used to provide control of the continuous pouring line that makes brake rotors for automobiles. The sand molds for the brake rotors are made by an automatic molding machine that sets each finished mold on a conveyor line.

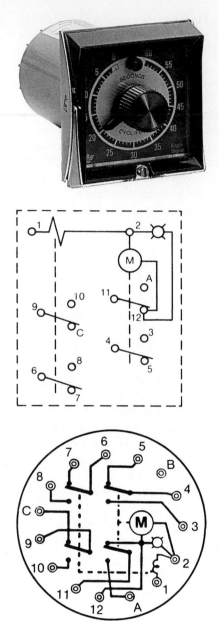

FIGURE 2-28 Synchronous motor-driven timer with a diagram of the internal electric contacts circuit. (Courtesy of Danaher Controls.)

The molds move along the conveyor line until they reach the pouring station where they will trip a photoelectric switch that controls a timer. You can see that the motor starter for the conveyor motor is connected in series with the 4-5 NC delay contacts and the 6-7 NC instantaneous contacts.

When the mold moves into place under the pouring system, the photoelectric switch will activate the timer and the NC 6-7 instantaneous contacts will open and de-energize the conveyor motor starter, which stops the conveyor immediately. The instantaneous 6-7 contacts will switch to the 6-8 position, which will energize the pouring system solenoid. The pouring system solenoid controls a large gear box that tips a ladle into position so that gray iron fills the mold. The amount of time delay is adjusted so that the sand mold is filled to the top. When the amount of

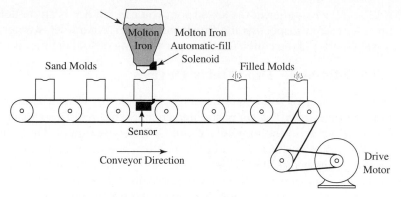

Hardware for Automatic Iron Pouring Line

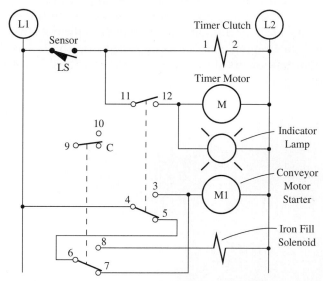

Electrical Diagram of Timer Motor for Automatic Pouring Line

FIGURE 2-29 Example application of a motor-driven timer used to control pouring gray iron into sand molds passing by on a conveyor line.

time delay expires, the timed 4-5-3 contacts will change from the normally closed 4-5 position to the 4-3 position, which will energize the conveyor motor starter again.

When the motor starter is energized again, the conveyor will begin to move the mold so that it no longer breaks the photoelectric beam. The conveyor continues to move until the next mold on it blocks the photoelectric switch light beam, which activates the process again. Some automatic pouring lines use laser detectors to monitor sprue holes or vent holes in the top of the sand mold to ensure the mold fills completely with molten gray iron during the pouring process. When molten iron appears in the vent hole, the mold is completely filled and the laser control can stop the pouring operation before the time delay elapses. The laser control provides a safety backup to the timer in case the molds fill too quickly.

2.7.5.2 Off-Delay Timers ■ The motor-driven timer cannot provide both on-delay and off-delay functions from the same device. If an off-delay function is required, a separate timer must be used. This timer is called a *reverse-acting* timer. This means that you must determine the type of timer you need before you purchase

it. If you have a timer in the stockroom you can tell if it is an on-delay or off-delay timer by closely inspecting the clutch mechanism. An on-delay timer has a normally open clutch, and the clutch for the reverse-acting off-delay timer is normally closed.

2.7.6 Solid-State Time-Delay Devices

A wide variety of solid-state timers is available for industrial applications. These timers range from the simplistic and inexpensive resistive-capacitive (*RC*) time-constant timers to the more elaborate CMOS-type timers. These timers provide a set of terminals where an external resistor (potentiometer) can be connected to provide a means of adjusting the time delay remotely. The amount of resistance in the *RC* time-constant timers will determine the amount of time delay. Some of the timers will provide a knob directly on the timer that allows the amount of time delay to be set locally, while other inexpensive timers provide a fixed amount of time delay that must be determined prior to purchase.

Fig. 2-30 shows examples of several solid-state timers. The solid-state timer is also available as an add-on element similar to the pneumatic element. This allows the solid-state timing element to be added directly to a traditional control relay. The solid-state timers shown in Fig. 2-30 are available with a variety of outputs.

FIGURE 2-30 Solid-state time-delay devices. The inexpensive solid-state devices use *RC* time-constant technology, and the more expensive types use CMOS technology. (Courtesy of National Controls Corporation.)

FIGURE 2-31 Microprocessor-based programmable-type timer. This device can be programmed to provide on-delay, off-delay, and counting functions. This allows one device to be stocked that can be programmed at the time of installation to provide mulitple functions. (Courtesy of Danaher Controls.)

Since these devices are solid state, they can provide outputs that include delay-on make (on-delay), delay-on brake (off-delay), repeat cycle, interval cycle, and single shot. Other types of solid-state time-delay devices provide thumbwheels on the front of the device that allow the amount of time delay to be set as a value.

2.7.7 Programmable Timers

The newest types of time-delay devices are microprocessor-based programmable timers. These devices utilize all of the features of a microprocessor, which include a keypad for input and light-emitting diode (LED) display. The amount of time delay and the type of outputs are all programmable through the keypad. The operation of the timer function such as pulse cycle, repeat cycle, or single shot is also programmable. Fig. 2-31 shows an example of the microprocessor-based programmable timer.

Since the function of counting is very similar to timing, the programmable-type timer can be programmed to operate as a time-delay device or as a counter. This feature provides industry the ability to purchase and stock one device for use as a timer or counter.

SOLUTION TO JOB ASSIGNMENT◄····················

Your solution for the job assignment should include using a clocked flip-flop similar to the one used to control the two air compressors in Fig. 2-6. Your solution should also include a set of flip-flops that provides counting capability like the ones in Figs. 2-8, 2-9, 2-17, and 2-18. If you choose to use an up/down counter, the good parts sensor should be connected to the up-count input, and the bad parts sensor should be connected to the down-count input. Each time the good parts sensor transitions, the count is incremented (1 is added to the total), and each time the bad parts sensor is transitioned, the count is decremented (1 is subtracted from the total). A self-contained industrial counter could be used for this part of the application.

If your solution includes a shift register to track the automotive bodies, it should look like the example in Fig. 2-11 or 2-12. A sensor should detect whether or not a body is present and this information is loaded into the shift register. The robot welders would take the place of the spray-painting nozzles as outputs. If a FIFO register is used to keep track of alarm conditions, it should look like the example in Fig. 2-15 or 2-16. You will need to determine the alarm conditions that your alarms will monitor.

QUESTIONS

1. Explain how NAND logic gates can be used to make an *RS* latch.
2. Provide an example of an industrial application that uses an *RS* latch and explain the operation of the application.
3. Explain the difference between a latch and a flip-flop.
4. Provide an example of an industrial application that uses a FIFO.
5. Explain the operation of a bit shift register.
6. Provide an example of an industrial application that uses a bit shift register.
7. Explain the difference between a bit shift register and a word shift register.
8. Provide an example of an industrial application that uses a word shift register.
9. Explain the operation of a ring counter.
10. Explain the operation of a Johnson counter.

TRUE OR FALSE

1. _____ An *RS* latch will have an output at *Q* any time the *S* input is HI.

2. _____ An edge-triggered flip-flop will only change state when a pulse has transitioned the clock input.

3. _____ A shift register stores the value at the data input and shifts it to the right or left one bit each time a clock pulse is received.

4. _____ An edge-triggered flip-flop changes its output only when a clock pulse occurs at the clock input.

5. _____ The gray code is used in encoders because it has more bits than the binary code, which means it can count larger numbers.

MULTIPLE CHOICE

1. The *RS* latch will provide an output at its *Q* output when _____.
   a. *S* is HI and *R* is HI.
   b. *S* is HI and *R* is LO.
   c. *S* is LO and *R* is HI.

2. The edge-triggered *JK* flip-flop shown in Fig. 2-3 will have its *Q* output HI when _____.

  a. *J* is HI, *K* is LO, and the clock pulse transitions from the off to the on state.

  b. *J* is HI, *K* is HI, and the clock pulse transitions from the on to the off state.

  c. *J* is LO, *K* is HI, and the clock pulse transitions from the off to the on state.

3. The ring counter uses _____.

  a. four flip-flops to count values 0–9.

  b. ten flip-flops to count values 0–9.

  c. four flip-flops to count values 0–7.

  d. ten flip-flops to count values 0–7

4. The motor starter coil for compressor 1 connected to the *Q* output of the edge-triggered *JK* flip-flop in Fig. 2-4 _____.

  a. energizes every other time the pressure switch PS1 closes.

  b. energizes every time the pressure switch PS1 closes.

  c. energizes every time safety switch PS2 or PS3 closes.

5. A one-shot timer _____.

  a. provides a series of short output pulses every 0.1 second as long as it receives a HI enable signal.

  b. provides a single output pulse of short duration regardless of the duration of the input pulse.

  c. provides a time delay-on function between 1–10 seconds.

PROBLEMS

1. Sketch the diagram of an *RS* latch and identify its input and output terminals.

2. Sketch the truth table for an *RS* latch.

3. Draw a 555 timer used as a monostable vibrator and identify its inputs and outputs.

4. Sketch the diagram of an edge-triggered *RS* flip-flop and identify its input and output terminals.

5. Draw the symbol for on-delay timer contacts (delay on) and off-delay timer contacts (delay off) and explain how each operates.

REFERENCES

1. Floyd, Thomas L., *Digital Fundamentals,* 5th ed. New York: Macmillian Publishing Company, 1994.

2. Kissell, Thomas E., *Modern Industrial/Electrical Motor Controls.* Englewood Cliffs, NJ: Prentice-Hall Inc., 1990.

3. Maloney, Timothy J., *Industrial Solid-State Electronics,* 2nd ed. Englewood Cliffs, NJ: Prentice-Hall Inc., 1986.

4. Texas Instruments Data Books: Programmable Logic Data Book; TTL Data Book, 1981, 2nd Ed.; TTL Data Book, 1976; Low Voltage Logic Data Book, 1993; Advanced Logic Data Book, 1991; Linear Circuits Data Book, 1992.

3

PROGRAMMABLE CONTROLLERS

After reading this chapter, you will be able to:

1. Describe the four basic parts of any programmable logic controller (PLC).
2. Explain how a PLC program is different from a BASIC or Fortran program.
3. Explain the four things that occur when the PLC processor scans its program.
4. Compare how a PLC controls a conveyor sorting system to the way a typical solid-state logic circuit would control it.
5. Explain the function of an input module and describe the circuitry used to complete this function.
6. Explain the function of an output module and describe the circuitry used to complete this function.
7. Explain the function of an internal control relay.
8. Discuss the classifications of PLCs.

9. Describe what happens when a PLC is in run mode.
10. Explain the function of the address of an input or output to the PLC processor and to the technician who must troubleshoot the PLC.
11. Identify the input and output instructions for a PLC.
12. Explain what a mnemonic is.
13. Read and interpret a PLC program.
14. Explain the operation of a latch and unlatch coil.
15. Explain the operation of a PLC on-delay timer.
16. Explain the operation of a PLC up counter and down counter.
17. Explain how a PLC timer or counter can reach values above 9999.
18. Describe the operation of a PLC sequencer.

You are a technician who has been assigned the job of converting a machine that is presently controlled by relay logic and motor-driven timers so that it will be controlled by a programmable logic controller (PLC). The machine also has a drum switch with a number of cams used to energize a group of solenoids in a machine sequence. The drum switch should be replaced by a sequencer instruction in the PLC. Your assignment includes providing a printout of the ladder logic diagram and a full set of drawings including the switches and outputs connected to input and output modules. Your supervisor will tell you the brand of PLC you will be using and the addresses for the inputs and outputs you should use. Your supervisor will also provide the original relay logic diagram.

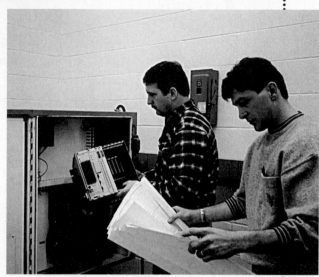

3.1 OVERVIEW OF PROGRAMMABLE CONTROLLERS

The programmable controller has become the most powerful change to occur in the electronics world for factory automation. The programmable controller (P/C) is also called the programmable logic controller (PLC). Since the personal computer is called a PC, the programmable controller is referred to as a PLC to prevent confusion. As an electronics technician, you will run into the PLC in a number of places such as on the factory floor, in a repair facility, or you might work for a company that manufactures components and boards that are used to interface with PLCs.

In this chapter you are going to study PLCs from two different perspectives. First, you will see how the PLC can be programmed to perform the same logic functions that the IC logic chips and relays in Chapters 1 and 2 performed. Second, you will see the large amount of electronics involved in the input and output modules that the controller uses for interfacing a variety of industrial voltages. These modules contain a variety of electronic circuits to convert 110 volts and 220 volts AC signals to the low-voltage signals required in the PLC. From this information you will see that the PLC is as important to the industrial automation world as the personal computer is to the business world.

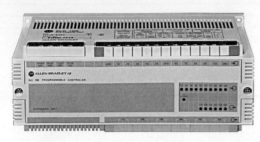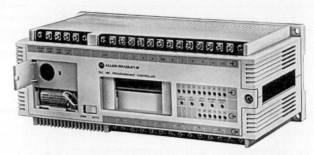

FIGURE 3-1 Typical programmable controllers. (Courtesy of Rockwell Automation's Allen Bradley Business.)

Today in industry you may find PLCs as stand-alone controls or as part of a complex computer integrated manufacturing (CIM) system. In these large integrated manufacturing systems the PLC will control individual machines or groups of machines. PLCs may also provide the interface between machines and robots, or machines and color graphics systems that are called man-machine interfaces (MMI).

In the 1970s and 1980s companies were able to hire both an electrician and an electronics technician to install, interface, program, and repair PLCs. Since the late 1980s the number of PLCs has grown so large that many companies have found that it is better to hire one individual with both electronic and electrical experience. This change will provide a large number of employment opportunities for electronics technicians who are willing to learn about industrial electrical applications such as the PLC. Other employment opportunities exist in companies that manufacture electronic boards and interface devices for PLCs. If you have never heard of a PLC or if you have had a minor introduction to them, this chapter will provide you with all of the information necessary to work successfully with them.

A picture of typical types of PLCs is shown in Fig. 3-1. This chapter will use generic PLC addresses where possible. When specific applications are provided, the Allen-Bradley SLC 100 and SLC 500 will be used as example systems. The early examples in this chapter will not be specific to any brand of PLC so that you may understand the functions that are generic to all controllers.

3.1.1 The Generic Programmable Logic Controller

The programmable logic controller (PLC) is a computer that is designed to solve logic (AND, OR, NOT) that specifically controls industrial devices such as motors and switches and allows other control devices of varied voltages to be easily interfaced to provide simple or complex machine control. The PLC is an enhancement of the simple logic circuits that were shown in Chapters 1 and 2, but it is different from other computers in that it is designed to have industrial-type switches, such as 110 volts push-button and 110 volts limit switches, connected to it through an optically isolated interface called an input module. The PLC can also control 480 volts three-phase motors through 120 volts motor starter coils with a similar interface called an output module.

The computer part of the PLC, which is called a central processing unit (CPU), allows a program to be entered into its memory that will represent the logic functions. The program in the PLC is not a normal computer programming language like BASIC, Fortran, or C. Instead the PLC program uses contact and

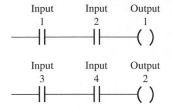

FIGURE 3-2 Example of a typical programmable controller program called ladder logic. This program uses the generic addresses of input 1, input 2, output 1, output 2, and so on.

coil symbols to indicate which switches should control which outputs. These symbols look similar to a typical relay ladder diagram shown in Chapter 1. In most systems the program can be displayed on a computer screen where the actions of switches and outputs are animated as they turn motors and other outputs on or off. The animation includes highlighting the input and output symbols in the program when they are energized. Other systems use light-emitting diodes (LEDs) on a hand-held programmer that illuminate when an input or output is energized to show its status.

In short, the PLC executes logic programs like logic gates found on solid-state ICs, but its programs are designed to look exactly like relay ladder diagrams that electricians have used since earliest electrical controls were introduced. The PLC was designed to provide the technology of logic control with the simplicity of an electrical diagram so that its inputs and outputs are easy to troubleshoot. Since the logic is programmable, it is easily changed. The logic can also be easily stored on a disk so that it can be loaded into a PLC on the factory floor anytime the program logic for a controller needs to be changed as different parts are made or when the machine function needs to be changed.

When PLCs were first introduced, they allowed the factory electrician to use them to help troubleshoot large automated systems. As PLCs have evolved, they now provide many of the more complex logic functions such as timing, up/down counting, shift registers, and first-in, first-out (FIFO) functions.

Fig. 3-2 shows an example of a PLC program. You should notice that the program looks exactly like an *electrical relay diagram.* As the program (diagram) gets larger, additional lines will be added. This gives the program the appearance of a wooden ladder that has multiple rungs. For this reason the program is called a *ladder diagram,* or it may be referred to as *ladder logic.*

The original function of the PLC was to provide a substitute for the large number of electromechanical relays that were used in industrial control circuits in the 1960s. Early automation used large numbers of relays which were difficult to troubleshoot. The operation of the relays was slow and they were expensive to rewire when changes were needed in the control circuit. In 1969 the automotive industry designed a specification for a reprogrammable electronic controller that could replace the relays. This original PLC was actually a sequencer-type device that executed each line of the ladder diagram in a precise sequence. Later in the 1970s the PLC evolved into the microprocessor-type controller used today. Several major companies such as Allen-Bradley, Texas Instruments, and MODICON produced the earliest versions. These early PLCs allowed a program to be written and stored in memory, and when changes were required to the control circuit, changes were made to the program rather than changing the electrical wiring. It also became feasible for the first time to design one machine to do multiple tasks by simply changing the program in its controller.

3.1.2 Basic Parts of a Simple Programmable Controller

All PLCs have four basic major parts: power supply, processor, input modules, and output modules. A fifth part, a programming device, is not considered a basic part since some PLCs will not have one if its program is loaded from an EPROM (erasable programmable read-only memory) chip. Fig. 3-3 shows a block diagram of the typical PLC. From this diagram you can see that input devices such as push-button switches are wired directly to the input module, and the motor starter or solenoids are connected directly to the output module. Before the PLC was invented, each switch would be directly connected to the motor starter or solenoid it was controlling. In the PLC the switches and output devices are connected to the modules, and the program in the PLC will determine which switch will control which output. In this way, the physical electrical wiring needs only to be connected one time during the installation process, and the control circuitry can be changed unlimited times through simple changes to the ladder logic program.

3.1.3 The Programming Panel

At the bottom of the diagram in Fig. 3-3 you can also see a programming panel. A programming device is necessary to program the PLC, but it is not considered to be one of the parts of a PLC because the programming panel or device can be disconnected after the program is loaded and the PLC will run by itself. The programming device is used so humans can make changes to the program, trouble-shoot the inputs and outputs by viewing the status of contacts and coils to see if they are energized or de-energized, and for saving programs to disk, or loading programs from disk.

The programming panel can be a dedicated device or it can be a personal or portable computer with PLC programming software loaded on it. The ladder logic program is able to be displayed on the programming device where it can become animated. This feature is unique to the PLC and it helps the technician troubleshoot very large logic circuits that control complex equipment. When a switch that is connected to the PLC input module is turned on, the program display can show this on the screen by highlighting the switch symbol everywhere it shows up in the program. The display can also highlight each output when it becomes energized by the ladder logic program. When the on/off state of a switch on the display is compared with the electrical state of the switch in the "real world," a technician can quickly decide where the problem exists.

FIGURE 3-3 Block diagram showing the four major parts of a programmable controller. The programming panel is the fifth part of the system, but it is not considered a basic part of the PLC since it can be disconnected when it is not needed.

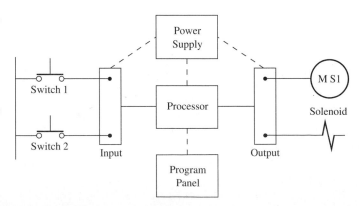

Another feature that makes the PLC so desirable is the fact that it has most of the logic functions found in the machine language program of any microprocessor chip. This means that functions such as timing and counting can be executed by the PLC rather than using electromechanical or electronic timers and counters. The advantage to using timer and counter functions in the PLC is that they do not involve any electronic parts that could fail, and their preset times and counts are easily altered through changes in the program. Additional PLC instructions provide a complete set of mathematical functions as well as manipulation of variables in memory files. Modern PLCs such as Allen-Bradley PLC5 and Siemens TI545 have the computing power of many small mainframe computers as well as the ability to control several hundred inputs and outputs.

3.2 AN EXAMPLE PROGRAMMABLE CONTROLLER APPLICATION

It may be easier to understand the operation of a PLC when you see it used in a simple application of sorting boxes according to height as they move along a conveyor. It is also important to understand that the addressing schemes for inputs and outputs of some PLCs may be difficult to understand when you are first exposed to them. For this reason the first examples of the PLC presented in this section will use generic numbering that does not represent any particular brand name. This will allow the basic functions of the PLC that are used by all brand-name of PLCs to be examined in a simplified manner. Later sections of this chapter will go into specific numbering systems for several major brands of PLCs.

Fig. 3-4 shows a pictorial diagram as a top view of the conveyor sorting system. The locations of the photoelectric switches that detect the boxes and the pneumatic cylinders used to push the boxes off the conveyor are also shown in this diagram. Each air cylinder is controlled by a separate solenoid. The conveyor is powered by a three-phase electrical motor that is connected to a motor starter. Start and stop push buttons are used with the photoelectric switches to energize the coil of the motor starter that turns the conveyor motor on and off. Fig. 3-5 shows all of the

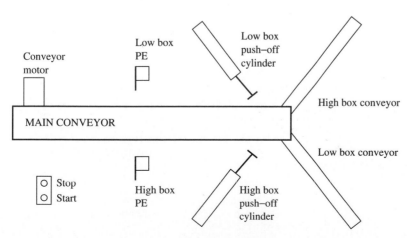

FIGURE 3-4 Diagram of a conveyor sorting system that is controlled by a generic programmable controller.

FIGURE 3-5 Input and output diagram that shows all switches that are connected to the PLC input module and motor starters that are connected to the PLC output modules. The input and output numbering is generic and does not represent any brand-name programmable controller.

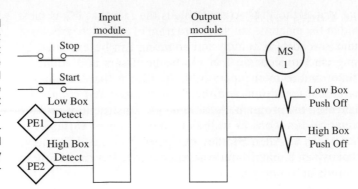

switches connected to an input module and solenoids and motor starter connected to an output module. The inputs are generically numbered input 1, input 2, and so on. The outputs are numbered output 10, output 11, and so on. The ladder logic program that determines the operation of the system is shown in Fig. 3-6. The diagram of the switches and outputs connected to the input and output modules uses the standard electrical symbol for each switch and output device.

In the ladder diagram you will see that each switch is represented as a set of open contacts or closed contacts (--] [-- or --]\[--). Each motor starter or solenoid in the hardware diagram is represented by an output symbol --( )--. The numbers that are assigned to the switches, motor starters, and solenoids in the hardware diagram are used in the logic diagram so that you can determine which switch is being used.

FIGURE 3-6 Ladder logic diagram showing the logic control for the conveyor sorting system. The diagram is generic and does not represent any particular PLC brand.

```
                                                           Motor run
        Start              Stop                              CR1
1   --------] [-----------] [----------------------( )-
          |  CR1  |
          |--] [--|
                        High box      Low box       Conveyor
                        detected      detected     motor starter
        CR1              CR2           CR3             MS1
2   --------] [----------]\[--------]\[----------( )-

        PE 2                              High box
        High box                        control relay
3       sensor                              CR2
    -------] [----------------------------------( )-

        PE 2           PE 1              Low box
        High box       Low box         control relay
4       sensor         sensor              CR3
    -------]/[----------] [-------------------( )-

                                          High box
        CR2                                solenoid
5   -------] [-------------------------------------( )-

                                          Low box
        CR3                                solenoid
6   -------] [-------------------------------------( )-
```

3.2.1 Operation of the Conveyor Sorting System

When the conveyor system is ready for operation, the start push button must be depressed. Since the stop push button is wired as an NC switch, power will pass through its contacts in the program to the start push-button contacts. When the start push button is depressed, the motor run condition will be satisfied in the first rung of the program and the motor run control relay in the program will be energized. The PLC can have hundreds of control relays in its program. The advantage of having control relays in a PLC program is that they provide the logic functions of a control relay, which can have multiple sets of NO and NC contacts, yet they only exist inside the PLC so they can never fail or break.

A set of NO contacts from CR1 is connected in parallel with the start push button to seal it in because the start push button is only momentarily closed. The photoelectric switches are used to energize and de-energize the conveyor motor starter in the second line of the logic. Line 2 of the diagram is designed so the start/stop circuit can turn the motor starter on or off without affecting the start/stop circuit. The photoelectric switches send out a beam of light that is focused on a reflector mounted on the opposite side of the conveyor. As long as a box does not interrupt the beam of light, the photoelectric switches will be energized.

When the motor run control relay coil (CR1) in the first rung of the program is energized, its NO contacts in the second rung of the program will become "logically" closed. Since no boxes are being sensed by the photoelectric switches, they will be energized and their contacts in the program, low box PE and high box PE, will pass power to the output in the second rung called motor starter. This directly controls the coil of motor starter 1. When motor starter 1 is energized, the conveyor motor is energized and the conveyor belt begins to move.

As boxes are placed on the conveyor, they will travel past the photoelectric switches. If the box is a low box, photoelectric switch PE1, which is mounted to detect the low box, will become energized. This will cause three things to happen. First, the NO contacts of PE1 low box sensor in rung 4 will close. Since the box is not a high box, the NC contacts of PE2 high box sensor in rung 4 will remain closed to pass power to PE1 low box sensor contacts (that are now closed) and on to the coil of CR3 (low box control relay). Second, the coil of CR3 is energized and all of the contacts in the program that are identified as CR3 will change state. This means the NC CR3 contacts in rung 2 will open and de-energize MS1, the conveyor motor starter. When MS1 is de-energized, the motor starter will become de-energized and the conveyor motor will stop. Third, when the coil of CR3 is energized by PE1 low box sensor, the output for the low box solenoid is energized and the rod of the low box push-off cylinder is extended. The low box is pushed off the main conveyor onto the low box conveyor.

After the low box is pushed off the main conveyor, it will no longer activate PE1, and the coil of CR3 will become de-energized in rung 3. When the coil of CR3 is de-energized, the NC CR3 contacts in rung 2 will return to their closed state, which will energize the output for MS1 and start the conveyor motor again.

When a high box moves into position on the conveyor, it will block both the high box photoelectric switch PE2 and the low box photoelectric switch PE1. This could create a problem of having both the low box control relay CR3 and the high box control relay CR2 energized at the same time if it were not for the logic in rung 4 that is specifically designed to prevent this. Since the contacts of the high box sensor PE2 are programmed normally closed in rung 4, they will open when the high box is detected and not allow power to pass to the coil of CR3. This type of logic is called an *exclusion,* since it prevents both control relay coils from energizing at the same time even though both photoelectric switches are activated by the high box.

When the high box is detected by PE2, the NC PE2 contacts in rung 3 will close and energize the coil of the high box control relay CR2. When the coil of CR2 is energized, all of the contacts in the program that are identified as CR2 will change state. The NC CR2 contacts in rung 2 will open and de-energize the coil of MS1, the conveyor motor starter. The NO CR2 contacts in rung 5 will close at this time and energize the high box solenoid and air will be directed to the high box pneumatic cylinder. Then the high box will be pushed off the conveyor.

After the box has been pushed off the conveyor, the photoelectric switch that detected the high box will return to its normal state, and high box contacts in rung 2 will return to their normally closed logic state. The motor starter will become energized again and start the conveyor motor. The high box contacts in rung 3 will return to their normally open logic state and the push-off cylinder will be retracted. This allows the conveyor to return to its normal operating condition.

3.3 SCANNING A PLC PROGRAM WHEN IT IS IN THE RUN MODE

The PLC processor examines its program line by line, which is the way it solves its logic. The processor in the PLC actually performs several additional functions when it is in the *run mode.* These functions include reading the status of all inputs, solving logic, and writing the results of the logic to the outputs. When the processor is performing all of these functions, it is said to be *scanning* its program. Fig. 3-7 shows an example of the program scan.

The PLC provides a means to determine when the switches that are connected to its input module are energized or de-energized. When the contacts of a switch are closed, they allow voltage to be sent to the input module circuit. The electronic circuit in the input module uses a light-emitting diode and a phototransistor to take the 110 volts ac signal and reduce it to the small-voltage signal used by the processor.

The contacts in the program that represent the switch will change state in the program when the module receives power. If you use the programming panel to

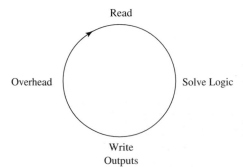

FIGURE 3-7 Example of a typical programmable controller scan. The processor continually reads the status of all inputs, solves logic, and writes the results of the logic to all outputs. The fourth part of the scan cycle is called overhead and it includes the functions of communications with the programming panel, printer, color screens, and so on.

examine the contacts in the PLC program, they will highlight when the switch they represent is energized. The PLC will transition the state of the programmed contacts at this time. For example, if the contacts are programmed normally open, when the real switch closes and the module is energized, the PLC will cause the programmed contact to transition from open to closed. This means that the programmed contacts will appear to pass power through them to the next set of contacts. If both sets of contacts are passing power in line 1 of the program, the output will become energized.

3.3.1 Image Registers

One of the least understood parts of the PLC is the *image register*. The image register is a memory section that is specifically designed to keep track of the status (on or off) of each input. This means that the image register is 1 bit wide. Fig. 3-8 shows an example of the image register. From this diagram you can see that all of the inputs and outputs for the PLC each have one memory location where the processor will place a 1 or 0 to indicate whether the input or output connected to that address is energized or de-energized. The processor determines the status of all inputs during each scan cycle by checking each one to determine if the switch connected to it is energized. If the switch is energized and the input module sends a signal to the processor, the processor will place a binary 1 in the memory location in the image register that represents that input. If the switch is off, the processor will place a binary 0 in the memory location for the image register. This means the image register will have an exact copy of the status of all its inputs during the *read* part of the scan. Some PLCs also call the image register the *image table*.

In the second part of the PLC scan the processor uses the values in the image register to solve the logic for that particular rung of logic. For example, if input 1 is ANDed with input 2 to turn on output 12, the processor looks at the image register address for input 1 and input 2 and solves the AND logic. If both switches are high, and a 1 is stored in each of the image register addresses, the processor will solve the logic and determine that output 12 should be energized.

After the processor solves the logic and determines the output that should be energized, the processor will move to the third part of the scan cycle and write a 1 in the image register address for output 12. This is called the *write cycle* of the processor scan. When the image register address for output 12 receives the 1, it immediately sends a signal to the output module address. This energizes the electronic circuit in the output module, which sends power to the device such as the motor starter that is connected to it.

It is important to understand that the PLC processor updates all of the addresses in the image register during the read cycle even if only one or two switch addresses are used in the program. When the processor is in the *solve logic* part of the scan, it will solve the logic of each rung of the program from left to right. This means that if a line of logic has several contacts that are ANDed with several more contacts that are connected in an OR condition, each of these logic functions is solved as the processor looks at the logic in the line from left to right. The processor also solves each line of logic in the program from top to bottom. This means that the results of logic in the first rung are determined before the processor moves down to solve the logic on the second rung. Since the processor continues the logic scan on a continual basis approximately every 20–40 msec, each line of logic will appear to be solved at the same time.

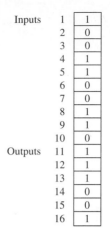

Inputs	1	1
	2	0
	3	0
	4	1
	5	1
	6	0
	7	0
	8	1
	9	1
	10	0
Outputs	11	1
	12	1
	13	1
	14	0
	15	0
	16	1

FIGURE 3-8 Example of an image register for a PLC. The image register is 1 bit wide and has a memory location for each input and output in the PLC.

The only time the order of solving the logic becomes important is when the contacts from a control relay in line 1 are used to energize an output in line 2. In this case the processor will actually take two complete scan cycles to read the switch in line 1 and then write the output in line 2. In the first scan, the processor will read the image register and determine the start push button has closed. When the logic for rung 1 is solved, the control relay will be energized. The processor will not detect the change in the control relay contacts in the second line until the second scan when it reads the image register again. The output in the second line will finally be energized during the write cycle of the second scan and power will be sent to the motor starter coil that is connected to the output address. It is also important to understand that the processor will write the condition of every output in the processor during the write part of the scan even if there is only one or two outputs used in the program.

The image register is also useful to the troubleshooting technician, since the contents of each address can be displayed on a CRT. The technician can compare the image register with the input and output modules and the actual switches and solenoids that are hard wired to the modules.

A separate image register is also maintained in the processor for the control relays. In some systems these relays are called *memory coils* or *internal coils.* A control relay has one coil and one or more sets of contacts that can be programmed normally open or normally closed. When the coil of a control relay is energized, all of the contacts for the relay will change state. Since control relays reside only in the program of the PLC and do not use any hardware, they cannot have problems such as dirty contacts or an open circuit in their coil. This means that if the contacts of a control relay will not change from open to closed, you would not suspect the coil of having a malfunction. Instead you would look at the contacts that control power to the control relay coil in program, and one or more of them will be open, which causes the coil to remain de-energized.

When the processor is in the run mode, it updates its input, output, and control relay image registers during the write-I/O part of every scan cycle. Since the image registers are continually updated approximately every 20–40 msec, it is possible to send copies of them to printers or color graphic terminals at any time to indicate the status of critical switches for the machine process. This allows the machine operator to do minor troubleshooting when a problem occurs before a technician is called. For example, if an operator is running a press or machine that must have a door or gate closed before it will begin its cycle, a fault monitor on a color graphic display can show the status of the door switch. If the door becomes ajar and the cycle will not start, the operator can look at the fault monitor and the copy of the input image register from the PLC would indicate that the door switch was open. The operator could take the appropriate action instead of calling a technician. In this manner, the PLC not only controls the machine operation, but it also helps in troubleshooting problems.

3.3.2 The Run Mode and the Program Mode

When the PLC is in the *run mode,* it is continually executing its scan cycle. This means that it monitors its inputs, solves its logic, and updates its outputs. When the PLC is in the *program mode,* it does not execute its scan cycle. The name *program mode* was originally used because older PLCs could not have their programming changed while executing their scan cycle. Modern PLCs have the ability to have their program edited and changed while the processor is in the run mode. At first glance, changing the PLC program while it is in the run mode looks dangerous, especially when the machine the PLC is controlling is in automatic operation. But you will find in larger control applications like pouring glass continually or continous steel rolling mills, it is not practical to stop the machine process to make minor program changes. In these types of applications the changes are made while the PLC is in the run mode. As a rule, you should switch the PLC to the program mode if possible when any program changes are being made.

3.3.3 On-Line and Off-Line Programming

The programming software for a PLC may allow you to write the ladder logic program in a personal computer and later download the program from the personal computer to the PLC. If you are connected directly to a PLC and you are writing the ladder logic program in the PLC memory, it is called *on-line programming.* If you are writing the ladder logic program in a personal computer or programming panel that is not connected to a PLC, it is called *off-line programming.* Most modern PLC programming software allows the program to be written on a personal computer without being connected to the PLC. This allows new programs or program changes to be written at a location away from the PLC. For example, you may write a program in Detroit, save the changes on a floppy disk, and mail the disk to Chicago where the PLC is connected to automated machinery. If you are using a *hand-held programmer,* it must be connected directly to a PLC so that you are writing the program in the PLC memory. It is also important to understand that some small PLCs like the Allen-Bradley SLC 100 do not have a microprocessor chip, so you cannot write the PLC program on line. Since the SLC 100 does not have a microprocessor chip, all programming must be completed off line, and then you must download the program changes to the PLC memory.

3.4 EQUATING THE PLC WITH TRADITIONAL SOLID-STATE LOGIC

It may be easier to equate the operation of a PLC controlled program with the same program using solid-state logic (AND, OR, and NOT) functions. Fig. 3-9 shows a diagram of the conveyor box sorting system using solid-state logic functions. You can see that the start and stop push buttons are inputs to the first AND gate (G1). The output of G1 is sent to the three-input AND gate (G2) along with the inverted signal from the high box photoelectric switch and low box photoelectric switch. The output of G2 is sent to the motor starter coil. AND gate (G3) is used to determine that the box is a low box by using a normal input from low box PE and an inverted input from high box PE. The output of G3 is used to energize the low box push-off solenoid. The high box PE signal is also used to energize the high box push-off solenoid.

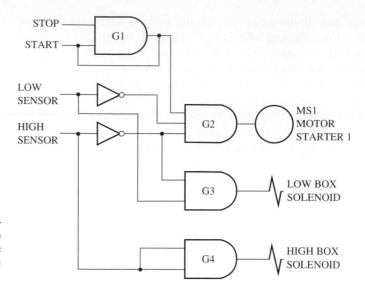

FIGURE 3-9 Solid-state logic gates used to provide the same logic as the PLC box sorting program.

As you know, the input switches must be sent through a signal conditioner because they are powered by 110 volts ac and they interface with the low-voltage logic chips. The outputs must also be sent to a signal conditioner (amplifier) to increase the low voltage logic output signal to 110 volts ac to be usable to energize the motor starter and solenoids. The input and output modules of the PLC provide the same function of the signal conditioners and amplifiers in the solid-state logic circuit.

FIGURE 3-10 Five basic ways to program contacts in a PLC: in series (AND), in parallel (OR), normally closed contacts (NOT), series normally closed contacts (NAND), and parallel normally closed contacts (NOR).

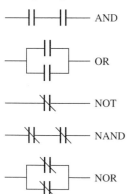

3.5 FEATURES OF THE PROGRAMMABLE CONTROLLER

The examples in Figs. 3-4 through 3-6 show several unique features of the PLC. First, you should notice that the ladder diagram allows you to analyze the program in the sequence that it will occur. Second, the PLC allows you to program more than one set of contacts in the program for each switch. For example, the photoelectric switch may be in the program both normally open and normally closed as many times as the logic requires. The PLC gives you the capability of solving AND, OR, and NOT logic by connecting contacts in series and parallel, and using NO and NC contacts. Fig. 3-10 shows examples of the five basic logic circuits as you would see them in ladder logic. As complex as some ladder logic programs look, you must remember that the logic program can only consist of combinations of these five basic functions.

Another feature of the PLC is the control relay or internal coils or memory relays. The relay has one coil and it can have any number of NO or NC contacts. When the coil is activated, all the contacts in the program that have the same address number as the coil will change state. The control relay exists only inside the PLC, which means it can never break or malfunction physically. Also since control relays exist only in the PLC memory, you can have as many of them as the memory size can accommodate. This means you can add control relays to accommodate changes to the machine control logic as long as the PLC has space remaining in its memory.

It may help to see a case where a control relay is used in an actual circuit. For example, in Fig. 3-6, you can see control relay (CR3) is used to indicate the logic has determined that a low box has been detected. The control relay in this rung of the program does nothing more than represent the conditions that indicate a low box has been detected. These conditions consist of the low box PE1 being on and the high box PE2 being off. Anytime CR3 is energized, a low box has been detected, and it will change the state of its contacts throughout the remainder of the program to represent the low box has been detected.

In many cases each control relay in a program represents a *condition* for the machine operation, such as "ready to go to auto," "all safety doors closed," or "ready to jog." If you think of the condition the control relay represents each time you see the control relay contacts, you will find it easier to read and interpret the program.

Another feature of the PLC that makes complex factory automation easier to troubleshoot is the way that each input switch or each output load is electrically connected to its interface module. In Fig. 3-5 you can see that when a switch is connected to an input module only two wires are involved. The outputs also use only two wires to connect the load device like the coil of a motor starter to the module. This means that when there is a malfunction in the electrical part of the circuit, the troubleshooter only has the switch and two wires to check for any individual input and the coil and two wires for any output. When you compare this type of wiring to a complex series-parallel circuit in an older-style relay panel, you can see it is much easier to troubleshoot two wires and a switch, or two wires and a solenoid. In the older-style relay panels a typical signal may pass through 10 to 15 relay contacts and switches and all the wiring that connects them together before it gets to a solenoid. This means that an open could occur in this complex circuit when any set of contacts opens, any switch opens, or if a wire has a fault where it becomes open.

To make troubleshooting the switch and two wires that connect it to an input module even easier, a small indicator light called a *status indicator* is connected to each input circuit. It is mounted on the face of the module so it is easy to see. The troubleshooter can look at the status indicator and quickly determine if power is passing through the contacts or not. If the switch contacts are closed but the status indicator is not illuminated, the troubleshooter needs to check only the two wires and the switch for that circuit to see where the problem is. This feature is particularly valuable when the automated system the PLC is controlling has several hundred switches that could cause a problem. A status indicator is also connected to each output circuit and mounted on the face of the output module. Anytime the PLC energizes an output, the status indicator for that circuit will be illuminated.

3.5.1 Classifications of Programmable Controllers

PLCs are generally classified by size. The small-sized system costs from $500 to $1000 and has room for a limited number of inputs and outputs. As a general rule, the small PLCs have less than 100 inputs and outputs with approximately 20 inputs and 12 outputs mounted locally with the processor. Additional inputs and outputs can be added through remote I/O racks to accommodate the remaining inputs and outputs. These PLCs generally have 2K to 10K of memory that can be used to store the user's logic program.

Medium-sized PLCs cost from $1000 to $3000 and have extended instruction sets that include math functions, file functions, and PID process control. These PLCs are capable of having up to 4000 to 8000 inputs and outputs. They also support a wide variety of specialty modules such as ASCII communication modules, BASIC programming modules, 16-bit multiplexing modules, analog input and output modules that allow interfaces to both analog voltages and currents, and communication modules or ports that allow the PLC to be connected to a local area network (LAN).

Large-sized PLCs were very popular in the early 1980s before networks were perfected. The concept for the large-sized PLC was to provide enough user memory and input and output modules to control a complete factory. Problems occurred when minor failures in the system brought the complete factory to a halt.

The advent of local area networks brought about the concept known as distributive control, where small and medium size PLCs are connected together through the network. In this way, the entire factory is brought under the control of a number of PLCs, but a failure in one system will not disturb any other system. The sizes of the majority of PLCs that you will encounter today will be small and medium systems that are used as stand-alone controls or as part of a network.

3.6 OPERATION OF PROGRAMMABLE CONTROLLERS

Now that you have a better understanding how the basic PLC operates you are ready to understand more complex operations.

3.6.1 Immediate I/O Updates

In some applications the PLC scan time is too long. One way to lower the scan cycle time to ensure that an input is read more quickly is called an *immediate I/O update.* The switches that require immediate update must be connected to a module that has special high-speed solid-state devices that can detect the transition of a signal in the *nanosecond* time frame. The high-speed module has access to a special part of the processor memory. When the processor receives a signal from one of its immediate inputs, it will drop out of its normal scan and immediately service the program logic the switch is controlling. This means that the normal scan cycle for the PLC may be 20 msec, and when the switch that is connected to the high-speed module transitions, the processor will acknowledge the request, and interrupt its primary scan to *actuate* or *call* the secondary scan called the *immediate interrupt scan.* The immediate interrupt scan operates independently of the primary scan. This means that the immediate interrupt scan can operate at a much higher speed than the normal scan.

When the input to the immediate interrupt module is transitioned and the

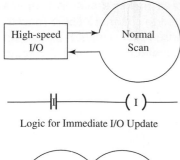

Logic for Immediate I/O Update

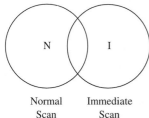

Normal   Immediate
Scan   Scan

FIGURE 3-11 Ladder logic for the heat-treating section of the carbon brush manufacturing system.

signal is sensed, the immediate interrupt scan will execute the logic where the input is used, and write the results of the logic for that line to the correct output as quickly as possible. To ensure the output signal is actuated as quickly as possible, the output module must also be an immediate interrupt output. An example of an immediate interrupt signal is a *dud detector* for a printing process for soft-drink cans. When soft-drink cans are manufactured, they pass through a high-speed inking process at speeds of over 1000 cans per minute. If one can is not inked correctly, it must be detected as a dud and ejected from the line while the cans pass at high speed. The sensor for the dud detector is connected to an immediate interrupt module. When it identifies a can that is a dud, an immediate interrupt signal is sent to the PLC processor. The PLC processor interrupts its normal scan and activates its immediate scan where the logic is activated and an immediate output signal is produced. The output signal is sent to the immediate interrupt output module that produces the dc signal that is sent to the high-speed solenoid. When the solenoid is energized, it activates a pneumatic high-speed push-off cylinder, which knocks only the dud can off the line as the cans pass by without stopping or slowing the line. Prior to high-speed immediate interrupt modules, the dud detector circuit would be controlled with solid-state logic components instead of a PLC.

Fig. 3-11 shows an example of the logic in a PLC program that uses the immediate update instruction. This diagram also shows how the immediate update scan cycle is called from the normal scan cycle to execute high-speed inputs and outputs.

3.6.2 The Need for High-Speed Inputs and Outputs

Prior to the immediate I/O update capability, there were many industrial applications that the PLC was too slow to control. In these cases, traditional solid-state ICs were used. For example, when a PLC is used for counting functions such as making aluminum beverage cans, the maximum line speed was limited to approximately 600 cans per minute if the scan cycle for the PLC was 10 msec. The immediate I/O update function allows this time to increase by approximately five times.

Another way to read inputs and write outputs at higher speeds is to use a module that has a microprocessor specifically dedicated to counting. These specialized modules called *high-speed input modules* can sense input signals at speeds up to 50 kHz and energize one or two outputs without interacting with the main PLC program.

3.7 ADDRESSES FOR INPUTS AND OUTPUTS FOR THE SLC 100

Up to this point in our discussion we have identified inputs and outputs in generic terms. In reality the processor must keep track of each input and output by an address or number. Each brand-name PLC has devised its own numbering and addressing schemes, which generally fall into one of two categories: sequentially numbering each input and output, or numbering according to location of input and output modules in the rack. The sequential numbering system is used in most small-sized PLCs and about half of the medium-sized systems. Allen-Bradley, one of the most popular brands, uses two different addressing systems. In its medium-sized PLCs (PLC2 and PLC5) the address for an input or output uses numbers that indicate the location of the input or output module in the rack, and it uses a sequential numbering system for addresses in its small-sized SLC 100 PLCs.

From this point on, the applications will use the address and numbering scheme of the Allen-Bradley SLC 100. If you are using a different brand-name PLC, you will need to convert the addresses to your system if you wish to try the program examples. Fig. 3-12 shows an example of the numbering system used in the Allen-Bradley SLC 100 small-sized PLC. (The SLC 150 is a similar system that is a little larger in that it has a combination of 32 inputs/outputs (I/O)'s directly attached to its processor, while the SLC 100 has a combination of 16 inputs /outputs attached to its processor.) From this example you can see that for the SLC 100 the input addresses start with address 001 and continue through address 010. Output addresses start with address 011 and continue through address 016. The first group of I/O is mounted locally on the processor. It is very important to understand that each I/O point address refers to a specific circuit in the hardware module and to a specific address in the processor's memory. The addresses in memory are assigned to *memory blocks*. The diagram that displays all of the addresses in a memory block is called a *memory map*.

FIGURE 3-12 Input and output addresses for the Allen-Bradley SLC 100. (Courtesy of Rockwell Automation's Allen Bradley Business.)

Address Block Number	Input Addresses (10)	Output Addresses (6)	Relay-Type Instruction Addresses	Associated Address Block Number
1	1–10	11–16	17, 18	1
2	101–110	111–116	117, 118	2
3	201–210	211–216	217, 218	3
4	301–310	311–316	317, 318	4
5	401–410	411–416	417, 418	5
6	501–510	511–516	517, 518	6
7	601–610	611–616	617, 618	7

If expansion modules are added to the processor, the first one would have input addresses 101–110 and output addresses 111–116. Up to six expansion modules can be added to the system so that the highest input addresses would be 601–610 and the highest output addresses would be 611–616.

This figure also shows the addresses of internal relays. As you know, the internal relays act like a control relay in that both perform logic but do not directly energize an output. From the figure you can see the internal relays are numbered 17 and 18 in the first block and 117 and 118 in the next block. If six additional expansion modules are used, the addresses can go to 617 and 618.

3.8 EXAMPLE INPUT AND OUTPUT INSTRUCTIONS FOR THE SLC 100

Fig. 3-13 shows an example of all of the input and output instructions the SLC 100 uses. From this figure you can see that normally open and normally closed contacts are provided for input instructions. The SLC 100 uses the name *examine on* for the NO contacts --] [--, and *examine off* for the NC contacts --]\[--. The term

Condition Instructions --] [--

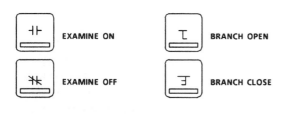

FIGURE 3-13 Example input and output instructions available in the Allen-Bradley SLC 100. (Courtesy of Rockwell Automation's Allen Bradley Business.)

Output Instructions --( )--

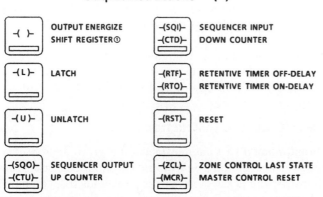

Instruction set of the SLC controller. Symbols are representations of the pocket programmer keys.

① The Shift Register symbol is --(SR)--. It does not appear on the programmer key.

examine was chosen to remind you that the processor looks at (examines) the electronic circuit in the input module to see if it is energized or de-energized. The processor will keep track of the status of the circuit by placing a 1 in the image register address if it is energized and a 0 if it is de-energized. The word *examine* is also used to remind you to look at the status light on the front of the module so that you can also determine if the input circuit is energized or de-energized. The branch instructions are used to allow contacts to be connected in parallel to each other in the ladder diagram.

The output instruction in this figure includes the traditional *coil,* which uses the symbol -( )-. This instruction is named *output energize.* Specialized outputs called the *latch* --(L)-- and the *unlatch* --(U)-- are also provided. The latch and unlatch coils act similarly to the flip-flops discussed in Chapter 2. The latch output will maintain its energized state after it is energized by an input even if the input condition becomes de-energized. The unlatch output must be energized to toggle the output back to its reset (off) state. The latch coil and unlatch coil must have the same address to operate as a pair.

Latch coils are used in industrial applications where a condition should be maintained even after the input that energized the coil returns to its de-energized state. It is important to understand that the latch coil bit will remain HI even if power to the system goes off and comes back on. This feature is important in several applications such as fault detection. Many times a system has a fault such as an overtemperature or overcurrent condition. After the fault is detected, the machine is automatically shut down by the logic and by the time the technician or operator gets to the machine, the condition is back to near normal and it is difficult to determine why the system shut down. If a latch coil is used in the detection logic, the fault bit will remain HI until the fault reset switch energizes the un-latch instruction.

Another application for the latch coil is for feed water pumps or fans that are located in remote locations in the building such as in the ceiling. A latch coil is used in these circuits because the motors must return to their energized condition automatically after a power loss. In some parts of the country, it is common for power to be lost for several seconds during severe lightning storms. If a regular type of coil is used in the program, a technician would have to go to each fan or pump and depress the start push button to start the motor again. If a latch coil is used, the coil will remain in the state it was in when the power went off. This means that if the latch coil was energized and the motor was running when power goes off, the latch coil will remain energized when power is returned and the motor will begin to run again automatically.

Other output instructions have a rather specific function that is beyond traditional relay logic. These instructions include the timer, counter, sequencer, and zone control. Each of these instructions will be covered in detail.

3.8.1 Mnemonics: Abbreviations for Instructions

In the SLC 100 as in other brand-name PLCs each instruction is given an abbreviation or name. The abbreviation is also called a *mnemonic* (pronounced new-mon-ic). The mnemonic for each instruction is selected because it sounds like the name of the function that the instruction provides. For example, the mnemonic for up counter is CTU, while the mnemonic for the down counter is CTD. The mnemonic is also the letters the PLC's microprocessor recognizes as instructions. You will need to learn the mnemonic for each instruction so that you can read and write

PLC programs. The mnemonics for different PLC brand names may be slightly different but they will always describe the instruction they represent.

3.9 USING THE SLC 100 TO CONTROL A BATCH-MIXING AND HEAT-TREATING PROCESS: MAKING CARBON BRUSHES

The best way to understand all of the instructions in the SLC 100 is to study an industrial application that uses a large variety of its instructions. Most programming manuals that come with the PLC explain the instructions without linking them to an actual application. The application in this example will help you understand each instruction.

The batch-mixing and heat-treating system shown in Fig. 3-14 uses contacts, coils, timers, counters, and a sequencer to mix the ingredients to make carbon brushes. One of the largest uses of carbon brushes is in dc motors such as the starter motors in automobiles. After the ingredients are mixed and the brushes are pressed out of an extruder, they have to be sent through a heat-treating process to cure them. A diagram of the process machinery for this application is shown in Fig. 3-15.

3.10 INPUTS AND OUTPUTS FOR THE CARBON BRUSH PROCESS SYSTEM

The PLC program is shown in Fig. 3-14. The controls for the carbon brush system have nine input switches connected to the input module and six devices connected to the output module of the SLC 100 PLC. The input/output (I/O) diagram in Fig. 3-16 shows these components connected to their respective input and output modules. You should also notice that the address for each device is provided. The I/O diagram will also show the electronic symbol for each switch and it will show if the switch contacts are wired normally open or normally closed. You will need to refer to the PLC program and the I/O diagram as the different parts of the machine are explained.

3.11 OPERATION OF THE HEAT-TREATING PART OF THE SYSTEM

Since the heat-treating part of the brush-making system consists of simple switches and outputs, we will begin our discussion at this point. The first three rungs of the program show examples of NO (examine on) contacts and NC (examine off)

FIGURE 3-14 Ladder logic for the heat-treating section of the carbon brush manufacturing system.

```
            START           STOP           START/STOP
            001             002            PERMISSIVE CR
                                               017
1  - - - - -] [- - - - - - - -] [- - - - - - - - - - - - -( )
                 :
            017  :
   - - - - -] [- -:

            START                          FAN MOTOR
            PERMISS.       LS1      LS2        MS
            017            003      004        011
2  - - - - -] [- - - - - - - -]\[- - - - - -]\[- - - - -( )

            FLOW           TEMP.          STAGE 1
            SWITCH         SWITCH         HEATING
            005            006            012
3  - - - - -] [- - - - - - - -] [- - - - - - - - - - - - -( )

            STAGE 1
            HEATING
            012                        901        20 SECOND
4  - - - - -] [- - - - - - - - - - - - - - - - - - -(RTO)   TIME DELAY

            STAGE 1
            HEATING
            012                        901
5  - - - - -]\[- - - - - - - - - - - - - - - - - - -(RST)   TIMER RESET

            TIMER                      STAGE 2
            CONTACTS                   HEATING
            901                        013
6  - - - - -] [- - - - - - - - - - - - - - - - - - -( )

            START          TEMP.          CONVEYOR     CONVEYOR
            PERMISS.       SWITCH         RUN MS       MOTOR
            017            006            014
7  - - - - -] [- - - - - - - -] [- - - - - - - - - - - - -( )

            TEMP.                      ALARM
            SWITCH                     HORN         LATCHES
            006                        015          ALARM HORN
8  - - - - -] [- - - - - - - - - - - - - - - - - - -(L)

            ALARM          TEMP.          ALARM        UNLATCHES
            RESET          SWITCH         HORN         ALARM HORN
            007            006            015
9  - - - - -] [- - - - - - - -]\[- - - - - - - - - - - - -(U)

            GOOD PART                  GOOD PARTS
            SENSOR                     COUNTER
            008                        902
10 - - - - -] [- - - - - - - - - - - - - - - - - - - - - -(CTU)

            BAD PART                   BAD PARTS
            SENSOR                     COUNTER
            009                        902
11 - - - - -] [- - - - - - - - - - - - - - - - - - - - - -(CTD)

            START          TIMER          LUBE
            PERMISS.       CONTACT        TIMER
            017            903            903
12 - - - - -] [- - - - - - - -]\[- - - - - - - - - - - - -(RTO)

            TIMER                      LUBE
            CONTACTS                   SOLENOID
            903                        016
13 - - - - -] [- - - - - - - - - - - - - - - - - - - - - -( )

            TIMER                      TIMER
            CONTACTS                   RESET
            903                        903
```

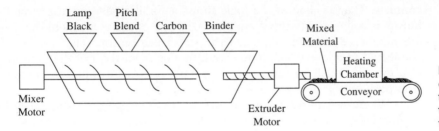

FIGURE 3-15 Diagram of the mixing and heat-treating equipment for the carbon brush manufacturing system.

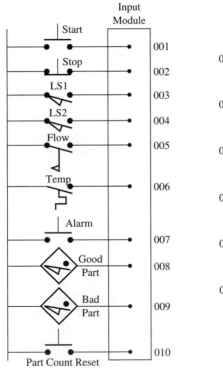

(a)

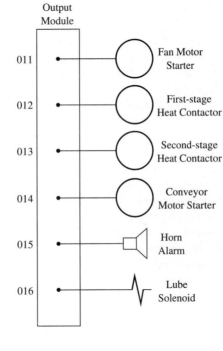

FIGURE 3-16 (a) Diagram showing the symbol and address for each input and output for the brush manufacturing mixing and heat-treating system. (b) List of input switches and output devices for the mixing and heat-treating system.

INPUTS		OUTPUTS	
START PB	001	011	FAN MOTOR STARTER
STOP PB	002	012	STAGE-1 HEATING CONTACTOR
LS1	003	013	STAGE-2 HEATING CONTACTOR
LS2	004	014	CONVEYOR MOTOR STARTER
FLOW SW.	005	015	HORN ALARM
TEMP. SW.	006	016	LUBE SOLENOID
ALARM RESET	007		
GOOD PART SENS	008		
BAD PART SENS	009		
PART CNT RESET	010		

(b)

contacts. The first line of the logic shows a traditional start/stop circuit. The start button is connected to input address 001 so all of the contacts in the program that are numbered 001 will represent the start push button. The stop push button is connected to address 002. You should notice from the I/O diagram in Fig. 3-16a that the start push button is wired as a set of NO push button contacts, and the stop push button is wired as a set of NC contacts.

When the operator is ready to start the machine, the start push button will be depressed. You will notice in the ladder diagram that both the start and stop switches use NO contacts. The stop push button that is hard wired to the input module must use NC contacts as a fail-safe condition. This means that the switch will stop the system if the switch becomes faulty or if one of the wires connected to the switch comes loose from its terminals. Since the real stop switch uses NC contacts, its programmed contacts must be normally open for the circuit to operate correctly. At first glance this seems contradictory, but since the stop push-button contacts are held closed by a spring, they will send a signal to the input module at all times unless the push button is depressed. This means the NO contacts (002) in the program will change state and pass power to the start push-button contacts (001) as long as the contacts in the hardware part of the stop push button remain closed. Since contacts 002 (stop push button) will continually pass power, output 017 in line 1 will become energized as soon as the start push button is depressed.

You should also notice in rung 1 that the 001 start contacts have a set of NO 017 contacts connected in parallel with them. The parallel contacts have the number 017, which means they are controlled by the output 017. These contacts will act as a seal-in circuit around the start contacts. When the start push button is depressed, its hardware contacts are closed momentarily. Then when the operator's finger is removed from the button, the contacts will open again because of the spring pressure. Since the programmed start contacts 001 will only be closed momentarily, the output would become de-energized as soon as the start button is released. The seal-in contacts from output 017 will provide an alternate path around the start contacts. Anytime the stop push button is depressed, its contacts 002 will open and de-energize output 017.

Output 017 has been chosen as the output in this rung because it is an *internal relay*. Since this rung represents the *permissive* start-circuit condition, it means that output 017 does not need to be connected directly to an output module address, since it will not provide power directly to any of the output devices. A second set of 017 NO contacts is used in rung 2 with two door limit switches to provide a signal to energize the fan motor starter coil. If both doors on the oven in the heat-treating section are closed, their limit switches will pass power and the fan motor starter will be energized as soon as the start/stop permissive circuit (output 017) becomes energized. If either of the doors are opened while the oven is in use, the limit switch for the door will open and de-energize the motor starter. Since the limit switches are in rung 2, their operation will not affect the start/stop circuit in rung 1.

The start/stop permissive contacts (017) are also used to energize the coil for the conveyor motor starter (014) in rung 7. The motor starter for the conveyor motor can be de-energized by the set of NC 006 contacts from the temperature switch which will open anytime the temperature exceeds 180°F, or by depressing the stop push button, which will de-energize the 017 permissive contacts.

3.11.1 Using Latch and Unlatch Coils

Rungs 8 and 9 use a special type of coil to ensure that an alarm light and horn are energized any time the temperature exceeds 180°F. The temperature switch 006 is used to activate the alarm. A problem could arise if the temperature returns to a safe value before the operator determines that a problem has occurred. The latch coil -(L)- 015 in rung 8 will operate like a flip-flop in that it will remain energized even if contacts 006 become de-energized. The alarm light and horn will remain energized until the unlatch coil 015 in rung 9 is energized. The operator must depress the alarm acknowledge button 007 and the NC 006 contacts in rung 9 must be closed to indicate the temperature is below 180°F. The latch and unlatch coils must both have the same address number to operate as a pair.

3.12 USING TIMERS TO STAGE THE ELECTRIC HEAT

The electric heating elements for the heat-treating part of this machine are energized in two stages. The logic for staging the heating is shown in rungs 3, 4, 5, and 6 in the diagram in Fig. 3-14. A time delay is needed to allow a 20-second delay between the time the first-stage heating element is energized and when the second-stage heating element is energized. The time delay is necessary to prevent the oven from becoming too hot too quickly. Since the first element is allowed to energize 20 seconds prior to the second element, the oven will begin to heat up slowly prior to having full heat applied.

In rung 3 the contactor for the first-stage heating element is energized if two conditions are met. First, contacts 005 from the air flow switch must be energized, which indicate the fan motor is energized and proper air flow has been achieved. (The air switch is mounted near the fan so that it will be activated when air is moving.) Second, the oven temperature must be less than 180°F, since the temperature switch 006 is set to open its contacts at this temperature.

The control logic for this system involves electrical control and physical function. This means that even though there is no electrical connection between the start/stop circuit and the electric heat, they are connected by the physical interaction between them. The electric heat is controlled indirectly by the start/stop circuit, since the coil for the first-stage heating contactor in rung 3 cannot become energized unless the fan motor is running.

3.12.1 RTO Timers

The SLC 100 uses the mnemonic RTO to represent the *retentive timer on-delay*. The retentive timer will operate like the traditional hardware motor-driven timer discussed in Chapter 2. Fig. 3-17 shows more information about the timer instruction. Here you can see that the timers use addresses 901–932. Each timer has a *preset value* (PR value) and an *accumulative value* (AC value). The preset value is the amount of time delay the timer will provide and the accumulative value is the actual time delay that has accumulated since the timer has been energized. The timer

Timer and counter addresses 901–932

Time base 0.1 Seconds   Range 0.1 Seconds–999.9 Seconds

(a)

On-delay timer

```
        001                        903
-------]  [------------- (RTO)
                                   PR 2.0


        005                        903
-------]  [------------- (RST)
```

(b)

Nonretentive timer

```
        002                        901
-------]  [------------- (RTO)
                                   PR 2.0


        002                        901
-------]\ [------------- (RST)
```

(c)

FIGURE 3-17 (a) Information about timer address or counter addresses, the time base for timers in the SLC 100, and range for timer preset values. **(b)** RTO timer instruction used as on-delay timer. **(c)** Making a nonretentive timer from an RTO instruction.

accumulative value increments at a rate determined by the time base. The time base for the SLC 100 timers is 0.1 second. When the timer AC value equals the PR value, the timer times out, and all of the contacts under its control (contacts that have the same address as the timer) will become activated and change state.

Another important part of each timer is the reset function. Each timer must be reset at some point to be used for the next timing cycle. The timer is reset by the RST instruction. The RST instruction must have the same address as the timer to provide the reset function for that timer. For example, if the timer has an address of 903, its RST instruction must also have the address 903. When RST 903 is energized, the AC value of RTO 903 is reset to 0.

The RTO timer is retentive since its AC value will be frozen and retained in the memory address anytime power to the timer instruction is interrupted. When power is applied to the timer, it will pick up the time delay exactly where it left off and resume the time delay. Fig. 3-17b shows a method to make the RTO timer act like a nonretentive timer. This diagram shows a set of NO 012 contacts energizing the RTO instruction, and a set of NC 012 contacts connected to the RST for the timer. Anytime the 012 contacts are opened to stop the timer, the NC set connected to the RST will cause the timer to reset and return its AC value to 0.

The operation of the RTO timer in the heat-treating system is shown in rungs 4, 5, and 6. After the coil of the first-stage heating contactor (output 012) is energized in rung 3, its contacts are used again in rung 4 to energize timer 901. Since NO contacts 901 are used in rung 6, they will pass power to the second-stage heating contactor (output 013) after timer 901 times out.

The SLC 100 also provides an off-delay timer with the RTF instruction. The RTF operates exactly like a motorized off-delay timer. The accumulative time will not begin to increment until power has been applied to the instruction and then de-energized. The RTF instruction begins its operation when the transition from on to off occurs. The contacts for the off-delay timer can be normally open or normally closed.

3.12.2 Using the RTO As an Automatic Resetting Timer for Lubrication

Rungs 12, 13, and 14 use an RTO timer that automatically resets itself with a lubrication system solenoid. The automatic resetting timer function is used in this application to provide automatic lubrication at a fixed time period. The automatic reset part of this circuit is accomplished by using an RTO timer instruction and placing a set of the NC timer contacts in series with it, and placing a set of NO timer contacts in series with the timer reset instruction. Since the RTO 903 instruction has a set of NC contacts to energize it, it will begin the cycle by receiving power. Every 60 seconds the timer will close the NO 903 contacts in line 13 of the program to energize the lube solenoid and close the NO 903 contacts in rung 14 that provide a signal to the timer RST instruction. The RST instruction will cause the timer to reset and start its timing cycle again automatically. It is important to understand that the lube system only needs its solenoid pulsed to activate the lubrication system.

3.13 COUNTER OPERATION IN A PLC

The counter instruction operates similarly to the timer instruction in the PLC. The major difference between the counter and the timer is the timer instruction will continually increment its accumulative value at a rate determined by the time base when its contacts that enable power to the instruction are closed. On the other hand, the counter must see a complete contact transition from open to close each time it increments the accumulative value. This means that the contacts must also return to their open state before they can transition for a second time. It is important to note that the counter does not care how long the contacts stay closed once they transition; it only looks for the transition. Fig. 3-18 shows the instruction for a counter that includes reset operation.

The counter has preset and accumulative values like the timer. The preset value is the desired value for the counter and may represent how large a value the counter is expected to count. The maximum preset value for the counter is 9999. The counter also has several outputs under its control. When the AC value equals the PR value, the *status bit* goes HI. The status bit controls all NO and NC contacts that have the same number as the counter. The counter also has a bit that indicates if the counter is in an overflow state or an underflow state. The counter is in an overflow state if the AC count tries to exceed 9999, and the underflow state occurs if the AC count tries to go below 0. The overflow and underflow conditions both set the same bit in the counter. The bit address is determined by adding 50 to the counter address. For example, if the counter address is 902, the overflow bit would control any contacts that have the address of 952.

The counter has a reset instruction like the timer, which will cause the PR value to return to its original value. When the up-counter instruction is used, the PR value will start at zero, and when the down-counter instruction is used, the PR value will start at its PR value and end at zero. The counter status bit will energize when the PR value equals the AC value for an up counter and when the AC value equals zero for the down counter.

```
        003                  902
- - - - - -]  [- - - - - - - - - - (CTU)
                             PR 1000
                         * MAX PRESET 9999


             NO
        Counter Status
             Bit
             902                 014
- - - - - -]  [- - - - - - - - - - (    )


             NC
        Counter Status
             Bit
             902                 015
- - - - - -] \ [- - - - - - - - - - (    )
           Counter
          Over/Under
          Flow Bit
             952 *               016
- - - - - -]  [- - - - - - - - - - (    )
```

*Overflow and underflow bit address is counter address + 50. Example: 902 + 50 = 952

```
        005                  902
- - - - - -]  [- - - - - - - - - - (RST)
```

FIGURE 3-18 Example of a typical counter instruction and its reset. NO and NC status bits are shown with overflow and underflow bits.

3.13.1 Using an Up/Down Counter to Count Good and Bad Parts

Counters are used to count up, count down, and to provide reset functions. The Allen-Bradley SLC 100 provides separate up-counter (CTU) and down-counter (CTD) instructions. The reset instruction operates like the reset for the timer instruction. The counter instructions share the same memory location in the PLC as the timers, which means that they share addresses 901–932. Another important point to remember is that since timers and counters reside in the PLC memory, they never break down. When a timer in a PLC will not time, or if a counter will not count, you should suspect the inputs to the timer or counter are not transitioning, or the reset line is held HI, but you should not suspect that the timer or counter instruction is broken. You may also find the PLC has lost all or part of its program from its memory and the program will need to be reloaded.

The up counter and down counter both use the PR and AC values like the timer. The up-counter instruction will add 1 to the AC value in the counter each time the instruction is transitioned, and the down-counter instruction will subtract 1 from the AC value.

3.13.2 Using Counters in the Carbon Brush Application

The application program for the carbon brush process system uses up-counter and down-counter instructions that are addressed to the same counter. A photoelectric switch at address 008 detects all of the brushes as they pass into the oven. These contacts are used as the input to the up counter in rung 10 and they count each brush. All of the brushes are weighed as they pass out of the oven, and any that are too light or too heavy are detected by the solid-state scales. The contacts for the solid-state scales are numbered 009, which are used as the input to the down counter in rung 11. Since the up counter counts all parts and the down counter subtracts all of the bad ones, the AC value in counter 902 will represent the good parts. Anytime the counter needs to be reset, a manual push button at address 010 in rung 16 must be depressed by someone responsible for quality control. Push button 010 could be a key-lock switch if security is needed.

3.13.3 Using a Resetting Counter Application

Another useful function that counters provide in industrial applications is the ability to keep track of steps in a sequential operation and reset automatically on the final step. An example of this type of application is counting the brushes as they exit the process oven and grouping them to be packaged four per pack. Although this counter is not part of the main program (see Fig. 3-19), it could easily be added if a packaging function was required. In this example counter 905 has a preset value of 4, which represents the four brushes that will be grouped and wrapped in a plastic wrapper. Each time a brush is counted, contacts 101 are closed momentarily and then opened, which causes the counter to increment. When the fourth brush has been counted, the contacts transition for the fourth time and the counter AC value will reach its PR value. This means the counter is done and its status contacts 903 in the second rung will be closed and energize the wrapper solenoid. This causes the wrapping machine to wrap the four brushes. The NO 903 contacts in the third rung will energize the reset (RST) instruction for the counter and the counter AC value will return to zero and the counter will be ready to count the next four brushes. Since the wrapper solenoid in the second rung will only be energized for one scan cycle (approximately 20 msec), the wrapper machine must have a seal-in feature that allows it to continue its operation after its input is strobed.

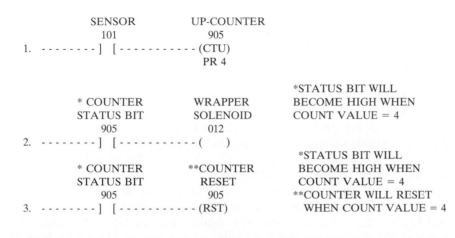

FIGURE 3-19 An example program of a counter used to count four parts for packaging. When the four parts are counted, an output pulse is sent to the wrapper machine.

3.14 TIMING AND COUNTING VALUES BEYOND 9999

One problem with counters and timers in PLCs is that their preset values are generally restricted to 9999 or less in the smaller units and 65,535 in the larger units. In modern manufacturing facilities, it is not uncommon for machines to produce over 100,000 products or parts each day, and if a timer is used to keep track of operating hours on a machine, the value can exceed 6000 hours per year. If the PLC is restricted to the smaller amounts, you can see that the preset values for counts in the counter and seconds in the timer will easily be exceeded. It is easy to add a second counter to the timer or production counter to allow the PLC to time or count an unlimited number.

Fig. 3-20a shows an example of a timer that has a counter added to it to provide the extended values. Fig. 3-20b shows two counters that work together to provide counts up to 20,000, and Fig. 3-20c shows an example of two counters that can count up to 300,000. When timers and counters use extra counters to extend their capacity, the extra counter is called a *cascaded counter.*

In the example in Fig. 3-20a, the first rung shows a normal RTO timer with the address of 904. A set of enable contacts is connected in series with timer RTO 904. The preset value of this timer is set to 600 to represent 60 seconds. When the timer times out, its status contacts 904 in the second rung will close and add 1 count to counter 905. The preset value of counter 905 is set at 60 to represent 60 minutes per hour. Each time the 904 timer times out, 60 seconds (1 minute) has passed and its 904 contacts in the third rung will reset the timer so it will start its timing cycle again. The AC value in the counter will always be the number of hours the timer has operated. Additional counters can be added to count 24 hours per day, 7 days per week, and 52 weeks per year. You can see that any number of counters can be cascaded by simply using the status contacts of the counter before it to enable it.

Fig. 3-20b shows counter 906 being enabled by contacts 007. Each time the contacts transition they will add 1 to the AC value in counter 906. When the AC value in counter 906 reaches 9999, it will close its overflow contacts 956 in the second rung. The 956 contacts will remain closed, which will allow enable contacts 007 to increment the second counter 907. When the second counter AC value reaches 9999, both counters can be reset by the third and fourth rungs. A manual reset could also be added to prevent the AC value in both counters from being lost. The grand total of both counters would be the sum of counters 906 and 907.

Figure 3-20c shows a way to use two counters to count values up to 99,999,999. This is accomplished by using the status contacts of the first counter to provide a pulse to the second counter each time the first counter's AC value reaches the PR value. In this example, the PR value is set at 9999. Each time the first counter has 9999, its status bit 908 in the second rung will close. The 908 status contacts in the third rung will also close. This will reset the 908 counter and allow it to start counting again. The 909 counter can be pulsed 9999 times so this means that the two counters can count up to 99,999,999 together.

If the PR value for each counter is set at 1000, the total count can be read by placing the AC value of counter 909 to the left of the AC value of counter 908 and reading the values as a six-digit number. For example, if counter 909 has a value of 102, and counter 908 has a value of 358, the grand total for both counters

```
            ENABLE                  MAIN
            SWITCH                  TIMER
             001                     904
1  --------] [----------- (TOF)
                                   PR 600

            CLOSES
       ONCE EVERY TIME            CASCADE
       TIMER TIMES OUT            COUNTER
             904                     905
2  --------] [----------- (CTU)
                                   PR 60

            CLOSES                RESETS
       ONCE EVERY TIME            MAIN
       TIMER TIMES OUT            TIMER
             904                     904
3  --------] [----------- (RST)
```

(a)

```
            ENABLE                  MAIN
            SWITCH                 COUNTER
             007                     906
1  --------] [----------- (CTU)
                                  PR 9999

            CLOSES
          AFTER MAIN
          COUNTER    ENABLE      CASCADE
          IS FULL    SWITCH      COUNTER
             956       007          907
2  --------] [------] [----- (CTU)
                                  PR 9999

    CLOSES AFTER CASCADE      RESETS MAIN
      COUNTER IS FULL           COUNTER
             957                   906
3  --------] [----------- (RST)

    CLOSES AFTER CASCADE      RESETS CASCADE
      COUNTER IS FULL           COUNTER
             957                   907
4  --------] [----------- (RST)
```

(b)

```
            ENABLE                  MAIN
            SWITCH                 COUNTER
             008                     908
1  --------] [----------- (CTU)
                                  PR 9999

      CLOSES AFTER MAIN
          COUNTER               CASCADE
       REACHES 9999             COUNTER
             908                   909
2  --------] [----------- (CTU)
                                  PR 9999

    CLOSES AFTER CASCADE      RESETS MAIN
      COUNTER IS FULL           COUNTER
             908                   908
3  --------] [----------- (RST)
```

(c)

FIGURE 3-20 **(a)** Example of a counter added (cascaded) to a timer to provide timer values above 9999 seconds. **(b)** Two counters cascaded to count up to 20,000. **(c)** Two counters cascaded to count up to 99,999,999.

would be 102,358. If you place 1000 in each counter PR value, the capacity of each counter is diminished, but it is much easier to total the count quickly.

3.15 SEQUENCERS USED TO CONTROL THE BATCH MIXING FOR THE CARBON BRUSH APPLICATION

The carbon brush manufacturing process must control the mixing of several ingredients. The mixing and blending process occurs in a very specific sequence. Since this operation is sequential, it will be easier to control it with the SLC 100 sequencer instruction (see Fig 3-21). From this example you can see that the sequencer instruction looks like a timer in the program except its mnemonic is SQO. The sequencer also uses a reset (RST) like the timer and counter. To fully understand the sequencer, you need to see that it controls its outputs like a matrix that is made of columns and rows (see Fig. 3-22).

3.15.1 Sequencer Application for the Carbon Brush Mixing System

The sequencer instruction in Fig. 3-22 is filled in to represent the industrial mixing application of the carbon brush manufacturing process. Since carbon brushes are made from four raw materials (lamp black, pitch blend, carbon, and binder), the system has four solenoids that control dispensing these materials. The mixing system also has two major motors: the mixer/blender and the extruder motor. Each of these six outputs is identified across the top of the matrix.

Seven steps in the process are identified down the left side of the matrix. The process begins with the zero step, which is used to reset and start the process. You should notice that all the outputs are turned off at this point by placing a zero in each column. When the sequence begins, an internal timer is used to drive the sequencer and ensure that each step lasts for 3 minutes. Step 1 is the start of the process and the lamp black dispensing solenoid is energized for 3 minutes.

At the end of 3 minutes, the timer times out and sends a signal to the sequencer that advances it to step 2. During step 2 you will notice that the lamp black solenoid remains on, and that the pitch blend dispensing solenoid and the mixer motor starter are energized to start mixing the product. The amount of lamp black in the product requires that its solenoid remains on for 6 minutes (two of the sequencer steps). If it was required to be on for a longer period of time, additional steps would be inserted to ensure it remained on for the correct amount of time. The

```
    ENABLE
    SWITCH      SEQUENCER
     001                    908
- - - - - - ] [ - - - - - - - - - - - (SQO)

    MANUAL       SEQUENCER
      PB           RESET
     003                    908
- - - - - - ] [ - - - - - - - - - - (RST)
```

FIGURE 3-21 Example of the sequencer instruction for the SLC 100.

STEPS	DISPENSE LAMP BLACK	DISPENSE PITCH BLEND	DISPENSE CARBON	DISPENSE BINDER	MIXER-BLENDER MOTOR	EXTRUDER MOTOR
0	0	0	0	0	0	0
1	1	0	0	0	0	0
2	1	1	0	0	1	0
3	0	0	1	0	1	0
4	0	1	0	1	1	0
5	0	0	0	0	1	0
6	0	0	0	0	0	1

FIGURE 3-22 Example of the matrix for a sequencer instruction. The outputs represent the solenoids and motor starters for the mixing part of the carbon brush manufacturing system. (Courtesy of Rockwell Automation's Allen Bradley Business.)

amount of time delay that is used in the timer is the lowest unit of time that any one step requires.

When the timer times out and moves the sequencer to step 3, you will notice that the lamp black and the pitch blend solenoids are turned off, but the mixer motor remains energized. The carbon solenoid is also energized during this time, which allows carbon to be dispensed while the mixing is occurring. This causes the product to be mixed and blended thoroughly.

When the sequencer moves to step 4, you will notice that the pitch blend solenoid is energized again. Now the binder solenoid is energized so that binder is blended with the product as it is continually mixed. At the end of this 3-minute period, the timer advances the sequencer to step 5 where all of the solenoids except the mixer blender motor are de-energized. At this step the mixer continues to be energized, so the product is continually mixed for an additional 3 minutes.

In the final step, the extruder motor starter is energized and it begins to turn a large fluted screw, which forces the material out through a die. This process is called extrusion and it ensures that all of the brushes are the same width. A knife blade continually cycles so that each carbon brush is cut to the same size. The carbon brushes drop onto a metal conveyor belt where they are moved into the oven for the curing process. You should remember that all of the logic in the controller prior to the sequencer is used to control the conveyor and oven. When the sequencer ends step 6, the sequencer is reset to step 0 where all the outputs are de-energized until the process is ready to start again.

3.15.2 Programming the Sequencer Matrix into the SLC 100

The matrix is divided into three major sections that are identified across its top row: *bit address data, program code,* and *preset values.* Each of these sections is discussed separately. Fig. 3-23a shows each of these sections for the sequencer instruction. Fig. 3-23b shows a sequencer instruction completely filled in.

In batch-mixing applications, the sequencer will control six outputs 111–116 as bit addresses. These are the addresses to which the solenoids and motor starters are connected. The sequencer knows that the outputs are going to be addresses 111–116 because the person who enters the program uses the table in Fig. 3-24 to locate the output addresses. You can see from the table that the output addresses are placed into groups of six real outputs and two internal outputs. It is important

CLASSIFICATION □ –(SQ1)– □ –(SQ0)–   ADDRESS: _____   □ TIME DRIVEN □ EVENT DRIVEN   GROUP NUMBER: _____

BIT ADDRESS DATA									PROGRAM CODE		PRESET VALUES
	B				A				Data B	Data A	
Bit Addresses →											
Mask Data →											
Step Data → 0											¦ ¦ ¦
1											¦ ¦ ¦
2											¦ ¦ ¦
3											¦ ¦ ¦

FIGURE 3-23 **(a)** Diagram of the sequencer instruction with the three major sections highlighted. The major sections are the bit address data, the program code, and the preset values. **(b)** Example of the sequencer instruction completely filled in for the carbon brush mixing operation. (Courtesy of Rockwell Automation's Allen Bradley Business.)

CLASSIFICATION □ –(SQ1)– ☒ –(SQ0)–   ADDRESS: _904_   □ TIME DRIVEN □ EVENT DRIVEN   GROUP NUMBER: _1_

BIT ADDRESS DATA									PROGRAM CODE		PRESET VALUES
	B				A				Data B	Data A	
Bit Addresses →	118	117	116	115	114	113	112	111	Data B	Data A	
Mask Data →	0	0	1	1	1	1	1	1	3	F	
Step Data → 0	0	0	0	0	0	0	0	0	0	0	1 ¦ 8 ¦ 0
1	0	0	1	0	0	0	0	0	2	0	1 ¦ 8 ¦ 0
2	0	0	1	1	0	0	1	0	3	2	1 ¦ 8 ¦ 0
3	0	0	0	0	1	0	1	0	0	6	1 ¦ 8 ¦ 0

to use the addresses of only one group in any one sequencer instruction and to use them in the order of least significant bit (LSB) address in the rightmost column.

In the example in Fig. 3-23b you can see the programmer entered addresses 111–116 in the matrix instruction under the column of Bit Address Data, and in the row marked Bit Addresses. If the outputs were to be connected to addresses

External Output Bit Addresses and Internal Bit Addresses (SQO Sequencers)			
Bit Addresses	Group Number	Bit Addresses	Group Number
011–016 (output) 017–018 (internal)	0	411–416 (output) 417–418 (internal)	4
111–116 (output) 117–118 (internal)	1	511–516 (output) 517–518 (internal)	5
211–216 (output) 217–218 (internal)	2	611–616 (output) 617–618 (internal)	6
311–316 (output) 317–318 (internal)	3		

FIGURE 3-24 A table showing the output addresses assigned to each group number for a sequencer. (Courtesy of Rockwell Automation's Allen Bradley Business.)

211–216, the programmer would have entered these numbers in this part of the sequencer matrix. Since this application only requires six outputs to control its solenoids and motor starter, only one group needs to be selected. If more than eight outputs were needed, a second group would have to be selected and it would be controlled by a second sequencer.

3.15.3 Selecting a Mask for the Sequencer

The next step of the programming process is to select a *mask* for the outputs. The mask is a set of 1's and 0's that is placed in the matrix in the row directly below the Bit Address row, which indicates the addresses you are placing under control of the sequencer. If you only needed four of the outputs to be controlled by the sequencer, you could place a 0 in the mask of the other outputs. Then they would not be controlled by the sequencer and they could be used in the ladder logic as needed. The 1's and 0's in the mask are actually ANDed logically with the number addresses listed above it. This means that if a 1 is placed in the mask directly under address 111, output 111 will be controlled by the sequencer. If a 0 was placed in the mask directly below address 112, it would mean that address 112 would not be used by the sequencer. (*Note:* In the SLC 100 the mask is placed in the sequencer during the programming phase of the project and it is normally not changed.)

The mask is determined by a four-digit hexadecimal number shown in Fig. 3-25. From this table you can see that the numbers use binary 1's and 0's to represent numbers 0–15. Since the hexadecimal system uses a single value to represent all of the numbers, the letters A–F are used to represent the two digit numbers 10–15. Once you have determined the 1's and 0's to correspond with the outputs you want to have energized and de-energized, you can use the table to convert this value to its hexadecimal equivalent. The hexadecimal number is then placed in the sequencer matrix in the row marked Mask Data under the column marked Program Code (Data B and Data A). You should notice that since addresses 111–114 are listed under column A, and addresses 115–116 are listed under column B, the mask

Hex Mask and Hex Data Codes
(for personal computer software data form)

FIGURE 3-25 Hexadecimal numbers used to represent the mask for the sequencer. (Courtesy of Rockwell Automation's Allen Bradley Business.)

Binary Value	Hex Value
0000	0
0001	1
0010	2
0011	3
0100	4
0101	5
0110	6
0111	7
1000	8
1001	9
1010	A
1011	B
1100	C
1101	D
1110	E
1111	F

for addresses 111–114 should be placed in Data A column and the mask for addresses 115–116 should be placed in Data B column.

3.15.4 Setting the Time for Each Step of the Sequencer

In our example for mixing and blending the products to make the carbon brushes, we set each step to last for 3 minutes. The value for the time delay is placed in the column in the matrix that is identified as Preset Values. The time must be listed in seconds. Since the example needed 3 minutes, the value 180 would be entered into the column for each step to represent the 3 minutes of time required. The timer that keeps track of this time delay is integrated as an internal part of the sequencer instruction and no additional external timer is required. The time delay for each step does not need to be the same amount of time.

3.16 OTHER SEQUENCER FUNCTIONS

The sequencer instructions can be cascaded together to provide complex machine sequence control. The sequencers can be driven by time or by events that occur in the machine operation. It is also possible to drive sequencers backward or forward to specific steps for troubleshooting or other purposes.

3.17 CIRCUITS FOR INPUT AND OUTPUT MODULES

At times you will need to connect a wide variety of switches to input modules and a wide variety of output devices to output modules. It will be necessary to understand the electronic circuits inside these modules that provide the signal conversion and the devices that provide circuit isolation. Fig. 3-26 shows the electronic circuits for an ac voltage input module, an ac voltage output module, and the relay output module. The circuit for the ac voltage input module is shown in Fig. 3-26a. From this diagram you can see that the circuit represents a single circuit in the module. Typical modules will have 2, 4, 8, 16, or 32 of these circuits that will all be the same. The circuit has two terminals for the input signal to be connected: the line *ac hot* and the *ac common*. The switch that is being monitored by the module would be connected in series with the ac hot terminal. This means that L1 voltage from the ac power supply would have to be connected to the switch so that when it closes, it would pass this voltage to the ac hot terminal of the module. The L2 voltage from the ac power supply would be connected directly to the ac common terminal on the module. If the module has multiple circuits, the common voltage can be jumpered from common terminal to common terminal for each circuit.

When the switch that is connected to the ac hot terminal is closed and passes voltage to the input part of the module circuit, the voltage passes through the resistive network that will drop the 110 volts to a lower voltage. The lower voltage flows through the bridge rectifier where it is turned into pulsing dc voltage. The

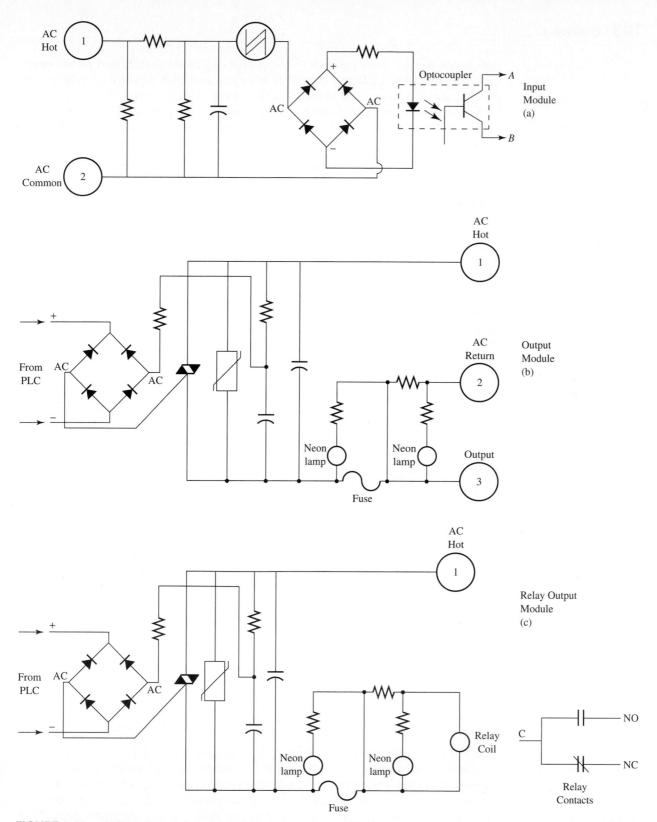

FIGURE 3-26   **(a)** Electronic circuit for a 110 volt ac input module. **(b)** Electronic circuit for 110 volt ac output module. **(c)** Electronic circuit for relay output module.

pulsing dc voltage flows through the light-emitting diode (LED) part of the opto-coupler. When the LED has current flowing through it, it produces a light that is focused directly on the base of the phototransistor mounted inside the optocoupler. The phototransistor will go into conduction when light is focused on it and signal current will flow through its emitter-collector circuit into the input bus of the PLC. When the PLC receives the signal, a 1 is written into the image register during the next scan cycle.

3.17.1 Electronic Circuit for a 110 ac Output Module

Fig. 3-26b shows the electronic circuit for a 110 volt ac output module. From this diagram you can see that each circuit has three terminals: the ac hot and ac return (common) that supply voltage to power the module, and the terminal for the output signal. The circuit for the output module uses a bridge rectifier and a triac to convert the low voltage dc signal from the PLC bus to control an ac output voltage signal of 110 volts. This is a unique circuit because the bridge rectifier is generally thought of as a device that converts ac voltage to pulsing dc voltage. In this circuit the bridge rectifier is going to act as a switch for the ac current. When the ac hot and ac return are applied to the circuit, voltage will flow completely through the circuit until it reaches the triac. The triac is an electronic device that will act like an open switch until it receives a pulse on its gate. (Chapter 4 explains the operation of the triac in detail.) When the triac receives the gate pulse, it will go into conduction and pass the voltage signal on to the output terminal of the circuit. The pulse for the triac gate is supplied by ac current flowing through the bridge rectifier. If the dc voltage side of the rectifier does not have current flow, the ac side will not have current flow. When the PLC sends a signal from its output bus to the module, it will allow current to flow on the dc side of the bridge, which will allow the ac voltage to flow through the bridge and provide a signal to the gate of the triac.

When voltage flows through the triac, it will move on to the output terminal of the module. The output section of this circuit has a neon indicator lamp connected so that it will illuminate anytime voltage is present at the output terminal. This indicator lamp is called the status indicator since the troubleshooting technician can look at it and quickly determine if the output circuit is receiving power. Some modules use an LED instead of a neon indicator lamp. You should also notice a fuse is connected in series with the output signal. A second indicator lamp is connected to this part of the circuit to indicate the fuse has blown. If the fuse is good and providing a current path, the neon indicator will not receive enough voltage to illuminate, and if the fuse blows (opens) a small amount of current will flow through the indicator lamp and it will illuminate to indicate a fuse has blown.

3.17.2 Electronic Circuit for the Relay Output Module

The electronic circuit for the relay output module is shown in Fig. 3-26c. From this diagram you can see that the circuit for this module is very similar to the regular ac output module except the relay output module has a small relay mounted in the circuit board where its coil can be connected directly to the output circuit. The relay provides a set of NO and NC contacts with a single common point. The contacts provide complete isolation from all voltages in the PLC and in the module. This means that the output signal from the PLC will act to open or close the relay contacts.

The relay contacts are used quite frequently in applications where the PLC must provide signals to a variety of other electronic devices such as input signals to robots. It is possible to have one robot that uses a current sourcing signal as an input, while two others use current sinking and low-voltage dc. This means that three different types of output modules would have to be used or several separate signal conditioning circuits would have to be made to accommodate all of these signals. The isolated contacts of the relay output module allow each robot to use its own power supply to send voltage to one side of the relay contacts, and when the contacts close, the signal is sent back to the robot as an input. This means each robot will receive the correct voltage at the correct polarity without interfering with other devices that are also connected to the PLC. Since the relay output module has both open and closed contacts available, differences in signal logic can also be accounted for, since some electronic devices look for the input signal to transition from on to off, while normally they look for the transition from off to on.

3.18 WIRING INPUT SWITCHES AND OUTPUT DEVICES TO THE PLC

All of the inputs and outputs that are used in the PLC system must be connected to the input or output module hardware. Fig. 3-27 shows an example of wiring a push-button switch to an input address and a motor starter coil connected to an output address. From this diagram you can see the location of the screws on the SLC 100 and where the wires from the switch and coil for the motor starter are connected.

The hard-wired circuit of a PLC must also be protected by a real relay called a *master control relay.* This relay will be identified as CRM (control relay master) in the diagram shown in Fig. 3-28. In this diagram you can see that all of the voltage to the input and output modules is controlled by the contacts of the CMR. The coil of the CMR is connected to a start/stop push-button station. This allows the power to be shut off to the inputs and outputs by depressing the stop button. You should notice from this diagram that power is connected directly to the processor.

3.19 ANALOG INPUT MODULES

The SLC 100 has an analog input module that allows analog values to be used as alarm inputs. Analog signals vary voltage or current from a minimum value to a maximum value. The analog voltage signal usually is a 0–10 volt signal and the current signal is 4–20 mA. The analog input module for the PLC has four separate channels to read four signals. Each signal has two setpoints that are programmable. This means that the voltage signal can come from a sensor, such as a level sensor in the bins for the carbon brush manufacturing process system. The sensor would send 10 volts when the bin is completely full, 5 volts when the bin is half full, and 0 volts when the bin is empty. The main feature of the analog signal is that every voltage between 0–10 volts will be available for the sensor to send to indicate the level of the raw material in the bin. The analog system generally has resolution of approximately 0.1 volt.

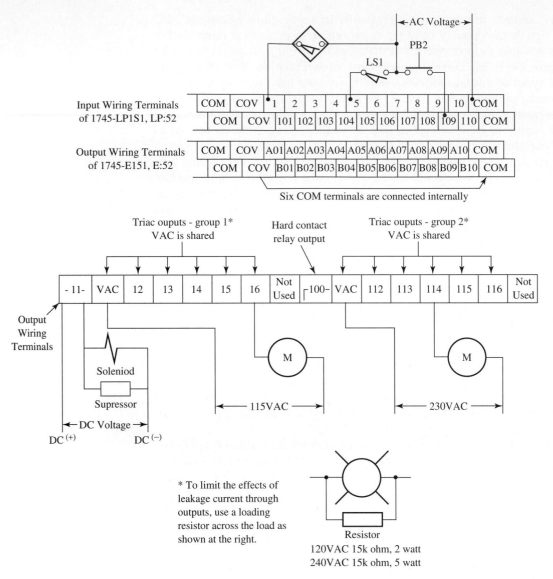

Input Wiring Terminals
of 1745-LP1S1, LP:52

Output Wiring Terminals
of 1745-E151, E:52

Six COM terminals are connected internally

Triac ouputs - group 1*
VAC is shared

Hard contact
relay output

Triac ouputs - group 2*
VAC is shared

Output
Wiring
Terminals

Soleniod

Supressor

115VAC

230VAC

DC Voltage

DC ⁽⁺⁾   DC ⁽⁻⁾

* To limit the effects of
leakage current through
outputs, use a loading
resistor across the load as
shown at the right.

Resistor
120VAC 15k ohm, 2 watt
240VAC 15k ohm, 5 watt

FIGURE 3-27 Wiring a push-button switch to an input module and wiring a motor starter to an output module of an SLC 100. (Courtesy of Rockwell Automation's Allen Bradley Business.)

The programmable values for the analog input will act as alarm points. For example, if it is important to know when the bin is 90% full so the bin would not overfill, the setpoint would be set for that level and a set of contacts in the analog module would transition when the bin reached 90% or above. The second setpoint could be set at 20% to transition a set of contacts to indicate anytime the bin was below 20% full. The alarms would indicate if the bin was too full or too empty. Since the SLC 100 does not have the capability to work with numbers, no other math function such as averaging can be accomplished. In larger PLCs, a full set of math instructions is available to perform math and compare functions. Also output analog features would be available to allow an analog output signal to be sent to motor drives or valves to operate them from 0–100% in smooth increments of 0.1 volt. The SLC does not provide analog output signals.

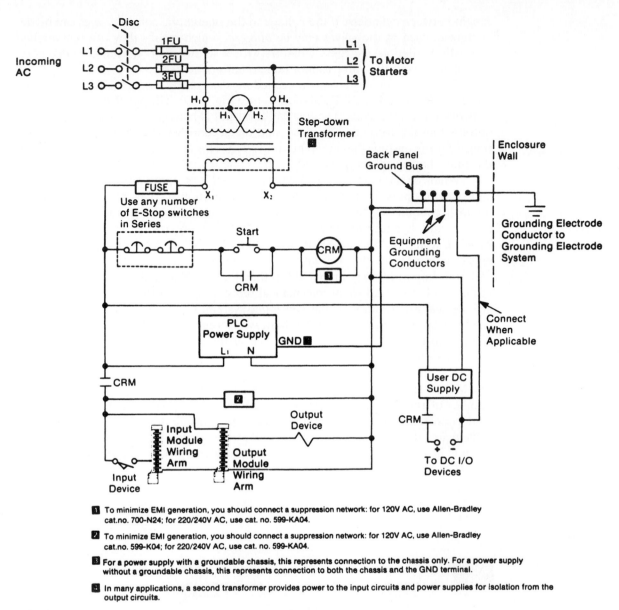

FIGURE 3-28 Electrical diagram of the master control relay that is used to disconnect power to all inputs and outputs. Notice the relay in the diagram is identified as CRM (control relay master). (Courtesy of Rockwell Automation's Allen Bradley Business.)

1 To minimize EMI generation, you should connect a suppression network: for 120V AC, use Allen-Bradley cat.no. 700-N24; for 220/240V AC, use cat. no. 599-KA04.

2 To minimize EMI generation, you should connect a suppression network: for 120V AC, use Allen-Bradley cat.no. 599-K04; for 220/240V AC, use cat. no. 599-KA04.

3 For a power supply with a groundable chassis, this represents connection to the chassis only. For a power supply without a groundable chassis, this represents connection to both the chassis and the GND terminal.

4 In many applications, a second transformer provides power to the input circuits and power supplies for isolation from the output circuits.

3.20 MASTER CONTROL RESET AND ZONE CONTROL LAST STATE

There are times when the PLC controlled system needs to have a select number of its outputs de-energized immediately regardless of the condition of the logic for the rung. For example, if a system has a number of pneumatic cylinders that are used to move boxes onto a conveyor, at times the boxes have a tendency to get caught between cylinders. If the voltage to the pneumatic control valves can be de-energized, each of the valves may be *plugged,* which means they can be switched manually by depressing their actuators with a screwdriver. If voltage remains on the solenoid coil, the valve cannot be manually switched.

The PLC provides a function called *master control reset* (MCR) that has the power to de-energize a group of outputs (see Fig. 3-29). The MCR function consists of two MCR output coils. The first MCR output is placed in the rung ahead of the outputs that are to be de-energized, and the second MCR output is placed in the rung after the outputs that are to be controlled. This builds a *fence* or *barrier* around the rungs having the outputs that are to be de-energized by the MCR.

When the contacts in series with the first MCR apply power to it, the MCR allows the outputs to operate in the normal condition so that they will respond to the logic condition of the rung they are in. If the contacts that enable the first MCR to open and de-energize the MCR, it will immediately de-energize all output coils between it and the second MCR instruction. The MCR acts like a programmable master control relay that is used for emergency stop conditions.

It is important to point out at this time that the National Electric Code (NEC) requires a hard-wired electromechanical relay to be used to de-energize power to all output as in a traditional relay controlled system. This means the MCR instruction can be used in addition to the real hard-wired relay, but it cannot be used instead of the real relay. A table is also provided in the figure to indicate how the MCR affects all other instructions.

The zone control last state (ZCL) instruction provides a similar function to the MCR. The ZCL uses two output instructions to set up a barrier around a number of rungs that are to be controlled. When the first ZCL rung is false, the zone control instruction *freezes* all of the outputs inside the zone to their last state. This means that all outputs in the zone that are on will remain energized, and all outputs in the zone that are off will remain de-energized. One function of using the ZCL instruction is to provide a means to use outputs more than once in a program. This programming feature is especially useful where the program has multiple functions such as a pallatizing system that can pack 4, 8, or 12 boxes per layer.

USING THE HAND-HELD PROGRAMMER

After you have determined the program that you want to enter into the PLC, you will need a hand-held programmer or programming software that resides on a personal computer to enter the program into the processor's memory. After the program is developed and entered into the processor's memory, it can be stored on EPROM (erasable programmable read-only memory) or on a floppy disk so that a copy is always available if anything happens to the copy in the PLC's memory. Fig. 3-30 shows a picture of the hand-held programmer and Fig. 3-31 shows a diagram of its keypad. The programming software also provides a means of displaying the program on the computer screen in a format that shows each input instruction as NO or NC relay contacts, and output instructions are shown as relay coils.

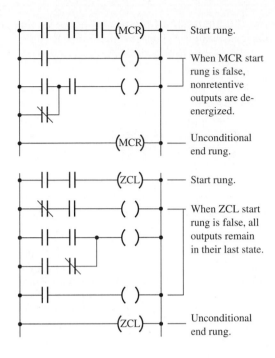

Start rung.

When MCR start rung is false, nonretentive outputs are de-energized.

Unconditional end rung.

Start rung.

When ZCL start rung is false, all outputs remain in their last state.

Unconditional end rung.

How MCR Affects Instructions

Instruction	MCR Zone Start Rung FALSE	ZCL Zone Start Rung FALSE
Nonretentive Outputs	De-energized	Remain in last state
Latch/Unlatch	Remain in last state	
Counters, RTO Timers	AC value stops incrementing, value retained. Status, overflow, and underflow bits remain in last state.	
RTF Timers*	AC value increments	AC value stops incrementing, value retained. Staus, overflow, and underflow bits remain in last state.
Sequencers	AC value stops incrementing, value retained. Stop number retained. Bit addresses remain in last state.	
Reset	Remains in last state	
Shift Register	Outputs disabled. Shifting disabled.	Outputs remain in last state. Shifting disabled.

*CAUTION Since the RTF timer runs inside of a FALSE MCR zone, its accumulator, status, and overflow bits could change state while an MCR zone is false. This could affect other outputs in your program.

FIGURE 3-29 Example of master control reset (MCR) and zone control last state (ZCL) instructions. The table indicates the condition of all outputs that fall under control of these instructions. (Courtesy of Rockwell Automation's Allen Bradley Business.)

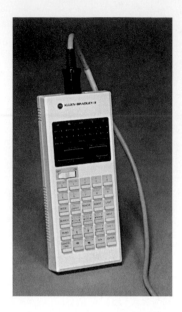

FIGURE 3-30 A hand-held programmer for the SLC 100. The hand-held programmer is also called the pocket programmer. (Courtesy of Rockwell Automation's Allen Bradley Business.)

FIGURE 3-31 Diagram of the hand-held programmer's keys. A list of the abbreviations and symbols for each function is also provided. (Courtesy of Rockwell Automation's Allen Bradley Business.)

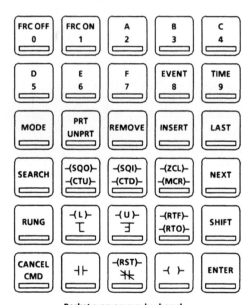

Pocket programmer keyboard.

Abbreviations and Symbols			
FRC OFF	Force OFF	–(RTF)–	Retentive Timer Off-Delay
FRC ON	Force ON	–(RTO)–	Retentive Timer On-Delay
PRT	Protect	CANCEL CMD	Cancel Command
UNPRT	Not Protect		
–(SQO)–	Sequencer Output	–(RST)–	Reset
–(SQI)–	Sequencer Input	⊥	Branch Open
–(CTU)–	Up Counter	⊐	Branch Close
–(CTD)–	Down Counter		
–(ZCL)–	Zone Control Last State	–│├–	Examine ON
–(MCR)–	Master Control Reset	╫	Examine OFF
–(L)–	Latch	–( )–	Output Energize
–(U)–	Unlatch		Shift Register (use shift key)

3.21.1 Force and Unforce Functions

The diagram of the hand-held programmer shows all of the keys that are used for programming any of the logic functions discussed so far. Other features such as *search* and *force off/force on* are also provided. The search function allows you to scan through the program to find specific instructions or rungs. The force-on and force-off instructions allow any input or output address to be energized or de-energized regardless of the logic. This means an output that controls the mixer motor on the carbon brush mixing control could be forced on to test the motor during troubleshooting.

It is important to understand that when an input or output is force on or off, it is no longer controlled by the program logic. This may result in unsafe conditions, so you must be aware of the results of forcing any inputs or outputs. It is also important to understand that if you force an input on and it is used in several places in the program that have outputs that are latched or sealed, the latch or seal condition may remain energized even after you de-energize the force instruction.

3.22 ADVANCED INSTRUCTIONS FOUND IN LARGER PLCs

In larger PLCs a wide range of advanced instructions is provided. These instructions include the complete set of math functions (addition, subtraction, multiplication, division, square root, and trigonometry functions). The numbers used in these applications can also have several different formats such as integers (whole numbers), floating-point decimals (numbers with decimal points), exponential numbers, binary, hexadecimal, and binary coded decimal. Each of the number formats can be converted to other formats. Another set of functions provided in these PLCs is the compare functions, including equal, less than, greater than, and all of the complements to these.

The larger PLCs also provide a wide range of file functions. A file is a group of memory locations used to store numbers. The functions include loading a file, unloading a file, moving all addresses from one file to another, word-shift registers, word-FIFO registers, and file logic instructions. Logic instructions, for example, allow all of the memory locations of one file to be ANDed with all of the memory functions of a second file. All of the other logic functions such as NOT, OR, NAND, NOR, and EXOR are also available.

Larger PLCs also provide analog capabilities for analog input and analog output modules. The analog modules are A/D or D/A modules that can convert voltages or currents to digital numbers or vice versa. The digital values can completely interface with all of the instructions inside the PLC and they can take advantage of closed-loop control functions. The closed-loop control functions in the PLC are provided as a menu where the setpoint, process variable, and output are entered as addresses and the range for each is set. The menu also accepts values for proportional gain, integral, and derivative (PID) values that will set the rate of reaction for the loop. All of the analog input and output signals are scaled to match the process variable in the system.

The important part of these larger systems is that they provide all of the machine language functions of the microprocessor, and they provide the original capability of the controlling input and output signals directly. In fact many of these PLCs cost less than $2000 and they provide the same functions as many minicomputers.

3.23 TYPICAL PROBLEMS A TECHNICIAN WILL ENCOUNTER WITH PLCs

As a technician you may encounter problems with PLCs at either the software or hardware level. At the software level, you may be required to lay out and enter the original ladder logic program. You may be also involved at the startup of a machine when a program has been written for the PLC, and a technician is required to test the program against the machine operation. In other cases, you may be involved in using a program to troubleshoot the system when a fault occurs. Sometimes you will be requested to make changes in the documentation that describes the operation of the system and identifies all the inputs and outputs. In each of these cases, you may need additional information to satisfactorily accomplish these jobs. This additional information may be provided in manufacturers' manuals, or in short 4–5-day training courses that are provided on specific PLCs.

You may also be involved with the hardware modules and I/O devices that are connected to PLCs. In some systems you will be requested to read the specification sheets for the signals that each module sends or receives so that you can select the proper types of electronic devices to interface with the PLC. Sometimes you will be requested to analyze a signal to see that it meets the proper criteria for data transmission. If system component failure occurs, you will be required to use all the knowledge you have about electronics to test the hardware and make decisions. You now have a better idea of how electronics have been blended with the programmable controller to provide industrial automation.

3.24 ADDRESSING OTHER ALLEN-BRADLEY PLCs

As a technician, you may find it necessary to work on other PLCs such as the Allen-Bradley PLC2. This mid-sized PLC is very common in industrial applications. When you troubleshoot a limit switch that is connected to a terminal on one of the input modules, or a motor starter that is connected to a terminal on one of the output modules, you will need to know how to decode its address to find the correct status indicator on the correct module. The address for each input and output is defined by the location of inputs and outputs in the permanent internal memory of the PLC2. A memory map for the permanent memory for the PLC2 is shown in Fig. 3-32.

From the diagram in this figure, you can see the first section of memory is called Processor Work Area 1, and this area is used only by the processor. The output image table is located at address 010 to 017 (octal). The input image table is located at address 110 to 177 (octal). These addresses will also be assigned to the coils and contacts in the program. The remainder of the memory map will be

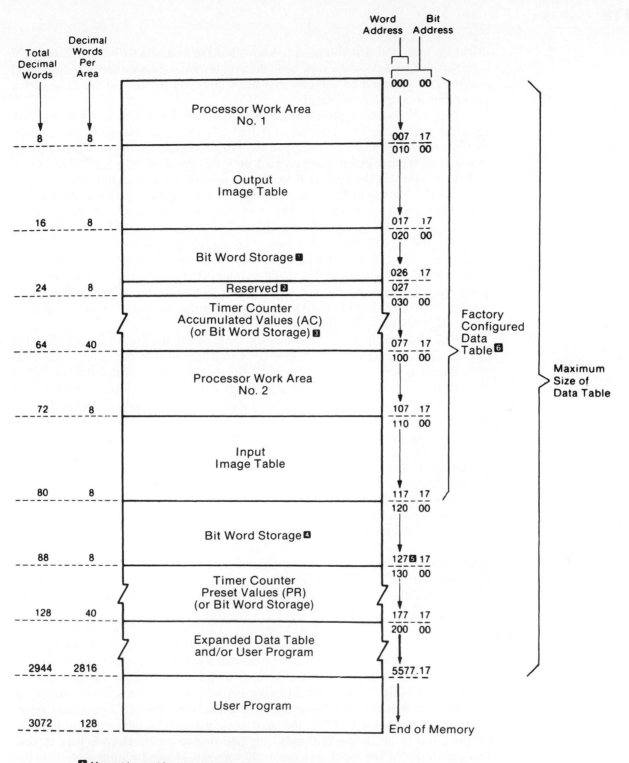

Total Decimal Words	Decimal Words Per Area		Word Address	Bit Address
			000	00
		Processor Work Area No. 1		
8	8		007	17
			010	00
		Output Image Table		
16	8		017	17
			020	00
		Bit Word Storage ❶		
24	8		026	17
		Reserved ❷	027	
		Timer Counter Accumulated Values (AC) (or Bit Word Storage) ❸	030	00
64	40		077	17
			100	00
		Processor Work Area No. 2		
72	8		107	17
			110	00
		Input Image Table		
80	8		117	17
			120	00
		Bit Word Storage ❹		
88	8		127❺	17
			130	00
		Timer Counter Preset Values (PR) (or Bit Word Storage)		
128	40		177	17
			200	00
		Expanded Data Table and/or User Program		
2944	2816		5577.17	
		User Program		
3072	128		End of Memory	

Factory Configured Data Table ❻

Maximum Size of Data Table

❶ May not be used for accumulated values.
❷ Not available for bit/word storage. Bits in this word are used by the processor.
❸ Unused timer/counter memory words can reduce data table size and increase user program area.
❹ May not be used for preset values.
❺ Do not use word 127 for block transfer data storage.
❻ Can be decreased to 48 words.

FIGURE 3-32 Example of Allen-Bradley PLC2 memory and data table. (Courtesy of Rockwell Automation's Allen Bradley Business.)

discussed later, but first you need to know that the numbers 010–077 and 110–177 are also used to decode the rack, module, and terminal location for each input and output. This information will also indicate where you would find the field wiring connected to a specific terminal.

Fig. 3-33 shows a typical address for a limit switch connected to terminal 110/03 and a motor starter connected to output 011/06. From this figure you can see that the first three digits of the input are 110. The first 1 in the address indicates that it is an input address. The second 1 indicates the address is located in rack 1, and the 0 indicates the module is located in module group 0 in the rack.

Notice that the PLC2 processor is mounted in the far left slot of the rack. In the addressing scheme, the first slot is dedicated to the processor and it is not counted as one of the rack addresses. The remaining slots are paired for addressing purposes and are called *module groups*. The module groups are numbered 0–7 and their numbers are indicated across the top of the rack. The number 03 in the input address indicates the limit switch would be connected to terminal 03 on the module. Since the addresses use the octal numbering system, terminal 03 is actually the fourth terminal on the module. An arrow indicates the exact location for the address of the limit switch.

The example in Fig. 3-33 shows the address for the output in this circuit is 011/06. The same addressing scheme is used. The first 0 in the address indicates the address is for an output. The first 1 in the address indicates the module is located in the second module group, which is number 1. (Remember that the addresses for an octal numbering system start at 0.) The terminal number for this output address is 06, which is the seventh terminal on the module.

To fully understand all of the addresses in the PLC2 memory map and in the addresses of input and outputs, you need to know that the PLC2 can have a maximum of seven racks that will be numbered 1–7. The racks are available in several sizes for 2 slots, 4 slots, 8 slots, and 16 slots. This means the system has the possibility for module groups for the largest rack and the groups will be numbered 0–7. Since each module group in the rack has two slots which will each accommodate one module, the numbering system must indicate whether the terminal number in the address is for the module in the left or right slot of the module group. The numbers for the terminals on the module that is in the left slot of a module group are 0–7, and the numbers for the terminals on the module in the right slot are 10–17. Since the PLC2 uses the octal numbering system, numbers 8 and 9 are not used.

Now when you refer to the memory map for the PLC2, you will see that the addresses for the outputs and inputs are an exact copy of the rack, module group, and terminal addresses used to determine the physical location of the module in the rack. In fact address 010 in the memory map is called Output Image Table for Rack 1, and address 020 is called Output Image Table for Rack 2. It is important to remember that the term *image table* is used because the PLC2 processor keeps an updated copy of a 1 or 0 in this area of its memory to indicate if an input or output is on or off. If the PLC uses only one or two racks, which is very typical, the remainder of the input addresses 130 through 170 and output addresses 030 and 070 can be used for other purposes such as storing the preset and accumulative values for timers and counters. Additional memory locations are available at addresses above 200. The total number of memory addresses above 200 will depend on the size and amount of memory that is purchased and installed in the PLC. The user program consisting of contacts and coils that determine the program logic is stored in the user program area at the bottom of the memory map.

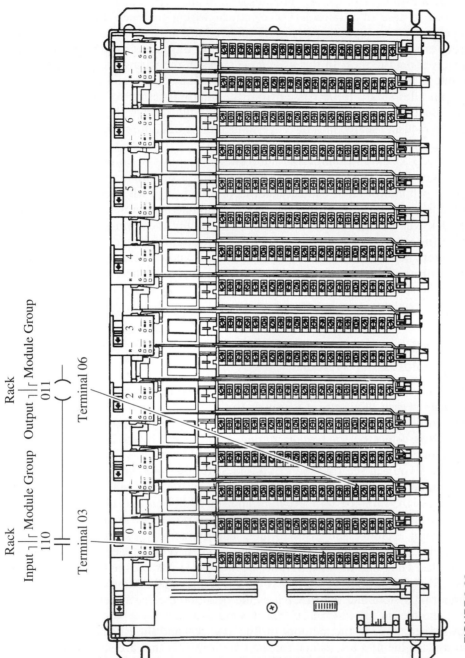

FIGURE 3-33 Allen-Bradley PLC2 addressing system for an input and an output. (Courtesy of Rockwell Automation's Allen Bradley Business.)

119

3.24.1 Allen-Bradley PLC2 Addressing System

The addressing system for the PLC2 is more complex because this system can have multiple I/O racks and a variety of input or output modules in each rack. The racks provide a means of mounting modules so that they can quickly be removed and replaced in case of a failure. The electrical wiring for each input or output is connected to a wiring arm that makes connection on the front of the module. The wiring arm can swing out of the way to allow the module to be removed and replaced from the rack. In this way, no changes in the wiring are required when a module is replaced. The rack also provides a means to connect each module to a bus that allows data to be passed from the microprocessor to each module. The bus connection is on the back edge of each module. When the module is placed into the rack and seated into the edge connector, it will automatically be connected to the processor bus.

Fig. 3-33 shows examples of addressing an input and an output. You should remember from the discussion in the previous section that the numbers in the address will indicate four distinct features of each signal: whether the signal represented is an input or an output, the rack the module is mounted in, the slot within the rack where the module is mounted, and finally the terminal on the module where the device is connected. All of this information is necessary for both the computer and the technician. The computer must know where the input signals are coming from and where output signals are going. The technician must also know where the specific input or output module is physically located in the rack to take a voltage reading.

3.24.2 Allen-Bradley SLC 500 and PLC5 Addressing Systems

The Allen-Bradley SLC 500 and PLC5 use a memory structure that is nearly identical to each other and it is similar to the PLC2 system. The main difference between the PLC2 and the PLC5 and SLC 500 memory maps is that more of the memory is dedicated for specific purposes in the PLC5 and SLC 500. Fig. 3-34 shows an example of the memory map for the PLC5 and SLC 500 systems. From this figure you can see that the PLC5 and SLC 500 can have a maximum of 1000 files numbered 0–999. The first eight files are dedicated, and the remaining memory can be assigned for the user program and for additional locations for any of the eight types of files except inputs and outputs. This provides a great deal of flexibility in assigning memory addresses to files. The memory types are: outputs, inputs, status files, bit files, timer files, counter files, control files, integer files, and floating-point files. The other major difference between file addresses in the memory of the PLC5/SLC 500 and the PLC2 is that the PLC5/SLC 500 addresses use the decimal numbering system and the PLC2 uses the octal numbering system.

The files in the PLC5/SLC 500 also are assigned file-type numbers. Fig. 3-35 shows an example of the memory for the SLC 500 and the PLC5. From the figure you can see the files are named for the type of values that are stored in them. For example, the output file is number 0 and it contains all of the output image register addresses. These are the addresses for all of the hardware output modules. The input image register is stored in file 1 and it provides a similar function for all of the input addresses.

The first nine files are assigned by the processor and will always have the assigned number. When an address is used in the program a key letter is used with

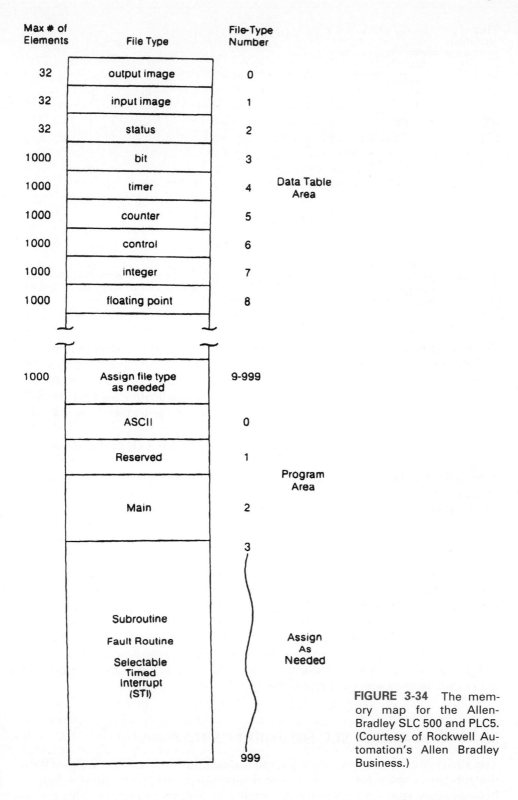

Max # of Elements	File Type	File-Type Number	
32	output image	0	
32	input image	1	
32	status	2	
1000	bit	3	Data Table Area
1000	timer	4	
1000	counter	5	
1000	control	6	
1000	integer	7	
1000	floating point	8	
1000	Assign file type as needed	9-999	
	ASCII	0	
	Reserved	1	Program Area
	Main	2	
		3	
	Subroutine / Fault Routine / Selectable Timed Interrupt (STI)		Assign As Needed
		999	

FIGURE 3-34 The memory map for the Allen-Bradley SLC 500 and PLC5. (Courtesy of Rockwell Automation's Allen Bradley Business.)

FILE NUMBER	FILE TYPE
0	OUTPUTS
1	INPUTS
2	STATUS
3	BIT
4	TIMER
5	COUNTER
6	CONTROL
7	INTEGER
8	FLOATING POINT
9–999	ASSIGNED AS NEEDED

FIGURE 3-35 Example of Allen-Bradley PLC5 and SLC 500 data files. (Courtesy of Rockwell Automation's Allen Bradley Business.)

the file type to define the addresses within the file. For example, all timer functions that use a timer file address would use the key letter T and the number 4 as in T4. The remainder of the key letters are as follows: output=O, input=I, status=S, bit=B, timer=T, counter=C, control=R, integer=N, and floating point=F. If additional files are required, the key letter is used and any file number above 10 can be assigned.

Fig. 3-36 shows an example of addressing inputs and outputs for the PLC5. This addressing system is similar to the previous systems except that the letter I is used as the first part of an input's address, and the letter O is used for the first part of an output's address. The remainder of the SLC 500 and PLC5 address that describes the location of the rack, module group, and terminal number is similar to the PLC2. The main difference in the PLC5 is that two digits are reserved for rack addresses since the PLC5 can have a maximum of 64 racks numbered 0–77 (octal).

You should notice that the programming areas of the PLC's memory are also shown in this diagram. The programming areas of memory include program 0, which is the area for ASCII, program 1, which is reserved by the processor, program 2 where the main program is stored, and the remainder of the memory where the subroutines and fault routines are stored. The PLC5 and SLC 500 can have a number of separate programs stored in the main program area, but you should remember that the processor executes all of the programs in the main program area of memory as one large program.

3.24.3 Allen-Bradley SLC 500 and PLC5 Hardware

The PLC5 is used in many industrial applications. In modern industrial facilities, the amount of space for electrical control cabinets on the factory floor is limited because more space is needed for production or process equipment. This means that the size of PLC controllers and modules must be as small as possible. The SLC 500 is much smaller than the PLC5 since it uses hardware that is similar in

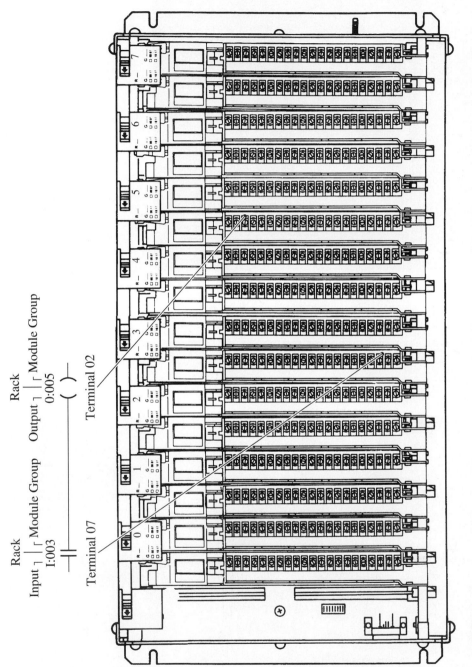

Rack

Rack

Input ┐ ┌ Module Group   Output ┐ ┌ Module Group
   I:003   O:005
     ─┤ ├─   ─( )─

Terminal 07

Terminal 02

FIGURE 3-36 Example of a PLC5 addressing system for an input and an output. (Courtesy of Rockwell Automation's Allen Bradley Business.)

FIGURE 3-37 A PLC5 processor in a rack with I/O modules. (Courtesy of Rockwell Automation's Allen Bradley Business.)

size to the SLC 100. Fig. 3-37 shows a typical PLC5 system installed in a rack, and Fig. 3-38 shows a picture of the SLC 500 system. Both systems can use software that resides on a personal computer and the SLC 500 can use a hand-held programmer for making changes to its program. The hand-held programmer is also shown in Fig. 3-38.

FIGURE 3-38 A variety of Allen-Bradley SLC 500 programmable controllers with input and output modules. A hand-held programmer is also shown. (Courtesy of Rockwell Automation's Allen Bradley Business.)

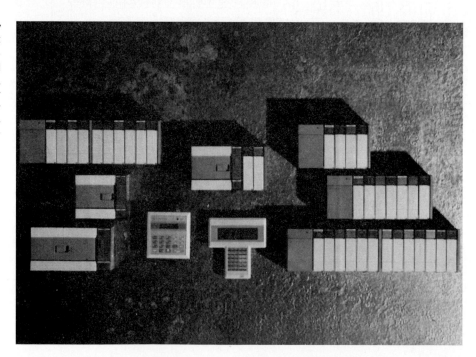

THIS RUNG IS USED TO START THE PUMP MOTOR. DEPRESS
THE START PUSH BUTTON TO START THE PUMP MOTOR.

```
             START          STOP          PUMP MOTOR MS
             001            002            017
  1  - - - - -] [- - - - - - - - -] [- - - - - - - - - - - - - -( )

             017
     - - - - -] [- -
```

THIS RUNG CONTROLS THE HIGH-PRESSURE ALARM. IF THE
PRESSURE AT THE PUMP MANIFOLD EXCEEDS 2000 PSI, THE
SYSTEM IS IN A HIGH-PRESSURE CONDITION AND PRESSURE
SWITCH PS1 (003) WILL CLOSE AND SET THE ALARM.

```
             PUMP
             MOTOR MS       PS1           ALARM
             017            003            011
  2  - - - - -] [- - - - - - - - -] [- - - - - - - - - - - - - -( )
```

THIS RUNG CONTROLS THE EXTEND CYLINDER SOLENOID.
WHEN THE EXTEND SOLENOID PUSH BUTTON 006 IS DEPRESSED
THE EXTEND SOLENOID IS ENERGIZED.

```
             PUMP           CYLINDER       EXTEND
             MOTOR MS       EXTEND         SOLENOID
             005            006            012
  3  - - - - -] [- - - - - - - - -] [- - - - - - - - - - - - - -( )
```

FIGURE 3-39 Example PLC program with tags and comments that are useful for understanding the ladder logic and for troubleshooting.

3.25 DOCUMENTING A PLC PROGRAM WITH SOFTWARE

Most PLC systems have programming software that is available from the company that made the PLC or from second-party software writers. This software will reside on a personal computer or portable computer and it will provide both on-line and off-line programming capabilities. This software will also provide a means of documenting the program by giving names (*tags*) to switches and outputs as well as their address numbers. The tags can be displayed while the program is being used to troubleshoot the machine. This means the troubleshooting technician will have the name of each switch displayed so that it is easy to identify.

Another feature provided by the software is the ability to add comments to each rung of the program. The comments allow the programmer to add a description of what the line of logic is doing and it can also explain the operation of the rung while various inputs and outputs in that rung are energized. The size of the comment can be as large as memory permits. It is important to understand that the tags and comments do not reside in the PLC; rather they are stored in the portable computer on the hard drive or on a floppy disk. Fig. 3-39 shows an example of a program that has both tags and comments.

────────────────

▶ S O L U T I O N   T O   J O B   A S S I G N M E N T

You were assigned to convert a machine that is operating on relay control to be controlled by a PLC. Your teacher should supply the original diagram. The first step of your solution should be to identify each input switch and identify its type. The inputs should then be assigned an address and a diagram similar to the one in Fig. 3-5 or the diagram in Fig. 3-16 should be prepared that shows all inputs and outputs that the system will need. The next step of your solution should include a step-by-step conversion of the relay logic to PLC ladder logic. Since the original machine used relay logic, the conversion should be rather simple. If the machine has timers or counters, you will need to assign a number for each PLC timer and counter that you will need. The preset value of each timer and counter should also be determined, and the condition that will be used to reset each should be identified. When you are ready to program the timer and counter into the program, use the examples in Figs. 3-17, 3-18, and 3-19. If your solution uses a sequencer, you will need to list all of the outputs for the sequence in the order that they need to be energized. Figs. 3-22 and 3-23 can be used as examples for your sequencer instruction. Be sure to include all of the drawings and documentation for your program with your solution.

Q U E S T I O N S

1. Describe the four basic parts of any programmable controller.
2. Explain the operation of a PLC up counter and down counter.
3. Discuss the differences between a program in a programmable controller and a BASIC or Fortran program.
4. Explain the four things that occur when the PLC processor scans its program.
5. Discuss the advantages a PLC controlled system would have over a hard-wired relay-type control system if you wanted to add one more photoelectric switch and output to detect a medium-sized box in the box sorting system described in Section 3.2.
6. Explain the function of an input module and how it completes this function. (Include the type of circuit you would find in an input module.)
7. The input module has a status indicator for each input circuit. Explain how the status indicator is used in troubleshooting.
8. Explain the function of an output module and how it completes this function. (Include the type of circuit you would find in an output module.)
9. Explain the function of an internal control relay.
10. Explain how a set of PLC contacts can be used more than once in a program.
11. Discuss the classifications of PLCs according to size.

TRUE OR FALSE

1. _____ One advantage of all PLCs is that you can use contacts with the same address more than once in the program.
2. _____ When the PLC processor is in the run mode, it does not execute its scan cycle.
3. _____ A mnemonic is a timer whose time delay is controlled by air pressure.
4. _____ When a timer in a PLC program will not time, you should suspect the timer is broken and you will need to program a replacement timer.
5. _____ A control relay (memory coil) in a PLC program is different from an output instruction in that its signal does not control a circuit in an output module and it is used mainly to determine logic conditions.

MULTIPLE CHOICE

1. A latch coil _____.
   a. maintains its state when power is interrupted.
   b. is only sealed in and returns to reset when power is interrupted.
   c. requires an unlatch coil with the same address.
   d. both a and c.
2. A seal-in circuit such as the one used for a start/stop circuit _____.
   a. maintains its state when power is interrupted.
   b. does not maintain its state when power is interrupted and must be energized after power has been returned by depressing the normally open push button.
   c. requires an unlatch coil to de-energize the output.
3. When the input contact that enables a retentive timer (RTO) is opened, the accumulated value in the timer will _____.
   a. reset (go to zero).
   b. freeze (remain at its present value).
   c. go to an undetermined value and the timer must be reset manually.
4. A nonretentive timer has its accumulative value reset _____.
   a. any time the enable contacts are opened.
   b. only when the reset for the timer is HI.
   c. any time the accumulative value in the timer exceeds 99.
5. When the PLC processor is in the program mode _____.
   a. it executes its scan cycle.
   b. it does not execute its scan cycle.
   c. it is impossible to tell whether the processor is executing its scan cycle because you are not on line.

PROBLEMS

1. Enter the conveyor box sorting program in Fig. 3-6 in a programmable controller and execute the inputs to watch the program operate. Describe the operation of each output.
2. Enter an on-delay timer and an off-delay timer in your PLC and describe the following: preset time, accumulative time, what must occur for the timer to run, what must occur for the timer to time out, and which contacts change when the timer times out.
3. Enter an up counter and a down counter in your PLC and describe the following: preset value, accumulative value, what must occur for each counter to count, what occurs when the counter's preset and accumulative values equal, what happens when the counter's accumulative value reaches 9999, which contacts change when the counter reaches its preset value, and which contacts change when the counter reaches 9999.

4. Program a timer and cascade the counters that will keep track of seconds, minutes, hours, and days.

5. Program a sequencer into your PLC and describe its operation. Your teacher may give you an example program or use the sequencer for the carbon brush application.

REFERENCES

1. *Allen-Bradley SLC 100, SLC 150, SLC 500, PLC5 Manuals.* Milwaukee, Wisconsin, 1989–1994.

2. Kissell, Thomas E., *Modern Industrial/Electrical Motor Controls.* Englewood Cliffs, NJ: Prentice-Hall Inc., 1990.

3. Kissell, Thomas E., *Understanding and Using Programmable Controllers.* Englewood Cliffs, NJ: Prentice-Hall Inc., 1986.

4. Maloney, Timothy J., *Industrial Solid-State Electronics,* 2nd ed. Englewood Cliffs, NJ: Prentice-Hall Inc., 1986.

4

SOLID-STATE DEVICES USED TO CONTROL POWER:
SCRs, TRIACS, AND POWER TRANSISTORS

OBJECTIVES

After reading this chapter, you will be able to:

1. Identify the electronic symbol of a silicon controlled rectifier (SCR) and name its terminals.
2. Draw the two-transistor equivalent circuit of the SCR and explain its operation.
3. Explain how an SCR uses an ac voltage sine wave or pulsing dc voltage to turn off (commutate).
4. Explain the difference between the firing angle and the conduction angle of the SCR.
5. Explain how an SCR is turned on by its gate.
6. Explain why it is important to use a plastic-case or isolated-case oscilloscope to make measurements when SCRs or other solid-state components are used in circuits that do not have a neutral wire, such as three-phase circuits.
7. Draw the waveforms of an SCR that is connected to a resistive load and uses an ac power supply.
8. Identify two ways an SCR may turn on from unwanted conditions.
9. Identify three solid-state devices that are used to trigger an SCR.
10. Draw an RL resonant circuit that is used to commutate an SCR.
11. Explain the difference between thyristors and transistors.
12. Identify the electronic symbol for a triac and name its terminals.
13. Explain the operation of the triac using the two-SCR model.
14. Identify the polarity of the terminals and gate of a triac in each of the firing quadrants.
15. Explain how a triac is commutated.
16. Identify the electronic symbol for a darlington transistor and name its terminals.
17. Explain the advantages of using a darlington transistor.
18. Identify the electronic symbols of the silicon controlled switch, gate turn-off device, insulated gate bipolar transistor, and light-activated SCR.
19. Explain the advantages provided by a high-voltage transistor.
20. Use the table provided to compare all of the solid-state devices used to control power.

You are working on a machine with a 120 volt dc motor and a 120 volt ac motor that are each $\frac{1}{4}$ hp. These motors have been started and stopped with single-pole switches. Your supervisor would like you to select the appropriate solid-state controller for each motor to turn it on and off and provide speed control. Since your supervisor does not really understand electronics, your assignment is to explain the type of solid-state devices that would be used to control the dc motor and the ac motor. You must also explain how these devices control current.

4.1 OVERVIEW OF SCRs, TRIACS, AND TRANSISTORS IN INDUSTRIAL APPLICATIONS

Silicon controlled rectifiers (SCRs), triacs, and high-powered transistors are used in many types of circuits to control larger voltages and currents. Many of these use 480 volt ac three-phase circuits and they may control over 50 A. Control of these large voltages and currents by solid-state devices was difficult until the last few years when larger solid-state devices were developed. Prior to this, engineers designed systems using several smaller devices in parallel to provide the capacity to control the larger currents. The advent of these larger devices has allowed control circuits for general-purpose power supplies, ac and dc variable-speed motor drives, stepper motor controls, servo motor controls, welding power supplies, high-frequency power supplies, and other types of large amplifiers. SCRs, triacs, and other solid-state devices used for switching larger voltages and currents on and off are commonly called *thyristors*. Thyristors control switching in an on-off manner, similar to a light switch which is different from a transistor that can vary the amount of current in its emitter-collector circuit by changing the bias on its base. The amount of current that flows through a thyristor must be controlled by adjusting the point in a sine wave where the device is turned on.

Prior to thyristors, the most common way to control large amounts of voltage was to use one or more transformers. Large currents were controlled by placing a large variable resistor called a rheostat in series with the load that was being controlled. The main problems with the transformers and rheostats is they are expensive and they frequently required maintenance. They also wasted energy, which is converted to heat that has to be removed and takes additional energy and lowers the efficiency of the devices. The solid-state devices provide a means to control the current in a variable fashion like a dimmer switch that is used to control the level of light. These devices can now safely control voltages over 1000 volts and currents up to 1000 A.

4.2 SILICON CONTROLLED RECTIFIERS (SCRs)

Silicon controlled rectifiers (SCRs) were first used in control circuits in the early 1960s. Fig. 4-1 shows a symbol for the SCR and identifies the anode, cathode, and gate terminals. (Notice that the cathode is identified by the letter C or the letter K.) This figure also shows pictures of several types of SCRs. The SCR acts like a solid-state switch in that current will pass through its anode-cathode circuit to a load if a signal is received at its gate. The SCR is different from a traditional switch in that the SCR will change ac voltage to dc voltage (rectify) if ac voltage is used as the power supply. The SCR is also different from a switch in that the amount of time the SCR conducts can be varied so that the amount of current provided to the load will be varied from near zero to maximum of the power supply. Fig. 4-2 shows an SCR connected to a resistive load.

The SCR will operate differently with various types of power supplies. If the power supply in Fig. 4-2 is dc voltage with no ripple, the SCR will stay on once a

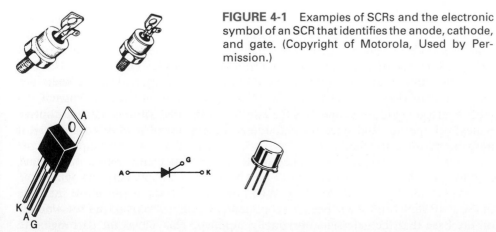

FIGURE 4-1 Examples of SCRs and the electronic symbol of an SCR that identifies the anode, cathode, and gate. (Copyright of Motorola, Used by Permission.)

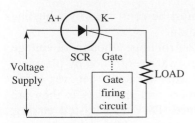

FIGURE 4-2 SCR connected to a resistive load and power supply.

positive voltage signal is applied to its gate. The only way to turn power off to the SCR is to turn off the power supply. You can see that this is not very useful. If the power supply to the SCR is pulsing dc or ac, the SCR will turn off (*commutate*) naturally at the end of each half-cycle when the voltage goes to zero. If the power supply is ac voltage, the SCR will only conduct during the positive half cycle of the waveform.

The SCR can vary the amount of current that is allowed to flow to the resistive load by varying the point in the positive half-cycle where the gate signal is applied. If the SCR is turned on immediately, it will conduct full voltage and current for the half-cycle (180°). If the turn-on point is delayed to the 90° point in the half-cycle waveform, the SCR will conduct approximately half of the voltage and current to the load. If the turn-on point is delayed to the 175° point in the half-cycle, the SCR will conduct less than 10% of the power supply voltage and current to the load, since the half-cycle will automatically turn off the SCR at the 180° point. This means that the gate of the SCR can be used to control the amount of voltage and current the SCR will conduct from near zero to maximum.

4.2.1 Anode, Cathode, and Gate

The SCR symbol in Fig. 4-1 shows an arrow in a circle like a junction diode. The *anode* and *cathode* of the SCR are identified exactly like the junction diode. The arrow is the anode, and the opposite terminal is the cathode. The third terminal of the SCR is called the *gate*. The gate is pulsed to turn on the SCR and allow current to flow through the anode–cathode circuit.

4.2.2 Two-Transistor Model of an SCR

The proper name for the SCR is the *reverse blocking triode thyristor*. The name is derived from the fact that the SCR is a four-layer thyristor made of PNPN material. Fig. 4-3a shows the four-layer PNPN material. Fig. 4-3b shows the PNPN material split apart as two transistors, a PNP and an NPN. Fig. 4-3c shows the SCR as two transistors. These figures will help you understand how the operation of the SCR can be explained by the four-layer (two-transistor) model.

The anode is at the emitter of the PNP transistor (T2), and the cathode is at the emitter of the NPN transistor (T1). The gate is connected to the base of the NPN transistor. Since the anode is the emitter of the PNP, it must have a positive voltage to operate, and since the cathode is the emitter of the NPN transistor, it must be negative to operate.

When a positive pulse is applied to the gate, it will cause collector current I_C to flow through the NPN transistor (T1). This current will provide bias voltage to the base of the PNP transistor (T2). When the bias voltage is applied to the base of the PNP transistor, it will begin to conduct I_C, which will replace the bias voltage on the base that the gate signal originally supplied. This allows the gate signal to

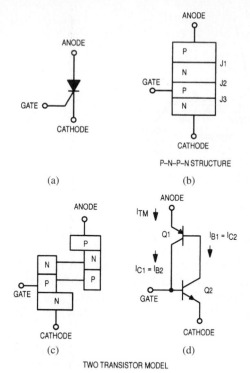

FIGURE 4-3   **(a)** Symbols of SCR. **(b)** SCR as a four-layer PNPN device. **(c)** shows the PNPN layers split apart as a PNP transistor and NPN transistors. **(d)** shows the diode operation using the two transistors. (Copyright of Motorola, Used by Permission.)

be a pulse, which is then removed since the current through the SCR anode to cathode will flow and replace the base bias on transistor T1.

Since the SCR is the equivalent of the two-transistor model, you can see that it will block reverse current just like a junction diode. If reverse voltage is applied to the anode–cathode circuit, both of the transistors will fail to conduct.

4.2.3 Static Characteristic Curve and Waveforms for the SCR

Fig. 4-4 shows the static characteristic curve for the SCR. From this figure you can see that the reverse voltage and forward voltage characteristics are similar to a junction diode. The main difference is that the SCR must be set into the conduction mode before it will begin to conduct. The main operating area of the SCR is the upper right quadrant of the graph where the SCR is conducting in the forward-bias mode. You can see that if reverse voltage is applied, the current flow will be blocked just like a junction diode. In the forward-bias mode you should notice that the current will be zero until the SCR is set into conduction. The point where the SCR goes into conduction is identified by the knee.

Fig. 4-5 shows the waveform for an SCR that is used to control current in a circuit that has ac voltage applied. The ac sine wave is shown in the top part of this diagram. The SCR current is shown in the diagram below the ac sine wave. In this example the SCR is set into conduction (*triggered*) at the 135° point of the ac sine wave. The SCR waveform shows no current flowing until the SCR is triggered at the 135° point. At that point the SCR begins to conduct and continues to conduct until the ac sine wave passes through the 180° point when its voltage goes to zero and then begins to reverse. Both of these conditions will cause the SCR to stop conducting and turn off. When the SCR is turned off, it is called *commutation*.

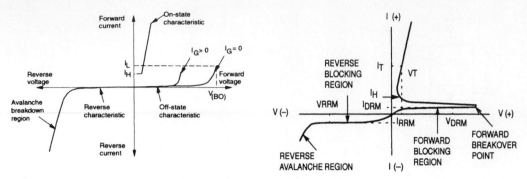

FIGURE 4-4 Static characteristic curve for the SCR. (Courtesy of Philips Semiconductors.)

The point in the ac sine wave where the SCR is triggered into conduction is called the *firing angle*. You should notice that the smaller the firing angle for the SCR, the earlier it will turn on during the ac sine wave. This means that the SCR will start into conduction earlier and, therefore, conduct more current. Since the SCR has commutation at exactly the same point in the ac sine wave (at the 180° point), the amount of current the SCR will conduct is determined by how early in the ac sine wave the SCR is triggered.

The amount of time the SCR is conducting current is also expressed in degrees (°) and it is called the *conduction angle*. This means that the firing angle and the conduction angle are the complement of each other, since their total is always 180° when the SCR is used in an ac circuit. For example, if the SCR has a firing angle of 45°, it would turn on at the 45° point and remain in conduction for the remainder of the 180°, which is three-fourths of the sine wave. This means the conduction angle would be 135°. You must be very careful when the SCR is being discussed to determine if its operation is described in terms of the firing angle or the conduction angle.

4.2.4 Waveforms of the SCR and the Load

Confusion with waveforms may also arise when you use an oscilloscope to display the waveforms across the SCR and the load resistor. Since the SCR will exhibit characteristics like a switch, the voltage will be measured across the SCR when it

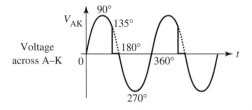

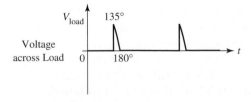

FIGURE 4-5 Waveforms of an ac sine wave applied to an SCR circuit. The top diagram shows the waveform of voltage that is measured across the anode and cathode of the SCR. The bottom waveform shows the voltage measured across the load. Notice the SCR is fired at 135°.

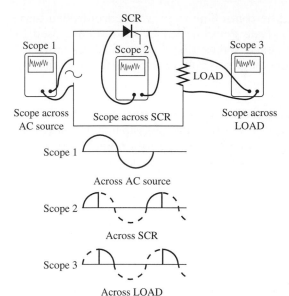

Scope 1
Across AC source

Scope 2
Across SCR

Scope 3
Across LOAD

FIGURE 4-6 Diagram of an SCR controlling voltage to a resistive load. The diagrams below the SCR circuit show the waveforms that you would see if you place an oscilloscope across the AC source, across the SCR, and across the load resistor.

is off, and the voltage will be measured across the load when the SCR is in conduction. This means that the oscilloscope will show the waveform of the firing angle when it is across the SCR, and it will show the conduction angle when it is across the load. Fig. 4-6 shows these two waveforms.

4.2.5 Methods of Turning on an SCR

The SCR is normally turned on by a pulse to its gate. It can also be turned on by three alternative methods that include exceeding the forward breakover voltage, by excessive heat that allows leakage current, or by exceeding the *dv/dt* level (allowable voltage change per time change) across the junction. The three alternative methods of turning on an SCR generally cause conditions which should be controlled to prevent the SCR from being turned on when this is not wanted.

4.2.5.1 Turning on an SCR by Exceeding the Forward Breakover Voltage ■ Each SCR has a forward-bias voltage level that should never be exceeded because it will immediately go into conduction, even when a gate signal is not applied. If this forward voltage level is exceeded, it is called *forward breakover voltage* (V_{BO}). When excessive forward voltage is applied, the SCR goes into conduction and the current that flows through the anode will *latch* the SCR in conduction even if the excessive forward voltage returns to normal levels. This means that a sudden uncontrolled pulse of excessive forward voltage could turn the SCR on and it would remain on until it was commutated. SCRs must be sized so that the V_{BO} rating is sufficiently

▼ IMPORTANT SAFETY NOTICE...

It is very important to note that you must use an isolated-case or plastic-case oscilloscope to make measurements in circuits that are not grounded. This means that you must use the plastic-case oscilloscope for all measurements in industrial circuits that are powered with 240 volt ac single-phase or any three-phase circuit. The problem of using a grounded-case oscilloscope is that one of the leads is grounded, and if you touched any component or test point in the circuit that is not grounded with this lead, you would create a direct short circuit. It is also important to understand that if you remove the grounding terminal of the power cord of a metal-case oscilloscope and use the oscilloscope to measure voltages in three-phase circuits, the metal case of the oscilloscope will be the same potential as L1, L2, or L3 and you will receive a severe electrical shock if you come into contact with the case.

larger than the circuit voltage it will be controlling to prevent uncontrolled turn-on. In some circuits additional solid-state devices such as varistors are used to prevent any undesirable forward-voltage spikes that may exceed the forward breakover level. The varistor will shunt the excess voltage to prevent it from turning on the SCR

4.2.5.2 Turning on an SCR by Leakage Current ■ The junction of the SCR is generally subjected to a buildup of heat. If this heat is not dissipated or controlled, it will allow the temperature of the junction to increase to a point where it will allow leakage current to flow through the junction. If the leakage current reaches a sufficient level, the SCR will turn on and allow forward conduction. This also allows the forward current to latch the SCR in conduction until it is commutated. The maximum junction temperature will be identified in the SCR specifications by the term $T_{j(max)}$. The temperature of the SCR junction is controlled by providing methods of moving the heat away from the SCR. This can be accomplished by providing heat sinks or fans to move the heat.

4.2.5.3 Turning on an SCR by *dv/dt* ■ The PN junction in the SCR has inherent capacitance, as do all PN junctions. The capacitance is proportionate to the size of the junction. When voltage is applied to the anode-cathode of the SCR, a small current will begin to flow, which allows the capacitance in the junction to begin to charge. If the charge is allowed to grow large enough to exceed the *dv/dt* rating, the SCR will go into conduction and latch up. The *dv/dt* is the change of voltage (Δv) over the change of time (Δt). This current can be calculated by the following formula.

$$i_C = C\ \Delta v/\Delta t$$

4.2.6 Turning on the SCR by Gate Triggering

When a positive pulse is applied to the gate of the SCR, it must be large enough to provide sufficient current to the first junction (the base terminal of transistor T1 in the two-transistor model in Fig. 4-3). If the current level of the pulse is sufficient, the first junction will go into conduction and the current flow through it will cause the second junction (transistor T2) to go into conduction. The current through the second junction will be sufficient to latch up the SCR by supplying an alternative source for the gate current. This means that the current to the gate can be removed and the SCR will remain in conduction. The SCR will commutate when the power supply it is connected to returns to the *zero voltage* level at 180° or when ac voltage is in reverse polarity (181° to 360°). If the pulse of current to the gate is too small or is not long enough in duration, the SCR will not turn on.

If you look at the SCR as a three-part device (anode, cathode, and gate), the positive pulse of gate current (i_G) is applied to the gate terminal and it will flow through the cathode where it leaves the device. The timing of the pulse is very critical if the SCR is being used to control the current proportionally. Since the current is being controlled from zero to maximum, the amount of resolution will be determined by the accuracy of the gate pulse timing.

4.2.7 Characteristics of the SCR Gate Signal

Since the most common way to turn on an SCR is through a positive pulse, it is important to understand the characteristics of this signal. The gate pulse must be positive in respect to the cathode and its current should be 0.1–50 mA. The minimum gate current (I_{GT}) and the minimum gate voltage (V_{GT}) will vary slightly as the amount required for turn-on will decrease as the temperature of the junction increases. After the SCR is turned on by the pulse, the gate will receive its current from sustained minimum *holding current* (I_{HO}) of the anode–cathode circuit. If the anode–cathode current (i_{AK}) drops below the minimum holding current, the SCR will commutate. The anode–cathode current will automatically go below the minimum when the ac voltage sine wave passes the 180° point and reverses its polarity.

A specification sheet that shows all of the typical characteristics for Motorola SCRs is provided in Fig. 4-7. Use this type of specification sheet to determine the voltage and current characteristics of an SCR you are working with, or you can use the sheet to select a replacement part. The first characteristic that is shown indicates the maximum peak repetitive forward or reverse blocking voltage for the SCRs listed on the sheet. The SCRs are listed in two groups: The first group is for blocking voltages when the gate is open, and the second group is for blocking voltages when gate current is less than 5 mA. You should notice that if you need an SCR that will block 100 volts with its gate open, you would select an SCR that is part number 2N3870 or 2N6171. If you need an SCR with a maximum blocking voltage of 600 volts with its gate open, you would select 2N3873 or 2N6174. If you need the maximum blocking voltage of 700 volts with less than 5 mA of gate current, you would select 2N3873 or 2N6174.

The third row on the page indicates average on-state current for the SCRs listed. You should notice that two values are listed for all the SCRs shown on this sheet. The first value indicates the maximum on-state current for these SCRs is 22 A if the temperature is controlled between −40°C to +85°C. If the temperature is above +85°C, the maximum current is derated to 11 A. The fourth row of this sheet shows the peak nonrepetitive surge current is 360 A, and the fifth row shows the circuit fusing should be 510 A. The remainder of the sheet shows design characteristics of all the SCRs listed on this page. These characteristics include peak repetitive forward or reverse current, forward on voltage, gate trigger current, holding current, turn-on time, and turn-off time. The top of the sheet shows general characteristics and a pin outline for all of the SCRs on this sheet and a symbol of an SCR with its anode, cathode, and gate identified.

4.2.8 Basic Gate Circuit

Fig. 4-8 shows a simple circuit that provides a gate pulse during each positive half-cycle of the ac sine wave. The fixed resistor and the adjustable resistor provide a voltage drop that sets the amount of gate voltage. Fig. 4-9 shows two sets of diagrams of the ac sine wave, the gate signal, the waveform across the SCR, and the waveform across the load. The minimum gate current I_{GT} is shown as a dotted line in the diagram of the gate signal. In Fig. 4-9a the gate current becomes strong enough at the peak of the ac cycle at the 90° point. The waveform for the SCR and the load shows the SCR turning on at the 90° point and staying on to the 180° point where the ac reverses its polarity. In Fig. 4-9b, the variable resistor has been adjusted so that the amount of voltage for the gate signal has increased significantly. This increase in voltage provides an increase in gate current so that the minimum gate current I_{GT} is exceeded at the 30° point. This means that the SCR is in conduction for 150° (30° to 180°).

This method of gate control is rather simplistic since it depends on the gate current exceeding the minimum current requirement to turn on the SCR. If the variable resistor is adjusted to a large value, the minimum amount of current

Silicon Controlled Rectifiers
Reverse Blocking Triode Thyristors

... designed for industrial and consumer applications such as power supplies; battery chargers; temperature, motor, light and welder controls.

- Economical for a Wide Range of Uses
- High Surge Current — I_{TSM} = 350 Amp
- Practical Level Triggering and Holding Characteristics —
  4 and 5.2 mA (Typ) @ T_C = 25°C
- Rugged Construction in Either Pressfit or Isolated Stud Package

2N3870
thru
2N3873
2N6171
thru
2N6174

SCRs
35 AMPERES RMS
100 thru 600 VOLTS

CASE 174-04
(TO-203AA)
STYLE 1
2N3870 thru 2N3873

CASE 311-02
STYLE 1
(Stud Isolated)
2N6171 thru 2N6174

MAXIMUM RATINGS (T_J = 25°C unless otherwise noted.)

Rating	Symbol	Value	Unit
*Peak Repetitive Forward or Reverse Blocking Voltage, Note 1 (T_J = −40 to +100°C, 1/2 Sine Wave, 50 to 400 Hz, Gate Open)	V_{RRM} or V_{DRM}		Volts
2N3870, 2N6171		100	
2N3871, 2N6172		200	
2N3872, 2N6173		400	
2N3873, 2N6174		600	
*Peak Non-Repetitive Forward or Reverse Blocking Voltage (t ≤ 5 ms)	V_{RSM} or V_{DSM}		Volts
2N3870, 2N6171		150	
2N3871, 2N6172		330	
2N3872, 2N6173		660	
2N3873, 2N6174		700	
*Average On-State Current, Note 2	$I_{T(AV)}$		Amps
(T_C = −40 to +65°C)		22	
(T_C = +85°C)		11	
*Peak Non-Repetitive Surge Current (One cycle, 60 Hz) (T_C = +65°C)	I_{TSM}	350	Amps
Circuit Fusing (t = 8.3 ms)	I^2t	510	A^2s

*Indicates JEDEC Registered Data.

Note 1. V_{DRM} and V_{RRM} for all types can be applied on a continuous basis. Ratings apply for zero or negative gate voltage; however, positive gate voltage shall not be applied concurrent with negative potential on the anode. Blocking voltages shall not be tested with a constant current source such that the voltage ratings of the devices are exceeded.

FIGURE 4-7 Typical specification sheet for an SCR. (Copyright of Motorola, Used by Permission.)

required for turn-on will not be achieved and the SCR will be in the off-state. Actual control is usually a total of about 90° between the 0° point and the 90° point. If a larger range of control is required, a triggering device such as a diac or unijunction transistor (UJT) will be required.

4.2.9 Using Unijunction Transistors and Diacs to Trigger SCRs

In most industrial electronic circuits where SCRs are pulsed, two devices called *unijunction transistors* and *diacs* are used to control the firing angle. Fig. 4-10a shows a circuit that uses a unijunction transistor (UJT) to trigger an SCR, and Fig.

2N3870 thru 2N3873 ● 2N6171 thru 2N6174

MAXIMUM RATINGS (T_C = 25°C unless otherwise noted.)

Rating	Symbol	Value	Unit
*Peak Gate Power	P_{GM}	20	Watts
*Average Gate Power	$P_{G(AV)}$	0.5	Watt
*Peak Forward Gate Current	I_{GM}	2	Amps
Peak Gate Voltage	V_{GM}	10	Volts
*Operating Junction Temperature Range	T_J	−40 to +100	°C
*Storage Temperature Range	T_{stg}	−40 to +150	°C
Stud Torque	—	30	in. lb.

* Indicates JEDEC Registered Data.

*THERMAL CHARACTERISTICS

Characteristic	Symbol	Max	Unit

Thermal Resistance, Junction to Case
    2N3870 thru 2N3873
    2N6171 thru 2N6174
	$R_{\theta JC}$	
		0.9
		1
			°C/W

* Indicates JEDEC Registered Data.

ELECTRICAL CHARACTERISTICS (T_C = 25°C unless otherwise noted.)

Characteristic	Symbol	Min	Typ	Max	Unit

*Peak Repetitive Forward or Reverse Blocking Current
(V_{AK} = Rated V_{DRM} or V_{RRM}, gate open, T_C = 100°C)
    2N3870, 2N6171
    2N3871, 2N6172
    2N3872, 2N6173
    2N3873, 2N6174
(Rated V_{DRM} or V_{RRM}, gate open, T_C = 25°C)
    All Devices
	I_{DRM}, I_{RRM}	
		
		—
		—
		—
		—
		
		—
			
			1
			1
			1
			1
			
			—
				
				2
				2.5
				3
				4
				
				10
					mA
					
					μA

*Peak On-State Voltage (I_{TM} = 69 A Peak)	V_{TM}	—	1.5	1.85	Volts
*Gate Trigger Current (Continuous dc) *T_C = −40°C (V_D = 12 V, R_L = 24 ohms) T_C = 25°C	I_{GT}	— —	9 4	80 40	mA
*Gate Trigger Voltage (Continuous dc) *T_C = −40°C (V_D = q2 V, R_L = 24 ohms) T_C = 25°C	V_{GT}	— —	0.9 0.69	3 1.6	Volts
*Holding Current (Gate Open) *T_C = −40°C (V_D = 12 V, I_{TM} = 200 mA) T_C = 25°C	I_H	— —	14 5.2	90 50	mA
*Gate Controlled Turn-On Time (t_d + t_r) (I_{TM} = 41 Adc, V_D = rated V_{DRM}, I_{GT} = 40 mAdc, Rise Time ≤ 0.05 μs, Pulse Width = 10 μs)	t_{gt}	—	—	1.5	μs
Circuit Commutated Turn-Off Time (I_{TM} = 10 A, I_R = 10 A) (I_{TM} = 10 A, I_R = 10 A, T_C = 100°C)	t_q	 — —	 25 35	 — —	μs
Forward Voltage Application Rate (T_C = 100°C, V_D = Rated V_{DRM})	dv/dt	—	50	—	V/μs

*Indicates JEDEC Registered Data.

FIGURE 4-7 (Continued)

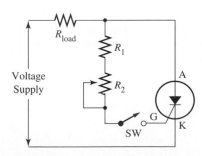

FIGURE 4-8 A fixed resistor and adjustable resistor provide a pulse during each positive half cycle of the AC input voltage.

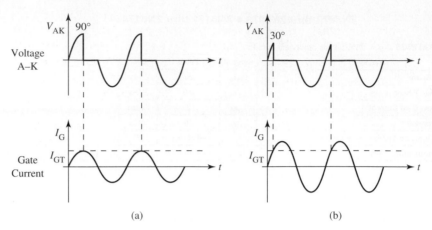

FIGURE 4-9 **(a)** The top diagram shows the waveform of ac supply voltage measured across the A–K circuit of an SCR when the SCR is turned on at the 90° point. The waveform in the bottom diagram shows the amount of gate current required to cause the SCR to go into conduction at this point. **(b)** The top diagram shows the waveform of an SCR that goes into conduction at the 30° point. Notice in the bottom diagram that the level of gate current is increased, which causes it to reach the gate current threshold at the 30° point instead of the 90° point.

4-10b shows a diac used to trigger an SCR. The UJT and diac are solid-state devices that provide a sharp pulse with sufficient current to cause the SCR to go into conduction. The pulse has a very sharp rise in current over a short time duration. The resistor and capacitor in each circuit provide an *RC* time constant that causes the time delay for the pulse. Since the resistor is variable, the larger the resistance is, the later in the sine wave the UJT or diac fires to provide the gate pulse. In some circuits, the SCR is so large that a smaller separate SCR is used to provide the gate signal for the larger one. These smaller SCRs are called *gaters*. Chapter 5 will cover UJTs, diacs, and other triggering devices in detail.

4.2.10 Methods of Commutating SCRs

Once an SCR is turned on, it will continue to conduct until it is *commutated* (turned off). Commutation will occur in an SCR only if the overall current gain drops below unity (1). This means that the current in the anode–cathode circuit must drop below the minimum (near zero) or a current of reverse polarity must be applied to the anode-cathode. Since the ac sine wave provides both of these conditions near the 180° point in the wave, the main method to commutate an SCR is to use ac voltage as the supply voltage. In an ac circuit, the voltage will drop to zero and cross over to the reverse direction at the 180° point during each sine wave. This

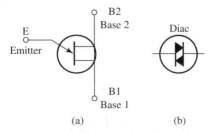

FIGURE 4-10 **(a)** Electronic symbol of a UJT. **(b)** Electronic symbol of a diac. These devices are used to fire an SCR.

means that if the supply voltage is 60 Hz, this will happen every 16 msec. Each time the SCR is commutated, it can be triggered at a different point along the firing angle, which will provide the ability of the SCR to control the ac power between 0° to 180°. The main problem with using ac voltage to commutate the SCR arises when higher-frequency voltages are used as the supply voltage. You should keep in mind the SCR requires approximately 3–4 μsec to turn off; therefore, the maximum frequency is dependent on the turn-off time.

If the SCR is used in a dc voltage circuit, similar commutation methods can be achieved if the supply voltage is pulsing dc that returns to near 0 volts at the end of each cycle. Pulsing dc voltage would be available from rectified ac voltage before it is filtered. In some cases, the SCR is used to rectify the ac voltage instead of junction diodes, which would provide the ability to control current as well as rectify the voltage.

If the SCR is used in a pure dc voltage circuit, a means of commutating it must be devised. Fig. 4-11 shows several of these types of commutation circuits. In Fig. 4-11a the SCR is in a dc circuit and a switch is placed in series with the anode–cathode circuit. This may seem redundant since the SCR acts like a switch, but you should remember that the SCR has the ability to turn on in such a way to prevent contact bounce or arcing at contacts. In this circuit, the switch is only used

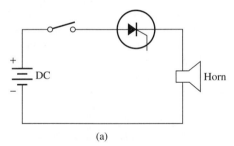

(a)

FIGURE 4-11 **(a)** A switch is used to commutate the SCR in a dc circuit by interrupting current flow. This type of circuit is used to provide control in alarms or emergency dc voltage lighting circuits. **(b)** A series *RL* resonant circuit use to commutate an SCR. **(c)** A parallel *RL* resonant circuit used to commutate an SCR.

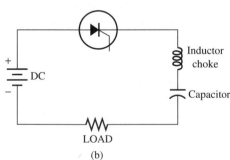

(b)

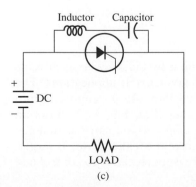

(c)

to turn off the SCR. This type of circuit is usable in a simple alarm circuit. The SCR acts like a relay in that the anode–cathode circuit will provide the function of the contacts, and the gate will provide the function of the coil. Since the circuit operates well with dc voltage, it can be used in battery-operated systems when the power is interrupted in the factory. The SCR could control a horn for the alarm, or it could switch on emergency lights. When the alarm is acknowledged, the switch would be opened and the SCR would turn off. The switch would need to remain open for 0.1 msec for the hold in current to drop below the minimum.

The circuit in Fig. 4-11b shows an *LC* commutation method. In this circuit an inductor and capacitor are connected in series with the SCR and load. When the SCR is triggered and current begins to flow, the *LC* circuit will become active. The time it takes to charge the capacitor will be a function of the resonant cycle. When the capacitor is fully charged at the halfway point in the resonant cycle, it will begin to reverse the current to the anode, which will commutate the SCR. This means the SCR conduction cycle will be half the resonant cycle. The capacitor and inductor can be sized to set the resonant cycle.

Fig. 4-11c shows a parallel resonant circuit. In this circuit, the *LC* circuit is connected in parallel with the SCR. In this circuit, the capacitor is charged as soon as power is applied. When the SCR is turned on, it provides a path for the capacitor to discharge. When the capacitor becomes fully discharged, it will begin to charge again. Since the SCR is still in conduction, the current charging the capacitor will be reverse biased to the current in the SCR. As soon as the current in the capacitor is larger than that flowing in the SCR, the SCR will be commutated. This means commutation will take place after approximately half the resonant cycle.

4.2.11 SCR Turn-Off Time and Reverse Recovery Time

The *turn-off time* is defined as the time between the point where the current starts to reverse and the point in time when the SCR can block the next cycle of forward voltage. This time characteristic is important to determine the maximum frequency at which the SCR can operate. The turn-off time will increase if the junction temperature increases, as the forward current amplitude increases, as the blocking voltage amplitude increases, as the rate of fall of the forward current, and as the rate of rise of the forward blocking voltage. The turn-off time will be shown on the SCR specification sheet.

The *reverse recovery time* is defined as the time when forward current stops and reverse recovery current drops below 10%. The reverse recovery current is the current that flows for a short period immediately after the SCR is commutated. The smaller this time is the better the SCR will operate in rectifier circuits or in inverter circuits. The reverse recovery time is also shown on the specification sheet.

4.2.12 Example Applications Using SCRs

SCRs are used in a wide variety of industrial applications that include large currents up to 2000 A at 4000 volts as well as a variety of low-current applications. The larger SCRs are usually packaged as modules so that they can fit compactly into motor frequency drives and other types of power supplies. Fig. 4-12 shows an example of SCRs in modular packages. From this diagram you can see that the terminals for these devices are larger because the current for the device is larger. Since these SCRs are in a package, their temperature is much easier to control and replacement of the complete assembly is also easier.

Some applications such as rectifier circuits may use SCRs in place of diodes. Fig. 4-13 shows a variety of combinations of SCRs that provide single-phase and three-phase half-wave and full-wave rectifier bridges. These circuits provide a means of combining the rectification part of the SCR with its ability to control the amount of current that flows in the circuit. You should also notice that one combination of two SCRs that are connected in inverse parallel is called an *ac switch*. The two SCRs are specifically chosen as a matched set and are mounted in the package together. Since the package is encapsulated in plastic like the ones shown in the previous figure, you do not have to worry about a mismatch of current or voltage characteristics between the SCRs. This type of package for the SCR is commonly used in motor drive circuits, welding, and power supply circuits that use larger currents where a large triac is not available. You will see later in this chapter that the triac is essentially two SCRs connected inverse and parallel to each other.

The second set of diagrams shows a combination of SCRs and diodes to provide a variety of circuits such as doublers, common cathode devices, common anode devices, and a three-phase full-wave bridge made from a combination of SCRs and diodes. The diode is used for part of the circuit because it is less expensive than the SCR, yet it will provide rectification like the SCR. Since the bridge rectifier is basically a series circuit for the SCR and diode, only one device needs to have the ability to control the amount of current flow in the circuit.

Fig. 4-14 shows a diagram of a typical ac variable-frequency drive. The circuit for this drive shows SCRs used in the rectification part of the circuit. The SCRs

FIGURE 4-12 Typical SCR packages. Notice that the terminals are larger for larger-sized wire required to carry the currents. If more than one SCR is mounted in a package, they are connected internally as pairs so the module is easy to install. (Courtesy of Darrah Electric Company, Cleveland, Ohio.)

THYRISTOR/SCR CIRCUITS

CIRCUIT TYPE	CIRCUIT DESIGNATION	CIRCUIT SCHEMATICS
HALF-WAVE SCR	PTA	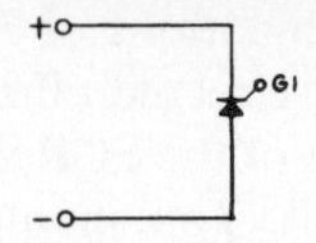
SCR DOUBLER	PTD	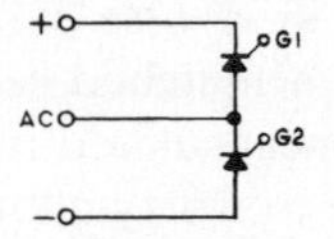
AC SWITCH	PAA	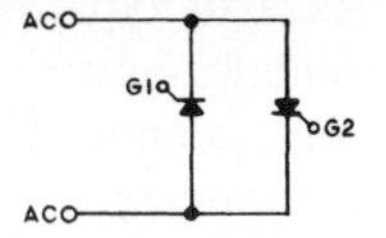
FULL SCR BRIDGE 1 PHASE	USE 2 PTD ASSEMBLIES	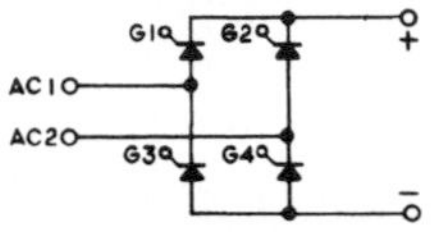
FULL SCR BRIDGE 3 PHASE	USE 3 PTD ASSEMBLIES	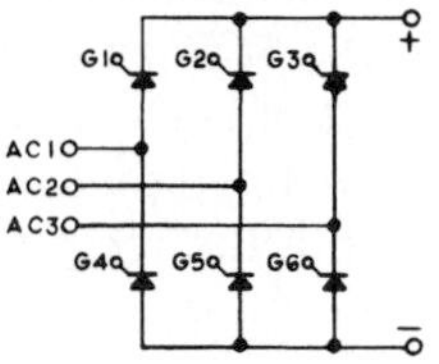

HYBRID CIRCUITS DIODES/SCR

CIRCUIT TYPE	CIRCUIT DESIGNATION	CIRCUIT SCHEMATICS
HYBRID DOUBLER	PHD	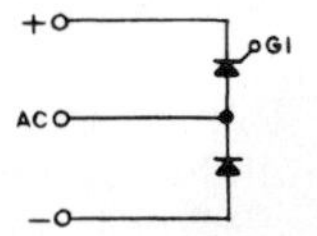
HYBRID DOUBLER	PHA	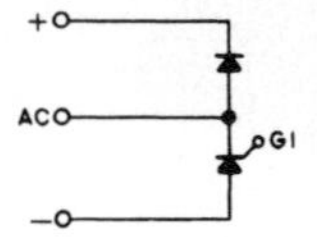
HYBRID BRIDGE COMMON CATHODE SCRS	USE 2 PHD ASSEMBLIES	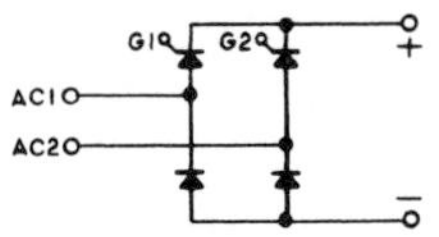
HYBRID BRIDGE COMMON ANODE SCRS	USE 2 PHA ASSEMBLIES	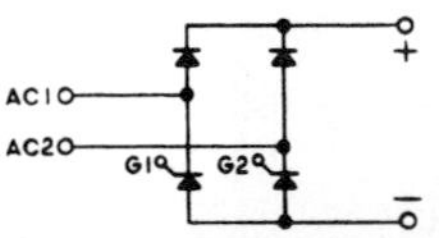
3∅ HYBRID BRIDGE	USE 3 PHD ASSEMBLIES	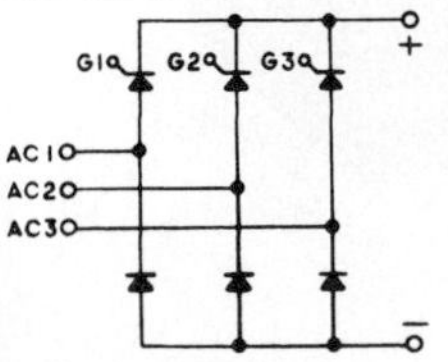

FIGURE 4-13 Diagrams of SCRs used in half-wave and full-wave rectifier applications. (Courtesy of Darrah Electric Company, Cleveland, Ohio.)

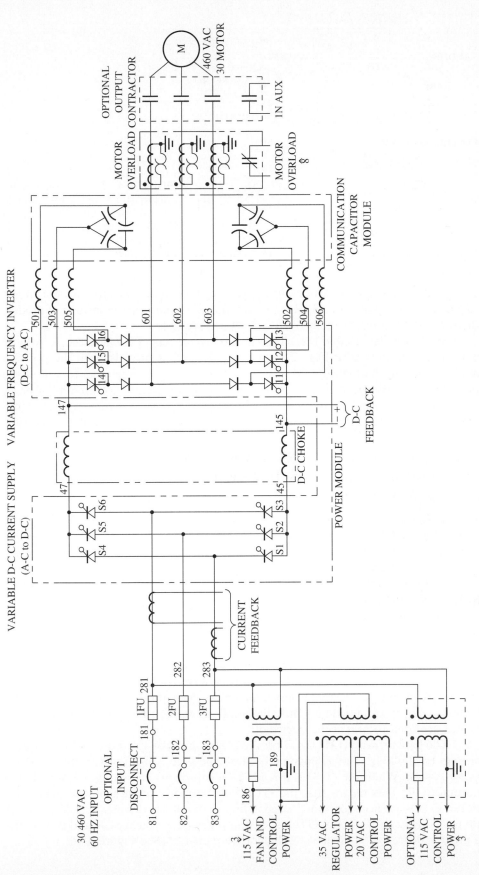

FIGURE 4-14 SCRs used as bridge rectifiers to control the ac voltage to dc voltage conversion for a variable-frequency drive. Pairs of SCRs are also used in the inverter section where dc voltage is changed back to ac voltage for the drive. (Courtesy of Rockwell Automation's Allen Bradley Business.)

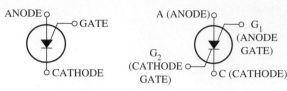

Complementary SCR   **SCS**

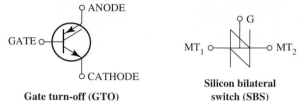

Gate turn-off (GTO)   **Silicon bilateral switch (SBS)**

FIGURE 4-15 Electronic symbols for the complementary SCR, the silicon controlled switch (SCS), and the gate turn-off device (GTO). (Copyright of Motorola, Used by Permission.)

provide the ability to convert the three-phase voltage that is the input for the circuit to pulsing dc voltage. Since SCRs are used in the rectifier section instead of diodes, the input voltage can be controlled as well as rectified.

SCRs are also used in this circuit in the inverter section where the dc voltage is turned back into ac voltage. Since the devices must provide both the positive and the negative half-cycles, a diode is connected in inverse parallel to a diode to provide the hybrid ac switch. This combination of devices is not used as often in newer drives because a variety of larger triacs and power transistors is available that can do the job better. This type of circuit was very popular in the mid-1980s.

4.2.13 Complementary SCRs, Silicon Controlled Switches, Gate Turn-Off Devices, and Other Thyristors

Several variations of the SCR have been produced to correct deficiencies in the SCR. These solid-state devices provide functions that allow the thyristor to control current somewhat differently than the SCR. The devices include the *complementary SCR, silicon controlled switch* (SCS), *gate turn-off devices* (GTOs), and *light-activated SCRs* (LASCRs). Fig. 4-15 shows the electronic symbols for the complimentary SCR, SCS, and GTO.

The complementary SCR uses a negative gate pulse instead of a positive pulse. This feature is useful in circuits where the gate must be pulsed by the cathode side of the circuit. In some cases it is important to control the turn-off feature of the device, so devices like the silicon controlled switch (SCS) and the gate turn-off device (GTO) are used. The SCS is a low-powered SCR with two gate terminals. If the anode–cathode current is less than 4 mA, the cathode gate (G_C) is used, and if the current is greater than 4 mA, the anode gate (G_A) is used. The gates in this device are used to switch the device off.

The gate turn-off (GTO) devices are useful in circuits that use higher frequencies up to 100 kHz. This is possible because the gate is used to turn the SCR off as well as on. When the GTO is forced off, it interrupts current flow faster than a normal SCR. This means that the GTO will be ready to turn on the next cycle faster. GTOs are now commonly used in motor frequency drives where frequencies up to 120 Hz are used. The GTO is used in place of a typical SCR and it provides a greater degree of control at the higher frequencies. (See Figure 4-16.)

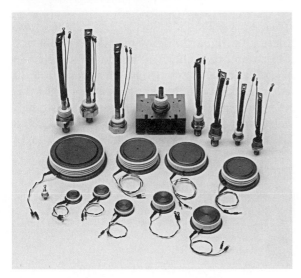

FIGURE 4-16 Typical packages for GTOs and SCRs. (Courtesy of POWEREX Inc.)

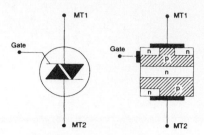

FIGURE 4-17 Electronic symbol of the triac and a diagram of its pn structure. (Courtesy of Philips Semiconductors.)

The light-activated SCR (LASCR) is a specialized SCR that can be turned on by light as well as by a gate pulse. The device has a window that allows light to shine on a light-sensitive silicon wafer. The amount of light required to turn the SCR on is variable since additional bias can be added to the gate. If the gate is biased with positive voltage to a level just below the turn-on, a very small amount of light will be needed to make the SCR go into conduction. If no bias voltage is used, a large amount of light will be needed to turn on the SCR.

Another feature of the LASCR is that the light pulse provides isolation. This is extremely important where the SCR may be turned on inadvertently by voltage spikes or unwanted circulating voltages that get inside shielded cables. Voltage spikes and unwanted circulating currents are common in high-voltage switching circuits in welding or other similar industrial applications. The LASCR is commonly used as part of optocouplers. The optocoupler is a device that has a light-emitting diode (LED) encapsulated so that its light shines directly on a light-sensitive switching device. When an input signal is sent to the LED, it produces a light that turns on the LASCR. If the signal becomes electrically contaminated by unwanted noise, the light will not transmit these signals to the LASCR side of the circuit. Section 5-6 in Chapter 5 explains these types of devices in detail.

4.3 TRIACS

The *bidirectional triode thyristor* (triac) is a solid-state device that acts like two SCRs that have been connected in parallel with each other (inversely) so that one SCR will conduct the positive half-cycle and the other will conduct the negative half-cycle. This means that the triac can be used for control in ac circuits. Before the triac was designed as a single component, two SCRs were actually used for this purpose. Fig. 4-17 shows the symbol for the triac, and its pn structure. The terminals of the triac are identified as *main terminal 1* (MT1), *main terminal 2* (MT2), and *gate*. The multiple pn structure is actually a combination of two four-layer (pnpn) junctions.

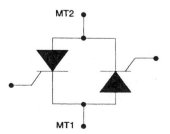

FIGURE 4-18 Two SCRs connected in inverse parallel configuration to provide the equivalent circuit for a triac.

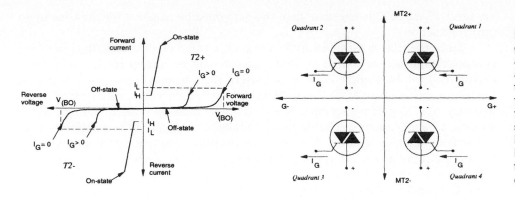

FIGURE 4-19 (a) A graph of the positive and negative voltage and current characteristics of a triac. **(b)** The four quadrants identified for the graph to describe the voltage polarity of the triac when it is in conduction. (Courtesy of Philips Semiconductors.)

The triac is required in circuits where ac voltage and current need to be controlled like the SCR controls dc current. Another difference between the triac and SCR is that the triac can be turned on by either a positive or negative gate pulse. The gate pulse need only be momentary and the triac will remain in conduction until the conditions for commutation are satisfied.

4.3.1 Operation of the Triac

The operation of the triac may best be explained by the two-SCR model in Fig. 4-18. From this figure you can see that the SCRs are connected in an *inverse parallel* configuration. One of the SCRs will conduct positive voltage and the other will conduct negative voltage. Unlike the two SCRs, the triac is triggered by a single gate. This prevents problems of one SCR not firing at the correct time and overloading the other. In the 1960s and 1970s when triacs were not available or were too small, two SCRs were actually connected together and used as a device to control current in an ac circuit.

The firing of the triac can be described by the diagram in Fig. 4-19. In this figure you can see that the triac can conduct both positive and negative current. The graph uses typical identification for its four quadrants. Voltage is shown along the horizontal *x*-axis, and current is shown along the vertical *y*-axis. This diagram also shows a second graph with the four quadrants identified. These quadrants will be used to explain the operation of the triac as polarity to its MT1, MT2, and gate changes.

When polarity is discussed, it is important to determine the point of reference. The polarity of a triac is described in *reference to* or *respect to* MT1. This means that if the gate is positive, it would be in respect to MT1. The triac can fire and go into its on-state in each of the four quadrants if the polarity of its gate and main terminals is correct. Fig. 4-20 shows a table with the correct polarity for firing the triac in each of the four quadrants. Quadrants 1 and 3 are the most used quadrants for the triac. If the triac uses the polarities of quadrant 1 for conduction where MT2 is positive, the gate must also be positive, and if the polarities of quadrant 3

Quadrant	Polarity of MT2 wrt MT1	Gate polarity
1 (1+)	MT2+	G+
2 (1-)	MT2+	G-
3 (3-)	MT2-	G-
4 (3+)	MT2-	G+

FIGURE 4-20 Polarity of terminal MT2 (with respect to terminal MT1) and the polarity of the gate for each of firing quadrants for a triac.

are used where MT2 is negative, then the gate must be negative for the triac to go into conduction.

Since the triac is manufactured from multilayered pn junctions, the operating characteristics such as holding current (I_H), gate trigger current (I_{GT}), and latching current (I_L) will be slightly different in each of the four quadrants. For example, if the triac operates in quadrant 2, its holding current is slightly higher, and if it operates in quadrant 4, its gate trigger current will be higher than the other quadrants. Some specialized applications will be designed to take advantages of these differences in triacs. It should also be noted that some manufacturers make a variety of thyristors having the same power ratings but with varying gate sensitivities.

4.3.2 Characteristics of Triacs

Specification sheets for triacs will list several useful characteristics in circuit design and troubleshooting. These characteristics include *latching current (I_L)*, *holding current (I_H)*, *gate trigger current (I_{GT})*, *main terminal RMS current ($I_{T(RMS)}$)*, *reverse current (I_R)*, *breakover voltage (V_{DROM})*, *gate to cathode voltage* also called *gate voltage (V_{GT})*, *maximum peak gate power ($P_{GM(max)}$)*, *maximum average gate power ($P_{G(AV)}$)*, and on-state voltage across MT1 and MT2 (V_{TH}), which is also called *forward voltage (V_T)*.

The latching current is the amount of current the triac gate needs to stay in conduction after it has been initially turned on. The holding current is the main terminal current required after the triac has been turned on. The gate trigger current is the amount of current that is necessary to go into conduction initially. It is important to remember that since the gate must be a specific polarity for the triac to fire in a specific quadrant, it is possible to apply bias voltage of similar polarity to help the triac trigger with smaller gate currents. Or bias voltage of the opposite polarity can be applied, which requires the application of a stronger gate current get the triac into conduction.

The other two specifications pertaining to the gate refer to the average amount of power the gate requires to put the triac into conduction ($P_{G(AV)}$) and the maximum amount of power the gate can handle ($P_{GM(max)}$). If the maximum gate power begins to exceed the limit in the specifications, the triac will begin to have thermal problems and erratic triggering will occur.

The main terminal rms current rating is the maximum amount of current that the triac can control. This value is listed as an rms value because the triac is primarily used in ac circuits. The breakover voltage is not listed as forward breakover like the SCR because the triac will go into conduction if the breakover voltage in either direction exceeds the specified level. Like the SCR, the breakover voltage rating of the triac should exceed the maximum peak circuit voltage, so it does not go into conduction if a spike of voltage occurs. Triacs are available with typical breakover voltage ratings (V_{DROM}) that range from 100–1000 volts in 100 volt increments. The on-state voltage is the voltage measured across MT1 and MT2 when the triac is in conduction. A specification sheet for the triac is provided in Fig. 4-21.

4.3.3 Switching on the Triac

The simplest way to control a triac is by applying current to its gate. When the triac is switched on by gate current, the current in the MT1–MT2 circuit does not begin to flow immediately because the anode current in the pn junction is delayed slightly. This time delay is not a problem when the triac is used in lower frequencies, but it will become a concern in higher-frequency circuits.

Triacs
Silicon Bidirectional Thyristors

... designed for full-wave ac power control applications, and specifically designed to be used in conjunction with MOC30XX opto couplers in circuits similar to that shown on page 3-189.

- Blocking Voltages to 400 Volts
- Load Current Controlled Up to 40 A
- Glass Passivated Junctions for Greater Parameter Uniformity and Stability
- Gate Triggering Guaranteed in Four Modes
- Designed for Use with MOC Series Optoisolators Having Triac Driver Outputs
- MAC3010/MAC3030 Are Recommended For Use With MOC3010/MOC3030 Optoisolators
- MAC3020/MAC3040 Are Recommended For Use With MOC3020/MOC3040 Optoisolators

**MAC3010
MAC3020
MAC3030
MAC3040
Series**

**TRIACs
4, 8, 15, 25 and 40
AMPERES RMS
250 thru 400 VOLTS**

CASE 77-07
(TO-225AA)
STYLE 5
-4

CASE 221A-04
(TO-220AB)
STYLE 4
-8, -15, -25

CASE 263-04
STYLE 2
-40

CASE 311-02
STYLE 2
-40I

MAXIMUM RATINGS (T_J = 25°C unless otherwise noted.)

Rating	Symbol	Current Ratings					Unit
		-4	-8	-15	-25	-40 -40I	
On-State RMS Current (see Figure 1) (Full Cycle Sine Wave 50 to 60 Hz)	$I_{T(RMS)}$	4	8	15	25	40	Amps
Peak Non-Repetitive Surge Current (One Full Cycle, 60 Hz, T_J = 110°C)	I_{TSM}	30	80	150	250	300	Amps
Circuit Fusing Considerations (T_J = −40 to +110°C, t = 8.3 ms)	I^2t	3.6	26	90	260	370	A^2s
Peak Gate Voltage (t ≤ 2 μs)	V_{GM}	±5	±10	±10	±10	±10	Volts
Peak Gate Power (t ≤ 2 μs)	P_{GM}	10	20	20	20	20	Watts
Average Gate Power (T_C = 80°C, t ≤ 8.3 ms)	$P_{G(AV)}$	0.5	0.5	0.5	0.5	0.5	Watts
Peak Gate Current (t ≤ 2 μs)	I_{GM}	11	12	12	12	12	Amps
Operating Junction Temperature Range	T_J	*	−40 to +125			*	°C
Storage Temperature Range	T_{stg}	−40 to +150					°C
Mounting Torque	—	6	8	8	8	30	in. lb.
MAC3010/MAC3030, Note 1 (T_J = 25 to 125°C) MAC3020/MAC3040	V_{DRM}	250 400	250 400	250 400	250 400	250 400	Volts

Note 1. V_{DRM} for all types can be applied on a continuous basis. Blocking voltages shall not be tested with a constant current source such that the voltage ratings of the devices are exceeded.

FIGURE 4-21 Specification sheet for a typical triac. (Copyright of Motorola, Used by Permission.)

Several conditions must be accounted for when using gate current to switch on the triac. These include maximum power and average power. The average power is the minimum power the triac gate requires to switch on the junction, and the maximum gate power is the maximum power the gate can handle without overheating. Minimum power can be achieved by applying a series of pulses instead of a continuous gate signal. Since the gate signal is a pulse, it can be stronger than the

MAC3010, MAC3020, MAC3030, MAC3040 Series

-8, -15, -25 CURRENT RATINGS

ELECTRICAL CHARACTERISTICS (T$_C$ = 25°C, and Either Polarity of MT2 to MT1 Voltage unless otherwise noted.)

Characteristic	Symbol	Min	Typ	Max	Unit
Peak Forward or Reverse Blocking Current (Rated V$_{DRM}$ or V$_{RRM}$) T$_J$ = 25°C T$_J$ = 125°C	I$_{DRM}$, I$_{RRM}$	— —	— —	10 2	μA mA

Peak On-State Voltage
(I$_{TM}$ = $\sqrt{2}$ I$_{T(RMS)}$ A Peak; Pulse Width ≤ 2 ms, Duty Cycle ≤ 2%)
           MAC3030-8
           MAC3030-15
           MAC3030-25
	V$_{TM}$	
		
		—
		—
		—
			
			—
			—
			—
				
				1.6
				1.6
				1.85
					Volts

Gate Trigger Current (Continuous dc) (V$_D$ = 12 V, R$_L$ = 100 Ohms) MT2(+), G(+); MT2(−), G(−) All Types	I$_{GT}$	—	—	40	mA
Gate Trigger Voltage (Continuous dc) (V$_D$ = 12 V, R$_L$ = 100 Ohms) MT2(+), G(+); MT2(−), G(−) All Types (T$_J$ = 125°C, R$_L$ = 10 k Ohms) MT2(+), G(+); MT1(−), G(−) All Types	V$_{GT}$	 — 0.2	 — —	 2 —	Volts
Holding Current (V$_D$ = 12 V, I$_{TM}$ = 200 mA, Gate Open)	I$_H$	—	—	40	mA
Gate Controlled Turn-On Time (I$_{TM}$ = 2 I$_{T(RMS)}$ A Peak, I$_G$ = 100 mA)	t$_{gt}$	—	1.5	—	μs
Critical Rate of Rise of Commutation Voltage (I$_{TM}$ = $\sqrt{2}$ I$_{T(RMS)}$ A Peak, Commutating di/dt = 0.36 I$_{TM}$ A/ms, Gate Unenergized, T$_C$ = 80°C)	dv/dt(c)	—	5	—	V/μs
Critical Rate of Rise of Off-State Voltage (Exponential Waveform, T$_C$ = 125°C)	dv/dt	—	40	—	V/μs

FIGURE 4-21 (Continued)

continuous signal since it spends half of its time in the off-state. Since the gate is pulsed, it will not add to thermal buildup, which causes other triggering problems.

Another way the triac can go into conduction is if the off-state voltage increases (rises) too quickly. This value of voltage is specified as *dv/dt* (delta voltage over delta time) and the circuit should be designed to include an *RC* snubber circuit to limit the rise time. If the change (Δ) of voltage increases too quickly, the triac can be triggered into unwanted conduction and erratic circuit operation may occur. The *RC* snubber circuit provides a capacitor that will charge, which will delay the rise time slightly. The maximum rise time and the values of the resistor and capacitor will vary from device to device and they can be determined from the manufacturer's specification sheet for each device.

4.3.4 Circuit Commutation for the Triac

The triac is slightly different from the SCR in that it must be commutated for two different functions. First, it must commutate after each half-cycle so that it can go into conduction in the opposite direction immediately when the polarity of the ac voltage changes. The second type of commutation occurs when the triac is switched to the off-state, where current flow is interrupted completely. When the triac has been in conduction in the forward direction and then it is turned off completely, its junctions cannot go into the blocking state immediately because the minority carriers have to be cleared out of the junction. If they remain, they will try to recombine, which will not completely block the off-state current.

If the triac is used to block current in an off-state for a resistive circuit, the voltage and current waveforms will remain similar, since they are *in phase* with

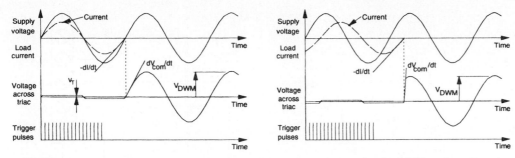

FIGURE 4-22 **(a)** Waveform for triac commutation in a resistive circuit. **(b)** Waveform for triac commutation in an inductive circuit. (Courtesy of Philips Semiconductors.)

each other. If the triac is used in an inductive circuit such as controlling a motor, the current waveform will be *out of phase* (lagging) with the voltage waveform. This causes a problem when the triac is commutated to remain in the off-state. If the current waveform is lagging, it will continue to have some amount of current available when the voltage waveform reaches the off-state. If the commutation condition does not continue for a longer time than the time the current is lagging the voltage, the triac will immediately go back into conduction because of the remaining current. The ratio of delta voltage over time ($\Delta v_{com}/\Delta t$) and the ratio of delta current over time ($\Delta i/\Delta t$) can be calculated from the following formulas;

$$di/dt = 2\pi f \sqrt{2I_{T(RMS)}}$$

and

$$dv_{com}/dt = 2\pi f \sqrt{2V_{(RMS)}}$$

The waveforms that show the effects of commutation in resistive and inductive circuits are shown in Fig. 4-22.

4.4 POWER TRANSISTORS

Transistors have become more widely used in ac variable-frequency motor speed drives and other power applications than in previous years because today larger transistors are available that can control voltage over 1000 volts and current over 1000 A. Transistors that are manufactured for these higher-power applications function exactly like the smaller-signal PNP and NPN transistors you have used in your electronic laboratory experiments, which will make understanding their operation much easier. When available, transistors are preferred over thyristors because they are much more versatile and can react to higher frequencies. One of the problems with SCRs, triacs, and other thyristors is that they must depend on reverse-bias voltages to help turn them off once they have been fired. This is not a problem with the transistor since it can easily be controlled by controlling the current to its base.

Typically you will find power transistors mounted in modules where their internal connections have been made during the manufacturing process. This ensures that the transistors used in pairs are matched and that critical connections will endure through all types of rugged operations. Cooling and other protection

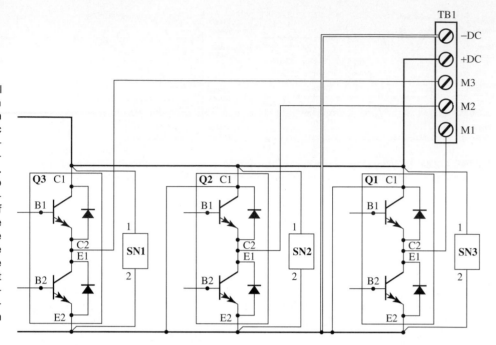

FIGURE 4-23 Typical transistor modules in the output section of a variable-frequency ac motor drive. The transistor modules are identified as Q1, Q2, and Q3. Each module has two transistors, one to provide the positive part of the ac wave, and the second to provide the negative part. The large bold lines are from the dc bus voltage that feeds the output section. (Courtesy of Rockwell Automation's Allen Bradley Business.)

issues are also controlled when the transistors are packaged as a module. You will typically find transistors as power transistors, darlington pairs, or as specialized transistors called *insulated gate bipolar transistors* (IGBTs). These types of transistors will be discussed in the remainder of this chapter.

4.4.1 Power Transistor Applications

Power transistors are now available with current ratings over 1000 A and voltage ratings over 1000 volts. These larger ratings allow transistors to be used easily in industrial applications such as variable-frequency drives and in power supplies. Fig. 4-23 shows a typical diagram of these transistors used in the output stage of an ac variable-frequency drive.

From this diagram you can see that the transistors in the output stage of the variable-frequency drive are connected in pairs. The pairs of transistors are mounted in modules identified as Q1, Q2, and Q3. These modules mount into the drive so that they can dissipate heat and be easily connected to the other components of the drive. The two large bold lines that are shown directly above and below each transistor module represent the dc bus voltage. The dc bus voltage is filtered pure dc and it may be as much as 500–900 volts. The base of each transistor is controlled by a *firing circuit* or *driven circuit* that receives its signal from the microprocessor in the drive. The microprocessor controls each transistor so that it can produce an output signal to the motor that looks like three-phase ac.

The point where each transistor pair comes together is connected to a terminal marked M1 for the first pair, M2 for the second pair, and M3 for the third pair. The M terminals are the connections for T1, T2, and T3 of the three-phase motor. You should also notice that terminal board TB1 in the drive where these M terminals are provided also provides a +DC and −DC terminal. These terminals are useful for connecting other dc-powered devices for the drive such as dc breaking and also for use as a test point. *Snubber circuits* (SN1), (SN2), and (SN3) are provided as

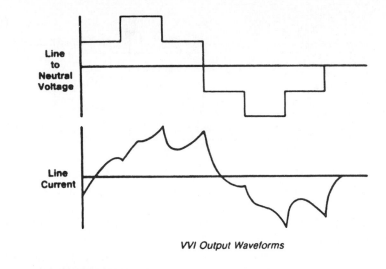

VVI Output Waveforms

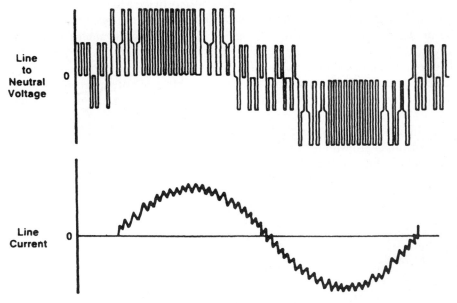

PWM Output Waveforms

FIGURE 4-24 Output waveform for transistors used in three-phase variable-frequency drive. The top diagram is for variable voltage input (VVI) drives, and the bottom diagram is for pulse-width modulation (PWM) drives. (Courtesy of Rockwell Automation's Allen Bradley Business.)

a module for each transistor. The snubber circuit may contain *metal-oxide varistors* (MOVs) or some combination of MOVs and capacitors depending on the exact transistor configuration for the output stage. The snubber prevents unwanted voltage surges from damaging the transistors.

A complete diagram of the drive will be studied later in Chapter 11. The input voltage shown in the diagram as the dark bold lines across the top and bottom of the circuit is pure dc voltage. The top transistor in each pair is pulsed (ramped) on to provide the positive half-cycle and the bottom is pulsed on to provide the negative half-cycle for one phase of the output. A diagram of the output is shown in Fig. 4-24.

Each pair of transistors will produce one phase of the three-phase output. The base of each transistor is controlled by a microprocessor that provides the timing for each part of the circuit. In some drives the current requirements are

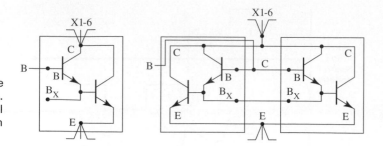

FIGURE 4-25 Example transistor modules. (Courtesy of Rockwell Automation's Allen Bradley Business.)

larger, so two or more transistors must be connected in parallel to control these larger currents. Fig. 4-25 shows a diagram of two different methods of connecting additional transistors to provide large-current control. These transistors are available in modules so that technicians do not need to worry about internal connections. The modules provide terminals to connect the transistors with other parts of the motor drive circuitry. These terminal points also provide a location for troubleshooting the PN junction of each transistor. A standard *front-to-back* resistance test can be made for the top transistor in Fig. 4-25a by placing the meter leads on the B and C terminals to test the base–collector junction, and by using terminals B and BX to test the base–emitter junction. The second transistor in the set can also be tested by using the E, BX, and C terminals. Most maintenance manuals for the drives will provide information that specifies the proper amount of resistance when the junction is forward and reverse biased during the test.

Fig. 4-26 shows a picture of a typical transistor module used in motor drives. The terminals shown on this module will be identified exactly like the ones in the diagrams in Figs. 4-23 and 4-25. These terminals will provide easy field wiring in the drive. The terminal connections will also make it simple to remove wires for

FIGURE 4-26 Transistor modules that show the terminal layout. The terminals in this module match the diagram shown in Figs. 4-23 and 4-25. (Courtesy of POWEREX Inc.)

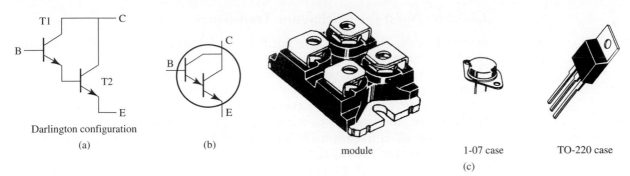

FIGURE 4-27 **(a)** Internal configuration for a darlington transistor pair. This diagram shows transistor T1 as the driver of transistor T2, which is the output for the darlington. **(b)** Electronic symbol for the darlington transistor. **(c)** Typical packages for the darlington transistor including the module, 1-07 case, and the TO-220 case. (Courtesy of Philips Semiconductors.)

testing and troubleshooting. If the module must be changed in the field, the technician can simply remove the wires and change the module.

4.4.2 Darlington Transistors

Darlington transistors are sometimes called *darlington pairs* because two bipolar transistors are packaged together to provide better operation for high-power and high-frequency applications in motor drives and power supplies. Fig. 4-27 shows the diagram of two transistors that have been manufactured as a darlington pair. From this figure you can see that the first transistor T1 is the driver of the second transistor T2. Transistor T2 is called the output for the darlington pair. The input signal is sent to the base of T1, which will act like a typical bipolar transistor. The larger the base current becomes, the larger the collector current will be up to the point of saturation. The emitter of T1 is used to provide base current to the output transistor T2. The base current of T2 will be used to drive the collector current, which will be the output current for the darlington.

In some darlington packages, a speedup diode is added between the bases of T1 and T2. The anode of the speedup diode is connected to the base of T2, which means it is used in reverse bias when the driver transistor is turned on. The diode is used to help speed up the turn-off of the transistors when the signal to the base is removed. The diode will not show up in the darlington symbol.

Since the darlington configuration uses two transistors, together they provide a higher gain than a single transistor. The darlington also has a higher $V_{CE(SAT)}$. The higher gain and higher saturation means the darlington can be used in high-voltage applications. This configuration also provides a faster switching speed with less power loss during the switching process, which saves energy. This feature makes the darlington useful in high-frequency applications up to 20 kHz.

Fig. 4-27b shows the electronic symbol for the darlington pair. You can see from the symbol that the terminals of the transistor are identified as B (base), C (collector), and E (emitter). Even though the darlington pair consists of two transistors, it will be considered one transistor when it is shown in the circuit diagram. Fig. 4-27c shows the typical packages for the darlington pairs. The first package is a module that is commonly used in motor drives and other power circuits. The second package is a 1-07 case, which utilizes a metal case to help dissipate heat. The third case is the standard TO-220 case, which provides a metal tab for mounting and dissipating heat.

4.4.3 The Need for Darlington Transistors

In recent years, electronics have been integrated into motor speed drives and a variety of switching-type power supplies. This meant that standard discrete components needed to be altered to provide better characteristics. The need for the darlington pair grew from the limitations of SCRs and triac-type thyristors. Thyristors control current by delaying the turn-on time. The later the pulse is applied to turn them on, the smaller the amount of current they will conduct during each cycle. On the other hand, a transistor uses variable current (0 to saturation), which provides an output current that will be a duplication of the input. This means the transistors will produce an analog signal when an analog signal is provided to its base. The simple bipolar transistor has several limitations including slow switching speeds, low gains, and larger power losses due to the switching process. A family of high-gain transistors called *metal-oxide semiconductor field effect transistors* (MOSFETs) were produced to address the gain problem, but they did not have the capability of controlling larger currents, so the darlington pair was designed. The darlington pair can actually be two discrete transistors that are connected in the driver/output configuration, or they can be a single device that has the two transistors internally connected at the point where it was manufactured as a single package.

4.5 INSULATED GATE BIPOLAR TRANSISTORS

Another transistor-type device is a combination of a *metal-oxide semiconductor* (MOS) transistor and a *bipolar transistor*. This device is called an *insulated gate bipolar transistor* (IGBT). Combining these two devices provides the high-voltage gain feature of the MOS device, and the high-current gain feature of the bipolar transistor. The combination of these two devices also provides better speed characteristics. The MOS part of the device provides low on-state losses, which make it easier to control at higher voltages. The device also exhibits switching speeds up to 20 kHz. Fig. 4-28 shows the electronic symbol for the devices. The IGBT is used in circuit applications up to 1000 volts.

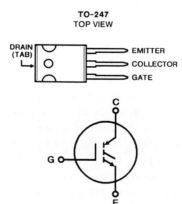

FIGURE 4-28 Electronic symbol and typical package for the insulated gate bipolar transistor (IGBT). (Courtesy of Harris Semiconductors.)

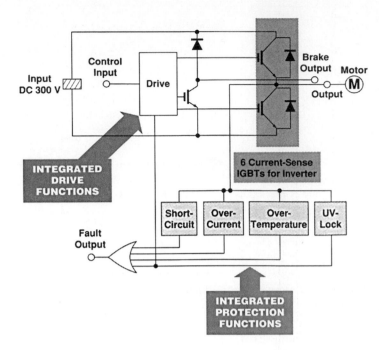

FIGURE 4-29 Block diagram for a circuit that provides IGTBs to sense current in a dc motor drive circuit. The output signal from the IGBTs is sent to protect circuits in the drive. The IGBTs also provide signals that are used for dc braking. (Courtesy of POWEREX Inc.)

Fig. 4-29 shows a typical application for an IGBT. In this application several IGBTs are used in a motor drive circuit to monitor current for the drive. The output from the current-sensing IGBTs is sent to four protection circuits that check for short-circuit faults, exceeding maximum overcurrent, exceeding maximum temperature, and undervoltage (UV). The IGBTs are also used in this application to provide an output signal that can be used to provide dc braking for the motor.

4.6 HIGH-VOLTAGE BIPOLAR TRANSISTORS

The *high-voltage bipolar transistor* (HVT) has become very useful in semiconductor controls where voltages of 850–1500 volts are controlled. These types of transistors are specifically designed for the higher voltages and they are normally used as electronic switches. Fig. 4-30 shows a cut-away diagram of a high-voltage transistor and a cross-sectional diagram. From the cut-away diagram you can see that the base and emitter leads are connected with wires to the integrated circuit part of the device. The collector is common to the case of the package, which provides better thermal transfer properties. In the cross-sectional diagram, you can see that the more n-material that is used in the device, the higher the voltage specifications for the device will be.

4.7 JUNCTION FIELD EFFECT TRANSISTOR (J-FETS)

J-FETs are used in power electronic circuits to control high voltages at lower currents. The maximum current for J-FETs that are used in power control circuits

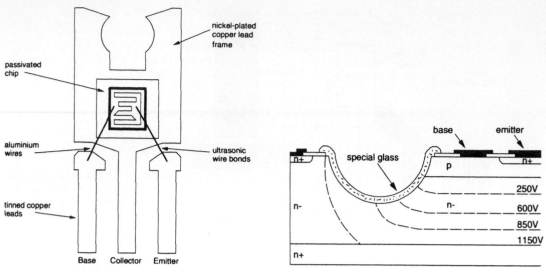

FIGURE 4-30 Cut-away diagram and cross-sectional diagram for a high-voltage transistor (HVT). (Courtesy of Philips Semiconductors.)

is approximately 10 A. Fig. 4-31 shows the symbols for an n-channel J-FET and a p-channel J-FET and a diagram that shows the arrangement of the three layers of the n-channel transistor. The J-FET has three terminals that are called the source, drain, and gate. If you compare the J-FET to an NPN or PNP transistor, the source provides a function similar to the transistor emitter, the drain is similar to the

FIGURE 4-31 The electronic symbols for n-channel and p-channel J-FET. The location of the three layers is also shown. (Courtesy of Philips Semiconductors.)

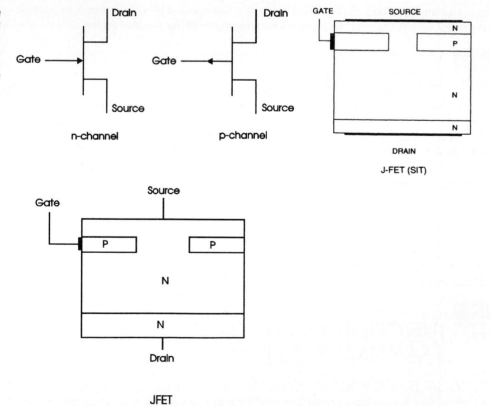

collector and the gate is similar to base. The n-channel has its source and drain made of n-material, and the p-channel has its source and drain made of p-material. Even though the J-FET cannot control larger currents like a thyristor, it is used in circuits because it provides fast reaction times and high impedance.

4.8 COMPARISON OF POWER SEMICONDUCTORS

As a technician you will find all of the devices that have been previously discussed in a wide variety of electronic circuits. They have been selected for their particular application because of their special properties. These properties are presented in a table in Fig. 4-32 so that you can compare them. From this table you can see that thyristors can control voltages up to 5000 volts, where IGBT, MOS devices, and J-FET devices can handle voltages up to 1000 volts. The maximum current ranges from 10 A for J-FET devices to 5000 A for thyristors. Other information is compared such as turn-off requirements, drive circuit complexity, device protection, and switching losses. The column on the far right side of the table describes the comparison.

FIGURE 4-32 A comparison of the power devices. The following abbreviations are used: HVT (high-voltage transistor), J-FET (J-type field effect transistor), MOS (metal-oxide semiconductor), THY (thyristor), GTO (gate turn-off device), and IGBT (insulated gate bipolar transistor). (Courtesy of Philips Semiconductors.)

	HVT	J-FET	MOS	THY	GTO	IGBT slow	IGBT fast	Unit
V(ON)	1	10	5	1.5	3	2	4	V
Positive Drive Requirement	-	+	+	+	+	+	+	+ = Simple to implement
Turn-Off requirement	-	-	+	(--)	-	+	+	+ = Simple to implement
Drive circuit complexity	-	.	+	(-)	.	+	+	- = complex
Technology Complexity	+	.	.	+	-	-	-	- = complex
Device Protection	-	.	+	+	-	-	-	+ = Simple to implement
Delay time (ts, tq)	2	0.1	0.1	5	1	2	0.5	μs
Switching Losses	.	++	++	--	-	-	.	+ = good
Current Density	50	12	20	200	100	50	50	A/cm^2
Max dv/dt (Vin = 0)	3	20	10	0.5	1.5	3	10	V/ns
dI/dt	1	10	10	1	0.3	10	10	A/ns
Vmax	1500	1000	1000	5000	4000	1000	1000	V
Imax	1000	10	100	5000	3000	400	400	A
Over Current factor	5	3	5	15	10	3	3	

········▶ **S O L U T I O N   T O   J O B   A S S I G N M E N T**

Your solution should include the selection of an SCR-type controller for the dc motor drive and an SCR, triac, or transistor-type controller for the ac variable-frequency drive. Use the material in Figs. 4-3–4-9 to explain to your supervisor how the SCR operates. You should include how the SCR is turned on and what the firing angle and conduction angle are. You should also explain that the SCR needs a pulsing input voltage like a sine wave from ac or pulsing dc to ensure that it is turned off. This is how the SCR provides the degree of control.

When you explain the devices that could be used for the ac drive, you should include the SCR since it can be used in pairs like the ones in Fig. 4-18. The diagram in Fig. 4-14 shows pairs of SCRs used to control ac motors.

If triacs are your choice, use the information in Figs. 4-17–4-21 to explain their operation. You may include the fact that a triac actually operates like two inverse parallel SCRs. Be sure to explain that the triac can control voltage in both the positive and negative half-cycles of the ac sine wave.

If you choose to use transistors to control the ac motor, use the information in Figs. 4-23–4-27 that explains the operation of this type of control. Fig. 4-23 and the accompanying diagrams in Fig. 4-24 will help you explain this type of transistor control. You can use the table in Fig. 4-32 to show the comparison of all of these devices.

········▶

Q U E S T I O N S

1. Explain how an SCR uses an ac voltage sine wave or pulsing dc voltage to turn off (commutate).
2. Explain how an SCR is turned on by its gate. Use the terms *minimum gate current, minimum gate voltage,* and *holding current.*
3. Explain why it is important to use a plastic-case or isolated-case oscilloscope to make measurements when SCRs or other solid-state components used in circuits do not have a neutral or ground wire, such as three-phase circuits.
4. Identify two ways an SCR may turn on from unwanted conditions.
5. Explain how the circuit for the SCR may be modified to prevent unwanted turn-on.
6. Explain the difference between the firing angle and the conduction angle of the SCR.
7. Explain the operation of the triac using the two-SCR model.
8. Identify the polarity of the terminals and gate of a triac in each of the firing quadrants.
9. Explain how a triac can be commutated.
10. Identify the electronic symbol for a darlington transistor and name its terminals.
11. Explain the advantages of using a darlington transistor over a single bipolar-type transistor.
12. Explain how a GTO is different from an SCR.

TRUE OR FALSE

1. _____ A triac can go into conduction with either a positive or a negative pulse if the polarity of the MT1 and MT2 terminals is correct.

2. _____ When working with a metal-case oscilloscope, you can safely measure voltages in three-phase circuits if you cut the ground terminal off of the power cord, or if you use a two-terminal to three-terminal adapter on the power cord.

3. _____ A single SCR is usually used in a circuit to control DC loads, and a triac is used to control AC loads.

4. _____ Darlington pair transistors are used in place of a single bipolar transistor because they can switch power faster than a single bipolar transistor.

5. _____ When the term *commutation* is used in reference to an SCR circuit, it means the method the SCR uses to turn on and go into conduction.

MULTIPLE CHOICE

1. When the SCR goes into conduction _____.
   a. its anode must have positive voltage applied to it and the gate must have a negative pulse.
   b. its anode must have negative voltage applied to it and the gate must have a negative pulse.
   c. its anode must have positive voltage applied to it and the gate must have a positive pulse.

2. When a triac is fired in the first quadrant _____.
   a. MT2 must be positive and the gate must be negative.
   b. MT2 must be positive and the gate must be positive.
   c. MT2 must be negative and the gate must be positive.

3. Using the data specification sheet in Fig. 4-7, SCR 2N6172 has a maximum peak repetitive voltage rating of _____ and reverse blocking voltage rating of _____.
   a. 100 volts and 150 volts.
   b. 200 volts and 330 volts.
   c. 400 volts and 660 volts.

4. Darlington pairs consist of _____ that are encapsulated to act as a single device.
   a. two bipolar transistors
   b. two SCRs
   c. two triacs

5. Referring to the table in Fig. 4-31, _____ are available that have a maximum of 5000 volts and 5000 A.
   a. GTOs
   b. IGBTs
   c. thyristors

PROBLEMS

1. Draw an *RL* resonant circuit that is used to commutate an SCR.

2. Draw the electrical symbols for the following devices and identify their terminals: SCR, triac, GTO, IGBT, complementary SCR, silicon controlled switch, and a silicon bilateral switch.

3. Draw the two-transistor equivalent circuit of the SCR and explain its operation.

4. Determine the number of degrees the SCR in Fig. 4-6 would be in conduction if the gate of the SCR receives a pulse that causes the SCR to go into conduction at the 40° point.

5. Draw the waveforms that you would see if you connected an oscilloscope across an SCR that is connected to a resistive load and uses an ac power supply. Compare this waveform to the one you would see if you connected the oscilloscope across the load.

REFERENCES

1. *Discrete Semiconductor Products Databook,* 1989, 2900 Semiconductor Drive, P.O. Box 58090, Santa Clara, CA 95052-8090.

2. Maloney, Timothy J., *Industrial Solid-State Electronics,* 2nd ed. Englewood Cliffs, NJ: Prentice-Hall Inc., 1986.

3. *Philips Power Semiconductor Applications,* Philips Semiconductors Strategic Accounts International Sales, 811 East Arques Ave., Sunnyvale, CA 94089, 1991.

4. *Rectifier Applications Handbook,* 3rd ed. Motorola Literature Distribution, P.O. Box 20912, Phoenix, AZ 85036.

5. *Rectifier Device Data,* REV 1, Motorola Literature Distribution, P.O. Box 20912. Phoenix, AZ 85036.

6. *Small Signal Transistors, FETs and Diode Devices Handbook,* REV 4, Motorola Literature Distribution, P.O. Box 20912, Phoenix, AZ 85036.

7. *Thyristor Device Data,* REV 5, Motorola Literature Distribution, P.O. Box 20912, Phoenix, AZ 85036.

8. *TVS (Transient Voltage Suppression)/Zener Device Data,* REV 1, Motorola Literature Distribution, P.O. Box 20912, Phoenix, AZ 85036.

5

SOLID-STATE DEVICES USED FOR FIRING CIRCUITS

OBJECTIVES

After reading this chapter, you will be able to:

1. Identify the electronic symbol for the unijunction transistor (UJT) and name its terminals.
2. Explain the operation of the UJT and include the function of the intrinsic standoff ratio.
3. Draw the diagram for a typical capacitor waveform that is used as an input signal to the UJT and show the resulting output waveform.
4. Calculate the intrinsic standoff ratio for a UJT.
5. Explain the operation of a relaxation oscillator and how it could be used to provide a timing pulse for a UJT.
6. Calculate the frequency of the relaxation oscillator.
7. Calculate the minimum and maximum sizes for resistors in a UJT firing circuit.
8. Calculate the peak voltage that the UJT will provide.
9. Explain the operation of all the components in a UJT circuit that are used to provide the firing pulse for an SCR.
10. Describe the waveform for each of the components in the UJT circuit that provides the firing pulse for the SCR.
11. Identify the symbol for the programmable unijunction transistor (PUJT, or PUT) and name its terminals.
12. Draw the diagram for the output pulse of a PUT.
13. Identify the symbol for a diac and name its terminals.
14. Draw the diagram for the output pulse for the diac.
15. Identify the symbol for a complementary SCR and name its terminals.
16. Draw the diagram for the output pulse for the complementary SCR.
17. Identify the symbol for a silicon controlled switch (SCS) and name its terminals.
18. Draw the diagram for the output pulse for the SCS.

165

19. Identify the symbol for a silicon unilateral switch (SUS) and name its terminals.
20. Draw the diagram for the output pulse for the SUS.
21. Identify the symbol for a silicon bilateral switch (SBS) and name its terminals.
22. Draw the diagram for the output pulse for the SBS.
23. Identify the symbol for a SIDAC and name its terminals.
24. Draw the diagram for the output pulse for the SIDAC.

JOB ASSIGNMENT

You have been asked to test a motor drive that is used to control the speed of a pump. Your supervisor thinks the problem is in the electronic components in the drive. You need to identify the components in the drive from the diagram that is provided. The second part of your job assignment is to connect an oscilloscope at various points in the circuit and compare the actual waveforms to the predicted waveforms for the components in the firing circuit and the thyristors that control the large currents.

5.1 OVERVIEW OF FIRING DEVICES

Each of the thyristors discussed in Chapter 4 requires a pulse to turn on. A family of solid-state devices has been specifically designed to produce the pulse signal that provides sufficient voltage and current. These devices also make the adjustment of time easy to manage so that the firing angle can be adjusted throughout the operating range of the thyristor. The *unijunction transistor* (UJT), *programmable unijunction transistor* (PUT), and the *diac* are the most common devices that provide pulses for this purpose. Other devices such as the *silicon bilateral switch* (SBS), *silicon unilateral switch* (SUS), and the *complementary unijunction transistor* (CUJT) are less frequently used to provide triggering pulses. The operation of each of these devices will be provided and example application circuits will be explained. The discussion of firing devices has been separated from the individual thyristors because it is not uncommon to find any one of the firing devices used in all of the thyristor circuits.

5.2 UNIJUNCTION TRANSISTOR (UJT)

The unijunction transistor (UJT) is a solid-state device that has been specifically designed to provide a sharp pulse when its breakover voltage level is reached. The UJT is also called a breakover voltage switch. The UJT has three terminals called *base 2* (B2), *base 1* (B1), and *emitter* (E). The functions of the bases and the emitter are different than the base and emitter in a bipolar transistor in that the unijunction transistor has only one pn junction and the bipolar transistor has two junctions. Fig. 5-1a shows the pn junction diagram of the UJT that identifies the basic parts. From this diagram you can see that the UJT is a large section of n-material that has base 1 and base 2 terminals attached at opposite ends. The emitter is made from a small section of p-material that is joined directly to the n-material to make the one junction. The point where the emitter is joined to the n-material is offset toward the base 2 end of the junction.

Fig. 5-1b shows the UJT as a function of its internal resistance between terminals B2 and B1. The point where the p-material is connected at the junction will act like a normal p–n junction or diode. Fig. 5-1c shows the electronic symbol of the UJT and it also identifies the B2, B1, and emitter terminals.

5.2.1 Theory of Operation of the UJT

The function of the UJT is to provide a pulse that turns on rapidly and provides sufficient voltage and current to trigger an SCR or triac. Fig. 5-2 shows an example of the waveforms for the input and output signals of a UJT. The top waveform shows the type of signal that a relaxation oscillator provides from the output of an *RC* (resistor capacitor) circuit. This gradually sloping waveform is the same waveform you would see when a capacitor charges and it is used as an input to the UJT. The UJT's output voltage will *break over* at some point and produce the pulse that looks like the lower waveform in the diagram. You should notice that the diagrams

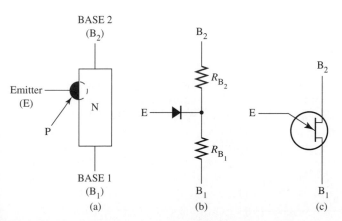

FIGURE 5-1 **(a)** Diagram of unijunction transistor (UJT) p and n junction. **(b)** Diagram of internal resistance and diode equivalent circuit. **(c)** Electronic symbol for (UJT).

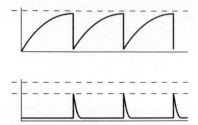

FIGURE 5-2 The top waveform shows the output signal of the relaxation oscillator that is sent to the UJT as the input signal. The bottom waveform shows the sharp pulse output signal from the UJT.

of the two waveforms show the relative time frame for each signal. The UJT will not provide an output until the voltage in the oscillator circuit charges the capacitor to a point where the UJT reaches breakover. When the UJT reaches breakover, it will produce a pulse that rises to its maximum voltage very quickly. This is the type of pulse that is needed to trigger a thyristor.

5.2.2 Firing Characteristics of the UJT

When the UJT reaches the breakover voltage, a complex set of reactions occurs. These reactions can be explained by applying a variable voltage (V_S) to the emitter to base 1 (E–B1) terminals and a fixed voltage (V_{BB}) to the base 2 to base 1 (B2–B1) circuit. Fig. 5-3 shows a diagram for this circuit. The UJT in this diagram is shown as its equivalent circuit with the n-material as two resistors identified as r_{B2} and r_{B1}. The base is shown in this diagram as a diode where it makes the junction between the p-material and the n-material.

When V_S is at a minimal level, the diode will not be forward biased and the current flow from the B2 terminal to the B1 terminal will be near zero. The actual amount of current can be calculated by dividing V_{BB} by the equivalent resistance in r_{B2} and r_{B1}. The actual voltage across each resistor could then be calculated by multiplying the current times the resistance. This voltage is important because the ratio of the voltage across r_{B2} when compared to the supply voltage V_{BB} is known as the *intrinsic standoff ratio*, which is used to calculate several characteristics such as peak voltage (V_P) and voltage at specific terminals of the UJT. Manufacturers' specification sheets will provide either the intrinsic standoff ratio or the data required to calculate it. The intrinsic standoff ratio is represented by the Greek letter η (eta). The following formula shows the relationship of the base 1 and base 2 voltages.

$$(\eta) = \frac{V_{B1}}{V_{B2}}$$

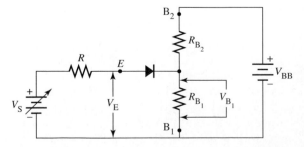

FIGURE 5-3 Diagram of the equivalent circuit for the UJT with bias voltage applied.

The intrinsic standoff ratio can also be calculated as the relationship of the internal resistance by the following formulas: Note that the *total internal resistance* is identified as r_{BB}, which is also called *interbase resistance.*

$$\eta = \frac{r_{B1}}{r_{B1} + r_{B2}}$$

· ·

PROBLEM 5-1

If a UJT has an internal resistance of 5.8 kΩ at r_{B1} and 1.9 kΩ at r_{B2}, what is its intrinsic standoff ratio?

SOLUTION

$$\eta = \frac{r_{B1}}{r_{B1} + r_{B2}} = \frac{5.8 \text{ k}\Omega}{5.8 \text{ k}\Omega + 1.9 \text{ k}\Omega} = 0.75$$

The amount of V_S at the emitter that is required to turn on the diode is called the peak voltage (V_P). When this voltage amount is exceeded, the UJT will go into conduction and provide the pulse. The amount of V_S that is required to cause this can be calculated by the following formula: $V_{PEAK} = \eta V_{BB} + V_E$. The value of V_E is the typical diode junction voltage drop of approximately 0.6 volt.

· ·

PROBLEM 5-2

If the V_{BB} voltage is 24 volts, and the intrinsic standoff ratio was calculated in Problem 5-1 as 0.75, what is the peak voltage V_P?

SOLUTION

$$V_P = 0.75 \ (24 \text{ V}) + 0.6 = \textbf{18.6 V}$$

· ·

5.2.3 Voltage and Current Characteristics When the UJT Fires

When the source voltage to the emitter rises to a point that it exceeds the peak voltage rating, the high internal resistance of base 1 drops to near zero and the flow of majority carriers increases to the point of full current flow from emitter (E) to base 1 (B1). The internal resistance of base 2 remains fixed, which means the current flow will cause a voltage to occur at base 2 as it passes through it. The change in the internal resistance of base 1 is so rapid that the voltage waveform quickly rises to its peak value and then gradually drops back to near zero. If the supply voltage is provided by a capacitor, it will quickly discharge during this period. The time between pulses will be determined by the frequency of the *RC* oscillator.

Fig. 5-4 shows the waveform of the characteristic curve for V_E versus I_E for the UJT. This waveform is also what the output pulse from the UJT will look like. The peak voltage (V_P), peak current (I_P), valley voltage (V_v), and valley current (I_V) are shown on the waveform. It is important to notice how quickly the voltage of the output pulse rises from off-state to peak. Since the rise time for this pulse is nearly instantaneous, it is the characteristic of the UJT that makes it so useful

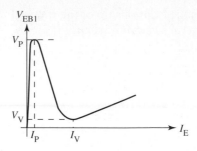

FIGURE 5-4 Waveform of output pulse from UJT.

in providing triggering pulses. The time it takes the pulse voltage to return to the valley voltage level will be a limiting factor when using the UJT high-frequency circuits.

5.2.4 Relaxation Oscillator

The relaxation oscillator is the basic timing circuit for a number of triggering devices including the UJT. The main components in the oscillator are the resistor and capacitor that make a charging RC time-constant circuit. In most oscillators, the capacitor is a fixed value and the resistor is a potentiometer. This means the larger the resistance, the slower the oscillator. Fig. 5-5 shows a variable resistor (50 kΩ potentiometer) set at 35 kΩ and 0.1μF capacitor connected to the base of the UJT. The UJT has an external resistor R_2 that is 330 Ω connected in series with its B2 terminal and a second external resistor R_1 that is 33 Ω connected to its B1 terminal. **It is important to note that R_1 and R_2 are real fixed resistors that are external to the UJT, and they should not be confused with the internal resistance (R_{B1} and R_{B2}) of the UJT.**

5.2.5 Voltage and Current in the Relaxation Oscillator

When the dc voltage is applied to the circuit shown in Fig. 5-5, the 0.1μF capacitor begins to charge at a rate determined by the potentiometer that is set for 33 kΩ.

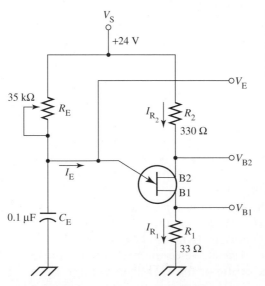

FIGURE 5-5 Diagram of relaxation oscillator connected to a UJT. The voltage measurements of V_E, V_{B2}, and V_{B1} are all made in reference to ground. Arrows are provided to show the current paths.

The timing of the *RC* circuit sets the frequency of the oscillator. The following formula is used to calculate the approximate frequency of the oscillator:

$$f = \frac{1}{T} = \frac{1}{R_E C_E}$$

Since the capacitor will be approximately 63 percent fully charged at the end of the first time constant, you can use a *rough rule of thumb* to calculate the time to charge the capacitor to the firing level by the formula:

$$t = R_E C_E \qquad \text{or} \qquad T = RC \ln \frac{1}{1-n}$$

When the capacitor charges to 63 percent of full charge during the first time constant, the amount of voltage will generally be sufficient to exceed the V_P level of the UJT, which will cause it to go into conduction and provide its pulse. When the supply voltage V_S charges the capacitor to the level where it exceeds the V_P level of the UJT, the internal resistance of r_{B2} inside the UJT will go to zero, which allows current I_E to flow from the emitter through the device out to terminal B1 and through external resistor R_1 as current I_{R1}. When current I_{R1} flows through resistor R_1, a voltage is developed across the resistor. This voltage is called V_{B1} and it is measured from terminal B1 to ground. Fig. 5-2 shows the voltage waveform of the capacitor in the top diagram, and the waveform of the voltage pulse that occurs from the current flow through resistor R_1 is shown in the bottom diagram.

On closer inspection of the diagram in Fig. 5-2 you should notice several important points. First, the waveform in the top diagram that represents V_E, which is the voltage from the capacitor that is sent to the emitter, never reaches 0 volts. This small amount of voltage is the valley voltage that was also defined in Fig. 5-2.

The second point you should notice is that when the diagram is viewed as a timing diagram, the start of the pulse from the UJT is shown occurring at the point where the capacitor voltage (V_E) exceeds the V_P of the UJT. The waveform rises quickly so that it looks almost like a vertical line, which indicates very little time has passed. The waveform is shown falling at the point where the capacitor is discharging. The amount of time for the capacitor to discharge will be a function of the internal resistance in r_{B1} inside the UJT and the RB1 external resistor. These two resistances are in the current path when the capacitor discharges through the emitter–base 1 circuit of the UJT.

One problem may arise with the oscillator circuit if care is not taken when the size of the R_E resistor is determined. Since R_E is a variable resistor, it can be adjusted from near zero to the maximum resistance of the potentiometer. If R_E gets too large, it will limit the amount of current that can flow to the emitter when the capacitor reaches full charge. When the capacitor reaches V_P voltage level of the UJT, the UJT will fire. When the UJT fires, a minimum current must continue to flow through R_E to keep the UJT in conduction long enough to provide the output pulse. If R_E is sized too large, this current will be limited and the UJT will not provide the correct pulse. R_E can be calculated by applying Ohm's law, which has been incorporated into the following formula. (V_P and I_P are provided in specification sheets, which can be found in the appendix.)

$$R_{E\text{max}} = \frac{V_S - V_P}{I_P}$$

It is also important that R_E is not sized too small. Since R_E is variable, the minimum amount of resistance (a fixed resistor) is normally added in series with the potentiometer so that if it is set to zero, the minimum amount of resistance will remain in the circuit. If R_E is too small, it will allow too much current to continue to flow after the capacitor is completely discharged. If this happens, the UJT will remain in conduction and will never return to a point where it can deliver another pulse. The minimum value for R_E can also be calculated by applying Ohm's law, which has been incorporated into the following formula. (V_S and V_V are provided in the specification sheet.)

$$R_{Emin} = \frac{V_S - V_V}{I_V}$$

· ·

PROBLEM 5-3

You can now try all of the formulas presented thus far to calculate all of the important parameters about the UJT firing circuit. You will need to calculate the following:

1. Calculate η, the intrinsic standoff ratio.
2. Calculate the frequency of the pulses.
3. Calculate V_P.
4. Calculate the minimum and maximum size range for R_E.
5. Is the value of R_E at 7.5 kΩ within the minimum to maximum range?

The following characteristics are provided on the specification sheet for the UJT:

$$r_{BB} = 8.5 \text{ k}\Omega \qquad r_{B2} = 3.2 \text{ k}\Omega \qquad r_{B1} = 5.3 \text{ k}\Omega$$
$$I_P = 5.2 \text{ }\mu\text{A} \qquad V_V = 1.5 \text{ V} \qquad I_V = 3.2 \text{ mA}$$

The following information is provided about the components in the oscillator circuit for this problem.

$$R_E = 7.5 \text{ k}\Omega \qquad C = 0.1 \text{ }\mu\text{F} \qquad V_S = 20 \text{ V}$$
$$R_{B1} = 33 \text{ }\Omega \qquad R_{B2} = 330 \text{ }\Omega$$

SOLUTION

1. The intrinsic standoff ratio (η) can be calculated by the formula:

$$\eta = \frac{r_{B1}}{r_{BB}} = \frac{5.3 \text{ k}\Omega}{8.5 \text{ k}\Omega} = 0.62$$

2. The frequency of the oscillator (f) can be calculated by the formula:

$$f = \frac{1}{(R_E)(C)} = \frac{1}{(7.5 \text{ k}\Omega)(0.1 \text{ }\mu\text{F})} = 1.33 \text{ kHz}$$

3. V_P can be calculated by the formula:

$$V_P = \eta(V_{B2B1}) + 0.6 \text{ V}$$

It will be necessary to calculate V_{B1B2}, which can be determined by using the following ratio:

$$\frac{V_{B2B1}}{V_S} = \frac{r_{BB}}{R_2 + r_{BB} + R_1} \qquad \text{restated} \qquad \frac{V_{B2B1}}{20 \text{ V}} = \frac{8.5 \text{ k}\Omega}{330 \text{ }\Omega + 8.5 \text{ k}\Omega + 33 \text{ }\Omega}$$

Therefore,

$$V_{B2B1} = \mathbf{19.2\ V}$$

You can now calculate

$$V_P = (0.62)(19.2\ V) + 0.6 = \mathbf{12.5\ V}$$

4. The minimum and maximum range for R_E can be calculated by the formulas:

$$R_{Emin} = \frac{V_S - V_V}{I_V} = \frac{20\ V - 1.5\ V}{3.2\ mA} = 5.78\ k\Omega$$

$$R_{Emax} = \frac{V_S - V_P}{I_P} = \frac{20\ V - 12.5\ V}{5.2\ \mu A} = 1.44\ M\Omega$$

Since R_S is 7.5 kΩ, it is within the R_{Emin} and R_{Emax}.

5.2.6 Synchronizing a UJT Pulse with AC Line Voltage

One of the easiest ways to synchronize the UJT pulse with the phase angle of the supply voltage is to use an ac voltage, or a pulsing dc voltage as the power source. Fig. 5-6 shows an application where an SCR is used to control power to an industrial resistive heating load. The power source for this circuit is 60 Hz ac voltage. The D1 diode in the circuit rectifies the ac voltage to pulsing half-wave dc voltage. The pulsing half-wave dc voltage is satisfactory for this circuit because the SCR can only operate in the forward-bias direction. Fig. 5-7 shows waveforms at various points in the circuit. These waveforms would be as you would see them if you used an oscilloscope to measure them.

The first waveform at the top of the figure is the sine wave from the ac input voltage. Notice that the peak voltage (163 volts ac) is calculated from the 115 volts rms.

The pulsing dc voltage shown in the second waveform

▼ IMPORTANT SAFETY NOTICE...

It is important to remember from the previous chapter that if you use an oscilloscope to make measurements in a circuit that has single- or three-phase voltages, you should always use a plastic-case or isolation-type oscilloscope that is not grounded.

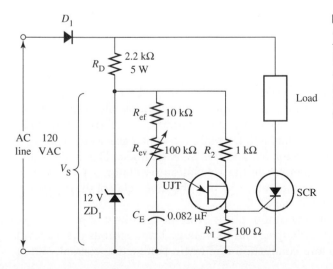

FIGURE 5-6 A circuit diagram of an SCR used to control an industrial electrical heating element. The relaxation oscillator provides an input signal to the UJT, which provides a pulse to the SCR gate.

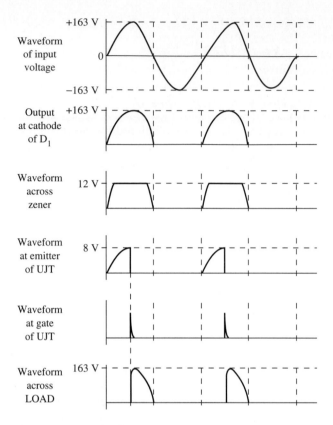

Waveform of input voltage

+163 V

0

−163 V

Output at cathode of D_1

+163 V

Waveform across zener

12 V

Waveform at emitter of UJT

8 V

Waveform at gate of UJT

Waveform across LOAD

163 V

FIGURE 5-7 Waveforms from oscilloscope measurements of various points in the SCR circuit. The first waveform is the ac input voltage. The second is the pulsing dc voltage after the diode. The third voltage is from the zener diode. The fourth voltage is the capacitor signal to the UJT emitter. The fifth is the pulse from the UJT that is sent to the gate of the SCR, and the sixth is the voltage across the load resistor.

is what you would see on the oscilloscope if you placed the leads at the cathode of diode D_1 and terminal L2 of the input voltage. The diode in this application is acting as a *rectifier* since it converts the ac voltage sine wave to pulsing dc voltage.

The third waveform is the signal that occurs across the anode–cathode terminals of zener diode ZD_1. You should notice that this signal looks similar to the pulsing dc signal, but in this case, the zener diode *clips* the top of the waveform at the 12 volt level so that it appears flat. The 12 volt zener diode acts like a voltage regulator at this point in that it will maintain 12 volts to the oscillator as long as the voltage supply to it exceeds 12 volts. The circuit is designed so that voltage supplied to the zener is well in excess of 12 volts. If the 115 volt supply fluctuates and drops up to 10%, it will not affect the operation of the firing circuit because the zener diode will continue to provide a consistent 12 volt signal. Even though the signal from the zener is regulated, it will still maintain the half wave characteristics so that it will retain phase relationship with the original sine wave.

The fourth waveform shows the signal the oscilloscope would display when its leads are connected across the capacitor. The oscilloscope leads are actually connected to the UJT emitter terminal and terminal L2, which acts as the common to the zener diode, capacitor, resistor R_1, and the SCR cathode. This is the same waveform shown in previous examples of the relaxation oscillator. As you know, the adjustment of the 100 kΩ potentiometer will vary the time delay of the oscillator, which will vary the *firing angle* of the SCR.

The fifth waveform shows the output pulse from the UJT as it appears across resistor R_1. When the capacitor charges, and its voltage level exceeds the V_P level of the UJT, the UJT will go into conduction and begin conducting current. When this current flows through the UJT, it will also flow through resistor R_1, since it is connected in series with the UJT. The size of the voltage pulse will be determined by voltage drop. This can be calculated from Ohm's law by multiplying the current times the resistance. The larger the current flow, the larger the voltage drop across the resistor. The gate of the SCR is connected to the point between the UJT B1 terminal and the top of R_1. This means that the gate of the SCR will receive the voltage pulse that is measured across R_1. The timing of this waveform will control the firing angle of the SCR. In this waveform you should notice that the firing angle is set at 85° which means the *conduction angle* will be 95° (180° − 85°).

The sixth waveform shows the voltage measured across the load. This waveform will have the outline of the original pulsing dc voltage. The actual point where the voltage turns on in the 180° half-wave will be determined by the time delay that is set by the relationship of the resistor and capacitor in the relaxation oscillator. You should notice that the *turn-off point* in the half wave is always at the 180° point, and the *turn-on point,* which is also called the firing angle, will vary. If the gate pulse is sent to the SCR at the 1° point, the SCR will turn on early in the half-cycle and remain in conduction for 179°, which will provide maximum voltage to the load. If the pulse is delayed and the SCR is triggered later at the 45° or 90° point, the load will receive a smaller amount of voltage. The waveform of the voltage that is measured across the load shows the voltage turned off from the start of the pulse to the 95° point. At the 95° point, the UJT provides a gate pulse to the SCR and it goes into conduction, allowing voltage to flow to the resistance heating load.

You should also notice that when the SCR turns off (commutates), it will remain off for 180° while the negative part of the ordinal sine wave for the ac supply voltage occurs. This will show up as "no voltage" because the diode is rectifying at this time. The SCR can only provide control for the positive half-cycle of the original ac input voltage. If a triac is used in place of the SCR, it could control both the positive and negative half-cycles of the input voltage.

5.2.7 Isolating the UJT Firing Circuit from the SCR Load Circuit

There will be times when the relaxation oscillator and UJT firing circuit must be isolated from the SCR and its load. The main reason for this is that the load may use high-voltage ac as a supply, and the UJT may use low voltage. Since the two circuits are isolated, the UJT must provide a series of pulses so that it can become synchronized with the ac voltage that supplies the load. Fig. 5-8 shows the UJT and SCR circuit diagram, and Fig. 5-9 shows the waveforms for the UJT and the load.

Fig. 5-8 shows the traditional relaxation oscillator connected to the emitter of the UJT. The major difference in this circuit is that the primary side of a *pulse transformer* is connected to the B1 terminal of the UJT instead of a resistor. This means that when current flows through the UJT, it will also flow through the primary side of the transformer. The transformer will provide isolation as it produces a pulse in its secondary coil.

The secondary coil of the pulse transformer is connected to the gate of the SCR. The SCR is connected so that it controls current flow through a bridge rectifier. This circuit is designed so that the SCR will see only pulsing dc voltage, and the load will see the complete ac voltage sine wave. Since the SCR sees the positive and negative half-waves as rectified positive pulses, it can control the full 360° of the original ac supply voltage.

FIGURE 5-8 Electrical diagram of a UJT providing a firing pulse to an SCR. The UJT uses a pulse transformer to provide isolation from the large ac voltage in the load circuit. The other unique part of this circuit is that the SCR can control both halves of the ac sine wave by controlling dc current in a bridge rectifier.

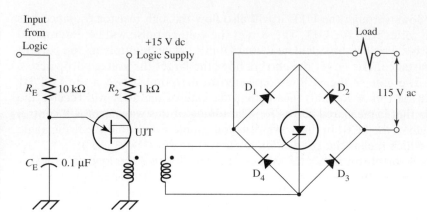

The circuit uses the feature of a bridge rectifier that current will flow in the ac side of the circuit anytime current is flowing in the dc side of the bridge rectifier. Since the SCR can easily control the dc current, it will indirectly have control over the current in the ac side of the circuit. This type of circuit was popular before larger triacs were manufactured. Now that triacs are available in larger sizes, they would be used in this type of control circuit.

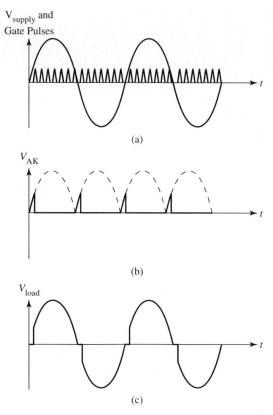

FIGURE 5-9 Waveforms for the isolated UJT firing circuit that uses a pulse transformer. Notice that the oscillator produces a *train of pulses* because the UJT triggering circuit is not synchronized with the SCR load circuit. **(a)** ac voltage supply and gate pulses. **(b)** Voltage anode to cathode (A–K). **(c)** Voltage across the load. (Courtesy of Philips Semiconductors.)

5.3 PROGRAMMABLE UNIJUNCTION TRANSISTORS

The *programmable unijunction transistor* (PUT) is similar to the UJT in that it has the ability to provide a pulse that is used to trigger SCRs and other thyristors. The main difference between the PUT and the UJT is that the peak voltage level (V_P) is determined in the PUT by external circuitry whereas V_P in the UJT is determined by the intrinsic standoff ratio. Since the peak voltage of the PUT can be adjusted by changing the circuit, designers like to use PUTs instead of traditional UJTs. Fig. 5-10 shows the electronic symbol for the PUT. From the figure you can see that the PUT has an anode, cathode, and gate. These terminals have the same names as in the SCR, but they perform a different function in the PUT.

5.3.1 Operation of the Programmable Unijunction Transistor

Since the PUT functions like a UJT, it will be easier to show the similarities of each. The cathode of the PUT provides the same function as base 1 in the UJT. When the PUT fires and goes into conduction, current flows out of the device through the cathode terminal. All measurements of voltage for the PUT will use the cathode as the reference terminal.

The anode of the PUT provides a similar function as the emitter of the UJT. When the anode voltage increases to the point called V_{PEAK}, the PUT will go into conduction. The level of peak voltage for the PUT will be determined by external voltage applied to the gate. This means that the peak voltage can be *programmed* or set to any level. Fig. 5-10 also shows the characteristic curve for the PUT. You can see that the curve has a *knee* in the positive quadrant of its graph. The knee indicates the point where V_{PEAK}, the anode voltage *breakover point* is.

The gate of the PUT provides a similar function as the base 2 terminal of the UJT. This relationship is not as close as the other two terminals, but it still has some similarities. The gate receives the external voltage, which becomes the basis for the peak voltage V_P. The formula for determining V_P is:

$$V_P = V_G + 0.6 \text{ V}$$

This is also the junction voltage. The actual value of V_P can be varied by changing the value of the voltage sent to the gate.

Fig. 5-11 shows a diagram of a PUT in a relaxation oscillator. This circuit will demonstrate how the operation of the PUT is different from the operation of the

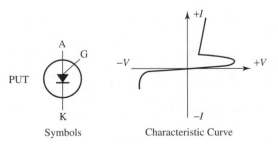

Symbols   Characteristic Curve

FIGURE 5-10 Electronic symbol and characteristic curve for the programmable unijunction transistor (PUT). (Copyright of Motorola, Used by Permission.)

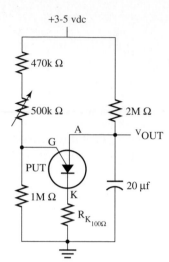

FIGURE 5-11 Diagram of a programmable unijunction transistor used in a relaxation oscillator circuit.

UJT. You should remember that the *RC* time constant determined the frequency of the oscillator for the UJT. In the PUT circuit, the peak voltage (V_P) will determine the frequency of the oscillator, which will show up in the timing of the PUT output pulses. This occurs because the *RC* part of the circuit will provide a consistent ramped voltage from the capacitor charging, and the gate voltage will set the voltage level (V_P) where the PUT will fire. For example, if the capacitor charged from 0–8 volts, and the gate voltage on the PUT is 7 volts, the PUT will fire when the capacitor voltage charges to the 7.6 volt level. If the amount of variable resistance is changed so that the gate voltage to the PUT is changed to 5 volts, the PUT would then fire when the capacitor charges to 5.6 volts, which would occur earlier in the time line. As the voltage at the gate of the PUT is lowered, the frequency of the PUT will increase. This function will be quite useful because it is easy to change the voltage level to the gate of the PUT with something as simple as a variable resistor. Or a digital to analog (D/A) converter that is controlled by a programmable logic controller (PLC) could be used.

Fig. 5-12 shows a complete circuit where a triac provides current control to a universal motor. The universal motor can use ac voltage or dc voltage. In this

FIGURE 5-12 Programmable unijunction (PUT) used to fire a transformer that fires a triac for controlling current to a motor. (Courtesy of Philips Semiconductors.)

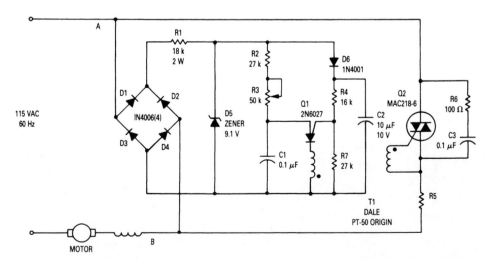

circuit the motor uses ac voltage. The triac is controlled by a programmable unijunction (PUT) that is labeled Q1 in this diagram. The PUT fires the primary side of a transformer, and the secondary side is connected to the gate of the triac. You should recognize the relaxation oscillator that is made from the resistors R_2 and R_3 and capacitor C_1. When the resistance of the potentiometer R_3 is varied, the oscillator frequency will change and the triac is fired earlier or later in its cycle, which will change the amount of current going to the motor. You should also notice that the triac is connected in the 115 volt ac part of the circuit, and the PUT uses dc voltage that has been rectified by the bridge diodes, and regulated by the 9.1 volt zener.

5.4 USING A DIAC TO PROVIDE A PULSE FOR THYRISTORS

The diac is also known as the *bidirectional trigger diode* or as the *symmetrical trigger diode.* Fig. 5-13 shows the electronic symbol of the diac and the waveform for the pulse that it produces. From this figure you can see that the symbol looks like two diodes connected in inverse parallel to each other. An alternate symbol is also shown that looks like a transistor with two emitters and no base. The terminals of the diac are called *anode 1* and *anode 2*. The diagram of the p–n-material shows that the major part of the diac is a pnp device. A slice of n-material is placed directly into both sections of the p-material. The anodes are connected to this slice of n-material.

5.4.1 Theory of Operation of the Diac

Since the diac is a bidirectional device, it can be connected in a circuit without regard to polarity. This means that it does not matter if anode 1 or anode 2 is connected to the trigger lead of the thyristor it is providing the pulse to. When the supply voltage to the diac is below the *forward breakover level* called $+V_{BO}$, the diode has virtually no current flow. You can see the characteristic curve for the diac that is shown in Fig. 5-13. After the supply voltage breaks the $+V_{BO}$ level, the junction in the diac breaks down and presents a very low resistance, which allows current to flow. This part of the characteristic curve is called the *knee,* and it is utilizing some part of *breakover voltage* to make the device provide a pulse.

When reverse voltage is applied to the diac, it will block current until its negative breakover voltage level $(-V_{BO})$ is exceeded. When the $-V_{BO}$ is exceeded, the resistance in the reverse junction drops to near zero, and the diac conducts

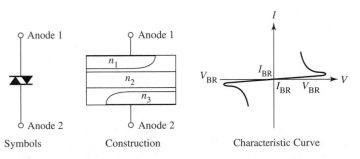

Symbols   Construction   Characteristic Curve

FIGURE   5-13 Electronic symbol for a diac. A diagram of the equivalent p–n junction and the waveform for the characteristic curve of the pulse are also provided. (Courtesy of Philips Semiconductors.)

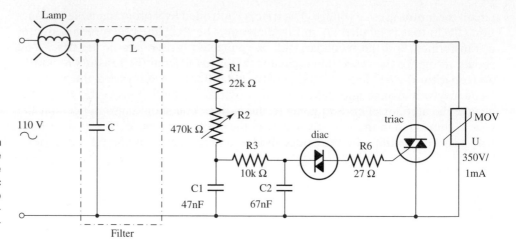

FIGURE 5-14 Diagram of diac used to provide a positive and negative pulse to control a triac that is used in a lamp dimmer circuit. (Courtesy of Philips Semiconductors.)

negative current. Since the actions in both the positive and negative directions are similar, the waves for each are nearly identical.

Diacs are available in a variety of V_{BO} ratings so that their operation can be matched to any specific circuit application. For example if a 32 volt diac is used, the capacitor in the oscillator will need to provide a voltage in excess of the diac V_{BO} rating. Since the diac can provide a positive and a negative pulse, it will generally be used in circuits where the thyristor is switching ac voltage. For example, the diac is usually used to provide the pulse for a triac.

Fig. 5-14 shows a diagram of a circuit that has a diac providing a pulse for a triac. In this circuit you can see that a variable resistor R_2 and capacitor C_1 provide the oscillator function that produces a ramped voltage signal sent to the diac. Since the capacitor voltage will increase to a value that exceeds the V_{BO} rating of the diac, the diac will produce a triggering pulse during each oscillation of the capacitor circuit. It is important to understand in this circuit that the supply voltage is ac. This means that the capacitor will charge with positive voltage and provide a positive pulse, and then charge with negative voltage to provide the negative pulse. Since the same ac voltage is used to power the triac, the positive and negative pulses from the diac will be synchronized with the firing quadrants of the triac. When the resistance of the potentiometer increases, the frequency of the oscillator will slow down and the triggering pulse to the triac will occur later in the ac cycle, which will cause the triac to send less current to the lamp. If the frequency of the oscillator is increased, the triac will fire earlier in the cycle and more current will be sent to the lamp. You should also notice that a metal-oxide varistor (MOV) is connected in parallel across the triac to protect it against voltage transients and surges.

5.5 OTHER SOLID-STATE DEVICES USED AS TRIGGERS

A variety of other solid-state devices is used to provide pulses to thyristors. Some of these devices are smaller thyristors while others are simple solid-state devices. Each of these devices will be presented and discussed in the following section. The

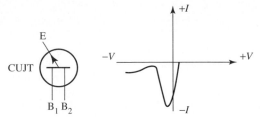

FIGURE 5-15 Electronic symbol and characteristic curve for a complementary unijunction transistor (CUJT). (Copyright of Motorola, Used by Permission.)

symbol for each device and its characteristic curve will be used to explain the type of pulse the device will produce.

5.5.1 Small SCRs and Complementary UJTs

In the early 1970s the majority of the triggering devices available today had not been designed yet. This meant that devices such as small silicon controlled rectifiers (SCRs) were used to provide a current pulse to the larger SCRs to control current. These smaller SCRs were sometimes referred to as *gaters* because they provided the gate signal for the main SCRs.

As the applications for large-voltage and large-current thyristors increased, newer devices were designed specifically to provide the control pulses. If a triggering device names its terminals *anode, cathode,* and *gate,* you will see that it will have some of the original operating characteristics of the SCR and may use the feature of forward breakover voltage. You will also notice that most of the newer devices have had enhancements added to make them more useful for a specific purpose.

Some devices will be an enhancement of other useful triggering devices. For example, a device that grew from the original UJT is the *complementary unijunction transistor* (CUJT). Fig. 5-15 shows the electronic symbol and the characteristic curve for the CUJT. From this diagram you can see that the CUJT provides a pulse that is identical to the original UJT except the pulse is negative. This pulse is useful in circuits where the supply voltage and the pulse voltage are negative, which is the third firing quadrant of the triac.

5.5.2 Silicon Controlled Switch (SCS)

The *silicon controlled switch* (SCS) provides both a positive and a negative pulse. The electronic symbol and the characteristic curve are shown in Fig. 5-16. From this diagram you can see that the characteristics of this device make it look similar to an SCR. In fact the SCS is a type of SCR with a second gate added that turns the device off. The first gate (G_1) is used to turn the SCS on, and the second gate

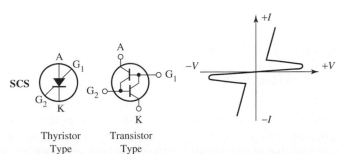

FIGURE 5-16 Electronic symbol and characteristic curve for a silicon controlled switch (SCS). (Copyright of Motorola, Used by Permission.)

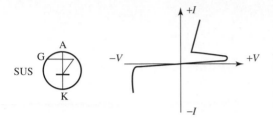

FIGURE 5-17 Electronic symbol and characteristic curve of the silicon unilateral switch (SUS). (Copyright of Motorola, Used by Permission.)

(G_2) is used to turn it off. The SCS is used in circuits that require higher frequencies where the device must be turned off quickly so that it is ready to turn on again. When traditional triggering devices are used, they do not turn off fast enough to be able to turn on again at higher frequencies. If the device that provides the firing pulse to the main thyristor does not turn off fast enough, it tends to keep the thyristor in the conduction state at all times.

5.5.3 Silicon Unilateral Switch (SUS)

The *silicon unilateral switch* (SUS) is a solid-state device that uses breakover voltage to go into conduction and provides a positive pulse. The electronic symbol and characteristic curve are shown in Fig. 5-17. From the diagram you can see that the SUS has an anode, cathode, and gate like an SCR. In fact this device is essentially a miniature SCR, but unlike the SCR, the SUS is generally set into conduction by causing the anode–cathode voltage to rise beyond the breakover point (V_{BO}). You should remember from the explanation of the SCR that it is generally put into conduction by a pulse to its gate. In the SUS, a small bias voltage is usually added to the gate to lower the V_{BO} level. Since the characteristic curve shows a knee in the positive quadrant of the graph, you will see that the SUS depends primarily on breakover voltage for its normal operation.

5.5.4 Silicon Bilateral Switch

The *silicon bilateral switch* (SBS) is basically equivalent to two SUSs. The electronic symbol, equivalent circuit, and characteristic curve of the SBS are shown in Fig. 5-18. From the symbol in the figure you can see that the SBS has three terminals called *anode 1, anode 2,* and *gate.* The characteristic curve of the SBS shows that the device can provide a pulse in both the positive and negative quadrants of its graph. This means that the device can provide both a positive and a negative pulse for devices such as the triac. The term *bilateral* in the name of this device means it has the ability to provide both a positive and a negative pulse. It is also possible to use only the positive pulse for SCR applications.

The SBS is actually a simple integrated circuit that consists of transistors, diodes, and resistors that are connected as two antiparallel regenerative switches. The components of this integrated circuit are matched so that the signal is asymmetrical. The voltages for the positive and negative pulses have less than a 0.5 volt difference.

5.5.5 SIDAC

The SIDAC is a newer triggering device that is a high-voltage bilateral trigger. This device will switch from the blocking state to the conducting state anytime the applied voltage exceeds the breakover voltage $V_{(BO)}$ level of the device in either the positive or the negative direction. Typical breakover voltages range from 104–

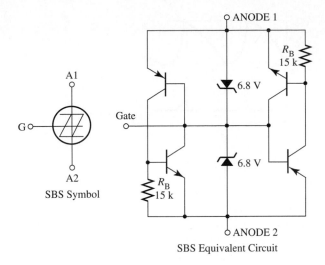

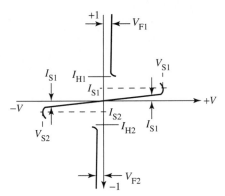

FIGURE 5-18 Electronic symbol, equivalent circuit, and characteristic curve of a silicon bilateral switch (SBS). (Copyright of Motorola, Used by Permission.)

280 volts. The electronic symbol for the SIDAC is shown in Fig. 5-19. From the symbol you can see that the terminals are named MT1 and MT2. A characteristic voltage and current curve is also shown for the device. From this curve you can see a typical knee for both the forward and reverse voltage areas. The knee indicates that the device will go into conduction any time the voltage level is exceeded.

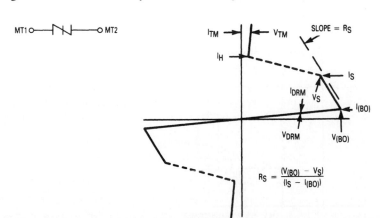

FIGURE 5-19 Electronic symbol and characteristic voltage and current curve for the SIDAC. (Copyright of Motorola, Used by Permission.)

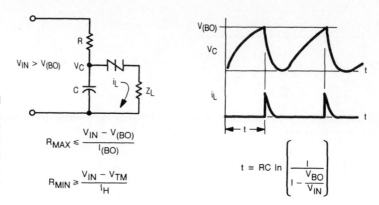

The SIDAC can be connected to an oscillator circuit just like the other trigger devices. The components in the oscillator must be sized so that the voltage supplied to the SIDAC will exceed its forward breakover rating so that the SIDAC will go into conduction and produce a very sharp high-voltage pulse to be used as a trigger. Fig. 5-20 shows an example of the oscillator circuit and the input and output pulses for the SIDAC.

The SIDAC is usable as a high-voltage, high-current trigger in electronic circuits. Fig. 5-21 shows an example of the SIDAC being used in a Xenon flasher circuit. In this circuit, the SIDAC is rated for 125 volts forward voltage. When the oscillator voltage exceeds 125 volts, the SIDAC goes into conduction and produces a high-voltage pulse to the primary side of the pulse transformer. The secondary side of the pulse transformer provides the voltage and current for the Xenon flasher. The smaller devices like an SBS or SUS will not have sufficient voltage and current pulse for this application.

5.5.6 Silicon Asymmetrical Switch (SAS)

The *silicon asymmetrical switch* (SAS) is similar to the silicon bilateral switch (SBS). The major difference is that the SAS has a different firing voltage in the positive quadrant than it does in the negative quadrant. This feature is useful in circuits that are affected by hysteresis, such as circuits with larger inductors. These circuits include most motor circuits and transformer circuits.

You should remember that when an inductor (coil) charges with current, it will become magnetized at a specific rate. When the current flow is stopped or reversed, such as at the end of each half-cycle when ac voltage is used, the rate at which the original magnetic field collapses and the reverse magnetic field builds will not be the same rate as the original signal. This effect is called *hysteresis* and it may carry over to the electronic components used to provide the firing pulse to the

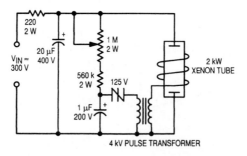

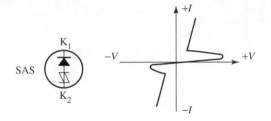

FIGURE 5-22 Electronic symbol and characteristic curve for the silicon asymmetrical switch (SAS). (Copyright of Motorola. Used by Permission.)

thyristors that are controlling the current flow. When hysteresis occurs in electronic devices, it will cause them to heat up more than normal. The asymmetrical characteristic of the SAS will help limit the effects of hysteresis. Fig. 5-22 shows the electronic symbol and characteristic curve for the SAS. From this diagram you can see that the terminals on the SAS are identified as cathode 1 (K_1) and cathode 2 (K_2).

5.5.7 Asymmetrical Silicon Bilateral Switch (ASBS)

The *asymmetrical silicon bilateral switch* (ASBS) is similar to the silicon bilateral switch (SBS) except it has two different firing levels. This feature is similar to the silicon asymmetrical switch (SAS) and provides the same function. The ASBS provides positive and negative pulses that are nearly identical except for the difference in voltages. Fig. 5-23 shows the electronic symbol and characteristic curve of the ASBS. From this diagram you can see the terminals for the ASBS are also called cathode 1 (K_1) and cathode 2 (K_2).

5.5.8 Shockley Diodes

The Shockley diode, also called the four-layer diode, is a unidirectional diac, which is basically an SUS without a gate terminal. The terminals are named *anode* and *cathode* like a regular junction diode. The Shockley diode will break over when the forward-bias voltage increases beyond the forward breakover voltage (V_{BO}) level. Fig. 5-24 shows an electrical symbol and the characteristic curve for the Shockley diode. This diode is mainly used as a trigger in SCR circuits.

5.5.9 Schottky Diode

The Schottky diode is shown in Fig. 5-25. From the electronic symbol in the figure you can see that the Schottky diode has *anode* and *cathode* terminals like a conventional junction diode. The symbol for this diode is also similar to the junction diode symbol except the letter S is added to the anode side of the symbol to indicate it is a Schottky diode. The characteristic curve in this figure shows that the Schottky diode has a better forward current-carrying capability when compared to a normal

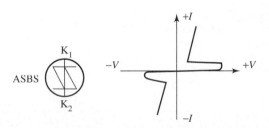

FIGURE 5-23 Electronic symbol and characteristic curve for the asymmetrical silicon bilateral switch. (Copyright of Motorola. Used by Permission.)

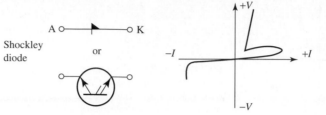

FIGURE 5-24 Electronic symbol and characteristic curve for a Shockley diode. (Copyright of Motorola. Used by Permission.)

junction diode. This type of diode is a four-layer diode like the Shockley diode, which is used primarily to provide a positive pulse to SCRs.

5.6 LIGHT-ACTIVATED SOLID-STATE DEVICES

A variety of light-activated and light-sensitive devices has been designed to provide isolation between the firing circuit and the thyristor devices. This has been especially important since thyristors are commonly used in industrial circuits where three-phase voltages up to 480 volts ac and 1000 A are controlled. These circuits tend to have uncontrolled transient voltage spikes that may be three to five times the original supply voltage. If these transient voltages make their way into the firing circuit, they will cause severe damage. In traditional circuits these transient spikes will travel from the load circuit back into the firing circuit through common connections or common ground points.

If light is used to *couple* or *connect* the firing circuit to the load circuit, these transients will not have a path to get back into the firing circuit. A complete family of light-sensitive devices has been designed. Fig. 5-26 shows the electronic symbols for the *photodiode, phototransistor, light-activated SCR* (LASCR), *light-activated programmable unijunction transistor* (LAPUT), *light-activated switch* (LAS), and the *light-activated SCS* (LASCS).

From the figure you can see that the symbol for each of these devices has two arrows pointing in toward the device. The arrows indicate that the device is activated by light or is sensitive to light. Some devices like the photodiode and phototransistor will vary their junction resistance with the amount of light that is focused on the device. Other devices such as the light-activated SCR and light-activated switch will go into conduction when a pulse of light is directed on the device, and the device will remain in conduction even after the light is removed.

FIGURE 5-25 Electronic symbol and characteristic curve for a Schottky diode. (Copyright of Motorola, Used by Permission.)

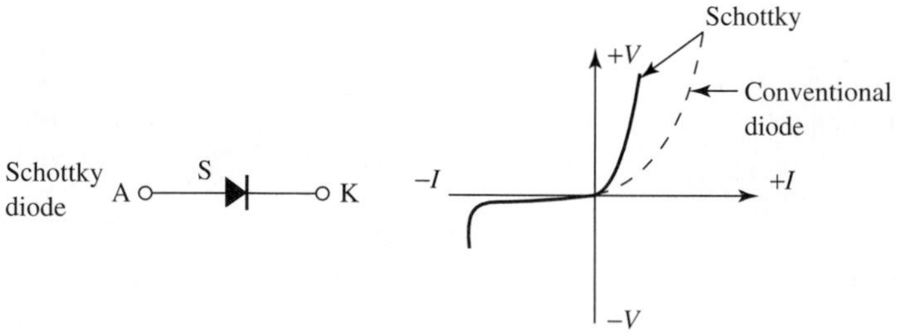

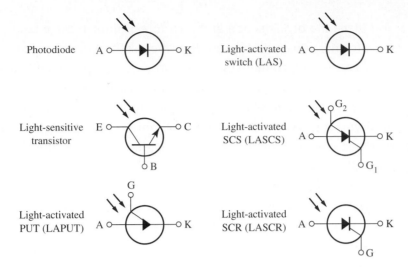

FIGURE 5-26 Electronic symbol for the photodiode, phototransistor, light-activated SCR (LASCR), light-activated PUT (LAPUT), light-activated switch (LAS), and the light-activated SCS (LASCS). (Copyright of Motorola, Used by Permission.)

REVIEW

As a technician you will find a wide variety of solid-state devices used as triggering devices that provide pulses for thyristors. Depending on the age of the circuits, you may find rather simple circuits like oscillators that provide a pulse to a UJT, or you may find the more complex solid-state circuitry of an SBS. You should remember that the function of each of these devices is similar in that they are trying to provide a pulse with sufficient voltage and current and the proper polarity to energize a wide range of thyristors. The main feature of each device is that it must be able to go into conduction and provide the pulse instantaneously. The timing of the pulse will be varied to change the output of the thyristor. You may need to refer to specific examples in this chapter to determine symbols, their terminals, and the waveform each device should produce.

SOLUTION TO JOB ASSIGNMENT ◀ ⋯⋯⋯⋯⋯⋯⋯⋯

The first step in solving the job assignment is to review the diagram of the circuit for the pump drive. The main thyristor and the triggering devices must be identified. If the motor is a dc motor and its circuit has an SCR for the thyristor control and a UJT as the firing device, you can use the circuit shown in Fig. 5-6 as a guide. The waveforms in Fig. 5-7 can be used to explain the typical waveforms for this type of circuit.

If the pump is an ac motor you can use the circuit shown in Fig. 5-12 as an example. The firing device may be a diac or SIDAC and their output signals will be similar to the one shown in the second part of Fig. 5-20. When you place the oscilloscope at the various points in the circuit to verify these waveforms, you must be sure that the oscilloscope is a plastic-case or isolated-case scope so that you do not provide a short circuit in the motor drive. You can start at the power supply and work toward the motor end of the circuit, or you can start at the motor and move back toward the power supply until you find where the circuit problem is located. You should remember that each section of these circuits should produce the signals that have been discussed.

It is also important to determine the phase relationship between the relaxation oscillator and the supply voltage so that you can determine the firing angle for each circuit. If the correct voltage, current, and frequency are provided to the device as an input signal, that device should produce an output or the device is defective. It will also be possible to test the pn junctions of some of the devices and determine if they are good or bad.

QUESTIONS

1. Explain the operation of the unijunction transistor (UJT) and include the function of the intrinsic standoff ratio.
2. Explain the operation of a relaxation oscillator and how it could be used to provide a timing pulse for a UJT.
3. Identify all the components in the UJT shown in Fig. 5-6 and explain the function of each.
4. Describe the waveform for each of the components in the UJT circuit that provides the firing pulse for the SCR. Use Fig. 5-6 as your example circuit.
5. Explain the main advantage the SIDAC provides over other triggering devices.
6. Describe the difference between a UJT and a PUT.
7. Explain why a diac is used instead of a UJT to trigger a triac.

TRUE OR FALSE

1. _____ The intrinsic standoff ratio for a UJT is determined by the amount of resistance provided by circuit resistors R_1 and R_2. (Use Fig. 5-5 as a reference.)
2. _____ The size of the variable resistor and capacitor that are connected to the emitter of a UJT in a relaxation oscillator will determine the frequency of the output pulse of the UJT.
3. _____ The unique characteristic of a diac is that its output pulse is only positive.
4. _____ The SIDAC is useful to trigger larger thyristors because it can produce high-voltage and high-current pulses as firing signals.
5. _____ The Shockley diode and the Schottky diode are primarily used to trigger SCRs.

MULTIPLE CHOICE

1. The UJT will produce an output pulse when _____.
   a. the voltage across B2–E exeeds the breakover voltage level.
   b. the voltage across B1–E exceeds the breakover voltage level.
   c. the voltage across B1–B2 exceeds the breakover voltage level.

2. If the gate of an SCR requires a larger amount of voltage than a UJT can provide, _____.

   a. you will need to use a smaller SCR.

   b. you can use a pulse transformer like the one in Fig. 5- 8 to increase the strength of the pulse from the UJT.

   c. you can connect the UJT in reverse bias to double its output pulse.

3. The difference between the PUT and the UJT is that _____.

   a. the peak voltage of the PUT is determined by external components connected to it.

   b. the intrinsic standoff ratio of the UJT can be varied by external resistance connected to it.

   c. the frequency of the output pulse of the UJT can be varied and the frequency of the output pulse of the PUT is fixed.

4. The diac _____

   a. produces only a positive pulse.

   b. produces only a negative pulse.

   c. produces both a positive and a negative pulse.

5. The SIDAC _____

   a. produces only a positive high-voltage pulse.

   b. produces only a negative high-voltage pulse.

   c. produces both a positive and a negative high-voltage pulse.

PROBLEMS

1. Calculate the intrinsic standoff ratio of a UJT that has an internal resistance of 6.4 kΩ at r_{B1} and 1.7 kΩ at r_{B2}.

2. Use the intrinsic standoff ratio that you calculated in problem 1 to determine the peak voltage for the UJT if its V_{BB} voltage is 24 volts.

3. Calculate the frequency of a relaxation oscillator in Fig. 5-5 if the potentiometer is set at 19 kΩ.

4. Calculate the intrinic standoff ratio, frequency, V_p, minimum and maximum sizes of R_E for an oscillator circuit with a unijunction transistor that has the following characteristics.

$$r_{BB} = 7.4 \text{ k}\Omega \qquad r_{B2} = 2.9 \text{ k}\Omega \qquad r_{B1} = 5.1 \text{ k}\Omega$$
$$I_P = 4.9 \text{ } \mu\text{A} \qquad V_V = 1.5 \text{ V} \qquad I_V = 3.0 \text{ mA}$$

The circuit components have the following ratings.

$$R_E = 7.8 \text{ k}\Omega \qquad C = 0.1 \text{ } \mu\text{F} \qquad V_S = 20 \text{ V}$$
$$R_{B1} = 32 \text{ } \Omega \qquad R_{B2} = 320 \text{ } \Omega$$

5. Draw the diagram for a typical capacitor waveform that is used as an input signal to the unijunction transistor (UJT) and show the resulting output waveform.

6. Draw the symbols for the following components and identify their terminals: UJT, diac, SBS, SUS, SCS, SAS, ASBS, CUJT, PUT, zener diode, metal-oxide varistor, Shockley diode, Schottky diode, complementary UJT, SIDAC, photodiode, light-activated switch, light-sensitive transistor, light-activated PUT, light-activated switch (LAS), light-activated SCS (LASCS), and light-activated SCR (LASCR).

REFERENCES

1. *Discrete Semiconductor Products Databook,* 1989, 2900 Semiconductor Drive, P.O. Box 58090, Santa Clara, CA 95052-8090.

2. Maloney, Timothy J., *Industrial Solid-State Electronics,* 2nd ed. Englewood Cliffs, NJ: Prentice-Hall Inc., 1986.

3. *Philips Power Semiconductor Applications,* Philips Semiconductors Strategic Accounts International Sales, 811 East Arques Ave., Sunnyvale, CA 94089, 1991.

4. *Rectifier Applications Handbook,* 3rd ed. Motorola Literature Distribution, P.O. Box 20912, Phoenix, AZ 85036.

5. *Rectifier Device Data,* REV 1. Motorola Literature Distribution, P.O. Box 20912, Phoenix, AZ 85036.

6. *Small Signal Transistors, FETs and Diode Devices Handbook,* REV 4. Motorola Literature Distribution, P.O. Box 20912, Phoenix, AZ 85036.

7. *Thyristor Device Data,* REV 5, Motorola Literature Distribution, P.O. Box 20912, Phoenix, AZ 85036.

8. *TVS (Transient Voltage Suppression)/Zener Device Data,* REV 1. Motorola Literature Distribution, P.O. Box 20912, Phoenix, AZ 85036.

6

PHOTOELECTRONICS, LASERS, AND FIBEROPTICS

OBJECTIVES

When you have completed this chapter, you will be able to:

1. Explain the difference between infrared, ultraviolet, and incandescent light.
2. Describe the operation of photovoltaic cells.
3. Describe the operation of photoconductive cells (cadmium sulfide).
4. Explain the operation of retroflective, through-beam, and diffuse scan photoelectric switches.
5. Explain why photoelectric switches use modulated light sources.
6. Provide the symbol for a light-emitting diode and explain its operation.
7. Explain the operation of a photodiode.
8. Explain the operation of a phototransistor.
9. Explain the operation of optoisolation that uses light-emitting diodes and phototransistors.
10. Provide examples of optoisolation in industrial electronic controls.
11. Explain the operation of light-activated SCRs.
12. Explain the operation of photoelectric devices used for solid-state displays.
13. Explain the operation of a bar-code system.
14. Explain the operation of a fiberoptic system used to transmit and receive signals in industrial applications.
15. Discuss industrial applications where fiberoptics would be necessary.

You are requested to help interface a computer to a machine to monitor a sensor that indicates (counts) the number of good parts that are accepted and bad parts that are rejected. The sensor for this system is rated at 110 volts AC and is an integral part of the machine. The signal that you send to the computer is a low-voltage TTL signal that must be completely isolated so that noise in the 110 volt system does not get into the computer.

After you have interfaced the sensor signal to the computer, you find that there is too much electrical noise and interference on the factory floor and you must move the computer to a nearby room. The wiring that was originally used to transmit the signal between the sensor and the room is receiving interference from the machines on the floor. Your job is to isolate the signal where it is connected to the computer and transmit the signal so that it is immune to interference. Since you had to move the computer away from the machine, you must also provide a means to display the data the computer is compiling at a location near the machine.

After you have interfaced the counter and display with the computer, you are requested to look into the possibility of using a bar-code reader to track each of the good and bad parts. You are requested to recommend the types of bar-code equipment that may be used for this application.

6.1 ANALYSIS OF LIGHT

If you are to fully understand how solid-state devices utilize light, you must first understand some basic concepts about light. Light travels in the form of waves and it is identified by its wavelength. Light is actually one form of *electromagnetic energy* (EME). In 1871 Heinrich Hertz, a German scientist. studied several phenomena about the wavelength and frequency of light and other forms of EME. The unit of frequency, hertz (Hz), has been named in his honor. Fig. 6-1 shows the frequency of various sources of EME from low-frequency sound waves to high-frequency cosmic rays. Even though light waves oscillate at different frequencies, all EME

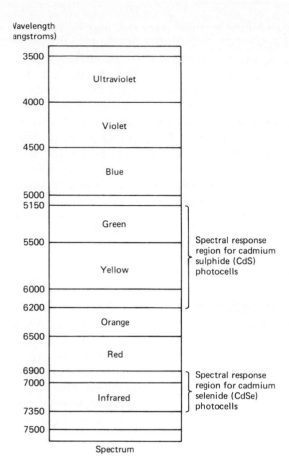

Vavelength
angstroms)

3500

Ultraviolet

4000

Violet

4500

Blue

5000
5150

Green

Spectral response
region for cadmium
sulphide (CdS)
photocells

5500

Yellow

6000
6200

Orange

6500

Red

6900
7000

Spectral response
region for cadmium
selenide (CdSe)
photocells

Infrared

7350

7500

Spectrum

FIGURE 6-1 Spectrum that shows the wavelength of various light sources. (Courtesy of Honeywell's Micro Switch Division.)

travels at 186,000 miles per second or 10^{10} meters per second, which we call the speed of light.

Hertz found in his studies that the higher the frequency of a wave, the shorter its wavelength will be. Photoelectric and solid-state optic devices for industrial controls use primarily infrared, visible, and ultraviolet light sources. If we apply Hertz's theory, we find that ultraviolet light has the shortest wavelength, and infrared light has the longest wavelength. Ultraviolet light and infrared light are not visible to the human eye, whereas other light waves between violet and red are called *visible light.* The human eye can detect a wide variety of colors and hues from the primary frequencies from 10^{14} to 10^{16} hertz.

You will gain a better understanding of light as you see how solid-state devices are designed to take advantage of the superior qualities of each type of light. For example, solid-state devices such as the *light-emitting diode* (LED) can be designed to emit light in the visible range to be used as indicator lights, or they can be designed to emit infrared light, which must be detected by infrared detectors, and are not affected by natural light.

The graph in Fig. 6-1 also shows the spectral response region for *cadmium sulphide* (CdS) photocells and for *cadmium selenide* (CdSe) photocells. CdS photocells are responsive to green and yellow light sources and CdSe photocells respond to light sources in the infrared light range. This information is useful when photoelectric controls are selected for use in industry. For example, if the sensor is used where background light may interfere, the cadmium selenide photocell may be used because it is only sensitive to infrared light.

This chapter will explain the operation of various types of photoelectric controls that use photovoltaic cells and photoconductive cells. This chapter will also explain the operation of light-emitting devices and light-activated devices such as photodiodes, phototransistors, and light-activated SCRs. The role of optoisolation will be explained as light-emitting devices are coupled with light-activated devices. Various types of solid-state displays are also covered. The last part of the chapter will explain newer optic technology such as lasers and fiberoptics. Lasers have become a vital tool for cutting and industrial processes as well as the main part of fiberoptic transmission and receiving systems. Fiberoptics has been fully integrated into industrial communications to provide immunity from interference caused by electromagnetic fields.

6.2 PHOTODIODES

Silicon photodiodes can operate in either photovoltaic or photoconductive mode. In the photovoltaic mode the photodiode produces a small amount of voltage when it is exposed to light and its junction is unbiased. This small voltage can be amplified to operate controls in a variety of applications including 110 volt circuits. When multiple cells are used together to produce larger amounts of voltage they are called solar cells because they can produce usable power from sunlight. This power can be used to recharge batteries, or provide power to operate sensors and transmitters on remote pipelines as they pass through deserts or other areas where AC voltage is not available.

Fig. 6-2 shows a photovoltaic cell and its symbol. A graph that shows the amount of voltage that a typical photocell produces for a given amount of light is also provided in this figure. The symbol shows two arrows indicating light pointing into the battery. The arrow running through the battery indicates the amount of voltage produced is variable. The graph shows that the amount of light that strikes the surface of the photovoltaic cell must increase exponentially to provide additional voltage. This means that a typical cell may produce approximately 0.17 volt with light intensity of 10 foot candles, and the voltage increases to only 0.2 volt when the light intensity doubles to 20 foot candles.

FIGURE 6-2 **(a)** Photovoltaic cell. **(b)** Electronic symbol for photovoltaic cell. **(c)** Graph of the amount of open circuit voltage vs. amount of light. (Courtesy of EG&G VACTEC, EG&G Optoelectronics Group.)

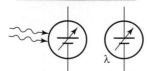

(b)

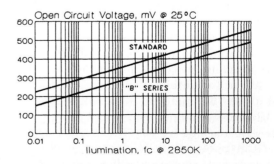

OPEN CIRCUIT VOLTAGE VS ILLUMINATION

When the photodiode is used in the photoconductive mode a reverse bias voltage is applied to its junction. In this mode the resistance of the junction changes as light strikes it which causes the photodiode to function as a current source. Other cells that are typically made of cadmium sulfide or cadmium selenide are also used as photoconductive cells.

6.2.1 Photovoltaic Cells

Photovoltaic cells are used as a light detector because of their ability to produce a small voltage when light strikes their surface. Since the voltage produced by the cell is small, it must be amplified to be useful as a light activated photoelectric switch. A small amount of bias voltage can be added to this type of circuit to compensate for background light. If the bias voltage is positive, less voltage would be required from the photovoltaic cell to cause the transistor to conduct, thus the light switch would be more sensitive to small changes in light. If the bias were negative, the photovoltaic cell would need to produce more voltage to cause the transistor to conduct and energize the switch. This would indicate that a powerful beam of light is striking the cell.

6.2.2 Theory of Operation of Photovoltaic Cells

Photovoltaic cells are similar to the P–N junction used in diodes. For this reason they are sometimes called *photoconductive diodes*. The major difference from the junction diode is that the junction in the photovoltaic cell is made by placing layers of selenide material on a metal base. The selenide material becomes the negative terminal of the cell and the metal base becomes the positive terminal. An opening is provided so that light can strike the junction. The photocell junction region is made thin so light (photons) striking the surface can liberate free electrons into the conduction-band region and allow them to diffuse across the junction. When the region is saturated, a differential voltage is set up across the terminals. The process becomes continuous as additional photons are absorbed from the light and cause new electrons to become excited and move into the conduction-band region of the junction. Typically a photovoltaic cell can produce 3.5 mA per square inch from a light source of 1000 foot candles.

6.2.3 Applications for Photoelectric Controls

Photoelectric controls are used in a variety of applications to indicate the presence of parts, the position of parts, and the level of liquid in a tank. These controls are also used in numerous other applications that include counting products, detecting missing caps or labels on containers, and detecting products in shrink-wrap packages. Fig. 6-3 and Fig. 6-4 show two applications of photoelectric switches. The photoelectric switches have photovoltaic sensors. You can see in the examples that photoelectric switches use a photoemitter, which sends a beam of light that strikes a receiver. When something breaks the beam of light, the photoelectric switch is activated. In Fig. 6-3 the photoelectric sensor is used to detect the presence of a cap and label on each bottle as it passes the inspection point. If both light beams are broken, the package has both a cap and label. If a cap is missing, the top beam of light will strike the receiver, and if the label is missing, the lower beam of light will be detected. A third photoelectric sensor could be set up to detect (count) each package as it passes. The combination of all three sensors could be sent to a logic circuit so that the number of good and bad parts are calculated.

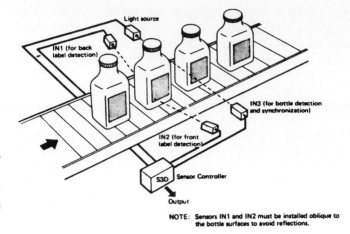

FIGURE 6-3 Applications that use photoelectric sensors to check for caps and labels on bottles. (Courtesy of Omron Electronics.)

A second application is shown in Fig. 6-4 where the photoelectric sensors are mounted so their light beams are directed through a clear plastic pipe that acts as a site glass for a tank that holds a liquid. If the level of the liquid is high enough in the tank, the beams of light will be broken. Any number of sensors can be mounted in this manner to detect the level in the tank at various heights.

6.3 PHOTOCONDUCTIVE CELLS

The second type of photoelectric device is the *photoconductive cell,* which changes its internal resistance when light strikes its surface. The cell is made from cadmium sulfide (CdS) or from cadmium selenide (CdSe). Fig. 6-5 shows the electrical symbol and a picture of a cad cell. This figure also shows a graph of the amount of resistance versus the amount of light for the cad cell. Cadmium sulfide is responsive to green or yellow light, which is in the range of 5150 to 6200 angstroms (Å). Under conditions

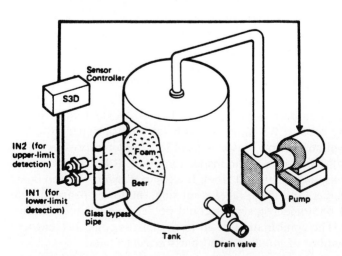

FIGURE 6-4 Wavelengths of various sources of light. (Courtesy of Omron Electronics.)

FIGURE 6-5 Typical cad cells and the electronic symbol of a Cad cell. (Courtesy of EG&G VACTEC, EG&G Optoelectronics Group.)

of no or little light, the resistance of the cell is 500 kΩ to 20 MΩ. When light strikes the surface of the cell, the resistance drops from 4 kΩ to 20 kΩ. This feature allows the cad cell to be used as the detector part of a photoelectric switch by being connected to an amplifier. One typical application uses a cad cell to activate the base of a transistor amplifier. When light strikes the cad cell, its resistance drops and current increases to the base of the transistor. When the base current increases, the emitter–collector current increases, the relay coil becomes energized, and its contacts close to allow current flow to a motor. The sensitivity of this device can also be adjusted by applying a small amount of bias voltage to the transistor.

Since the cadmium sulfide cell operates in the visible light spectrum, it has an inherent problem in that it has trouble distinguishing control light and ambient or background light. This is a problem in most industrial areas, since background light is ever present from a variety of sources: overhead lighting used to light work areas, reflected light from shiny surfaces, and light from processes such as pouring molten steel or arc welding.

In other applications the photoelectric switch may use these sources of light to detect the presence of molten iron or the flame in a furnace. Fig. 6-6 shows a photoelectric switch being used on an automatic pouring line where brake rotors are made in an iron foundry. The automatic control senses that the sand mold is in place and the ladle is tipped to begin pouring the iron. When the iron fills the mold to the top, a vent hole will also fill with molten iron. The photoelectric element is focused on the vent, and when the molten iron appears, the light it emits will be sufficient to activate the switch, which stops the ladle-pouring cycle.

Fig. 6-7 shows the electrical diagram of a fuel oil burner control that uses a cadmium cell in a safety circuit to detect the presence of a flame. In this application the cadmium cell, called a flame-detection circuit, is positioned where it can receive light from the flame of a fuel oil burner that is part of a residential or light industrial boiler. When the thermostat closes the R–W contacts, it will provide voltage for the burner circuit. This voltage will provide current through the coil of the 1K relay

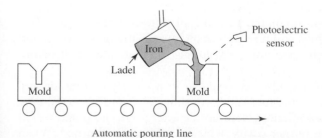

Automatic pouring line

FIGURE 6-6 Cad cell in a photoelectric sensor is used to determine when a sand mold is poured full of molten cast iron. The photoelectric sensor is part of a foundry automated pouring system where automotive brake parts are made.

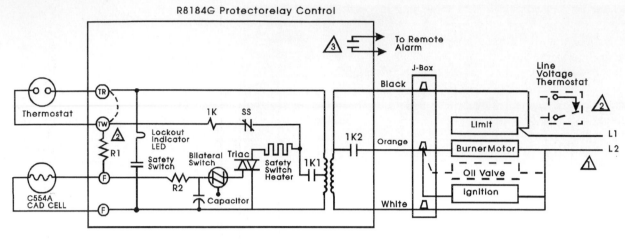

R8184G Protectorelay Control

△1 Power supply. Provide disconnect means and overload protection as required.

△2 To use R8184C with line voltage controller, jumper T-T terminals and connect line voltage thermostat in series with limit controller.

△3 Remote alarm contacts (available on some models); make to turn on alarm when safety switch locks out.

FIGURE 6-7 Electrical circuit diagram of cadmium cell used to detect the presence of a flame for a fuel oil burner control. (Courtesy of Honeywell Tradeline Controls.)

coil, through the SS (safety switch heater coil), through the triac back to the common side of the transformer. The current is sufficient to energize the 1K relay coil, which will pull in the 1K1 and 1K2 NO contacts. The 1K2 contacts will close to energize the burner motor, the oil solenoid valve, and the ignition module that provides the spark for ignition. The 1K1 contacts provide a path to the center tap of the transformer so that the 1K coil remains energized.

When the ignition process occurs, the flame should appear and create enough light to cause the resistance in the cadmium sulfide cell to be reduced, which will create a short circuit in parallel with the resistor-capacitor oscillator. This turns off the silicon bilateral switch (SBS), which will turn off the triac. The triac acts like a solid-state switch. When it is turned off, it will stop current flow through the heater of the SS switch.

If a flame is not detected, the resistance in the cadmium cell will remain high and the capacitor will continually charge and discharge, which produces a pulse to the SBS. The SBS will provide a pulse to the triac, which keeps it in conduction. When the triac is in conduction, it will allow current to pass through the heater of the SS switch. If the heater of the SS switch has current pass through it for 30 seconds, its NC SS contacts will open and turn off all power to the circuit. This will de-energize the 1K coil, which will turn off the oil burner, the oil solenoid, and the ignition. A reset button will "pop out," which must be manually reset before the circuit can operate again.

The SS switch in the boiler control is available in a variety of time delays. Anytime the flame has a problem, the fuel valve will shut off according to the time set by the time delay. If the fuel solenoid were not de-energized, the pump would continue to pump fuel into the burner cavity, which would not be burned. When the burner is started the next time, this fuel would tend to ignite and burn out of control.

All residential and commercial fuel oil furnaces have a similar safety feature. Larger industrial heating systems for gas and oil systems also have some type of safety feature. The main problem with the cadmium cell in the flame-detection circuits is that it will get a coating of soot or smoke on the lens and the amount of light reaching the cell surface will be diminished, which will cause the system to trip out incorrectly at some time.

6.3.1 Cadmium Selenide Cell

An alternate control design which eliminates the problem of ambient light causing erratic sensing uses a cadmium selenide (CdSe) cell. The CdSe cell is responsive to light in the range of 6900–7350 Å, which is in the infrared spectrum, and all other sources of light in the range of 3500–6800 Å will be ignored. This provides a degree of rejection to unwanted light. The CdSe cells can be used where all types of sources of ambient light are present, but since they are of the wrong wavelength, they will not cause the cell to activate.

6.3.2 Modulated Light Source (MLS)

The majority of photoelectric devices used today operate on the principal known as modulated light source (MLS). These devices use a light-emitting diode (LED) as a light source and a phototransistor as the receiver. The operation of the LED and phototransistor will be discussed in detail later in this chapter in Section 6.5. In the MLS-type switch, the phototransistor detects the presence of light from the LED and it provides the amplification for the circuit. The LED is modulated (turned on and off) at a specific frequency by an oscillator circuit. The phototransistor in the receiver is connected to a circuit that is tuned to respond to the specific frequency of the LED, and it filters out or rejects light sources that have other frequencies. Since all sources of light have a specific frequency, this method provides simple filtering, which is called *light rejection*.

Pulsing the LED also allows a more intense light source with less power consumption to be used. Less power is consumed in this application because the light is pulsed rather being continuously illuminated. Heat from the LED is also dissipated more easily when it is pulsed. The actual on-time for the LED may only be several msec, while the off-time may be over 100 msec. This provides an off-to-on duty cycle of better than 10:1. It also means that the pulse may have the strength of several hundred mA when the LED is pulsed on, and yet it consumes an average of less than 30 mA.

6.4 INDUSTRIAL PHOTOELECTRIC DEVICES

Cadmium sulfide cells are used in some photoelectric devices in conjunction with incandescent light sources. The cadmium sulfide cells are used with LEDs that emit visible light and use modulating light source technology, where the LED is pulsed at a specific frequency. All of these photoelectric devices are designed around several principals for detection applications. In industry, the photoelectric switch is generally thought of as a limit switch that can detect motion, position, level, or the presence of a part without being required to come into physical contact with what it is detecting. Normally a limit switch has two drawbacks in industrial applications. First, the sensing cam on the limit switch must come into contact with what it is sensing, which eventually wears the cam down. Second, it must be mounted close enough to the product it is sensing and it becomes easily damaged. In contrast, photoelectric devices have the advantage that they can sense motion, position, level, or the presence of a part without coming into contact with what they are sensing, and they can be mounted in a safe location.

Modern photoelectric devices use three principles of transmitting and receiving light for detection. These are the *through-beam scan, retroflective scan,* and the *diffuse scan.* Fig. 6-8 shows examples of these three types of applications, and Fig. 6-9 shows pictures of typical photoelectric switches.

6.4.1 Through-Beam Photoelectric Devices

The through-beam photoelectric device may also be called the *direct scan* or *separate* type of control because the emitter sends a beam of light directly to the receiver, which is mounted across from it. The emitter and receiver circuits are mounted in separate housings. The through-beam is used in applications where the sensing range is a rather long distance. In this type of application the light only needs to travel the distance in one direction. In other devices such as the retroflective scan, the same light that is emitted must also travel back to the receiver after it is reflected.

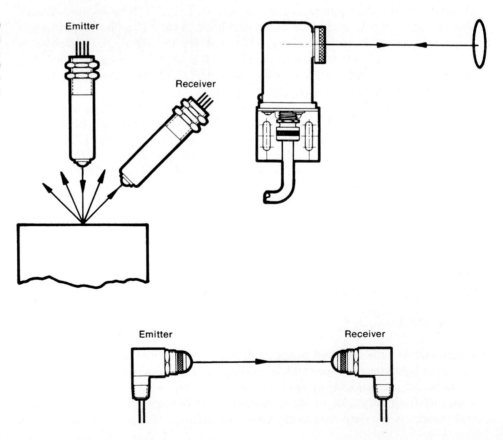

FIGURE 6-8 Three types of photoelectric switches: through-beam, retroflective, and diffuse scans. (Courtesy of Honeywell's Micro-Switch Division.)

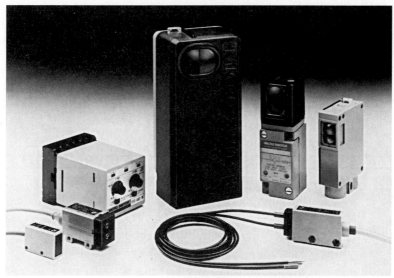

FIGURE 6-9 Typical photoelectric switches used in industrial applications, (Courtesy of Honeywell's Micro-Switch Division.)

The through-beam device is used in industrial production applications as well as nonproduction systems like material handling. One application of material handling for the through-beam sensor is to detect forklifts that must pass throughout a factory. The sensors are used to open doors automatically when the forklift is ready to pass through the doorway. Another application would be to sense when semitrailers are in the correct position on loading docks. This is important when forklifts must move in and out of the trailers for loading and unloading.

Through-beam sensors are also used as light curtains around robots or other dangerous machinery. Light curtains use multiple beams of light that are spaced at varying heights across the entrance of the work location. If anyone tries to enter the work area and breaks any of the beams of light, the sensor will detect the intrusion and shut down the system.

Fig. 6-10 shows the light-curtain sensor set up to protect a robot work area, a press work area, and a palletizing area. It is very dangerous if people get too

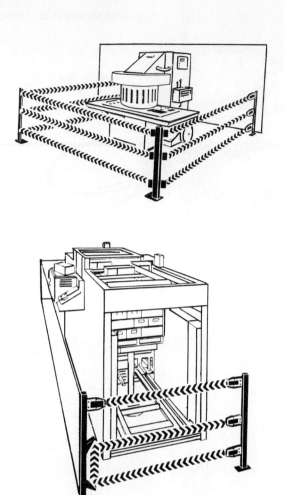

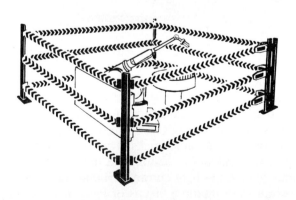

FIGURE 6-10 Light curtains used to protect the area around a robot work cell, a turret punch press, and palletizing and depalletizing equipment. (Courtesy of Tapeswitch Corporation.)

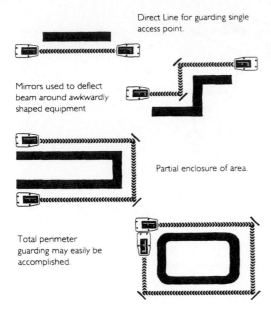

Direct Line for guarding single access point.

Mirrors used to deflect beam around awkwardly shaped equipment

Partial enclosure of area.

Total perimeter guarding may easily be accomplished.

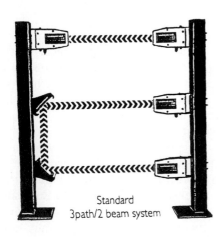

Standard
3path/2 beam system

FIGURE 6-11 Light-curtain photoelectric sensor used to provide a safety curtain around equipment that may pose a hazard to humans. The top diagram of this figure shows a single line of light used to provide the barrier. The middle diagrams show several ways to use mirrors to direct the light around corners, and the bottom diagram shows how several lights can be made to protect at various heights in a doorway. (Courtesy of Tapeswitch Corporation.)

close to the operating machinery in each of these work areas. The light curtain will trip a switch anytime the light is broken.

Fig. 6-11 shows a picture of the light-curtain photoelectric sensor that is used to keep humans out of the robot work area. As you can see, the light can be focused around corners by using mirrors. If two mirrors are used, the light can be directed around three sides of the work area, and if three mirrors are used, the light can be directed to protect all four sides of the work area.

Another application for the through-beam sensor is to detect the position of tensioning arms on a textile machine or a paper roll. The tensioning arm is moved up or down to provide the correct amount of tension on the roll material. The photoelectric switches can be used to determine if the tension bar is at the highest or lowest point or if it is in the middle position. This information can be sent to the motor drive. If the tensioning arm is in the middle position, the motor speed is correct. If the tensioning bar is at its lowest position, it means that too much slack is in the roll and the motor should speed up to take out some of the slack.

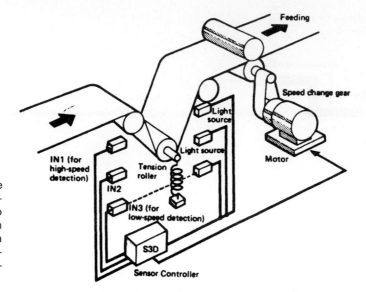

FIGURE 6-12 Example of through-beam photoelectric sensors used to determine the position of a tensioning bar in a paper or fabric roll process. (Courtesy of Omron Electronics.)

If only the top photoelectric switch is activated, it means the tensioning bar is at the top and the roll does not have sufficient slack so the motor should be slowed down. Fig. 6-12 shows an example of this type of application.

6.4.2 Retroflective Scan Photoelectric Devices

The retroflective scan photoelectric device has the sender and receiver circuits mounted in the same housing. A typical application for this device is sensing boxes on a conveyor line and counting them. The device sends the modulated light source aimed at a reflector. When no boxes are on the conveyor and the path of light is clear, the beam of light will strike a reflector mounted on the opposite side of the conveyor and return the light to the receiver in the device. When boxes come down the conveyor and block the light beam, the photoelectric switch will sense this change in light. The electronic circuit in the switch allows the device to be set for *light sensing* or *dark sensing*. If the switch is set for light sensing, the output circuit in the device will be energized when light is sensed, and if it is set for dark sensing, the output will be energized when light is blocked. Since the light must travel out to the reflector and back to the receiver, this type of sensor is generally used for medium-range or short-range sensing.

Another method of using the retroflective scan sensor uses a beam of light from the emitter that is aimed so it is focused directly on the part that is being sensed. When the part is present, the surface of the part will reflect the light back to the receiver. The part that is sensed generally must have a shiny surface or some degree of light-reflecting capability. The switch is also set to be more sensitive than in instances where a reflector is used. For example, the switch can sense the presence of a tin or aluminum can very easily. It can also detect the presence of tools such as drill bits or milling cutters and indicate when they are missing or broken. More advanced applications use the sensor to detect the presence of containers or the edges of parts where the material being sensed will contrast with the background.

6.4.3 Diffuse Scan Photoelectric Switches

The *diffuse scan* photoelectric switch is also called the *specular scan* switch. This type of control is also shown in Fig. 6-8 as one of the three main methods of using

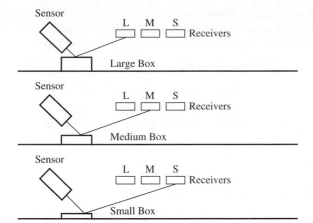

FIGURE 6-13 Example of a diffuse scan photoelectric sensor with one emitter and several receivers to detect boxes of different heights.

photoelectric controls. The diffuse scan switch uses the concept that light will reflect off the surface of an object at a specific angle, which depends on the angle that the emitted light strikes the surface and the angle of the surface. For example, if a part is present on a moving conveyor line, an emitter and receiver can be set so that light can be reflected off the surface of the part when it is detected. When the part is not present on the conveyor, the light will deflect a different angle and not strike the receiver. This concept can also be used to detect multiple parts of different heights that are all placed on the same conveyor. One emitter can be used for this application and several different receivers are mounted at different locations to detect the reflected light from the part it is trying to detect. Since the angle of reflection will be different for each part, they can each be detected separately. Fig. 6-13 shows an example of this type of application.

6.4.4 Electronic Circuits for Photoelectric Switches

The through-beam, retroflective, and diffuse scan photoelectric switches all have similar electronic circuits. The electrical block diagram for a retroflective control is provided in Fig. 6-14. The circuit has two distinct parts: the emitter and the receiver. A power supply, voltage regulator, and spike limiter are shared with both

FIGURE 6-14 Electrical block diagram of a retroflective photoelectric sensor. (Courtesy of Rockwell Automation's Allen Bradley Business.)

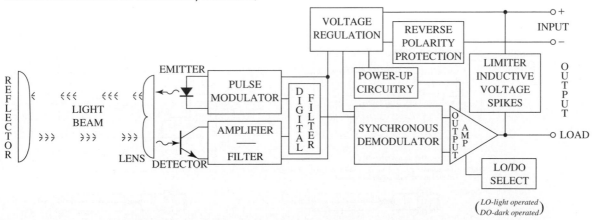

circuits. The emitter uses a pulse modulator (oscillator) circuit to pulse an LED at a specific frequency. The receiver uses a phototransistor that sends its signal to an amplifier. A digital filter is used to compare the received signal to the emitted signal. If the signal is the correct frequency, the amplified signal is sent to the synchronous demodulator and on to the output amplifier. The amplifier output is determined by the LO/DO selector, which determines if the photoelectric switch should be enabled as a *light-operated* or *dark-operated* switch. The output signal from the photoelectric control can be selected when the device is purchased and installed by choosing a plug-in output module.

6.4.5 Output Signals for Photoelectric Controls

Photoelectric controls must be capable of being interfaced to numerous types of electronic and electrical control systems that operate as part of complex machines or as stand-alone controls. These controls provide output circuits to interface to 110 or 220 volts AC, 24 volts DC, or small TTL voltage levels for either PNP or NPN circuits. Photoelectric controls also provide a relay output that has NO and NC contacts. The contacts are isolated so that nearly any voltage can be switched on or off through them. Fig. 6-15 shows examples of the electronic-type output circuits.

The output circuit for AC voltages is designed to operate at 110 or 220 volts. It uses an SCR or triac to switch the AC voltage to the load. You will notice that the AC source voltage is also used to provide voltage for the power supply of the photoelectric switch and its internal circuitry. This is known as a *three-wire* configuration. If the relay-type output and the isolated contacts are used, the control will require two wires for the power supply and two wires for the switched output. Thus, the switch is called a *four-wire* control. You should also remember that since the electronic circuits that control the emitter and receiver require DC voltage, a bridge rectifier must be used to convert the AC voltage to DC. The zener diode is

FIGURE 6-15 Electrical diagrams of the output stages of photoelectric sensors. The top two diagrams show an SCR used with a bridge to provide an output for AC voltages and the bottom two diagrams show a PNP and an NPN transistor used to provide an output for DC voltages. (Courtesy of Omron Electronics.)

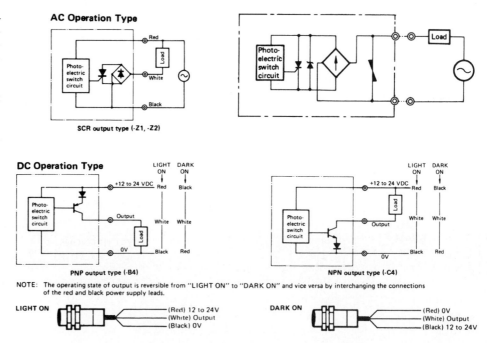

used as a regulator and a varistor is used to protect the circuit from excessive incoming voltage.

The DC voltage output is similar to the AC voltage output except that it does not require the bridge rectifier, and it can use an NPN transistor to provide a *current-sinking* circuit or a PNP transistor to provide a *current-sourcing* circuit. You can see in the diagram that a diode is connected in series with the transistor to ensure correct polarity. The detector and amplifier section of the photoelectric device provides a small dc voltage signal to the base of the transistor anytime the switch has been activated. The signal at the base of the transistor causes current flow in the transistor's emitter–collector circuit. Since the load is connected in series with the collector, it will be energized anytime current flows.

Circuits for interfacing the photoelectric output to TTL and CMOS circuits are also available. The voltage for these types of circuits must be supplied separately from the power supply voltage for the photoelectric switch and it must also be totally isolated to protect them. These types of outputs are useful when a photoelectric device must be interfaced to a microprocessor.

6.5 OPTOELECTRONIC DEVICES

In previous examples of photodiodes used as photovoltaic and photoresistive devices, light was absorbed to make a change in the energy level of the P–N junction. A transistor can also be manufactured to allow light to strike its junction and this process can be modified to supplement the action of base current. This type of transistor is called a *phototransistor.* In a normal transistor the amount and polarity of current injected into its base will affect the current flowing in its emitter–collector circuit. In the phototransistor the amount of light striking the junction will affect the amount of current in the emitter–collector circuit. A similar process can be used to make the SCR light activate. The main advantage of the light-activated devices is that light is immune to signal noise. This means that anytime a device is activated by light, it will provide a degree of isolation.

If the process is reversed and energy is put into the P–N junction, light can be emitted in the infrared and visible light ranges. The electronic device specifically designed as a solid-state lamp is called a light-emitting diode (LED). The LED can be manufactured to emit light in the visible light range or the infrared light range. LEDs can be mounted in a chip so the emitted light is focused directly on a phototransistor. The combination of these two components provides optical isolation, and the device is called an optocoupler.

6.5.1 Phototransistors, Light-Activated SCRs, and Phototriacs

Transistors can be manufactured with the junction exposed so that light can strike it. These devices typically have a glass cover to allow light to strike the collector junction. When light strikes the junction, it causes free electrons to move across it and current flow down through the emitter junction will also increase. The amount of collector current will change as the amount of light striking the junction changes. The effect of the light is similar to applying bias voltage to the base of a junction transistor. If voltage is applied to the base of the phototransistor, the amount of light required to make a change in current can be offset to make the device more or less sensitive. Fig. 6-15 shows the electronic symbol for the phototransistor and other photoelectronic devices.

6.5.1.1 Photodarlington Transistors ■ The phototransistor generally uses a junction-type transistor. It can be more specialized for higher gain and faster speed by using a darlington pair. The photodarlington provides higher gain because the collector circuit of the transistor that receives the light source is connected to the base of the second transistor. This arrangement also allows for high-speed switching, since the first transistor desaturates the output of the second transistor and lowers switching losses. The specialized optical devices are also available as a photo Schmitt trigger for high-speed switching circuits.

6.5.1.2 Light-Activated SCRs (LASCRs) and Phototriac Drivers ■ Other solid-state devices available as light-activated devices include the *light-activated SCR* (LASCR) and *phototriac drivers*. These devices provide the same type of optical isolation to the power thyristors. The light source provides the means to activate the SCR into conduction just like a pulse of positive voltage to its gate. The phototriac provides a similar function in that light causes it to go into conduction much the way a gate pulse would. Once the LASCR is turned on, its anode–cathode circuit operates like a traditional SCR. The same is true for the phototriac in that once it is energized by light, its MT1 and MT2 circuit operates like a traditional triac.

Light-activated devices are used as stand-alone devices to indicate the presence or absence of light, or they may be manufactured in conjunction with a light source and mounted in an encapsulated package to form an optoisolation device.

6.5.1.3 Light-Activated PUT (LAPUT) ■ The light-activated programmable unijunction transistor (LAPUT) provides an output signal like the PUT anytime light strikes it. The light provides a degree of isolation. The regular PUT receives its input signal from a low-voltage source that is very susceptible to induced voltages or other sources of electrical noise. The LAPUT receives its input signal from a light source that is not bothered by electrical noise.

6.5.2 Light-Emitting Diodes (LEDs)

The light-emitting diode is a solid-state lamp that can emit light from a P–N junction instead of using a filament like a traditional lamp. Fig. 6-16 shows a drawing, typical dimensions. and the electronic symbol of an LED. The symbol indicates that the LED still has an anode and a cathode like a traditional diode. The cathode terminal is near the flat edge of the diode.

The LED is manufactured with different materials to give off different colored light. The P–N junction of these materials is designed so that it is exposed to allow light to be emitted when forward-bias voltage is applied. When forward-bias voltage is applied, the majority carriers are injected into the junction where they recombine and release energy. In this device the energy is released in the form of light energy. If the junction is made of gallium, aluminum, and arsenide (GaAlAs) on gallium arsenide (GaAs), the LED will emit infrared light that must be detected by infrared sensors, since it cannot be seen by the human eye. If gallium arsenide and phosphorus (GaAsP) on gallium phosphorus (GaP) is used, the LED will emit a red light. Green light can be produced by LEDs using gallium and phosphorus or gallium phosphide and zinc. It is also possible to provide yellow and orange light-emitting

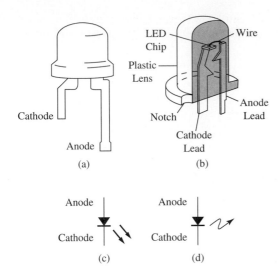

FIGURE 6-16 **(a)** Typical light-emitting diode (LED) with anode and cathode identified. **(b)** Cut-away diagram of LED. **(c)** The symbol for the LED that shows a diode with two arrows pointing away from it indicating it gives off light. **(d)** The symbol for the LED that shows a lightning bolt pointing away from it indicating it gives off light.

diodes by starting with the material used to produce red light and varying the amount of doping of each material.

6.5.2.1 Identifying Anode and Cathode Leads of LEDs ■ At times you will need to identify the anode and cathode leads of an LED. In Figs. 6-16a and 6-16b you can see that if the terminals of the LED are different lengths, the shorter lead is the cathode, and the longer lead is the anode. You should also notice that if the LED has a notch in its plastic lens, the notch will be near the cathode lead. You should always verify the lead identification with a front-to-back test of the PN junctions for the LED.

6.5.2.2 Applications for LEDs ■ Light-emitting diodes are typically used as indicators like the filament-type lamp of previous years. The major difference is that the LED does not have a filament that must heat up to produce the light. The filament of the older-style lamps created problems such as requiring large amounts of power, the need to get rid of excess heat, and the problem of short life expectancy because of filament failure. The LED remedies these problems and is usable in high-speed applications, such as indicating transmission of data and in applications with photoelectric switches where the light source is modulated at high frequencies.

LEDs are available in a variety of case designs so that they can be soldered directly into printed circuit boards or mounted on the front of devices, such as computer disk drives and in programmable controller modules to indicate power is applied. Another application for LEDs is to be combined with phototransistors to make optoisolators.

6.6 OPTOISOLATORS AND OPTOINTERRUPTERS

As more electronic devices were used to form circuits, the demand to connect circuits with differing voltage potentials and differing impedances together became more prevalent. For example, as computers and programmable controllers became

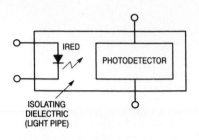

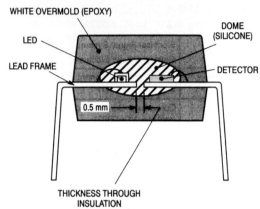

FIGURE 6-17 Electrical block diagram and physical layout of a typical optocoupler. The optocoupler is also called an optoisolator and it is usually packaged as a six-pin IC chip. (Copyright of Motorola, Used by Permission.)

more usable on the factory floor, it became evident that some type of interface would be needed that could isolate the 220 volt AC and 110 volt AC signals that most machinery used from the small DC bus voltages found in computers. Isolation is also a problem when larger AC and DC voltages need to be interfaced to TTL logic circuits.

The simple solution to this problem is to combine an LED with a phototransistor. The new device is totally encapsulated so that the light from the LED is focused directly on the opening in the phototransistor, and no other forms of light could be detected. The input signal is connected to the LED and the output signal is connected to the transistor. The device is called an *optocoupler* or *optoisolator*. Fig. 6-17 shows a block diagram of an optocoupler that shows an LED shining light directly on a photodector, which is usually a phototransistor. The second diagram in the figure shows how the LED is located so its light is focused directly on the phototransistor.

Fig. 6-18 shows the six-pin IC package for an optocoupler and the electronic diagram of its pin outline. The IC package may also be called an IC or a *chip*. From this diagram you can see that the anode of the LED is pin 1 and the cathode is pin 2. The emitter of the phototransistor is pin 4, the collector is pin 5, and the base is pin 6. It is important to note that each type of optocoupler may use different

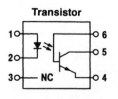

FIGURE 6-18 Pin outline for an optocoupler for a six-pin IC. A sketch of a six-pin IC is also shown. (Copyright of Motorola, Used by Permission.)

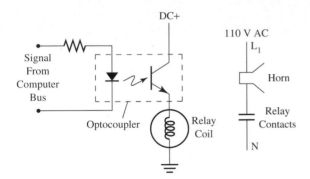

FIGURE 6-19 Electrical diagram of an optocoupler used to interface an annunciator horn to a computer. The relay coil is connected to the output stage of the optocoupler.

pin assignments, so you must be sure to check the manufacturer's pin outline diagrams.

6.6.1 Applications for Optocouplers

Fig. 6-19 shows an optocoupler interfacing a computer output signal to a relay coil and the contacts of the relay are used to energize a 110 volt alarm lamp and annunciator. This circuit allows the small-voltage signal from the computer to safely energize the high-voltage lamp and horn without the fear of allowing any high-voltage spikes to get back into the computer. Optocouplers are also used in programmable logic controller (PLC) 110 volt input and output modules to provide isolation between the 110 volt signals and PLC bus. In industrial applications a limit switch on a machine is wired for 110 volt AC so that it is not bothered by induced electrical noise. The 110 volt AC signal is connected to the programmable controller input module circuit consisting of a bridge rectifier that converts the AC signal to DC, a resistor, and the LED for the optocoupler. The transistor side of the optocoupler is connected to the input bus of the PLC. Since the signal emitted by the LED is transferred by light, the high and low voltages of the circuit are isolated.

6.6.2 Theory of Operation of the Optoisolated Circuits

In Fig. 6-19 the computer sends out a low-voltage signal from its bus to the LED side of the optocoupler. The forward-bias voltage passes through the resistor, which sets the voltage drop so that the LED receives approximately 5–16 mA depending on the type of optocoupler. The forward-bias voltage causes the LED to emit light to the phototransistor. When the phototransistor receives the light, it goes into conduction and current begins to flow through its emitter–collector circuit and the relay coil becomes fully energized, which pulls its NO contacts closed. When the NO contacts close, 110 volts AC passes through them to the annunciator (horn) and it begins to sound an alarm tone. You can see that the signal that comes from the computer bus is isolated from the dc voltage in the relay coil by the optocoupler, and it is further isolated from the 110 volts AC in the annunciator by a relay. This ensures that the computer bus is protected from unwanted voltage spikes or noise that may be generated when the relay coil and horn are de-energized.

In Fig. 6-20 a circuit is shown where the optocoupler is used in a PLC input module. The LED side of the optocoupler is connected to the ac voltage side of the circuit. In this application, the ac voltage is sent to the optocoupler when a limit switch closes. The ac voltage is rectified to dc voltage when it passes through

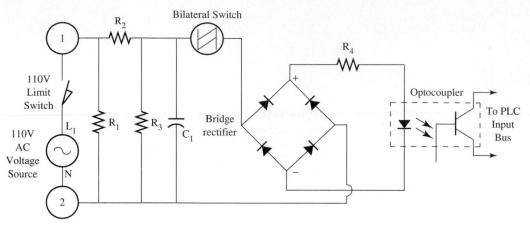

FIGURE 6-20 Optocoupler used in a PLC input module. When the switch closes, 110 volts AC is provided to terminal 1. The phototransistor in the optocoupler is connected to the PLC input bus.

the diodes in the bridge rectifier. The dropping resistor ensures that the proper amounts of voltage and current reach the LED. When the forward-bias voltage reaches the LED, it emits light to the phototransistor, which goes into conduction and allows current to flow in its emitter–collector circuit. This dc voltage is sized specifically to energize the programmable controller bus. Again you can see that the optocoupler provides isolation between the high voltage at the limit switch and the low-voltage dc signal that is sent to the programmable controller bus. You must remember that the limit switch uses 110 volts AC because it may be mounted in the machine at a distance of 150–200 feet from the programmable controller. If the low-voltage signal was sent out to the limit switch, it would be susceptible to noise and other induced signals.

6.6.3 Optoisolation Relays (Solid-State Relays)

The industrial applications that require optoisolation circuits are so prevalent that several companies make plug-in and stand-alone optoisolation circuits called solid-state relays (SSRs). The SSR provides the optocoupler circuit in an encapsulated module that has larger terminals available so that it can be used in industrial circuits and requires 3–32 volts dc to turn it on. The LED section of the optocoupler acts like the coil of a traditional relay. This part of the SSR requires dc voltage because the LED must be forward biased to produce light. The phototransistor section of the optocoupler inside the SSR is equivalent to the contacts in a relay.

If a traditional phototransistor is used, the SSR will be rated for dc voltages. If the SSR is rated for ac voltages, it will use photosensitive solid-state devices to trigger other devices such as triacs or two inverse parallel SCRs for switching, or it can trigger the phototriac directly. Fig. 6-21 shows examples of SSRs used in conjunction with several types of transistor circuits to provide interface capabilities with TTL circuits. The internal diagram of the SSR is shown in Fig. 6-21c. In this diagram you can see that the SSR is an optocoupler that uses an LED and a phototransistor. A 1000 Ω resistor is connected internally in series with the LED so that the user does not need to worry about needing additional resistance to prevent excessive current. It should be noted that if the voltage of the input signal is too low, there may be insufficient current to properly illuminate the LED.

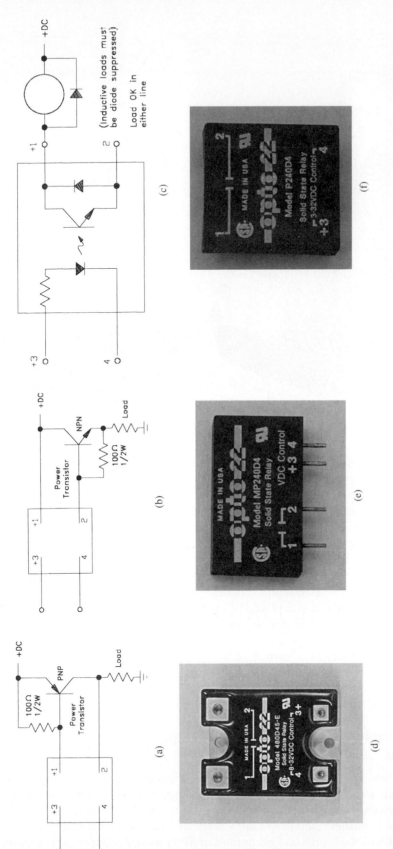

FIGURE 6-21 (a–f) Examples of solid-state relays and their electrical diagrams. The diagrams show pnp and npn transistors used in the output stage. Figure 6-21c shows a typical load connected to the relay. (Courtesy of Opto 22, Remecula, CA.)

213

FIGURE 6-22 Typical rack with solid-state relays mounted in it. A wide variety of relays is available to provide interfaces to dc, ac, and analog signals. (Courtesy of Opto 22, Remecula, CA.)

A diode is connected internally in reverse bias across the emitter–collector terminals of the transistor to prevent large transient currents from getting into the transistor if inductive loads such as relay coils are connected to the output. It is also suggested that a second diode be connected externally across the terminals of any coils or inductive loads that are used with dc voltages. The diode will provide a path for the reverse-voltage transients that occur when voltage is turned off and the magnetic field in the coil collapses.

Since the SSRs are available as stand-alone or plug-in devices, they provide the advantage of being removed and replaced very quickly. If they are the plug-in type, they can be removed and replaced by someone with minimal technical knowledge, since the wiring for this type is connected to the socket which is soldered directly to a printed circuit board.

You will find SSRs in a variety of applications, such as in microprocessor controlled systems like the high-speed weighing system and the single-point temperature controllers where they provide a simple interface for alarms and other outputs. Fig. 6-22 shows the relays plugged into a module board.

6.6.4 Optocouplers Used in Input and Output Module Circuits

Optocouplers are also commonly used to provide isolation for input and output modules that are used to interface between programmable logic controllers (PLCs) or for other computer-type systems. The companies that make the SSRs also provide generic input and output circuits so that a designer can interface a wide variety of ac and dc circuits to the inputs and output bus of a common desktop-type computer. This allows the computer to be used to run a variety of software and still have the

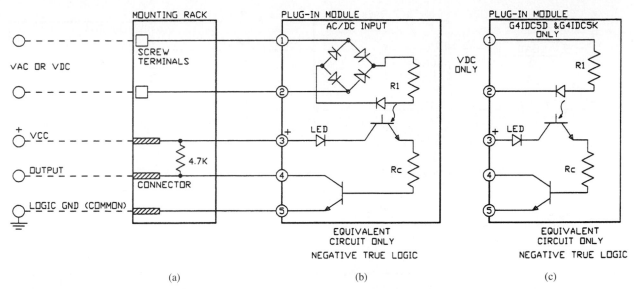

FIGURE 6-23 **(a)** Electrical diagram of typical solid-state relay used to interface input signals. The diagram shows the terminals in the rack allow for either ac or dc signals to be connected. If an ac signal is used, an ac relay must be installed in the rack, and if a dc signal is used, a dc relay must be used. **(b)** The diagram for an ac relay. **(c)** The diagram for a dc relay. (Courtesy of Opto 22, Remecula, CA.)

ability to read inputs and write outputs (turn them on or off) through its parallel or serial port. Fig. 6-23 shows the typical circuit for an input module. This circuit is similar to a typical PLC input module.

Fig. 6-23a shows that the two terminal connections are used for the input signal. You should notice that this module can use either ac or dc signals as the input. Since the input signal generally comes from the machine operation, a common switch device such as a limit switch or push-button switch could be used to switch the input signal voltage on or off. The bottom three terminals provide the signal to the computer bus. You should notice that this circuit expects to receive V_{CC} from the computer and then it will be switched on or off by the transistor in the opto-coupler circuit and returned to the computer.

Fig. 6-23b shows the optocoupler part of the circuit. Since this circuit can use either an ac or a dc input signal, a diode bridge rectifier is used. The bridge rectifier will change any ac signal to dc so that it is compatible with the LED. If the input signal is dc, it will come through two of the diodes in the bridge and also will provide dc voltage to the LED. An internal resistor R_1 is used to provide the proper amount of current for the LED. The phototransistor side of the optocoupler has an LED that is used as an indicator to show when the circuit is on or off.

Fig. 6-23c shows a circuit that is used if the signal will be only dc. In this diagram you can see that the bridge rectifier has been eliminated since the incoming signal will always be compatible with the LED. The transistor side of this circuit is identical to the circuit in Fig. 6-23b.

The optocoupler circuit is also used in output modules to provide a high degree of isolation between the computer that originates the low-voltage output signal and the industrial load such as a motor starter coil or solenoid. Fig. 6-24 shows an example of this type of circuit. From this diagram you can see that the output signal from the computer bus comes into the circuit on the bottom two terminals. The V_{CC} voltage for this circuit is generated by the computer bus. The low-voltage signal is sent to the LED part of the optocoupler. A second LED is

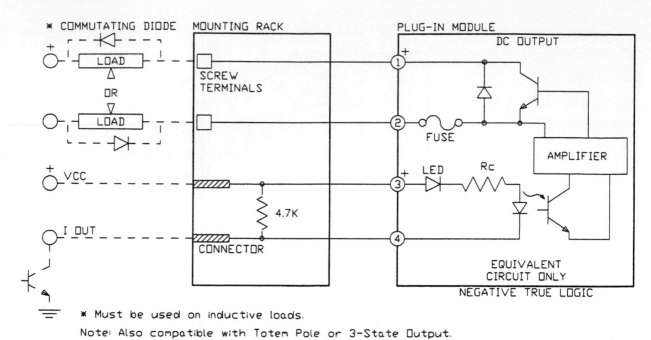

FIGURE 6-24 Electrical diagram of the solid-state relay used as an output module. Notice that the output circuit includes an optocoupler and a transistor that is used as an amplifier. (Courtesy of Opto 22, Remecula, CA.)

used as an indicator that is mounted on the face of the module. This LED will be illuminated anytime the computer sends the bus signal to the optocoupler.

The second part of this circuit is the traditional phototransistor. You should notice that a second transistor is used to provide amplification of the signal to get it to a useful voltage. A diode is connected across the emitter–collector circuit in reverse bias to absorb any transient voltages that may occur if an inductive load such as a relay coil is connected as the load device. Any inductive device that is powered by dc voltage will send large transient voltages when its magnetic field collapses as its voltage supply is turned off. You should also notice that a fuse is used in series with the load in the output side of the transistor circuit. This provides additional protection and it is replaceable.

6.6.5 Specialty Types of Optocouplers

A large variety of optocouplers has been designed to meet the demands of numerous applications. For example, optocouplers are available that are specifically designed for high-gain signals that use darlington pairs and for high-speed switching where Schmitt triggers are used. Other conditions such as common-mode rejection, ac/dc voltage to logic-level signal interfaces, low-current applications, TTL applications, high-gain applications, and for multiplexing data applications require special types of optocouplers.

Fig. 6-25 shows example circuits of the low-input current logic gate optocoupler. Fig. 6-26 shows examples of specialty types of optocouplers that use transistors with a base terminal where bias can be added, darlington pair transistors that are used for higher gain, and Schmitt triggers that are used for high-speed switching. Fig. 6-27 shows the diagram for a shunt drive circuit for optocoupler interface between TTL and CMOS signals, and Fig. 6-28 shows an optocoupler that allows ac or dc voltage as the input and the output is converted to a logic-level signal.

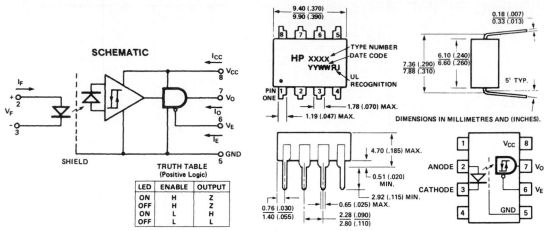

FIGURE 6-25 Schematic diagram and pin outline for a low-power optocoupler. This type of device is used where the input signal is a low-power signal. (Copyright of Motorola, Used by Permission.)

In Fig. 6-26 you can see the diagram of the low-input power logic gate opto-coupler. This optocoupler combines a GaAsP LED with an integrated high-gain photon detector. The detector portion of the device provides a three-state output stage and has a detector threshold with hysteresis. The need for pull-up resistors is negated by the three-state output. The hysteresis provides differential-mode noise immunity and prevents the possibility of chatter in the output signal. Chatter may occur if the contacts of the input device bounce during closure. The contact bounce may appear as more than one signal transistion, and the hysteresis in the circuit ensures the input signal only represents the intial contact closure. This optocoupler is specifically designed to switch at small current thresholds as low as 1.6–2.2 mA. A truth table is also provided for this circuit.

Fig. 6-27 shows the diagram for a shunt drive circuit that uses an optocoupler to provide an interface between TTL/LSTTL/CMOS logic circuits. The LED in this circuit can be enabled by as little as 0.5 mA at a frequency of 5 megabaud (5 million pulses per second). This makes the circuit usable as a logic-level translator or for microprocessor I/O isolation. This circuit also eliminates several problems and increases common-mode rejection, since the path for leak current in the LED is eliminated.

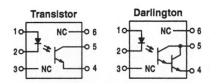

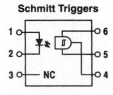

FIGURE 6-26 Diagrams of optocouplers that use transistors, darlington transistors, and Schmitt triggers for their output stage. (Courtesy of Hewlett Packard Company.)

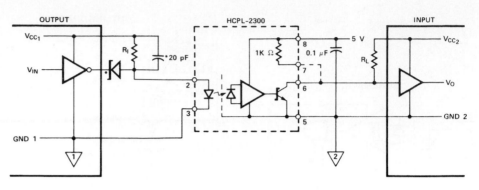

OUTPUT   HCPL-2300   INPUT

V_{IN} V_{DC}	V_{CC1} V_{DC}	R_I kΩ	R_L kΩ	V_{CC2} V_{DC}
5	5	6.19	1 (INTERNAL)	5
10	10	14.7	2.37	10
15	15	21.5	3.16	15

*SCHOTTKY DIODE (HP 5082-2800, OR EQUIVALENT) AND 20 pF CAPACITOR
ARE NOT REQUIRED FOR UNITS WITH OPEN COLLECTOR OUTPUT.

**AC Input
Transistor Output**

**AC Input
Resistor–Darlington
Output**

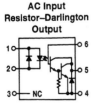

The optocoupler circuit that will accept either ac or dc voltage signals and interfaces them to logic-level signals is shown in Figs. 6-28 and 6-29. In Fig. 6-27 two similar optocouplers are shown that have a junction diode connected in inverse parallel to the LED on the input side of the circuit. This design protects the LED from reverse breakdown when ac voltage is applied. The output side of the first optocoupler uses a transistor with surge protection. The second circuit shows the output stage of the optocoupler using a darlington pair with surge protection to provide increased amplification and speed of switching.

The optocoupler in Fig. 6-29 shows the input side of the circuit utilizing a bridge rectifier so that either ac or dc voltage can be sensed. A clamping diode is also provided to prevent damage due to overvoltage conditions. A buffer is provided in the input stage of the circuit to control the threshold voltages over a wider range for input signals. Signals as low as 1.2 mA at 3.7 volts can be detected. The output stage of this device provides an open collector circuit so it is compatible with TTL saturation voltages and CMOS breakdown voltages.

Other types of specialty optocouplers have been developed to handle problems that occur when optocouplers are used in ac circuits. One circuit is shown in the first part of Fig. 6-30 where a phototriac is used instead of a phototransistor. Since the triac is used, ac voltages can be controlled directly. The second diagram in this figure shows a triac connected to a *zero-crossing circuit*. The zero-crossing circuit is used to ensure that the triac switches ac voltage on and off exactly when the ac sine wave is at 0 volts. This means that the triac is only turned on when the sine wave is at 0° or at 180°, which means that voltage and current are minimal when the triac allows current to flow to the remainder of the circuit. This allows circuit components such as lamp filaments to last much longer, since they are not subjected to high-voltage transients from switching ac voltage and current when the sine wave is at a peak.

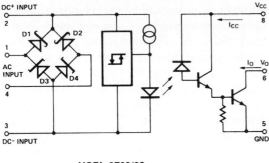

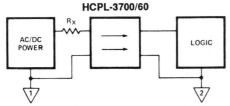

HCPL-3700/60

FIGURE 6-29 Electrical diagram of an optocoupler specifically designed to accept an AC or DC voltage input. (Courtesy of Hewlett Packard Company.)

6.6.6 Adding Bias to the Phototransistor of the Optocoupler

The optocoupler is also useful in circuits where the bias of the phototransistor is changed. When bias voltage is added to the base of the phototransistor in an optocoupler, it can make the optocoupler more or less sensitive. If the bias voltage to the base of an NPN transistor is slightly positive, the optocoupler will become more sensitive because the LED does not need to produce as much light to make the phototransistor begin to conduct. If the bias voltage is slightly negative, the optocoupler will become less sensitive and the LED must have more current applied before it can produce enough light to overcome the bias on the phototransistor. The optocoupler must have a base terminal for the transistor brought out to a pin so that it is usable to add bias to it. The optocouplers in Fig. 6-28 show transistors with their base brought out to pin 6.

6.6.7 Using an Optocoupler to Convert a 4–20 mA Signal to a Variable-Voltage Signal

The optocoupler can also be used to convert a 4–20mA signal to a variable-voltage signal. A 4–20 mA signal is common in process control applications, such as level sensors or flow sensors in food processing. The 4–20 mA signal is used for two

Random Phase Triac Driver

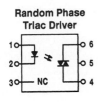

Zero Crossing Triac Driver

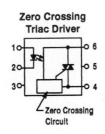

FIGURE 6-30 Electrical diagram of optocouplers specifically designed with triac and zero-crossing triac drivers as the output circuit. (Copyright of Motorola, Used by Permission.)

reasons. First, the milliamp signal is a current loop-type signal that is more resistant to noise than a voltage signal; and second, the minimum value for the 4–20mA signal is 4 mA which is *offset* above 0 mA. Since the minimum signal value is offset 4 mA from 0 mA, a broken wire can be detected if the current value falls to 0 mA.

In some applications the amplifier is designed for the input signal to be a voltage signal rather than a milliamp signal. For example, in a circuit that controls the speed of a dc motor that pumps a beverage into a holding tank, the original input signal may come from a potentiometer that is adjusted manually by the equipment operator who determines the pump speed. If the circuit is automated so the input signal comes from a flow sensor that uses a 4–20 mA signal, an op-amp circuit can be used to convert the 4–20 mA signal to a 0–5 or 0–10 volts dc signal. Fig. 6-31 shows an example of this circuit. The optocoupler replaces a fixed potentiometer that is part of a relaxation oscillator used to adjust the firing angle of an SCR amplifier that controls the speed of a DC motor. From the figure you can see that several resistors have been added to the optocoupler circuit. The 100 k pot is the zero pot, and the 500 k pot is the span pot. Together these resistors provide bias to the base of the phototransistor. When the 4–20 mA signal is applied to the LED side of the optocoupler, it will change the amount of light that is emitted to the phototransistor. The power source for the phototransistor circuit is coming from the firing circuit for the SCR drive. The pots must be adjusted so that enough negative bias is applied to the base of the phototransistor so that when the 4 mA signal is applied, the output of the transistor is zero. You should remember that 4 mA input signal must look like the 0 voltage signal that the original fixed pot supplied. The negative bias from the zero pot will ensure that the amount of light the LED is emitting when 4 mA is applied will not be sufficient to start the transistor into conduction. The 500 k pot is used to provide enough positive bias to the phototransistor so that it is in full saturation when the light from the LED peaks when the 20 mA signal is applied. The process of setting the potentiometers to ensure that the minimum and maximum signal voltages is produced is called *setting zero and span*.

6.6.8 Photo IC Interrupter

Another specialty application for the optocoupler technology is an integrated circuit called a *photointerrupter* that is specifically designed for high-speed oscillation of the LED and detection circuit. The photointerrupter is also called an *optointerrupter*.

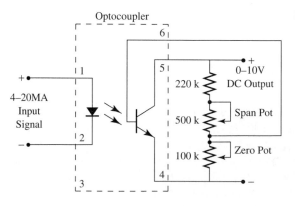

FIGURE 6-31 Electrical diagram of an optocoupler that is used to convert a 4–20 mA analog signal to a 0–10 volt dc analog signal.

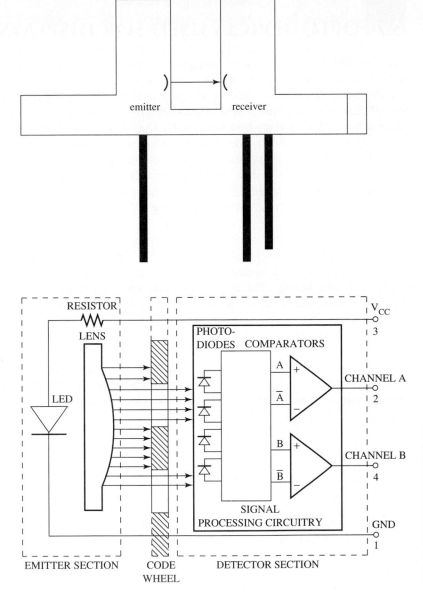

FIGURE 6-32 Drawing and electrical diagram of a photointerrupter used to detect data from punch cards or other similar media as it moves through the slot in the head where the sender and receiver are mounted. (Courtesy of Hewlett Packard Company.)

Fig. 6-32 shows a typical optointerrupter and a circuit for the device. The device is designed with a slot so data material such as punch cards and punch tape can be passed through to be read at high speed.

In this device the traditional phototransistor is replaced with a Schmitt trigger circuit, which can turn on and off at much higher frequencies that provide switching at speeds of 3 μsec. The LED is mounted on one side of the slot, and the Schmitt trigger is mounted on the opposite side. The Schmitt trigger circuits have either an *open collector* interface or the traditional *pull-up resistor* interface, which allows the circuit to produce the active HI or active LO output.

6.7 OPTODEVICES USED FOR DISPLAYS

The simplest type of display, known as the *seven-segment* display and shown in Fig. 6-33a, is so named because it uses seven LED segments to produce all of the numbers 0–9. You can see from the figure that by lighting the different segments, the outline of each number will appear. A driver circuit can be connected to the display to convert binary numbers or BCD (binary coded decimal) numbers to values 0–9. More advanced displays provide a decimal point and + and − signs.

The technology has evolved to provide four-character and eight-character displays (see Fig. 6-33b) that can show numbers 0–9 and all the alpha characters needed to make a message. Fig. 6-34 provides a matrix that shows the ASCII code for the symbols, alpha characters, and numbers for a typical four-character or eight-character display.

Since each display can only show one number or character at a time, as many as 20 or 30 of these displays can be put together as a message board to form words and sentences that scroll across them. A driver circuit must be used with the message board that can decode ASCII messages and time the display so the letters scroll from character to character to make the sentence appear to move across the display. These displays are especially useful for microprocessor controlled equipment and are generally used in conjunction with a keypad so information can be entered and read. Some of the applications in Chapter 13 utilizes this type of display to show operating parameters and diagnostic information. Programmable controllers can also use this type of display to show the values from counters and timers right at the machine operator's station. This provides information in an inexpensive format.

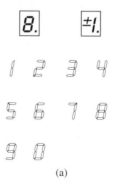

(a)

FIGURE 6-33 **(a)** Seven-segment display that shows how numbers 0–9 and characters such as + and − can be displayed with an example of how each number 0–9 will utilize the seven segments. **(b)** A CMOS-type display.

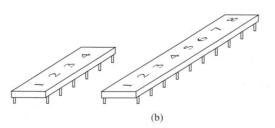

(b)

6.7.1 Liquid Crystal Displays (LCD) and Plasma Display Screens

When the amount of information that needs to be displayed increases, a message display may not be able to display data effectively. In these applications, an LCD screen or plasma display screen can display information on a screen that looks like a television or computer monitor. Since these types of screens can be connected directly to computers, microprocessors, or programmable controllers, they can provide the operator with sufficient information to start up, operate, and troubleshoot a complex machine. A keyboard or keypad can be used in conjunction with the display screen so that the operator or supervisor can enter data as well as read information from the machine. Several companies manufacture a combination display and keypad that can be used as a *hand-held teach pendant* for inputting data to robots or other computer controlled equipment. Another feature of these displays

FIGURE 6-34 A matrix that shows how letters and characters can be generated on a panel display. You can see that each character and number is assigned an ASCII code value from the matrix. The columns in the matrix are identified by the first four bits (D0–D3) and the rows are identified by the last three bits (D4–D6). Hexidecimal values could also be used. The columns use values 0–A, and the rows use values 0–7. (Courtesy of Hewlett Packard Company.)

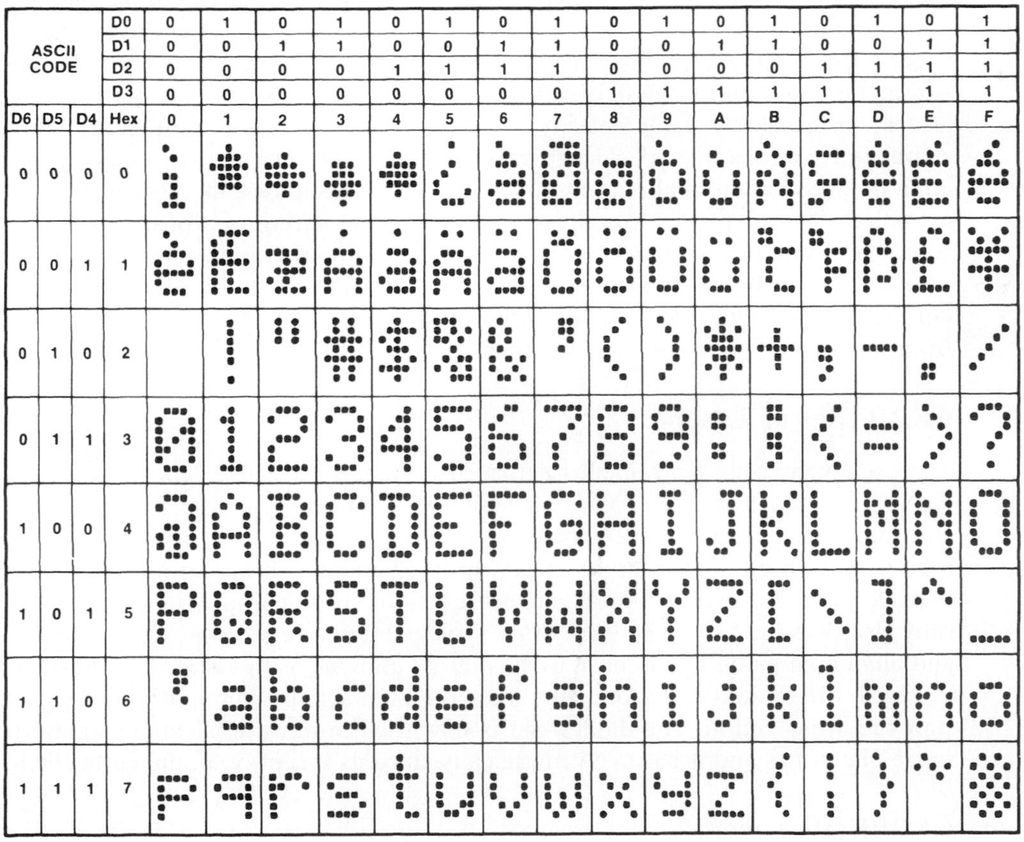

Notes: 1. High = 1 level.
2. Low = 0 level.

and teach pendants is that they can utilize microprocessor technology and be connected to any network where other computers, microprocessor-based equipment, or programmable controllers are connected and read or write information to or from any machine.

6.8 LASERS

Laser is an acronym for *light amplification stimulated emission of radiation*. The laser produces a single coherent beam of light that is generally one color so that only a single frequency is used. The light is useful because it can be directed for many applications such as measurements, cutting, melting, surface hardening, welding, and for use in fiberoptics. Lasers can be classified as gas, liquid, solid, or high- or low-power lasers.

6.8.1 Theory of Laser Operation

The coherent light frequencies of lasers can range in the electromagnetic energy spectrum from infrared to ultraviolet. The major difference between laser light and normal light is how the light originates. The laser light begins with the excitation process that creates a population inversion of photons. Photons (from molecules and atoms) are emitted in all directions and in turn stimulate other photons that are in phase, have the same polarity, and travel in the same direction. Some of the stimulated photons are specifically reflected back into the active region of the atom to ensure additional photons are stimulated. This allows the light to become amplified. Since lasing is actually a process of stimulating photons, it will only continue to produce a beam of light as long as the population inversion continues at a level (threshold) strong enough to sustain stimulation. In comparison, the normal types of light that have been discussed previously in this chapter originate from spontaneous movement of photons.

6.8.2 Types of Lasers

There are many types of available lasers such as the ruby, dye, CO_2 (carbon dioxide), HeNe (helium-neon), Nd-YAG (neodymium-yttrium, aluminum, garnet), Nd-glass, and the semiconductor-diode type. The HeNe-type laser is the most commonly used laser with outputs of up to 25 mW/m laser length. These types of lasers are used in bar-code readers and alignment functions in industrial applications. The ruby laser was the first to be built. It uses ionized chromium as the active material and aluminum oxide as the host. Since a large amount of the chromium must be raised to the excited state, the ruby laser requires a larger amount of energy to start the lasing action. The diode laser is now used in most applications of short-range fiberoptics and it has evolved radically since its inception in the early 1960s.

6.8.3 Diode Lasers

The original diode lasers were made from simple *homojunction* PN devices. There are over 20 compounds made from semiconductor material that can produce the laser effect. The most commonly used material is AlGa (aluminum and gallium), which is also used in LEDs. In the case of diode lasers, the diode junction is forward biased, which results in a large injection current and the population inversion necessary to start the lasing effect. In a typical LED diode junction, light is emitted from all areas of the surface and hence it is not focused. In the laser diode, the light emission is controlled to occur only from one edge of the junction, so it can be sharply focused.

Newer versions of the diode laser are made by adding additional junctions to the original PN junction. These junctions are called single or double heterojunctions. Several additional p-type layers may be added to the original single PN junction that provide differing energy levels as different types of material are used. The differing energy levels cause electrons and holes to reflect back into the active region, which help sustain the laser effect. This provides laser power in the range of 10–20 W of pulsed peak power. This type of power is necessary for fiberoptics applications.

6.8.4 Industrial Laser Applications

Lasers have become a vital part of industrial applications such as fiberoptics, precision measurement, bar-code reading, cutting, finishing, and manufacturing on both metal and nonmetal materials. For example, a laser can be combined with a robot to perform very close tolerance and intricate cutting processes. The robot can provide the accurate location repeatedly and the laser can provide accurate cutting features. Fig. 6-35 shows a robot with a laser that is used in a cutting application. In other applications, the laser can be used in heat treatment to harden or soften metal in combination with other manufacturing processes such as cutting, surface finishing, or face hardening.

In measurement applications, the laser can be used to determine if parts coming from a die or press still have the required tolerance. Since the laser can make measurements that can detect very small differences, the condition of the die

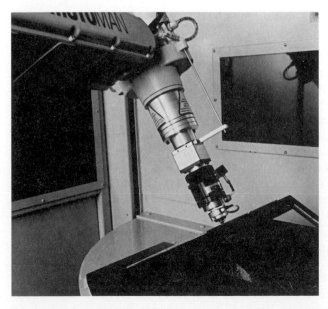

FIGURE 6-35 A robot that is using lasers for welding and cutting applications. (Courtesy of Motoman.)

FIGURE 6-36 An industrial laser that is enclosed for the safety of personnel who must work near it. This type of application ensures that workers will not be exposed to the laser light. (Courtesy of Motoman.)

can be determined and repairs can be made if it is worn. The speed at which the laser can make these measurements allows for every part to be tested as it comes from the manufacturing process.

One problem with using lasers in the factory is that humans working in the vicinity of the machine must be protected against the harmful laser or light rays that are present. Since these laser or light rays will cause harm to eyes, most industrial applications are completely enclosed. Fig. 6-36 shows an enclosed laser system that will prevent the laser from harming the eyes of humans who must work in the area of the laser.

 # 6.9 FIBEROPTICS

Another use of light technology is fiberoptics. In the late 1970s scientists found that a glass fiber could be used in place of traditional wire to transmit a signal. The major difference between fiberoptic systems and traditional wire data systems is that a beam of light is used to transmit the signal through the glass fiber instead of a wire. The signal was encoded into a beam of light by pulsing a light on and off to indicate the 1's and 0's of a binary code just as an electronic signal voltage pulses to provide 1's and 0's. Since the signal is sent through the fibers on a beam of light, it cannot be influenced by electrical signals that are nearby, or by induced electrical currents and noise that are common in modern industry. Another advantage of fiberoptics is that the signal can be sent over great distances without the signal losses associated with copper wire. Fiberoptic systems are compatible with traditional RS232 and RS422 systems for sending data information.

6.9.1 Fiberoptic Transmitter and Receiver

The light that travels through the fiberoptic system must be sent from a transmitter module and received by a receiver module. Fig. 6-37 shows a typical transmitter and receiver for a fiberoptic system. The emitter in the transmitter is a GaALAS device (LED) that can send an optical signal up to distances of 10,000 m at data rates of 1000 Mbaud. Low-cost products are limited to distances of approximately 100 m.

From the diagram in Fig. 6-37a you can see that the transmitter receives a traditional electronic data signal on its input line. V_{CC} is also applied to the transmitter to provide power for the circuit. The input signal is amplified to provide enough power for the LED. The pulse rate of the electronic signal becomes the pulsed light signal that is sent through the fiber cable. The receiver module in Fig. 6-37b has a photodiode (PD) mounted so that the beam of light is focused directly on it. The diode will turn on and off at the same rate as the light pulses so that the electronic data signal sent to the transmitter is duplicated exactly in the receiver voltage. The output of the receiver is a traditional electronic signal. The fiberoptic cable simply provides a means to transmit the data signal over a great distance without the chance of interference.

The diagram in Fig. 6-37c shows how all of this circuit is encapsulated inside the cable-end hardware. An additional diagram is provided in Fig. 6-38 that shows the receiver circuit. You can see that these circuits utilize op-amps and other transistors to send and receive the signals.

6.9.2 Types of Fiberoptic Cables

Fiberoptic cables are available in a large variety of combinations. Figs. 6-39 and 6-40 show a picture of typical cables and diagrams of the way cables are packaged. You can see in the picture that some of the cables come with preformed ends, while others must be cut and spliced to length in the field where the connections are made. The diagram in Fig. 6-40 shows four combinations of fiberoptic cables.

FIGURE 6-37 **(a)** Cross-sectional view of a typical fiberoptic fitting. **(b)** An electrical diagram of a fiberoptic transmitter. **(c)** An electrical diagram of a fiberoptic receiver. (Courtesy of Hewlett Packard Company.)

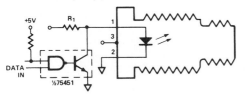

HFBR-1201 TRANSMITTER

(a)

HFBR-2201 RECEIVER

(b)

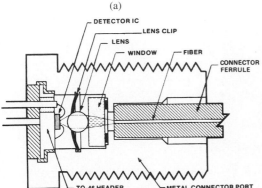

(c)

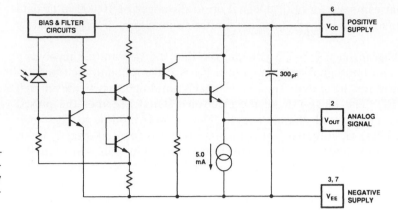

FIGURE 6-38 Electrical diagram of a fiber-optic receiver. (Courtesy of Hewlett Packard Company.)

The first diagram shows two optic fibers with two reinforcement segments combined to make one cable. The reinforcement segments are necessary to provide additional strength to the cable when it is installed in applications where it is not fully supported along its entire length. Two additional diagrams show a cable that has 12 fibers with a variety of reinforcement members. These cables provide the additional fibers so more signals can be sent through a single cable. The last diagram in this figure shows a 24-fiber cable. This cable combines 24 fibers with reinforcement members to provide a larger number of fibers to send or receive data.

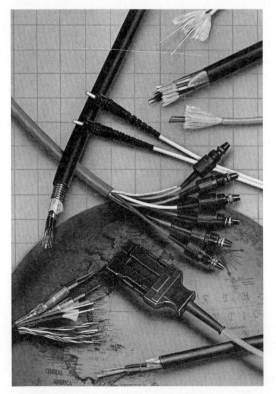

FIGURE 6-39 A variety of fiberoptic cables used in typical industrial applications. (Courtesy of Belden Wire & Cable Company.)

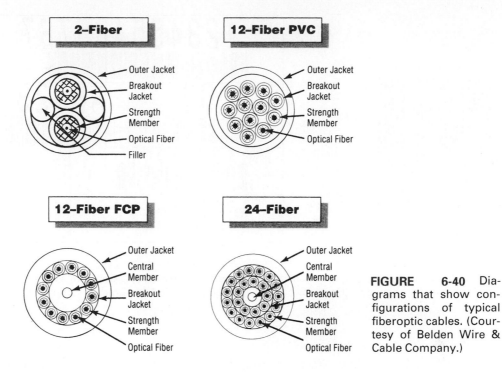

FIGURE 6-40 Diagrams that show configurations of typical fiberoptic cables. (Courtesy of Belden Wire & Cable Company.)

6.10 BAR-CODE EQUIPMENT

Another application of fiberoptics is used in bar-code readers. You have probably noticed bar codes on all of the products you purchase in a department store or grocery store. Bar codes are the black stripes of ink on the package label that indicate what the product is and other important data that are necessary for controlling stock and to determine the price of each item. The bar code is a symbol that may contain letters, numbers, or symbols to identify the product. Several different formats have been developed for the bar codes. The most common format is called *universal product code* (UPC). Other formats called *Interleave 2 of 5*, and *Code 39* are also used. Today UPC has become a subset of the *European Article Number* (*EAN*). A typical UPC symbol usually has the UPC type represented by one character, the UPC manufacturer or vendor ID number represented by five characters, the UPC item number represented by five characters, and finally, one character that is used as a check digit. Fig. 6-41a shows an example of a typical UPC bar-code symbol.

Another format for a bar-code symbol is shown in Fig. 6-41b. This format is the Automotive Industry Action Group (AIAG), which is used in the automotive industry to identify parts that are shipped between the parts manufacturer and the assembly plant. In this figure you can see that the format includes five separate bar-code symbols that are made up of numbers and letters.

A wide variety of additional formats has been designed over the past few years for each special area of technology, such as the American Gas Association,

FIGURE 6-41 **(a)** Example of universal product code (UPC) format. **(b)** Automotive Industrial Action Group (AIAG) format for bar codes. (Courtesy of Rockwell Automation's Allen Bradley Business.)

the American Paper Institute, the health industry, the Grocery Manufacturers of America, the National Welding Association, and many others. These formats allow the bar-code symbols to represent information that is useful to these applications.

6.10.1 Bar-Code Scanner and Decoder

A bar-code system is made of two parts: the printer and the scanner/decoder. Printers are available to print labels or to print directly on products such as cardboard boxes and packages. The ink must contrast with the background so that the symbols can be easily read by the scanner. The second part of the system consists of a scanner that reads the code symbols, and a decoder that interprets the code.

The scanner can be portable (hand held) or stationary such as the readers in a grocery store. The basic parts of the scanner are shown in Fig. 6-42. In this figure you can see the block diagram that shows a light source focused on the bar code. This light must be reflected back to the photodetector part of the scanner. Since the reflected light diffuses (spreads out), the window that receives the reflected light must be aligned correctly so that the scanner can read the code. After the signal is received, it is amplified, conditioned, and then passed on to the decoder section of the system. The decoder uses a microprocessor to decode the data and compare them to data in the database. The decoded data can be stored or displayed so that the humans using the system can read the data.

The light sources that are used to illuminate the bar code can be from an LED, a visible laser diode, an infrared laser diode, or a helium-neon (HeNe) laser. The type of light source depends on the type of ink that is used in the bar code and the location where the codes will be read. Since some bar codes are read in an industrial environment, it may be difficult to get light sources to reflect the bar code correctly, so one of the invisible light sources such as infrared laser or HeNe laser is used.

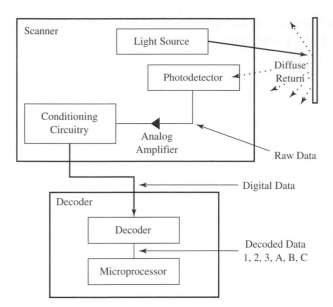

FIGURE 6-42 Block diagram of a bar-code scanner and decoder. (Courtesy of Rockwell Automation's Allen Bradley Business.)

6.10.2 Types of Scanners

Scanners come in a variety of forms, with the *hand-held, pencil wand,* and *slot types* being the most popular in industrial applications. A variety of scanners are shown in Fig. 6-43. The hand scanner looks similar to a gun and it has a large window

FIGURE 6-43 (a–c) Typical bar-code equipment. (Courtesy of Rockwell Automation's Allen Bradley Business.)

that allows the light source to be directed on the bar code. This type of scanner is useful for inventory control where packages are stationary or where packages are too large to be moved across the scanner. The pencil or wand-type scanner is also useful where bar codes are read on smaller packages. The slot type is useful for smaller packages that are all uniform in size and can be automatically passed through the reader.

6.10.3 Typical Bar-Code Applications in Industry

Bar-code readers have made inventory control and shipping much more efficient in industrial applications. In most heavy manufacturing industries large quantities of raw material and parts must be used every day and replaced so that the operation can continue the next day. A system called *just-in-time delivery* (JIT) has been implemented over the past ten years. This system operates with only a one-day or half-day inventory at the manufacturing site. This means that the parts that will be used later in the day or tomorrow are just being made and shipped so that they will arrive *just in time* to be used. In the years prior to the 1980s, companies would keep two to three weeks' worth of raw material in a warehouse at the manufacturing plant, and assembly plants would keep a similar amount of stock at their sites. All of this additional stock costs millions of dollars to keep in inventory since it has to be paid for when it is delivered, and the company cannot receive any money in return for it until the parts are assembled, shipped, and sold.

Since most all companies have limited the amount of stock that is on hand, it is important to be able to track all inventory from the time it is consumed to the point where it is reordered and received. You can start to appreciate this task when you consider the number of parts that are needed at an automobile assembly plant where several thousand cars are produced daily, or a bottling plant where millions of soft drinks are bottled every day.

Fig. 6-44 shows a diagram of a typical industrial site that uses bar codes to control inventory and parts as they move through the plant. You can see in the upper right corner of this diagram that one system may be set up to read packages as they are palletized. Another system in the lower part of the diagram shows a bar-code reader reading cans as they pass a canning and filling line. A third system shown on the left side of the diagram is checking labels on a bottling line.

Each of the systems can be set up to operate locally and also report to a main computer or a PLC. The PLC may be used to operate diverter and sorting equipment. It may also be used to control the palletizing equipment. The majority of this equipment may be stationary equipment with additional hand-held systems to catch parts that are not on the material-handling lines.

As a technician you may be called to check the interface between the bar-code readers and the PLCs or computers. The bar-code systems are fully compatible with the PLC and you will only need to provide a power source and the cables that connect the equipment to the bar-code reader module. The bar-code module is designed to interface directly to the bar-code equipment on the factory floor and then reside in the rack like any other PLC module where data can be easily read on the bus. The data that are brought into the PLC are compared to the database to indicate the product that is being read. After the PLC identifies the product, it can manipulate the automation along the line and direct the parts to the correct section of the shipping system.

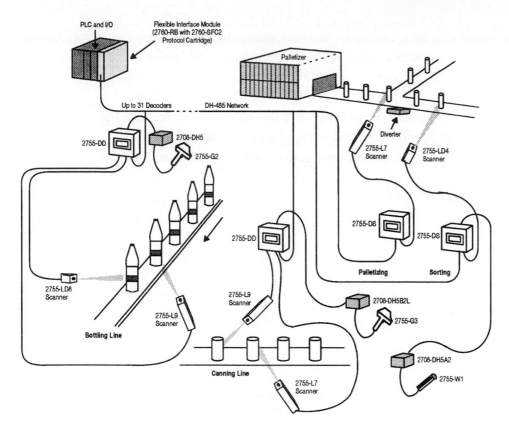

FIGURE 6-44 Applications showing bar-code equipment used in a typical industry. (Courtesy of Rockwell Automation's Allen Bradley Business.)

SOLUTION TO JOB ASSIGNMENT

Your solution to the job assignment should include some type of isolation to get the signal from the machine to the computer. Since the counter uses a photoelectric device like the ones in Fig. 6-8 to detect and count the good and bad parts, its signal must be sent to one of the optocouplers shown in Figs. 6-17, 6-18, 6-19, 6-24, 6-26, 6-27, 6-28, 6-29, or 6-30. These optocouplers provide a variety of interfaces to any computer signal. Your solution may use a solid-state relay (SSR) found in Fig. 6-21 instead of an optocoupler.

The solution for moving the computer to an office and rerouting the signal should include a fiberoptic cable with a transmitter and receiver. The hardware should fit directly in the back of the computer and into the display that is the source of the original signal. Since you are using a fiberoptic cable, you do not need to be concerned about interference, and the signal can be sent up to 15,000 feet without secondary amplification. A second display could also be set up near the office computer if necessary. You should select equipment with a display that uses characters like the ones in Fig. 6-33. This type of display can show multiple characters, and the computer or a driver can be used to send the messages that indicate the count.

The solution to suggesting a bar-code system to track the parts can be found in Section 6.10 and in Figs. 6-41 to 6-43.

QUESTIONS

1. Use the frequencies of infrared, ultraviolet, and incandescent light to explain the differences among them.
2. Describe the operation of photovoltaic cells.
3. Provide an example where a retroflective photoelectric switch would be used in industry.
4. Provide an example where a through-beam photoelectric switch would be used in industry.
5. You are asked to provide a light curtain around an irregular-shaped robot work cell to protect against personnel entering the work cell while the robot is running. Discuss several methods of providing this protection.
6. Explain the operation of a bar-code system. Be sure to include the function of the photodetector, conditioning circuitry, amplifier, decoder, and microprocessor.
7. Provide an example of an industrial application where bar codes are used. Be sure to explain the type of bar-code format and type of readers that are used.
8. Explain the operation of a fiberoptic system used to transmit and receive signals in industrial applications.
9. Discuss industrial applications where fiberoptics would be necessary.
10. List four applications where lasers are used in industry today.

TRUE OR FALSE

1. _____ Photoelectric switches use modulated light sources so that the switch is not activated by ambient light.
2. _____ A diffuse scan photoelectric switch can reflect the beam of light off a product at such an angle it will strike the receiver.
3. _____ The AIS, UPS, and EAS are typical bar-code formats used in industrial applications today.
4. _____ Fiberoptic cable can have more than one optic fiber.
5. _____ Lasers are mainly used in weapons and do not have many industrial applications.

MUTLIPLE CHOICE

1. The optocoupler uses _____ to receive the light emitted by the LED.
   a. phototransistors
   b. darlington transistors
   c. Schmitt triggers
   d. all the above

2. The _____ photoelectric switch can use one beam of light and three receivers to detect three different sizes of boxes as they pass on a conveyor.

   a. through-beam

   b. diffuse scan

   c. retroflective

3. The input switch should be connected to terminals _____ and the load should be connected to terminals _____ in a solid-state relay application.

   a. 3 and 4, 1 and 2

   b. 1 and 2, 3 and 4

   c. It doesn't matter which terminals are used for input and output.

4. A photoelectric switch that is wired as a light-operated switch will be activated _____.

   a. when light strikes the receiver.

   b. when light does not strike the receiver.

   c. at the point where the light beam transitions from light to dark.

5. A photointerrupter is used to _____.

   a. detect light beams from great distances.

   b. detect light beams at high speeds such as reading punch data tapes.

   c. interrupt light beams so that photoelectric switches will detect correctly in industrial locations where ambient light is a problem.

   d. all of the above.

PROBLEMS

1. Select one of the diagrams from Fig. 6-21 that could be used to control a dc load. Explain your choice.

2. Select one of the diagrams from Fig. 6-21 that could be used to control an ac load. Explain your choice.

3. Use the graph in Fig. 6-2 to calculate the amount of voltage a photovoltaic cell produces at 100 foot candles of light.

4. Draw a sketch of retroflective, through-beam, and diffuse scan photoelectric switches and explain their operation.

5. Draw the diagram (pin outline) of an IC optocoupler that uses an LED and phototransistor.

6. Provide example circuits of optoisolation that use darlington transistors, triacs, and Schmitt triggers in industrial electronic controls.

REFERENCES

1. *Generation 4 Digital I/O Family Data Book.* OPTO 22, 43044 Business Park Drive, Remecula, CA 92590-3665.

2. Kissell, Thomas, *Modern Industrial Electrical Motor Controls.* Englewood Cliffs, NJ: Prentice-Hall, Inc., 1992.

3. Kissell, Thomas, *Understanding and Using Programmable Controllers.* Englewood Cliffs, NJ: Prentice-Hall, Inc., 1986.

4. *Optical Devices Data Sheet Pack.* 1993. NEC Electronics Inc., 475 Ellis Street, P.O. Box 7241, Mountain View, CA 94039.

5. *Optoelectronics Designer's Catalog,* 1991–1992. Hewlett Packard Company.

6. *SSR Family Data Book,* OPTO 22, 43044 Business Park Drive, Remecula, CA 92590-3665.

7

INDUSTRIAL POWER SUPPLIES, INVERTERS, AND CONVERTERS

When you have completed this chapter, you will be able to:

1. Explain the operation of a single-diode half-wave rectifier and draw its input and output waveforms.
2. Draw the output waveform for a two-diode center-tapped rectifier that produces a full-wave output and explain its operation.
3. Explain the operation of a four-diode full-wave bridge rectifier and draw its input and output waveforms.
4. Calculate the average dc output voltage from the input rms voltage for the single-diode half wave rectifier, two-diode center-tapped rectifier, and the four-diode full-wave rectifier.
5. Show how the capacitor and inductor are connected in a power supply and explain their function.
6. Explain the function of a zener diode in a power supply.
7. Explain the operation of a metal-oxide varistor (MOV) as it is used in a power supply.
8. Explain the operation of six-phase rectifiers.

9. Compare the differences between half-wave and full-wave rectifiers.
10. Explain the operation of a crowbar circuit.
11. Discuss the operation of an uninterruptible power supply.
12. Draw the diagram of an inverter and explain its operation.
13. Identify three types of inverter circuits (variable-voltage input, pulse-width modulation, and current-source input) and explain their operation.
14. Explain the operation of a cycloconverter.
15. Draw the diagram for a linear power supply and explain its operation.
16. Explain the operation of a switch-mode power supply.
17. Explain the operation of the buck converter, the boost converter, and the buck-boost converter.
18. Identify typical voltages that would be found in power supplies used in industrial electronic circuits.
19. Provide examples of power supplies used in industrial applications.

236

You are working on a process oven that is used to heat automobile seats to remove wrinkles in seat covers. The process oven has a conveyor that is powered by a three-phase ac motor. The conveyor's speed is not adjustable at this time. Your supervisor informs you that due to changes in quality standards, the speed of the conveyor must now be made adjustable. You are also informed that the lab technicians are adding a small computer to take quality control data such as temperature, cook time, and conveyor speed. Your supervisor reminds you that it is important to protect the computers that gather the test data against possible power losses. Your assignment (Courtesy of Fostoria Industries.) is to select equipment to control the speed of a three-phase ac motor and to protect against the loss of voltage to the computers. Include a report that provides an explanation of the basic theory of operation for all of the equipment you have selected.

7.1 OVERVIEW

Solid-state electronics have become integrated into all aspects of industrial power supplies, converters, inverters, and choppers. Industrial power supplies are used in applications where a variety of voltages is required, such as power for PLC processors and their analog modules and other specialty modules. Power supplies are also used in all types of digital displays and cathode ray tube (CRT) color displays. You can assume that you will find a power supply in every electrical device on the factory floor. Any equipment that has electronic circuits in it must have a dc supply voltage available. Since all power in the factory originates as ac voltage, converters must be used in the power supplies to convert ac power to dc power. These circuits called converters will be similar to the diode rectifier circuits you have seen in the basic electronic courses. Larger industrial systems such as motor drives, welding equipment, and battery chargers will require larger three-phase rectifier circuits.

Some industrial applications such as ac variable-frequency motor drives and welders will have a section where ac voltage is converted to dc and another section where the dc voltage must be converted back to ac voltage. The circuit that changes dc voltage to ac voltage is called an *inverter*. The inverter circuit is needed to change the dc voltage back to ac voltage because the frequency of the incoming ac voltage is fixed at 60 Hz and the output section of the drive needs to be able to provide the frequency of the output voltage between 0–120 Hz.

Choppers are special inverter circuits that take the dc voltage and convert it to a variety of other dc voltages. This is a useful circuit in dc motor drives where the dc voltage must be manipulated to provide changes in current and voltage for changing motor speed and torque characteristics. Chopper circuits are also used in welding power supplies and dc power supplies for a variety of other solid-state devices. This chapter will help you understand how solid-state devices are used in industrial power supplies to provide the correct voltages and frequencies through converters, inverters, and choppers.

7.2 INDUSTRIAL RECTIFIER CIRCUITS: AC-TO-DC CONVERSION

One of the most widely used electronic circuits to convert voltage is called the *rectifier* circuit. Since the rectifier circuit uses diodes to convert ac voltage to dc, it is also called a *converter* circuit. All power that is supplied to a modern factory is alternating current (ac), so it is important to have circuits that can convert the ac power to dc power since most solid-state devices require a source of dc power to operate. Several single-phase rectifier circuits will be provided to help review the basic principles that are necessary to understand the more complex three-phase converter circuits.

7.2.1 Single-Phase Rectifiers

Single-phase rectifier circuits have been used since the advent of vacuum-tube diodes. When vacuum tubes were first introduced to control voltage and current, they required a variety of dc power supplies. Since the dc power supplies originated from ac voltage, vacuum-tube diode rectifiers were used to convert ac voltage to dc voltage. When solid-state devices were developed, the first uses for solid-state diodes were to provide rectification of ac voltages to provide the necessary dc voltages.

Fig. 7-1 shows an example of a power supply that uses a single solid-state diode rectifier. From this figure you can see that this type of power supply uses a transformer to increase or decrease the voltage from the 110 volt ac supply voltage. When ac voltage is applied to the transformer primary circuit, its secondary will supply voltage to the diode rectifier. A resistor is shown to indicate a typical load.

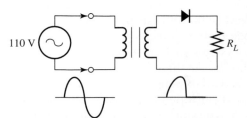

FIGURE 7-1 Electrical circuit diagram of a single-diode rectifier power supply. The waveforms show ac voltage supply and half-wave dc at the load resistor.

The diagram also shows a typical waveform for the input voltage and the waveform for the rectified voltage. You would see the full sine wave if you placed an oscilloscope across the two transformer leads that are the input of the diode rectifier, and you would see the single half-wave if you placed the oscilloscope leads across the load resistor. You should notice that since only one diode is used, the half-wave in the output waveform occurs when the ac input voltage is positive between 0° and 180°. No output voltage will be supplied during the point from 180° to 360° where the ac voltage is negative. (Note: The ac line voltage in North America may be generated from any value between 110–125 volts, and the amount of voltage available at the electrical outlet may be slightly higher or lower at various times of the day. To avoid confusion, the value 110 volts will be used for discussion of all ac line voltage in this text.)

It is important to be able to calculate the amount of *dc average voltage* (dc half-wave) that will be available at the load. The average dc voltage read with a dc voltmeter is called the dc voltage for the rectifier. The formula for the dc average voltage for a single-diode rectifier is:

$$V_{dc\,av} = \frac{V_P}{\pi}$$

where $\pi = 3.1416$.

This formula can be converted by dividing V_p by π so that the new formula is:

$$V_{dc\,av} = 0.318\ V_P$$

The *peak voltage* can be calculated from the *root-mean-square* (rms) value of the ac supply voltage. Peak voltage is the value of voltage that is measured from the 0 volt line to the peak of the sine wave and it must be measured with an oscilloscope or with a peak-reading voltmeter. The rms voltage is the voltage you would read if you were using an ac voltmeter. For example, the 110 volts you would find in a wall receptacle is actually 110 volts rms. The formula for calculating peak voltage from rms is:

$$\text{peak volts} = \text{rms volts} \times 1.414$$

or

$$V_P = V_{rms} \times 1.414$$

It is also important to remember that the diode in this circuit will drop approximately 0.7 volt since it is a silicon diode. (It should be noted that in larger-power diodes, the drop may be as much as 2 volts.) This voltage drop would be calculated if the information you are determining needs to be more precise. The equation for $V_{dc\,av}$ with the voltage drop for the diode is:

$$V_{dc\,av} = \frac{V_P - 0.7}{\pi}$$

∙∙∙

EXAMPLE 7-1

Calculate the $V_{dc\,av}$ voltage for a single-diode half-wave rectifier that has an input voltage of 110 volts ac rms (including the 0.7 volt drop for the diode).

SOLUTION

The first step of this calculation involves finding the peak ac voltage from the rms voltage.
From the equation

$$V_P = V_{rms} \times 1.414$$

$$110V_{rms} \times 1.414 = 155.54V_{dc\,peak}$$

From the equation

$$V_{dc\,av} = \frac{V_P - 0.7}{\pi}$$

$$\frac{155.54V_P = 0.7}{\pi} = 49.29V_{dc\,av}$$

. . . .

7.2.2 Two-Diode Full-Wave Single-Phase Rectifiers

A drawback of the single-diode half-wave rectifier is that it only produces a half-wave dc output. If a second diode is added to this circuit and a center-tapped transformer is used, the output waveform will be two positive half-waves. The first diode provides an output half-wave when the supply voltage is between 0° and 180°, and second half-wave output is provided by the second diode when the supply voltage is between 180° and 360°. Fig. 7-2 shows the electrical diagram of the two-diode full-wave bridge circuit. This diagram also shows the sine wave for the single-phase input voltage and the waveform of the two positive half-waves for the output.

The transformer for this circuit is the same transformer from the previous circuit and it will produce output voltage from a two-diode rectifier with a center-tap transformer that will be approximately the same voltage as the output of the single-diode rectifier. At first it may appear that the circuit with the center tap transformer will provide more output voltage since it has two diodes, but you should remember that since the secondary windings of each transformer are equal, the voltage from each half of the center-tap transformer (X1 to CT) will be half the voltage between X1–X2. This means that the voltage from each half of the transformer will be added together to provide approximately the same dc average voltage at the output of the circuit as the output voltage from the single-diode rectifier.

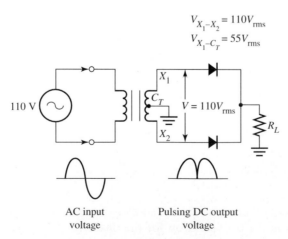

$$V_{X_1-X_2} = 110V_{rms}$$
$$V_{X_1-C_T} = 55V_{rms}$$

$$V = 110V_{rms}$$

AC input
voltage

Pulsing DC output
voltage

FIGURE 7-2 Electrical diagram of a two-diode bridge rectifier circuit. The waveforms for the ac sine wave at the input and the two half-waves at the output are also shown.

The equation for calculating the dc average voltage for the two-diode full-wave bridge rectifier is found in two steps after the peak secondary voltage is determined. Notice that since this is a center-tap transformer, the $V_{\text{out peak}}$ can be calculated either of two ways using X1–X2 as the full secondary voltage divided by 2, or using X1–CT as the amount of voltage for half the transformer. Be sure the voltage drop for each diode is subtracted from secondary voltage.

$$V_{\text{out (peak)}} = 1.414 \times \left(\frac{V_{\text{secondary}_{(X1-X2)}}}{2} \right) - 0.7 \, V_{\text{secondary}(X1-X2)}$$

or

$$V_{\text{out (peak)}} = 1.414 \times (V_{\text{secondary}_{(X1-CT)}}) - 0.7 \, V$$

Next calculate the dc average voltage from the equation. Note that since there are now two diodes in the circuit, the output voltage will include voltage from both half-waves, so the new formula will show two times the peak voltage.

$$V_{\text{dc av}} = \frac{2V_P}{\pi} \qquad \text{(where } V_P \text{ is } V_{\text{out (peak)}} \text{ from either of the previous equations)}$$

. . . .

EXAMPLE 7-2

Calculate the $V_{\text{dc av}}$ voltage for a two-diode full-wave rectifier that uses a center-tap transformer. The secondary voltage is 110 volts ac rms as measured between X1 and X2 just like the single-diode rectifier circuit, and the voltage from X1 to the CT is 55 volts. (Be sure to include the 0.7 volt drop for each of the diodes.)

SOLUTION

The secondary peak voltage $V_P = 155.54$ (X1–X2) needs to be divided by 2 to get the V_P for the value of voltage from one line to the center tap (X1–CT). $V_P = 77.77$ volts and this value can be used directly in the equation. When you are finished with the calculation, you should notice the answer 49.06 volts is approximately the same voltage as the 49.29 volts from the single-diode rectifier. The slight difference in voltage comes from the voltage drop of the second diode.
From the equation

$$V_{\text{dc av}} = \frac{2(V_{P(X1-CT)} - 0.7)}{\pi}$$

$$\frac{2(77.77V_P - 0.7)}{\pi} = 49.06 \, V_{\text{dc}}$$

. . . .

7.2.3 Four-Diode Full-Wave Bridge Rectifier

Another circuit that provides a full-wave output uses four diodes and a regular transformer without the center tap. This circuit uses two diodes at a time to rectify each half of the sine wave. Fig. 7-3 shows an example of this type of circuit. The input sine wave and the output full wave (two half-cycles) are also shown. You should notice in this circuit that ac voltage from the bottom terminal of the transformer is applied to the bridge where the cathode of diode 1 and the anode of

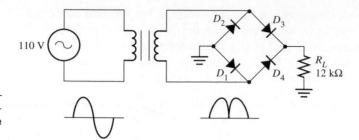

FIGURE 7-3 The electrical diagram of a four-diode full-wave bridge rectifier.

diode 4 are connected, and from the top terminal of the transformer where the cathode of diode 2 and the anode of diode 3 are connected. This means that the ac voltage is connected where the anode of one diode is connected to the cathode of the second diode.

The output for the bridge circuit will have its positive dc voltage terminal at the point where the cathode of diode 3 and diode 4 are connected, and the negative point of the circuit will be where the anode of diode 1 and diode 2 are connected. This point is also grounded.

When ac voltage is applied to the four-diode full-wave bridge rectifier, the positive half of the sine wave will be rectified by diodes 1 and 3. The negative half of the sine wave is rectified by diodes 2 and 4. From the top circuit in Fig. 7-4 you can see that the positive half-cycle of the ac is shaded, and the first half-wave is shaded to indicate the output for this part of the circuit. The bottom circuit shows the negative half of the sine wave being rectified. The path the electrons would travel through the bridge is also shown. Notice that electron flow is always against the arrows of the diodes.

In some industrial power supplies the four-diode full-wave bridge rectifier is drawn slightly differently, even though it operates exactly like the previous circuit. Fig. 7-5 shows an example of the full-wave bridge drawn with the diode bridge turned on its side so they look like a square rather than a diamond. The bridge is illustrated this way because the six-diode three-phase bridge rectifier uses a similar pattern.

You should begin to notice the four-diode full-wave bridge uses two diodes at a time to rectify each half-cycle of the ac sine wave. If one diode of either of

FIGURE 7-4 Electrical diagram that shows the current path of the positive and negative half-cycles of the sine wave as it is rectified through the bridge.

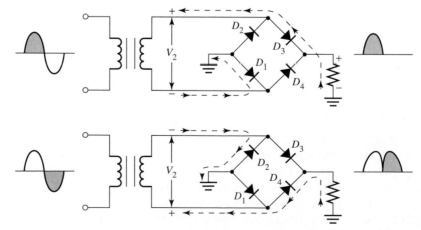

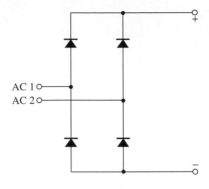

FIGURE 7-5 Electrical diagram of the four-diode bridge where the diodes are shown in a box formation rather than a diamond. The circuit function is exactly like the bridge shown in Fig. 7-4.

the two diode sets is faulty, an open will occur for that half-cycle and the output voltage of the bridge will drop approximately in half. If one diode from each set develops an open, the output from the bridge will be zero.

The equations for calculating the dc average voltage at the output of the four-diode bridge are similar to the two-diode full-wave bridge except the equations for the four-diode bridge must account for a 0.7 volt drop in each of the two diodes used to rectify the positive and negative half-cycles. The equation to determine the $V_{dc\,av}$ must be calculated in two steps. The first step will use an equation to determine the peak voltage (less the 1.4 volt drop for the two diodes):

$$V_P = (V_{secondary\,rms} \times 1.414) - 1.4 \text{ V}$$

The second equation is:

$$V_{dc\,av} = \frac{2V_P}{\pi}$$

....

EXAMPLE 7-3

Calculate the $V_{dc\,av}$ for a four-diode full-wave rectifier that is connected to the secondary of a transformer that provides 110 volts ac rms. (Be sure to include the 1.4 volt drop through the diodes.)

SOLUTION

From the equation

$$V_P = (V_{sec.rms} \times 1.414) - 1.4 \text{ V}$$
$$V_P = (110 \times 1.414) - 1.4 \text{ V}$$
$$V_P = 154.14 \text{ V}$$

From the equation

$$V_{dc\,av} = \frac{2V_P}{\pi}$$

$$V_{dc\,av} = \frac{308.28 V_P}{\pi}$$

$$V_{dc\,av} = 98.12 \text{ V}$$

....

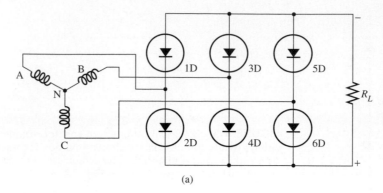

(a)

FIGURE 7-6 **(a)** Electrical diagram of the three-phase bridge rectifier that is connected to the secondary winding of a three-phase transformer. **(b)** Three-phase input sine waves. **(c)** Six half-waves for the dc output.

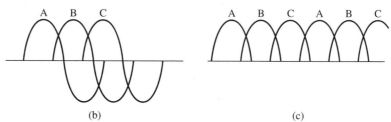

(b)   (c)

7.2.4 Three-Phase Full-Wave Rectifier

Most industrial power supplies for motor drives and welding applications use three-phase ac voltage. This means that the rectifier for these circuits must use a three-phase bridge, which has six diodes to provide full-wave rectification (two diodes for each line of the three phases). Fig. 7-6 shows the electrical diagram for a three-phase bridge rectifier. From this diagram you can see that the secondary winding of a three-phase transformer is shown connected to the diode rectifier. Phase A of the three-phase voltage from the transformer is connected to the point where the cathode of diode 1D is connected to the anode of diode 2D. Phase B is connected to the point where the cathode of diode 3D is connected to the anode of diode 4D, and phase C is connected to the point where the cathode of diode 5D is connected to the anode of diode 6D. The anodes of diodes 1D, 3D, and 5D are connected together to provide a common point for the dc negative terminal of the output power. The cathodes of diodes 2D, 4D, and 6D are connected to provide a common point for the dc positive terminal of the output power.

A good rule of thumb for determining the connections on diode rectifiers is that the ac input voltage will be connected to the bridge where the anode and cathode of any two diodes are joined. Since this occurs at two points in the bridge, in a four diode bridge the two ac lines will be connected there without respect to polarity since the incoming ac voltage does not have a specific polarity. The positive terminal for the power supply will be connected to the bridge where the two cathodes of the diodes are joined, and the negative terminal will be connected to the bridge where the two anodes of the diodes are joined.

The diagram in this figure also shows the waveforms for the three-phase sine waves that supply power to the bridge, and for the six half-waves of the output pulsing dc voltage. You should notice that since the six half-waves overlap, the dc voltage does not have a chance to get to the zero voltage point; thus, the average dc output voltage is very high.

The three-phase full-wave bridge rectifier is used where the required amount of dc power is high and the transformer efficiency must be high. Since the output waveforms of the half-waves overlap, they provide a *low ripple percentage.* In this circuit, the output ripple is six times the input frequency. Since the ripple percentage is low, the output dc voltage is usable without much filtering. This type of rectifier is compatible with transformers that are wye or delta connected.

7.2.5 Other Types of Three-Phase Rectifiers

Several other variations of the three-phase rectifier are used in some industrial power supplies because they provide an advantage of less power being converted by each individual diode, which means smaller diodes can be used to provide the same voltage and current as another rectifier circuit. Two of the more usable types of alternative rectifier circuits are shown in Figs. 7-7 and 7-8. A three-phase double-wye rectifier with an interphase transformer is shown in Fig. 7-7. This rectifier circuit is sometimes used instead of the normal three-phase bridge because each diode in this circuit must rectify only one-sixth of the total dc load, whereas each diode in the normal bridge circuit must contribute one-third of the dc load since it takes a pair of diodes to rectify each phase.

This circuit consists of the secondary windings of two three-phase wye-connected transformers that have their neutral points connected by an *interphase transformer.* The polarity of the windings of the first secondary must be reversed with respect to the polarity of the windings in the second transformer. This means

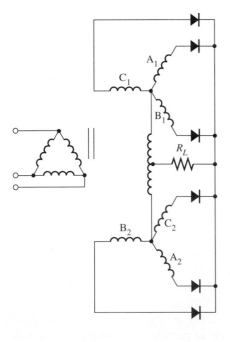

FIGURE 7-7   Electrical diagram of a three-phase double-wye rectifier with an interphase transformer. The output waveform for the rectifier is also shown.

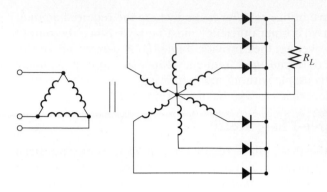

FIGURE 7-8 Electrical diagram of a six-phase star rectifier circuit. This rectifier circuit utilizes a secondary transformer winding that has all six windings connected at a center tap.

that the secondary windings of the two transformers are actually wired in inverse parallel to each other. This allows the diodes to all be connected the same way, with their cathodes all connected together to provide the positive dc power. The interphase transformer connects the two neutral points of the wye secondary transformers and it has a center tap, which becomes the negative terminal of the dc power supply. The output waveform for the rectifier is also shown, and you can see that the amount of ripple is the same as the three-phase bridge rectifier.

You should remember the output waveform for the half-waves created by the three-phase sine waves will show two half-waves for the sine-wave input of each phase. This means that the output waveform will have six half-waves that will overlap because the sine waves of the three-phase input voltage overlap at 120° intervals. The overlap of the six half-waves in the output section results in a higher average dc voltage and less ripple that needs to be filtered.

Another method of connecting the secondary windings of the transformer for the three-phase rectifier is shown in Fig. 7-8. In this diagram you can see that the secondary windings of the transformer consist of six separate windings. All six of the windings are connected at one end to form a center point for the *star configuration,* which is actually a type of wye-connected transformer. The cathodes of each diode in this rectifier are connected to provide the positive terminal of the dc power supply. The center point of the star is the negative terminal of the dc power supply. This circuit is used where it is important that all of the diodes in the circuit have a common connection for their cathodes.

7.2.6 Six-Phase Full-Wave Bridge Circuits

In the 1970s and 1980s power supplies required higher currents and voltages than the individual diodes could provide in the basic four-diode bridge. If the amount of current or voltage the power supply requires is larger than the individual diode can provide, the diodes can be connected in parallel to provide the extra current, and they can be connected in series to meet the higher voltage specification. Two types of circuits are generally used to provide these configurations. Fig. 7-9a shows an example using 12 diodes that are connected in parallel as a six-phase full-wave bridge to provide extra current. Fig. 7-9b shows examples of 12 diodes connected in series to allow the bridge to be used in a circuit where the system voltage is higher than the specification for any of the individual diodes.

Since both of these types of circuits use 12 diodes, the amount of ripple in the output section is reduced to approximately 1% since the output frequency will be 12 times the input frequency. In the circuit in Fig. 7-9a where the diodes are

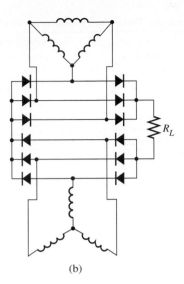

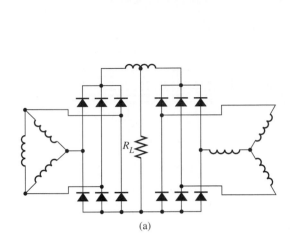

FIGURE 7-9 **(a)** Electrical diagram of 12 diodes connected in parallel as a six-phase full-wave bridge rectifier. This circuit can provide larger current that exceeds the specification of each individual diode. **(b)** Electrical diagram of 12 diodes connected in series as a six-phase full-wave bridge rectifier that provides a voltage that exceeds the specification of each individual diode.

connected as two parallel power supplies that are interconnected with equalizing reactors, each of the bridge circuits provide half of the load. In the circuit in Fig. 7-9b, each phase of the input voltage will pass through four diodes instead of two so the voltage drop is shared four ways.

7.2.7 A Comparison of the Different Types of Rectifier Circuits

It is important to compare all of the different types of rectifier circuits. You will find each of these different types of circuits when you troubleshoot modern electronic controls such as ac and dc motor drives, welding power supplies, uninterruptible power supplies (UPS), and other industrial systems that require dc voltage. Fig. 7-10 provides a table that shows all of the types of rectifier circuits that have been previously discussed and each of their characteristics are compared.

In the first line of the table you can see a comparison of the average current that each diode will carry. In the four-diode bridge and in the six-phase rectifier circuit where the diodes are connected in series, each diode must carry 0.333 of $I_{F(AV)}$, and in all of the other circuits, each diode must carry 0.167. This means that if the current requirements for the output of the power supply are larger than the current rating of any individual diode, the circuit where the diode carries 0.167 of the full-load current should be used.

You should also notice in this table that as more diodes are used in a rectifier circuit, the ripple factor (percentage of total rms ripple) becomes better (lower). This table also shows that the conversion efficiency of each of the three-phase rectifiers is above 96%. This means that each of these types of three-phase rectifiers will produce approximately the same amount of dc voltage for a given input ac voltage.

In the examples of the single-phase power supplies the amount of dc average voltage was calculated using several equations. The equations for solving the amount of dc average voltage for three-phase rectifiers become more complex because the instantaneous values of each phase of the supply voltage must be accounted for. Thus, it is easier to use a rectification ratio for each type of three-phase power supply, which is shown in the bottom line of the table.

Summary of Significant Three-Phase Rectifier Circuit Characteristics for Resistive Loads

	Half-wave Star	Bridge	Double Wye with Interphase Transformer	Full-wave Star	Wye-Delta Connections	
					Parallel	Series
Rectifier Circuit Connection						
Load Voltage and Current Waveshape Characteristic One Cycle						
Average Current through Diode, $I_{F(AV)}/I_{L(DC)}$	0.333	0.333	0.167	0.167	0.167	0.333
Peak Current through Diode, $I_{FM}/I_{F(AV)}$	3.63	3.14	3.15	6.30	6.30	6.30
Form Factor of Current through Diode, $I_{F(RMS)}/I_{F(AV)}$	1.76	1.74	1.76	2.46	2.46	2.46
RMS Current through Diode, $I_{F(RMS)}/I_{L(DC)}$	0.587	0.579	0.293	0.409	0.409	0.818
RMS Input Voltage per Transformer Leg, $V_I/V_{L(DC)}$	0.855	0.428	0.855	0.741	0.715	0.37
Diode Peak Inverse Voltage (P.I.V.), $V_{RRM}/V_{L(DC)}$	2.09	1.05	2.42	2.09	1.05	1.05
Transformer Primary Rating, VA/P_{DC}	1.23	1.05	1.06	1.28	1.01	1.01
Transformer Secondary Rating, VA/P_{DC}	1.50	1.05	1.49	1.81	1.05	1.05
Total RMS Ripple, %	18.2	4.2	4.2	4.2	1.0	1.0
Lowest Ripple Frequency, f_r/f_i	3	6	6	6	12	12
Rectification Ratio (Conversion Efficiency), %	96.8	99.8	99.8	99.8	100	100

FIGURE 7-10   A table that shows the comparison of all of the features of three-phase power supplies. (Copyright of Motorola, Used by Permission.)

7.2.8 Using Capacitors and Inductors as Filters for Power Supplies

Most power supplies found in industrial electronic circuits have capacitors and inductors used as filters. A filter on the power supply circuit will reduce the amount of ripple to a point where the output dc voltage is nearly a straight line, or pure dc. It is important in some circuits where the dc voltage is converted back to ac voltage that all traces of the original frequency of the input voltage is removed.

Fig. 7-11 shows a diagram of a typical capacitor and inductor in the power supply circuit. The capacitor is connected in parallel with the load, and the inductor is connected in series with the dc voltage terminals. You should remember from

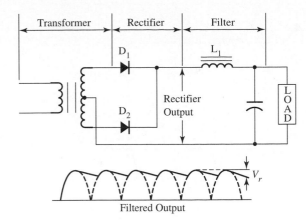

FIGURE 7-11 Electronic circuit that shows a capacitor connected in parallel and an inductor connected in series with the dc output part of the rectifier circuit. (Copyright of Motorola. Used by Permission.)

the basic electrical courses that you have studied that the capacitor will charge when voltage is supplied to it. It will then discharge the stored voltage when the supply voltage is less than the stored charge. The effect of the capacitor charging and discharging is to smooth out the area between the peaks of the full-wave dc output voltage. The waveform in Fig. 7-11 shows the effect of the capacitor filter. (It should be noted at this time that it is customary to show only the voltage waveform since it can easily be seen from an oscilloscope. The current waveform exists but it is difficult to view it directly.) The inductor provides essentially the same function to the current waveform as it stores energy in its magnetic field and releases it back into the output circuit. The effect of the inductor storing and releasing energy into the output circuit is to provide a slight phase shift, which smooths the area between the current peaks. Together the capacitor and inductor filter the dc full-wave output voltage and current to a smoother, near pure supply of dc power. The capacitor can be increased in size or several capacitors can be used together in parallel to increase the filtering capability of the circuit.

The inductor that is used for filtering is generally called a *choke* and it looks very similar to a small transformer except it will have two wires instead of four. In the rectifier circuit for larger motor drives, the capacitors in the filter will have a precharge circuit that limits the rate that voltage is supplied to the capacitors when power is initially applied. The capacitors also have a discharge resistor to ensure that all of the stored potential is removed from the capacitor when power is turned off. It is important to remember that these filter capacitors store a large amount of energy and it will take several seconds for them to discharge after power is removed.

7.2.9 Using a Zener Diode for Voltage Regulation

Most industrial power supplies require the dc output voltage to have some type of regulation to keep the output voltage level constant when the input voltage fluctuates. The ac voltage that supplies power to industry today will fluctuate up to 10% of the supply voltage specification. This means that in the hottest days of the summer it may not be uncommon for the three-phase 208 volts that is supplied to a machine to drop below 200 volts. When this occurs, the dc output voltage of all of the rectifier circuits will also drop. When the dc voltage drops, the circuit may become unreliable, so a zener diode is generally used in the output section of the power supply to provide voltage regulation. Fig. 7-12 shows a circuit with the zener diode connected in parallel with the load.

FIGURE 7-12 **(a)** Electrical diagram that shows a zener diode connected to the output side of the rectifier circuit so that it can supply a regulated voltage to the load. **(b)** The waveform of full-wave rectification voltage at the output of the rectifier. **(c)** The waveform of the regulated voltage measured across the load after it passes the zener diode.

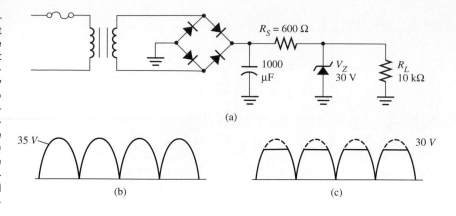

The zener diode must be rated for the same voltage the dc load requires. For example, if the dc load needs 20 volts dc, the zener will be rated for 20 volts. The operation of the zener diode in the regulation part of the circuit can easily be explained using the diagram in Fig. 7-12. In this circuit you can see that the zener diode is connected in parallel with the load resistor and it is connected in reverse bias with the power supply. In normal conditions, the output voltage from the rectifier is 15% to 20% larger than the zener level. The zener diode will go into conduction when the voltage reaches or exceeds its rating. In the circuit in the figure, the zener diode is rated for 30 volts dc, and the full-wave bridge voltage that is supplied to it is 35 volts. Since the 35 volts exceeds the zener rating, the zener diode will go into conduction and create a voltage drop across its terminals of exactly 30 volts. Since the load resistor is connected in parallel with the zener, the 30 volts across the zener terminals will be the same voltage supplied to the load resistor. Anytime the voltage is less than the zener rating, the zener will present high resistance and have no effect on the circuit.

If the ac voltage that supplies the diode section of the rectifier drops 2% or 3%, the peak half-wave dc voltage at the output will drop 1–2 volts. This means that the original 35 volts will become 33 or 34 volts. Since this voltage still remains larger than the zener diode rating, the zener diode will continue to stay in conduction and provide the 30 volt drop and continue to supply 30 volts to the load. As you can see, the incoming ac voltage will need to *drop* more than 10% to cause the dc half-wave voltage to become lower than the zener diode rating, and cause it to stop regulating.

It is important to understand that as the dc half-wave voltage becomes larger, the zener must shunt more current. The term *shunt* means that the zener diode will provide a parallel path for this excess voltage and current so that the load resistor will always be supplied with a constant 30 volts. Zener diodes come in a variety of voltage and current ratings for all types of industrial power supplies up to several hundred volts.

If you use an oscilloscope to view the waveforms in a power supply, you will notice that the voltage from the terminals of the rectifier will be complete full-wave rectification like the waveform shown in Fig. 7-12b. The waveform that would be observed at the load is shown in Fig. 7-12c and it shows the top parts of the rectified full-wave as missing because the zener diode shunted the excess voltage so that the load would see the regulated 30 volts.

7.2.10 Surge Protection for Rectifier Circuits

The rectifier circuits in power supplies are subject to a large variety of transient voltages from lightning or from inductive loads such as motor and coils that produce a spike when they are de-energized. When these transients occur, they may have voltage levels that are two to five times the original supply voltage. The fuses in a circuit are designed to protect a circuit against overcurrent, but they cannot detect overvoltage.

A solid-state device called a *metal oxide varistor* (MOV) is designed to be connected in parallel across the input power supply. Fig. 7-13 shows three diagrams of MOVs connected in power supply circuits. In Fig. 7-13a, you can see that this circuit uses two MOVs. MOV1 is connected in parallel across the incoming voltage lines immediately after the mainline fuse. The MOV exhibits high resistance until its voltage rating is exceeded. This means that if the incoming voltage to the power supply circuit that the MOV is protecting is normal, the MOV will have very high resistance and act like an open circuit. If the incoming voltage increases to a point that exceeds the MOV's voltage rating, the internal resistance of the MOV is quickly reduced to near zero, which will allow maximum current to flow through it. Since MOV1 is connected directly across the two wires of the incoming voltage immedi-

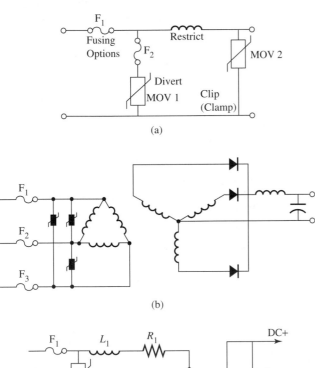

(a)

(b)

(c)

FIGURE 7-13 (a) Electronic diagram that shows two metal-oxide varistors (MOVs). MOV1 is for protection against overvoltage conditions from incoming voltage, and MOV2 provides surge protection against transient voltages that may be produced in an inductive load. (b) Electronic diagram that shows three MOVs connected in a three-phase power supply. (c) Electronic diagram that shows two MOVs connected in a single-phase power supply. (Copyright of Motorola. Used by Permission.)

ately after the main circuit fuse F1, the excess current that is flowing through the MOV will cause fuse F1 to open.

In some applications the F1 fuse must be sized as a time delay fuse so that inductive loads such as motors can draw locked rotor amperage (LRA) for several seconds when the motor is started. In these cases a second fuse (F2) that is faster acting is added in series with the MOV to protect it. When MOV1 goes into conduction, the excessive current it draws will be sufficient to cause both fuses to open. Since fuse F2 is specifically designed to react fast enough to protect the MOV, the MOV will not be damaged by the overcurrent.

MOV2 is connected after the choke in this circuit so that it is parallel to the load. If excessive voltage occurs, the MOV will cause a short circuit that results in excessive current flowing through the choke. The choke will absorb some of this excessive current and protect the MOV until fuse F1 opens and clears the fault.

Fig. 7-13b shows a three-phase rectifier circuit with three MOVs. One MOV is connected across each incoming phase. If the voltage of any phase becomes excessive and exceeds the level of any MOV, the MOV will begin conducting current and cause sufficient current to flow so that the main fuses in the circuit will open.

Fig. 7-13c shows two MOVs connected across the single-phase lines for a full-wave bridge rectifier. If the voltage that is supplied to the bridge rectifier becomes excessive, the MOVs will go into conduction and cause excessive current to flow, which will cause the main fuses to open.

7.2.11 Crowbar Protection Against Overvoltage

Another way to protect power supply circuits against overvoltage conditions is with a *crowbar* circuit.

Fig. 7-14 shows two examples of a crowbar circuit. In these circuits, an SCR is used to sense the overvoltage condition and go into conduction. The SCR is strategically located in the circuit to cause a short circuit of sufficient size to cause

FIGURE 7-14 Electronic diagrams of two types of crowbar circuits that use an SCR to provide overvoltage protection in power supplies. (Copyright of Motorola. Used by Permission.)

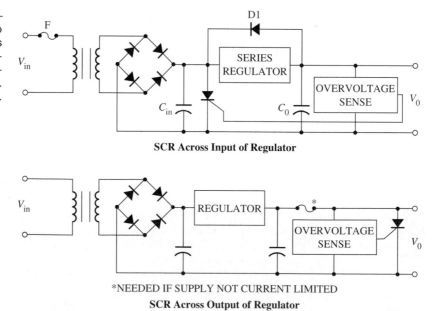

SCR Across Input of Regulator

*NEEDED IF SUPPLY NOT CURRENT LIMITED
SCR Across Output of Regulator

the fuse or circuit breaker to open and protect the circuit. The circuit in Fig. 7-14a uses an SCR that is connected in parallel across the dc terminals of the bridge rectifier. The fuse in this circuit is located in the primary circuit of the transformer. When the overvoltage-sensing circuit at the output terminals of the power supply detects the voltage has exceeded the setpoint, it will provide a pulse to the gate of the SCR, which will cause it to go into conduction. When the SCR goes into conduction, it will create a short circuit in the dc circuit, which will cause the primary side of the transformer to draw excessive current and open the fuse. The SCR is used because it is large enough and fast enough to withstand the short-circuit energy that is developed when it goes into conduction.

The crowbar circuit in Fig. 7-14b shows the SCR connected across the dc output terminals. In this circuit the fuse is connected in the dc side of the circuit prior to the SCR. When the overvoltage condition is sensed, a pulse is sent to the gate of the SCR and the SCR goes into conduction. When this occurs, a short circuit is developed in the secondary of the power supply and the dc current will increase sufficiently to cause the fuse to open. These types of circuits could also use circuit breakers instead of fuses.

7.3 APPLICATIONS FOR INDUSTRIAL POWER SUPPLIES

The rectifier sections are used in all types of industrial applications. Virtually every piece of industrial equipment that has electronic circuity must have a means of converting ac voltage to dc voltage. These systems use the rectifier circuits discussed earlier to provide the dc power. The rectifier circuits will include capacitors and inductors for filtering, and the combined circuit will be called a power supply. Industrial power supplies will be slightly different from the ones used in consumer products such as radios, televisions, and other electronic components in that their supply will be single-phase and three-phase high voltage. This means that the voltage level will normally be 208 or 480 volts.

It will be easier to understand the operation of these types of power supplies if you examine several specific applications. The next sections will explain the operation of the power supply circuit in two common industrial applications: the ac variable-frequency motor drive and a welding system. This section will also explain the operation of an uninterruptible power supply (UPS) that provides backup voltage to computers and PLCs in case of a power outage.

7.3.1 Power Supply for a Variable-Frequency Motor Drive

You can get a better idea of how the diodes in the rectifier and the devices in the filter and regulator sections of a circuit all work together when you see a complete electronic circuit for an operational system. Fig. 7-15 shows the electrical diagram of a variable frequency motor drive that is commonly used in industrial applications. From this diagram you can see that the drive circuit uses three-phase supply voltage, so a three-phase full-wave bridge rectifier is used. A picture of this type of bridge is shown in Chapter 4 and you should remember that it is encapsulated so you do

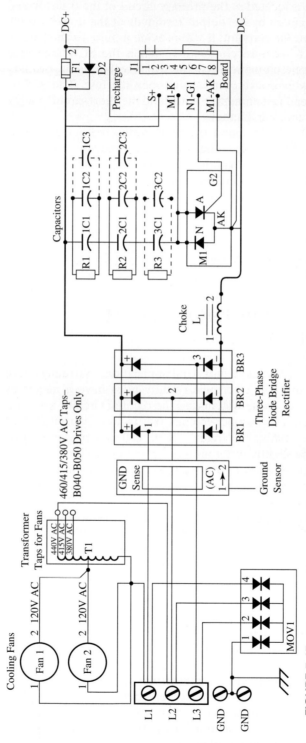

FIGURE 7-15 Electronic diagram of the power supply for a variable-frequency drive. (Courtesy of Rockwell Automation's Allen Bradley Business.)

not see each of the diodes. Rather you would find the terminals for the three-input terminals and the two dc output terminals.

This circuit also uses metal-oxide varistors (MOVs) to check for an overvoltage condition at the incoming supply. You should notice that the electronic symbol that looks like two diodes connected end to end is used instead of the box symbol that was used for MOVs in the previous figures. The MOVs are connected between each of the three phases of incoming power and ground to protect the electronic parts from excessive supply voltage. The fuses in the system only check for overcurrent, and they are not aware of an overvoltage condition. If an overvoltage condition occurs, the MOVs will go into conduction (lower their resistance) and cause an overcurrent condition that will make the fuse open. The fuses that protect the supply voltage are not shown in this diagram. They would be located in series with the L1, L2, and L3 terminals of the supply voltage.

The next components in the dc voltage section of the drive are the choke (inductor) and capacitors (filter). The inductor in this circuit is labeled L1 for inductor 1 and it is connected in series with the dc negative voltage line and the capacitors that are connected in parallel across the dc lines. The positive dc and negative dc line will be called the *dc bus* because they supply all dc power to the inverter section of the drive. More than one capacitor is used on larger drives to provide the correct amount of filtering, which will make the pulsing half-wave dc nearly pure dc.

The capacitors will have resistors connected across their terminals. The function of these resistors is to provide a means to bleed the capacitors down when power is de-energized. You should always treat the capacitors in a power supply as though they are fully charged, so you do not get hurt from the high voltage they store even when the main power is turned off. It is more important to understand that these capacitors will constantly charge and discharge with every half-wave when power is applied, so they can continually discharge high voltage to you if you come into contact with them. The capacitor precharge circuit is connected to the bank of capacitors used in the filter. The precharge circuit ensures that when voltage is initially applied to the capacitors, it is ramped up slowly so that the capacitors do not absorb maximum voltage. This circuit will help prevent the capacitors from charging too quickly and becoming damaged.

7.3.2 Welding Power Supply

The next application of industrial equipment that uses a large power supply is a welding system. This welding system is specifically designed for dc arc welding. This means that it will get its supply power from an ac power source that must be converted to dc voltage. Fig. 7-16 shows the diagram of the power supply for this welding system. You can see that the supply voltage in this system is single-phase (two wires) ac voltage, which can be 220, 380, or 440 volts. The main transformer has multiple taps to accommodate each of these supply voltages. All the technician needs to do when supply power is connected during the installation process is to measure the supply voltage and connect the lines to the appropriate terminals on the main transformer. The taps on the transformer are clearly identified so that the connection can be made for each different type of voltage.

The main transformer consists of two single-phase primary windings that are connected in parallel. This means that each of the primary windings will receive

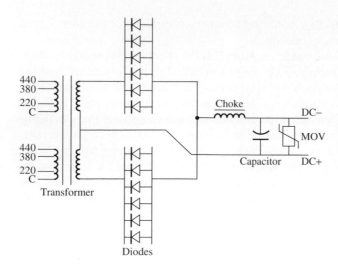

FIGURE 7-16 Electrical diagram of the power supply for a six dc arc-welding system. Notice that this power supply uses six diodes in each section of a center-tapped transformer full-wave bridge rectifier. The diodes are connected in parallel to supply larger currents.

the same voltage. The secondary windings of this transformer and the diodes are connected like a center-tapped full-wave bridge rectifier. The main difference in this circuit is that each section of the rectifier uses six diodes that are connected in parallel so that the power supply can provide current that is larger than the rating of each individual diode. In fact the current rating of this power supply will be 12 times the size of each individual diode because this configuration of rectifier uses a center-tap transformer. This means that the current rating of each diode can be 100 A, and the power supply can provide over 1000 A. This type of circuit was very popular in the 1980s when the size of diodes was restricted to 100–150 A. Today larger diodes are available to provide current in excess of 1500 A.

The output section of this power supply has a filter with a choke (inductor) that is connected in series with the dc negative line, and a capacitor that is connected in parallel across the positive and negative dc lines. A metal-oxide varistor is connected in parallel with the capacitor and the load. You should remember that the MOV is in the circuit to protect against overvoltage conditions. The MOV would cause a fuse in the transformer secondary to open anytime the voltage exceeds its rating.

7.3.3 Uninterruptible Power Supplies

Another popular type of power supply used in industrial applications is called an *uninterruptible power supply* (UPS). The UPS has become important in industrial and commercial power supplies because it provides a means of supplying power to computers and programmable logic controllers (PLCs) in applications where a power failure cannot be tolerated. In most parts of the United States, weather conditions such as lightning storms and ice storms may cause the power company to loose power for a period of time. The amount of time the power is disrupted may last from 10 seconds to several hours. If the power outage occurs while a computer or PLC is running, it will cease operating and may loose the information in active memory. The outage condition will also cause the systems to be restarted, which may take additional time.

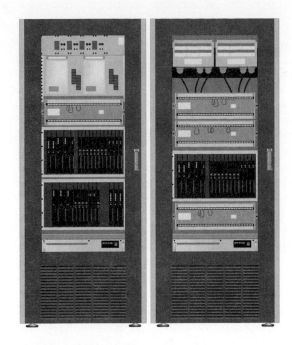

FIGURE 7-17 An uninterruptible power supply (UPS) in an enclosure with several PLC racks. (Courtesy of Liebert Inc.)

The UPS combines a power supply with a battery to provide a circuit that can provide output power while the power company's incoming power is down. This is accomplished through a rectifier section in the system that converts the ac voltage into dc voltage. The major difference in the UPS is that the power from the rectifier section is used to charge a battery. The battery acts as a buffer for the voltage because the battery supplies dc voltage to the inverter part of the system. Here dc voltage is turned back into ac voltage with a set frequency that matches the power system of the equipment connected to it. This means that some of the dc voltage may go through the battery to keep it charged. The battery is sized so that it is large enough to continue providing dc voltage to the inverter even when the ac incoming voltage has been interrupted. The battery is rated for the amount of time the UPS system is supposed to provide the voltage during an outage. Since most computers and PLCs do not use a large amount of voltage, even very small UPS systems can provide dc voltage for up to an hour. Larger UPS systems are designed to provide backup power to complete buildings such as hospitals or for control rooms for nuclear power plants. Fig. 7-17 shows a picture and Fig. 7-18 shows an electronic block diagram of a UPS.

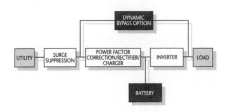

FIGURE 7-18 An electronic block diagram of an uninterruptible powersupply (UPS). (Courtesy of Liebert Inc.)

7.4 INVERTERS: CHANGING DC VOLTAGE TO AC VOLTAGE

Inverters are circuits specifically designed to change dc voltage to ac voltage. As you know systems such as variable-frequency motor drives and uninterruptible power supplies (UPS) convert ac power to dc and then convert the dc back to ac. This may sound like a strange way to provide an ac output voltage if ac voltage is the original supply, but in the case of the variable-frequency motor drive, the frequency of the supply voltage will be 50 or 60 Hz and the output ac voltage needs the possibility of frequencies between 1–120 Hz. In the case of the UPS, the ac supply voltage needs to be changed to dc so it can be stored in a battery for later use if the power supply is interrupted. Since the voltage is changed to dc and is stored in a battery, it must be changed back to ac to be usable. In the UPS, the output frequency will be a constant 60 Hz.

The earliest use of converting dc voltage to ac voltage was in circuits specifically designed for providing variable-frequency ac voltage for single- and three-phase ac motors. Today inverter circuits similar to the original ones are used to provide three-phase voltage from single-phase voltage sources, to provide three-phase voltage with variable-frequency and variable voltage from a fixed three-phase power source, and to provide isolation by using batteries as buffers or storage. Modern industrial circuits use one of three types of inverters: variable-voltage input (VVI), pulse-width modulation (PWM), and current-source input (CSI).

It is vitally important that you understand that inverter circuits may be found in equipment as one part of the total system or they may be a stand-alone circuit. For example, in an ac variable-frequency motor drive the inverter is only a part of the total circuit, and its job is to change dc voltage back to ac voltage.

At this point you may not fully understand why all the equipment and circuits are needed to change voltages and frequencies in industrial applications, but you should understand that in factories today a large variety of expensive equipment exists that may come from different parts of the world. Therefore, requirements for voltages and frequencies may not match what is available in the factory. In some cases this equipment is one of a kind and must be installed and used as is, which requires circuits that can change voltage and frequency easily.

7.4.1 Single-Phase Inverters

The simplest inverter to understand is the single-phase inverter, which takes a dc input voltage and converts it to single-phase ac voltage. The main components of the inverter can be either four silicon controlled rectifiers (SCRs) or four transistors. Fig. 7-19 shows a typical inverter circuit that uses four SCRs, and Fig. 7-20 shows a typical inverter circuit that uses four transistors. Originally called a *dc-link converter,* now it is simply called an *inverter.*

The diagram in Fig. 7-19 shows four SCRs used in the inverter circuit. In this circuit SCR_1 and SCR_4 are fired into conduction at the same time to provide the positive part of the ac waveform and SCR_2 and SCR_3 are fired into conduction at the same time to provide the negative part of the ac waveform. The waveform for the ac output voltage is shown in this figure, and you can see that it is an ac square wave. A phase-angle control circuit is used to determine the firing angle, which provides the timing for turning each SCR on so that they provide the ac square

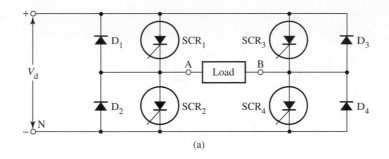

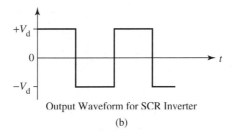

(a)

(b)

Output Waveform for SCR Inverter

FIGURE 7-19 **(a)** Electrical diagram of a typical inverter circuit that uses four silicon controlled rectifiers (SCRs). **(b)** Output waveform for SCR inverter.

FIGURE 7-20 Electronic diagram of a transistor inverter with the output waveforms for the ac voltage. (Courtesy of Rockwell Automation's Allen Bradley Business.)

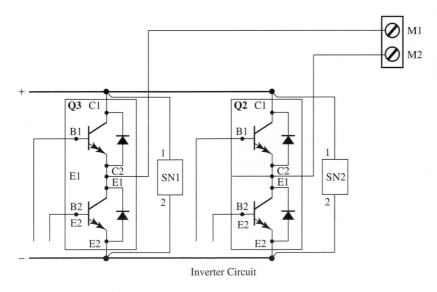

Inverter Circuit

Output Waveform

wave. Examples of these types of firing-angle circuits are shown in Chapter 5. The load is attached to the two terminals where the ac square wave voltage is supplied.

7.4.2 Using Transistors for a Six-Step Inverter

Fig. 7-20 shows the electrical diagram of an inverter that uses four transistors instead of four SCRs. Since the transistors can be biased to any voltage between saturation and zero, the waveform of this type of inverter can be more complex to look more like the traditional ac sine wave. The waveform shown in this figure is a *six-step* ac sine wave. Two of the transistor will be used to produce the top (positive) part of the sine wave, and the remaining two transistors will be used to produce the bottom (negative) part of the sine wave.

When the positive part of the sine wave is being produced, the transistors connected to the positive dc bus voltage are biased in three distinct steps. During the first step, the transistors are biased to approximately half-voltage for one-third of the period of the positive half-cycle. Then these transistors are biased to full voltage for the second third of the period of the positive half-cycle. The transistors are again biased at the half-voltage for the remaining third of the period. This sequence is repeated for the negative half-cycle. This means that the transistors that are connected to the negative dc bus are energized in three steps that are identical to the steps used to make the positive half-cycle.

Since six steps are required to make the positive and negative half-cycles of the ac sine wave, this type of inverter is called a *six-step inverter*. The ac voltage for this inverter will be available at the terminals marked M1 and M2. Even though the ac sine wave from this inverter is developed from six steps, the motor or other loads see this voltage and react to it as though it was a traditional smooth ac sine wave. The timing for each sine wave is set so that the period of each is 16 msec which means it will have a frequency of 60 Hz. The frequency can be adjusted by adjusting the period for each group of six steps.

7.4.3 Three-Phase Inverters

Three-phase inverters are much more efficient for industrial applications where large amounts of voltage and current are required. The basic circuits and theory of operation are similar to the single-phase transistor inverter. Fig. 7-21 shows the diagram of a three-phase inverter with three pairs of transistors. Each pair of transistors operates like the pairs in the single-phase six-step inverter. This means that the transistor of each pair that is connected to the positive dc bus voltage will conduct to produce the positive half-cycle, and the transistor that is connected to the negative dc bus voltage will conduct to produce the negative half-cycle.

The timing for these transistors is much more critical since they must be biased at just the right time to produce the six steps of each sine wave, and they must be synchronized with the biasing of the pairs for the other two phases so that all three phases will be produced in the correct sequence with the proper number of degrees between each phase.

7.4.4 Variable-Voltage Inverters (VVIs)

A variable-voltage inverter (VVI) is basically a six-step, single-phase or three-phase inverter. The need to vary the amount of voltage to the load became necessary when these inverter circuits were used in ac variable-frequency motor drives and welding circuits. Originally these circuits provided a limited voltage and limited

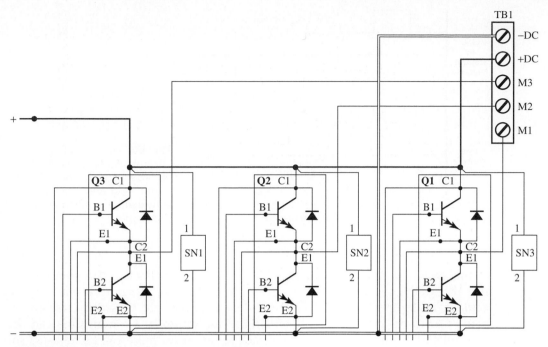

FIGURE 7-21 Electrical diagram of a three-phase inverter that uses six transistors. (Courtesy of Rockwell Automation's Allen Bradley Business.)

variable-frequency adjustments because oscillators were used to control the biasing circuits. Also many of the early VVI inverters used thyristor technology, which meant that groups of SCRs were used with chopper circuits to create the six-step waveform. After microprocessors became inexpensive and widely used, they were used to control the biasing circuits for transistor-type inverters to give these six-step inverter circuits the ability to adjust the amount of voltage and the frequency through a much wider range. Motors needed the adjustable frequency to increase or decrease their speeds from their rating that was determined by the number of poles the motor has when it is manufactured. The voltage of the drive needed to be constantly adjusted as the frequency was adjusted so that the motor received a constant ratio of voltage to hertz to keep the torque constant. This became a problem at very low speeds where motors tended to loose torque.

Fig. 7-22 shows a diagram of the voltage and current waveform for the VVI inverter. From this diagram you can see that the voltage is developed in six steps and that the resulting current looks like an ac sine wave. These are the waveforms that you would see if you placed an oscilloscope across any two terminals of this type of inverter.

7.4.5 Pulse-Width Modulation Inverters

Another method of providing variable-voltage and variable-frequency control for inverters is to use pulse-width modulation (PWM) control. This type of control uses transistors that are turned on and off at a variety of frequencies. This provides a unique waveform that makes multiple square wave cycles that are turned on and off at specific times to give the overall appearance of a sine wave. The outline of the waveform actually looks very similar to the six-step inverter signal. An example

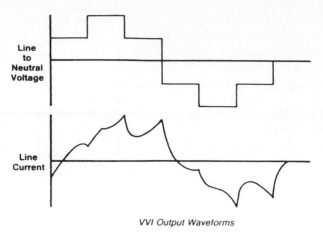

Line
to
Neutral
Voltage

Line
Current

VVI Output Waveforms

FIGURE 7-22 Voltage and current waveforms for the variable-voltage input (VVI) inverter. (Courtesy of Rockwell Automation's Allen Bradley Business.)

of this type of waveform is provided in Fig. 7-23. From this diagram you can see that the overall appearance of the waveform is an ac sine wave. Each sine wave is actually made up of multiple square wave pulses that are caused by transistors being turned on and off very rapidly. Since the bias of these transistors can be controlled, the amount of voltage for each square wave pulse can be adjusted so that the entire group of square waves has the overall appearance of the sine wave. If you look at the voltage waveform for the PWM inverter, you will notice that the outline of the ac sine wave still looks like the six-step sine wave originally used in the VVI inverters. The height of the steps of the ac sine wave is also increased when the voltage of the individual pulses are increased. This increases the total voltage of the sine wave that the PWM inverter supplies.

The width (timing) of each square wave pulse can also be adjusted to change the period of the group of pulses that makes up each individual ac sine wave. When

FIGURE 7-23 Voltage and current waveforms for the pulse-width modulation (PWM) inverter. Notice that the overall appearance of each waveform is an ac six-step sine wave and that it is actually made of a number of square wave pulses. (Courtesy of Rockwell Automation's Allen Bradley Business.)

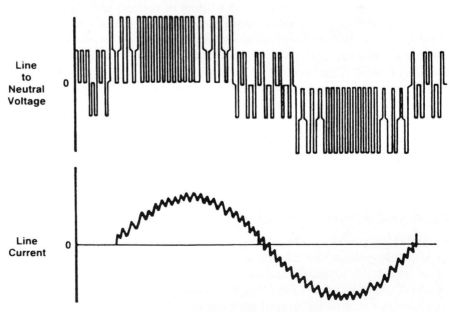

Line
to
Neutral
Voltage

0

Line
Current

0

PWM Output Waveforms

the width of the sine wave changes, it also changes the period for the sine wave. This means that the frequency is also changed and is controlled for the PWM inverter by adjusting the timing of each individual pulse. Since adjusting the voltage and frequency is fairly complex, the PWM inverter uses a microprocessor to control the biasing of each transistor. If thyristors are used as in SCRs, the microprocessor will control the phase angle for the firing circuit.

Early PWM circuits used thyristors such as SCRs to produce the square wave pulses. The control circuit included triangular carrier waves to keep the circuit synchronized. This sawtooth waveform was sent to the oscillator circuit that controlled the firing angle for each thyristor. Today the PWM inverters mainly use transistors because of their ability to be biased from zero to saturation and back to zero at much higher frequencies. Modern circuits will more than likely use transistors for these circuits because they are now manufactured to handle larger currents that are well in excess of 1500 A.

7.4.6 Current-Source Input (CSI) Inverters

The current-source input (CSI) inverter produces a voltage waveform that looks more like an ac sine wave and current waveform that looks similar to the original on/off square wave of the earliest inverters that cycled SCRs on and off in sequence. This type of inverter uses transistors to control the output voltage and current. The on-time and off-time of the transistor are adjusted to create a change in frequency for the inverter. The amplitude of each wave can also be adjusted to change the amount of voltage at the output. This means that the CSI inverter like the previous inverters can adjust voltage and frequency usable in variable-frequency motor drive applications or other applications that require variable voltage and frequency. Fig. 7-24 shows the voltage and current waveform for the CSI inverter.

7.4.7 Cycloconverters

A *cycloconverter* is a circuit designed to convert the frequency of ac voltage directly to another frequency of ac voltage without first converting the voltage to dc voltage. The history of this circuit dates back to the 1930s when mercury arc rectifiers were used to control the frequency of railroad engines in Germany. The supply voltage for these original circuits was a fixed 50 Hz ac sine wave common in Europe. The train engines used low frequency (16.6 Hz) so their electric motors would turn

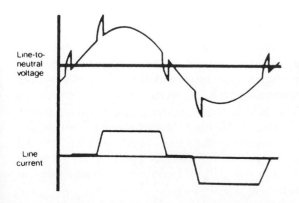

FIGURE 7-24 Voltage and current waveform for the current-source input (CSI) inverter. (Courtesy of Rockwell Automation's Allen Bradley Business.)

Line-to-neutral voltage

Line current

slowly, creating a tremendous amount of torque. These earliest rectifiers were rather large tube thyristors. The input circuit for the cycloconverter used a large transformer, and the output section used the thyristors to adjust the timing of the output stage, which allowed frequency to be changed.

Since modern electronic technology provides many ways to control voltage and frequency, the cycloconverter circuit is no longer useful. In fact many advantages are provided by converting the ac voltage to dc before the frequency is converted back to ac and adjusted for the output section. For example, when the ac input voltage is rectified to dc and filtered, all transient signals and voltage spikes are removed so that when the dc is converted back to ac, the output circuit is effectively isolated from the input.

7.4.8 Applications for Inverters

Inverters are seldom found as stand-alone circuits. You will normally find them used in conjunction with other circuits such as rectifiers and filter circuits in power supplies that will provide a source for the dc voltage the inverter needs. You may also find the inverter as an integral part of the dc to dc converter circuitry used in many types of dc power supplies. The major use of inverters in industry today is for variable-frequency ac motor drives and high-frequency power supplies for welding applications. Chapter 11 will explain motor drives and other applications in detail.

7.5 DC-TO-DC CONTROL (CONVERTERS AND CHOPPERS)

Originally in the 1970s and 1980s, the conversion of dc voltage to a different value of dc voltage was accomplished with a circuit called a *chopper*. The chopper was originally specifically designed to convert a fixed dc voltage into variable dc voltages primarily used to control the speed of dc motors. Since dc voltage is not readily available as a supply source in industry, these circuits must rely on a rectifier circuit to change ac supply voltage to dc. In this section we will treat these circuits from the point where dc voltage is supplied to them from the rectifier circuits. In this sense, they will be classified as a dc-to-dc converter. When troubleshooting these circuits, you must test the rectifier circuits to ensure a sufficient supply of dc voltage is available, even though the circuit diagram does not include the ac-to-dc rectification.

7.5.1 Overview of DC-to-DC Voltage Conversion

As electronic devices have become larger and faster, circuits have evolved and ac motor control has become more popular than dc motor control for industrial applications. Today dc-to-dc voltage conversion is more widely used in power supply circuits because every piece of equipment that has an electronic board in it requires a wide variety of dc voltage supplies. Each voltage must be supplied from a power supply. This means that computers, PLCs, and all other electronic equipment require a dc power supply. Today the older chopper circuits have been modified into the new power supply technology with newer types of circuits and they are all more commonly called *converters*. Most often you will find converter circuits today in *switch-mode power supplies* (SMPS).

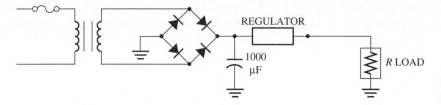

FIGURE 7-25 Electronic diagram of a linear power supply.

7.5.2 Linear Power Supplies

Linear power supplies have been popular since the beginning of vacuum-tube electronics. Their operation is simple, but their efficiency is quite poor in the range of 30% to 40%. Fig. 7-25 shows the electronic diagram for a typical linear power supply. From this diagram you can see that the first part of the power supply is exactly like the rectifier sections presented earlier in this chapter. The power supply uses a transformer and a four-diode full-wave bridge rectifier to produce the pulsing dc output waveforms. A capacitor is used as a filter to smooth out the dc. The remainder of the circuit contains a regulator and the load. The regulator is the part of the circuit that makes a linear power supply different from the newer switch-mode power supplies.

The regulator in this circuit acts as a voltage divider between the regulator and the load. To understand this operation you need to think of the load as a fixed resistance and the regulator as a variable resistance. As you know, when two resistances are connected in series, such as the regulator and the load resistance, the amount of voltage supplied to them will be shared. The amount of voltage will be split by the ratio of the resistance. For example, if the ratio of the resistance is 2 : 1, the regulator will have twice as much voltage measured across it than the load. If the power supply delivered 30 volts, the regulator would have 20 volts dropped across it, and the load would have 10 volts measured across it.

If the resistance of the regulator was changed so that the ratio of resistance with the load was 1 : 1, the voltage across the regulator would drop to 15 volts, and the voltage measured across the load terminals would be increased to 15 volts. This means that the voltage to the load terminals will change anytime the resistance in the regulator changes. This type of circuit is simple to operate, which makes the linear power supply easy to manufacture and troubleshoot. The problem with this type of power supply is all of the voltage that is dropped across the regulator is wasted energy. If the regulator drops 10% of the voltage and the load gets 90%, the power supply is operating somewhat efficiently. If the regulator drops 90% and the load gets 10%, you can see that this is a tremendous amount of wasted energy. The other drawback of the linear power supply is that since it must drop a portion of the supply voltage through the regulator, the components in the regulator must be sized large enough to handle the excess heat that is generated. This tends to make the linear power supply up to two times larger (and heavier) than the new switch-mode power supplies.

7.5.3 Switching Power Supplies

Switch-mode power supplies (SMPS), also called switching power supplies, have become more popular than linear power supplies in the past ten years because they provide a regulated voltage with more efficiency and they do not require the larger transformers and filtering devices that the linear power supplies require. For

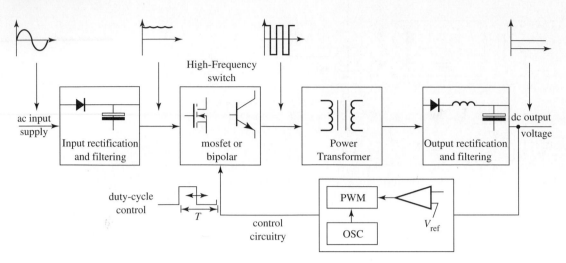

FIGURE 7-26 Electronic block diagram of a switch-mode power supply (SMPS). (Courtesy of Philips Semiconductors.)

example, the linear power supplies generally have average conversion efficiencies of 30%, while the SMPS have efficiencies up to 80%. Since the SMPS do not need the larger components, they are more usable in modern circuits where cabinet space and board space is at a premium. Designers are continually trying to reduce the size and weight of electronic controls, and one easy way has been to change to SMPS.

Fig. 7-26 shows an electronic block diagram of a switch-mode power supply. This diagram will help you understand how the SMPS converts a dc input voltage to a new value of dc voltage that is filtered and regulated. The first block of the power supply is called the *rectifier and filter section* and it is shown in the diagram as a diode and capacitor, indicating the ac voltage is rectified to pulsing dc and then filtered to reduce the amount of ripple. The second block in the diagram shows the symbols of a MOSFET and bipolar transistor. This section is called the *high-frequency switching section* and it uses either MOSFETs or bipolar transistors to convert the dc voltage to a high-frequency ac square wave. The high-frequency ac square wave can be 20–100 kHz. The incoming ac voltage is rectified to dc and then the high-frequency switching section changes it back to ac for several reasons. First, the incoming voltage is always fluctuating and it is full of transient voltages that can be damaging to solid-state components if they are allowed to reach them. The two-step conversion helps to isolate these fluctuations and transients. The second reason is that higher frequencies allow for higher conversion efficiencies.

The next section of the SMPS is the *power transformer section*. The power transformer will isolate the circuits and step up or step down the voltage to the level required by the dc voltage. The output of the transformer is sent to a second rectifier section. Since the first rectifier section was for the input voltage, it is called the *input rectifier,* and since the second rectifier is used for supplying output voltage, it is called the *output rectifier section.* The output rectification section is different from the input rectifier in that the frequency of the voltage in the second section will be very high (20–100 kHz). This means that the output ripple of the high-frequency voltage will be nearly filtered naturally because of the number of overlaps between each individual output pulse. Since the ripple is very small, the actual capacitors in the filter section will be rather small.

The final section of the SMPS is the *control and feedback block,* which contains circuitry that provides pulse-width modulated output signal. The pulse-width modulation provides a duty cycle that can vary pulse by pulse to provide an accurate dc output voltage. This block of the power supply uses an operational amplifier to compare the output voltage to a reference voltage and continually make adjustments to the output voltage. The oscillator in this circuit provides the frequency for the duty cycle.

The final three blocks (power transformer, output rectification, and control/ feedback) can have different circuitry called *topology* that is simpler or more complex than this example. Each of the topologies have advantages and disadvantages. The more common topologies will be presented in the next section.

7.5.4 The Buck Converter

The buck converter circuit is the basis for several other similar circuits called *forward converters.* The buck converter circuit and the input and output voltages for this circuit are shown in Fig. 7-27. This circuit would be connected directly after the power transformer. From the diagram you can see that this circuit is fairly simple in that it consists of a transistor, inductor, diode, and capacitor. When the transistor is turned on, power will flow directly to the output terminals. This voltage must also pass through the inductor, which will cause current to build up in it in much the same way that a capacitor charges. When the transistor is switched off, the stored current in the inductor will cause the diode to become forward bias, which will let it freewheel and allow the current to be delivered to the load that is connected to the output terminals.

The waveforms at the bottom of the diagram show the square wave of the input voltage in the top line. The waveform on the bottom line shows the effects of the inductor putting the stored current back into the circuit to the load. This

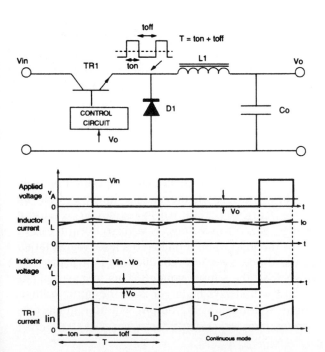

FIGURE 7-27 Electronic diagram and waveforms for a buck converter circuitry for a switch-mode power supply. (Courtesy of Philips Semiconductors.)

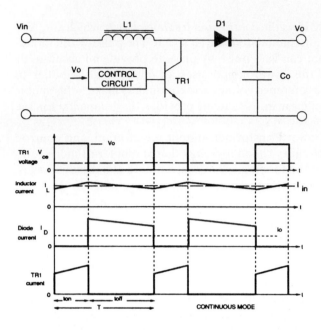

FIGURE 7-28 Electronic diagram and waveforms for a boost regulator for a switch-mode power supply. (Courtesy of Philips Semiconductors.)

current is shown as the dashed line that occurs during the square wave off-cycle. This type of circuit is called a *step-down buck converter* because the output voltage will always be smaller than the input. Voltage regulation for this circuit is controlled by duty cycle. If the off-time for the duty cycle is lengthened, the average voltage to the output will be lower, and if the off-time for the duty cycle is shortened, the average power will increase.

7.5.5 The Boost Regulator

The *boost regulator* is a second type of fundamental regulator circuit for the switch-mode power supply. The electronic diagram and waveforms for this type of converter are shown in Fig. 7-28. From this diagram you can see that the transistor has been moved to a point after the inductor and it is now connected directly across the positive and negative lines of the output. The *freewheeling diode* is connected in series and reverse bias with the inductor. The capacitor remains in parallel with the output voltage terminals to provide filtering. It is important to remember that this circuit is also connected directly to the secondary windings of the power transformer, just like the buck regulator.

When the transistor is switched on, current flows in this circuit and builds up in the inductor. When the transistor is turned off, the voltage that was built up across the inductor due to the stored current is returned to the circuit because it is reverse bias to the applied voltage. Since this voltage is reverse bias, it will be allowed to pass through the diode to the load when it builds to a level that is larger than the applied voltage. When the diode goes into conduction, it will pass the power stored in the inductor along with the supply voltage. This means that the output voltage of the boost converter will always be larger than the input voltage, hence, the name *boost*. The amount of voltage at the output will be regulated by adjusting the duty cycle of the circuit.

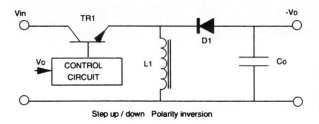

Step up / down Polarity inversion

FIGURE 7-29 Electronic diagram of a buck-boost regulator. (Courtesy of Philips Semiconductors.)

7.5.6 The Buck-Boost Regulator

A combination of the buck regulator and boost regulator is shown in Fig. 7-29. This combination circuit is called the *buck-boost regulator* and it utilizes the strong points of both of the previous regulators. From the diagram of this circuit you can see that the transistor is connected in series like the buck converter and the inductor has been moved to a position where it is connected in parallel with the output terminals. The freewheeling diode is connected as it was in the boost regulator.

The transistor controls the voltage to the output in this circuit. When it is turned on, the inductor will store energy. When the transistor is turned off, the stored energy will be large enough to forward bias the diode and pass voltage to the output terminals. Since this circuit has the basic operation of both the buck and boost regulators, it means that the output voltage can be regulated both above and below the input-voltage level. For this reason the buck-boost regulator is more popular. The waveforms for this type of circuit are similar to the boost regulator.

7.5.7 The Forward Converter

The forward converter is basically a buck converter with a transformer and a second diode added to allow energy to be delivered directly to the output through the inductor during the transistor on-time. Fig. 7-30 shows the electronic diagram and waveforms for the forward converter. In this diagram you can see that the transistor is connected in series with the primary of the additional transformer. The second transformer provides a phase shift that causes the polarity of its voltage to be such that it will flow to the output while the transistor is in conduction. This allows a better flow of energy to the output through the transistor. This means that the output current is continuous with the result that the ripple will be minimal. This also means that the filtering capacitors can be smaller, which makes the entire power supply smaller.

7.5.8 The Push-Pull Converter

As the switch-mode power supply has evolved, additional adjustments to the original circuits have been made to get more power from smaller components. This means that the efficiency for the system must be increased. One simple way to do this is to use a center-tapped transformer that utilizes both the top and bottom half-cycles. Fig. 7-31 shows a diagram of the push-pull converter in which the push-pull converter utilizes a center-tapped transformer for both the primary and secondary windings. The primary winding is controlled by two transistors, which allow one of them to conduct during each half-cycle, so the output is receiving voltage directly through one of them at all times. This means that the efficiency of this configuration is

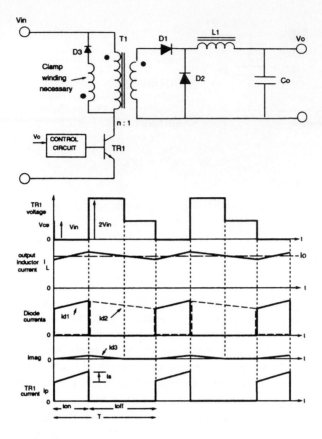

FIGURE 7-30 Electronic diagram of a forward converter and waveforms. (Courtesy of Philips Semiconductors.)

approximately 90%. This allows the overall size for the power supply to be smaller for a comparable power supply whose efficiency is 75% to 80%.

7.5.9 Half-Bridge Converter

One of the problems with the push-pull converter is that the flux in the two sections of the center-tapped transformer primary and secondary windings can become unbalanced and cause heating problems. Another problem is that each transistor must block twice the amount of voltage than other converters block. The half-bridge converter provides several advantages over the push-pull converter. Fig. 7-32 shows the electronic diagram for the half-bridge converter. From this diagram you can see that the more expensive center-tapped transformer is replaced with a

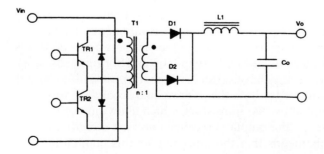

FIGURE 7-31 Electronic diagram of a push-pull converter. (Courtesy of Philips Semiconductors.)

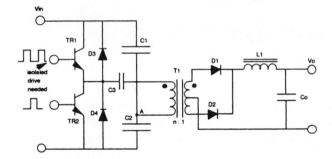

FIGURE 7-32 Electronic diagram of a dc half-bridge converter. (Courtesy of Philips Semiconductors.)

traditional transformer. This circuit still uses two transistors and two sets of diodes like the push-pull circuit. The main difference of the half-bridge converter is that it utilizes two large bulk capacitors (C1 and C2). These capacitors are connected so that each one is in series with one of the transistors. This means that power can be transferred to the output during the on-time for each transistor, which increases efficiencies to the 90% range. Since center-tapped transformers are not used, the problem with flux unbalance is also eliminated. These advantages also allow this type of converter to be utilized in power supplies up to 1000 W.

7.5.10 The Full-Bridge Converter

The full-bridge converter adds two additional transistors to the half-bridge converter. This means that four transistors are available to provide power to the output section, so this type of converter is used in power supplies in excess of 1000 W. Fig. 7-33 shows the electronic diagram of the full-bridge converter. Each transistor has a clamping diode connected across its collector–emitter terminals and they are driven alternately in pairs. Transistors T1 and T3 are energized together for one half-cycle, and transistors T2 and T4 are energized together for the other half-cycle. One advantage of the full-bridge converter is that it only requires one capacitor for smoothing the output voltage, whereas the half-bridge converter required two. The full-bridge converter will be used in larger power supplies usually over 1000 W. Since this type of converter is more complex, it is normally used in the largest types of power supplies.

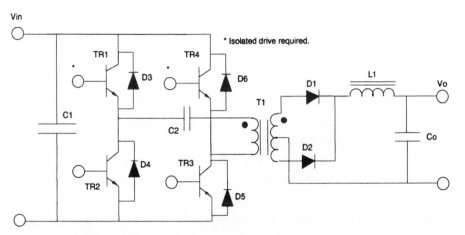

FIGURE 7-33 Electronic diagram of a full-bridge converter.

7.6 WHAT YOU MAY FIND WHEN YOU WORK ON A POWER SUPPLY

When you are asked to work on a system that uses a power supply, you should remember that the power supply is required to provide one or more of the following functions: ac-to-dc voltage conversion, ac-to-dc to ac voltage conversion, voltage conversion with battery backup, voltage conversion with a variety of dc voltages available, or ac voltage conversion with variable frequency. The equipment and circuits for each of these functions will be similar in that they will have converter circuits that include a variety of single-phase or three-phase rectifiers, and they may have inverter circuits that provide dc- to ac-voltage conversion.

If the equipment is designed to provide backup power capability, it will have one or more batteries for storage of dc power, which can later be converted by the inverter section back to ac voltage. When you are requested to work on any of these types of power supplies, you will need to review the electronic diagram and compare it to the diagrams provided in this chapter to better understand the function of the circuit. You can also use the waveforms provided in the chapter diagrams to compare to the waveforms that you will measure with an oscilloscope when you connect it to the circuit.

▶ SOLUTION TO JOB ASSIGNMENT

Your solution to the job assignment should include a variable-frequency ac motor drive to control the speed of the three-phase conveyor motor. An uninterruptible power supply (UPS) should be part of your solution to provide backup power to the computer and PLC.

The theory of operation for the variable-frequency drive should include a three-phase rectifier section that is used as the converter to produce full-wave pulsing dc voltage. A capacitor is used to filter the voltage and an inductor is used to filter the current. The inverter for the motor drive will use pulse-width modulation (PWM) or some other type of inverter circuit to change the dc voltage back to three-phase ac voltage. The inverter will also determine the frequency of the output ac voltage, which will change the speed of the conveyor motor.

The theory of operation for the UPS uses rectifier and filter sections similar to the variable-frequency drive. The UPS uses a battery connected in parallel with the dc bus to store dc power in case the incoming ac power is interrupted. The inverter is also connected in parallel to the dc bus, and it will take power directly from the rectifier section if the ac incoming power is being supplied. When the ac power is interrupted, the inverter will receive power from the battery. The inverter for the UPS may be the PWM type or any of the other types that were discussed.

QUESTIONS

1. Explain the operation of a single-diode half-wave rectifier and draw its input and output waveforms.
2. Explain the operation of a two-diode center-tapped rectifier that produces a full-wave output and draw its input and output waveforms.
3. Explain the operation of a four-diode full-wave bridge rectifier and draw its input and output waveforms.
4. Explain the function of the capacitor and inductor in a power supply.
5. Explain the function of a zener diode in a power supply.
6. The fuse in a circuit can only protect against an overcurrent. Explain what components can be added to the circuit to protect it against overvoltage conditions.
7. Explain the operation of an uninterruptible power supply.
8. Explain the operation of an inverter.
9. List three types of inverter circuits and explain the output waveform of each.
10. Explain the difference between a linear power supply and a switch-mode power supply.

TRUE OR FALSE

1. _____ The output of a single-phase half-wave rectifier is two positive pulses for each sine wave input.
2. _____ The rms value of voltage will always be smaller than the peak value for the same voltage.
3. _____ The three-phase full-wave rectifier uses six diodes and has six half-wave outputs.
4. _____ An inverter changes DC voltage to AC voltage.
5. _____ A converter changes DC voltage to AC voltage.

MULTIPLE CHOICE

1. Power goes through an uninterruptible power supply and travel throughout the components to the load in the following order: _____.
   a. surge protection section, rectifier section, battery, and then to inverter section.
   b. surge protection section, inverter section, battery, and then to rectifier section.
   c. surge protection section, battery, rectifier section, and then to inverter section.
2. The crowbar circuit uses _____.
   a. an overvoltage-sensing circuit and triac to cause the circuit fuse to open when an overcurrent condition is sensed.
   b. an overvoltage-sensing circuit and an MOV to cause the circuit fuse to open when an overvoltage condition is sensed.
   c. an overvoltage-sensing circuit and an SCR to cause the circuit fuse to open when an overvoltage condition is sensed.

3. A pulse-width modulation (PWM) inverter provides _____.
   a. an output waveform that can have its voltage, current, and frequency varied.
   b. an output waveform that can have only its voltage varied.
   c. an output waveform that can have only its current and frequency varied.
4. A chopper is _____.
   a. a converter circuit that changes ac to dc.
   b. an inverter circuit that changes dc to dc.
   c. a converter circuit that changes dc to dc.
5. A switch-mode power supply (SMPS) consists of the following circuits: _____.
   a. an input inverter circuit, a high-frequency switch, a power transformer, and an output inverter circuit.
   b. an input rectification circuit, a high-frequency switch, a power transformer, and an output rectification circuit.
   c. an input rectifier circuit, an input voltage regulator, a high-frequency switch, an output voltage regulator, and an output rectification circuit.

PROBLEMS

1. Calculate the average dc output voltage from the input rms voltage for the single-diode half-wave rectifier if the voltage at the transformer secondary is 208 volts rms.
2. Calculate the average dc output voltage for a two-diode center-tapped rectifier if the input voltage at the secondary of the transformer is 230 volts rms.
3. Calculate the average dc output voltage for a four-diode full-wave bridge rectifier if the input voltage at the secondary of the transformer is 230 volts rms.
4. Show two ways that MOVs can be connected in a power supply to protect it from over-voltage.
5. Draw the circuit for a linear power supply, explain the points you would test, and describe the waveform of the voltage at each point.
6. Draw two ways SCRs can be used in crowbar circuits and explain the operation of each of the circuits.
7. Draw the block diagram circuit for an uninterruptible power supply and identify its components.

REFERENCES

1. *Bulletin 1336 Adjustable Frequency AC Drive Maintenance Manual,* Allen-Bradley, 1201 South Second Street, Milwaukee, WI, 53204.
2. *Bulletin 1336 Adjustable Frequency AC Drive User Manual.* Allen-Bradley, 1201 South Second Street, Milwaukee, WI, 53204.
3. Kissell, Thomas E., *Modern Industrial/Electrical Motor Controls.* Englewood Cliffs, NJ: Prentice-Hall Inc., 1990.
4. Maloney, Timothy J., *Industrial Solid-State Electronics,* 2nd ed. Englewood Cliffs, NJ: Prentice-Hall Inc., 1986.
5. *Power Supply Reference Guide.* Power One Inc., 740 Calle Plano, Camarillo, CA 93012.
6. *UPStation D Series UPS.* Liebert Corporation, 1050 Dearborn Drive, P.O. Box 29186, Columbus, OH 43229.

8

OPERATIONAL AMPLIFIERS

OBJECTIVES

When you have completed this chapter, you will be able to:

1. Identify the inverting, noninverting, + supply voltage, − supply voltage, and output terminals on a diagram of an op amp.
2. Use the diagram of a comparator to explain its operation.
3. Explain the terms *amplification* (gain) and *saturation*.
4. Describe the ideal amplifier.
5. Determine the output voltage for an inverting amplifier.
6. Determine the output voltage for a noninverting amplifier.
7. Explain the polarity of the output for an inverting and noninverting op amp when input voltage is positive and when it is negative.
8. Explain the two types of power supplies that are used with op amps.

9. Use the diagram of a voltage follower to explain its operation.
10. Use the diagram of a current-to-voltage converter to explain its operation.
11. Use the diagram of a voltage-to-current converter to explain its operation.
12. Determine the amount of output voltage for a summing amp.
13. Use the diagram of a summing amp to explain its operation.
14. Explain the operation of an integrator.
15. Explain the operation of a differentiator.
16. Use the diagram of a window detector to explain its operation.
17. Explain how to adjust zero and span.
18. Use the diagram of a differential amplifier to explain its operation.
19. Use the diagram of an instrumentation amplifier to explain its operation.

You have been assigned to design a simple over-temperature alarm for a cabinet that has a motor drive mounted in it. The alarm should use a thermocouple as the sensor and cause a light to come on when the overtemperature occurs. Your supervisor also tells you it would be nice to have a display to indicate the actual temperature the sensor is measuring. To save money, you are given an LED display that has been salvaged from an older piece of equipment. The LED display requires a 4–20 mA signal. You will need to design a circuit that can convert the millivolt signal from the thermocouple to a 4–20 mA signal that can be sent to the LED display.

Your supervisor is impressed with your ability and asks if it is also possible to design a simple circuit that could test the incoming voltage to the motor drive and set an alarm if the voltage is 5% too low or 5% too high.

8.1 OVERVIEW OF OPERATIONAL AMPLIFIERS

The operational amplifier (op amp) has been used as a linear amplifier since it was first introduced in the 1940s as a tube-type amplifier. Later in the 1950s when discrete solid transistors were being mass-produced, op amps were created by placing individual transistors together on circuit boards with capacitors and resistors to provide specific gain patterns. Many of these circuits were used in control circuits for rockets in the space program that first landed a man on the moon. Op amps were also used in missiles for the military as well as in aircraft circuits.

Later when integrated circuits (ICs) allowed multiple transistors to be manufactured right on a chip of silicon, the op amp was one of the first circuits developed. When the op amp was manufactured as an IC chip, it allowed the necessary internal resistors, capacitors, and transistors to be matched and produced on the chip. The IC is mounted in a housing that makes it easy to install in any circuit board application.

Today op amps are available with multiple circuits on one chip. For example, quad op amps are available that have four op amps mounted on one IC, and dual op amps are available that have two op amps mounted on the same IC. Op amps today are available with a variety of transistors such as the traditional PNP or NPN bipolar transistors, metal-oxide semiconductor field effect transistors (MOSFETs), junction field effect transistors (JFETs), complementary metal-oxide transistors (CMOS), and a combination of bipolar and JFETs (BIFETs).

Op amps are also designed for specific purposes such as instrumentation amplifiers, preamplifiers, comparators, and differentiators. This makes it easier to use op amps in these types of industrial circuits. This chapter will make it apparent how many industrial electronic circuits utilize op amps. The fundamental concept of each type of amplifier circuit will be explained so that you will be able to understand how they operate and what test will work when troubleshooting a circuit that has op amps.

8.1.1 What Op Amps Do

The easiest way to think about an op amp circuit is to think of a sensor signal like the millivoltage signal that comes from a thermocouple when it is heated. This voltage signal is so small that it is nearly useless. If the signal is sent through an op amp, its voltage can be increased to a value up to 20 volts and its current can be increased up to 50 mA, which would be large enough to energize the coil of a relay.

When a small signal is increased to a larger value, it is *amplified*. Amplification is necessary because many of the sensors that are used in industrial applications produce signals in the microvolt range (one-millionth of a volt) or in the millivolt range (one-thousandth of a volt). Many of these sensors such as the thermocouple were discovered in the late 1800s but they were not usable at that time because of the small amount of voltage they produced. After op amps were invented, a large variety of sensors became useful in industrial and medical systems.

The signal that is used as an input to the op amp can be ac or dc in nature, and the output will simply be a *larger* copy of the input signal. These signals will generally be linear, which means that the output can also be linear so that the entire range of signal voltage from minimum to maximum can be used.

8.2 BASIC OP AMP CIRCUIT

The schematic symbol for an op amp is shown in Fig. 8-1. From this diagram you can see that the basic outline of the op amp is a triangle that is placed on its side. The inputs are connected to the flat side of the triangle, and the output is connected to the point of the triangle.

The op amp needs to have a power supply because in reality the amplifier needs the power source to be able to increase the input signal to the strength that

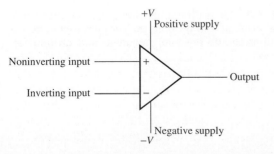

FIGURE 8-1 The schematic symbol for an op amp.

is necessary for the output signal to be useful. The power supply terminals will be identified on the op amp as the $+V$ and $-V$ terminals. All op amps need some type of power supply, but some diagrams will not show the power supply terminals as it is assumed that they are always there.

The input terminals on the op amp are identified by a + sign and a − sign. The + and − signs do not infer polarity, rather they indicate what will happen to the input signal when it is amplified and sent to the output terminal. For example, the − signal is called the *inverting input terminal*. If a positive signal is sent to the inverting input signal, the output would become inverted and it would be negative. Conversely, if a negative signal is sent to the inverting input, the output would be inverted and it would be positive.

The + terminal is called the *noninverting terminal*. This means that if a positive signal is sent to the noninverting signal, the output signal would not be inverted and it would remain positive. Conversely, if a negative signal is sent to the noninverting input, the output will also be negative.

8.2.1 The Op Amp Power Supply

The power supply for the op amp will be determined by the type of output signal the op amp is required to produce. For example, if the output signal needs to produce both positive and negative voltages (bipolar), then the power supply will need to have both positive and negative voltages and a ground available. If the op amp needs to produce only positive voltages, then the power supply can be a traditional positive and ground-type power supply.

The differential power supply provides a positive voltage, negative voltage, and ground used with op amps when the output signal is bipolar. This arrangement is basically two batteries that are connected in series. The point where the two batteries are connected together is the *ground point* that becomes the *reference*. This type of power supply was predominantly used when op amps were first introduced. The major advantage to this power supply is that it is possible for the output signal to be positive or negative with respect to ground. The main drawback of a differential power supply is that it is more expensive than simpler power supplies.

Recently applications that require only positive voltage with respect to ground have become more prevalent. When the output voltage is positive with respect to ground, the power supply for the op amp can be a traditional power supply that has a positive terminal and a negative terminal that is grounded. This means that the output voltage can be linear, but its polarity will always be positive in reference to ground. Fig. 8-2 shows these two types of power supplies. The traditional power supply with only a positive and negative terminal is shown in Fig. 8-2a, and the differential power supply that has a positive, negative, and ground is shown in Fig. 8-2b. The power supply voltage can be virtually any voltage between 10–30 volts.

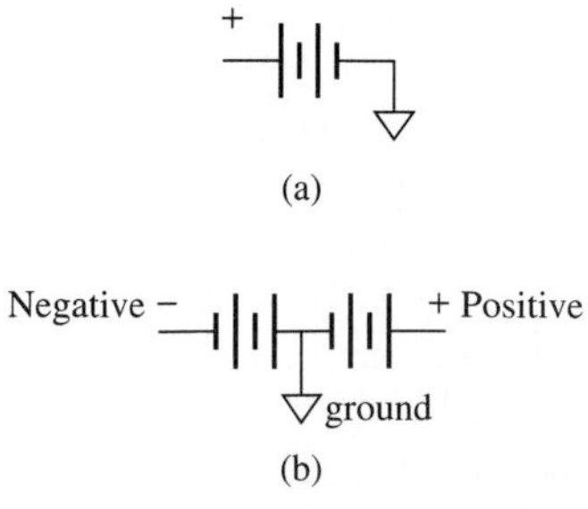

(a)

(b)

FIGURE 8-2 **(a)** A traditional power supply that has a positive terminal and ground. **(b)** The differential power supply that shows positive, negative, and ground terminals.

The more typical power supply voltages are $+/-$ 12 volts, $+/-15$ volts, and $+/-18$ volts. You should also notice the symbol for ground ($\downarrow$) is provided with the power supplies and it will be used in all the diagrams in this chapter to identify ground.

8.2.2 The Ideal Amplifier

The best way to understand the electronic operation of an op amp is to think of an ideal amplifier. The *ideal amplifier* would have infinite (very high) input impedance, infinite gain, and very low output impedance. Infinite input impedance is desirable because this means that the amplifier will not require the input sensor signal to have a large current, and it will not *load down* the sensor circuit. This enables sensors that produce very small microvolt signals to be used with the op amp.

Infinite gain is desirable so that the amplifier can use the full range of the power supply. This is important where the application requires the output to be linear and be able to reach all values from the minimum to saturation voltage, which is the maximum value of the power supply. We will see that in some circuits the actual amount of gain the op amp provides to the circuit can be controlled by external resistors that are connected to the op amp.

Low output impedance is necessary so that the op amp does not become loaded down when it provides voltage and current to the output device. This means that the op amp can act like a power supply and not affect the circuit that it is controlling.

8.3 THE COMPARATOR

The op amp can have components such as resistors and capacitors connected to its terminals to provide a wide variety of circuit functions. The simplest op amp circuit is called a *comparator circuit*. When the op amp has one input sensor signal connected to its inverting input terminal and a second input sensor signal connected to its noninverting input terminal, it will compare the voltage level of the two inputs and the output will be the amount of difference between the two signals multiplied by the gain. Since the gain for the op amp in this configuration is infinite, the slightest difference in voltage between the two inputs will cause the output to be driven to its maximum value. This maximum value is called *saturation*, and it will be approximately the full amount of the power supply voltage.

If the two sensor signals that are connected to the input terminals both produce a positive voltage, and the noninverting signal is larger, the output signal will be *positive saturation*. The amount of difference between the two signal voltages will not matter because the difference will be multiplied by gain, which is nearly infinite. If the inverting input is larger, the output will be *negative saturation*.

At first glance, the comparator circuit does not appear to be too useful because the output signal is not linear and it is always at saturation. This means that the output signal will always be maximum positive voltage, or maximum negative voltage, depending on which input is larger. Fig. 8-3 shows several modifications that can be made to the op amp circuit to make the comparator a useful circuit. For example, a resistor is connected in series with a cadmium sulfide cell (CdS cell) that is a light sensor. The sensor and the fixed resistor provide a voltage-drop circuit that is connected to the noninverting input, and a reference voltage is applied to

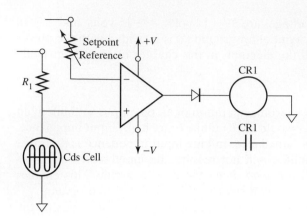

FIGURE 8-3 Electrical diagram of an op amp used as a simple comparator circuit. The Cds light sensor cell in the circuit will change resistance and alter the amount of voltage drop to the noninverting terminal.

the noninverting input through a variable resistor (potentiometer). A diode and the coil of a control relay are connected in series with the output terminal. The diode will ensure that only positive voltage will pass to the coil to energize it.

The potentiometer is adjusted to provide a reference voltage that will become the setpoint for the light sensor. For example, if the potentiometer is set so the reference is providing 10 volts to the inverting input, the comparator's output will be negative as long as the voltage supplied by the sensor to the noninverting terminal is less than 10 volts. As long as the op amp's output voltage is negative, the diode will be reversed biased and no current will flow to the relay coil. If the light conditions change and the sensor's resistance changes so that the voltage-drop circuit produces a voltage larger than the 10 volt reference voltage, the comparator's output will become positive and the diode will be forward biased. This will allow current to flow to the coil of the control relay. When current flows through the control relay coil, it becomes magnetized and closes the normally open contacts. An outdoor light can be connected to the control relay contacts, which means a variable resistor can be set to determine how dark it becomes before the light is turned on. If the reference voltage is increased, the resistance in the light sensor must increase to provide a larger voltage to the noninverting input to get the comparator to see it as the larger voltage and send positive saturation voltage to the output terminal.

This type of circuit could also be used for a variety of industrial applications such as determining when a heating or cooling system should be energized. The sensor in this type of circuit would be changed to a resistance-type temperature sensor such as a thermistor or resistive temperature detector (RTD). Fig. 8-4 shows an example of this type of circuit. You should also notice that two diodes are added to the output circuit. One is connected in forward bias so that positive voltage will cause the heating relay to be energized, and the other is connected in reverse bias so that the cooling relay will be energized when the op amp's output is negative. The contacts of the heating relay will energize electric heating coils and a fan. The contacts of the cooling relay will energize a motor starter, which will energize an air-conditioning system that includes a compressor and fan. You should also notice in both of the comparator circuit applications that the devices connected to the output terminal are relay coils, so they do care that the output signal is always at saturation. Since relay coils are essentially *on/off* devices, the saturation output is ideal.

When the RTD sensor is exposed to a temperature increase, its resistance

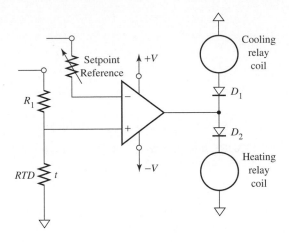

FIGURE 8-4 Electronic diagram of an op amp comparator circuit that has a heating and cooling relay coil connected to the output.

will increase. When the temperature increases and its resistance increases, the voltage across it will become larger. Since this voltage is supplied to the noninverting input and it is larger than the voltage that is connected to the inverting input, the output will be negative. This will forward bias the diode for the cooling system and the cooling relay coil will become energized.

If the temperature decreases, the resistance will decrease and the reference voltage at the inverting input will be the larger voltage and the output of the op amp comparator would be positive. The positive voltage will forward bias the diode for the heating system, which will allow current to flow through the heating system relay coil. The voltage supplied at the reference potentiometer will actually be considered a setpoint, and it will represent the temperature that will determine if the heating or cooling relay is energized.

When the input from the noninverting input is the larger voltage, the output will be positive saturation and the actual saturation voltage will be approximately 1 volt less than the supply voltage that is measured across $+V$ and reference (ground). If the input from the inverting input is larger, the output will be negative saturation. This time the actual saturation voltage for this configuration will be approximately 1 volt larger than the supply voltage measured across $-V$ and reference (ground).

8.4 THE VOLTAGE FOLLOWER

The op amp voltage follower connects its output directly back to its inverting input, which produces an output that is a copy of the input in intensity and polarity. This means that the voltage-follower circuit actually provides isolation from the input to the output. Since the output voltage is a copy of the input voltage, the gain will always be 1, which is also called *unity*. For this reason the voltage-follower circuit is sometimes called the *isolation amplifier* or *unity gain amplifier*. Fig. 8-5 shows the diagram for the voltage-follower op amp.

The heart of the op amp IC is the differential amplifier and its output is the differences between the voltages present at the inputs. Since the output of the voltage follower is the feedback signal to the inverting input, the difference between

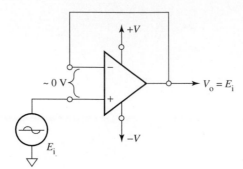

FIGURE 8-5 Electronic diagram of the voltage-follower op amp circuit.

the two inputs will be zero. This means that if a 5 volt signal is applied to the noninverting input, there will also be 5 volts at the inverting input coming from the feedback signal.

8.5 THE INVERTING LINEAR AMPLIFIER

When the op amp is used as a comparator, its output has two states: full on (saturation) and full off. If one resistor is added to the inverting input terminal of the op amp and a second resistor is added between the output terminal and the inverting input terminal to provide feedback, the output signal of the op amp will become *linear* and it will be *proportional* to the input signal. The size of the proportion will be called *gain*. The amount of gain for the amplifier will be determined by the ratio of the size of the input resistor R_i and the size of the feedback resistor R_f. This type of op amp circuit is called an *inverting amplifier*. It should also be noted that the inputs of the op amp have very high impedance, which is usually in the megohm range. This means that they sink virtually no current, and the input current is equal to the output current. The open-loop gain is also extremely high (approximately 100,000 Ω). This causes the difference between the voltages at the input pins to be less than a millivolt. This also means that if the noninverting input is connected to ground, the inverting input will be at *virtual ground*.

The formula for determining the gain of an inverting amplifier is:

$$A_{CL} = \frac{-R_f}{R_i}$$

where A_{CL} is the closed-loop gain.

The gain can be used to determine the size of the output voltage for any given input signal from the following formula.

$$V_o = \left(V_i \times \frac{-R_f}{R_i} \right)$$

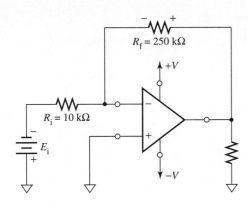

FIGURE 8-6 Electronic diagram of an inverting op amp.

Other useful formulas include:

$$V_v = V_i - I_i R_i \qquad V_v = V_o - I_i(-R_f) \qquad V_v = 0$$

$$V_i = I_i R_i \qquad V_o = I_i(-R_f)$$

$$\frac{V_o}{V_i} = \frac{I_i(-R_f)}{I_i R_i} \qquad \frac{V_o}{V_i} = \frac{-R_f}{R_i} = A \text{ (gain)}$$

where

V_o is the voltage at the op amp output.
V_i is the voltage at the inverting input of the op amp.
V_v is virtual ground at inverting input.
R_i is the input resistor.
R_f is the feedback resistor.
I_i is the input current.

Fig. 8-6 shows a diagram of the inverting op amp. From this figure you can see that R_i is 10 kΩ and R_f is 250 kΩ. It might be a good time to discuss some conventions used with op amps. The voltage on an op amp diagram may be identified with the letter V or the letter E. If the voltage is an input voltage, it will be identified as V_i or V_{in} and if it is an output voltage, it will be identified as V_o or V_{out}. Resistors will be identified as input resistors (R_i or R_{in}), feedback resistors (R_f), or load resistors (R_L).

· ·

EXAMPLE 8-1

Calculate the gain for the inverting op amp shown in Fig. 8-6. Also calculate the voltage at the output terminal if the input voltage is 0.1 volt.

SOLUTION

Calculate the gain using the following formula:

$$A = \frac{-R_f}{R_i} = \frac{-250 \text{ k}\Omega}{10 \text{ k}\Omega} = -25$$

Calculate the voltage at the output terminal using the following formula. (Remember the output voltage will be inverted because the input signal is received on the inverting input terminal of the op amp.)

$$V_o = \left(V_i \times \frac{-R_f}{R_f} \right)$$

$$V_o = (0.1 \times -25) = -2.5 \text{ volts}$$

8.5.1 Operation of the Inverting Amplifier

The operation of the inverting amplifier causes the output voltage signal to be an inverted amplified copy of the input signal. If the gain is 25 and the input voltage is 0.2 volt, the output signal will be −5 volts. The ratio of the input voltage to the output voltage will continue until the output voltage reaches saturation. Since the output signal is linear, it will be an exact copy of the input signal (except it will be amplified and its polarity will be reversed). The inverting op amp is very useful in amplifying small signals from sensors.

8.6 THE NONINVERTING OP AMP

The noninverting op amp is similar to the inverting op amp except that the inverting input is grounded and the noninverting input is not grounded. Fig. 8-7a shows an example of a noninverting op amp, and Fig. 8-7b shows a derivation of this op amp circuit. The input signal is applied to the noninverting input. The noninverting op amp provides a linear output signal like the inverting amplifier, and the ratio of the input resistance and the feedback resistance will determine the gain.

The inverting and the noninverting inputs have extremely high input impedance in the megohm range. This means that all of the output current will flow through R_{in} and R_f to ground. Since the op amp has extremely high open-loop gain in the range of 100000, the voltage on the inverting input pin and the noninverting input pin on the op amp will be within a millivolt or two of each other. This also means that the V_v at the inverting input will be equal to the voltage at the noninverting input. The formulas for gain and output voltage can be derived from this information, and you can see that these formulas are slightly different than the formula for the inverting op amp.

$$A_{CL} = 1 + \frac{R_f}{R_i}$$

$$V_o = V_i \times 1 + \frac{R_f}{R_i}$$

where
      A_{CL} is the closed-loop gain.
      V_o is the voltage at the op amp output.
      V_i is the voltage at the inverting input of the op amp.
      V_v is the point where voltage is at virtual ground.
      R_i is the input resistor.
      R_f is the feedback resistor.
      I_o is the output current.

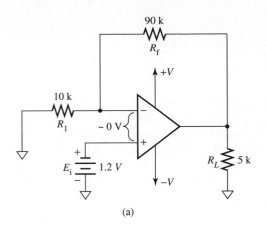

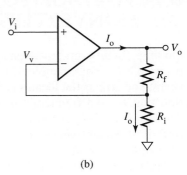

(b)

FIGURE 8-7 **(a)** Electrical diagram for a noninverting op amp. **(b)** Derivation of the noninverting op amp in Fig. 8-7a.

The diagram for the noninverting op amp is shown rearranged in Fig. 8-7b. This diagram is the exact circuit shown in Fig. 8-7a, but the R_i and R_f resistors are now shown so that their series circuit relationship can be better seen. When you think of R_i and R_f in series with each other, you get a better idea of how the formula for determining the gain of this amplifier is derived. The following formula shows how the final formula for gain is derived.

$$V_o = I_o(R_f + R_i) \qquad V_v = I_o R_i$$

$$A = \frac{V_o}{V_i} \qquad A = \frac{I_o(R_f + R_i)}{I_o R_i} \qquad A = \frac{(R_f + R_i)}{R_i} \qquad A = \frac{R_f}{R_i} + 1$$

EXAMPLE 8-2

Calculate the gain for the noninverting op amp shown in Fig. 8-7a. Also calculate the voltage at the output terminal if the input voltage is 1.2 volts.

SOLUTION

Calculate the gain using the following formula:

$$A_{CL} = 1 + \frac{R_f}{R_i} = 1 + \frac{90\,k\Omega}{10\,k\Omega} = 10$$

Calculate the voltage at the output terminal using the following formula. (Remember the output voltage will not be inverted because the input signal is received on the noninverting input terminal of the op amp.)

$$V_o = V_i \times \left(1 + \frac{R_f}{R_i}\right)$$

$$V_o = 1.2 \times 10 = 12\text{ V}$$

8.7 VOLTAGE-TO-CURRENT CONVERTERS

Op amp circuits can be designed for specific functions by adding resistors at appropriate places. For example, the major difference between the inverting and noninverting op amp is the location of the resistors and the input signal. Another application that is used frequently in industrial electronic circuits is a voltage-to-current converter. This type of circuit is useful because the input voltage can be converted to output current where the output current remains linear with the input voltage. This op amp circuit is especially useful in applications where the sensors produce a variable voltage and the control circuit requires a changing current such as a 4–20 mA signal.

The 4–20 mA signal is very popular in industrial process controls for several reasons. First, the signal is a current signal, which is rather strong and not subject to transients that are continually occurring in the area. Second, since 4 mA is the low end for the signal, it is above zero current. This means that anytime the current is less than 4 mA, the circuit is malfunctioning. For example, if the current is 0 mA, it can be presumed that the circuit has an open. If the circuit used 0–20 mA as the span, it would be difficult to determine if 0 mA was showing the lowest signal, or if the circuit had an open.

Fig. 8-8 shows the diagram for a voltage-to-current amplifier circuit. In this diagram you can see that the load is connected in the feedback loop and the input voltage signal is applied to the inverting input. This means that the load resistance is also the feedback resistance. In this configuration the load current will also flow through the input resistor that is connected to the noninverting input. This will cause the voltage drop at the noninverting input to change with the change in the input voltage signal at the inverting input. This basically keeps the output current linear with the input voltage signal.

The formula for calculating the current through the load is determined in the following steps.

First,

$$I_{R_i} = \frac{V_{in}}{R_i}$$

Second,

$$I_{R_i} = I_{load}$$

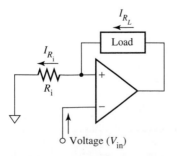

FIGURE 8-8 The electronic diagram of a voltage-to-current converting op amp circuit.

Therefore,

$$I_{\text{load}} = \frac{V_{\text{in}}}{R_{\text{i}}}$$

8.7.1 Applications for the Voltage-to-Current Converter

One of the frequent applications for the voltage-to-current converter is in a thermocouple transmitter for instrumentation circuits. In this application, the thermocouple will produce a signal that is between 0–70 mV when it is heated. This small-voltage signal will be connected to the noninverting input terminal where it is used as the input signal. The output for this circuit will be 4–20 mA. When the thermocouple is heated, its voltage will change and the output current will also change between 4–20 mA. This allows the thermocouple signal that is generally too small to be used as the sensor to provide feedback to instrumentation heating control systems.

Another useful application for this type of circuit is in sensors that use piezo-electric elements to produce small amounts of microvolts when they have force (pressure) exerted on them. These types of sensors are used in accelerometers and other applications to measure a variety of torques.

8.8 CURRENT-TO-VOLTAGE AMPLIFIERS

A similar type of circuit to the voltage-to-current converter amplifier is the current-to-voltage amplifier. In some applications, the sensor's signal will be a current-type signal and it will need to be interfaced with a circuit that is expecting a variable-voltage signal. The current-to-voltage amplifier is used to provide this linear conversion. Fig. 8-9 shows the electrical diagram for this type of op amp circuit. To understand the operation of the current-to-voltage converter, you must remember that the inverting input has extremely high impedance so all of the current from the current source will flow through the feedback resistor R_{f}. The noninverting input for this circuit is tied to ground and because it has high open-loop gain, the inverting input is virtually grounded. Even though the inverting input is grounded, it cannot sink any current. This means the output voltage across the load resistor is equal to the input current times the feedback resistor ($V_{\text{o}} = I_{\text{i}} \times R_{\text{f}}$), which causes the output voltage to be directly proportional to the input current.

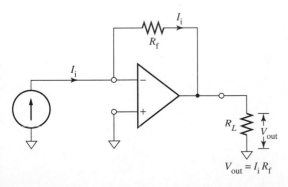

FIGURE 8-9 Electrical diagram of an op amp used as a current-to-voltage converter.

$$V_{\text{out}} = I_{\text{i}} R_{\text{f}}$$

8.9 SUMMING AMPLIFIERS

Since many control applications can be solved by mathematics, it is natural for the electrical op amp circuit to duplicate the mathematical functions. For example, the function of gain is sometimes called *proportional control*. This means that the mathematical function of *multiplication* is used to determine the output signal. Another useful mathematical function used to provide control is called *summing* (adding). The summing op amp circuit can take a number of voltage inputs and *add* them together to use as a single input voltage. Fig. 8-10 shows a typical summing amp circuit. From this diagram you can see that three signals represented by three voltage sources are connected in parallel to the inverting input. The op amp adds (sums) the voltage of all of these input signals and treats them as a single voltage input. The formulas for calculating the output for the summing amp are determined in the following steps. (Note that all of the R_i resistors are equal and identified as R.)

First,

$$I_1 = \frac{E_1}{R} \qquad I_2 = \frac{E_2}{R} \qquad I_3 = \frac{E_3}{R}$$

Second,

$$V_o = -(I_1 + I_2 + I_3) \times R_f$$

If the feedback resistor R_f is made equal to all of the input resistors (R), the formula can be written as:

$$V_o = -\left(\frac{E_1}{R} + \frac{E_2}{R} + \frac{E_3}{R}\right) \times R$$

All of the R's can be factored out, so the new formula is:

$$V_o = -(E_1 + E_2 + E_3)$$

This means that if the input voltage signals are 0.3 volt, 0.5 volt, and 0.8 volt, the output voltage for the summing amp would be -1.6 volts $-(0.3 + 0.5 + 0.8)$.

FIGURE 8-10 Electronic diagram of a summing amp.

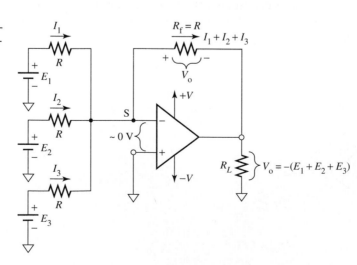

(Note: The polarity of the output signal is negative because the input voltage is applied to the inverting input.) This type of circuit is useful in applications where multiple sensors each produce a voltage that must be added to get the total effect of the process change.

8.10 OP AMPS USED AS INTEGRATORS

Op amp integrators have a variety of industrial applications such as replacing the early pneumatic (air) controlled process valves and actuators. It may be easier to understand how the op amp provides integration if you can understand the need for integral control. As process control applications became more complex, it became necessary to use integration to get the controls to provide more accurate regulation. Integration is a calculus function that measures the amount of the error and uses time to change the output signal to reduce the error. *Error* is the difference between the desired value (setpoint) and the actual value measured by the sensor. For example, if the temperature of an industrial heating process with a setpoint of 280° is controlled by a proportional steam valve that is controlled 0–100%, the integral control would use the actual temperature from the sensor and compare it to the setpoint of 280°. If the output signal to the steam valve is 40%, the actual temperature is 260°, and the setpoint is 280°, the integral controller would determine that the output signal should be increased to provide more heat. The integral controller would *regulate* how fast the change should occur over a given time.

When the op amp is designed to use *time* to determine the amount of change in the output, *capacitors* that take time to store a charge must be added to the circuit. The exact amount of time for the capacitor to reach full charge is determined by the size of the capacitor and the resistance connected to it. The resistor and capacitor make up a traditional time constant. Fig. 8-11 shows a diagram of the op amp integrator where a capacitor is added to the feedback part of the op amp circuit. As the input signal changes, the capacitor will be forced to change its charge to this new value, and the time it takes for the capacitor to change will be the *integral time*. The smaller the integral time, the faster the output will change.

More specifically, you can see from the diagram that the noninverting input is grounded. This means that the inverting input is virtually grounded. If the input voltage remains fixed, the input current will be constant. (The input usually presented to an integrator is a rectangular voltage pulse, with constant amplitude, for a time *t*.) From the definition of a capacitor (charge stored per unit of voltage),

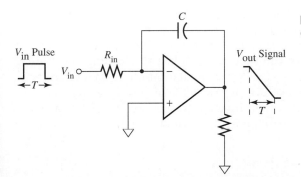

FIGURE 8-11 Electronic diagram of an op amp integrator circuit.

you can see the output voltage can be derived directly from the integrator in terms of the input current and time the input voltage is applied. The output voltage is linearly proportional to the time the input voltage is applied and the output is a negatively sloped linear voltage ramp.

The formula for calculating the output for an op amp integrator is:

$$E_o = - \frac{1}{R_1 \times C_1} \int E_I \, dt$$

where

E_o is the output of the op amp integrator.
E_I is the input voltage of the op amp integrator.
$R \times C$ is the time constant for the circuit.
dt is the change (delta) in time.

It should be noted that if a sine wave is used as the input signal to an integrator, the output will have a 90° phase shift from the input.

Other useful formulas:

$$I_{in} = \frac{V_{in}}{R_{in}} \qquad C = \frac{Q}{V}$$

$$V = \frac{Q}{C} \qquad \frac{dV}{dt} = \frac{1}{C}\frac{dQ}{dt} \; (C = \text{capacitance} = \text{constant})$$

$$\frac{dQ}{dt} = I_{in} = \text{constant} \qquad \frac{dV}{dt} = \frac{1}{C} I_{in}$$

$$\int dV = \frac{1}{C} I_{in} \int_0^t dt \qquad V = \frac{1}{C}\frac{I_{in}}{C}(T - 0) \qquad V = \frac{I_{in}}{C} T$$

8.11 OP AMPS USED AS DIFFERENTIATORS

A differentiator is a circuit that performs the mathematical operation of differentiation. The differentiator circuit provides a function that is inverse to the integrator. This means that since the output of the integrator is proportional to the length of time the input signal has been present, the output of the differentiator will be the inverse of the time the signal has been present at the input. Another way to say this is that the output of the differentiator is proportional to the rate of change for the input signal. The differentiator can be used to detect the leading and trailing edges of a rectangular pulse, or to produce a rectangular output from a ramp input.

The differentiator op amp is shown in Fig. 8-12. In this figure you can see that a capacitor is used to provide a time function to the circuit, so the capacitor is connected in series with the inverting input. The formula for the differentiator is:

$$E_o = R_F C \frac{dV_i}{dt}$$

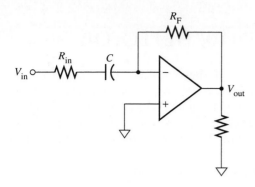

FIGURE 8-12 Electronic diagram of the op amp differentiator.

where

E_o is the output from the op amp.
R_f is feedback resistance.
C is capacitance.
dV_i is the change (delta) in input voltage.
dt is the change (delta) in time.

It should be noted that if a sine wave is used as the input signal to a differentiator, the output will have a 90° phase shift from the input.

A better understanding of the operation of the differentiator can be gained from the following information.

At the end of the ramp, the capacitor voltage is

$$V = Q/C$$

Divide both sides by the ramp time.

$$\frac{V}{T} = \frac{Q/T}{C} \qquad \frac{V}{T} = \frac{I}{C} \qquad \left(\frac{Q}{T} = I\right)$$

Solve for current.

$$I = \frac{CV}{T}$$

where

I = capacitor current
C = capacitance
V = voltage at end of ramp
T = time between start and end of ramp

The output voltage can now be calculated:

$$V_{out} = -IR$$

If we again look at the capacitance voltage,

$$V = \frac{Q}{C} \qquad \frac{dV}{dt} = \frac{1}{C}\frac{dQ}{dt} \qquad \frac{dV}{dt} \qquad \frac{dV}{dt} = \frac{1}{C}i$$

From the input voltage we see that its slope is constant; therefore, the capacitance voltage is constant. Remember that the noninverting input is grounded and, therefore, the inverting input is virtually grounded.

8.12 THE OP AMP WINDOW DETECTOR

When two or more op amps are used in a circuit, they provide a wide range of new circuit functions. One widely used application is called a *window detector*. This circuit is useful to monitor power supplies to determine if their voltage goes above or below a predetermined upper and lower limit (window). For example, it is very important that TTL power supplies maintain their voltage at a level that is no greater than 5.5 volts and no less than 4.5 volts.

The window detector circuit uses two op amps connected as comparators to test for voltage that exceeds an upper and lower level. Fig. 8-13 shows a diagram of the window detector circuit designed to detect if the power supply voltage for a TTL power supply exceeds the upper limit of 5.5 volts or becomes less than the lower limit of 4.5 volts. The op amp in the top of the diagram is connected to detect voltage that exceeds the 5.5 volt limit, and the op amp in the bottom part of the circuit is connected to detect when voltage falls below the 4.5 volt limit. The power supply voltage is identified in this circuit as E_1. It is connected to the inverting input of the op amp that tests for the upper limit, and it is connected to the noninverting input of the op amp that tests for the lower limit.

8.12.1 Operation of the Window Detector (Comparator)

If the power supply voltage is between 4.5–5.5 volts, the output of the two op amps will be added together, which will be sufficient to energize the output lamp. If an alarm is needed to indicate when the voltage is out of the upper and lower range, a relay coil could be used in place of the lamp, and an alarm could be wired through the normally closed contacts of the relay.

The easiest way to get a better understanding of this circuit is to show its operation with different voltages applied. For example, if the power supply voltage is 5 volts, this value will be less than the setpoint voltage of 5.5 volts for the upper limit that is applied to the noninverting input of the top op amp. Since the setpoint

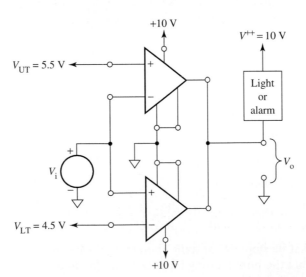

FIGURE 8-13 The electronic circuit diagram for a window detector using two op amps that are connected as comparators.

voltage at the noninverting input is larger, its output voltage will be positive saturation. This same 5 volts is applied to the noninverting input of the bottom op amp. Since the larger voltage is the 5 volts from the power supply that is applied to the noninverting input, its output will also be positive saturation.

If the power supply voltage drops below 4.5 volts, the setpoint voltage applied to the inverting input of the op amp at the bottom will be the larger voltage, and the output will become negative saturation. Since the power supply voltage is still less than 5.5 volts, the output of the top op amp will remain positive saturation. The positive saturation voltage from the top op amp and the negative saturation voltage from the bottom op amp will cancel each other and their total will not be sufficient to keep the lamp on or to keep a relay coil energized. The alarm would be sounded because it is wired through the normally closed contacts of the relay.

If the power supply voltage increased above the 5.5 volt limit, the op amp testing the lower limit would provide a positive saturation voltage for its output, and the exceeding voltage at the top op amp would cause the inverting input to be the larger voltage, which would cause its output to become negative saturation. Since the positive and negative saturation voltages are of equal value, they would cancel each other and the output would not be sufficient to keep the lamp illuminated. Or if a relay is used, its coil would not remain energized and the alarm would become powered.

8.13 ADDING ZERO AND SPAN TO AN OP AMP CIRCUIT

When an op amp is used as a signal transmitter or converter, many times the signal must be zeroed and spanned to match existing signals. For example, when a signal transmitter is used to convert the millivolt signal from a thermocouple so that it is compatible with the 4–20 mA signal for an instrumentation circuit, the potentiometer on the op amp must be adjusted so that the lowest thermocouple temperature causes the op amp output to be exactly 4 mA, and the highest thermocouple temperature causes the op amp output to be exactly 20 mA.

When the op amp is adjusted so that it puts out the 4mA output signal at its lowest value, it is called *setting zero*. Even though 4 mA is not 0 mA, it is still considered zero for this circuit because it is the lowest setting. Since 4 mA is above ground, it will also be called the *offset* for the output signal. The top value for the output signal is set for 20 mA, and it is called the *span setting*. In actuality, the span for this transmitter circuit is 16 mA, which can be determined by subtracting the lowest value of 4 mA from the highest value of 20 mA.

The locations of the zero and span potentiometer in the op amp circuit are shown in Fig. 8-14. From the op amp in this figure you can see that the potentiometer at the noninverting input is connected to a voltage supply to provide the offset voltage. In this case the offset voltage will also set the *zero* point for this circuit. The potentiometer is also used to set the feedback resistance for this circuit, which will set the amount of gain for the circuit. This in turn sets the *span*, which is the upper level for the output signal.

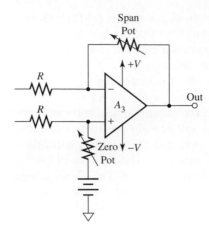

FIGURE 8-14 Electronic diagram of a zero and span potentiometer added to an op amp circuit so the op amp can be calibrated for zero and span.

8.13.1 Setting Zero and Span

At times you will be requested to adjust the zero and span potentiometers for a transmitter. It is important to understand that you need to be able to manipulate the input signal so that it can actually be set at its highest and lowest values. Since this is not always practical or possible, a power supply that is rated for calibration duty can be used to duplicate the highest and lowest voltages the input signal would provide.

The first step of the zero and span process is to set the power supply to provide the lowest voltage the input signal would provide. When this input is applied to the op amp, the zero potentiometer (pot) should be adjusted so that the output is exactly 4 mA. Next the calibration power supply should be set to apply the largest voltage the input sensor can produce. When this voltage is applied to the input of the op amp, the span pot should be adjusted so that the op amp output is 20 mA.

When the span pot is adjusted, it will generally raise or lower the zero voltage slightly, which means that you will need to repeat the steps for setting the zero level again. Each time the zero pot is adjusted, it will alter the span setting slightly, so the span adjustment will need to be completed again. Each time the zero and span pots are adjusted, they will each affect the other so the process must be completed several times until all error is removed. This can usually be accomplished in two to three cycles.

8.14 THE DIFFERENTIAL AMPLIFIER

The differential amplifier uses a single op amp to measure and amplify the difference between the inputs. This type of basic op amp circuit is used in conjunction with other amplifier circuits to provide more complex functions such as instrumentation amplifiers. Since this amplifier must measure the difference between two input voltages, it is important that the input resistors are precision-matched resistors and that the feedback resistor R_f is matched with resistor R_D.

The formulas for calculating the gain of this amplifier is:

$$m = \frac{R_F}{R_i}$$

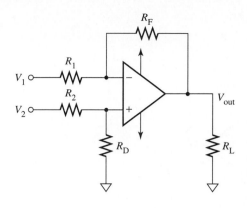

FIGURE 8-15 Electronic diagram of a basic op amp differential amplifier.

where m is differential gain, R_f is feedback resistor, and R_i is input resistors. (Note that both R_i resistors are the same value.)

The output voltage for the differential amplifier is found by determining the difference between the voltage at E_1 and E_2 and multiplying it by the gain. The formula for the output is:

$$V_o = m \times (E_1 - E_2)$$

where m is differential gain.

. . . .

EXAMPLE 8-3

Calculate the output signal for the op amp in Fig. 8-15 if the difference between voltage V_1 and voltage V_2 is +0.6 volt. The value of R_f is 90 kΩ and the value of each R_i input resistors is 10 kΩ.

SOLUTION

Step 1. Calculate the gain using the following formula.

$$m = \frac{R_F}{R_i} \qquad m = \frac{90 \text{ k}\Omega}{10 \text{ k}\Omega} \qquad m = 9$$

Step 2. Calculate the output voltage using the following formula.

$$V_o = m \times (E_1 - E_2) \qquad V_o = 9 \times (0.6 \text{ volt}) \qquad V_o = 5.4 \text{ volts}$$
. . . .

8.15 INSTRUMENTATION AMPLIFIERS

One of the largest growing applications for op amps is their use in instrumentation. The basic op amp circuits that use individual op amps can be modified slightly to combine several op amps to take advantage of each of their strengths. Fig. 8-16 shows a circuit where three op amps are connected to make a single instrumentation amplifier. Their individual output signals are also sent to the inputs of the third op amp so it will operate as a differential amplifier. A potentiometer is added to the third op amp at the noninverting input. This potentiometer provides a reference

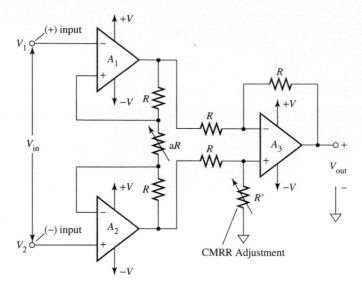

FIGURE 8-16 Electronic diagram of three op amps used to provide an instrumentation amplifier circuit.

directly to ground and will be used to balance out the *common-mode* voltage. This will provide a function that is similar to providing the offset for the zero and span amplifiers. The R_A potentiometer is used to balance the input impedances and gains for both the A_1 and A_2 amplifiers.

The instrumentation amplifier circuit is a differential amplifier optimized for high input impedance and a high common-mode rejection ratio (CMRR). From the diagram in the figure, you can see the inputs are voltage followers, and they have very high input impedance. This effectively isolates the input from the output. The second stage is a differential amplifier with a potentiometer to adjust the CMRR. Keep in mind the IC op amp has as its heart a differential amplifier. Noise picked up in the circuit is usually the same in both input lines. This is termed *common mode*. A differential amplifier will subtract out this noise. This means that the output of the instrumentation amplifier is both isolated and clean.

▶ S O L U T I O N   T O   J O B   A S S I G N M E N T

Your solution to the job assignment should include an op amp that can amplify the millivolt signal from the thermocouple to a voltage that is large enough to energize a relay coil. Since this is an alarm circuit, your solution should use a circuit like the comparator circuit in Fig. 8-3. Since the signal from the thermocouple is a voltage, it can be connected directly to the noninverting input.

Your solution to the second part of the assignment should include a voltage-to-current amplifier like the one shown in Fig. 8-8. You may also need to add the component to the circuit to provide a zero and span capability like the circuit shown in Fig. 8-14.

Your solution to the part of the assignment that requests an alarm circuit be made to determine if the drive voltage is 5% too high or 5% too low should include a window detector circuit as shown in Fig. 8-13. You will need to use a transformer to take a sample of the incoming voltage to the drive, since it is three-phase high voltage. Additional alarm circuits will be required if you are to monitor the voltage of all three phases. The secondary winding of the transformer will provide a voltage in the range of 5–10 volts that will be compatible with the op amp. The transformer will also provide isolation from the three-phase voltage. You will need to use a calibration power source to set the incoming voltage to the drive to 5% above and 5% below the normal voltage range so you can adjust the reference voltages on the window detector.

QUESTIONS

1. Sketch an op amp and identify the inverting input, noninverting input, $+V$ supply voltage, $-V$ supply voltage, and output terminals on the diagram and explain the function of each.
2. Explain the terms *amplification* (gain) and *saturation*.
3. Describe the ideal amplifier.
4. Use the diagram of a summing amp and explain its operation.
5. Use the diagram in Fig. 8-14 to explain the function of zero and span and how you would adjust each for a 0–10 volt output signal.
6. Use the diagram of a differential amplifier and explain its operation.

TRUE OR FALSE

1. _____ The polarity of the output for an inverting op amp will be positive when the input voltage signal is positive.
2. _____ If the output signal of the op amp is -12 volts to $+12$ volts, the power supply for the op amp will need to be a differential-type power supply.
3. _____ The ideal amplifier would have low input impedance, infinite gain, and high output impedance.
4. _____ In the window detector circuit shown in Fig. 8-13 the alarm will be sounded anytime the E_1 voltage becomes larger than 5.5 volts or smaller than 4.5 volts.
5. _____ The zero and span potentiometers shown in Fig. 8-14 provide a means to adjust the minimum and maximum values for the output voltage signal.

MULTIPLE CHOICE

1. If the voltage from $+V$ to ground on a differential power supply is $+14$ volts, you would expect to measure _____ volts from $-V$ to ground.
    a. $+28$ volts
    b. $+14$ volts
    c. -14 volts

2. If the signal to the inverting input terminal of the op amp is negative, the output signal will be _____.

   a. negative

   b. positive

   c. impossible to tell

3. If the setpoint voltage at the inverting input for the op amp in Fig. 8-4 is +1 volts and the signal from the temperature sensor provides +2.5 volts to the noninverting input, _____ relay will be energized.

   a. the heating

   b. the cooling

   c. impossible to tell which

4. If the voltage E_1 is 0.35 volt, E_2 is 0.15 volt, and E_3 is 0.25 volt, for the op amp in Fig. 8-10, the output signal is _____.

   a. +0.75 volt

   b. −0.75 volt

   c. impossible to calculate because the size of each resistor is not shown and the amount of gain is not provided.

5. If a differential amplifier has a feedback resistor R_f that is 80 kΩ and has input resistors R_i that are 10 kΩ each, the differential amplifier will have a _____ and if the differential voltage between the two inputs is 0.8 volts, the output signal will be _____.

   a. gain of 8, 8 volts

   b. gain of 6, 6.4 volts

   c. gain of 8, 6.4 volts

PROBLEMS

1. Calculate the output voltage for an inverting amplifier if the input voltage is 0.4 volt, R_f is 90 kΩ, and R_i is 10 kΩ.

2. Calculate the output voltage for a noninverting amplifier if the input voltage is 1.2 volts, R_f is 120 kΩ, and R_i is 8 kΩ.

3. Determine the output signal for a summing amp if it has 0.5 volt, 1.2 volts, and 1.6 volts input voltages applied to it.

4. Draw the diagram of an integrator and explain its operation.

5. Draw the diagram of a differentiator and explain its operation.

6. Draw the diagram of a window detector and explain its operation.

REFERENCES

1. Coughlin, Robert F., and Fredrick F. Driscoll, *Operational Amplifiers and Linear Integrated Circuits*, 2nd ed. Englewood Cliffs, NJ: Prentice-Hall, Inc., 1982.

2. *Linear Circuits Operational Amplifiers Data Book*. Texas Instruments Incorporated, Literature Center, P.O. Box 809066, Dallas, TX 75380-9066, 1992.

3. Maloney, Timothy J., *Industrial Solid-State Electronics*, 2nd ed. Englewood Cliffs, NJ: Prentice-Hall, Inc., 1986.

4. *Operational Amplifiers Databook*. National Semiconductor Corporation, 2900 Semiconductor Drive, P.O. Box 58090, Santa Clara, CA 95052-8090, 1993.

9

OPEN-LOOP AND CLOSED-LOOP FEEDBACK SYSTEMS

OBJECTIVES

When you have completed this chapter, you will be able to:

1. Explain the difference between open-loop and closed-loop systems.
2. Explain the terms *gain, reset,* and *rate.*
3. Explain the terms *proportional, integral,* and *derivative.*
4. Draw the block diagram for a typical servo (feedback) system and identify the setpoint (SP), process variable (PV), error, summing junction, controller amp, and output (final control element).
5. Discuss the difference between manual mode and automatic mode.
6. Explain the term *differential gap* (dead band) as it refers to an on-off controller.
7. Explain the term *bump.*

8. Discuss the term *bumpless transfer.*
9. Explain how a controller that uses proportional only (P only) will respond.
10. Explain how a controller that uses proportional, integral, and derivative (PID) will respond.
11. Explain the term *bias* (offset) and how it is used to help the P-only controller.
12. Explain the term *proportional band.*
13. Explain the operation of a ratio control system for an on-off heating contactor.
14. Explain how a loop is tuned.
15. Explain how you would determine the proper value of gain, reset, and rate for a new closed-loop system.
16. Explain the function of process alarms.
17. Explain the function of deviation alarms.

299

(Courtesy of ABB Instrumentation)

You are requested to convert several process systems that are presently using pneumatic controls to the newer digital-type controllers. After you have installed the new digital controllers, you are asked to determine what values of gain, reset, and rate (PID) should be placed in each controller. The first system you are working with is a batching tank used to store a product whose temperature is increased from 75°F to 150°F over a 20-minute period. This system uses a thermocouple as the sensor and a steam valve to control the amount of steam used to increase the temperature of the product in the tank.

The second system is used to control the level of product in the tank. A float is used as the level sensor and it controls a variable valve. Your job is to explain to your supervisor how the new system operates. You will also need to explain how deviation and fixed alarms function in this system and you will be expected to set these alarm points after the process engineer provides you with the data. You will need to explain to the process engineer how you will use either the open-loop or closed-loop method to determine what the gain and reset (proportional and integral) values should be for each of these new control systems.

9.1 OVERVIEW OF OPEN-LOOP AND CLOSED-LOOP SYSTEMS

Industrial control systems will be operated as open-loop or closed-loop systems. The first industrial open-loop and closed-loop control systems used pneumatic controllers. During the 1960s to 1980s, most controllers used op amps to provide the control functions. In the 1980s and 1990s microprocessors were incorporated with op amps to provide digital control. Since all these types of controllers are still in use today, you will need to know how to operate and troubleshoot them all. All these controllers use the functions of *gain, reset,* and *rate,* which is also called *proportional, integral,* and *derivative* (PID) to control the way the output changes.

This chapter will provide information about PID control and provide examples of open-loop and closed-loop systems.

9.2 PARTS OF A TYPICAL CONTROL SYSTEM

It is important to understand that the basic parts of any control system will have the same names and provide the same functions regardless if the controller is pneumatic, op amp, or a microprocessor-based system. Fig. 9-1 shows the basic parts of a control system shown in a block diagram. It will be easier to understand these basic parts of a control system if they are applied to a specific application. An electric heating system that is used to dry paint is shown in Fig. 9-2 and it will be used for this explanation. From the block diagram you can see that the system starts with a *setpoint* (SP) signal. The SP is the desired value or the desired temperature. For example, if we wanted the desired temperature of the heating system to be 280°F, the SP would be adjusted to 280°F.

The next part of the control system that we will examine is the *process variable* (PV), which is the signal that comes from the sensor. For example, in the heating system, the PV is the signal from the thermocouple which is the sensor for this system. The PV is also called the *feedback signal* and it is the present value or actual value of the temperature at the instant the sensor reading takes place. Since the sensor reading is continuous, the PV will change continually to indicate the changing temperature of the system.

The *summing junction* is the place in the control system where the SP is compared to the PV. This means that if the SP is 280°F and the PV signal indicates the actual temperature is 270°F, the difference is 10°F. The summing junction is identified by the Greek letter sigma (Σ).

The difference between the SP and the PV is called *error*. The error can be a positive value if the SP is larger than the PV, or it can be a negative value if the SP is smaller than the PV.

The *controller* will use gain, reset, and rate to adjust the output signal in response to the amount of error. Gain, reset, and rate are also called *proportional, integral,* and *derivative* (PID). The PID values can be adjusted to change the speed of response for the system. For example, a gain value can be used to make the output change the temperature at a given response rate of 1°F per minute, or it

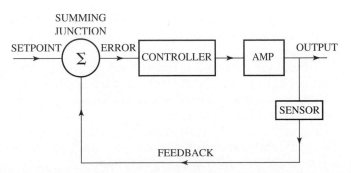

FIGURE 9-1 Block diagram of a control system.

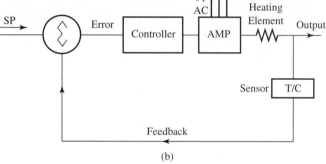

FIGURE 9-2 (a) An industrial electric heating system used to dry paint. This system is called a paint-drying oven. **(b)** The control diagram for the industrial paint-drying oven. (Courtesy of Fostoria Industries.)

can be set to provide a change of 3°F per minute. The function of gain, reset, and rate will be explained in detail later in this chapter.

9.2.1 An Electric Heating Application

The electric heating system in this application is a paint-drying oven. The paint-drying oven is shown in Fig. 9-2a and the control diagram of the oven is shown in Fig. 9-2b. In this application automotive trim parts are manufactured in a plastic press and then placed on a continuous overhead conveyor line. The conveyor line moves the parts through a paint booth where robots apply paint to the parts. The freshly painted parts are then moved through an electric paint-drying oven. The paint-drying oven uses electric heating coils to provide the proper amount of heat for the oven.

From the control diagram you can see the components for the parts of the basic control system. The controller for this application is a single-point electronic controller that has a keypad where the SP is entered by the operator. The sensor for this system is a thermocouple and it is also called the process variable. The SP is compared to the PV signal at the summing junction, which is inside the controller. The signal that comes from the summing junction is called the error signal, and it is sent to the controller part of the system.

The controller sends an output signal to an amplifier that controls the electric heating elements. The amplifier for this system is a three-phase current controller that adjusts the amperage to the electric heating elements from 0–100% with a 0–10 volt dc signal. The controller sends the 0–10 volt signal as an output signal to the amplifier.

9.3 A TYPICAL OPEN-LOOP SYSTEM

The block diagram for the control system shown in Fig. 9-1 is called a *loop* because a sensor is used to sample the temperature and send its PV signal from its sensor back to the summing junction where it can be compared to the SP. The sensor signal is also called the *feedback* signal because it gets fed back to the summing junction. When this happens, it is called *closing the loop,* and the system is called a closed-loop system.

If the feedback signal is not used, the system is called an *open-loop system.* In most systems it is possible to use a switch to control the PV (sensor) signal that is used for feedback. If the switch is closed, the sensor signal is used as feedback, and if the switch is open, the sensor signal is not used. This switch is used to determine if the loop is operated in open-loop or closed-loop mode. When the system is in open-loop mode, the output is adjusted manually by the operator. For example, in the paint-drying oven, the operator can set the output signal to 50%, which will provide 50% full energy to the heating element. The actual amount of heat the elements provide must be observed by the operator and the operator must decide if temperature adjustments are necessary. The drying oven is usually placed in open loop during the warm-up period for the oven or during system troubleshooting or calibration.

It is also important to understand at this point that the sensor signal has several names that are used interchangeably. These names are *process variable (PV) signal, feedback signal,* and *sensor signal.* One important distinction should be made at this time. The *process variable* is the temperature that is being changed, and the *process variable (PV) signal* is the signal that comes from the thermocouple that indicates what the temperature is.

Servo system is a name given to a wide variety of control systems that use feedback. These systems can usually be broken into the two broad categories of motion control and process control. Motion control servo systems can use linear motion devices such as a rack and pinion or a ball screw to provide linear motion, and they can use servomotors and gears to provide circular motion. Motion control systems are usually associated with robots, machine tools, and variable-speed motor drive applications. These types of systems will be covered in depth in a later chapter. This chapter will primarily use process control systems as examples because they are easy to visualize. Typical industrial process control systems such as controlling temperature and liquid levels will be used extensively in this chapter.

9.3.1 Manual Mode and Auto Mode

The terms *manual mode* and *auto mode* are sometimes used in place of open loop and closed loop. When the system control is placed in manual mode, the loop is

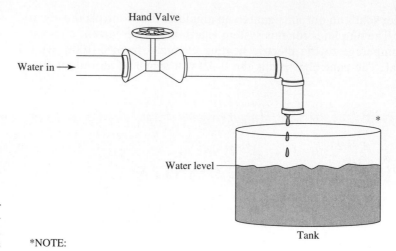

Hand Valve

Water in →

Water level

*

Tank

FIGURE 9-3 An example of an open-loop water-level control system. The operator controls the opening and closing of the water valve when the system is operated in manual mode.

*NOTE:
Operator must visually check water level when system is in manual control.

in open-loop operation, and when the system control is placed in auto mode, the loop is in closed-loop operation. This means that the switch that controls the feedback signal is also called the *auto/manual switch*. It may be easier to understand the operation of the loop if it is in manual mode (open loop) when you study an example.

An example of a typical open-loop system in industry would be an operator filling a tank with water. Fig. 9-3 shows an example of this type of system and you can see that when the operator manually opens a valve, water starts to fill the tank. When it is determined the water is at the correct depth, the operator turns the valve off. If additional water was needed, the operator would turn the water on again and put more water in the tank. This system is called an open-loop system for several reasons. First, the operator must *manually* open the water valve if more water is needed in the tank, and close the water valve when the proper amount of water has been added. Second, the operator *looks* into the tank to determine if more or less water is needed, and then opens or closes the water valve rather than the controller comparing the sensor signal to the setpoint. Open-loop systems tend to be very simple and used frequently in industrial applications because they are inexpensive to operate.

This same system can be operated as a closed-loop system if the sensor signal is sent to the controller and the *controller* determines the value for the output signal. The valve that controls the amount of water flowing in the tank needs to be an electrical or pneumatic valve that the output signal from the controller can adjust from open to closed.

9.3.2 An Example of a Sump Pump Control

Another rather simple system can be used to explain the major difference between open-loop and closed-loop control. In this control system a sump pump is turned on to pump out a sump when the level of the liquid in it gets too deep. Many industries have a sump, which is a large hole in the floor that collects waste water

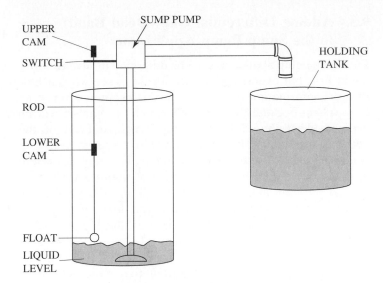

FIGURE 9-4 Diagram of a sump pump application.

or waste cutting oil. When the liquid level in the sump rises enough, the liquid must be pumped out of the sump into a holding tank. If this system was an open-loop system, someone would have to periodically check the level of the sump and turn the pump on when the level gets too high. When the pump lowers the level sufficiently, the pump would be manually turned off.

In a closed-loop system, a sensor would be connected to the switch mechanism to turn the pump on and off automatically. This sensor and control mechanism could be something as simple as a float that is mounted on the bottom of a rod. In Fig. 9-4 you can see that the rod is mounted on the sump pump and it activates the switch when the float moves the rod up or down with the liquid level. Two cams are attached to the rod at the point where the rod goes through the pump switch activator. One of the cams is attached to the rod just above the switch arm and the other cam is connected to the rod just below the switch arm.

When the liquid level rises in the sump, the float will move up with the level and the lower cam will force the switch arm up to turn the pump on. When the pump begins to pump the liquid into the holding tank, the level in the sump lowers, the float will move down, and the cam that is mounted on top of the switch arm will press the switch arm down to turn off the pump. When the liquid level rises again, the float will rise and turn on the switch again, and the process is repeated again. The float and rod provide the *process variable* (*feedback*) signal for this system.

9.3.3 On-Off Control

A traditional *on-off control* is used to control the level of water in the sump. This type of control can turn the output (final control element) on or off. In some cases the level where the pump is turned on and off is so close that the motor will begin to cycle on and off rapidly. This action is called *short cycling* and it can be harmful to the pump. The difference between the point where the pump is turned on and where the pump is turned off is called the *differential gap* and the sump pump system can be modified to allow the differential gap to be fully adjustable.

9.3.4 Adding Differential Gap (Dead Band) to the Sump Pump Application

The sump pump application can be described in several ways, which may be confusing at first, but eventually you will begin to see why each description is necessary as more information is provided in this chapter. First, it may be called a *closed-loop system* because the float and switch mechanism will turn the pump on automatically when the level in the sump is high, and turn off the pump when the level becomes low again. The system may also be referred to as a *simple on-off control system* because the pump motor can only be energized where it is running at 100%, or it can be de-energized where it is not running at all.

A differential gap can be designed into the sump pump system by modifying it so that it has two cams attached to the float rod. Fig. 9-5a shows an example of two cams set for a small differential gap and Fig. 9-5b shows an example of two cams set for a large differential gap. The points on the rod where the upper and lower cams are set are called the *setpoints*. If the upper cam is moved higher, the level of the liquid will be allowed to fill to a higher level before the pump is turned on. If the lower cam is set lower, the level in the sump will be allowed to go lower before the pump is turned off.

The distance between the upper and lower cam on the rod is called *differential gap* or *dead band*. If the distance between the upper cam and the lower cam is increased, the differential gap (dead band) will be larger. The larger the dead band, the larger the difference will be between the highest and the lowest levels the liquid. If the differential gap is large, the motor will not cycle as quickly and it will remain running longer once it is energized. The only drawback to a larger differential gap is that the level in the sump will vary from nearly full to nearly empty. This is not a problem with the sump pump application because the pump is trying to keep the sump empty. The differential gap would have to be smaller if the control system is used to control the level of water for a mixing application where the control system is used to measure the amount of water or level in the tank. In this type of control the amount of water needs to be controlled to a specific level, and a small differential gap would accomplish this.

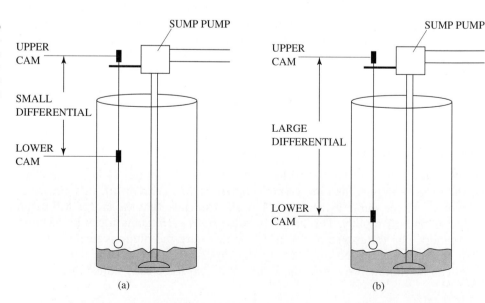

FIGURE 9-5 (a) Two cams set for a small differential for the sump pump application. **(b)** Two cams set for a large differential for the sump pump application.

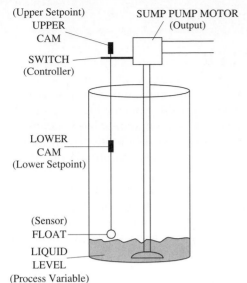

FIGURE 9-6 The sump pump application from Fig. 9-4 with its basic parts identified; the upper cam is the upper setpoint, the lower cam is the lower setpoint, the switch is the controller, the sump pump motor is the output device, the float is the sensor, and the water level is the process variable.

It is also important to be able to identify the basic parts of the process system such as setpoint (SP), process variable (PV), sensor, controller, and output for each system that you encounter. Fig. 9-6 shows these parts identified for the sump pump application. The upper cam is the upper SP, and the lower cam is the lower SP. The PV in this system is the liquid level and the float is the sensor that indicates the liquid level. Since the float sends the signal that indicates the level of the liquid in the sump to the switch, the sensor signal will be called the PV signal. The switch is the controller in this system and the pump is the output device or final control element that can change the liquid level by pumping the liquid into the holding tank and lowering the liquid level in the sump.

9.3.5 Automating the Sump Pump Application

The sump pump could become more automated if a microprocessor controller was used as the controller. A liquid-level sensor is used to indicate the level of the liquid in the sump instead of the float. The sensor sends a voltage signal to the controller to indicate the exact level of the liquid in the sump. This is the process variable (feedback) signal. An operator must enter a number into the controller to indicate the desired level (setpoint) for the liquid in the sump. The controller compares the setpoint and the process variable, and the difference between them is the error.

A process control diagram was previously presented in Fig. 9-1. In this figure you can see the setpoint (SP) and the process variable (PV) are compared in the controller at a point called the summing junction (Σ). The error signal comes from the summing junction and represents the difference between the SP and the PV. The error signal can be positive or negative depending on whether the actual level is higher or lower than the SP. The controller uses the error signal to determine when to turn on the output (pump motor). The controller calculates the error continually and when it determines the liquid level is above the SP, it will send a signal to turn on the pump. As the liquid level is pumped down, the controller calculation at the summing junction will determine that the level is below the SP and send a signal to turn the output off.

9.4 SINGLE-POINT PROCESS CONTROLLER

Prior to the advent of electronic devices such as op amps and microprocessors, closed-loop controllers were originally designed to control pneumatic (air) signals because pneumatic signals were easy to adjust between a minimum and maximum signal. After the op amp was perfected during the 1960s, most closed-loop controllers were designed as dedicated single-circuit controllers that used operational amplifiers to compare the SP and the PV.

During the 1980s, inexpensive microprocessors became readily available, so newer controllers now use op amp and microprocessor technology together and the controller is completely programmable. Fig. 9-7 shows a picture of this type of single-point process controller, which usually has two windows where numbers and letters can be displayed to show the amount of the SP and PV. These controllers also have a simple keypad with arrows that allow numbers to be changed up or down as the operator scrolls through a series of menu choices to enter setpoint, alarm points, and other operational values.

9.4.1 Operating a Controller in Manual and Auto Modes

Controllers are designed to allow the operator to switch the loop between open-loop control and closed-loop control. When the controller is set in manual mode, it is operating as an open-loop control system and the operator changes the output percentage to any value between 0% to 100% that is desired by pressing the up or down arrow keys on the face of the controller. The controller does not make any adjustments to the output on its own when it is in manual mode, even though the sensor continues to send a process variable (PV) signal, and the controller continues to display the PV. The operator may use the value displayed in the PV window to determine what the level is when adjusting the output percentage between 0–100%.

FIGURE 9-7 Typical single-point controller. (Courtesy of ABB Instrumentation.)

When the controller is set in auto mode, it becomes a closed-loop controller and the PV signal is compared to the setpoint (SP) to determine the error. The controller uses the error in closed-loop mode to continually adjust the output. In this manner, the controller will automatically change the output to satisfy the SP.

9.5 BUMPLESS TRANSFER

When a controller is running in manual mode and then switched to auto mode, the controller will immediately begin to compare the SP to the PV and determine the amount of error. When this occurs, it is important for the PV to be as near to the SP as possible so that the error is as small as possible. For example, in a temperature controller this means that the operator must be very careful and ensure that the temperature is increased in manual mode to a point where it is very near the SP. In some industries where the operator must tend to more than one machine, it is not practical for the operator to stay with the system until the process temperature is raised to the SP. In these systems where the controller must be placed in auto mode when the process temperature is well below the SP, a function called *bumpless transfer* is used to ensure that the SP and the PV are exactly the same when the controller is changed from manual to auto mode so that error is zero.

Bumpless transfer in its simplest form changes the SP to match the PV anytime the processor is changed from manual mode to auto mode. This ensures that the error is zero when the switch is made, and the controller will not be allowed to go into a condition that may cause the temperature to oscillate to extremes. For example, when a plastic injection molding machine is first turned on, its barrel will be at room temperature (approximately 75°F). The operating temperature for the barrel heaters is approximately 600°F. If the operator placed a setpoint of 600°F into the controller when the unit is first turned on and the barrel temperature is at room temperature, the amount of error would be 525°F ($E = \text{SP} - \text{PV}$). This would be a large error and it would cause the controller to have problems keeping the system under control until the barrel temperature reaches the SP.

The condition of excessive error can be controlled by bumpless transfer. When the operator first starts up the injection molding machine, the controller is set to manual mode and the output is set to 50%. This means the electric heating elements called band heaters receive 50% of the total possible energy and the temperature of the barrel begins to increase to approximately 400°F. At some point the operator changes the controller from manual mode to automatic mode. When this occurs, the controller changes the SP to match the actual temperature at that time. The actual temperature is provided to the controller from the thermocouple as a feedback signal.

Since the controller changes the SP to match the actual temperature from the sensor (PV) at the time when the controller is switched to auto mode, the SP will need to be adjusted later to the final control temperature. When the operator is ready, the SP can be changed to 600°F and the controller will add heat to increase the barrel temperature to 600°F, which is the normal operating temperature. If the controller would have been set in auto mode when the barrel temperature was room temperature (75°F), the 525°F error between room temperature and 600°F may have been too large and caused the controller to become unstable. If the process temperature was raised to 400°F while the controller is in manual mode, the error would be smaller and the controller stands a better chance of bringing the system to 600°F without going out of control.

9.6 VALVE POSITIONING SYSTEM TO CONTROL WATER LEVEL IN A TANK

The valve positioning system in Fig. 9-8 shows an electrical system that is combined with an op amp to provide a signal to a current to pressure (*I/P*) transmitter. The transmitter receives a 0–10 volt signal and sends a 3–15 psi air signal to open or close a water valve. When the valve is opened, additional water will flow into the tank and change the level of the float that is used as a sensor to indicate the water level in the tank. The water level in the tank is the PV for this system, and the valve is the output device.

The SP and PV (feedback) signal are both connected to movable arms on potentiometers. The voltage from these two potentiometers is used as the input for an op amp. The op amp is used as an error amp and its gain can be determined by the size of the feedback resistance and the input resistance (see Chapter 8 for formulas). The SP is entered for the system by adjusting the SP potentiometer. Since the voltage from that potentiometer will be different from the voltage that

FIGURE 9-8 Diagram of a servo system that controls the level of water in a tank.

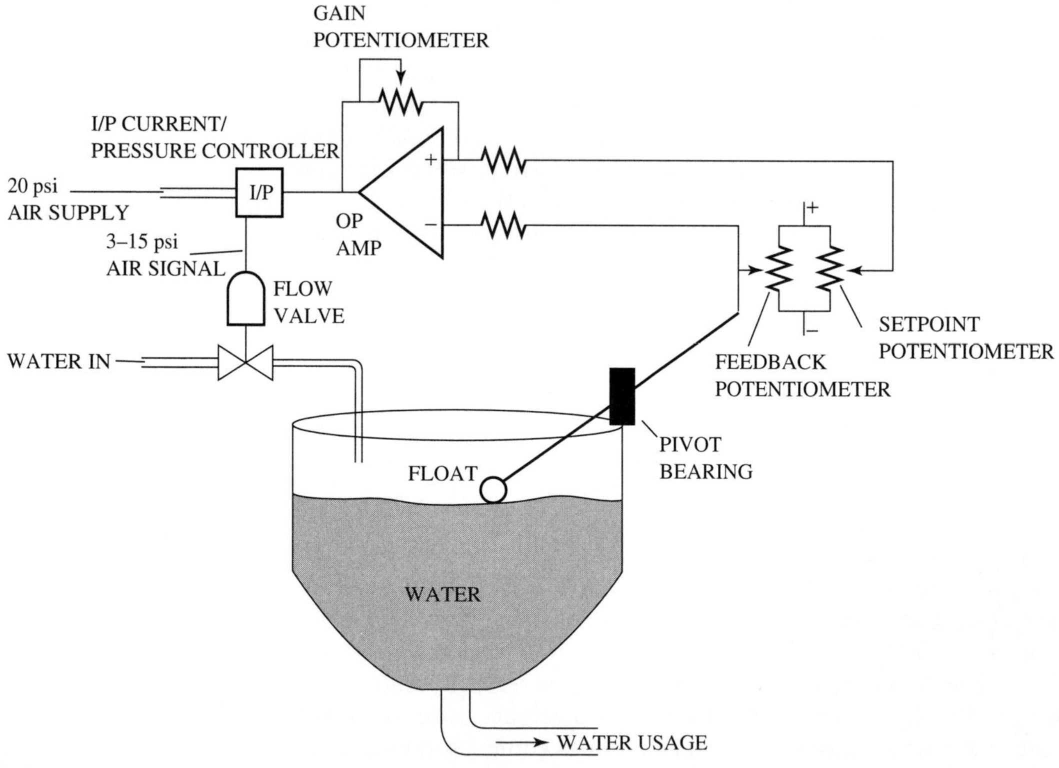

comes from the feedback potentiometer, the op amp will send a signal to the *I/P* transmitter that indicates the amount of error.

When the *I/P* transmitter receives the error signal, it will send a change of air pressure to the valve, opening farther to allow more water to flow into the tank, and the tank level will increase. When the tank level starts to increase, the float will rise and change the position on the feedback potentiometer. The voltage from the feedback potentiometer will increase, the amount of difference between the SP voltage and the feedback voltage will become smaller, and the output signal will decrease.

When the feedback signal becomes the same voltage as the SP signal, the error will be zero. Since the error is zero, the voltage to the valve would become zero. The valve for this application must hold its position when voltage is at zero, and it will increment (add to the position) or decrement (subtract from the position) anytime the valve position is increased or decreased.

The rate of response can be controlled by changing the amount of gain. This means that a variable resistor (potentiometer) is used as the feedback resistor on the op amp and it is called the *gain pot*. If the resistance of the potentiometer is increased, the gain will increase and the response will be faster.

9.7 HOW FAST CONTROLLERS WILL CHANGE THE OUTPUT

All controllers (pneumatic, op amp, and microprocessor) have some method of determining the speed at which they will adjust the output signal (between 0–100%) when some amount of error is detected. This is usually accomplished by circuits that control the *gain, reset,* and *rate* for the controller. In the older pneumatic controllers these factors were controlled by adding springs and levers to allow more or less air to be sent to the output device. If the full amount of air (15 psi) is sent to the output it will operate at 100%, and if the minimum amount of air (3 psi) is sent to the output, the output will operate at 0%. The op amp controller uses capacitors to provide some amount of time delay for the functions of gain, reset, and rate, which will determine the speed at which output changes.

Microprocessor controllers use mathematical calculations called *algorithms* to determine the amount of adjustment that must be made to the output signal when it receives a change in the error signal. The error signal is determined by comparing the SP to the PV signal from the sensor.

It may be easier to understand why a system must adjust the speed response of the output by using an example of a simple heating system. In this example the controller is trying to control the temperature in a laboratory testing furnace. The temperature may not vary too much unless someone opens the door to the furnace to add a new test specimen. If the furnace is operating at 450°F and the door is opened, the temperature may drop 15°F while the new test sample is placed inside. This drop in temperature will be seen as error and the controller will change the output accordingly to try and get the temperature back to 450°F. The technician can set the gain, reset, and rate variables to make the controller change the output immediately or gradually, depending on the type of sample that is being tested. Some types of materials may have limitations that require that their temperature may be increased at a rate of 1 degree per minute, so the controller response can be set by the gain value to regain the lost 15°F slowly over 15 minutes.

Another type of material may be able to absorb the change at a rate of 10°F per minute, and the gain can be changed to regain the lost temperature in 1.5 minutes. The nice thing about the microprocessor is that different gain values can be programmed in for each different type of sample that is being tested.

9.8 UNDERSTANDING GAIN, RESET, AND RATE

The terms *gain, reset,* and *rate* are functions that determine how fast the controller will change the output signal. They are also called *modes of operation* and their functions were developed with the early pneumatic controllers and vacuum-tube controllers and later refined with op amp controllers. These basic modes of controller response were based on mathematical formulas (algorithms) that have been used for many years to solve complex problems. These formulas are derived from the calculus functions *proportional, integral,* and *derivative* (PID). Today companies that make process control equipment will use the terms *gain, reset,* and *rate* interchangeably with the terms *proportional, integral,* and *derivative.* For the person trying to learn process control theory, this tends to make it more difficult. The table in Fig. 9-9 shows the relationship of these terms. The terms *gain, reset,* and *rate* originally referred to controller operation. *Proportional, integral,* and *derivative* refer to the math functions used to make the controller perform the actions of gain, reset, and rate.

Another problem with the terms *gain, reset,* and *rate* and the terms *proportional, integral,* and *derivative* has occurred with the advent of the microprocessor chip. Since the original calculation in the microprocessor uses formulas, these formulas can be changed slightly to provide different types of controller response when new models are produced each year. This means that you may find several different responses when using the gain function from different brand names of controllers or from different models of the same controller from year to year because a different formula to calculate the gain may have been used.

This may not be a problem for someone learning process control for the first time, but it is definitely a problem for someone who learned how a proportional controller or integral controller operated with op amps or pneumatic controllers. The function of proportional control is rather fixed and limited with the op amp or pneumatic controller and people became familiar with this type of response. In modern microprocessor controllers, the response the manufacturer calls proportional may have enhancements to the calculation the microprocessor performs so that the controller response reacts differently than traditional proportional control. The enhancements were added through the years to make the controller provide

GAIN   —   PROPORTIONAL

RESET   —   INTEGRAL

RATE   —   DERIVATIVE

FIGURE 9-9 Comparison of the PID terms *proportional, integral,* and *derivative* to the terms *gain, reset,* and *rate.*

better response and control. This means that you only need to understand the basic functions of gain, reset, and rate or proportional, integral, and derivative to determine how this response will react. From these basic functions you will be able to determine the specific detailed differences between brand names of controllers when you start to use them in the field.

9.8.1 Using Gain for Control

To make it easier to understand the function of gain, we will use a microprocessor controller for our examples. Later you will be able to apply the term to the op amp and pneumatic controllers.

The simplest type of control to understand is called *gain,* where the controller uses gain as a *multiplying factor.* The controller uses a mathematical calculation called the algorithm to compute the amount of output for each change of error. This means that the processor is continually checking (sampling) the value from its sensor, which is the PV and it compares this value to the SP. This comparison takes place at the summing junction and the result is called error. It should be pointed out at this time that the function of the summing junction is performed by a calculation and in older controllers this function was performed by op amps. The formula for error is:

$$\text{Error} = \text{Setpoint} - \text{Process Variable}$$

or

$$E = \text{SP} - \text{PV}$$

Note that some applications will use the formula:

$$E = \text{PV} - \text{SP}$$

The error can be a positive number or a negative number, depending on the amount of the values. For example, if the SP for a laboratory furnace is 450°F and the sensor indicates the actual temperature (PV) is 445°F, the error would be a positive 5°F. If the PV temperature is 455°F and the SP is 450°F, the error would be a negative 5°F. In this application the controller is designed to change the output so that heat is added to the furnace if the error is positive, and if the error is negative the controller is designed to turn off the heat source.

The controller takes the error from the summing junction and uses the algorithm to calculate the amount of output. The algorithm can be a complex formula but it can be simplified as shown.

$$\text{Output} = \text{Gain} \times \text{Error}$$

$$M_o = K_c \times E$$

In the formula, the output is referred to as M_o, the gain is referred to as K_c, which stands for controller (c) constant (K), and E is error. These terms have become standards for the Instrument Society of America (ISA). ISA was formed to standardize and promote the process control and instrumentation industry. Some companies that design process control and motion control equipment may use slightly different designations for the variables in these formulas.

The following example will help explain how the error and gain are used by the controller to change the output to respond to the changes to the system. At the start of this example, the temperature (PV) inside a laboratory furnace is room temperature (75°F) and the SP is 75°F. The amount of error at this point is zero,

SP	− PV	= Error	Error	× Gain	= Output
100°F	− 75°F	= 25°F	25	× 2	= 50%
100°F	− 80°F	= 20°F	20	× 2	= 40%
100°F	− 85°F	= 15°F	15	× 2	= 30%
100°F	− 90°F	= 10°F	10	× 2	= 20%

FIGURE 9-10 Table that shows the calculations the controller makes to continually change the output as the PV changes.

and the controller sets the output to zero. The technician starts the oven by entering a new SP at 100°F and sets the gain to a value of 2 in the controller. When the controller sees the new SP of 100°F, the error is calculated (100 − 75 = 25). The controller uses the error of 25 and multiplies it times the gain of 2 and the output is set to 50%. You should remember that the output can only be set to values between 0% and 100%.

When the output is set to 50%, the heating element begins to add heat to the furnace and the sensor continually sends the PV signal to the controller to indicate the increasing temperature. The controller is constantly recalculating the changing error and setting the new output. The following table shows a progression of these calculations as the controller samples the change in the PV and recalculates the output. The controller uses a *sample time* to determine how quickly to recalculate the output signal. The sample time may be fixed, or it may be adjustable in some controllers. Fig. 9-10 shows a table of the calculations the controller makes.

It should be noted that error is actually a percentage of the SP-PV over the total number of degrees (span) the system can control. For this example, the span is set for 100°F so the number of degrees of error is also the percentage of error.

From the calculations in Fig. 9-10 you can see that the controller continually reduces the percentage of output as the PV temperature inside the furnace increases and the temperature gets closer to the SP. The controller will need approximately 30% output to sustain the 85° temperature, and if the output is lowered below 30% the temperature will begin to decrease. This means that the controller will "level out" at some point and hold the output steady.

9.8.2 Problems with Using Only Gain for Control

The *gain-only* controller is also called the *proportional-only* controller. The calculations in the previous table show several inherent problems with gain-only control. First, the PV will never reach SP because the error is continually getting smaller as the temperature inside the furnace gets closer to the SP. This problem can be overcome somewhat by changing the amount of gain (multiplying factor). The next table in Fig. 9-11 shows the results if the gain factor is changed to 4.

When the gain is doubled from 2 to 4, the output is also doubled for each calculation. Remember, if the sample time for the controller is set for 1 second, the controller is making a new calculation every second. For simplicity, the values in the table only show the calculations at intervals of specific error points.

The numbers in the table show the progression of the changes in the output due to the action of gain. To start this process the technician makes the changes to the SP in the controller. This change in the SP is called a *bump*. The word *bump*

SP	− PV	= Error	Error	× Gain	= Output
100	− 75	= 25	25	× 4	= 100%
100	− 80	= 20	20	× 4	= 80%
100	− 85	= 15	15	× 4	= 60%
100	− 90	= 10	10	× 4	= 40%

FIGURE 9-11 Calculations for controller when gain has been changed to 4.

does not sound too technical but it is very descriptive since it refers to any change in the SP or PV. If the technician changes the SP, it is called a bump, and if the door to the furnace is opened and the temperature drops 10°F it is also called a bump. A bump may be thought of as any disturbance to the PV or the SP. When the technician bumps the SP from 75°F to 100°F, the controller will begin calculating the output using the new gain of 4. You can see that the output is initially set to 100% this time and the heating element will supply the full amount of heat. As the PV indicates the temperature is increasing, the output is also reduced more quickly because of the increase in gain. The result is still the same. The gain-only controller cannot get the temperature to the SP. Each time the gain is increased, the controller will get the PV closer to the SP, but because the output is always reduced when the error gets smaller, the controller will always tend to level out before the process temperature reaches the SP.

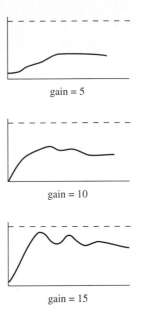

gain = 5

gain = 10

gain = 15

FIGURE 9-12 Three graphs that show the response curve for the heating system with gains of 5, 10, and 15.

If the gain is increased, the amount of output is increased for each change of error. This trend is useful to a point. Fig. 9-12 shows the result of increasing the gain in the form of several graphs. The first graph shows the original PV for the system and the response of the PV as the output is changed when the gain is 5. The second graph shows the response when the gain is increased to 10, and the third graph shows the response when the gain is increased to 15. From these graphs you can see that each time the gain is increased, the PV will get closer to the SP (dashed line). But you should also see that as the gain gets larger, the PV temperature begins to *oscillate*.

When the controller causes the system to oscillate, the PV temperature will actually go up and down. At first, as the gain is relatively small, the oscillations will be small, but eventually as the gain is increased, the system will begin to oscillate and the process temperature will surge above the SP and then well below the SP. Fig. 9-13 shows the results of too large a gain as the PV temperature oscillates out of control. Since the temperature at the top of the oscillation is well above the SP, and the temperature at the bottom of the oscillation is well below the SP, this type of control is not only unacceptable, it is also dangerous. The oscillations will continue to grow larger if the gain is increased or they will diminish if the gain is reduced and the PV will end up short of the SP.

The following four statements can be made about gain-only (proportional-only) control.

1. The PV will never reach the SP.
2. The larger the gain is, the closer the PV will get to the SP.
3. As gain is increased, the PV will oscillate more.
4. As gain is increased, the system will reach a point where the PV will oscillate out of control.

9.8.3 Using a Gain-Only Controller

Since the gain-only controller cannot get the PV to reach the SP, some adjustments must be made to use it successfully. One way to use a gain-only controller is to have a system that is repetitive. For example, if the technician started the process furnace every day and the SP was only bumped once each day, a gain and setpoint could be determined that would bring the furnace up to a steady-state temperature of 100°F. Since the steady-state temperature will be below the SP, the trick would be to choose the SP high enough above 100°F so that the steady-state temperature would settle out at 100°F. Since the system is always started from 75°F (room

FIGURE 9-13 Graph of large oscillations caused by the gain value becoming too large.

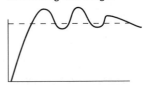

temperature), and the steady-state temperature is always 100°F, the SP might be 110°F so that the temperature would always *line out* at 100°F. Since the system is so consistent, the setpoint and gain could be adjusted slightly to ensure that the steady-state temperature of 100°F could be reached. When gain is changed, the rate of change for the output signal would also be changed, which would make the furnace gain temperature faster if it is increased, and slower if it is decreased.

9.9 ADDING OFFSET (BIAS)

Another way to make the gain-only controller usable is to add *bias* to the algorithm. Bias is also called *offset,* and it is some value between 0–100% that is added to the output after the gain calculation brings the controller to the steady-state mode. This will help overcome the problem of the gain-only controller not being able to get the PV to SP. The formula for determining the output (M_O) with bias (M_X) added to the original formula is:

$$(K_C \times E) + M_X = M_O$$

The bias is also called offset because it provides the function of adding offset value to the output device. For example, previously when the PV temperature in the laboratory furnace was 75°F and the SP was changed to 100°F, the gain factor in the formula could only get the process temperature to about 85–90°F. At this temperature, the controller reached a steady-state point where the output stabilized at approximately 35–40%. From trial runs in manual mode, the technician has determined that the furnace required approximately 48% output for the PV temperature to reach 100°F. The technician can begin adding bias to the controller which would add value to the output percentage and make it increase.

If a value of 10 was added as bias, the output would rise from 35% to 45% and the process temperature would rise to approximately 96°F. Since the PV temperature is closer to the SP, the value from the ($K_c \times E$) part of the formula will be reduced because the error is smaller. In fact when the PV temperature is exactly at the SP, the value from the ($K_c \times E$) part of the formula will be zero because the error is zero. At this point, the value that the controller sends the output must all come from the bias term. When the technician is originally setting up the controller for the furnace, bias could be continually added until the temperature is exactly 100°F. After the bias term has been determined, it can be written down to be used again. If the furnace was brought from room temperature to 100°F each time, the bias value could be left in the controller and the temperature would reach the SP each time the furnace was turned on.

9.9.1 Problems with Adding Bias Manually

Several problems occur when the bias is changed manually in the controller. One problem with leaving the bias value in the controller is that if a new SP such as 120°F is needed, the amount of bias will not be enough, and it will have to be increased. Conversely, if the SP was lowered to 90°F, the amount of bias would be too much and the temperature would overshoot. This means that the bias will need to be changed for every SP. Since the technician is present to enter the new SP, the bias can be changed at that time.

Another problem with putting bias in the controller and leaving it in is that it will be added to the output signal from the very start when the SP is changed. This would cause the output to change at a faster rate than it normally would.

9.10 EXAMPLE OF AN INDUSTRIAL SYSTEM THAT USES GAIN AND RESET

Before you begin the explanation of adding reset control to gain control, you should have an example of this type of control in mind. Fig. 9-14 shows a picture of a plastic injection molding machine. The injection molding machine is also called a plastic molding press. The temperature at the barrel of the press must be controlled within +/− 5°F at setpoints from 300°F to 600°F. The reset control helps the temperature controller hold the temperature to within +/− 5°F.

9.10.1 Adding Integral (Reset) to the Formula to Change Bias Automatically

When the integral is added to the original formula, the type of control is called PI (proportional and integral) control. *Integral* is a mathematical term that is used in controllers to determine the amount of error remaining anytime the PV does not equal the SP. In mathematics, you may have completed calculations using integration to determine the *amount of error under the curve.* The controller uses the integral in a similar manner. To be precise, the integral term is added to the original algorithm with the $K_c \times E$ term and its job is to *automatically change the bias* (offset) at a time period specified by the integral time. The formula for a proportional and integral (PI) controller can be slightly different from controller to controller. It is also important to understand that a formula for positioning systems (like a

FIGURE 9-14 A plastic injection molding machine that uses reset control with gain control to hold the temperature close to setpoint. (Courtesy of Cincinnati Milacron Plastic Machinery Group.)

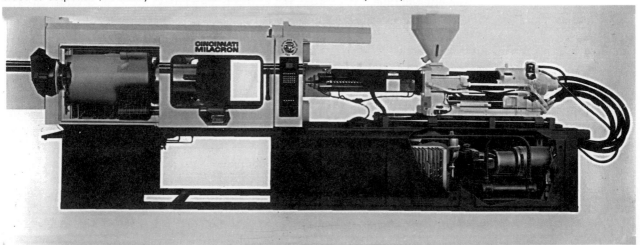

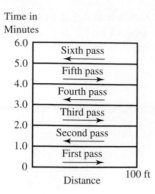

Time in
Minutes

FIGURE 9-15 Example of mowing a yard to show the function of integral action in a controller.

system that controls a valve positioner) would be different from a formula for motion systems (like the velocity controller for a robot drive). The formula for the valve positioning controller would allow the output to remain at any value between 0–100% when the SP is reached, while the motion controller system for the robot drive would set the output to 0 when the robot reached its final position (SP).

A simple way to understand the function of integral is to think about the way you mow the lawn. Fig. 9-15 shows a diagram of the way you would mow your lawn. You can see that you would mow from left to right on the first pass, and then turn around and repeat the next pass from right to left. You would continue walking back and forth across your lawn until all of the grass was cut. In this example we will only look at the first two passes. Let's say that it took 1 minute to make the first pass. If you change the time it takes to make the second pass, the speed that you must walk must also change. For example, if you mow the second pass in 30 seconds (0.5 minute) you would need to walk twice as fast, and if you took 120 seconds (2 minutes) to mow the second pass you would need to walk half as fast.

For this example we will equate the time it takes to mow the first pass to the response the controller gets from gain-only control. This means the gain action causes the first amount of change. The integral action then simply repeats the amount of change (mowing the second pass) in some given amount of time. The integral action continually adjusts (repeats) the output signal by the same amount of the original change caused by gain until the PV equals the SP. You can see that if the integral time is decreased, the repeat (second pass) must occur more quickly. This means the output signal would have to be increased to ensure that the process could move fast enough. If the integral time is increased, the time it takes to repeat the first pass would be slowed down. From this example you can now see that the integral part of the formula will continually adjust the output signal after gain causes the first amount of change in the system after the SP is changed. If the integral value in the controller is small, the amount of change caused by the action of gain will be repeated more rapidly and the controller will respond more rapidly. If the integral value in the controller is made larger, the amount of change caused by gain will be repeated more slowly.

You should notice that at this time we have not actually given any values to integral, rather we have used the terms *larger* and *smaller*. This refers to the fact that some value for integral is already placed in the controller, and this amount is either decreased to make the controller response faster or it is increased to make the response slower.

9.10.2 Using Values for Integral Action

Now that we know the action that the integral causes we can apply units to the values that represent it. Since integral action is the time it takes to repeat the action that was first caused by gain, we can describe an integral as the *number of minutes per repeat*. This can be shortened to *minutes/repeat*. In some controllers the time may be calculated in seconds rather than minutes. Since the time in these units refers to *real time,* it will be possible to use a strip chart recorder to graph the changes in the response for the system and determine the amount of change for the integral value.

In the lawn-mowing example, the first pass was completed in 1 minute. If the integral action completed the second pass in 1 minute, the response would be exactly the same as the first pass. If the second pass was completed in 2 minutes, the response would show that the speed on the second pass would be slower and it took twice as long to repeat the action. Fig. 9-16 shows a graphic representation of what would happen if the second pass took 30 seconds. From this graph you can see that the speed would be increased so that it would take half as long to repeat the original action. From this example, you can see that by specifying the amount of time to repeat the original action caused by gain, you can modify the total system response.

The effects of a large, medium, and small integral time can be shown in graph form. Fig. 9-17 shows examples of these three types of response. Again it should be noted that the relative terms of *small, medium,* and *large* are used so that you can see what will happen to the system if the integral is increased or decreased. In the graph at the top of the figure, you can see that since the integral time is larger, it will take longer for the system PV to reach SP. In the middle graph, the integral time has been reduced, and the system responds more quickly so it can reach SP more quickly. The graph at the bottom of the figure shows the response for the smallest integral, and you can see that the PV gets to SP very quickly.

By specifying the integral time, you can make the controller change the output more rapidly or more slowly to meet any conditions that the system requires. For example, if you were heating a furnace or the barrel of a plastic injection molding machine, you may be concerned that the temperature does not rise too quickly. You could specify a rather large integral time such as 2 minutes, and the amount of original change caused by the gain action could be added to the bias every 2

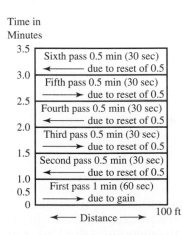

Time in
Minutes

FIGURE 9-16   Graph of changes caused by lowering the time for the second pass to 30 seconds.

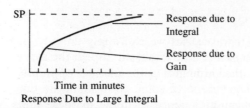

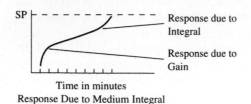

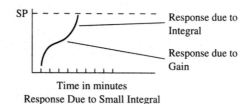

FIGURE 9-17 Three graphs showing the effects of large, medium, and small integrals.

minutes. This is why the units for integral action are referred to as *repeats per minute*. This means that when you place a value in the controller for integral action you must remember you are specifying the amount of time in minutes that you want the controller to repeat the original action caused by the gain calculation. (Please note that some controllers use seconds instead of minutes where faster response times are required.)

9.10.3 Reset Action

The term *reset* is often used when discussing the action that the integral term causes the controller to make. In original controllers that used pneumatic or solid-state amplifiers, the controller could achieve the effect of integral action without a calculation by simply resetting the amount of original change caused by the gain action. For this reason the term *reset* is still used in controllers. You will need to get used to referring to this part of the control as integral or reset action. The main point to remember about the integral is that the smaller the number that is used, the more quickly the system will respond, and the larger the number for the integral, the more slowly the system will respond.

9.11 ADDING DERIVATIVE (RATE) TO THE FORMULA

Derivative can be added to the formula to react with gain only as a PD (proportional, derivative) controller or it can be added with the PI (proportional, integral) controller to make a full three-mode controller called PID (proportional, integral, and derivative). The derivative has the inverse effect of integral. Instead of specifying the time it takes to repeat the action caused by gain, the derivative is specified as the *number of repeats in one minute.* You can see that for derivative action, time is kept constant and the number of repeats that you want to occur will be changed. (You should remember for integral action, the variable is time and the single repeat is the constant.) This means that if you want the controller response to become faster, you would specify a larger number for the derivative, which would be more repeats per minute.

For example, we can use the lawn-mowing example again to compare the action caused by the integral to the action caused by the derivative. We will show the results graphically in Fig. 9-18. In the original lawn-mowing example it took 1 minute to make the first pass, which we called the amount of change caused by gain. If the derivative was set to a 2, it would mean that after gain completed the first pass, we would want the controller to increase the output so that two passes could be made during the next minute. If the derivative was specified to be 0.5, the response rate would have been slowed so that only one-half pass would be mowed during the next minute, and it would take 2 full minutes to mow the second pass.

You can see that a derivative of 2 would provide a similar effect as an integral of 0.5, and a derivative of 0.5 would be similar to an integral of 2. When you think of an integral, you must think in terms of *minutes per repeat,* and when you think of a derivative you must think in terms of *repeats per minute.* It should be pointed out that technicians generally only need to understand the effects of increasing and decreasing gain, reset, and rate, and they do not necessarily need to know how to do the actual calculations.

9.11.1 Rate Action

In the early pneumatic and solid-state controllers, derivative action was referred to as *rate* because it controlled the rate of change for the controller by providing more or less time delay to the response. For solid-state, op amp controllers, this was accomplished by adding capacitance to the system. In pneumatic controllers the time delay was changed by adjusting small valves to bleed off or bypass air that the controller was sending to the output device.

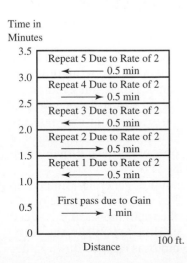

FIGURE 9-18 Graphic representation of adding rate to the control system. This diagram shows the effect of rate in the lawn-mowing example.

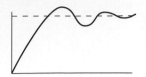

FIGURE 9-19 Graph of quarter-amplitude decay (QAD) response.

9.11.2 Other Effects of Derivatives

Derivatives affect the calculation differently than simply being the inverse of integrals. In a PI controller, the integral action works well over the long term, as the amount of error is continually checked and the output is changed to bring the PV closer to the SP. If the gain happens to be rather large, the amount of the original change will be large and each time the integral action repeats this amount of change the controller will change the output in larger increments. This has the effect of possibly causing the PV to overshoot the SP, as shown in Fig. 9-19. From this figure you can see that the PV overshoots the SP on the first oscillation but the controller reduces the amount of overshoot on each succeeding oscillation. This type of response is called *quarter-amplitude decay* because it reduces the error in four amplitudes. The reason the overshoot occurred is that in the PI formula, the controller does not know that the PV is larger than the SP until it happens. This would be similar to driving your car and looking out the side windows for stop signs. You would not know that you were at a stop sign until you were passing it.

Derivative action performs the function of *looking ahead* or *anticipating.* The derivative action causes the controller to determine that if the output continues at the same level, the controller response will cause the PV to overshoot the SP. Since the controller now knows (anticipates) that the present output will likely cause the PV to overshoot the SP, the derivative action will cause a small amount of value to be subtracted from the offset (bias term), which will cause the output to become smaller.

Using derivatives does have some drawbacks. If the sensor signal (PV) has any noise in it, this noise will show up on top of the normal signal. When this occurs, the derivative function sees the increase in the signal from the noise as normal and it tries to adjust the response accordingly. The result of this action is that each time noise is encountered, even if it is only for a millisecond, the derivative will adjust the output as though the noise on the signal was the original signal. This causes the output to be very rough and it means that if the derivative is used in a controller application, the sensor signal must be filtered.

9.12 PROPORTIONAL AND DERIVATIVE CONTROL (PD)

Most controllers use proportional-only (P-only) or proportional and integral (PI) control. Proportional and derivative (PD) control is generally used in applications where anticipation is required or in motion control applications. An example of motion control would be a dc servomotor that is connected to a ball screw so that a robot arm can move in a horizontal travel to put parts in several bins that are side by side. The ball screw is rotated by the motor and the platform that the robot sits on is moved in front of a conveyor where it can pick up parts and then move quickly to one of the bins in which the parts are dropped.

In this type of motion control application the motor is generally ramped up at some rate until the desired speed is achieved. Fig. 9-20 shows an example of this type of response. When the motor reaches full speed, the robot moves at this speed until it is in front of the bin in which the part is to be dropped. When the robot reaches the point where it is to stop, the speed of the motor is ramped down rather

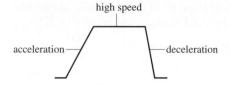

FIGURE 9-20 An example of proportional and derivative (PD) control.

abruptly so the ball screw stops the robot directly in front of the bin. The PD controller is useful in these types of applications because it allows the motor to ramp up and reach full speed rapidly and then move at high speed and ramp down quickly as it reaches the SP where the robot should stop and drop off the parts.

Since the PD control can respond quickly, it becomes a detriment in some process systems like controlling steam valves where this type of control may cause the valve to fluctuate too much, making it wear out.

9.13 PROPORTIONAL BAND

In the proportional-only controller the output will change from 0% to 100% depending on the amount of gain and the amount of error. For example, if you have a product such as soup that you would like to heat in a kettle that has a steam jacket, the amount of steam that is allowed to flow into the steam jacket will determine how hot the soup will become. Fig. 9-21 shows a graph that indicates the percentage the steam valve opens from 0% to 100% along the *x*-axis, and the temperature of the soup in the kettle along the *y*-axis of the graph. You should notice that the soup will be at 70°F when the valve is closed (0% open) and the soup will be 170°F when the valve is 100% open. This represents a 100°F span (70° to 170°).

Proportional band is used in some controllers instead of gain. This tends to be somewhat confusing because gain-type control is referred to as proportional

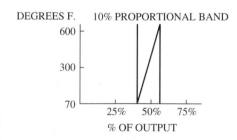

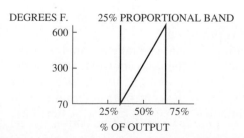

FIGURE 9-21 Graphs showing the effects of 10% proportional band and 25% proportional band.

control. *Proportional band* is the amount of change in error that will cause the output to go from full on to full off. The amount of change in error is calculated as a percentage of full-scale error. This can also be defined as $PB = \dfrac{100}{K_P}$.

Fig. 9-21 shows several examples of proportional band. From the first example you can see that if the proportional band is 10%, a change of 10% of the full-scale error will cause the output to swing 100%. In the second example you can see that if the proportional band is 25%, a change of 25% will cause the output to swing 100%.

Proportional band is also the inverse of proportional control (gain) and in modern microprocessor controllers it refers to the number of units such as degrees (°F) that the controller is trying to hold around the SP. If the proportional band is 10 and the controller is a temperature controller, it is trying to hold the temperature of the system within 10°F of the SP. This normally means the controller will not change the output if the PV temperature is not more than 5°F above or below the SP, which is a span (band) of 10°F.

Proportional band can be calculated by the formula:

$$P_B = \frac{1}{\text{gain}}$$

This formula will provide the amount of proportional band as a percent of full-scale error. If the gain is large, the proportional band will be small, and vice versa. For example, a gain of 10 will provide a proportional band of 0.1, and a gain of 50 will mean a proportional band of 0.02. Conversely, if the controller has a proportional band of 5, the gain would be 0.2.

9.14 RATIO CONTROL FOR ON/OFF HEATING CONTACTOR

In many applications such as electrical heating systems, the heating element is controlled by a contactor that has only two states (on or off). The contactor is used as the control device instead of an analog controller because of the expense. Since the output of the process controller is analog (0–100%) and the heating contactor is on/off, the controller must be set up to provide the output signal to the contactor as a ratio of time on versus time off. This type of control is also called *time proportioning*.

Fig. 9-22 shows an example of this type of control. From this figure you can see that the coil of the heating contactor is connected to the controller, and the heating element is connected to the contacts of the heating contactor. Anytime the contactor coil is energized, full voltage is sent to the heating element and it provides 500 W of heat to the furnace. Anytime the contactor coil is de-energized, no voltage is sent to the heating element and no heat is produced.

The on/off type of control for the heating element can take advantage of the 0–100% analog output of the controller by selecting the ratio control feature of the controller. The ratio control feature uses ratio of on-time to off-time to get the effects of the analog controller. For example, if the controller determined that the output should be 50%, the ratio of on-time to off-time will be 50%. A time base

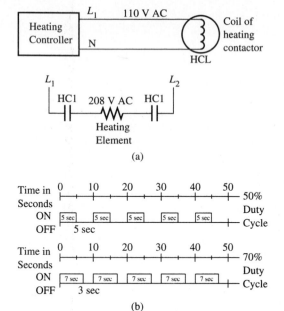

(a)

(b)

FIGURE 9-22 **(a)** Diagram of a 110 volt ac heating contractor coil (HC1) connected to a controller and its contacts controlling a 208 volt ac heating element. **(b)** The timing diagrams for a 50% duty cycle and a 70% duty cycle. Notice the on-time is 5 seconds and the off-time is 5 seconds for the 50% duty cycle, and the on-time is 7 seconds and the off-time is 3 seconds for the 70% duty cycle.

of 10 seconds is generally used, which means that for an output of 50%, the heating contactor would be energized for 5 seconds and de-energized for 5 seconds. If the controller determined the output should be 70%, the contactor would be energized for 7 seconds and de-energized for 3 seconds.

The contactor will be continually turned on and off during each 10-second cycle as determined by the amount of output percentage. The heating effect from the heating element is similar to that of an element controlled by the more expensive analog amplifier. This type of control is used extensively in the heating systems for plastic presses.

9.15 TYPES OF SYSTEM RESPONSE

The way a process control system responds is very important. It is necessary to understand the types of response so that you can adjust the controller. The system response will be described as an underdamped response, a critically damped response, and an overdamped response. Fig. 9-23, Fig. 9-24, and Fig. 9-25 show examples of these responses. The graphs display time (t) along the x-axis (left to right) and they show amount of PV along the y-axis (top to bottom). If the PV is above the SP, it is called *overshoot,* and if it is below the SP, it is called *undershoot.*

The normalized response is a medium-speed response, which does not allow the PV to overshoot. This is also called a first-order response and it would be useful in heating a product like the syrup used to make cola. If the syrup is overheated, it will burn and the cola will not taste good. Overheating the product will also affect the shelf life, which means the cola will lose its taste quickly while waiting to be sold.

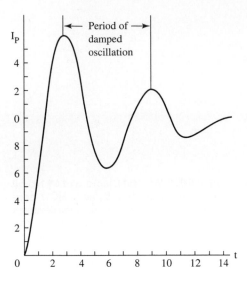

FIGURE 9-23 Graph that shows an underdamped system response.

The underdamped response is a little faster responding, and it allows the PV to oscillate slightly around the SP. Since the controller reduces the amount of overshoot and undershoot in four amplitudes, this type of response is called *quarter-amplitude decay* (QAD). This type of response is used to maintain the level in a filling bowl for a system where beverage cans are filled on an automated line. The response will occur when the level in the filler bowl gets low and fresh beverage is added to the bowl. The response to the low level needs to be fairly fast, and the system is not harmed if the level oscillates (QAD) around the SP as the bowl is being filled.

The overdamped response is a very slow response that is used when bringing systems up to temperature. For example, when a process cooker is first brought up to heat, it must be heated slowly so that it does not crack. The heat for the cooker comes from steam. The steam valve is controlled with an overdamped response so it opens slowly and allows the temperature to increase slowly. It may take over an hour to bring the temperature from room temperature to 200°F.

Loops that respond quickly are called critically damped loops. An example of this type of response would be a steam heating coil for a jacket-type product

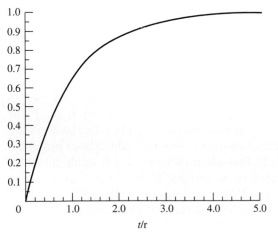

FIGURE 9-24 Graph of a critically damped system response.

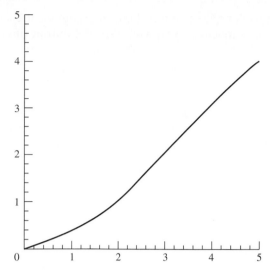

FIGURE 9-25 Graph of an overdamped system response.

heater. The jacket allows steam to circulate around a kettle and heat the product without touching it. Anytime the steam valve is opened or closed slightly, the temperature will change quickly.

9.16 TUNING THE CONTROLLER

Tuning is the process of adjusting the gain, reset, and rate (proportional, integral, and derivative) to make the system being controlled respond as desired. Usually a process engineer will provide specifications to indicate the type of response the system should have. Once the gain, reset, and rate are adjusted, the controller will automatically change the output to ensure the system responds correctly when the PV is above or below the SP.

Several methods can be used to determine what the original tuning variables for a new controller should be. The Ziegler-Nichols method includes setting the controller in manual mode, changing the output in percentage steps, and observing and recording the rate of change in the PV. A second way involves setting the controller in auto mode and causing the system to oscillate around the SP, and a third way is to use the auto-tuning function of the controller where the controller automatically performs the test described in the first method.

9.16.1 Determining Gain, Reset, and Rate (PID) in Manual Mode

One of the easiest ways to determine the value to use for gain, reset, and rate is to use the Ziegler-Nichols method. This method allows the process system to be operated under manual control while changes in the system are observed. The changes in the system will be used in a set of formulas that will yield P, I, and D values, which will cause the system to operate with a QAD response.

The first step of this method requires the controller to be placed in manual mode with the output set to 20%. This means that if you are trying to find the tuning variables for a temperature controller that is controlling the temperature for the barrel of a plastic injection molding machine, you would set the output of the controller to 20%. Since the controller can control the actual temperature from 70°F to 800°F, you would find that 20% of this span would provide a temperature of approximately 150°F. The controller would be allowed to operate at this percentage of output until the barrel reached this temperature and stayed there for 10 to 15 minutes. This is referred to as *allowing the temperature to line out,* which means that if the process variable was graphed on a recorder, the graph would show somewhat of a *straight line* after the system was allowed to run at a constant 20% output for 15 minutes.

The next step in the process is to increase the output 10% and measure the amount of change to the PV. The change in the PV is usually documented with a strip chart recorder so that a graph of the change will be available. The graph will show the amount of change to the PV and the amount of time over which this change occurred. Since the output started at 20%, a 10% change would increase the output to 30%. An example of the graph of the change in the temperature (PV) for this process is shown in Fig. 9-26.

The next step is to use the graph and determine the amount of process gain, which is the amount of change in the process variable (ΔPV) divided by the amount of percentage change to the output signal from 0–100% (Δ output).

$$K_P = \frac{\Delta PV}{\Delta Output}$$

(Note: It is important to understand that it is the amount of change of these variables and not the actual value that is used in the formula. For example, if the output signal changed from 40% to 60%, the amount of change is 20%, so the value of 20% would be used in the formula and not 60%.)

K_P is called process gain because it will reflect the inherent ability of the barrel to heat up some number of degrees when the output is changed by some percentage.

FIGURE 9-26 Graph of the temperature (PV) for the barrel of a plastic injection molding machine after the output signal is changed a total of 10%.

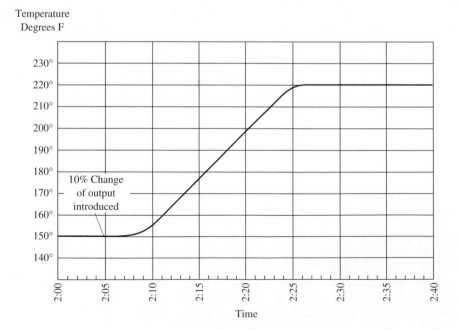

It is important at this time that you do not confuse process gain K_P with controller gain K_C that has been discussed as gain up to this point. From the graph in Fig. 9-27 you can see that the PV was at 150°F when the change was introduced to the output and the temperature increased to 220°F, which is a change (Δ) of 70°F. The change (Δ) in the output is 10%. The following formula will show the process gain K_P for these values is 7.

$$K_P = \frac{\Delta PV}{\Delta \text{Output}} = \frac{70}{10} = 7$$

The next step in this process is to determine the *dead time* for the system, which is the time it takes for the PV to respond (change) after the output is changed. Dead time is very important since it will be used to determine the amount of integral (reset) response that is needed. The dead time can be found on the graph by locating the point where the output was changed from 10% to 20% and determining how long it took for the PV to start changing. A tangent line is marked on the response graph to show where the change occurs. The bottom end of the tangent line touches the point where the change in PV starts (150°F) when the time is 2:10. The time of 2:10 will be the official point in time where the change begins. Since the 10% change was entered in the system at 2:05 and the system started to change at 2:10, the dead time is 5 minutes.

The third step in the Ziegler-Nichols method is to determine the *time constant* for the 10% change to the output. The time constant is the amount of time required for the PV to change 63% of the total change caused by changing the output 10%. Since the PV was 150°F at the start of the change and 220°F at the end of the change, the total change is 70°F, and 63% of 70° is 44°F.

The 44°F change is added to the original temperature of 150°F to determine that the point where 63% change has occurred is 194°F. A vertical line is drawn where 194°F occurs, and you can see it intersects the time axis at 2:18. Since the end of the dead time occurs at 2:10, and the point where the PV is changed 63%

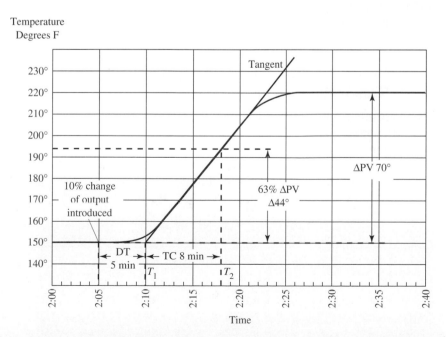

FIGURE 9-27 A graph of the PV response that shows the dead time (DT), time constant (TC), and the process gain (K_P) identified.

FIGURE 9-28 Formulas for figuring the value for gain (P), reset (I), and rate (D). These formulas will give the PID values to make the system response quarter-amplitude decay (QAD).

	Gain (P)	Reset (I)	Rate (D)
P only	$TC/(K_P*DT)$	----	-----
PI	$0.9TC/(K_P*DT)$	3 DT	-----
PID	$1.2TC/(K_P*DT)$	2 DT	.5DT

occurs at 2:18, the time constant time will officially be determined to be 8 minutes (2:18 − 2:10).

9.16.2 Determining PID Values from Process Gain, Dead Time, and Time Constant

The Ziegler-Nichols method uses a group of formulas to determine the P, I, and D values for a P-only controller, a PI controller, and for the three-mode PID controller. Remember that PID values for gain, reset, and rate would be put into a controller to get the system to provide a QAD response. Since this is a new system and we are not sure what values to use, we will use the values of process gain, dead time, and time constant that were determined when the system was tested in manual mode to help us determine the values the controller will use when it is in automatic mode. Fig. 9-28 shows these formulas.

Fig. 9-29 shows the formulas from Fig. 9-28 with the values from the graph filled in. This time the calculations are solved and the actual values for proportional (P), integral (I), and derivative (D) are shown. The proportional value is determined to be 0.22 if a gain-only controller is used. If the controller is to use gain and reset, the proportional value should be set at 0.20 and the integral value would be set to 15. If the controller is to be a three-mode controller, the gain should be set to 0.27, the integral should be set to 10, and the derivative should be set to 2.5. As a technician you would enter these values into the controller by using the keypad and the system response should be a quarter-amplitude decay (QAD). If you would like to change the response so that you have less overshoot, you could increase the integral value or decrease the gain value from the keypad.

9.16.3 Determining PID Values for a Slower Responding System

It is important to understand that the Ziegler-Nichols method will work to determine the PID values for any type of system. For example, if the system was a much slower responding system like a heating system with much smaller electrical heating elements, the graph would show that the system changed from 150°F to 162°F when the output was changed 10%. You should remember that the previous system

FIGURE 9-29 Formulas with the values for PID calculated for a quarter-amplitude decay response.

	Gain (P)	Reset (I)	Rate (D)
P only	TC/K_P*DT $8/(7 \times 5) = 0.22$	----	-----
PI	$0.9TC/K_P*DT$ $0.9 \times 8/(7 \times 5) = 0.20$	3 DT $3 \times 5 = 15$	-----
PID	$1.2TC/K_P*DT$ $1.2 \times 8/(7 \times 5) = 0.27$	2 DT $2 \times 5 = 10$	.5DT $.5 \times 5 = 2.5$

changed 140°F, so this system is slower responding. Since the output only changed 12°F when the 10% change of output was entered, the process gain (K_p) for this system is 1.2 instead of 7. The graph also indicates that the dead time also changes to 2.6 minutes, and the time constant is 1.2 minutes. When these numbers are entered into the Ziegler-Nichols formulas, the gain should be set at 6.4 for a proportional-only controller. If the controller is using proportional and integral (P and I), the gain should be set at 5.77, and the integral should be set at 7.8. If the controller is using proportional, integral, and derivative (PID), the gain should be set at 7.6, the integral should be set at 5.2, and the derivative should be set at 1.3.

9.16.4 Determining PID Values in Automatic (Closed-Loop) Mode

The values of PID for a new system can also be determined by placing the system in automatic (closed-loop) mode and using gain only. We will use the barrel heaters for the plastic injection press again, and the heating elements will be the smaller heating elements used in the previous example. The open-loop and closed-loop methods of determing PID values should produce similar results.

Since we want the system to operate in closed-loop mode and act as a proportional-only controller, the value of the integral will be set to infinity, and the value for the derivative will be set to zero. This effectively makes the system a proportional-only (gain-only) controller. Since we are trying to determine a safe value for gain, we should start testing the system with a small gain of 0.1. Next the SP will be adjusted to a value that represents 25% of the maximum PV the system can achieve. (This assumes that the system can safely operate at 25% of its maximum PV.) In the previous example of the barrel heaters for the plastic injection molding press, the maximum temperature the barrel could reach was 800°F, so 25% of 800°F would be a setpoint of 200°F.

The PV should be allowed to line out and then the SP be adjusted up 10°F to 210°F and allow the system to respond. We are now trying to make the PV temperature start oscillating. If the PV does not oscillate, you can turn the SP down 10°F to return to 200°F. After the temperature returns to 200°F the gain can be increased and the SP can be changed again to 210°F. The system should be checked to see if the temperature is beginning to oscillate. This process should be repeated until the oscillations become sustained, which means they are not growing any larger or any smaller. If the oscillations start to become larger, the gain should be reduced slightly. If the oscillations begin to decay, the gain should be increased slightly. The value of gain when the system is in sustained oscillation is called the *ultimate gain* (K_u). The period of one of the oscillations is also measured at this time and it is called the *ultimate period* (P_u). Fig. 9-30 shows a diagram of the sustained oscillation. You should notice that the ultimate gain is identified as a gain of 12 and the ultimate period (P_u) is 10 minutes.

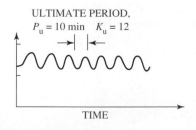

ULTIMATE PERIOD,
$P_u = 10$ min $K_u = 12$

TIME

FIGURE 9-30 A graph of a closed-loop system that is in sustained oscillation. Notice that the ultimate gain (K_u) is 12, and the ultimate period (P_u) is 10 minutes.

Type of Controller	Gain	Reset	Rate
P only	0.5 K_U	---	---
P and I	0.45 K_U	$P_U/1.2$	---
P I D	0.6 K_U	0.5 P_U	$P_U/8$

The formulas in Fig. 9-31 will be used to determine the amount of gain for a proportional-only (P-only) controller, the amount of gain and reset for a proportional and integral controller (PI controller), and the amount of gain, reset, and rate for a three-mode PID controller.

Since we know that the value of the ultimate gain (K_u) is 12, and the ultimate period (P_U) is 10 minutes, we can complete the calculations from the table and determine the values for gain, reset, and rate. These calculations are shown in the table in Fig. 9-32.

From the table in Fig. 9-32 you can see that the gain for a proportional-only controller should be set to the value of 6. If the system is to be operated as a proportional and integral (P and I) controller, gain should be set to a value of 5.4 and the integral should be set to 8.3 minutes. If the controller is operated as a PID controller, the gain should be set to a value of 7.2, the integral should be set to 5 minutes, and the derivative should be set to 1.25 minutes.

Since the open-loop method and closed-loop methods were both used to determine the values of gain, reset, and rate (PID), you should notice that both methods produced values that were similar. Most technicians prefer to use the open-loop method because they have more control over the output if the system starts to run away.

9.17 ADJUSTING GAIN, RESET, AND RATE AFTER A SYSTEM HAS BEEN RUNNING

More often adjustments to gain, reset, and rate are made after the system has been running for some time. The system response is determined to be too fast or too slow and you will be asked to make changes. This type of tuning is very easy to understand if you keep in mind the rules shown in Fig. 9-33.

Type of Controller	Gain	Reset	Rate
P only	0.5 K_U .5 × 12 = 6	---	---
P and I	0.45 K_U 0.45 × 12 = 5.4	$P_U/1.2$ 10/1.2 = 8.3	---
P I D	0.6 K_U .6 × 12 = 7.2	0.5 P_U .5 × 10 = 5	$P_U/8$ 10/8 = 1.25

To Make the Loop Respond Faster	
GAIN (P)	↑
RESET (I)	↓
RATE (D)	↑

FIGURE 9-33 Table that shows if gain, reset, and rate should be increased or decreased to make the loop response faster.

You should think of changing gain, reset, and rate like you would think about changing the speed of an automobile you are driving. You can slow down by letting off on the accelerator or by stepping on the brakes. Varying amounts of each can achieve the exact same speed. When changing the system response, you should remember that gain is used to set the first part of the response, and the integral is primarily used to make changes during the remaining cycle. This is important to understand if a system is controlling the temperature in an oven, which is brought up to temperature once a week on Monday morning and left to run at the SP for the remainder of the week. In this case the gain would be used when the system was brought up to temperature, and the integral should be used to make adjustments after it is up to temperature.

If the system requires the SP to be continually changed, or if the PV changes significantly, such as when a new batch is processed every hour, the gain will have a much greater effect on the response of the loop.

9.18 PROCESS ALARMS AND DEVIATION ALARMS

Both open-loop and closed-loop control systems need alarm points to alert the operators or supervisors of the condition of the system. Process alarms are used in a system to set the maximum and minimum limits that the system is allowed to reach. These limits are determined by the equipment manufacturer so that the equipment will not harm people who must work around the process and so that the equipment will not be damaged while operating. Fig. 9-34 shows examples of process alarms.

For example, the alarms on a heating system may be set at 600°F and 32°F. These values were determined because temperatures above 600°F will damage the furnace lining, and temperatures below 32°F will allow moisture to freeze and damage seals. If the system controls the level of liquid in a tank, the high and low alarms may be set as a percentage of the level of a full tank. The high-level alarm is set at 90% so the tank will not overfill, and the low-level alarm at 10% so the level does not go completely empty when a pump is running and cause the pump

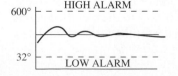

FIGURE 9-34 Example of high and low process alarms. Notice that the high alarm is set at 600°F and the low alarm is set at 32°F.

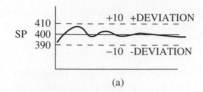

(a)

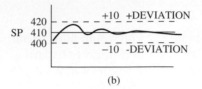

(b)

FIGURE 9-35 **(a)** Deviation alarm is set at 10° above and below the setpoint. **(b)** This graph shows a change in the setpoint from 400° to 410°. Notice that the deviation alarms remain at +/−10°, so now they will be set for 420° and 400°.

to cavitate and fail. Once the process alarms are set for a controller, they should not be changed.

Deviation alarms are used primarily for quality control. These alarms are set to be an amount that is added to and subtracted from the SP. For example, if the SP for a temperature controller is 400°F, and the quality will be affected if the temperature deviates 10°F above or below the SP, the deviation alarm would be set at +/−10°F.

The important point to remember about the deviation alarm is that if the SP is changed to 450°F, the alarms would remain plus 10°F and minus 10°F and they would move with the SP. The new alarm points would be 440°F and 460°F.

9.18.1 Inner and Outer Alarm Bands

The process and deviation alarms can be set so that two bands of alarm points are used. Fig. 9-35 shows an example of inner and outer alarms for the deviation alarms.

This figure shows the inner alarm point for the deviation alarm is set at +/−5°F, (395°F and 405°F) and the outer alarm is set at +/−10°F (390°F and 410°F). A warning light may indicate that the inner alarm point has been exceeded, and the operator can check the system to ensure the controller is trying to bring the process back into the correct operating range. If the condition continues to drive the process beyond the outer deviation point, additional controls may send the product to a holding area to be tested because the process exceeded the standards that are set for acceptable quality. The inner and outer deviation alarms provide a means to ensure that only top-quality products are produced.

The fixed process alarms can also use inner and outer alarm points. This gives the system the ability to warn the operator prior to the outer alarms being exceeded. For example, the fixed high inner alarms may be set at 550°F and the low inner alarm may be set at 50°F. If the temperature exceeds either of these alarm points, the operator will be warned by an audible or visual alarm so that the system can be checked before it starts to shut down. If the problem continues and the outer alarm point of 600°F or 32°F is exceeded, the system will go into fail-safe shutdown procedure. During the fail-safe shutdown procedure, safety devices will take control of the process equipment and bring the process safely under control. For example, if the outer high alarm on a heating system tripped, the power to the electric heating coils will be de-energized and an audible alarm such as a warning horn will indicate the system has a serious problem.

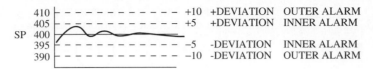

9.18.2 Statistical Process Control (SPC)

The closed-loop controller has built-in features such as the deviation alarms that allow it to aid in controlling product quality. Quality control personnel can determine the statistical process control limits, and these values can be loaded into the controller as inner and outer deviation alarm points. The process control signal from the controller can also be sent to a recorder and a complete and accurate record of all sensors can be recorded when the product is being manufactured. This information provides a method for the quality of the product to be monitored and improved.

9.19 ADVANCED CLOSED-LOOP SYSTEMS

In some applications, the variables in the closed-loop system function in a complex manner so that a simple closed-loop controller cannot control the loop accurately. For example, if a product such as soup is heated while it flows to the filler line, a tube and jacket-type steam heater must be used. The soup flows through the inside of the tubes in the heater and steam is injected around the outside of the tubes in the jacket to heat the soup. This system becomes complex for two reasons. First, the flow of soup through the heater is not at a constant rate, so the heater needs a varying amount of steam to heat the product to the same temperature. Second, the steam can change the temperature of the jacket much more quickly than the soup can change temperature. Since the soup and steam change temperature at different rates, a single controller would need two or three different gains, resets, and rates to be able to compensate for the differences in speed. It is easier to control this type of system with two loops that are connected to each other. This type of control is called *cascade control*.

9.19.1 Cascaded Loops

Fig. 9-37 shows the diagram of a cascade system that uses steam to heat a continuous flow of soup. A temperature controller is the primary control for the system. It is also called the outer control or the *master* and it has a thermocouple as its sensor to provide the PV signal. The thermocouple is identified in the diagram as the *temperature element* (TE). The output of the temperature controller is sent to the second controller that controls the steam valve, which is called the inner controller or the *slave*. The steam loop has a pressure sensor called a *pressure element* (PE) that sends the value of the steam pressure to the controller.

To operate the system, the operator enters the SP temperature in the master controller. For this example, the soup is heated to 180°F so the SP in the temperature

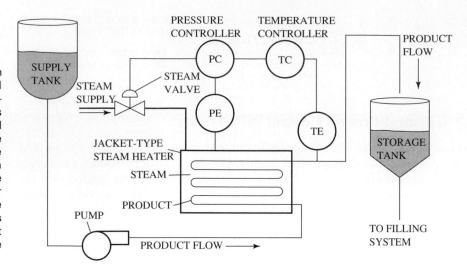

FIGURE 9-37 Diagram of a cascaded control system. The temperature controller samples the temperature and sends its output to the pressure controller. The pressure control system samples the pressure and the amount of error from the temperature control and determines the amount of output signal that is sent to the steam valve.

controller is set to 180°F. Since the system is a cascade system, the operator cannot adjust the SP in the pressure controller. The SP for the pressure controller is determined from the output signal from the master (temperature) controller. The output for the slave (pressure) controller is connected directly to the steam valve, which is opened when heat is needed and closed when the temperature is above the SP.

9.19.2 Operation of the Cascade Controller

The cascade loop control begins at the master controller. In this example the master controller is continually checking the soup temperature as it leaves the heater to the SP of 180°F. For this example we will assume the two controllers have the soup temperature at SP. When the flow of soup is increased, the additional soup flowing through the heat exchanger causes the soup leaving the heater to become colder than 180°F, and the master controller (temperature controller) begins to increase the amount of output signal that is sent to the slave. The output signal can vary from 0–100%. The slave controller has a setpoint range that is set from 0–50 psi, which is the range of the steam pressure. The master controller treats the SP range as a span of 0% to 100% even though the slave sees this value as 0 to 50 pressure. Since the two controllers are set for cascade mode, the increase in the output signal from the master will cause the SP to increase in the slave.

The slave controller controls the steam pressure like a typical single-loop controller, except its SP is changed only by the master controller rather than by the operator. At the beginning of the example, the output of the slave controller sets the steam valve to a position between 0–100% where the system was holding the soup temperature at 180°F. Since the soup temperature dropped, and the master controller increased its output, the SP of the slave increased, and the slave controller increased its output, which opened the steam valve farther. The increase of steam pressure will cause the temperature of the soup to increase. If the flow of soup slowed down, the soup temperature would rise above the SP, and the master would reduce its output signal, which would reduce the SP of the pressure controller. The pressure controller would then reduce its output signal, which would close down the steam valve and the temperature of the soup would drop.

Since the temperature of the soup changes rather slowly, and the temperature of the jacket which is caused by the steam can change rather quickly, the gain, reset, and rate in the temperature controller can be set for its response. The gain, reset, and rate can be set differently in the pressure controller for its response. In the temperature controller, rate (derivative) action will be useful because it can anticipate that the soup temperature is getting closer to SP and begin to compensate rather than wait for the temperature to reach the SP and then tell the slave to start turning back the steam. If anticipation is not used, the soup temperature will oscillate well above and below the SP because of the difference in response.

9.20 EXAMPLE SERVO APPLICATIONS

Closed-loop servo systems have changed during the past 15 years since the advent of the op amp and the microprocessor chip. In the 1960s and 1970s when op amp integrated circuits were introduced into control circuits on a mass scale, they provided a means to use analog sensors in systems that controlled analog output signals. For example, a load cell could be used to control a proportioning valve to provide an accurate filling and weighing system. A setpoint voltage that represented the desired weight could be set as one input to an op amp and the voltage from the load cell feedback signal could provide the voltage for the other input. The output signal is sent to a proportioning valve that opens from 0–100% to allow the proper amount of material to drop into the weighing hopper.

In the 1980s and 1990s when microprocessor chips became available, they were first integrated with the op amps. This type of control required analog-to-digital converters and digital-to-analog converters. Today a wide variety of sensors is available to provide a true digital input signal, and a wide variety of output devices is available that is controlled by digital signals. This means that a control system today may be all analog, or have analog inputs and outputs with digital microprocessor control, and still others may be a complete digital system where the input and output signals and the control are all digital.

9.20.1 A Simple Analog Servo Positioning System

One of the simplest types of closed-loop servo systems is an analog positioning system. Fig. 9-38 shows a diagram for this type of circuit. The input sensor for this system is a linear potentiometer. The resistance for the potentiometer is 1.5 kΩ, and its movable arm is connected to a tensioning arm that is used to apply pressure to a sealing bar for closing plastic bags in a packaging process. The sealing bar

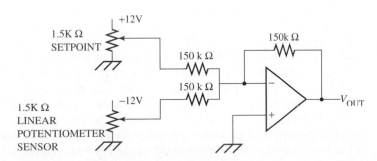

FIGURE 9-38 Example of an analog closed-loop servo system that indicates the position of a sealing bar used to close plastic bags in a packaging application.

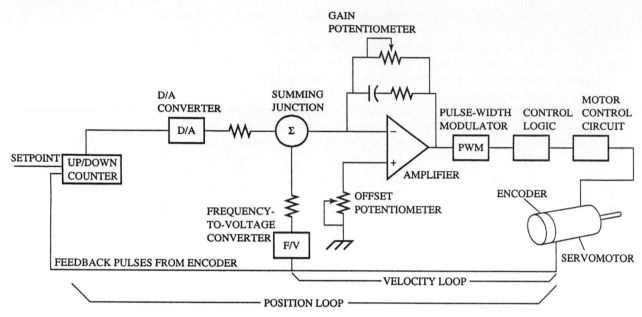

FIGURE 9-39 Block diagram of a servomotor used to control the position of an index table for a welding application. This diagram shows an encoder that is used to provide a velocity and position feedback signal. (Courtesy of Rockwell Automation's Allen Bradley Business.)

provides heat to the bag to cause the plastic to melt and seal. The amount of pressure the sealing bar applies to the plastic bag is set by an operator by adjusting the SP potentiometer. If more pressure is required, the SP is adjusted for more voltage, and if the pressure is too much the potentiometer is set to provide less voltage. The output of the op amp is sent to the coil of a magnetic pull bar that moves the sealing bar. The electromagnet in the pull bar applies more pressure to the sealing bar when the voltage becomes stronger. A spring provides a means to loosen pressure on the sealing bar when the magnetic strength is reduced. The arm of the linear pot is connected to the sealing bar and it is used to determine its position. This system is overly simplified so that you can understand each part.

9.20.2 Closed-Loop Velocity and Positioning Control of a Servomotor

A more complex application of a closed-loop control system is a motion control system that has a positioning loop and a velocity loop. Fig. 9-39 shows an example of this type of motion control loop used to drive a servomotor that moves a welding index table. The servomotor in this application has an encoder attached to its shaft that is used to provide velocity data as well as positioning data. The pulse train from the encoder is sent to the frequency-to-voltage converter, and the output voltage signal from the converter is sent to the summing junction. A position setpoint is set in the up/down counter, which is converted from a digital value to an analog voltage. The voltage is sent to the summing junction as a reference signal. The summing junction compares the position reference signal to the velocity feedback signal and produces an error signal that is sent to the op amp as the inverting input signal. If the error is negative, it means the motor has not moved the table into the correct position, and additional voltage will be sent to the motor from the output signal from the op amp. The output signal from the op amp must go through a pulse-width modulator (PWM) circuit, a control logic circuit, and a motor control circuit.

When the motor shaft moves the table, the encoder will indicate that the table is at a new position. The position loop compares the setpoint position (the desired position) in the up/down counter buffer with the actual position that is reported by the encoder. The amount of difference is converted from a digital signal to an analog voltage and sent to the summing junction. The velocity loop compares the amount of actual error to the amount of possible error and determines how close the indexing table is to its destination. The closer the table gets to its destination position, the smaller the amount of signal voltage is sent to the op amp and to the motor and it will slow the table. When the encoder indicates that the indexing table is at the correct position, the setpoint signal and the feedback signal will cancel each other and the resulting signal value will be zero, which means that no voltage is sent to the motor so it will stop at that position.

SOLUTION TO JOB ASSIGNMENT

Your solution to the job assignment should include installing two digital type controllers like the one shown in Fig. 9-3. The first is used to control the temperature of the product in this tank, and the second is used to control the level in the batching tank.

Your solution should include the explanation for your supervisor of the parts of the two systems that uses the diagram in Fig. 9-1. The sensor for the temperature control system should be some type of temperature sensor such as a thermocouple. The sensor will provide the process variable signal that is used as feedback. The output device should be a heating element, steam valve or some type of device that can increase the temperature of the product in the tank.

The sensor for the level control system should be a level sensor, which may have a float or a similar type of sensor. The output for the level loop should be some type of valve that will control the amount of product flowing into the tank.

Your solution should use Figs. 9-23, 9-24, and 9-25 to determine the type of response that should be used for each loop. It is important to remember that you would not be expected to determine the type of response, but you would be expected to understand the graph for an underdamped, overdamped, and critically damped response and be able to adjust gain, reset, and rate so the system provides the correct type of response.

Your solution should also include an explanation of either the open-loop or closed-loop method of determining the new tuning values of gain, reset, and rate for each system. Use the material in Section 9.16 to help your explanation and your solution should indicate why you selected the open-loop or the closed-loop method.

You should use the information in Section 9.18 to determine the types of alarms you will need. The types of alarms for your solution should include fixed alarms for the temperature system that protect the system from excessive temperature and from temperatures below 40°F. Since these alarms are adjustable, you do not have to specify the exact high temperature since you would get this information from the manufacturer's specifications. The fixed alarms for the tank level should be set at approximately 20% and 80% until your production supervisor would indicate otherwise. These values will sound an alarm if the tank becomes too full or too empty. Deviation alarms should be included, and you will need to depend on the quality control supervisor to indicate the exact values for these alarms.

QUESTIONS

1. Identify the parts of a closed-loop feedback system.
2. Explain the terms *proportional, integral,* and *derivative.*
3. Discuss four statements that can be made about a P-only controller.
4. Explain the term *bias* (offset) and how it is used to enhance the P-only controller.
5. Explain how you would determine the proper value of gain, reset, and rate for a new closed-loop system.
6. Explain what would happen to your process alarms if you changed the setpoint.
7. Explain the function of deviation alarms.
8. Explain what would happen to your deviation alarms if you increased the setpoint of the system from 150°F to 170°F.

TRUE OR FALSE

1. _____ When a system is in open loop the feedback signal is sent to the summing junction.
2. _____ Process alarms are fixed alarms used to protect the system.
3. _____ When a system is placed in manual mode, the output signal is changed by the controller.
4. _____ Deviation alarms are fixed alarms used to protect the system.
5. _____ A bump in a process control system is any change in setpoint or any change in the process variable.

MULTIPLE CHOICE

1. Bumpless transfer causes the _____.
   a. setpoint to be set equal to the output when the controller is changed from manual to automatic mode.
   b. setpoint to be equal to the process variable when the controller is changed from manual to automatic mode.
   c. process variable to be set equal to the setpoint when the controller is changed from manual mode to automatic mode.
2. The proportional band is the _____.
   a. amount of change in error that will cause the output to change from 0% to 100%.
   b. amount of error that can occur before the deviation alarm is activated.
   c. amount of error that can occur before the process alarm is activated.
3. A heating system that uses ratio control to control a heating contactor _____.
   a. turns the contactor on for a percentage of time equal to the percentage of the output signal.
   b. provides a voltage signal to the coil of the heating contactor that is between 0–100% of the applied voltage.
   c. provides an analog voltage signal directly to the heating element that is equal to the percentage of time the output signal is on.
4. The term *tuning a loop* refers to _____.
   a. adjusting the offset for the feedback signal.
   b. adjusting the gain, reset, or rate for a process control system.
   c. adjusting the setpoint for a process control system.
5. If the integral time of a process control system is decreased while the gain is held constant, the system will _____.
   a. respond more rapidly.
   b. respond more slowly.
   c. not change its response since the gain did not change.

PROBLEMS

1. Use the table in Fig. 9-28 to determine the value for gain, reset, and rate for a P-only loop, a PI loop, and a PID loop if the value for TC is 12 minutes, the value for DT is 2 minutes, and the process gain is 5.
2. Use the table in Fig. 9-31 to determine the values for gain, reset, and rate for a P-only loop, a PI loop, and a PID loop if the value for K_u is 15, and the value for P_u is 5 minutes.
3. Draw the block diagram for a typical servo system and identify the setpoint, process variable, error, summing junction, controller amp, and output (final control element).
4. Determine the proportional band if the gain is set to 1.2 for the system.
5. If the + deviation alarm is set to 15°F, and the − deviation alarm is set to −12°F, calculate the temperatures the PV must go above or below to actuate these alarms when the SP is set to 190°F.

REFERENCES

1. *Advanced Process Controls Training Manual PM-II.* Texas Instruments, Johnson City, TN, 1992.

2. Jacob, J. Michael, *Industrial Control Electronics.* Englewood Cliffs, NJ: Prentice Hall, 1988.

3. Johnson, Curtis, *Process Control Instrumentation Technology.* Englewood Cliffs, NJ: Regents Prentice Hall, 1993.

4. Maloney, Timothy J., *Industrial Solid-State Electronics,* 2nd ed. Englewood Cliffs, NJ: Prentice Hall, Inc., 1986.

10

INPUT DEVICES: SENSORS, TRANSDUCERS, AND TRANS-MITTERS FOR MEASUREMENT

OBJECTIVES

When you have completed this chapter, you will be able to:

1. Explain the term *sensor*.
2. Explain the term *transducer*.
3. Explain the operation of three types of temperature sensors: thermocouples, RTDs, and thermistors.
4. Describe two applications for temperature sensors.
5. Explain the operation of three types of pressure sensors: Bourdon tube, strain gauge, and piezoelectric.
6. Describe two applications for pressure sensors.
7. Explain the operation of three types of flow sensors: paddlewheel, pitot tube, and orifice plate.
8. Describe two applications for flow sensors.
9. Explain the operation of three types of level sensors: capacitive, sonic, and delta P.
10. Describe two applications for level sensors.
11. Explain the operation of four types of position sensors: LVDTs, rotary and linear en-coders, resolvers, and proximity switches.
12. Describe two applications for position sensors.
13. Use sketches to explain the operation of ball screw mechanisms and rack and pinion assembly used to change rotary motion into linear motion, and how linear motion can be turned into rotary motion.
14. Explain the operation of pH sensors, density sensors, and gas detectors.
15. Explain why a technician must understand the theory of operation of a sensor as well as the electronic signal that comes from the sensor.
16. Explain the operation of a Hall effect and explain applications where it can be used.
17. Calculate flow, temperature, level, and pressure.
18. Explain why the 4–20 mA and 0–10 volt signals are the most frequently used signals from sensors and transducers.
19. Explain how you would troubleshoot various sensors and transducers.

343

Photo Courtesy of Kistler-Morse.

You are requested to select and install several sensors for a new process application that involves mixing a liquid with two dry materials to make a polymer additive that is used in making rubber products. This application will involve a series of three 10,000-gal tanks that will hold raw materials. The tank that holds the liquid is heated with a steam jacket. A mixing tank is mounted on a rail and it moves under each tank to receive its 200 lb of each of the dry ingredients and 100 gal of the liquid ingredients that will be mixed together to form the polymer additive. The mixing tank must stop at exactly the correct position precisely below each tank so that it can receive the proper amount of each ingredient.

Your job is to research the sensors needed for this system and recommend which ones you would use and explain why you made your selection. A temperature sensor and level sensor are needed for the tank that holds the liquid. A pressure sensor is needed to indicate the amount of steam pressure in the steam jacket. A level sensor is needed on each tank that holds the dry ingredient, which is in powder form. A sensor is needed to measure the amount of the dry ingredient as it is dumped into the mixing tank. A flow meter is needed to measure the liquid ingredient as it is dumped into the mixing tank. Position sensors are needed to ensure the mixing tank is in the proper position before the ingredients are dumped into it. A density sensor is needed to determine when the mixture is the proper consistency.

You should also include all of the pertinent formulas that would help you determine if the sensor you are choosing for each application is the proper size or in the range for the amount of material that is being sensed.

344

10.1 OVERVIEW OF SENSORS, TRANSDUCERS, AND TRANSMITTERS

Sensors, transducers, and transmitters are perhaps the most important parts of industrial control systems. They are used in process control systems as well as motor control and motion control systems. You will find them in virtually every system because they provide feedback information about how well the system is doing.

A *sensor* is defined as a device that is sensitive to motion, heat, light, pressure, electrical, magnetic, and other types of energy. A *transducer* is defined as a device that can receive one type of energy and convert it to another type of energy. This means that a transducer may include a sensor to sense the amount of pressure, and a circuit to convert the amount of pressure to an electrical signal and transmit it to an electrical control system where it is used as the process variable (PV) or feedback signal.

Since transducers can convert one type of energy to another, it is important to be able to identify all the types of energy. It is also important to understand that the final form of energy is generally something that is compatible with electronic circuits. This means that the output of the transducer must be able to change voltage, current, resistance, frequency, capacitance, or inductance so it is compatible with electronic circuits.

The major forms of energy that sensors can detect can be classified as motion, temperature, light, pressure, electrical, magnetic, chemical, and nuclear. There are several variations of each of these and times when they overlap slightly such as when motion causes a change in pressure, or when light causes a change in temperature.

It is important to understand that a wide variety of sensors can measure temperature, but one type may be more useful to an electronic circuit because it converts its temperature measurement to an electrical signal compatible with electronic circuits. For example, the temperature sensor may be a thermocouple or a glass bulb containing mercury that can sense changes in temperature. Since the thermocouple converts the change in temperature to a change in millivolts, and the mercury bulb converts the change in temperature to the height of the column of mercury, the thermocouple will be more useful to an electronic circuit because the change of voltage can be easily transmitted as a feedback signal.

A transmitter is a device that can convert a very small signal to a more usable signal. The transmitters for the sensors used as industrial feedback signals must typically convert very small electrical signals such as microvolts (μV), millivolts (mV), milliamps (mA) or frequency into larger voltage and current signals such as 0–10 volts or 4–20 mA. The transmitter generally uses devices such as op amps to amplify and linearize the output signal. The transmitter may also provide a zero and span circuit that allows the signal to be calibrated with other parts of the electrical system such as the single-point controller discussed in the previous chapter.

It is very important to understand that all of the sensors and transducers will be grouped into a broad category called *instruments*. The number of sensors and transducers available for use in modern industrial systems seems almost unlimited. It is necessary to classify them in groups that represent the type of energy conversion that is used, such as temperature, motion, light, and so on. The second type of classification is defined by the type of electrical signal that the device produces, such as voltage, current, frequency, and so on.

TABLE 10-1

Type of Energy Source	Type of Sensor or Transducer That Produces Voltage or Current	Type of Sensor or Transducer That Produces a Change of Resistance or Impedance
Motion	Generator	Linear potentiometer
Light	Solar cell, photovoltaic	Cds Cell
Force (pressure)	Piezoelectric	Strain gauge
Temperature	Thermocouple	Resistive temperature detector (RTD) thermistor
Magnetic	Transformer hall effects	Magnetoresistive
Chemical	Batteries, fuel cells	Different concentrations

It will be easier to understand all of these devices and how they are used in electronic circuits in industrial applications if you think of them in groups that represent their theory of operation, and the type of signal that they produce. Since these two categories are very tightly defined, you will find that you only need to learn a few theories, and you can transfer this knowledge and be able to explain the theory of operation of virtually every sensor and transducer that you find. This will also be useful when you must troubleshoot these devices. Troubleshooting the electronic part of sensors and transducers is made easier since sensors produce limited types of electrical signals that you will be familiar with. Table 10-1 shows the type of energy source in the first column and the type of sensor or transducer that converts this form of energy to voltage or current. The third column shows the type of sensor or transducer that changes the original source of energy to a change of resistance or impedance.

10.1.1 Signal Types

Sensors are designed to detect the smallest amount of change in motion, light, pressure, temperature, magnetic force, or chemical reaction. The change of energy that the sensor detects must be converted into an electrical signal that is useful to an amplifier so that the final electrical signal is in one of the traditional formats. The more typical electrical signal formats are voltage, current, resistance, frequency, capacitance, and inductance. Some formats can be defined in a second form, such as resistance can be defined in terms of conductance ($1/R$), frequency can be defined in terms of the period ($1/P$), resistance, capacitance, and inductance can be defined in terms of impedance. The unit for conductance is siemen (S) or Mhos, the unit for frequency is hertz (H), and the unit for impedance (Z) is ohms (Ω).

In some cases a transmitter will provide additional signal conditioning, such as from voltage to current, or from current to voltage. Another common form of signal conditioning is from frequency to voltage. Over the years the electronic industry has standardized several signal types. These signal types are listed in the table in Fig. 10-1.

10.1.2 Analog and Digital Signals

Another important necessary conversion of signals is the conversion of analog signals to digital formats and the conversion of digital signals to analog formats. These signal conversions have become important in the last ten years because of the large increase in digital electronic circuits and devices. Digital signals are generally

Standard Types Of Electrical Signals
0 − 1000 millivolts
0 − 20 milliamps
4 − 20 milliamps
0 − 10 volts
2 − 10 volts
−10 to +10 volts

FIGURE 10-1 Types of standard electronic signals.

classified by the number of bits that are used. For example, if the signal uses 12 bits, it converts to the binary number 4096 so values from 0–4095 can be used; if 16-bit signals are used, it converts to the binary number 65566, so values from 0–65565 can be displayed. If the 16-bit number requires a +/− sign, one of the bit locations is used as the *sign bit*, so the numbers 0 to +32767 and 0 to −32766 are represented with the remaining 15 bits.

10.1.3 Basic Sensor Circuits

Since most sensors provide a very small signal, it must be amplified or compared before it is of any use. The bridge circuit is used to compare the signal to some setpoint (SP) value and the op amp circuit is used to amplify any sensor signal that is too small. Fig. 10-2 shows a temperature sensor that changes resistance with changes in temperature. This sensor is called a *resistive temperature detector* (RTD) and it is connected to a bridge circuit. The variable resistor in the bridge is adjusted to provide a setpoint. If the temperature changes, the resistance in the RTD will change in proportion and the voltage through the bridge will reflect this change. Since the temperature can increase or decrease, the voltage at the bridge can swing to positive or negative. It is important to understand that any resistive sensor can be used in a bridge circuit.

Another circuit that is widely used with sensors is the op amp circuit. Since sensors provide a voltage signal or a change in resistance, the op amp circuit must be able to amplify either voltage or resistive signals. Fig. 10-3 shows the op amp circuit with a photoconductive cell as its sensor. In this circuit, the sensor must also have a voltage supply. When light strikes the photoconductive cell, its resistance will change. Since this resistance is actually R_{in} for the op amp circuit, it will change the gain of the op amp when the resistance changes. When the gain of the op amp changes, it will change the output voltage of the op amp.

The op amp must also be able to amplify signals from sensors that change voltage rather than resistance. Fig. 10-4 shows a circuit with a solar cell (photovoltaic cell) connected to the op amp circuit. When light strikes the solar cell, it will change voltage, and the op amp will amplify the small-signal voltage. This is the typical circuit that will be used for all sensors that produce small voltages.

FIGURE 10-2 Diagram of a resistive temperature device (RTD) connected to a bridge circuit.

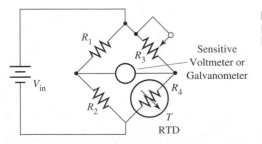

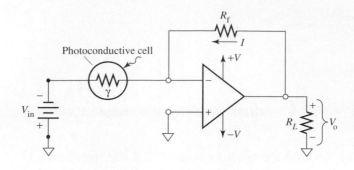

FIGURE 10-3 Diagram of a photocell that changes its resistance when light strikes it. The resistive sensors require a voltage supply.

10.1.4 Transmitters

A transmitter is a device containing a specialized circuit that can accept any of the signals from sensors such as millivolt, milliamp, voltage, or frequency pulses, and convert it to a 4–20 mA, 1–5 volt, 0–5 volt, or 0–10 volt signal. The electrical diagram of a typical transmitter is shown in Fig. 10-5. In this diagram you can see that the sensor signal is accepted as an input signal in the upper left corner of the circuit. The two-wire sensor signal is sent through a low-pass filter section and a signal amplifier, which is a circuit that contains one or more op amps to provide zero, span, and linearization capabilities. The signal is then sent through an isolation transformer and then to an output amplifier, which conditions the small signal to a more usable 4–20 mA signal. The output section of the transmitter is powered by an external voltage supply (usually 12–20 volts dc). You should notice that since the output signal for this type of transmitter is 4–20 mA, the output section will form a series circuit with the external power supply (V_S) and the system that will be accepting the signal. The system that accepts this signal as an input is identified as the load (R_L) in the transmitter circuit. You should also notice that the transmitter must have its own power supply of 12–50 volts dc to provide power for its electronic components and amplifiers. The diagram in Fig. 10-5a shows an example of an electrical diagram for a two-wire transmitter. The two-wire circuit receives a small-voltage input signal and the transmitter converts the signal to a 4–20 mA signal that is usable as the current input signal for recorders or controllers. This type of input signal is called a 4–20 mA loop.

The diagram in Fig. 10-5b shows the types of devices such as recorders and controllers that use the signal from the 4–20 mA loop. One of the main advantages of the current-loop signal is that the current is the same in all parts of the circuit, so each component in the circuit will receive the same signal. This means all the devices in the circuit such as the transmitter, ammeter, recorder, and controller are connected in series. Thus, they will all receive the same amount of current. This is

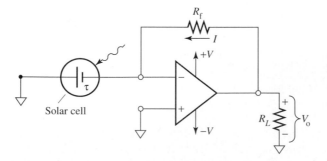

FIGURE 10-4 Diagram of a photoconductive cell that is connected to an op amp. Since the photoconductive cell produces its own voltage, it does not require an additional voltage source.

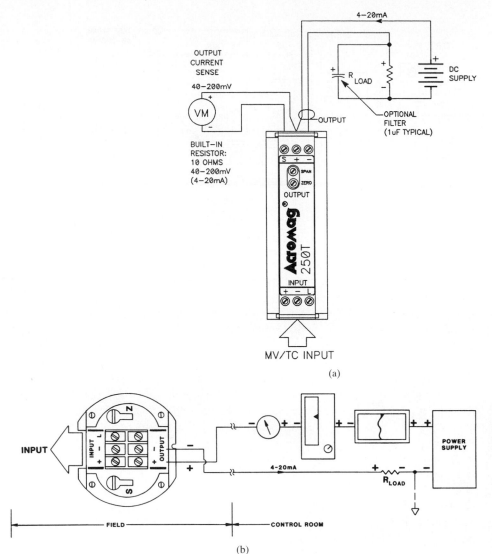

FIGURE 10-5 **(a)** Electrical diagram of a typical signal transmitter that converts a voltage signal to a 4–20 mA loop signal. **(b)** A diagram that shows an ammeter, recorder, and controller connected in series so that they all receive the same current. (Courtesy of Acromag Inc.)

important when all the devices in the circuit must be calibrated. It is also important that all of the devices in the circuit experience a loss of current if an open circuit occurs.

The other advantage of the 4–20 mA signal is that the 4 mA represents a *live zero*. This means that if the signal from the controller or transmitter ever drops to 0 mA, the circuit wiring or components have caused an open circuit to occur. If the controller or transmitter is trying to send a 0% output signal, the current loop will have 4 mA under normal conditions. Since the current loop can be used to send all components and devices connected in the loop the same amount of current, and the 4 mA offset can be used for a live zero, you will find that it is the most common type of signal used in industrial instruments and sensors.

Fig. 10-6a shows a diagram of a sensor connected to a filter, amplifier, and display as you would find in a typical industrial installation. From this diagram you can see that the sensor requires a power supply to amplify its signal. If a display is used, an additional power supply is required for it. The output signal format is

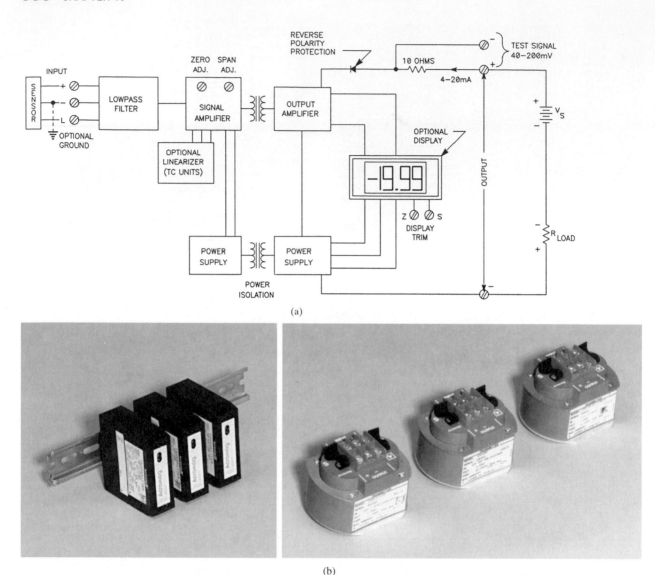

(a)

(b)

FIGURE 10-6 **(a)** Diagram of a typical sensor connected to filters, amplifiers, and displays. **(b)** Examples of signal transmitters that are mounted on DIN rails and three transmitters that can be mounted directly into any cabinet. (Courtesy of Acromag, Inc.)

4–20 mA, which means that the end user of the signal must be connected in series with the output amplifier. The end user of the signal in this diagram is indicated as the load (R_{LOAD}). Fig. 10-6b shows a picture of a transmitter mounted on a DIN rail and three thermocouple transmitters that can be mounted in a cabinet. A DIN rail is a mounting strip that is attached permanently to an electrical panel, and components such as transmitters can be mounted to the rail so that they can easily be removed and replaced if they need to be tested or changed.

10.1.5 Recorders

Another device used in many industrial applications with electronic sensors and controls is a recorder. The recorder is used extensively in the food processing and manufacturing applications because a record of critical process variables such as

temperature and time must be kept for government agencies. For example, the USDA (United States Department of Agriculture) requires records of temperature for any product that is cooked during its process to eliminate bacteria. The recorder may be a circular or strip chart type with the paper showing the readings over a specific period of time such as 24 hours. The recorder can accept 4–20 mA or 0–10 volt signals as an input. This means that recorders are fully compatible with the signal transmitters that were explained in the previous section. The recorder may have a display built in to show the amount of the PV that is recorded. Fig. 10-7 and Fig. 10-8 show pictures of several examples of recorders.

10.1.6 Alarm Systems

Another part of sensor and transducer systems can be an alarm package to indicate when the sensors are detecting values that exceed process alarms or deviation alarms. In Chapter 9 process alarms were identified as fixed alarm points indicating safety points that should not be exceeded. The deviation alarms were identified as

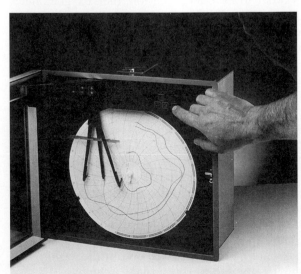

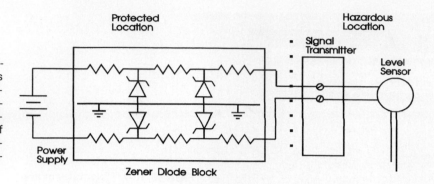

FIGURE 10-9 Electrical diagram that shows zener diodes and resistors used to limit voltage and current to provide explosion-proof sensors and transmitters for use in hazardous locations.

alarm points that moved with the setpoint to indicate quality conditions. Each type of alarm could have high and low settings with additional inner alarm points to signal when the alarm condition is approaching.

The alarm signals can be tied to a wide variety of signal lights or audible alarm horns to indicate when an alarm point has been exceeded. Many alarm packages include circuitry that records the exact value of the sensor signal as the alarm point is exceeded and the time and date that the event occurs.

10.1.7 Applications for Explosion-Proof Sensors

Sensors can be installed in all parts of an industrial application. Some applications such as food processing, mining, spray painting, and chemical processing require special enclosures or barriers for the sensors to protect them from explosions and contact with chemicals and acids. This can be accomplished by making the enclosure for the sensor so that it prevents the atmosphere from coming into contact with the sensor. In this type of application, the sensor is fully enclosed and sealed so that the explosive gases or chemicals cannot come into contact with the area where the sensor produces the source of ignition. Another way to enhance the seal is to pressurize the area inside the enclosure so that if small leaks do occur, the harmless gas that is used to pressurize the enclosure will leak out rather than allowing the explosive gas to leak in.

Another way to provide protection against explosion is to limit the amount of power the sensor will have. In order to have an explosion, fuel, oxygen, and a source of ignition must all be present. Since it can be assumed that the fuel and oxygen are always present in the area where explosion-proof sensors are needed, the goal is to limit the amount of energy available to the sensor or the part of the circuit that is located in the hazardous location. One way to limit the power to the area of the sensor that is in the exposed atmosphere is to use zener diodes to limit voltage, and resistors to limit the amount of current. Fig. 10-9 shows the block diagram for this type of circuit. From the circuit you can see that the zeners are located in the circuit between the power supply and the sensor so that the voltage to the sensor is limited. In the circuit in the bottom part of the diagram you can see that the zener-limiting circuit is used with a signal conditioner which would include the sensor. Each of these circuits meets ratings limiting the source of ignition so that these devices can be used in hazardous locations. These circuits can also be used with virtually any sensor.

The sensors that must come into contact with acids or other chemicals are provided with enclosures that are made of substances such as stainless steel to resist corrosive damage. These enclosures are also sealed to keep out the vapors and fumes that are present in these types of atmosphere. It is usually a common practice

Type	Enclosure
1	General purpose — indoor
2	Drip-proof — indoor
3	Dust-tight, rain-tight, sleet-tight — outdoor
3R	Rainproof, sleet resistant — outdoor
3S	Dust-tight, rain-tight, sleet-proof — outdoor
4	Water-tight, dust-tight, sleet-resistant — indoor
4x	Water-tight, dust-tight, sleet-resistant, — indoor/outdoor
5	Dust-tight — indoor
6	Submersible, water-tight, dust-tight, sleet-resistant — indoor/outdoor
7	Class I, group A, B, C or D hazardous locations, air-break — indoor
8	Class I, group A, B, C or D hazardous locations, oil immersed — indoor
9	Class II, group E, F, or G hazardous locations, air-break — indoor
10	Bureau of Mines
11	Corrosion-resistant and drip-proof, oil immersed — indoor
12	Industrial use, dust-tight and drip-tight — indoor
13	Oil-tight and dust-tight — indoor

FIGURE 10-10 Chart for the National Electric Manufacturers Association (NEMA) classification for enclosures. (Reprinted by permission of National Electric Manufacturer's Association.)

to pressurize the enclosures for these sensors in a similar manner as the explosion-proof enclosures. In other applications such as food processing, the equipment and sensors must be thoroughly washed down twice a day, so the sensor enclosures must protect against high levels of humidity and be able to withstand the direct spray from the wash down. These enclosures may be classified by the National Electric Manufacturers Association (NEMA). Fig. 10-10 shows the NEMA standards for enclosures. From this list you can see that the explosion-proof enclosures must have a 7, 8, or 9 rating, and if they are to be rated for wash down, they must be rated 4, 4x, or 6.

10.1.8 Overview

You can now see that even though a large variety of sensors is available for use in electronic circuits, they all have much in common, which makes it easier for industries to stock replacement parts. For this reason you will begin to see that electrically, all sensors will look similar since they tend to use either a 4–20 mA or a 0–10 volt signal. This also makes it easier to exchange sensor types and utilize a wide variety of sensors for any given application. Modern industrial electronic circuits and components provide the means to ensure that the wide range of sensors is compatible with industrial controllers through the use of transducers or transmitters. The major differences between sensors are the theories of operation they utilize to sense and amplify the signals. The remainder of this chapter will explain the different types of sensors and the theory of operation for each. You will need to thoroughly understand the theory of operation of each type of sensor so that you can troubleshoot them.

10.2 TEMPERATURE SENSORS

When you are working with systems where the temperature must be measured, you will have several choices for sensors. Let's say that you need a temperature

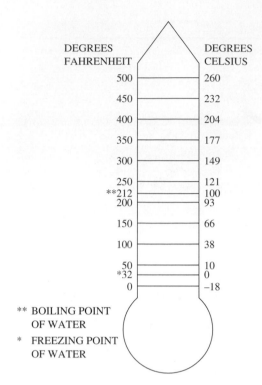

DEGREES
FAHRENHEIT

DEGREES
CELSIUS

500 — 260

450 — 232

400 — 204

350 — 177

300 — 149

250 — 121
**212 — 100
200 — 93

150 — 66

100 — 38

50 — 10
*32 — 0
0 — −18

** BOILING POINT
OF WATER

* FREEZING POINT
OF WATER

FIGURE 10-11 Conversion table for Fahrenheit and Celsius temperature scales.

sensor that produces a voltage and needs to measure temperatures between 75° and 600°. Then your choice would be a thermocouple. If the sensor signal needed to produce a change of resistance, then the choice would be a resistive temperature detector (RTD) or a thermistor. If the sensor signal needed to produce a proportional current, then the choice would be a solid-state temperature sensor.

You can start to see that each of the sensors provides a different type of output signal that will be compatible with the control circuit you are using. In addition, each type of sensor will have specific temperature ranges that it can operate in, which will also help to narrow down the choice. The remainder of this section will explain the operation of each type of sensor in greater detail.

10.2.1 Temperature Conversions

One problem for technicians is that temperature specifications will be listed in degrees Fahrenheit (°F) or in degrees Celsius (°C). Both systems are commonly used today in industry and the technician must be able to move accurately between the two systems. Fig. 10-11 shows a table that can be used to convert between the two systems. Since both of these standards are developed around the boiling point and freezing point of water, it is a good practice to memorize several temperatures such as the boiling point of water, the freezing point of water, and room temperature so that you can provide an estimate of a temperature reading from either system. For example, the boiling point of water is 212°F, and 100°C, and the freezing point of water is 32°F, and 0°C. If 72°F is considered room temperature, the equivalent temperature would be 22.2°C. One additional temperature that should be memorized is 50°C. Since this represents the halfway point in the Celsius system, it will give a good reference to estimate other temperatures (50°C is equal to 122°F). You should also notice that since the Fahrenheit system uses an offset of 32° for the

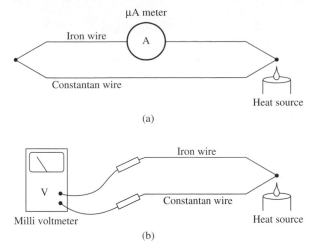

(a)

(b)

FIGURE 10-12 (a) The Seebeck effect shown by connecting two dissimilar metals at both ends to make a thermocouple junction. The Seebeck current will flow in this closed circuit. **(b)** Two dissimilar metals are joined at one end to make a thermocouple junction to produce a Seebeck voltage.

low temperature, you *cannot* simply divide the Fahrenheit scale in half to get the equivalent temperature for 50°C.

The formula for converting Celsius to Fahrenheit temperature scales is:

$$°C = \tfrac{5}{9} \, (°F - 32)$$

The formula for converting Fahrenheit to Celsius temperature scales is:

$$°F = \tfrac{9}{5} \, °C + 32$$

10.2.2 Thermocouples

The *thermocouple* is a temperature sensor that provides a small millivolt signal that ranges between 10–80 mV. Thermocouples are made by connecting two dissimilar metals at one end to form a junction. The theory of operation of the thermocouple can be explained by the Seebeck effect, which was discovered by Thomas Seebeck in 1821. The amount of voltage that the thermocouple produces depends on the two types of metal that are used to form the junction. When Seebeck made this discovery in 1821, the thermocouple was not useful because the small amount of voltage that was produced in the millivolt range could not be amplified at the time.

10.2.2.1 Seebeck Effect ■ Seebeck discovered that if two wires made from dissimilar metals are connected at both ends to make two junctions, when one end is heated, a small amount of current would flow through the circuit (see Fig. 10-12). In Fig. 10-12a you can see both ends of wires A and B are twisted together and one end is heated. Fig. 10-12b shows that if the circuit is altered so that the two wires are twisted together to make a junction at only one end and this junction is heated, the two wires will now produce an *open-circuit voltage* that is called the Seebeck voltage. The open-circuit voltage is proportional to the amount of heat added to the junction.

If a voltmeter is used to read the millivoltage that the thermocouple produced, a second junction of dissimilar metals is produced at the point where the thermocouple wire is connected to the copper wires of the voltmeter. This junction will produce a small amount of voltage in opposition to the original voltage produced at the thermocouple junction. One way to get around this problem is to extend the copper wire from the meter to point where a second junction can be produced in such a way that the voltage from this junction can be controlled.

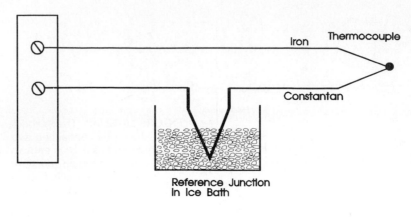

a

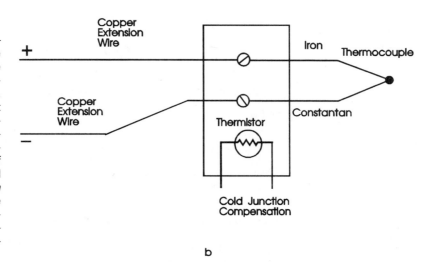

FIGURE 10-13 **(a)** Example of placing the second junction in an ice bath to provide cold-junction compensation. **(b)** Example circuit that shows a thermistor (thermal resistor) used to provide cold-junction compensation. This type of compensation is called *software compensation* because the reference voltage must be calculated with the hot-junction voltage to get an accurate reading.

b

10.2.2.2 Controlling the Reference Junction ■ The second junction explained in the previous section can be controlled so that it produces 0 volts. This is accomplished by forcing its temperature to 32°F or 0°C. In the 1800s this was accomplished by placing the junction into an ice bath, so this junction became known as the *cold junction*. Modern thermocouple circuits today can accomplish the same function with an electronic compensation circuit called *cold-junction compensation*. The cold-junction compensation circuit provides the equivalent voltage produced by the junction in the ice bath so that the actual voltage produced by the junction that is measuring the amount of heat can determine the temperature accurately. The original thermocouple junction used as the heat sensor is sometimes called the *hot junction*, and the compensation junction is sometimes called the *cold-junction* or *reference junction*. Fig. 10-13a shows a diagram of the reference junction that is placed in an ice bath, and a diagram of the cold-junction compensation circuit is shown in Fig. 10-13b.

In Figure 10-13b a special resistor that changes its resistance as the temperature changes is used to provide the cold-junction compensation circuit. This resistor is called a *thermistor* and it helps provide additional compensation that is needed if

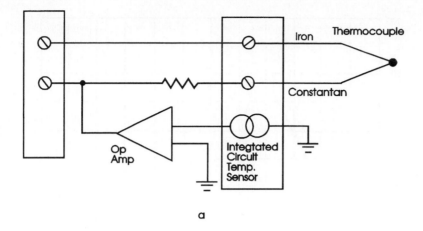

a

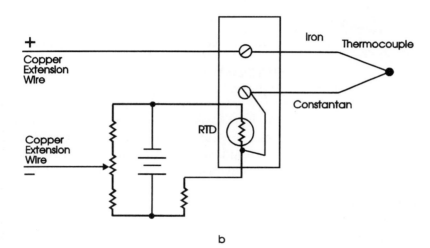

b

FIGURE 10-14 (a) An example of a temperature-sensing integrated circuit used to provide electronic ice point compensation. **(b)** An example of a battery used to provide hardware compensation to the thermocouple circuit.

the compensation junction is located where the temperature may change. For example, the temperature in a factory where the voltage reading part of the thermocouple circuit is located may be as warm as 98°F in much of the summer and be 70°F during the remainder of the year. This temperature difference would be sufficient to provide a constant error in the readings. Since the thermistor will automatically change its resistance as the temperature changes, it is an ideal component to use in the compensation circuit. The only drawback to using a thermistor for cold-junction compensation is that the effects of the compensation voltage must be calculated or computed with the hot-junction voltage to get an accurate reading. For this reason this type of circuit is called *software compensation* and is commonly used where the thermocouple is connected as a temperature sensor directly to a solid-state or electronic controller.

Cold-junction compensation could also be provided by using a small battery voltage that represents the reference junction temperature. This type of compensation is called *hardware compensation* and it is widely available on electronic temperature controllers and on the newer versions of digital voltmeters (DVMs). Since the small compensation voltage can be provided inside a digital voltmeter, it allows

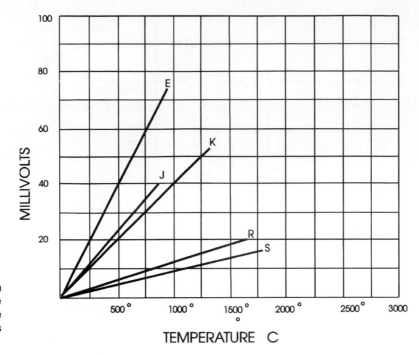

FIGURE 10-15 Graph of the temperature range and amount of voltage various thermocouples produce.

technicians to accurately check thermocouple temperatures with the voltmeter while troubleshooting. When better accuracy is required, an integrated circuit can be used in place of the battery to provide the hardware compensation voltage. The main component of this IC is a temperature-sensitive solid-state device called an *integrated circuit temperature sensor*. The electronic symbol for the temperature IC is two overlapping circles. The overlapping circles are generally shown inside a trianglc that is similar to the outline of an op amp. Since this type of compensation is provided by electronics, it is commonly called the *electronic ice point reference*. Fig. 10-14a shows an example of electronic compensation and Fig. 10-14b shows an example of battery hardware compensation.

10.2.3 Types of Thermocouples

One of the problems in designing a temperature sensor is trying to match the sensor to the range of temperatures that it is expected to measure. This means that no one sensor is able to accurately measure temperatures below 32°F and above 1000°F. For this reason a wide variety of thermocouples is needed to provide accurate readings over a wide range of temperatures. Fig. 10-15 provides a graph that shows the temperature range for five common types of thermocouples. Each type of thermocouple is identified by a single letter of the alphabet. The most common type is a *J-type* thermocouple made of iron and constantan. From the horizontal axis of the graph you can see that its temperature range is 0°C–750°C. This converts to 32°F–1382°F. The vertical axis of the graph shows that the J-type thermocouple produces approximately 40 mV at 750°C. A complete table for the J-type thermocouple is provided in the appendix.

This graph also shows that the S-type thermocouple has a range of 0°C–1450°C, which converts to 32°F–2642°F. You should also notice that the S type only produces approximately 17 mV at 1450°C.

EXAMPLE 10-1

Suppose you are asked to find the type of thermocouple that produces the most voltage for each degree of temperature. This thermocouple would provide the most resolution (mV/°C).

SOLUTION

From the graph in Fig. 10-15 you can see from the vertical axis that the E-type thermocouple produces the most voltage at over 70 mV. Its resolution would be determined by dividing 70 mV by 900°C.

$$\frac{70\,\text{mV}}{900°\text{C}} = 0.077\,\text{mV}/°\text{C}$$

10.2.3.1 Color Codes for Thermocouples ■ It is also important to understand that since the thermocouple produces a dc millivoltage, the two wires that are used as its terminals have polarity. This means that you must be able to determine which wire will produce the positive voltage and which will have negative polarity. Fig. 10-16 shows a complete list of 11 of the most common thermocouples and the polarity of each of the two wires. To make the polarity easier to identify, a standard color code has been established by the American National Standards Institute (ANSI) and it has been adopted by manufacturers. (It should be noted that the British, German, Japanese, and French have each adopted their own individual color codes, which are shown in the table.)

Type of T/C	American	British	German	Japanese	French
J Type	+ white − red black jacket	+ yellow − blue black jacket	+ red − blue black jacket	+ red − white yellow jacket	+ yellow − black black jacket
K Type	+ yellow − red yellow jacket	+ brown − blue red jacket	+ red − green green jacket	+ red − white blue jacket	+ yellow − purple yellow jacket
T Type	+ blue − red blue jacket	+ white − blue blue jacket	+ red − brown brown jacket	+ red − white brown jacket	+ yellow − blue blue jacket
E Type	+ purple − red purple jacket	+ brown − blue brown jacket	+ red − black black jacket	+ red − white purple jacket	+ yellow − purple purple jacket
N Type	+ orange − red orange jacket	* use American Standards	*	*	*

FIGURE 10-16 Thermocouple identification table showing the type of thermocouple by classification, the type of material that each thermocouple is made of, and the color code for the positive and negative wires. (©Copyright 1995 Omega Engineering, Inc. all rights reserved. Reproduced with the Permission of Omega Engineering, Inc., Stamford, CT 06907.)

In some applications where the thermocouple is mounted a long distance from the controller or display, it is not practical or cost-effective to use thermocouple wire to make these long runs. In these applications specialized extension wire is used because it is less expensive and easier to install since it is not as stiff as thermocouple wire. Extension wire is specifically designed so that it can transmit the millivoltage over a long distance and not be affected by temperature changes or voltage drops. Another important feature of extension wire is that it can be connected to thermocouple wire without making another junction that must have compensation.

•••

EXAMPLE 10-2

You are requested to identify the type and determine the polarity of a thermocouple that you suspect is not functioning correctly. The thermocouple is the temperature sensor for a piece of older equipment where you are working. Since this piece of equipment is not new, the documentation for the sensor is missing and the blueprint is severely faded so that you cannot read it, but you are able to determine that the machine was made in Milwaukee, Wisconsin, in 1975. You have determined that the thermocouple wire is white and red.

SOLUTION

Since you have determined the thermocouple wire is white and red, you can use the table in Fig. 10-16 and determine that this is a J-type thermocouple. Since you have determined this machine was built in the United States, you can assume that the thermocouple colors use the AN51 code. If your preliminary test indicates that this is not a J-type thermocouple, you may need to assume that this thermocouple may be following one of the foreign color codes.

A second test can help you determine if this is actually a J-type thermocouple and that it is following the ANSI color code. Since the table shows that the J-type thermocouple is made of iron and constantan, and the iron wire is the white wire, you can use a magnet to detect the iron wire. If the magnet shows that the white wire is magnetic, you can accurately say this is a J-type thermocouple. If the red wire is magnetic, this may further disprove your initial assumption and lead you to make further tests to determine which country this thermocouple comes from.

If you have determined this is a J-type thermocouple, you can use the table to determine that the temperature range for this thermocouple is 0°C–750°C and it will produce 0–42 mV over this temperature range.
•••

10.2.3.2 Color Codes for Foreign-Made Thermocouples ■ In recent years the amount of machinery manufactured outside of the United States has substantially increased. It is now possible to have a mixture of American- and foreign-made parts on the same machine. Much of this has occurred because multinational companies have purchased smaller companies worldwide that make a variety of parts and sensors, and the equipment builders will tend to use parts from their subsidiaries when assembling machinery.

The color code for foreign-made thermocouples is also shown in Fig. 10-16. It should be noted that foreign-made thermocouples use a color code for the jacket material that surrounds the thermocouple as well as a color code for the individual wires.

10.2.3.3 Troubleshooting and Calibrating Thermocouples and Transmitters ■ Since thermocouples produce small amounts of millivoltage for any given temperature, a precision millivoltmeter can be used to test the output signal from all types of thermocouples. The temperature at the tip of the thermocouple must be controlled while the millivoltage measurement is made. Since the thermocouple sends the transmitter a millivolt signal, a precision millivolt power supply could be used for transmitter calibration. Manufacturers have combined the voltmeter and power supply into one tester. This tester has the tables for millivolt signal strength and the cold-junction compensation built into it so the technician only needs to select the type of thermocouple and the temperature range, and the test circuit will provide an accurate reading from this. When the transmitter is being calibrated, the tester can send the voltage signal for the minimum and maximum ranges of each type of thermocouple.

10.2.3.4 Mounting Thermocouples in Thermocouple Wells ■ Thermocouples are used extensively in industrial applications to measure the temperature of liquids, surfaces, and enclosed spaces. For example, the thermocouple may be used to measure the temperature of liquids such as water, soup, cola, or any other liquid as it is heated during processing. As another example, surface temperatures of the barrel of a plastic injection molding machine or extruder must be constantly monitored to ensure that they remain at the proper temperature during processing. Enclosed areas such as the inside of a heat-treating oven or a drying oven for a spray-painting operation must also be constantly monitored. All of these applications and many others rely on the proper installation and positioning of the thermocouple for accurate temperature readings.

The operation of the thermocouple requires that its tip is placed directly in the liquid or space, or in direct contact with the surface that is being measured. Since many of these locations would destroy the tip of the thermocouple, a *thermocouple well* is used to provide a means of mounting the thermocouple without allowing it to be damaged. The thermocouple well is also called a *thermowell* and Fig. 10-17 shows a thermocouple well and probe. From this figure you can see that the well provides a cavity where the thermocouple can be easily inserted. When the thermocouple is completely inserted into the cavity, its tip will come into contact with the body of the well and heat can be transmitted between the two very easily so that the thermocouple can accurately measure the temperature.

All wells are mounted so that their tips are directly in the liquid or environment and they are secured so that a complete seal is provided. Since the outside of the thermocouple well provides the seal, the thermocouple can be quickly removed and replaced from the cavity in the inside of the well without breaking the seal between the well and the environment. The tip of the thermocouple must always be in contact with the well.

In Fig. 10-17 you can see that the thermowell is identified by the letter D, and the thermocouple element is identified by the letter C. The section identified by B is the extension assembly and these can be any length to ensure that the tip of the well is in the exact location, and the head of the assembly can be where it is easily accessible for maintenance and calibration. The *thermowell head* is identified by the letter A. The head has a cover that screws on or off to allow access to the thermocouple terminals. The cover provides protection for the head and the thermocouple terminals to keep them from becoming damaged or corroded by humidity or other caustic substances that may come into contact with the outside of the well.

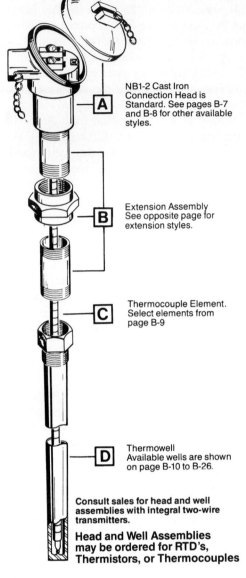

A — NB1-2 Cast Iron Connection Head is Standard. See pages B-7 and B-8 for other available styles.

B — Extension Assembly See opposite page for extension styles.

C — Thermocouple Element. Select elements from page B-9

D — Thermowell Available wells are shown on page B-10 to B-26.

Consult sales for head and well assemblies with integral two-wire transmitters.
Head and Well Assemblies may be ordered for RTD's, Thermistors, or Thermocouples

FIGURE 10-17 A thermowell that shows the thermocouple element inside. The extension section and sealed head are also shown. The letters indicate the parts of the thermowell. **A** is the sealed head, **B** is the extension section, **C** is the thermocouple element, and **D** is the thermowell. (©Copyright 1995 Omega Engineering, Inc. all rights reserved. Reproduced with the Permission of Omega Engineering, Inc., Stamford, CT 06907.)

Fig. 10-18 shows the thermocouple and the thermowell individually. The thermowell allows the thermocouple to be directly mounted in it by simply inserting the tip of the thermocouple completely into the well until it touches the bottom of the well. This type of well uses threads to provide a seal with the system. Thermowells are made from a variety of material such as stainless steel, Teflon (plastic), or Teflon-coated metal. The plastic-type thermowells are only usable in low-pressure applications with limited fluid flow, since they do not provide a seal that is as good as the metal types. The Teflon-coated and Teflon (plastic) types are useful in applications where the temperature of caustic acids, electroplating liquids, or other corrosives must be measured. You should be aware that many foods such as tomatoes are acidic and will severely corrode typical metals. Stainless steel and plastic both provide a degree of corrosion protection and they are the types of material generally used for these applications.

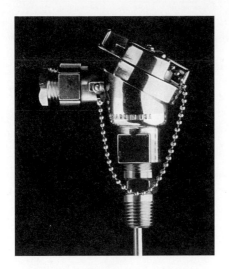

10.2.3.5 Thermocouple Transmitters ■ Thermocouple transmitters are used extensively in industrial electronic circuits because the raw signal from the thermocouple is in the millivolt range (up to 80 mV). This small amount of voltage is not very useful until it is amplified. The transmitter provides a circuit that converts the signal to a 4–20 mA signal or a 0–10 volts signal. The transmitter also provides the circuitry to *zero* and *span* the thermocouple signal. The zero and span circuit consists of potentiometers that the technician can adjust to ensure that the signal is exactly 4 mA when it is at its minimum temperature, and exactly 20 mA when it is at its maximum temperature.

10.2.4 Resistance Temperature Detector (RTDs)

In 1821 Sir Humphrey David discovered that the resistance of some metals changed when heated. This was approximately the same time that Thomas Seebeck discovered that voltage is produced when two different metals are joined to form a junction and the junction is heated. Since the amount of change in resistance was small and not uniform, Sir Humphrey David's discovery was not evaluated completely until 1871 when Sir William Siemens decided to use platinum as the metal. He found that platinum has a positive coefficient of resistance when heated. This means that the resistance of platinum increased as its temperature increased. He also found that this change was rather linear. In 1932 C. H. Meyers developed platinum into the sensor that we now know as the *resistance temperature detector* (RTD). Early versions of this sensor used a platinum wire that was wrapped around a ceramic stem. Today platinum wire is wound around a glass stem or ceramic bobbin to make the RTD. After the wire is wound onto the element, the entire sensor is sealed in glass in the shape of a bulb so that it is durable and can withstand higher temperatures. The glass bulb can be used as an exposed sensor or it can be shaped so that it will easily fit into a metal sleeve to make a probe. Fig. 10-19 shows examples of typical RTDs, which are also called *resistive temperature detectors*.

The latest RTDs use a platinum or metal glass slurry to create a thin film that is mounted on a flat ceramic substrate. The *thin film detector* (TFD) can be manufactured to be used as a small glass bead or it can be sealed in a metal element or in plastic such as Teflon. The main feature of these RTDs is that they are miniature in size, which makes them useful in temperature measurement applications where

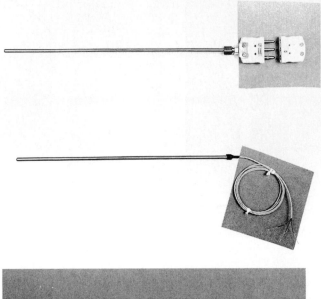

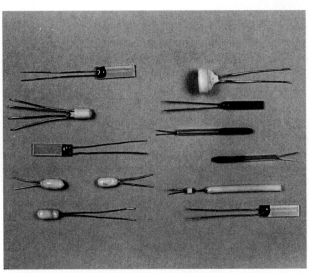

FIGURE 10-19 Examples of resistance temperature detectors (RTDs). (©Copyright 1995 Omega Engineering, Inc. all rights reserved. Reproduced with the Permission of Omega Engineering, Inc., Stamford, CT 06907.)

there is not a lot of room for a full-size sensor. Fig. 10-20 shows several examples of these smaller RTD sensors. It should be noted that since the material for the RTD is wound on a stem, the entire shaft of the RTD is temperature sensitive, whereas only the tip of a thermocouple is sensitive.

10.2.4.1 Types of RTDs ■ RTDs are available in a variety of metals with platinum, nickel, copper, and tungsten being the most common. Fig. 10-21 shows a table that indicates the temperature range and the resistance coefficient (α) for these RTDs. The temperature coefficient is the amount of resistance change you can expect for each °C of temperature change. For example, the resistance of a platinum RTD is 100 Ω at 0°C. Since its resistance coefficient is 0.0039, it must be multiplied by 100 Ω to determine the amount of resistance change per °C.

$$0.0039 \times 100 \ \Omega = 0.39 \ \Omega/°C$$

Since the platinum RTD will change 0.39 Ω for each °C, if the temperature changes 100°, the new resistance would change 39 Ω. This means that 39 Ω is added

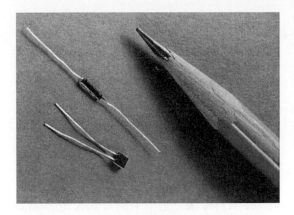

FIGURE 10-20 Examples of miniature RTD sensors made from thin film technology. You can see the relative size of the RTD sensor as compared to a standard pencil. (©Copyright 1995 Omega Engineering, Inc. all rights reserved. Reproduced with the Permission of Omega Engineering, Inc., Stamford, CT 06907.)

to the original 100 Ω resistance at 0°C and the new resistance for the RTD would be 139 Ω at 100°C. The following calculations show this.

$$0.39\ °C/\Omega \times 100° = 39\ \Omega$$
$$39\ \Omega + 100\ \Omega = 139\ \Omega \text{ at } 100°C$$

This table shows the temperature range for platinum as −184°C–815°C and for nickel as −73°C–149°C, which is the linear range for these RTDs. The extended range for platinum is −270°C–982°C, and for nickel the extended range is −101°C–315°C. These RTDs do not maintain a linear curve when they are measuring temperatures in their extended ranges.

10.2.4.2 Two-Wire, Three-Wire, and Four-Wire RTDs ■ Fig. 10-22 shows examples of two-wire, three-wire, and four-wire RTDs. The symbol for the RTD is a resistor with an arrow through it indicating it is a variable resistance. The resistor is also identified by R_t to indicate this resistor is temperature sensitive.

Fig. 10-22a shows a two-wire RTD; Fig. 10-22b shows a three-wire RTD with terminals 2 and 3 connected to a common point. The extra wire is used to provide temperature compensation and the typical circuit for this is shown in Fig. 10-23. Fig. 10-22c and 10-22d show two configurations for a four-wire RTD. Fig. 10-22c shows a second wire R_2 that is connected to R_1, and R_3 connected to R_4. Since terminals R_1 and R_4 have extra wires connected to them, they both can be used in the circuit to provide junction temperature compensation. Fig. 10-22d shows a second way to manufacture a four-wire RTD. In this application a two-wire RTD is used with two additional wires added as a lead resistance loop at terminals R_3 and R_4. The lead resistance loop is used in bridge circuits to provide temperature compensation.

Type of RTD	Temperature Range °C	Resistance Coefficient Alpha (α) OHMS/C°
Platinum	−184 to 815	0.0039
Nickel	−73 to 149	0.0067
Copper	−51 to 149	0.0042
Tungsten	−73 to 276	0.0045

FIGURE 10-21 A table for RTDs that shows the type of material, temperature range, and amount of resistance per °C. (©Copyright 1995 Omega Engineering, Inc. all rights reserved. Reproduced with the Permission of Omega Engineering, Inc., Stamford, CT 06907.)

FIGURE 10-22 a–d Examples of two-wire, three-wire, and four-wire RTDs. (©Copyright 1995 Omega Engineering, Inc. all rights reserved. Reproduced with the Permission of Omega Engineering, Inc., Stamford, CT 06907.)

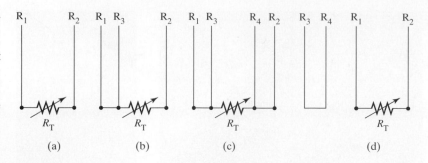

10.2.4.3 Electrical Circuits for RTDs

■ Since RTDs vary the amount of their resistance when they are heated, they must be connected to a resistive-sensitive circuit like a Wheatstone bridge to produce a usable signal as a sensor. Fig. 10-23 shows an example of a two-wire RTD connected to a bridge circuit. The two terminals in the middle of the bridge can be connected to a digital voltmeter (DVM) or they can be connected to a controller if the RTD is used as the temperature sensor for the system. Since the RTD can only change resistance, the bridge circuit must have an external power supply. In some applications the RTD must have additional wires connected to it to provide temperature compensation. Temperature compensation is necessary in some applications because of the difference in temperature between the point where the RTD is mounted and the location where the transmitter or controller is converting the value. When the two-wire RTD is connected in the Wheatstone bridge circuit as in Fig. 10-23, it becomes one of the "legs" of the bridge. In this configuration, the RTD does not have any means of providing temperature compensation. This configuration is the most common circuit for the RTD.

The circuit in Fig. 10-24 shows a three-wire RTD. In this circuit you can see that the third wire for the RTD is connected at the same terminal as one of the original two leads. This extra lead can be used in a wide variety of circuits to cancel the effects of unwanted temperatures so that all changes in resistance to the bridge come from the RTD sensor. Unwanted temperatures can come from the wire heating slightly due to current flowing through it, or from temperature changes of the air that is near the sensor and its wires. The compensation occurs because the same amount of current will flow in the compensation lead as in the original lead and the design of the circuit causes the voltage drop across each set of terminal wires to be the same, which effectively allows them to cancel each other.

Fig. 10-25 shows a diagram of a four-wire RTD. In this configuration, one additional wire is connected to each end of the original two-wire RTD. These additional wires provide another way to compensate for unwanted changes in resistance. The two-wire RTD is adequate for the vast majority of temperature sensors. The three-wire and four-wire RTDs are available for applications where the RTD must have greater accuracy.

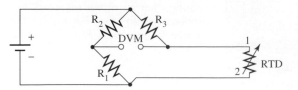

FIGURE 10-23 Electrical diagram of a two-wire RTD connected to a Wheatstone bridge circuit.

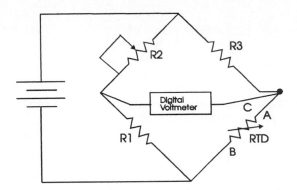

FIGURE 10-24 Diagram of a three-wire RTD connected to a Wheatstone bridge.

10.2.4.4 Using the RTD as a Motor-Starting Switch ■ Another way RTDs are used is to put them in series with the start winding of a single-phase motor to take the place of the start (centrifugal) switch. The start switch is used to energize the start winding of the single-phase motor for just a few seconds until the motor's armature comes up to approximately 75% speed. When the motor reaches 75% speed, the centrifugal switch is turning fast enough to open and disconnect the start winding until the next time the motor needs to start. When the motor is de-energized, it will slow down sufficiently to allow the contacts of the centrifugal switch to go back to the closed position. The RTD can duplicate the function of the centrifugal switch by being at low resistance when no current is flowing, and the resistance will increase to a point to stop current from flowing. The resistance of the RTD will increase as the amount of starting current increases and warms it up. This type of circuit is used predominantly on refrigerator compressors that only start and stop several times an hour. The RTD cannot be used to start motors that must turn on and off frequently because the RTD must have time to cool down before its resistance is low enough to allow full current flow to the start winding. Fig. 10-26 shows an example of this circuit.

10.2.4.5 Troubleshooting RTDs ■ The simplest way to troubleshoot the RTD is to use an ohmmeter. Since you know that the RTD must have some amount of original resistance at room temperature, you should be able to disconnnect and isolate its leads so you can measure its resistance. If the RTD is platinum, the amount of resistance should be approximately 110 Ω. The next step would be to heat the RTD and see if the resistance increases. Even though you may not know the actual temperature, this field test will indicate if the RTD can change resistance as its temperature changes. Remember, the resistance change may only be a few

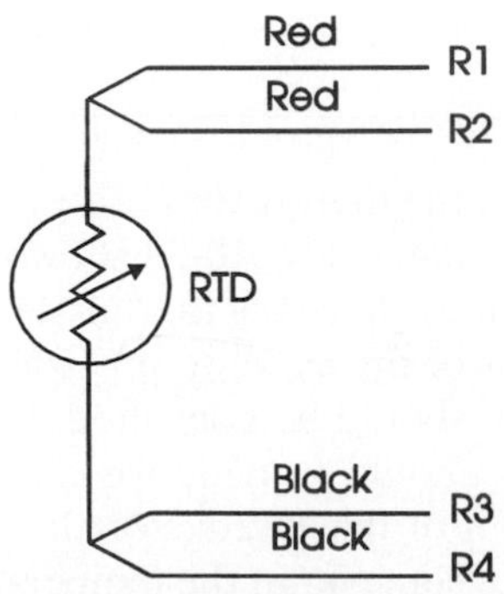

FIGURE 10-25 Electrical diagram of a four-wire RTD.

Red — R1
Red — R2

RTD

Black — R3
Black — R4

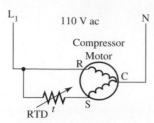

FIGURE 10-26 RTD used as the starting switch for a single-phase refrigeration compressor motor.

ohms. If the resistance does not change or if the resistance is infinity (∞), the RTD should be replaced.

The most common problem with RTDs is that the wire that is used for the sensor will break and cause an *open circuit*. If the original resistance reading of the RTD is infinity (∞), it indicates the RTD is open and it must be replaced. It is also possible for the RTD to become shorted, which would cause the resistance reading to be near zero and it would not change as the temperature changes.

After you have determined that the RTD is operational, its circuit may need to be calibrated. A calibration tester is available that is basically a precision Wheatstone bridge, which allows you to measure the amount of resistance the RTD provides for each temperature throughout its range. You must also be able to accurately control the temperature for each reading if the calibration is to be successful.

10.2.5 Thermistors

The thermistor is a temperature-sensitive resistor much like the RTD. The major difference between the thermistor and the RTD is that the thermistor generally has a negative temperature coefficient and the thermistor exhibits a greater change of resistance for each degree of temperature change. A negative temperature coefficient means that the resistance of the thermistor decreases as its temperature increases. It should be noted that some thermistors are available that have a positive temperature coefficient, but the vast majority used as temperature sensors will have a negative temperature coefficient. The main feature that makes the thermistor so useful is that it is far more sensitive than the thermocouple or the RTD. It is important to understand that the resistance of the thermistor is not linear.

Thermistors are generally manufactured in two broad categories. The first uses semiconductor material that is formed into a particular shape such as a beaded head, probe shape, or rod shape, and covered with glass. The second uses the semiconductor material that is manufactured into the shape of a wafer, disk, or washer and then is coated with a metal cover. The shape of these thermistors makes them useful in industrial applications where they may be mounted directly to nuts and bolts so that they are held in place. Fig. 10-27 shows examples of these types of thermistors.

10.2.5.1 Linearity of Thermistors ■ The response graph of the thermistor is different from the thermocouple and the RTD in that it is not linear. Fig. 10-28 shows the typical response for thermocouples, RTDs, and thermistors. Since the thermistor response is a curve, it is important to locate linear portions of the curve to provide the best accuracy and repeatability. In some cases a table should be consulted to determine the exact amount of resistance the thermistor should provide for any given temperature. You should also notice from the graph of the thermistor that the resistance decreases as the temperature increases, and at some point the temper-

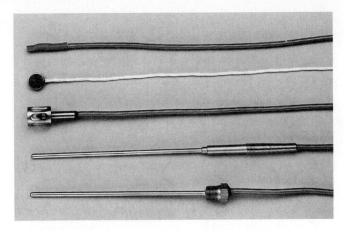

FIGURE 10-27 Examples of thermistors used as temperature sensors. (©Copyright 1995 Omega Engineering, Inc. all rights reserved. Reproduced with the Permission of Omega Engineering, Inc., Stamford, CT 06907.)

ature will increase sufficiently to make the resistance equal zero and the thermistor is no longer useful as a temperature sensor.

10.2.5.2 Applications for Thermistors ■ Thermistors are used in a variety of temperature-sensing applications that are similar to the RTD because they change the amount of resistance as the temperature changes. This means that thermistors will mainly use Wheatstone bridge circuits to provide the signal conditioning necessary for use as a sensor signal. You should notice from the examples in Fig. 10-29 that the symbol for the thermistor is similar to the RTD in that a variable resistor symbol is used and the word *thermistor* or R_t is used to indicate that the resistor is temperature sensitive. The first circuit example (Fig. 10-29a) shows the thermistor used as a very accurate temperature sensor that is connected as one leg of the bridge. Fig. 10-29b shows thermistors mounted in several locations that are all connected to one bridge circuit through a switch. The switch will make a connection to one thermistor at a time. Fig. 10-29c shows the thermistor providing temperature compensation for a precision meter movement. The precision meter movement will not be accurate if the ambient temperature where the meter is located increases or decreases and the internal resistance is changed. If the meter movement is subjected to higher currents, it will also tend to heat up. The thermistor will change its resistance and automatically compensate for the temperature differences.

Fig. 10-29d and Fig. 10-29e show examples of two or more thermistors connected to the same bridge circuit. In this configuration, two thermistors can be located some distance apart to provide two separate temperature readings that can

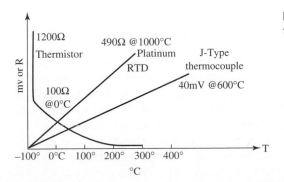

FIGURE 10-28 Response curves for thermistor, RTD, and thermocouple.

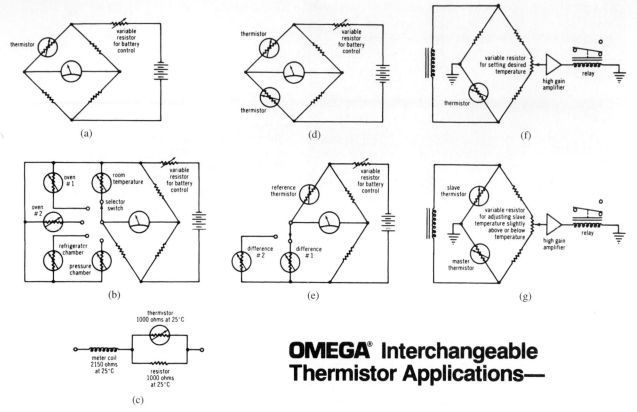

FIGURE 10-29 Seven applications circuits for thermistors. (©Copyright 1995 Omega Engineering, Inc. all rights reserved. Reproduced with the Permission of Omega Engineering, Inc., Stamford, CT 06907.)

be used as a differential reading. For example, one of the thermistors can be located at the inlet of a heat exchanger, and the other can be located at the outlet. The bridge circuit will show the temperature difference between the inlet and outlet. This type of configuration is also useful in air-conditioning system troubleshooting where the temperature differential of parts of the system such as expansion valves, evaporator coils, and condenser coils can be used to provide information for troubleshooting. Fig. 10-29f shows a thermistor used as a temperature sensor for an alarm circuit, or it can provide on/off control for a heating element. The output signal from the bridge is sent to an amplifier that is connected to the coil of a relay. If the temperature causes the resistance of the thermistor to increase or decrease and make the bridge unbalance, current will flow to the amplifier and it will provide current to the relay. The contacts of the relay can be used to switch an alarm horn or light on or off if the circuit is used as an alarm. Or the contacts can switch the heating element on if the temperature is low, and off if the temperature is high. In this type of application the thermistor will act like a sensor and the variable resistor is used to adjust the alarm setpoint temperature. Fig. 10-29g shows two thermistors used in a master/slave application. This type of circuit is used where two separate heating chambers or two separate liquid baths must be maintained at the same temperature. If the temperature of either chamber or bath becomes too hot or too cold, the bridge will become unbalanced and energize the output. The amplifier and relay part of this circuit are similar to the previous example.

10.2.5.3 Troubleshooting Thermistors ■ The thermistor is a temperature-sensitive resistor that usually has a negative temperature coefficient. This means that you can use an ohmmeter to test the thermistor if it is isolated or removed from its circuit and power source. As heat is added to the thermistor, its resistance should drop. You will need to have some data about the thermistor's temperature curve to know at what temperature the resistance becomes 0 Ω. If the temperature is in the middle of the thermistor's operating range, the amount of resistance change should be much larger than an RTD. For example, the resistance of one standard type of thermistor material would have 19.59 kΩ at 0°C and 407 Ω at 100°C.

If the resistance of the thermistor does not change, or if the value is always infinity (∞), the thermistor is defective and must be changed. The infinity reading indicates the thermistor has an open, but you must use a very high resistance scale on the meter and try to increase the temperature as much as possible to get the resistance to move lower.

10.2.6 Integrated Circuit (IC) Solid-State Temperature Sensors

When semiconductor material was first manufactured and used for diodes and transistors, it was found that the material also changed the amount of voltage across the PN junction when it is forward biased as the temperature changes. The early attempts at using the semiconductor material were not too successful because impurities in the material made its response nonlinear. Today with modern manufacturing technology, the solid-state material used in making ICs provides a linear output response. The semiconductor material also provides a compact sensor when it is manufactured into an IC. The IC sensors are very small and shaped like typical small transistors, which allows them to be placed in printed circuit boards and other electronic circuits. Fig. 10-30 shows examples of these types of solid-state IC sensors.

10.2.6.1 Applications for IC Temperature Sensors ■ Fig. 10-31 shows two example circuits for the IC temperature sensor. You should notice that the symbol for this device is two interlocking circles with a + and − used to indicate polarity. This type of sensor changes the amount of voltage across its PN junction as the temperature of the junction changes. In the first circuit, a bias voltage is applied to the circuit and a variable resistance is used to provide a setpoint. As the temperature changes, the amount of current will change proportionally because the voltage across the PN junction of the IC will change. This can provide very accurate temperature readings between −55°C–150°C. The IC sensor can be used on printed circuit boards to

FIGURE 10-30 Examples of solid-state IC temperature sensors. (©Copyright 1995 Omega Engineering, Inc. all rights reserved. Reproduced with the Permission of Omega Engineering, Inc., Stamford, CT 06907.)

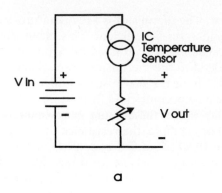

a

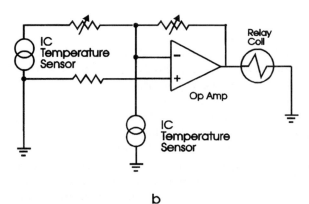

b

FIGURE 10-31 **(a)** Electrical diagram of an IC temperature sensor used in a circuit with a variable resistor to provide a setpoint setting. **(b)** Electrical diagram of two IC temperature sensors used with an op amp.

monitor the temperature of the heat sink or it can be integrated into a cold-junction compensation circuit for a thermocouple.

The second circuit shows two IC temperature sensors used with an op amp. The first IC is used as a temperature sensor and the second is used to provide temperature compensation for the circuit. You can troubleshoot the IC temperature sensor while it is in circuit. The amount of voltage through the PN junction should change as the IC temperature is changed. If the IC is used in a circuit where this voltage change provides a voltage drop, the current in the circuit will also change so you can use a voltmeter or ammeter to test the circuit. You should also understand that the change in temperature for the IC can come from external sources in much the same way as other temperature sensors are used, or the temperature change can come from a change in the amount of current flowing through the junction. An increase in current will cause an increase in temperature.

10.2.7 Infrared Thermometry and Optical Pyrometers

Newer technology has provided electronic circuits and detectors that allow noncontact infrared sensors to be used to measure temperature. It is important to understand that all objects emit energy if their temperature is above absolute zero. The amount of energy increases as the temperature of the object increases. Fig. 10-32 shows a picture of a typical infrared temperature-measuring device. The sensor is mounted in a hand-held instrument whose physical appearance is similar to a gun.

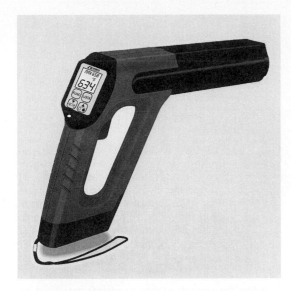

FIGURE 10-32 Hand-held infrared pyrometer. (©Copyright 1995 Omega Engineering, Inc. all rights reserved. Reproduced with the Permission of Omega Engineering, Inc., Stamford, CT 06907.)

The gun is used by pointing it at the heat or energy source and reading the amount of energy emitted by it. This type of temperature-measuring device is required where noncontact measurement is necessary because of moving parts or where the part may be contaminated if a temperature probe is placed directly on it.

Fig. 10-33 shows a block diagram of the operation of an infrared pyrometer. The pyrometer receives light waves that are emitted from a source whose temperature is being measured. When the lens of the pyrometer is focused on the part being measured, the emitted light waves are received by the detecting circuit. The signal from the detector is sent to an op amp where it is amplified. The output from the op amp can be sent directly to a meter movement or display to indicate the amount of heat, or the signal can be sent to a signal conditioner where it is converted to a 4–20 mA or 0–10 volt signal that is useful as the process variable signal for controllers and recorders.

10.2.7.1 Industrial Applications for Infrared Thermometry ■ Infrared thermometry is useful in noncontact temperature measurements such as determining the

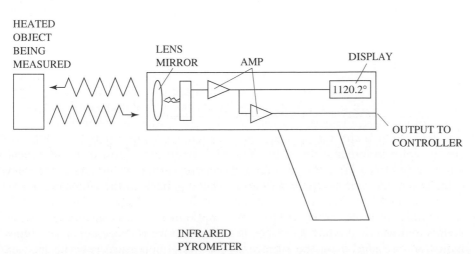

FIGURE 10-33 Block diagram of an infrared pyrometer.

FIGURE 10-34 Example of infrared thermometry used to provide noncontact temperature measurement of printed circuit boards on a production line. (©Copyright 1992 Omega Engineering, Inc. all rights reserved. Reproduced with the Permission of Omega Engineering, Inc., Stamford, CT 06907.)

temperature of a ladle of molten iron, or the temperature of a continuous sheet of glass as it moves from a curing oven. Smaller units are also useful on a manufacturing line to determine hot spots or overheated components on printed circuit boards during final quality control tests when boards are powered up for the first time as they are manufactured. Fig. 10-32 shows an example of a infrared pyrometer, and Fig. 10-34 shows an example of testing for warm components on a final inspection of a printed circuit board during the manufacturing process. The smaller sensors are called infrared thermocouples because they are approximately the same size as a thermocouple.

Another important application of noncontact temperature measurement is thermal photography, which is used to troubleshoot hot spots on industrial motors, switch gears, and transformers where the extra heat is an indication of a malfunctioning component or a loose connection. In this type of application, a camera with infrared-sensing capability is used to take a picture of the suspect part. The picture will indicate where temperatures are normal or abnormal. Once a hot spot is detected, appropriate action must be taken to prevent severe damage due to overheating.

10.2.7.2 Nonreversible Temperature Indicators ■ Another method to measure temperature is to use indicators and labels that are not reusable. These indicators are placed on the object whose temperature is being measured, and they will indicate the highest temperature the object was subjected to. Fig. 10-35 shows examples of these types of temperature sensors. The simplest sensor is a *single dot* of material that changes color when the specified temperature is exceeded. More complex indicators have four or more dots that will each change color at a different temperature. For example, the dots may change color at 5°F increments beginning at 100°F through 130°F. If the temperature has reached 110°F, the first three dots would change color to indicate this is the highest temperature. Most of these indicators must be discarded after they are used. These indicators use material that changes color at specific temperatures, and do not change back to the original color once the temperature is lowered.

Other examples of these types of temperature sensors include crayons and pellets that will melt when their specified temperature is exceeded. These types of indicators are placed on the surface where the temperature is to be measured.

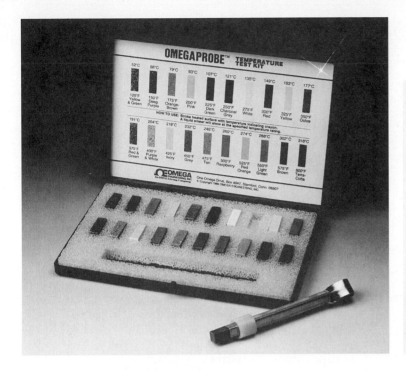

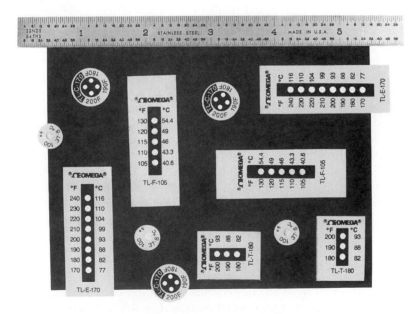

FIGURE 10-35 Examples of nonreversible temperature indicators that are used one time and discarded. (©Copyright 1995 Omega Engineering, Inc. all rights reserved. Reproduced with the Permission of Omega Engineering, Inc., Stamford, CT 06907.)

When the temperature reaches the specified value, the crayon or pellet will melt. It is not possible to determine how much the temperature has exceeded the melting point of the material, so a series of pellets or crayons are usually used so that one or more remain unmelted. This provides a means to determine that the temperature is in a range of not more than the temperature of the unmelted pellet but more than the range of the melted pellet.

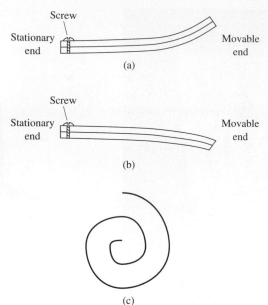

FIGURE 10-36 **(a)** A bimetal strip that is warped upward when it is heated. **(b)** A bimetal strip that is warped downward when it is cooled. **(c)** A bimetal strip in the shape of a coil to amplify the movement caused when the coil is heated.

10.2.8 Nonelectrical Temperature Sensors

Several types of nonelectrical temperature sensors use a variety of bimetal strips and filled-bulb temperature sensors. The bimetal temperature sensors work on the principle that two different types of metal are bonded (joined) together so that when they are heated one will expand faster than the other. This causes the bimetal strip to curl one direction when it is heated and the other direction when it is cooled. The bimetal effect can be amplified by shaping the sensor in the shape of a coil or helix. When the two types of metal begin to expand as they are heated, the end of the coil will move farther than a single piece of bimetal material. Fig. 10-36 shows an example of the bimetal strip when it is heated and when it is cooled and a bimetal strip in the shape of a coil. A different version of the bimetal sensor is the filled-bulb sensor. The bulb on this type of sensor is filled with alcohol or refrigerant that will change pressure when it is heated. When the fluid changes pressure, it will exert pressure on a Bourdon tube or helix and cause it to move or change shape. This movement can be used to activated a switch, so these types of devices can be used as thermostats or alarm switches.

10.2.9 Comparison of Temperature Sensors

At times you will need to compare the operation of a thermocouple, an RTD, a thermistor, and an IC temperature sensor. Fig. 10-37 shows a comparison of the thermocouple, RTD, thermistor, and IC sensor. From this chart you can see a graph of the output of each device as it is compared with a change in temperature. For example, the thermocouple provides a millivolt signal, the RTD provides a change of resistance that increases when temperature increases, the thermistor provides a change of resistance that decreases when temperature increases, and the IC sensor provides a voltage or current signal when temperature increases.

At this point we will review the advantages and disadvantages for each sensor. For example, some of the advantages of the thermocouple include that it is self-

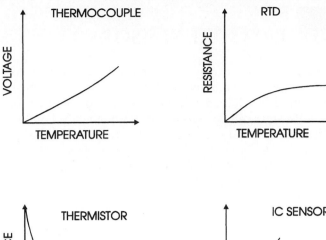

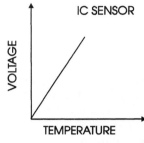

FIGURE 10-37 A comparison of thermocouples, RTDs, thermistors, and IC temperature sensors.

powered because it produces a millivolt signal which means it does not need a power supply, its two wire operation is simple to use, it is rugged and can be used in applications where the temperature is extreme and where it may get bumped frequently, and they have a wide range of devices available for different temperature ranges. The disadvantages of the thermocouple include that it is nonlinear in parts of its response curve, it requires a temperature compensation circuit (reference), it is the least stable, and it is the least sensitive.

The RTD is the most stable and most accurate of the temperature devices discussed in this chapter. Some disadvantages of the RTD are that it is more expensive than the other devices, it requires a current source such as a wheatstone bridge circuit with a power supply to provide a useful signal, it has a low absolute resistance, a small ΔR, and it is self-heating which means that it will add heat to itself as current flows through it which will cause errors in some readings.

The advantages of the thermistor are that it has a high output change of resistance for a change of temperature, it is faster reacting to a temperature change, and it can provide a signal from a two-wire configuration. The disadvantages of the thermistor are that it is not linear which means that only a small portion or its range can be used at a time, or it must be used in applications that do not require complete linearity such as a high/low alarm, it has a very limited temperature range, it is fragile and can be broken easily, a current source like the wheatstone bridge is required, and it is self-heating like the RTD.

The IC sensor is the most linear of these sensors. It has the highest output for the amount of temperature change, and it is by far the most inexpensive. Its disadvantages include its temperature range is less than 200°C, it requires a power supply like a traditional solid state circuit, it is slow to react to temperature changes, it is self-heating, and it has a limited number of configurations.

10.3 PRESSURE SENSORS

Pressure sensors are among the most useful sensors for industrial applications because pressure can be converted to determine the liquid level of a tank or the temperature in addition to measuring the amount of force that is placed on an object. Pressure can be used to determine the level of fluid in a tank by converting the amount of head pressure into the height of a column of the liquid. Pressure can also be used to determine temperature from conversion tables because there is a direct correlation between the temperature and the pressure of confined gases. Pressure can also be used to measure flow by calculating the pressure drop across an orifice plate. This means that you will see pressure sensors in a wide variety of sensor applications, but the basic theory of operation of the pressure sensor is all that is required to troubleshoot and repair all of these types of sensors. This section will explain the terminology of pressure, the operation of the different types of sensors, and applications of using pressure sensors in industrial measurements.

It may be helpful to start this section with several common problems you will face when you are working in industry. The first problem is that you are asked to select a sensor that can determine the amount of pressure that steam is exerting in a steam jacket that is used to heat the product your company is making. It is important to have a visual indication of the amount of steam and also have the sensor produce a 4–20 mA signal that can be used as the process variable for a closed-loop controller that controls the steam valve.

A second problem involves selecting a pressure sensor to determine the amount of vacuum (pressure below atmospheric pressure 14.7 psi) that exists in a product testing system that tests air-conditioning tubing. A third problem is to select a sensor that can measure the pressure of the hydraulic system on a plastic injection molding system that has pressures that reach 2000 psi. You will find that this section provides you with enough information to solve all of these problems.

10.3.1 Parameters of Pressure

It is important to understand the terms that are used to discuss pressure before you can begin to understand the theory of operation for different sensors. Pressure is defined as the *amount of force applied to an area*. This can be defined mathematically as pressure is equal to force per unit area:

$$P = \frac{F}{A}$$

The unit of pressure can be described in *pounds per square inch* (psi), which is the common standard for English units. To understand this concept, you need to understand all of the ways force can be used. In simplest terms *force* may be defined as any push or pull. If the force is produced by the shaft of a motor in a circular motion, it is defined as *torque*. Other important terms that refer to pressure are the terms *stress* and *strain*. *Stress* can be defined as the ratio of force over the cross-sectional area. These are basically the same terms that we used earlier to define pressure, so this means that the terms *stress* and *force* may be used in similar examples. The term *strain* refers to the change an object goes through when force is applied to it. For example, if four vertical support beams are used to hold up a tank of water, each of the beams will be deformed (compressed) slightly by the weight of the tank and water as the tank is filled. The deformation (change in length) can be divided by the original length to determine the amount of strain. In this section of the chapter, stress will be used to describe the force per unit area applied to a body, and strain will be used to describe the change in the length of the body. From the previous example, stress is the force that the weight of the water exerts on the beams, and strain is the amount each of the beams is compressed.

The next important area of pressure that must be understood is the units of measurement for pressure. Pressure measurement uses the amount of atmospheric pressure as a base value for most of the units of measure. For example, the amount of air pressure caused by the atmosphere at sea level is 14.7 psi. This means that each square inch of land has 14.7 pounds of force pushing down on it. This pressure is coming from the weight of the air around us. If you climb a mountain, the amount of air is less dense so the amount of pressure that it exerts will be less than 14.7 psi. Since the atmosphere is always around us, it is used as the starting point (zero reading) for most pressure measurements. For example, if you used a pressure gauge to measure the air in the tires of an automobile, the pressure gauge would read zero if you examined it before you placed it on the valve stem of the tire to measure the pressure. This means that the air pressure gauge is calibrated so that the 14.7 psi of atmospheric pressure that continually surrounds us is not added to the pressure of the air in the tire. If the atmospheric pressure was included in the pressure reading, it would make the value larger. You can now see why it will be important to indicate if the pressure reading includes atmospheric pressure or not. If atmospheric pressure is included in the reading, it is called *absolute pressure* and the units will be defined as *psia*. If the reading is from a gauge that does not include the original 14.7 psi from the atmosphere, it will be called *gauge pressure* and the units will be defined as *psig* or *psi*. In industry, if the term *absolute* is not specified with the pressure units, then the reading is designated as a gauge reading. It is also important to understand that since all readings are assumed to be gauge readings unless specified, the letter *g* is generally not used and the units are shortened to psi. This means that you are probably more familiar with the psig system than you realize because every time you used the term *psi* to describe the amount of air in a tire, or the amount of pressure in a hydraulic system, you were using the psig system. Fig. 10-38 shows a graph of absolute pressure and gauge pressure.

10.3.2 Vacuum: Pressure Readings Below Atmospheric Pressure

When the pressure is below atmospheric pressure, it is called a *vacuum*. The pressure range of vacuum is from 14.7–0 psi on the absolute scale (psia) or 0 to −14.7 psi if you are using the gauge scale (psig). (You should remember that most common pressure gauges read zero at atmospheric pressure so the common pressure gauge will need to read negative numbers to indicate pressures that are in the vacuum range.) This causes some problems when you are first introduced to pressure measurements since scientists and technicians will refer to the vacuum range as 14.7–0 psia, and you are trying to measure the vacuum with a gauge that reads the vacuum in values below 0.

Another problem involved in reading vacuum with a gauge is that most gauges that are designed to read over 100 psi have a difficult time accurately reading small changes in pressures below atmospheric pressure. This problem is overcome by using a column of water or a column of mercury to measure values in the vacuum range. Scientists found that if they placed water in a U-shaped tube, allowed one end of the tube to be open to the atmosphere, and placed a pump on the other

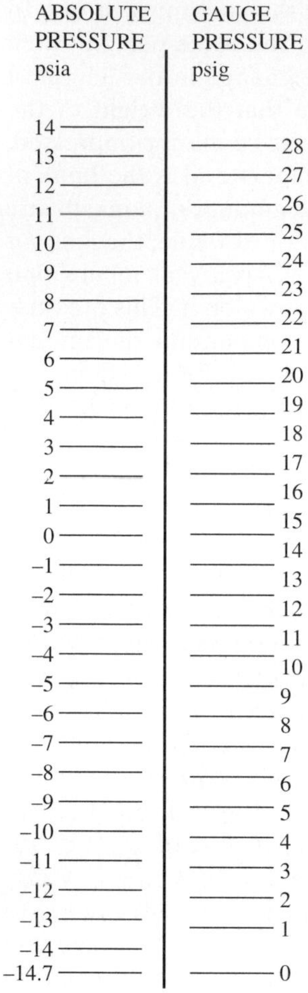

ABSOLUTE   GAUGE
PRESSURE   PRESSURE
psia   psig

```
14 ——————          —————— 28
13 ——————          —————— 27
12 ——————          —————— 26
11 ——————          —————— 25
10 ——————          —————— 24
 9 ——————          —————— 23
 8 ——————          —————— 22
 7 ——————          —————— 21
 6 ——————          —————— 20
 5 ——————          —————— 19
 4 ——————          —————— 18
 3 ——————          —————— 17
 2 ——————          —————— 16
 1 ——————          —————— 15
 0 ——————          —————— 14
-1 ——————          —————— 13
-2 ——————          —————— 12
-3 ——————          —————— 11
-4 ——————          —————— 10
-5 ——————          —————— 9
-6 ——————          —————— 8
-7 ——————          —————— 7
-8 ——————          —————— 6
-9 ——————          —————— 5
-10 —————          —————— 4
-11 —————          —————— 3
-12 —————          —————— 2
-13 —————          —————— 1
-14 —————
-14.7 ———          —————— 0
```

FIGURE 10-38 A comparison of absolute pressure (psia) and gauge pressure (psig).

end to begin to remove all of the air from one side of the tube, the water in the tube would rise on the side where the air was pumped out. The U-shaped tube is called a *manometer.* The water was actually being pushed up the column by the atmospheric pressure (14.7 psi) pushing down on the open side of the manometer. Scientists found that 14.7 psi had the ability to raise the column to a height of 33.9 ft. This also means that a complete vacuum could lift a column of water 33.9 ft. When this experiment was duplicated and mercury was used in place of water, it was found that a complete vacuum was able to raise a column of mercury to a height of 29.92 in. This number should sound familiar to you since weather forecasters use a mercury manometer to determine if the atmospheric pressure is slightly higher or lower than 14.7 psi. The mercury manometer is called a barometer. If the atmospheric pressure is lower than 14.7 psi, it is said the barometer is falling, which usually means storms are approaching. Or if pressure is slightly above atmospheric pressure, it is said the barometer is rising and good weather is expected. Fig. 10-39 shows one U-tube manometer with water in it and another with mercury. The figure shows the amount of vacuum in reference to feet of water column and inches of mercury. As the vacuum gets stronger, it is called a *deep vacuum* or a *complete vacuum.* The mercury manometer is the preferred instrument when deep vacuum or complete vacuum readings are measured because it is smaller (approximately

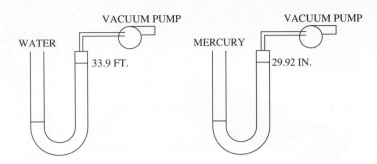

FIGURE 10-39 A U-tube manometer that shows a complete vacuum as 29.92 in. of mercury or 33.9 ft of water.

30 in. tall) than a water manometer (approximately 33.9 ft tall). A water manometer can provide better accuracy because each inch of water column represents 0.036 psi, whereas each inch of mercury represents 0.49 psi.

10.3.3 The Bourdon Tube as a Pressure Sensor

One of the most common sensors for pressure is a Bourdon tube. The simplest form of this sensor is made of a C-shaped metal tube. One end of the tube is sealed, and the other end is connected to the source of pressure that is being measured. The end that pressure is applied to is mounted in such a way that it cannot move. When pressure is applied to the inside of the tube, the sealed end of the tube will tend to straighten out, which will cause a small amount of movement at the sealed end of the tube. This movement can be amplified directly by gears and a pointer to make a direct-reading pressure gauge. Or the movement can be transferred to a linear potentiometer, which would let the signal convert pressure to a change of resistance. The potentiometer can be part of a bridge circuit so that the output signal can be represented by a change of voltage. Fig. 10-40 shows an example of a Bourdon tube sensor used in a pressure gauge.

The shape of the Bourdon tube can be modified so that a small amount of pressure will cause the tube to move farther. One modification involves bending the tube into a series of spirals and the second involves bending the tube into a helix. Fig. 10-41 shows examples of the spiral-type and helix-type Bourdon tube. The end of the spiral-type Bourdon tube will tend to move farther when pressure is applied since there is more tube exposed to the pressure. The helix is designed so that a long piece of tubing is flattened slightly and rolled into a series of continuous

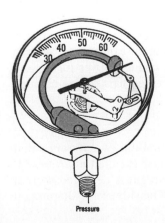

FIGURE 10-40 A Bourdon tube sensor that is used to measure pressure. (Courtesy of Instrumentation & Controls Systems.)

FIGURE 10-41 **(a)** Example of a spiral-type Bourdon tube. **(b)** Example of a helix-type Bourdon tube. (Courtesy of Instrumentation & Controls Systems.)

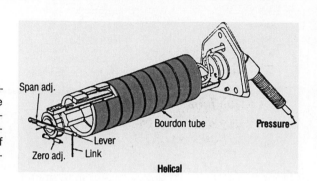

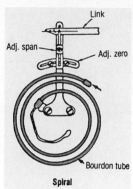

loops. The open end of the helix is firmly held in place and the sealed end is connected to a center shaft. When pressure is applied to the helix, the entire length of the tubing tries to open or unwind around the shaft, which causes the shaft to rotate. This motion can be connected to a pointer to provide a direct reading or to a potentiometer so that the amount of pressure can be converted to a change of voltage.

Bourdon tube sensors are used to measure pressure up to 100,000 psi. They are not useful for low pressures under 15 psi because the tubing is rigid and the amount of movement of the open end of the tube is very small. These types of sensors are used in pressure gauges that can indicate the amount of hydraulic or steam pressure. For example, a plastic injection molding machine will have several hydraulic gauges to indicate the amount of pressure the system uses to inject plastic into a mold, or the pressure that is available at the hydraulic pump.

10.3.4 Diaphragm Pressure Sensor

The diaphragm pressure sensor is designed to measure lower pressures in the range of 330 psi and vacuums to 29.9 in. of mercury (in. Hg). The diaphragm is made of a flexible membrane, which can be rubber for lower pressures and metal for pressures up to 330 psi. The diaphragm is mounted in the middle of the capsule so that it creates two chambers. One of the chambers is open to the atmosphere and the other is connected to the source of pressure being measured. When pressure is applied to the chamber, the diaphragm will expand slightly into the open chamber. The amount of movement will be proportional to the pressure being applied. The movement can be applied directly to a switch to create a low-pressure switch. Or it can be connected to a shaft that can amplify the amount of movement through gears to provide a direct reading on a gauge. The movement can also be applied to a potentiometer so the signal can be converted to a change of resistance and then to a change of voltage if the potentiometer is connected to a bridge circuit.

10.3.5 Differential Pressure Sensors

The diaphragm sensor can be modified to measure the amount of difference that exists between two sources of pressure. This type of sensor is called a *differential pressure sensor* and the chamber surrounding the diaphragm is designed so that pressure can be applied to both sides of the diaphragm. The side of the diaphragm

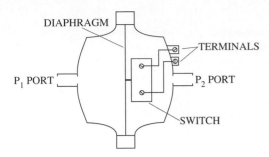

FIGURE 10-42 Example of a differential pressure sensor. Notice that this sensor uses the diaphragm to make two chambers. Pressure is applied to each chamber and the amount of difference between the two pressures can be sensed.

that has the most pressure applied to it will cause the diaphragm to move toward the side that has less pressure. The movement of the diaphragm can be detected and converted to linear motion to activate a switch or be used to move the arm of a linear potentiometer. Springs can be added to one or both sides of the diaphragm to adjust the amount of differential pressure required to make the diaphragm move. The differential pressure sensor can be manufactured to respond to a very small amount of pressure difference. Fig. 10-42 shows an example of a differential pressure sensor. The pressure ports for this type of switch are identified as P_1 and P_2.

A typical industrial application for a differential pressure switch includes testing the pressure difference across a fan or a water pump. If the pump is running, a high-pressure area will develop in front of the pump (P_1), and a low-pressure area (P_2) will develop behind the pump inlet. The amount of pressure difference will be called delta P (ΔP). (Note that Δ is the symbol for the Greek letter delta and it is used to indicate the amount of difference or change.) The differential pressure sensor can also be used to measure the pressure difference across an air filter. When the filter begins to become dirty, a pressure difference will develop across the filter and the differential pressure sensor will sense this difference. Fig. 10-43 shows diagrams of these examples.

10.3.6 Bellows Pressure Sensor

The bellows pressure sensor is made of a sealed chamber that has multiple ridges like the pleats of an accordion that are compressed slightly when the sensor is manufactured. When pressure is applied to the chamber, the chamber will try to expand and open the pleats. Fig. 10-44 shows an example of a bellows sensor, which uses a spring to oppose the movement of the bellows and provides a means to adjust the amount of travel the chamber will have when pressure is applied. In low-pressure bellows sensors, the spring is not required. The travel of the bellows can be converted to linear motion so that a switch can be activated, or it can be connected to a potentiometer. This type of sensor is used in low-pressure applications usually less than 30 psi. The bellows sensor is also used to make a differential pressure sensor. In this application two bellows are mounted in a housing so that the movement of each bellows opposes the other. This will cause the overall travel of the pair to be equal to the difference of pressure that is applied to them.

All of the foregoing pressure sensors provide a change of position or motion as pressure changes. This type of sensor limits the type of electrical interface to either an on/off switch or to a potentiometer. When a more accurate indication is required, other types of pressure sensors that produce a change of voltage or resistance will be used.

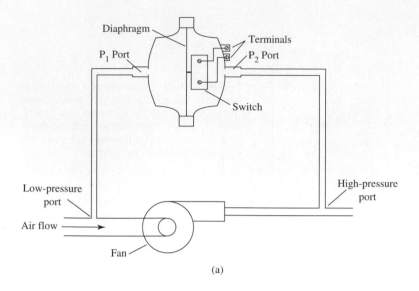

(a)

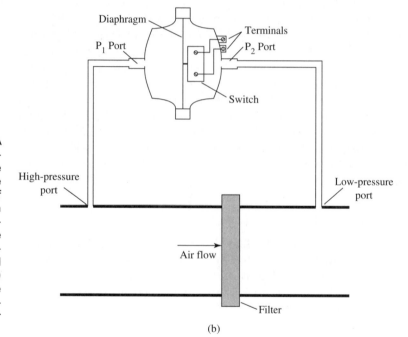

FIGURE 10-43 (a) A differential pressure sensor used to measure the pressure difference across a water pump. If the pump is running, a high pressure will develop in front of the pump and a lower pressure will develop behind the inlet of the pump. **(b)** A differential pressure sensor is used to determine how dirty an air filter is.

(b)

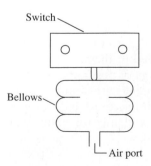

FIGURE 10-44 Example of the bellows pressure sensor. The movement of the bellows is used to activate a switch.

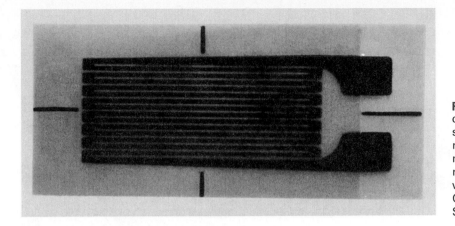

FIGURE 10-45 A typical bonded resistance strain gauge. (©Copyright 1995 Omega Engineering, Inc. all rights reserved. Reproduced with the Permission of Omega Engineering, Inc., Stamford, CT 06907.)

10.3.7 The Strain Gauge

A strain gauge consists of a grid of very fine wire that is bonded to a surface. When the surface moves due to changes in pressure, the resistance of the wire changes. The strain gauge must be connected in a circuit such as a Wheatstone bridge so that the change in resistance can be converted to a change in voltage. Fig. 10-45 shows an example of a typical strain gauge. In this figure you should notice that the strain gauge looks like a long piece of wire neatly laid out on a flat surface in multiple rows. The strain gauge has two terminals that are the ends of the long piece of wire. The material that the strain gauge is mounted to is called the *backing* or *carrier matrix*. This material is flexible like a thin piece of plastic tape but durable so that it can hold the strain gauge wire in its proper position. The backing also is the material that the adhesive is applied to so that the strain gauge can be glued into place. The overall size of the strain gauge will range from less than 1 in. long for the smaller sizes and up to 10 in. long for larger ones. When you purchase a strain gauge, it will come mounted on its backing ready to be glued to the surface of the material that will move or flex when pressure is applied to it. For example, a strain gauge could be glued to the end of the Bourdon tube or the bellows of the pressure gauges that were previously discussed. As pressure is applied to the Bourdon tube and bellows, they will deform or change and the strain gauge will monitor this change by changing its resistance. If the strain gauge is connected to a Wheatstone bridge, the amount of change in resistance can be converted to a change of voltage.

10.3.7.1 Types of Strain Gauges ■ Strain gauges come in a variety of shapes and material that provide more or less change in resistance as they are deformed due to pressure. The types of strain gauge material include piezoresistive or semiconductor gauge, carbon resistive gauge, bonded metallic wire, and foil resistance gauges. Fig. 10-46 shows examples of the variety of shapes of strain gauges. The most common type of strain gauge is the bonded wire strain gauge that is less than 1 in. long (Fig. 10-46a). Fig. 10-46b shows an example of longer bonded wire strain gauges, which are used to determine the amount of change of a larger surface.

The perpendicular strain gauge is shown in Fig. 10-46c. This type of gauge is used to measure strain in two directions such as when a surface is subjected to twisting forces. Fig. 10-46d shows a rosette strain gauge, which is used on surfaces where the stress and strain will concentrate at a point and fan out in any direction.

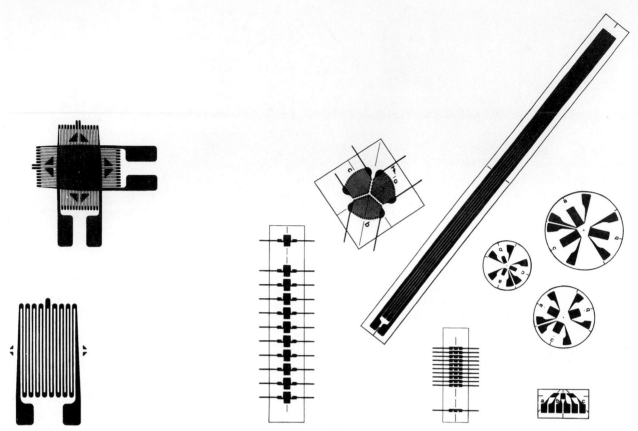

FIGURE 10-46 **(a)** Example of the short bonded wire strain gauge. **(b)** Example of a long bonded wire strain gauge. **(c)** Example of perpendicular strain gauge. **(d)** Example of a rosette strain gauge. **(e)** Example of a crack propagation gauge and a full-bridge diaphragm gauge that are used to monitor cracks. (©Copyright 1995 Omega Engineering, Inc. all rights reserved. Reproduced with the Permission of Omega Engineering, Inc., Stamford, CT 06907.)

This type of strain gauge is useful on surfaces that are rounded or curved like the domed surface of a tank. Fig. 10-46e shows a crack propagation gauge and a full-bridge diaphragm strain gauge used to monitor cracks. When a crack is detected in the surface of a vital piece of equipment, this type of strain gauge is mounted directly on the crack and any time the crack changes or enlarges, the amount of change can be monitored by the change in resistance of the strain gauge.

10.3.7.2 Connecting Strain Gauges in Circuits ■ The strain gauge must be connected to an electrical circuit for it to be useful. Fig. 10-47 shows an example of a two-wire strain gauge connected to a Wheatstone bridge circuit. From this example

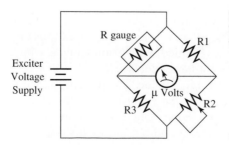

FIGURE 10-47 A strain gauge connected in a Wheatstone bridge circuit. Notice the symbol for the strain gauge is a resistor designated R_g.

you can see that the strain gauge acts like one of the legs of the bridge. The typical resistance for the strain gauge is 120 Ω. Resistor R_2 is used to balance all the resistance in the bridge. When this occurs and no force is applied to the surface where the strain gauge is mounted, the voltage-out signal will be zero. When a force is applied to the surface, it will deform the metal that the strain gauge is glued to. At this time, the wire in the strain gauge will also deform and the amount of resistance in the wire will change. The change in resistance is proportional to the deformation of the wire in the strain gauge, which means that the signal from the bridge circuit will be somewhat linear due to the change in pressure. You should notice that R_2 can be adjusted at any point to balance the bridge. If a small amount of pressure is applied to the strain gauge when the circuit is balanced, it is called a *preload*. The formula for calculating the voltage out from the bridge is:

$$V_{out} = V_{in}\left(\frac{R_3}{R_3 + R_g} - \frac{R_2}{R_1 + R_2}\right)$$

This means that $(R_1/R_2) = (R_g/R_3)$ when the bridge is balanced.

• •

PROBLEM 10-1

A strain gauge is connected to a bridge circuit where R_1 and R_3 are 150 Ω resistors. The resistance of the strain gauge with no weight applied is 120 Ω. To what value should R_2 be adjusted for the bridge to be balanced when no pressure is applied to the strain gauge?

SOLUTION

Since $(R_1/R_2) = (R_g/R_3)$, this formula can be manipulated to solve for R_2.

$$R_2 = (R_1) \times (R_g/R_3)$$
$$R_2 = (150\ \Omega) \times (120\ \Omega/150\ \Omega)$$
$$R_2 = 120\ \Omega$$

• •

PROBLEM 10-2

When a load of 10 psi is applied to a strain gauge in the previous problem, its resistance will change 20 Ω so that the total resistance of the strain gauge changes from 120 Ω to 140 Ω. If an input voltage of 10 volts is applied to the circuit, what will be the output voltage when the 10 psi load is applied to the strain gauge?

SOLUTION

Using the previous formula for voltage out:

$$V_{out} = V_{in}\left(\frac{R_3}{R_3 + R_g} - \frac{R_2}{R_1 + R_2}\right)$$

$R_1 = 150\ \Omega$ (given)

$R_2 = 120\ \Omega$ (calculated from prior problem)

$R_3 = 150\ \Omega$ (given)

$R_g = 140\ \Omega$ (increased 20 Ω because of load placed on strain gauge)

$$V_{in} = 10 \text{ volts}$$

$$V_{out} = 10 \text{ volts} \left(\frac{150 \, \Omega}{150 \, \Omega + 140 \, \Omega} - \frac{120 \, \Omega}{150 \, \Omega + 120 \, \Omega} \right)$$

$$V_{out} = 10 \text{ volts } (9.28)$$

$$V_{out} = 0.728 \text{ volt}$$

· ·

10.3.7.3 Gauge Factor ■ Gauge factor (GF) is the ratio of the amount of change in resistance to the change in the length (strain) along the axis of the gauge. The formula for calculating the gauge factor is:

$$GF = \frac{\Delta R/R}{\Delta L/L} = \frac{\Delta R/R}{\varepsilon}$$

(Notice that $\Delta L/L$ is identified as epsilon ε.) Gauge factor is a dimensionless value, and the larger the value the more sensitive the strain gauge is. The gauge factor is provided with the strain gauge; typical values are 2–2.5.

10.3.8 Load Cells

Many times it is not practical to use a single strain gauge because it will not provide meaningful data regarding total pressure because offset and compensation strain gauges would be required. A *load cell* is a sensor that uses strain gauges that are mounted in specific patterns to provide a meaningful value of the change in pressure or weight. One of the common applications of a pressure sensor in industrial applications is a scale used to weigh a product. Another application of load cells is to use several load cells that are permanently mounted under the legs of a bulk tank to measure the weight of the material that is placed into the tank. The load cells must be large enough to be able to determine the weight of the tank plus the weight of the material in the tank.

Load cells can be divided into two broad categories by type. The first category and most popular type is the *bending beam load cell*. This type of load cell uses a principle of measuring force on a cantilever beam. It will have shear stress and bending stress along the cross-sectional area of the beam. This stress can be measured by strain gauges and converted into an equivalent amount of pressure.

The second category of load cells is the *shear beam load cell*. This type of load cell uses an I-beam design because the shear stress is uniform at the web of the beam. The shear beam load cell is more expensive but it does have advantages such as returning to zero more quickly than the bending beam when the load is removed. It has a higher tolerance of dynamic forces and vibration, and it can be better sealed for environmental conditions such as the necessity of wash down in food processing.

Fig. 10-48 shows several types of load cells. Fig. 10-48a shows a compression load cell. This type of load cell can measure the amount of load that is placed on top of its load button. Fig. 10-48b shows a tension/compression load cell, which has a threaded stud that can be fixed to either or both of its surfaces. If a load is placed on top of the stud, it will measure compression forces, and if a load is connected to the bottom stud, the load cell will measure tension forces.

The load cell in Fig. 10-48c shows an S-shaped cell that can be used to weigh hoppers and tanks that are mounted to it. This type of load cell is commonly used in truck scales. In many industrial applications factory personnel are required to

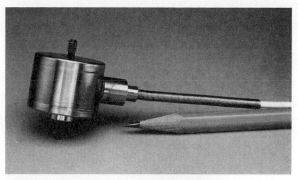

FIGURE 10-48 **(a)** Example of a compression load cell. **(b)** Example of a load cell that can measure both tension and compression. **(c)** Example of an S-shaped load cell. **(d)** Example of a double-ended shear beam load cell. **(e)** Example of a platform load cell. **(f)** Example of a reaction load cell that is used to measure rotary torque. (©Copyright 1995 Omega Engineering, Inc. all rights reserved. Reproduced with the Permission of Omega Engineering, Inc., Stamford, CT 06907.)

weigh the truck prior to unloading and after it is unloaded to determine the amount of product that was shipped. This provides a means to determine the inventory and verify billing amounts. Raw product must also be weighed when it is added to a recipe to make a finished product. Fig. 10-48d shows a double-ended shear beam load cell, which is used for weighing tanks because of its low profile. Fig. 10-48e

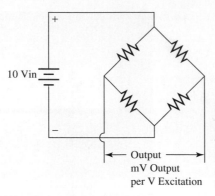

10 Vin

Output
mV Output
per V Excitation

FIGURE 10-49 Electrical diagram of a load cell. The load cell looks like a Wheatstone bridge, and its output voltage will increase as the load on the cell is increased. The output voltage is specified as mV/Volt of dc exciter voltage.

shows a platform load cell that is used in industrial weighing systems. This type of load cell can be used to weigh product in a stationary manner or it can measure product (on the fly) as it continually moves along a production and packaging line without stopping. These types of sensors are well suited for this type of application because of the accuracy (0.2 g) as well as the speed of the readings. Fig. 10-48f shows a load cell that is specifically designed to measure rotary torque.

10.3.8.1 Electrical Circuits for Load Cell Sensors ■ Load cells are made by combining a number of strain gauges in a common sensor that can measure changes in pressure. Since the individual strain gauges each change their resistance when they are deformed by strain, the load cell operation will look like a Wheatstone bridge circuit. When the pressure (load) is applied to the load cell, the resistance of the bridge will change the amount of output voltage from the sensor. Fig. 10-49 shows an example of the electrical diagram of a load cell. The voltage used to supply power to the load cell (bridge) is called *exciter voltage*. The voltage that comes from the load cell as an output signal will be measured in millivolts (mV). When a load is applied to the load cell sensor, the amount of millivolt output will increase. The total amount of output signal that the load cell can produce is specified in terms of millivolt of output per volt of exciter voltage (mV/V). Typical output voltage is 2 mV–3 mV per volt of excitation voltage. Typical excitation voltage is 5–10 volts dc. This means that most load cells are able to produce a maximum of 30 mV when fully loaded.

10.3.8.2 Troubleshooting Load Cells ■ Load cells are easy to troubleshoot because they act like any other resistive sensor. Since they use an exciter voltage like a Wheatstone bridge, you should start any troubleshooting procedure by measuring the supply voltage. If the exciter voltage is low or absent, you must take care of this problem before continuing the troubleshooting process.

The second part of the troubleshooting process involves applying a physical load to the cell. This means you will have to look at the specifications for the load cell and apply a load that weighs approximately 50% of the maximum allowable load. The 50% load should cause the load cell to produce approximately 50% of its rated output. If the load cell specification indicates the output voltage is 3 mV/ volt of excitation, and the exciter voltage is 10 volts, the maximum load is 200 lb, and the test weight is 100 lb, the following formula will indicate load cell output voltage at the test weight should be approximately 15 mV when a 100 lb test weight is applied:

$$V_{\text{out max}} = \text{nominal output (mV/V)} \times \text{excitation voltage}$$

$$V_{\text{out max}} = 3\,\text{mV/V} \times 10\,\text{V dc}$$

$$V_{\text{out max}} = 30\,\text{mV}$$

$$\text{Output voltage}_{\text{at test weight}} = \frac{\text{test weight}}{\text{max weight}} \times V_{\text{out max}}$$

$$\text{Output voltage}_{\text{at test weight}} = \frac{100\,\text{lb}}{200\,\text{lb}} \times 30\,\text{mV} = 15\,\text{mV}$$

• •

PROBLEM 10-3

You are asked to test a load cell that has an output of 2 mV/V and an excitation voltage of 5 volts dc. The load cell has a weight capacity of 500 lb, and the test weight is 125 lb. What should the output voltage be when the 125lb test weight is placed on the load cell?

SOLUTION

$$V_{\text{out max}} = \text{nominal output (mV/V)} \times \text{excitation voltage}$$

$$V_{\text{out max}} = 2\,\text{mV/V} \times 5\,\text{V dc}$$

$$V_{\text{out max}} = 10\,\text{mV}$$

$$\text{Output voltage}_{\text{at test weight}} = \frac{\text{test weight}}{\text{max weight}} \times V_{\text{out max}}$$

$$\text{Output voltage}_{\text{at test weight}} = \frac{125\,\text{lb}}{500\,\text{lb}} \times 10\,\text{mV} = 2.5\,\text{mV}$$

• •

10.3.9 Pressure Transducers and Transmitters

Load cells have been integrated directly into a number of pressure-sensitive devices such as transducers and transmitters. The *pressure transducer* is a device that can be threaded directly into a hydraulic or pneumatic line and read the pressure of the system. Fig. 10-50 shows an example of two pressure transducers. Fig. 10-50a shows a pressure transducer that has a capacity to measure pressure between 0–5000 psi. These sensors are available in pressure ranges from 0–50 psi through 0–1500 psi. The typical output for this type of sensor is rated like the load cell at 2 mV/V with exciter voltage of 10 volts. This makes these types of sensors easy to interface with input modules for programmable logic controllers (PLCs).

Fig. 10-50b shows a second type of sensor that uses strain gauge and load cell technology that integrates a load cell beam with a diaphragm pressure sensor. The sensor is mounted (threaded) in a pressure line where the force of the fluid changes the position of the diaphragm. A beam is mounted so that it is deflected when the diaphragm changes position. Strain gauges are mounted on the beam similar to a load cell where they can monitor the change of position of the beam as force from the diaphragm is transmitted to it. The main advantage of this type of sensor is that the chamber that measures the pressure does not need any O-rings or seals since a thin metal diaphragm is used. This also allows pressures up to 5000 psi to be accurately measured. The feature that makes this type of sensor so useful is that the exciter voltage can be 12–35 volts dc, which provides an output signal of 4–20

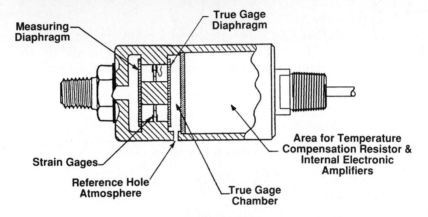

FIGURE 10-50 (a) A pressure transducer that measures pressure in the range of 0–5000 psi. **(b)** A pressure transducer that is used to measure high pressure. This type of sensor integrates a load cell with a pressure diaphragm. (Courtesy of Sensotec Inc.)

mA. The mA signal is preferred over the mV signal because it is not susceptible to transient or induced voltages. The 4–20 mA signal is easily interfaced to a PLC analog input module, display (meter), or recorder.

Fig. 10-51 shows an example of two pressure transducer electrical diagrams. The first electrical diagram shows a four-wire sensor with a 24 volt dc supply voltage and a 4–20 mA signal that is sent to a PLC analog input module. The second electrical diagram shows a three-wire sensor with an 18 volt dc supply voltage and a 0–6 volt dc output voltage. All of the pressure transducers and pressure transmit-

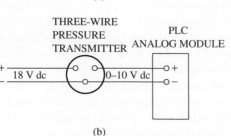

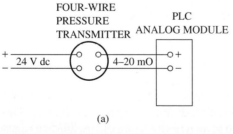

FIGURE 10-51 (a) A diagram of a four-wire pressure transducer. The transducer has a 24 volt dc power supply and the transducer produces a 4–20 mA signal for a programmable logic controller (PLC) analog module. **(b)** A diagram of a three-wire pressure transducer. The transducer has an 18 volt dc power supply and produces a 0–10 volt signal for a PLC analog module.

FIGURE 10-52 **(a)** An example of a truck scale used to weigh the amount of material the truck is hauling. **(b)** An example of a hopper scale that is used to accurately measure the amount of material in a hopper. This application is frequently used in batch processing. **(c)** An example of a tank scale designed to weigh the amount of liquid in a tank. (Courtesy of Fairbanks Scales.)

ters must be zeroed and spanned when they are installed so they accurately represent the minimum and maximum signals. The zero and span circuitry is provided through op amps that are used in the electronic part of the device.

10.3.10 Industrial Scales and Weighing Systems

Load cell sensors can be used in industrial scales and weighing systems. Fig. 10-52 shows several examples of industrial applications where accurate weights must be measured. The first example shows large scales where trucks can be weighed before and after they have been loaded. This provides an accurate measurement of the amount of material the truck is carrying. The weight is important to determine the amount of material that is transported so it can be paid for. Another important reason to have the accurate weight is so the truck can comply with road weight limits. If the truck is overloaded, it is subject to heavy fines as it travels through each state.

Fig. 10-52b shows a scale that is designed to weigh material as it is loaded into a hopper. The range of weight for this scale is from 0–50,000 lb. The material in the hopper is used as part of a recipe in a batch process and the amount of material must be accurately weighed if the recipe is to be followed correctly. Fig. 10-52c shows a tank scale that is designed specifically for weighing the contents of a cylindrical tank. This type of scale can weigh products up to 150,000 lb, and the controls for the scales can be used to weigh multiple batches that may be sent to

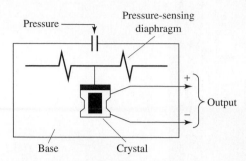

FIGURE 10-53 Example of a piezoelectric crystal used as a pressure sensor. The piezoelectric crystal will produce a small voltage that is proportional to the pressure. (Courtesy of PCB Piezotronics, Depew N.Y.)

a variety of blenders and hoppers. A wide variety of small accurate scales is also used in quality control applications in factories where the product must meet a weight standard.

10.3.11 Piezoelectric Pressure Sensors

In 1880 Pierre and Jacques Curie determined that a small amount of voltage could be produced by applying large amounts of pressure to certain crystals of elements. This phenomenon is called the *piezoelectric effect.* When pressure is applied to the crystal, it will deform and produce a small voltage. The amount of voltage is proportional to the amount of deformation. The best crystals that are used for this type of sensor come from ammonium dihydrogen phosphate and sintered ceramics. The amount of voltage that is produced is very small and the internal impedance of the crystal is very large, which makes the use of op amps a necessity to produce a usable signal.

When the piezoelectric effect is used in a pressure sensor, the sensor uses a diaphragm that deflects slightly when pressure is applied. A rod transfers this small amount of movement directly to the piezoelectric crystal. The pressure on the crystal causes a small voltage to be produced that is proportional to the pressure. The voltage is amplified to traditional voltage signal values (0–10 volts). Fig. 10-53 shows an example of a piezoelectric pressure sensor.

10.3.12 Capacitance Pressure Sensors

Pressure sensors can also use capacitance and reluctance to provide an electric signal that varies from minimum to maximum (0–10 volts). This type of sensor uses two plates that are separated by a dielectric to create the capacitance. In most cases the dielectric is a silicone oil filling. One of the plates is stationary, and the second plate is connected to a diaphragm so that when the diaphragm moves, the plate will move and the amount of capacitance will change. Since this type of circuit uses changing capacitance, it is most usable with ac voltage. When the capacitance changes in an ac circuit, the capacitive reactance changes and the amount of voltage drop across the capacitor can be detected.

The capacitive sensor can be enhanced so that two capacitors are used with two diaphragms called the sensing diaphragm and the isolation diaphragm. When the sensor is exposed to high pressure on one side and lower pressure on the opposite side, the sensing diaphragm moves. This moves the plate in the first capacitor and increases its capacitance. Since the two capacitors are linked, when the capacitance of one increases, the capacitance of the other decreases. The high and low pressures allow this sensor to measure the difference in pressure and the sensor is called a differential pressure sensor (ΔP). When pressure difference changes, the capacitance between the two capacitors changes and this change can be converted to a change in capacitive reactance in the ac circuit. The capacitive sensors can also be used to change the frequency of an oscillator circuit and the change in frequency can be used as an input to a variable-frequency circuit, which includes a frequency-to-voltage converter.

10.3.13 Variable Reluctance/Inductance Pressure Sensors

Reluctance is the opposition to current flow that is caused by magnetic circuits. Inductance is the property of a coil of wire to build magnetic flux lines. The amount of inductance in a coil depends on the number of turns of wire in the coil, the amount of current flowing in the wire, and the type of core that is placed inside the coil of wire or the position of a core when it is inside a coil. An easy way to change the amount of inductance in the coil is to change the position of a movable core inside the coil. The movable core can be attached to the diaphragm in a manner similar to the pressure sensors previously discussed. When the pressure moves the diaphragm, the core will move and change the amount of inductance in the coil of wire. When the amount of inductance changes, the amount of inductive reactance also changes. If the coil is placed in an ac circuit, the change of inductance can be used to change the voltage drop across the inductor or it can be used in an oscillator circuit to change the frequency of the circuit. These types of sensors are also called *variable reluctance,* since the opposition to inductance is called reluctance.

10.3.14 Differential Pressure Sensors (ΔP Sensors)

Differential pressure sensors take the difference between two readings and produce a signal based on the total amount of difference. The total amount of difference is identified by the Greek letter delta (Δ), and this is why these types of sensors are called *delta P sensors* or *ΔP sensors.* The operation of one type of differential pressure sensor was discussed in the capacitive pressure sensors. Other types of differential pressure sensors use a similar theory of operation in that they use a diaphragm that will move in proportion to the pressure movement instead of a capacitance sensor. The differential sensor is used to measure the amount of pressure drop across a hydraulic filter to indicate when it is dirty, or it can measure the pressure drop across an orifice plate to determine fluid flow. The amount of pressure differential may be very small even though the amount of each individual pressure P_1 and P_2 is relatively high. For example, it is possible to measure a 2 psi pressure drop across a filter even when the input pressure is 150 psi. This means that the inlet pressure is 150 psi and the outlet pressure is 148 psi. You will be able to recognize pressure differential sensors because they use two pressure lines instead of one. Since the P_1 and P_2 pressures are relatively high it is not possible to use the bellows low-pressure switches described earlier in this section.

10.3.15 Ultralow-Pressure Sensors

Ultralow-pressure sensors are useful in detecting extreme low-pressure readings. For example, when a small fan is used to exhaust gas from a small industrial furnace, the total amount of pressure is usually less than 1 psi above atmospheric pressure. The sensor to measure this low pressure usually uses a rubber diaphragm that is very sensitive. It is important that the total maximum pressure level is never exceeded or the sensor will be permanently damaged. Other very low pressures are used to set the amount of gas pressure in a regulator where natural gas or propane gas is used to heat a process oven. The total pressure of this gas must be approximately 10 in. of water column, which is about 0.36 psi. This small amount of pressure cannot be accurately measured with a traditional sensor, so the ultralow-pressure sensor or a U-tube manometer that is filled with water or mercury must be used.

10.3.16 Multiple Uses for Pressure Sensors

In the beginning of this section several problems were posed regarding the selection of a pressure sensor to measure steam pressure, hydraulic pressure, and vacuum. From the material presented in this section you can see that since steam pressure is low to medium pressure, a Bourdon tube sensor, bellows, or strain gauge (load cell) sensor could be used. Since hydraulic pressure tends to be up to 1500 psi, Bourdon tube sensors or strain gauge (load cell) sensors would probably be used. If vacuum is measured, a bellows or diaphragm sensor that can measure low pressures or a manometer could be used. You can see that a large variety of sensors is available to match the application that you will encounter. You will also begin to see that it will be possible to use a pressure sensor to measure high pressure and low pressure. It is also possible to use a pressure sensor for making temperature readings if a conversion chart is used. You will see in the next section how pressure sensors can be used to measure flow and level.

10.4 FLOW SENSORS

Many industrial applications require a measurement of flow. Sometimes the flow measurement must be very accurate such as the measurement of material that is used in processes or manufacturing so that total amount of raw material can be determined. These types of flow measurements are important when the amount of material being metered is for recipe purposes, or when material is loaded or transported and it must be measured so it can be paid for. In other applications like the hydraulic system on a large machine, it is important to determine the amount of fluid flowing in specific parts of a system to determine if the valves are working correctly or if a failure has occurred and fluid is being bypassed to the tank. In some applications the exact amount of flow is not important, and a more general measurement will be usable to indicate if a pump is operational, or if a drain line is functioning correctly.

This section will explain the different types of flow sensors and the theory of operation of each type of sensor. All of the sensors in this section will be classified as intrusive or nonintrusive. If the sensor disturbs the flow of fluid that it is measuring, it will be classified as an intrusive flow sensor. If the sensor can measure the fluid flow without disturbing the flow, it will be classified as a nonintrusive flow sensor.

Another way to classify flow sensors is by the way they measure the flow. For example, a *positive displacement flow meter* is very accurate because it physically measures every drop of fluid, whereas a *bypass meter* is not as accurate because it measures a small amount of the fluid flow and calculates the remainder. Differential flow meters place an obstruction in the path of the fluid flow, such as an orifice plate, and the amount of flow is determined from a calculation that is based on the amount of pressure drop the fluid has as it moves through the orifice plate. A velocity flow meter measures the velocity of fluid as it passes a turbine and calculates the total amount of flow based on the velocity of the fluid as it passes the turbine.

The fourth way to measure fluid flow is to use a mass flow sensor that calculates the total volume of material that passes through the sensor and determines the flow from this figure.

Another way to classify flow meters is to determine how the signal from the sensor will be used. For example, the signal can be used to determine the amount of flow at any given point in a system and indicate it on a meter or display. Or it can be used to totalize the flow to determine the volume of fluid that has passed the sensor in a given time. The flow signal can also be sent to a recording device or a transmitter where it is sent to a control system. In the control system, the sensor signal will act as the process variable where it is the feedback signal for the system. In this type of application, the flow sensor signal can be sent to a controller where it can be compared to a setpoint. The controller can increase or decrease its output signal based on this comparison. The flow sensor signal can also be used to control the system or report the flow for an alarm package that can be set to indicate when the flow is too high or too low.

10.4.1 Flow Technology

It is important to understand the terms relating to flow before you can fully understand flow sensors. *Fluid* is the term that describes any substance that flows. Liquids such as water or hydraulic fluid and gases such as oxygen or nitrogen are all considered fluids since they flow. The terms *liquid* and *gases* describe two of the three states (solid, liquid, gas) of any substance. The flow rate of a fluid as it flows through a pipe can be calculated by the formula $Q = V \times A$ where

Q = liquid flow through a pipe

V = average velocity of the flow

A = cross-sectional area of the pipe

It is important to understand this formula is a basic relationship for flow and does not take into account the density or viscosity of the fluid or the friction in the walls of the pipe.

• •

PROBLEM 10-4

Determine the flow of hydraulic fluid through a 2 in. diameter pipe that has an average velocity of 60 in. per second.

SOLUTION

1. First you must find the cross-sectional area of the pipe by the formula:

$$A = \pi r^2$$
$$A = 3.14 \times (1 \text{ in.})^2$$
$$A = 3.14 \text{ sq in.}$$

2. Use the formula $Q = V \times A$ to find flow:

$$Q = 60 \text{ psi} \times 3.14 \text{ in.}^2$$
$$Q = 118.4 \text{ cu in./sec}$$

• •

Other terms that are important to understand are *laminar flow* and *turbulent flow*. Fig. 10-54 provides three diagrams that show uniform laminar flow, nonuniform

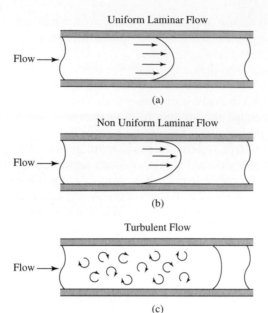

Uniform Laminar Flow

(a)

Non Uniform Laminar Flow

(b)

Turbulent Flow

(c)

FIGURE 10-54 **(a)** An example of uniform laminar flow. **(b)** An example of nonuniform laminar flow. **(c)** An example of turbulent flow.

laminar flow, and turbulent flow. Laminar flow means that the fluid flows parallel to the pipe and the flow is rather smooth. Turbulent flow is characterized by swirling action of the fluid inside the flow. Fig. 10-54a shows a diagram of uniform laminar flow. From this diagram you can see that the flow is very uniform and has the shape of a bullet, which indicates that the flow is streamline. Fig. 10-54b shows a diagram of nonuniform laminar flow. This diagram shows the overall shape of the flow is parallel, which indicates it is laminar, but the front nose of the flow is parabolic, which means the fluid has more friction on one side of the pipe than the other. Fig. 10-54c show an example of turbulent flow. From this diagram you can see that the fluid is swirling as it flows. The swirling action tends to create opposition to the flow.

Another important term pertaining to fluid flow is the *Reynolds number.* The Reynolds number is the ratio of a fluid's *inertial forces* to its *drag forces.* The flow rate and the specific gravity of a liquid are part of its inertial forces, and the pipe diameter and viscosity of the fluid make up its drag forces. The Reynolds number is dimensionless, but it can be calculated by the following formula:

$$R = \frac{3160 \times Q \times G_t}{D \times \mu}$$

where

R = Reynolds number

G_t = specific gravity of the liquid

Q = flow rate for the liquid in gallons per minute (gpm)

D = the inside diameter of the pipe in inches

μ = viscosity of the liquid in ft/sec

. . . .

PROBLEM 10-5

Determine the Reynolds number for a liquid flowing through a pipe with an inside diameter of 2 inches, whose specific gravity is 0.713, its flow rate is 0.2 gps (12 gpm), and its viscosity is 1.17 ft²/sec.

SOLUTION

Use the formula (remember to convert diameter to feet):

$$R = \frac{3160 \times Q \times G_t}{D \times \mu}$$

$$R = \frac{3160 \times 0.2 \text{ gps} \times 0.713}{.16 \text{ ft} \times 1.17 \text{ ft}^2/\text{sec}}$$

$$R = 2409.7$$

. . . .

10.4.2 Calculating Flow from a Pressure-Drop Measurement

One of the easiest ways to determine the amount of flow is to place an obstruction in the flow such as an orifice plate to create a pressure drop. These types of sensors are categorized as obtrusive flow sensors. Since they are mounted directly in the fluid flow, they tend to cause small amounts of disturbance to the flow. In most cases the amount of disturbance is not important, but in some fluids the disturbance will create problems that will affect the quality of the product. If this occurs, an alternate method of measuring flow may be needed.

Fig. 10-55 shows a diagram of the orifice plate that is mounted in a pipe. When fluid begins to flow in the pipe, a pressure is created on the side identified as P_1. When the fluid flows through the orifice, a lower pressure is created on the

Front view of
orifice plate

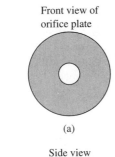

(a)

FIGURE 10-55   (a) An orifice plate that is removed from the piping so that the relative size of the hole (orifice) can be viewed. (b) An orifice plate mounted in the piping so that it can create a pressure drop that can be used to calculate flow.

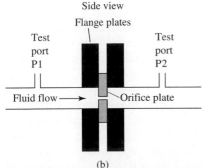

(b)

opposite side of the orifice plate identified as P_2. The simplified formula for flow using the pressure drop is:

$$Q = k\sqrt{P_1 - P_2}$$

where: Q = flow in gallons/minute (gpm)

k = is the constant that is determined by the orifice plate

P_1 = is higher pressure in front of the orifice

P_2 = is lower pressure behind the orifice

••

PROBLEM 10-6

If the pressure P_1 is 4.5 psi and pressure P_2 is 4.1 psi across an orifice plate and the value for k is 10.6, what is the total flow in gpm?

SOLUTION

Using the formula:

$$Q = k\sqrt{P_1 - P_2}$$
$$Q = 10.6\sqrt{0.4}$$
$$Q = 6.7 \text{ hpm}$$

••

10.4.3 Types of Flow Meters That Utilize a Pressure Drop

Several types of flow meters utilize a pressure drop to determine the flow of a liquid. In these types of flow meters, the method to create the pressure drop is all that is different in them. In fact these types of flow meters actually use a differential pressure sensor to produce the electric signal. The operation of differential pressure sensor was discussed in the previous section of this chapter.

One type of flow meter that uses a pressure drop is called an orifice plate flow meter because it uses an orifice plate to create the pressure drop. Fig. 10-55 shows an orifice plate that is outside of the piping so you can see the relative size of the hole. This figure also shows the way the orifice plate is mounted in the piping so a pressure drop is created when fluid flows through it. You should understand that the mounting apparatus for the orifice plate, called a *flange,* makes it easy to open the piping and inspect or change the orifice plate. It is also important to understand that orifice plates are available in a variety of sizes that are all accurately drilled to a specific size. The size of the hole in the orifice plate must be accurately gauged so it can be used in the flow rate calculation. Notice that two ports are located in the pipe, one on either side of the orifice plate, so that the pressure drop can be sampled by the differential pressure sensor. One problem of the orifice plate flow meter is that it causes a very high turbulence in the fluid as it passes through the orifice.

10.4.3.1 Venturi Flow Meter ■ A venturi is a point in a pipe that has been narrowed so that the flow is restricted slightly. In Fig. 10-56 notice that a high-pressure port is provided in front of the point where the venturi is narrowed down and a low-pressure port is provided directly after the point where the pipe is narrowed down. The high-pressure port is provided to sample the increased fluid pressure where it

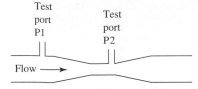

FIGURE 10-56 A venturi uses a slight restriction (a point where the pipe diameter is narrowed down) to create a pressure drop.

will increase slightly because of the restriction caused by the venturi, and the low-pressure port is provided to sample the pressure as it drops after it flows past the restriction caused by the venturi. Additional pressure ports are provided on both sides of the restriction in larger venturies to allow an average pressure to be measured. The venturi is widely used because it has no moving parts and the small amount of restriction it causes to create the pressure drop does not disturb the fluid flow too much. This means that the venturi can handle larger volumes of flow than most other types of flow meters.

10.4.3.2 Flow Tubes ■ The flow tube is similar to the venturi in that it places a restriction in the flow that creates a pressure drop. Fig. 10-57 shows an example of a flow tube that is located in a pipe. From this diagram you can see that the flow tube looks like a short piece of pipe with a thick wall. The front face of the pipe is rounded so that fluid is disturbed as little as possible when it enters the tube. The diameter of the flow tube is approximately half the diameter of the original pipe so that a pressure drop is created as the flow is diverted through the flow tube. A pressure tap is provided prior to the flow tube to measure the higher pressure and directly after the opening of the flow tube to measure the lower pressure. The difference in the pressures must be measured by a pressure differential sensor and the pressure drop is used to calculate the flow.

10.4.3.3 Pitot Tube ■ The *pitot tube* (pronounced like the words *pea toe*) is a device that has two tubes that are placed in a fluid flow to sense the impact pressure and the static pressure used to determine the amount of pressure drop. Fig. 10-58 shows two examples of pitot tubes. In the first example two tubes are connected side by side. One of the tubes has a hole in it that faces the fluid flow (impact tube), while the other tube has a hole in it that faces away from the fluid flow (static tube). The impact tube measures the higher pressure (impact pressure), and the static tube measures the lower pressure (static pressure) as fluid flows past the pitot tube.

The second example shows two tubes that are mounted one inside the other. The inside tube measures the impact pressure, and the outside tube measures the static pressure. In both examples the ends of the pitot tube that are outside the pipe are connected with plastic tubing to a sensor or instrument that can measure

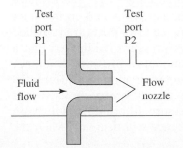

FIGURE 10-57 A flow tube that is mounted in a pipe to cause a pressure drop.

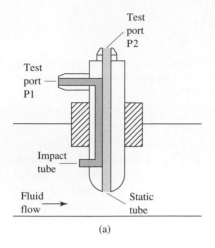

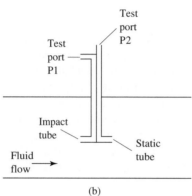

FIGURE 10-58 **(a)** Pitot tube with its two tubes mounted side by side. **(b)** Pitot tube with one tube inside the other. The inner tube senses the impact pressure and the outer tube senses the static pressure.

a very small pressure differential. The difference between the impact pressure and the static pressure is very small so a manometer or ultralow differential pressure sensor must be used to measure the pressure differential. The manometer can be used if only a visual indication is needed, and a differential pressure sensor or transducer is required if the pressure drop is converted to an electrical signal.

FIGURE 10-59 Pressure ports are located on the inside radius and outside radius of an elbow to measure the small amount of pressure drop that occurs when fluid flows through the elbow.

10.4.3.4 Elbow Meters ■ Another method to create a pressure drop in a fluid system is to use an existing elbow in the piping system. It has been determined that the pressure of fluid will show a slight pressure differential as it passes through an elbow. The fluid that flows near the inside radius of the elbow will have a slightly lower pressure than the fluid that flows on the outer radius of the elbow. Fig. 10-59 shows an example of this type of meter. You should also notice that the distance the fluid travels on the outer radius of the elbow will be longer than the distance on the inner radius of the elbow. Two low-pressure ports are provided on the inner side of the elbow to provide the lower-pressure reading, and two ports are provided on the outer side of the elbow to provide the higher-pressure reading so that a better average of each pressure is determined. The amount of pressure difference from this type of sensor is very small and a manometer or ultralow differential pressure sensor must be used to sense it. The pressure difference is used in a calculation to determine fluid flow. The amount of pressure difference will increase when the fluid flow increases, and the pressure difference will decrease when the fluid flow decreases.

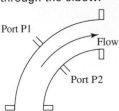

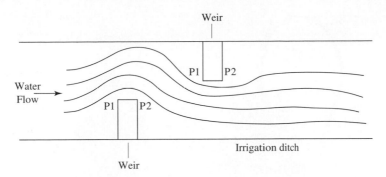

FIGURE 10-60 Weirs are used to create a pressure drop so flow can be calculated.

10.4.3.5 Weirs ■ Another way to create a pressure difference in a flow of fluid is to place a weir directly in the flow. A *weir* is a narrowing of an open channel to create an obstruction that is placed in the flow to cause a pressure drop as fluid flows through it (see Fig. 10-60).

When the fluid flow increases, the pressure drop through the weir will increase. One example of this type of application would be a weir placed in a canal used to move cooling water to a pond. Many industries use ponds for cooling water or to store waste water while it is being treated. It is important to get an estimate of the amount of water that is flowing into the pond if water-treatment chemicals are added, so a weir is used to measure flow. The pressure of the water when it is obstructed by the weir will cause the velocity of the flow to change slightly, which will cause a slight pressure increase. When the water exits the weir, its pressure will decrease slightly. This small amount of pressure difference can be used to calculate flow. Weirs are also used extensively in irrigation ditches to measure the amount of water flow. In these applications the amount of flow is totalized so the water can be sold.

10.4.4 Velocity Flow Meters

Velocity flow meters use the change in velocity that occurs when fluid flow changes to measure the amount of flow. Velocity meters can be used in systems that are not adversely affected by changes in viscosity. Examples of the velocity flow meter are the paddlewheel flow meter, the turbine flow meter, the vortex flow meter, the electromagnetic flow meter, and the ultrasonic (Doppler) flow meter.

10.4.4.1 Paddlewheel Flow Meters ■ A paddlewheel flow meter uses a completely different way to determine the amount of fluid flow than the pressure differential flow meter. This type of sensor has a paddlewheel that is placed directly in the fluid flow so that it can rotate freely as fluid passes it. The faster the fluid flows past it, the faster the wheel rotates. Each paddle or web of the paddlewheel has a magnet mounted in it so that a sinusoidal waveform is produced when it passes the detector mounted in the head of the sensor. A frequency-to-voltage converter is used to convert the sinusoidal signal to a variable-voltage signal. Fig. 10-61 shows a paddlewheel flow meter mounted in a pipe. It is important to understand that the paddlewheel flow meter is mounted in a straight run of piping where the flow will be laminar. The paddlewheel is a low-cost sensor used in applications that do not require a high degree of accuracy.

10.4.4.2 Turbine Flow Meters ■ Turbine flow meters are very similar to the paddlewheel flow meter except the turbine flow meter is much more accurate. From

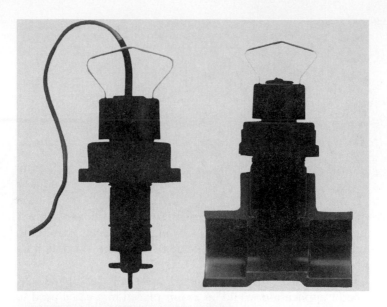

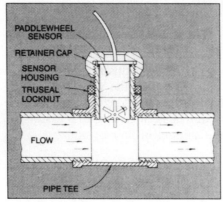

the diagram in Fig. 10-62 you can see that the turbine wheel is mounted so that it is directly in the flow. Each of the vanes of the turbine wheel has a magnet mounted in it so that when the vane spins under the magnetic pickup, an electric pulse is generated. When fluid starts to flow, the wheel will begin to rotate. When the flow increases, the wheel will rotate more quickly and the number of electric pulses will

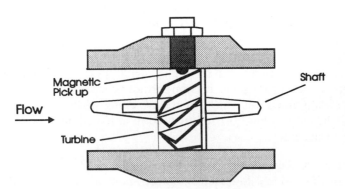

FIGURE 10-62 A turbine flow meter. The turbine flow meter has a magnet mounted in each of the vanes. When fluid flows past the turbine, it will spin and electric pulses are pro-duced.

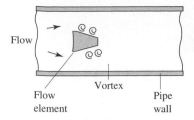

FIGURE 10-63   A vortex flow meter. Notice the blunt object placed in the flow stream that causes the flow to create vortices as the fluid flows around it. Each vortex is counted with a very sensitive electronic detector.

increase. The electric pulses can be averaged over time to provide a flow rate, or they can be totalized to determine the total flow.

10.4.4.3 Vortex Flow Meters ■ The *vortex flow meter* is also a velocity-type flow meter that places a flow element in the flow stream that is not streamline. When the flow stream strikes the flow element, a series of vortices is produced (shedded). For this reason, this type of flow meter is sometimes called a *vortex shedding flow meter.* When a vortex is produced, it causes the fluid to create a swirling motion as it moves. A very sensitive electronic detector can detect the presence of the vortices. The number of vortices that are produced is directly proportional to the flow rate. This type of sensor is very accurate and it can also be used when the fluid has suspended solids (slurries) moving in the flow. The number of vortices that are detected can be averaged to produce a flow rate, or they can be added to produce a total flow. A frequency-to-voltage circuit and amplifier must be used to increase the small-pulse signal from the electronic detector. It is also important to understand that the vortex-type flow meter works best in very turbulent flows. Fig. 10-63 shows the location of the flow element in a pipe, and Fig. 10-64 shows a picture of several vortex flow meters.

10.4.4.4 Electromagnetic Flow Meter ■ Another type of flow meter uses the operating principle of creating an electrical field in the fluid and then measures the strength of the field. This type of flow meter takes advantage of the principle of *electromagnetic induction.* As you know, a voltage can be induced in a conductor

FIGURE 10-64   Vortex meters used to measure flow. (©Copyright 1995 Omega Engineering, Inc. all rights reserved. Reproduced with the Permission of Omega Engineering, Inc., Stamford, CT 06907.)

when it is passed through a magnetic field. In this case the fluid must be electrically conductive so that it will accept the magnetic field and act as a conductor.

Fig. 10-65 shows an electromagnetic flow meter. The meter has two major parts. The first part is a *set of coils* that creates the magnetic field. When the fluid flows past these coils, it will act like an electrical conductor and will hold an electrical charge. The second part of the flow meter is a *set of electrodes* that is used as the detector. The detector acts like a voltmeter and measures the intensity of the electrical charge. The stronger the fluid flow, the stronger the electrical charge.

This type of flow meter is extremely useful in that it does not need to place any disturbance in the fluid flow, and it is also useful to measure the flow of very corrosive fluids. Early versions of this type of meter consumed large amounts of electrical power, but newer versions use pulse technology to limit the current supplied to the magnetizing coils so that the power consumption is reduced.

10.4.4.5 Ultrasonic Flow Meters ■ The ultrasonic flow meter uses Doppler meters to measure the shift of a frequency signal that is sent into the liquid flowing through a pipe. From the diagram in Fig. 10-66 you can see that the transmitting element injects a signal with a given frequency into the fluid flow. Bubbles in the fluid or any suspended solids in the fluid reflect the signal back to a receiver. The receiving element called a *Doppler meter* is placed a short distance downstream and it detects the frequency as the fluid flows past it. A special circuit called a *time-of-travel meter* measures the time delay or shift in the frequency that is caused by the fluid flow. The faster the fluid flows, the more the frequency is shifted. This type of flow meter is useful because it can measure the flow without creating any obstruction. It is important to understand that some amount of suspended solids or bubbles must be present in the flow to get the best reflection of the signal.

This type of meter is also used extensively as a portable flow meter. The portable flow meter is used in troubleshooting to detect flow in complex hydraulic systems. When a hydraulic system starts to fail, it generally leaks fluid past an inoperable valve back to the reservoir. Since the hydraulic piping is made of metal

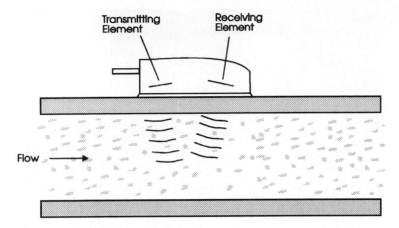

FIGURE 10-66 An ultrasonic flow meter that uses Doppler meters to detect the shift of frequency of an injected signal. The frequency signal is injected into the fluid and the flow will cause a shift in the signal as it is reflected to the Doppler meters.

or hoses, it is difficult to tell how much fluid flow is being bypassed, so an ultrasonic flow meter can be temporarily placed around each segment of the hydraulic piping and the amount of flow in the main lines and the return lines to the reservoir can be determined.

10.4.5 Positive Displacement Flow Meters

Another category of flow meters is called the positive displacement flow meter. This type of flow meter is the most accurate in that the flow is broken into segments and each segment is measured as the flow is moved. One example of a positive displacement flow meter uses a piston pump. The volume of the piston is known, and all of the fluid flows through the piston pump. This allows the total volume of fluid to be measured by counting the strokes of the piston. For example, if the piston has a volume of 0.1 gal, and it makes 200 strokes per minute, the flow would be 20 gpm. Another type of positive displacement flow meter uses a set of oval gears (lobes) that pumps a specific volume of fluid each time the gears mesh. The faster the flow moves, the faster the gears rotate. The number of rotations are counted and the flow rate can be calculated because the volume of fluid that is pumped during each revolution of the lobes is known. Fig. 10-67 shows a positive displacement flow meter.

The nutating disk is another type of positive displacement flow meter. The nutating disk is a movable disk that is offset so that it makes a concentric circular motion each time it rotates. The housing for the disk is perfectly round, so that each time the disk rotates, its oval path will trap a specific amount of fluid. Since the volume of trapped fluid is known, the number of revolutions the disk makes can be counted and the flow rate can be calculated.

10.4.6 Mass Flow Meters

A mass flow meter is the most accurate type of flowmeter designed to measure the flow of gases as well as other fluids. Two types of mass flow meters are commonly used. The *Coriolis mass flow meter* is so named because it uses the Coriolis phenomenon to measure mass flow instead of volumetric flow, and the other is called a *thermal mass flow meter*.

The Coriolis mass flow meter uses a U-shaped tube that is designed to vibrate up and down at its natural frequency while all of the fluid flows through it. A strong magnet is used to make the U-tube vibrate. Fig. 10-68 shows an example of the

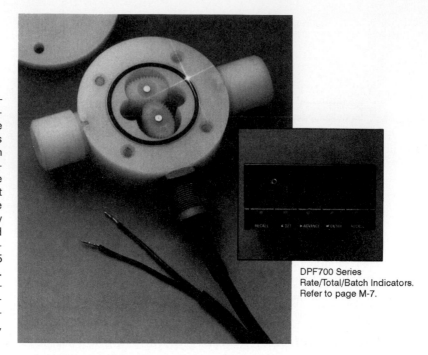

FIGURE 10-67 A gear-type positive displacement flow meter. The gears (lobes) rotate as fluid flows through them and the number of rotations are counted. Since the volume of fluid that is trapped between the gears is known, the flow rate can be calculated from the number of rotations. (©Copyright 1995 Omega Engineering, Inc. all rights reserved. Reproduced with the Permission of Omega Engineering, Inc., Stamford, CT 06907.)

DPF700 Series
Rate/Total/Batch Indicators.
Refer to page M-7.

Coriolis mass flow meter in three distinct stages. Fig. 10-68a shows the tube when it is not vibrated, and Fig. 10-68b shows the tube as it vibrates. These two diagrams show that when the U-tube vibrates it will naturally move up and down. When fluid is flowing through the tube, it will oppose the up-and-down movement, which will cause the U-tube to twist. The third part of the diagram, Fig. 10-68c, is an end view of the U-tube and it shows the effect of the twisting motion. The amount of twist will be directly proportional to the amount of flow. Sensors are located near the tube to detect the amount of twist and convert it to a usable variable-voltage

FIGURE 10-68 **(a)** The U-tube portion of a Coriolis mass flow meter. All fluid flow will travel through the U-tube portion of the flow meter. **(b)** This diagram shows the U-tube vibrating up and down. The up-and-down movement is caused by a magnet. **(c)** This diagram shows an end view of the U-tube. When fluid is flowing through the U-tube as it vibrates, the tube will tend to flex. (Courtesy of Micro Motion.)

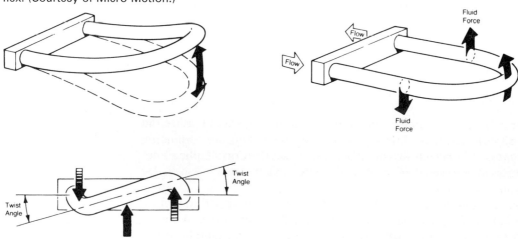

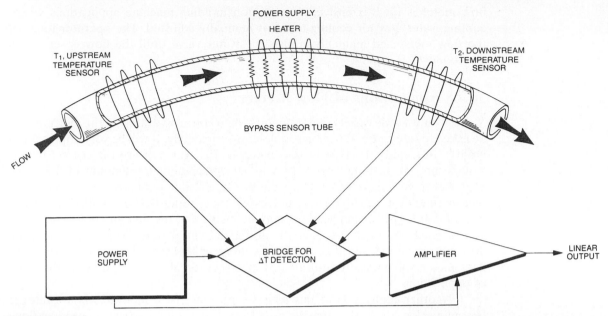

FIGURE 10-69   A thermal mass flow meter that shows the thermal element directly below the flow passage. Notice that the fluid flow does not come directly into contact with the thermal element, but it is close enough to provide heat transfer. (©Copyright 1995 Omega Engineering, Inc. all rights reserved. Reproduced with the Permission of Omega Engineering, Inc., Stamford, CT 06907.)

signal. This type of mass flow meter is useful to measure fluids whose viscosity continually changes because they do not need to have pressure or temperature compensation.

The thermal mass flow meter is usable for measuring the mass flow of gases. This type of mass flow meter uses a thermal element whose temperature changes as fluid flows past it. The amount of heat loss is directly proportional to the fluid flow. The thermal element is mounted close to the fluid flow but it does not come directly into contact with the fluid. This allows this type of mass flow meter to be used in virtually all types of applications where the density, pressure, and viscosity may change. The flow meter uses an electronic package that contains a flow analyzer, temperature compensator, and a signal conditioner to provide a linear output signal. Fig. 10-69 shows a diagram of a thermal mass flow meter.

10.4.7 Nonelectrical Flow Meters

In some cases the flow meter will be used by an operator as a visual indicator rather than provide an electrical signal. The most widely used visual flow meter is called a variable-area flow meter. This type of flow meter consists of a float that is weighted so that gravity pulls it down, and flow will push it up. The float is located in a tapered glass tube that has the flow rate indicated on it so when the float moves up or down the flow rate can be visually determined. This also means that it is important for this type of flow meter to be mounted in a vertical position.

When the flow through this type of flow meter is minimal, the float will rest at the bottom of the tap. When flow begins, the float will be lifted from the bottom of the flow meter and will be suspended in the flow. The stronger the flow, the higher the float is lifted. An indicator on the float is lined up with the scale that is printed or etched on the flow meter to identify the amount of flow. This type of flow meter is used extensively in injection molding machine applications where cooling-water flow for cooling molds is manually adjusted. The operator looks at the flow meter and manually adjusts the water valve until the flow is set at the desired level.

10.4.8 Applications of Flow Meters

You will find a wide variety of flow meters when you are working on the job. You will be asked to calibrate and troubleshoot them periodically. It will be very important to be able to identify the type of flow meter so that you can use the correct test to troubleshoot it. In some typical flow meter applications the amount of flow may be an average that is not 100% accurate. It will not matter because the rough measurement will be sufficient to determine that proper flow is established. Examples of these types of applications may be measuring the flow of cooling water to a large machine or process. Since the exact amount of water is not important, the flow meter will indicate that the flow is more or less than it was previously. The flow meter will basically indicate the presence or absence of flow in this type of application.

Another application of this type of flowmeter is to indicate when a filter is becoming clogged. A filter may be used to trap unwanted particles in product flow. The flow will drop off as the filter becomes clogged, and the signal from the sensor can be used to indicate a dirty filter.

In other applications, a mass-flow flow meter may be used to accurately measure the flow of gases or fluids that are used in processes. These types of flow meters are also used in biomedical applications where accurate measurements are required. Another application for the mass-flow flow meter is where an accurate amount of gas is added to a heating chamber when special metals or ceramics are processed. In this type of application the addition of certain gases will change the composition of the metal or ceramic material.

The positive displacement flow meter is used where accurate flow measurements must be made. For example, in a fiberglass manufacturing application, a number of chemicals are blended continuously to make the fiberglass resin. It is important to be able to monitor the flow of each of the chemicals so the exact amount of each can be added continually. Another application of the continuous flow sensing is where a catalyst is added to fiberglass as it is molded into automotive body parts. This process is called reactive injection molding (RIM) and it is important to accurately measure the amount of catalyst that is continually added to the fiberglass as it is injected into the mold.

Other applications will include measuring the flow of virtually every liquid product that is made such as beverages, juices, fertilizers, chemicals, and petroleum products. The flow meter may be used to meter the raw products prior to processing, or it may measure the flow of the finished product. Some flow meters are mounted in remote locations such as in the desert on petroleum pipelines or in Alaska and they use a solar energy converter to provide a supply of electricity. The signals from these remote flow meters are sent to a satellite and retransmitted to a control room that can be hundreds of miles away. This type of flow meter is called a *remote transmitting unit* (RTU). Fig. 10-70 shows an example of an RTU with its solar package and satellite link. This type of sensor is also used on a smaller scale in tank farm applications where large volumes of liquids are moved and stored in tanks over an area of several hundred acres. It would be very expensive to connect each sensor with wire, so the RTUs are used to send the flow meter measurements

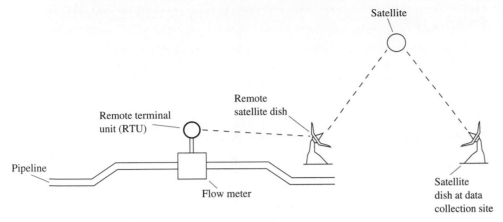

FIGURE 10-70 A remote terminal unit (RTU) with a satellite transmitter used to send flow data from a remote pipeline to a data collection site.

on radio carrier waves. A group of sensors similar to flow meters are density meters and viscosity meters. The next section will discuss these meters.

10.5 DENSITY AND VISCOSITY SENSORS

When product flow is measured, it will be affected by density, viscosity, and temperature of the product. Density is also measured when products such as tomato paste, pudding, paint, petrochemicals, and a wide variety of similar products are produced. Density is a measurement of mass per unit volume and it is also called *specific gravity*. Specific gravity is a measurement of the density of a product at a set temperature compared to the density of water. The specific gravity of water is considered 1. A simple way to discuss density is to define how "runny" something is at a set temperature when compared to water. The product will be more runny or less runny than water when the two are compared at the same temperature.

The formula for density is:

$$D = M/V$$

where M is the mass of the liquid and V is the volume.

The common units for density are grams per cubic centimeter and pounds per cubic foot. The formula for specific gravity is the density of a liquid at a set temperature divided by the density of water at the same temperature. It is important that the temperature of the liquid and the temperature of water during the comparison are identical.

$$\text{Sp Gr} = \frac{D_\text{L} \text{ at } T_1}{D_\text{W} \text{ at } T_1}$$

where D_L is the density of the liquid being measured.

T_1 is the control temperature for both liquids.

D_W is the density of water (approximately 62.4 lb/ft³)

PROBLEM 10-7

Calculate the density of a latex paint that has mass of 82 lb. and a volume of 1 ft³. Use the density to calculate the specific gravity of the paint at 75°F. The specific gravity of water at 75°F is 1, and the density of water is 62.4 lb/ft³.

SOLUTION

Use the formula $D = M/V$ to determine the density:

$$D = 82 \text{ lb/ft}^3$$

Use the formula $\text{Sp Gr} = \dfrac{D_L \text{ at } T_1}{D_W \text{ at } T_1}$ to calculate the specific gravity.

$$\text{Sp Gr} = \frac{82 \text{ lb/cu ft}}{62.4 \text{ lb/cu ft}}$$
$$\text{Sp Gr} = 1.3$$

10.5.1 Liquid Density Sensors

The density of a liquid can be measured as it flows through a piping system. Fig. 10-71 shows a picture and Fig. 10-72 shows a diagram of a typical liquid density sensor. From the diagram you can see that unlike a flow meter, the density sensor does not need to sample the entire flow. Rather it can be mounted in the piping to take a sample of the flow. A small portion of the flow will continually flow through a sensing chamber of the density sensor. The sensing chamber has a totally

FIGURE 10-71 A typical density sensor. The sensor is mounted in the piping system so that it can sample a part of the flow. (Courtesy of Princo Instrumentation, Inc.)

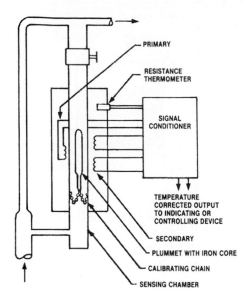

FIGURE 10-72 Diagram of a density sensor. This type of sensor takes a sample of the product as it flows past this sensor. (Courtesy of Princo Instrumentation, Inc.)

submerged *plummet* attached to a set of calibrating chains at several fixed reference points. The plummet is an object that will float in the liquid at the equilibrium which is calibrated to be near the midpoint of the density range for the liquid being measured. When the density of the liquid increases, it is thicker and the plummet will be raised slightly. When the density decreases, the liquid is thinner and the plummet will fall slightly. The change of distance the plummet moves is measured by a position sensor called a *linear variable differential transformer* (LVDT). The operation of the LVDT will be explained later in this chapter. An RTD is used to constantly measure the temperature of the product to adjust the calculation of the density.

The second part of the density sensor is the signal conditioner. The signal conditioner uses the position signal from the LVDT and the temperature signal from the RTD and converts them to 4–20 mA, 0–5 volts, or −5 volt to +5 volt signal that represents the density of the liquid being measured.

The density meter is calibrated and monitored continually. During the time the density meter is operating, a quality control technician will sample the product at the same point in the piping system as the density meter. The sample will be taken to a laboratory where a small amount is placed on a special testing paper that has been previously weighed. When the sample is placed on the paper, the paper and the liquid sample are weighed together. The sample is then heated in an oven for several minutes until most of the moisture is removed from the paper. At this point the sample is weighed again. The technician now has determined the weight of the product by subtracting the original weight of the sample paper from the weight of the original sample before it was heated and after it was heated. The difference between these weights is the amount of water that was in the product versus the amount of solids in the sample.

10.5.2 Viscosity Sensors

Viscosity is similar to the density of a product except viscosity will change continually as the temperature changes. For example, you may be familiar with motor oil. When motor oil is warm, it will flow like water, and when the temperature is cold

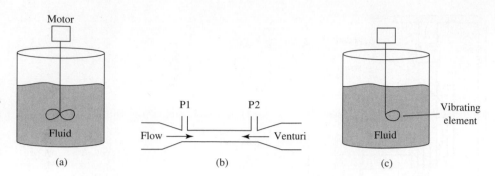

FIGURE 10-73 **(a)** A rotating disk viscosimeter. **(b)** A differential pressure viscosimeter. **(c)** A vibrating element viscosimeter.

such as in the winter, the motor oil will be so thick, it will barely flow. In fact, if thinning additives are not added to motor oil, it will have the consistency of grease if the temperature is lowered sufficiently. In large pipeline operations and in applications such as filling tomato paste into containers, the viscosity must be controlled so the amount of product flowing can be calculated accurately. Viscosity is controlled by adjusting the temperature or density of the material.

The instrument that measures the viscosity of a product is called a *viscosimeter* (see Fig. 10-73). Fig. 10-73a shows a rotating disk viscosimeter, Fig. 10-73b shows a differential pressure type, and Fig. 10-73c shows a vibrating element to measure the viscosity of a product. The rotating disk viscosimeter has an element or disk that is rotated at a specific speed. The amount of torque the motor requires to turn the disk will increase as the product gets thicker. The differential pressure viscosimeter uses a venturi to measure the pressure differential of product flowing through it. When the viscosity of the product changes, the pressure drop across the venturi will also change. The vibrating element viscosimeter uses an element that is vibrated at a specific frequency. The amplitude of the vibration will increase when the viscosity decreases.

10.6 LEVEL SENSORS

Level sensors are used to determine the amount of product in holding tanks and process tanks. In modern industry the raw materials used to produce all types of products and the finished product may be stored in a tank or bin. A means of measuring the level of the tank or bin is essential to determine the amount of product that is in them. This can be accomplished by a variety of sensors. This section will explain the operation of the sensors used to determine the level of products. Some sensors will use the pressure or weight of a product to determine its level. In these sensors, the internal transducers will be the same pressure sensors that were explained earlier in this chapter.

Level sensors can be categorized in several ways. One way is to determine if the level is to be measured at a given setpoint, or if it is to be measured continuously from minimum to maximum. The sensor that determines the level at a single point is called a *point-contact sensor,* and the sensor that measures the level from minimum to maximum is called a *continuous level sensor.*

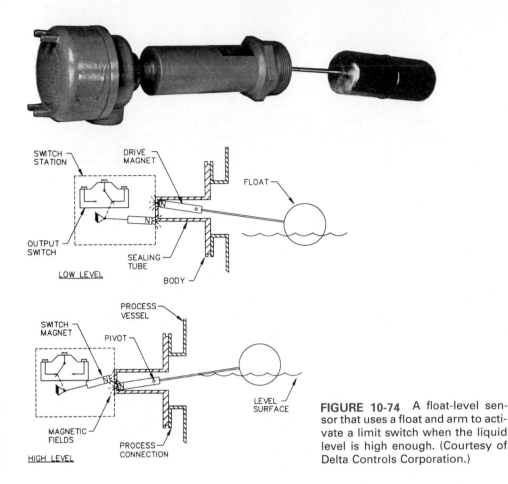

FIGURE 10-74 A float-level sensor that uses a float and arm to activate a limit switch when the liquid level is high enough. (Courtesy of Delta Controls Corporation.)

10.6.1 Point-Contact Level Sensors

Several point-contact level sensors are available. Most of these types of level sensors include a switch that is activated when the level reaches a specific point. Each of these types of level sensors will be discussed in this section.

10.6.1.1 Float-Level Sensor ■ The float-level sensor is the simplest level sensor to understand. From the diagram in Fig. 10-74 you can see that a float is connected to an arm, and the arm will activate a limit switch when the arm is raised. The float will be lifted when the level of the liquid is high enough. The limit switch can be adjusted so that the exact level where the switch is activated can be set.

Another type of float switch uses a magnetic-activated switch. From the two diagrams in Fig. 10-75 you can see the float in this type of switch has a rod connected to it just like the previous type float-level sensor. This type of sensor has a permanent magnet connected to the end of the rod. Fig. 10-75b shows that when the float raises with the liquid level, the magnet is moved into position so that it is near the magnetic-activated switch that is mounted in the head of the sensor. When the permanent magnet on the rod is in the correct position, it will pull the movable magnet that activates the switch. When the movable magnet is pulled to the magnet on the rod, the switch contacts close. You can see in Fig. 10-75a that when the liquid level drops and the float allows the rod to be lowered, the magnet will no

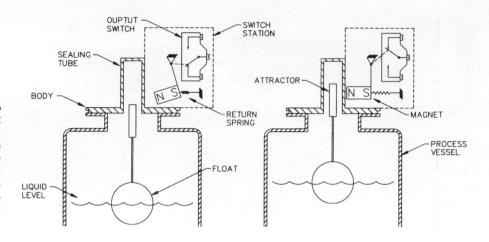

FIGURE 10-75 **(a)** A float-level sensor that uses a magnetic-actuated switch. This figure shows the sensor at low level. **(b)** The float-level sensor with the sensor at high level. (Courtesy of Delta Controls Corporation.)

longer have any attraction to the switch. Small springs will then cause the contacts to move to their normally open position. (Notice that the switch has a single-pole, double-throw configuration so that the common terminal can be connected for normally open or normally closed operation.) The switch is typically connected to a pump motor a solenoid valve that is used to change the amount of material entering or leaving the tank.

10.6.1.2 Multiple Float-Level Sensor ■ The multiple float-level sensor is similar to the single float-level sensor except the multiple float-level sensor provides the ability to have switches activated at more than one point. From Fig. 10-76 you can see that the liquid level moves the float up and down on a guide that keeps it aligned with the switches. The float has a magnet mounted inside it and the switches

FIGURE 10-76 **(a)** Multiple reed switches are mounted so that a magnet in the float will activate each switch when the magnet is near the switch. The float moves up and down a guide with the level of the liquid in the tank. **(b)** Close-up view of the float with the magnet encased in it, and the reed switch that is activated when the float moves past it. (Courtesy of **a.** Instrumentation & Controls Systems; **b.** ©Copyright 1995 Omega Engineering, Inc. all rights reserved. Reproduced with the Permission of Omega Engineering, Inc., Stamford, CT 06907.)

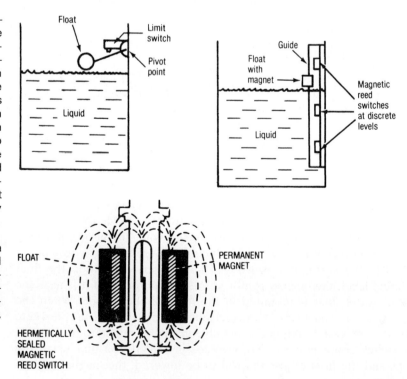

are magnetic-activated reed switches. When the magnet in the float comes near each switch, it turns it on, and when the magnet is not near the switch, the switch is turned off. Any number of switches can be added to the system so that the level at numerous points can be detected.

10.6.1.3 Displacer-Level Sensor ■

The displacer-level sensor is very similar to the float-level sensor in that it has a displacer element located directly in the liquid. The displacer element has a rod that connects it to a switch, and when the level increases, the displacer element rises with the level and moves the rod so that it activates the switch. The switch in this sensor is exactly like the magnetic switch explained in the float-level sensor. Fig. 10-77 shows a picture and Fig. 10-78 shows a diagram of this type of sensor and you can see that when the rod moves up, the magnet on the rod moves to where it is located close to the magnet that activates the switch. The magnet in the switch activator is movable, and when it is pulled toward the magnet on the end of the rod, it will activate its contacts. When the level drops and the displacer drops and the magnet on the rod drops, the magnet on the rod will no longer have any effect on the switch magnet. Small springs will cause the contacts to move back into their original position.

The major difference between the float and the displacer element is that the float is totally supported by the surface of the liquid, and the displacer is partially submerged. This is known as the *buoyancy principle*. The displacer element must be slightly more dense than the liquid that it is used in. Since the displacer element

FIGURE 10-77 A displacer-level sensor. (Courtesy of Delta Controls Corporation.)

FIGURE 10-78 **(a)** An example of a displacer-level sensor. The displacer element uses the buoyancy principle to allow the sensing element to be partially submerged so that action on the liquid surface does not interfere with the sensor's action. **(b)** The liquid-level sensor is measuring liquid level at an intermediate level. **(c)** The liquid-level sensor is measuring liquid level at a low level. (Courtesy of Delta Controls Corporation.)

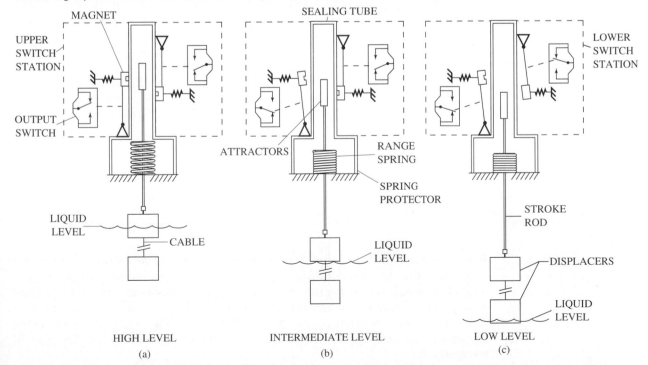

FIGURE 10-79 (a) An example of a paddlewheel-level sensor that uses a shaded pole motor to turn the shaft of the paddlewheel-level sensor very slowly. When the level of the material increases to a point where it reaches the paddlewheel, it stalls the motor. A sensor detects the slight increase of current when the motor stalls and activates a switch. (b) Application where paddlewheel sensor is mounted through the top of a tank to measure when the tank is full, and through the side of a tank to measure when the tank is half full. (©Copyright 1995 Omega Engineering, Inc. all rights reserved. Reproduced with the Permission of Omega Engineering, Inc., Stamford, CT 06907.)

is partially submerged, it is not subjected to the action on the surface of the liquid. Since pumps and agitators tend to make the surface very rough, a float sensor may be subjected to false actuation, or it may wear prematurely. If the surface is subject to floating debris or suspended solids, an additional displacer may be used that is submerged further below the surface so that it can activate safely.

10.6.1.4 Paddlewheel-Level Sensor ■ The paddlewheel-level sensor uses a paddlewheel that is similar to the paddlewheel flow sensor. In this application the paddlewheel is turned slowly by a small motor. When the level of the liquid or solid granules comes into contact with the paddlewheel, it will become stalled. When this occurs an amperage detector determines that the current has increased slightly, which indicates that the level has increased to a point where it stops the motor shaft from turning. The motor that is used for this type of sensor is a shaded pole motor. The operation of the shaded pole motor allows it to have its rotor stalled and not damage the motor. When the rotor stalls, the motor windings act similarly to a primary winding of a transformer, and a slight increase in current will occur. The operation of the shaded pole motor will be covered in a later chapter in this text.

Fig. 10-79 shows an example of the paddlewheel-level sensor. The paddlewheel can be mounted at any level on the side of the tank to indicate the level has been reached or exceeded. This type of sensor is used frequently to measure the level of granular plastic or other similar types of raw material that is stored in bins and hoppers.

10.6.1.5 Vibrating-Tines Level Sensor ■ The vibrating-tines level sensor uses a set of tines that acts like a tuning fork to determine when the level of material or liquid has exceeded the level setpoint. An electronic circuit makes the tines oscillate at a specific frequency. When the level of material rises and covers the tines, it stops them from oscillating. An electronic circuit detects the change in the oscillating frequency and activates a switch. Fig. 10-80 shows an example of this type of level sensor.

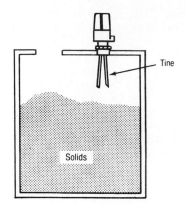

FIGURE 10-80 Example of vibrating tines used to determine the level of material. The tines are oscillated at a specific frequency. When the level of material reaches a point where it covers the tines, they will stop oscillating. The change of frequency is detected, which activates a switch. (Courtesy of Instrumentation & Controls Systems)

10.6.1.6 Beam-Breaker Level Sensor ■ A beam-breaker level sensor uses a light beam to detect the level of solids to determine when the setpoint level has been exceeded. From Fig. 10-81 you can see that two probes are mounted horizontally so that one can send a beam of light that is focused directly on the other. When the level of the material increases sufficiently, it will block the beam of light and the photoelectric detector will activate a switch. These types of sensors are useful to measure the level of granular solids. In some applications, multiple sensors are used to determine the level at several points.

10.6.1.7 Two-Wire, Conductance-Level Sensor ■ Another way to determine the level of certain liquids is to mount two wires at different heights in a tank where the level is measured. One wire is mounted near the bottom of the tank, and the second one is mounted at the level that is used as the setpoint. When the liquid level rises to a point where the second wire is covered, a small current is conducted through the liquid between the two wires. This small current is detected and the electronic circuit activates a switch. When the liquid level is below the second wire, the resistance of the air is too large and no current will flow (see Fig. 10-82). Some applications of this type of sensor use additional sets of wires that are mounted at several points along the side of the tank to detect the level of the liquid at more than one point. It is also important that the liquid has the properties that make it a conductor. Some liquids are dielectrics or insulators and this type of sensor would not be usable for them.

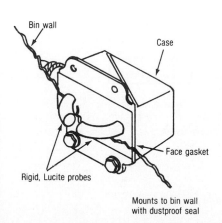

FIGURE 10-81 An example of a beam-breaker level sensor, which focuses a beam of light on a receiver. When the level of the material increases, it will block the beam of light, and a photoelectric sensing circuit will detect the difference and activate a switch. (Courtesy of Instrumentation & Controls Systems.)

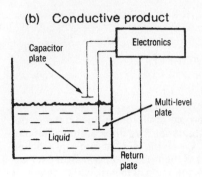

(b) Conductive product

Capacitor plate

Electronics

Multi-level plate

Liquid

Return plate

FIGURE 10-82 Example of two conducting wires used to detect the level of liquid. One wire is mounted near the bottom of a tank, and the other is mounted near the top. When the liquid level covers the top wire, a small current is conducted through the liquid. The small current is detected, which activates a switch. (Courtesy of Instrumentation & Controls Systems.)

Another version of this type of conductance-level sensor uses one or two probes instead of the wires. If one probe is used, the side of a metal tank is used as the other probe. When the liquid level increases to a point that it touches the tip of the probe, a small current is conducted between the probe and the side of the tank. If the sensor has two probes, a small current will flow through the liquid as it rises and touches both probes. Fig. 10-83 provides a diagram that shows examples of the probes being used as level sensors and for alarm points. You can see that one of the sensors has four separate probes of different lengths to indicate when the tank is one-quarter, one-half, three-quarters full, and completely full. Each probe is connected to an indicator lamp.

10.6.1.8 Thermistor-Level Sensor ■ The thermistor-level sensor operates from a principle that the temperature of the liquid being measured will be different from the temperature of the surrounding air. The thermistor is mounted in the tank at the specified level. Since it is not submerged in liquid, it will be measuring the temperature of the air. When the liquid level rises and covers the thermistor, it will be measuring the temperature of the liquid. The small change of temperature between the air and the liquid will cause the thermistor's resistance to change, which can be detected in the bridge circuit. The bridge circuit can be designed to activate a set of switch contacts.

FIGURE 10-83 Example of conductance probes being used to sense the level of liquid in a tank. The probes are shown as alarms to indicate high liquid levels and low liquid levels. (Courtesy of Delta Controls Corporation.)

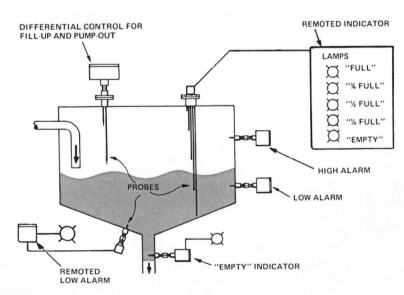

DIFFERENTIAL CONTROL FOR FILL-UP AND PUMP-OUT

REMOTED INDICATOR

LAMPS
- "FULL"
- "¾ FULL"
- "½ FULL"
- "¼ FULL"
- "EMPTY"

HIGH ALARM

PROBES

LOW ALARM

REMOTED LOW ALARM

"EMPTY" INDICATOR

10.6.2 Continuous-Level Sensors

The next group of level sensors is continuous-level sensors. These sensors provide a continuous reading of levels from minimum to maximum. This means that they provide an analog signal (0–10 volts dc, or 4–20 mA). These types of sensors are generally more expensive since they must provide a continuous signal. Several types of sensor elements are used to provide a continuous signal. These include the R.F. admittance (capacitance) type, the sonic type, conductive type, sounding type that uses a reel and tape, and the differential pressure sensor that can be used to determine the height of a liquid column. If the weight of a substance is known, a scales-type weighing sensor can be used to measure the total volume of the material being measured. This section will explain the operation of these types of sensors.

10.6.2.1 RF Admittance (Capacitance) Level Sensor ■ The *RF admittance-level sensor* uses pulsed radio frequency waves to determine when material or liquid is touching the end of its probe. Since this type of sensor uses the change in *dielectric* to determine the level of a liquid or granular solid material, it is also called a *capacitance-level sensor*. Fig. 10-84 shows an example of this type of level sensor and Fig. 10-85 shows the electronic circuit that produces the radio waves. From the first diagram you can see that this type of level sensor has a long probe. The circuity in the top of the sensor produces pulsed radio frequency waves that flow through the probe to the tank wall or the bottom of the tank, which provides a ground for the circuit. This means that the probe must be strategically located in close proximity to the outside wall or to the bottom of the tank so that the amount of dielectric in the air between the probe and the tank wall is known. If the bottom of the tank is used as the ground plane, the amount of capacitance between the probe and the level of the liquid or solid granular material will change as it rises and moves closer to the probe. The change of capacitance can continue to be measured as the material or liquid begins to cover the probe. If the side of the tank is used, the probe will use the side as the ground plane reference and it can also continue to determine

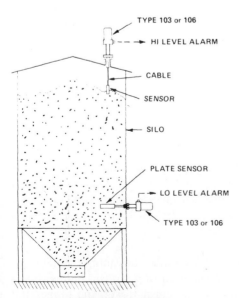

FIGURE 10-84 Example of an RF admittance (capacitance) level sensor. The probe can be mounted so that the side of the tank or the bottom of the tank can be used as the ground reference for the probe. (Courtesy of Delta Controls Corporation.)

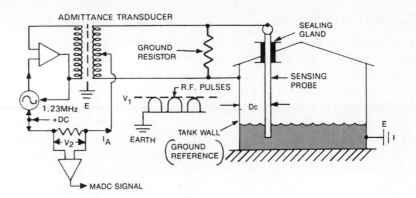

FIGURE 10-85 Electrical diagram of an RF admittance-level sensor. (Courtesy of Delta Controls Corporation.)

- HIGH FREQUENCY ADMITTANCE TECHNIQUES AVOID ERRORS DUE TO COATINGS, BUILDUP, AND MATERIAL CONDUCTIVITY VARIATIONS.

- GROUNDED PROBE PROVIDES A STABLE ZERO AND AVOIDS FAILURES DUE TO STATIC ELECTRICTIY.

the change of capacitance as the liquid moves up the probe. In both applications the dielectric of each type of liquid or material can be measured, and the probe can be calibrated to show the precise level of liquid or granular material as it gets closer to the probe or moves up on the probe.

The diagram in Fig. 10-86 shows an ac power source that is used to provide the radio frequency for the probe. An op amp is used in conjunction with a transformer to provide a series of pulses whose high frequency is in the radio wave range. The secondary of the transformer is center tapped and connected to an op amp that measures the amount of power used by the probe to transmit the energy wave. When the probe is in free air and no material or liquid is close to it, the amount of energy will be minimal. When the material or liquid begins to rise and changes the dielectric between the probe and the tank, the amount of energy to transmit the energy wave will change. The amount of change will be proportional to the level of the material or liquid. The output of this op amp will provide a 4–20 mA signal that is proportional to the level of the liquid. The sensor has a zero and span circuit to provide accurate calibration.

A two-probe version of this type of probe is also available for use in tanks that are nonmetallic. Some chemicals and other liquids must be stored in plastic tanks. In these types of applications, the tank wall or bottom cannot be used to create the ground plane. A two-probe capacitance-level sensor uses two probes that are located in close proximity to each other. If the entire level of the tank is to be measured, the two probes must be long enough to reach to the bottom of the tank. When the level in the tank is near empty, the probes are exposed to air, which allows the circuit to use the minimal amount of energy. When liquid level begins to rise and cover the probes, the capacitance will change because the dielectric between the probes is changing. The electronic circuitry for this type of sensor is very similar to the RF admittance-level sensor.

10.6.2.2 Sonic-Level Sensors ■ The sonic-level sensor takes advantage of the principle that the speed of sound waves traveling through air or a gas can be measured and timed. The longer the distance the sound waves must travel, the longer the time it will take for them to be sent and reflected. In this type of application, the

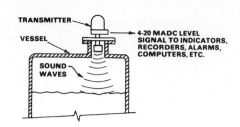

TRANSMITTER WITH INTEGRAL ELECTRONICS,
FLANGED VESSEL CONNECTION, AND E/P HOUSING

FIGURE 10-86 Ultrasonic-level sensor. The sensor has a transmitter and receiver in its head. Sound waves are transmitted to the liquid and the time it takes for them to be reflected and received by the sensor is detected. This time is converted to distance so the sensor can determine how far the level is from the probe. (Courtesy of Delta Controls Corporation.)

sonic transmitter is mounted in the top of a tank. The sensor has a transmitter and receiver mounted in the same head. An electrical pulse is applied to a crystal that causes it to vibrate against a diaphragm. When the diaphragm vibrates, sound waves are produced and they are directed at the level of the liquid or solid that is being measured. When the sound waves reach the liquid or solid, they will be reflected back to the sensor. The receiver in the sensor will detect the reflected sound waves and the electronic circuitry measures the amount of time between the instant the sound waves are sent and when they are received. The amount of time is accumulated digitally in the circuit in the form of a binary coded decimal (BCD) number. The number is used in a calculation to convert the amount of time to distance. The answer to the calculation is in a BCD format so a digital-to-analog converter is used to provide a 4–20 mA signal that represents the level of the liquid or material. The sonic type of level sensor is usable in applications for measuring material such as pitch blend and carbon black, which tend to be dirty, sticky, and gummy, or other materials that are corrosive or abrasive that may damage a probe if it came into contact with these substances. Some substances such as powders and substances that are foamy may not be suitable for this type of level probe because they tend to absorb sound waves rather than reflect them (see Fig. 10-86).

10.6.2.3 Sounding Tape and Reel-Type Level Sensors ■ The *sounding tape and reel-type level sensor* is sometimes called a *fishing reel* level sensor. The name of this sensor comes from the fishing term *sounding*.

In modern level sensors a weight is attached to a line and the line is lowered from the top of a tank. The weight will cause the line to have tension as long as it is suspended in air. When the weight comes into contact with the level of the material it is measuring, the amount of tension will be reduced and a sensor can detect this change. When the tension changes, a mechanism stops the line from lowering and begins to retrieve the line. A spool is used to reel the line in and the number of rotations the spool uses to reel the line completely to the top is measured and converted to a level measurement. Some of these sensors use a magnet in the spool and the number of times the magnet passes a detector is counted. In other applications a rotary potentiometer is connected to the spool and the number of rotations is converted directly into a change of resistance. In both types of sensors, the transducer signal is converted to a usable 4–20 mA output signal. If the spool uses a magnet and counter, the counter is reset when the tension changes and indicates the weight at its lowest point. When the reel begins to pull the line in, the counter will be ready to determine the amount of line that was lowered. Figs. 10-87, 10-88, and 10-89 show example applications using the sounding-type level sensor.

10.6.2.4 Differential Pressure Level Sensor ■ At first it is difficult to understand how pressure can be used to measure the level of a liquid, but if you understand that a column of liquid that is 1 inch square at its base will cause a pressure at the bottom of the column that is proportional to the height of the column, you can see how pressure can be converted to height. What this means is that the deeper a liquid becomes, the greater the pressure becomes at the bottom of the tank or vessel that holds the liquid. The pressure can be measured and formulas are available to calculate the depth or, if you prefer, the height of the liquid. This pressure is called *head pressure* or *hydrostatic head pressure.* The exact pressure at the bottom of the tank will be determined by the depth of the liquid, the specific gravity (density) of the liquid, and the atmospheric pressure pushing down on the liquid. The effects of atmospheric pressure can be equalized by using a differential pressure

FIGURE 10-87 **(a)** An application of a sounding-type level sensor that uses a weight and line to measure the level of granular material in a tank. **(b)** An application of a sounding-type level sensor that uses a weight and line to measure the level of liquid in a tank. The weight in this application will float on the liquid when it reaches its surface. (Courtesy of Delta Controls Corporation.)

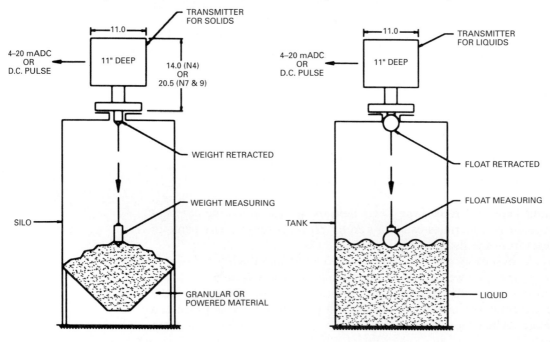

FIGURE 10-88 Example of a sounding-type level indicator shown mounted in place to indicate the level of a tank. (Courtesy of Delta Controls Corporation.)

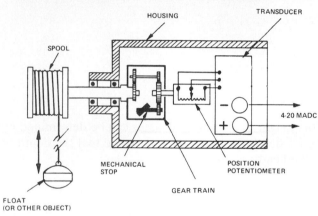

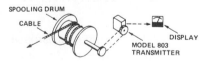

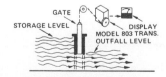

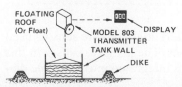

FIGURE 10-89 **(a)** An example of the spool mechanism used to measure the amount of line that has been lowered to determine the level of material. The output of this amplifier is 4–20 mA. **(b)** An example of the output of the level sensor sent to a display. **(c)** An example of a level sensor connected to a floating roof for a tank, and the output signal is sent to a transmiter and a display. (Courtesy of Delta Controls Corporation.)

sensor (ΔP sensor) as explained previously. Since one side of the ΔP sensor is open to atmospheric pressure, it will not matter what the atmospheric pressure is that pushes down on the liquid because the same pressure is pressing on the open side of the pressure differential sensor.

The concept of determining the level of liquid in a tank by calculating its head pressure may best be explained by calculating the pressure that a column of water exerts on the bottom of a tank when the level of water in the tank is 40 ft deep. From pressure conversion tables it is known that water exerts 0.434 psi per each foot of water column. This means that if the water is 1 ft deep in the tank, the pressure at the bottom of the tank would be 0.434 psi. The formula for determining the total amount of head pressure is:

$$P = \text{depth (ft)} \times \text{pressure of 1 ft of water (psi)}$$

. . . .

EXAMPLE 10-3

Find the pressure at the bottom of a tank that is 40 feet deep.

$$P_{\text{total}} = D \times P_{\text{at 1 ft}}$$
$$P = 40 \times 0.434$$
$$P = 17.36 \text{ psi}$$

SOLUTION

The formula for calculating the height of a column of water can be determined by changing the formula to solve for depth. It should be understood that the depth of the water would be equal to the height of the column of water.

$$D = \frac{P_{\text{total}}}{\text{pressure of 1 ft of water}}$$

. . . .

PROBLEM 10-8

Determine the level of water in a tank if the pressure at the bottom of a tank of water is 24.5 psi. (Do not account for atmospheric pressure because a ΔP sensor is being used to measure the pressure.)

SOLUTION

$$D = \frac{P_{\text{total}}}{\text{pressure of 1 ft of water}}$$
$$D = 24.5 \text{ psi}/0.434 \text{ psi per ft}$$
$$D = 56.45 \text{ ft}$$

. . . .

Figs. 10-90 and 10-91 show examples of the differential pressure sensors used to measure the liquid level in several tank applications. The pressure sensor in this type of instrument uses strain gauge technology that is integrated into a flexible diaphragm.

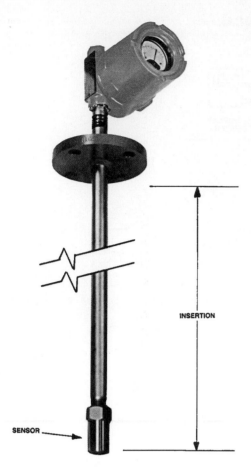

INSERTION

SENSOR

FIGURE 10-90 A pressure differential sensor that determines the level of liquid by using the pressure of a column of liquid to calculate the level. (Courtesy of Delta Controls Corporation.)

This is a good example of where this type of sensor is used to measure the level of liquid in a tank, but you must troubleshoot the sensor as though it is a pressure transducer and strain gauge. This is typical of many sensors that use one type of technology to measure one parameter and convert it to another.

10.6.2.5 Using Weighing Systems to Determine the Level ■ Another way to determine the level of liquid or granular solids in a tank is to weigh the tank when it is empty and when it is full and develop a scale. For example, if a tank weighs 1500 lb when it is empty and it weighs 2500 lb when it is full, the material in the tank can be calculated to weigh 1000 lb when the tank is full. If the tank has a uniform shape, the material in the tank should weigh 500 lb when the tank is half full, and 250 lb when the tank is a quarter full. A ratio can be developed by dividing the total weight of the material by 100%. This means that 1% equals 10 lb. If the tank is spherical or conical or other nonuniform shape, a table will need to be developed to determine the weight at each height of the tank because the values will not be linear.

The weighing system in this type of application takes advantage of the load cell technology to measure the weight. Fig. 10-92 shows an example of a weighing system used to determine the level of material in a hopper. You can see that this type of sensor depends on the weight of the material to determine the level in the tank. For this reason it is important that the material does not become wet because

BURIED HORIZONTAL CYLINDRICAL TANK VOLUME INDICATOR

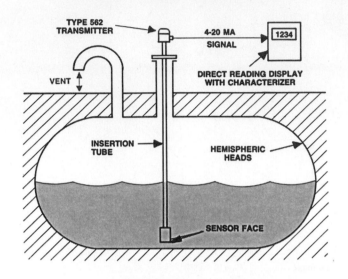

TYPE 562
TRANSMITTER

4-20 MA
SIGNAL

1234

DIRECT READING DISPLAY
WITH CHARACTERIZER

VENT

INSERTION
TUBE

HEMISPHERIC
HEADS

SENSOR FACE

UTILITY CATCH TANK
(SHOWING TWO MOUNTING
METHODS FOR THE 562)

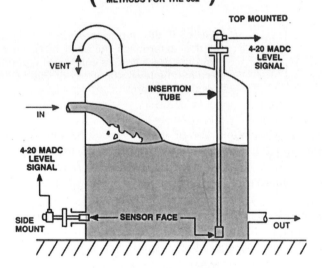

TOP MOUNTED

4-20 MADC
LEVEL
SIGNAL

VENT

INSERTION
TUBE

IN

4-20 MADC
LEVEL
SIGNAL

SIDE
MOUNT

SENSOR FACE

OUT

FIGURE 10-91 (a) A ΔP level sensor used to measure the level in a buried tank. **(b)** An application where the ΔP sensor can be mounted in the top of a tank or in the side of a tank to determine the level of liquid in the tank. (Courtesy of Delta Controls Corporation.)

its weight would not be the same as when dry and the instrument would think the tank was fuller than it actually is. Fig. 10-92 shows examples of these types of weighing systems for a variety of tanks.

10.7 POSITION SENSORS

Linear and rotary motion is used in many of the previous sensors to detect pressure, flow, level, and temperature. For this reason it is important to be able to measure linear and rotary position or distance moved. It is also important to be able to

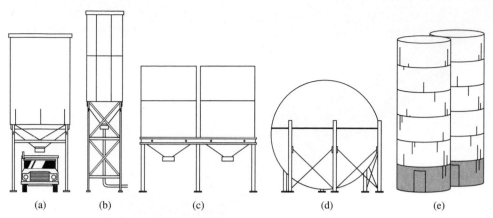

(a)   (b)   (c)   (d)   (e)

FIGURE 10-92 **(a)** Example of weighing system that measures the amount of product dispensed from the tank into a truck. **(b)** The load cells for this system are mounted in the feet of the tank to measure the total weight of the tank. **(c)** An example of two tanks that share load cells to determine the amount of product in each tank. **(d)** Example of load cells in the feet of a spherical tank to measure the amount of product in the tank. **(e)** Example of load cells in cylindrical tanks to measure the amount of product in the tanks. (Courtesy of Fairbanks Scales.)

measure linear and rotary motion and distance for many positioning and robotic applications. This section will discuss the operation of linear and rotary potentiometers, linear variable differential transformers (LVDTs), magnetostrictive sensors, encoders, and resolvers that are all used to measure distance or position. It is also important to understand that the two main types of motion, linear and rotary, are virtually interchangeable with the use of ball screw mechanisms and rack and pinion devices. This means that if the output motion of a sensor is linear, it can be changed to rotary motion with a rack and pinion, and if the output motion is rotary, it can be changed to linear motion with a rack and pinion or a ball screw mechanism. This simple conversion allows a larger variety of transducers to be used to measure distance or position.

An example of a rack and pinion and a ball screw mechanism is shown in Figs. 10-93a and 10-93b. The ball screw mechanism is shown in Fig. 10-93a. From this diagram you can see that the screw part of the ball screw looks like a threaded rod. The number of threads per inch will determine the amount of resolution the ball screw will have. When the ball screw is rotated, the ball bearings will rotate with the screw and cause the traveling portion of the mechanism to move in a linear motion. This is similar to having a threaded rod with a nut on it. If the nut is kept from turning and the rod is rotated, the nut will move back and forth along the threads of the rod.

Fig. 10-93b shows the rack and pinion mechanism. From the diagram you can see that if rotary motion is used to turn the pinion gear, the rack will change this motion into linear motion. If linear motion is input into the rack, the pinion gear would transfer this motion into rotary motion.

10.7.1 Linear Potentiometer

The potentiometer is perhaps the most common position sensor. This type of sensor is basically a fixed resistor with a movable tap that allows the amount of resistance

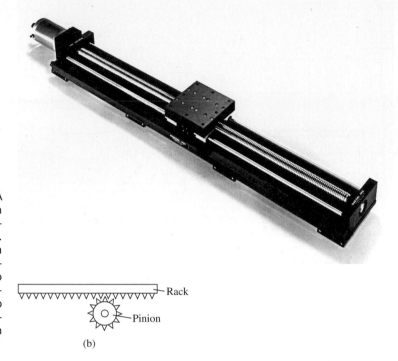

FIGURE 10-93 (a) A ball screw mechanism that converts rotary motion into linear motion. **(b)** A rack and pinion mechanism used to convert rotary motion into linear motion, or to convert linear motion into rotary motion. (Courtesy of Parker Hannifin Corp., Paedal Division.)

between the tap and either end of the resistor to provide a portion of the total resistance (0–100%). Fig. 10-94 shows an example of the linear potentiometer and you can see the main parts are the two terminals of the fixed resistor and the center tap. The center tap is also called the *wiper, slider,* or *tap.* Potentiometers are sized by the amount of resistance that is available between the two fixed terminals.

The reason the potentiometer is so usable is that it is fairly linear along its entire length. The resistor is usually made of fine resistance-type wire that is tightly wound around an insulating core. The slider makes contact with the wire and provides a means to alter the amount of the total resistor that is used. For example, if the total resistance is 1000 Ω, and the slider is positioned at a point that is the center of the resistor, the amount of resistance between either end and the slider will be 500 Ω. If the slider is moved 75% of the way toward one end, the amount of resistance between the slider and one fixed end will be 750 Ω and the amount of resistance between the slider and the other end will be 250 Ω.

The linear potentiometer is usable in applications where the amount of linear movement must be measured accurately. In many of the previous applications of sensors and transducers, such as the low-pressure sensor whose diaphragm moves a small amount, a sensor like the linear potentiometer is used to measure the movement. The linear potentiometer is useful with the linear part of the rack and pinion or ball screw mechanism. Another application for the linear potentiometer is on plastic injection molding machines where the travel of the movable section of the mold must be accurately measured and converted to a position. The linear potentiometer in this application may be over 48 in. long, since the amount of travel will be up to 48 in. In this application, the linear potentiometer is fixed and the movable center tap is connected to the movable part of the mold so that it will move with the mold as it opens and closes and continually indicates the position.

FIGURE 10-94 A linear potentiometer with a diagram of fixed ends and wiper.

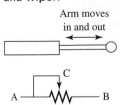

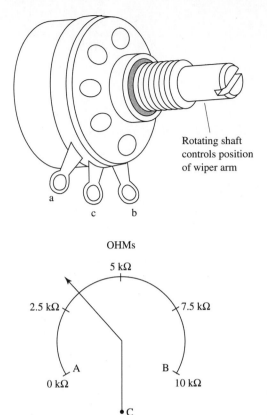

FIGURE 10-95 A rotary potentiometer is shown with its electrical diagram.

10.7.2 Rotary Potentiometer

Fig. 10-95 shows an example and a diagram of a rotary potentiometer. Electrically the linear and the rotary potentiometer operate the same. The main difference is that the shaft of the rotary potentiometer converts rotary motion into a change of resistance. Some rotary potentiometers are the single-turn type while others are the multiple-turn type. The single-turn potentiometer allows the rotating shaft to move the wiper from one end of the fixed resistor to the other in one shaft rotation. It is common in this type of application to refer to the rotation of the shaft in degrees (0–360°). This means that the amount of resistance for each degree can be determined, and then the amount of measured resistance can be converted to angular displacement. For example, if the total resistance of the potentiometer is 1000 Ω, the shaft would be measuring 180° if the resistance of the wiper to one end was 500 Ω.

The rotary potentiometer is used in applications such as indicating the position of welding turntables. A turntable is used to mount multiple parts that are to be welded. The table is indexed a specific amount so that each part is presented to a welding robot at the same location each time. The position of the turntable can be measured anywhere within 360°, and each welding position can be calculated in degrees.

10.7.2.1 Electrical Circuits for Potentiometer Position Sensors ■ Both the linear and the rotary potentiometer must be connected to an electrical circuit for their output to be useful. Fig. 10-96 shows examples of two of the most common circuits

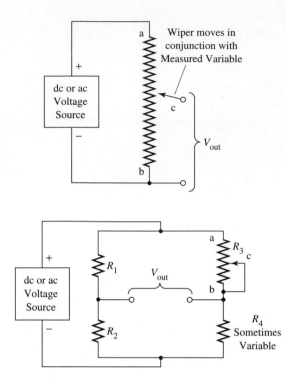

FIGURE 10-96 Circuits for the potentiometer that are used to convert the change in the wiper into a voltage signal.

the potentiometer uses to convert position into a voltage or current. The first diagram shows the potentiometer connected across a power supply. The two fixed ends of the potentiometer are connected directly to the power supply, and the wiper can be used to show a variable voltage in reference to either the positive or negative power supply terminal. When the wiper moves and changes the resistance, the amount of voltage will also change. The supply voltage for this application could also be an ac voltage.

The second type of circuit for the potentiometer is the Wheatstone bridge circuit. The bridge allows the change of resistance to change the output voltage of the bridge. The potentiometer can comprise one side of the bridge if it is used with two fixed resistors, or it can be one leg of the bridge if it is connected to three fixed resistors.

10.7.3 Linear Variable Differential Transformers (LVDTs)

The *linear variable differential transformer* (LVDT) is a position sensor that uses a primary transformer winding and two identical secondary transformer windings. These windings are wound around a hollow tube which provides a cavity for a movable core. The movable core is attached to the part of the system whose position or motion is being measured. The movable core provides a magnetic linkage between the primary and secondary windings. As you know, when ac voltage is applied to the primary side of the transformer, the amount of secondary voltage for a transformer is determined by the number of windings and the type of core material. When the movable core is near the secondary winding, it will produce the maximum amount of voltage, and when the core is not near the secondary winding, the secondary winding will produce the least amount of voltage because the transformer essentially has an air core at that time.

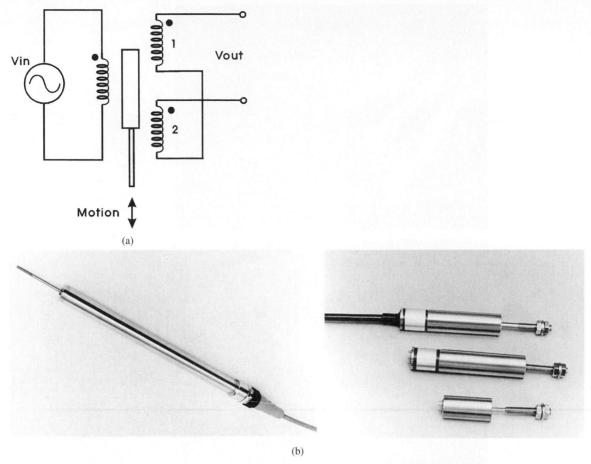

FIGURE 10-97 **(a)** Electrical diagram of the primary and two secondary windings of an LVDT sensor. **(b)** Examples of operating LVDT positioning sensors. (Courtesy of Sensotec.)

Fig. 10-97 shows an example of an LVDT and its electronic circuitry. From the electronic circuit you can see that the two secondary windings of the transformer are connected as series opposing transformers. The output of each transformer is connected to an op amp and the remainder of the circuit changes the ac output to dc. When ac voltage is applied to the primary winding of the transformer, the output of each secondary will be 180° out of phase with the primary. If the movable core is positioned near the right end, the voltage from that secondary will be greater than the other. When the core is positioned near the left end, the voltage from that secondary will be greater. If the core is an equal distance between the two ends, the voltage will be zero because an equal amount of positive and negative voltage will be produced. The amount of voltage will be zero for two reasons when the core is an equal distance between the two secondaries. First, the amount of voltage produced by the positive secondary will be equal to the amount produced by the negative secondary, which means the op amp will see their sum as zero. The second point to remember is that since the core is an equal distance between the two secondaries, it will have a minimum amount of core in each winding, which means the amount of voltage produced by each secondary will be minimal. When the LVDT's output is zero, it is called the null point for the sensor.

Since the remainder of the circuit is used to convert the output signal to a dc

(a)

(b)

FIGURE 10-98 **(a)** Examples of two different types of magnetostrictive positioning sensor. Three sensors are shown with movable rods, and the fourth sensor is shown with a fixed rod and movable magnet. **(b)** An example of a magnetostrictive type positioning sensors mounted on an injection molding machine. (Courtesy of MTS Systems Corporation.)

differential voltage, the output voltage will be positive with respect to ground when the movable core is near one end. When it is near the other end, the output voltage will be negative with respect to ground.

The LVDT is used to sense position because it is very accurate and it has very little friction when movement occurs. The LVDT is easy to troubleshoot since it is essentially a transformer. The first measurement should be the amount of primary voltage. If primary voltage is present, the core can be moved from one end to the other and the output voltage should change from positive dc to negative dc.

10.7.4 Magnetostrictive-Type Position Sensor

The accuracy of a linear potentiometer is limited to its resolution, which is determined by how tight the wires for the resistor can be wound. Other linear position sensors have been developed that provide a greater degree of accuracy. The *magnetostrictive-type* position sensor uses a magnetic field that is distorted as a waveguide

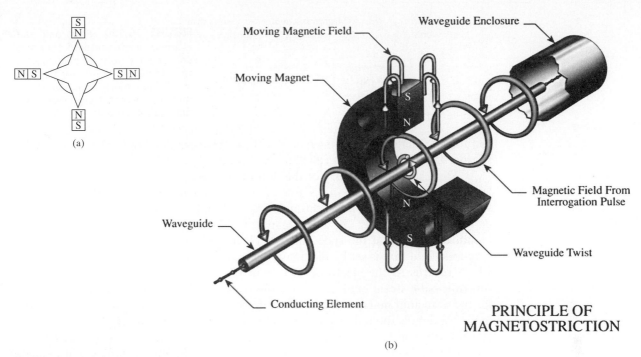

(a)

(b)

FIGURE 10-99 **(a)** The location of the magnets in the doughnut. Notice that four magnets are mounted so their north and south poles alternate, which makes the magnetic field S shaped or star shaped. **(b)** Cut-away diagram that shows the magnets mounted in the sensor. Notice that the magnetic field produced by the doughnut magnets is shown around the rod, which is called a waveguide. A second circular magnetic field is shown around the waveguide (rod). This magnetic field is developed from a high-current pulse sent through the waveguide. (Courtesy of MTS Systems Corporation.)

is moved through the field to determine position. Fig. 10-98 shows an example of these sensors. One type has a magnetostrictive sensor with a movable rod. For this type of sensor the body is mounted on the fixed part of a machine, and the rod is mounted to the movable part of the machine. When the movable part of the machine moves back and forth, the rod is extended and retracted in the sensor body.

The second type of magnetostrictive positioning sensor has the rod fixed to the body. The body is mounted to the stationary part of the machine and a set of four magnets shaped like a doughnut are mounted to the movable part of the machine. When this type of sensor is mounted on the machine, the rod is inserted through the center of the magnetic doughnut. When the movable part of the machine moves back and forth, the magnet will move back and forth over the rod. This is the same effect as the sensor in the first example where the rod moves back and forth through the doughnut.

10.7.4.1 Theory of Operation for the Magnetostrictive Positioning Sensor ■ The theory of operation of the magnetostrictive positioning sensor can be best explained with several diagrams. Fig. 10-99a shows an example of the four magnets that are mounted in the shape of a doughnut. From this diagram you can see that the two magnets that are mounted opposite each other have the same pole mounted toward the interior of the sensor. This means that every other magnet has the opposite pole mounted toward the interior of the sensor. This causes an S-shaped magnetic field moving from magnet to magnet. Fig. 10-99b shows the magnets as you would see them mounted in the sensor. This example shows the sensor with a movable

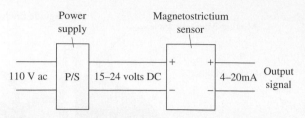

FIGURE 10-100 Electronic diagram of a magnetostrictive sensor. The 110 volts ac is converted to a dc power source for the sensor. When the magnet or rod is moved, the output signal will vary between 4–20 mA.

rod that slides through the magnetic field as it moves in and out of the body of the sensor. The movable rod is called a *waveguide.*

The electronic circuitry of this sensor produces a second magnetic field that envelopes the waveguide in a series of circles. This magnetic field is produced by sending a high-current pulse through the waveguide. This magnetic field is called an *interrogation pulse.* The interrogation pulse (magnetic field) travels down the length of the rod at the speed of light. When the interrogation pulse meets the magnetic field that is set by the doughnut magnets, a *torsion strain pulse* is created which returns to the electronic circuit in the sensor's head. The velocity of the return torsion strain pulse is predictable so it can be measured to determine how far the doughnut magnets are from the sensor head. This means that if the machine moves and pulls the rod farther out of the sensor body, the interrogation pulse will travel farther and it will take longer for the return torsion pulse to get back to the sensor head. When the sensor electronics calculate the distance, they will indicate the rod has moved to the new position.

The electronic circuit for this sensor usually produces a 4–20 mA signal that is easily interfaced with other sensors and controls on the machine. The 4–20 mA signal can easily be converted to a 2–10 volt signal by using a 500 Ω resistor in the circuit. Electronic circuitry is also available to provide the output signal as a binary or binary coded decimal (BCD) signal. Zero and span potentiometers are provided to allow the exact position of the rod and magnets to be aligned with zero points and maximum travel on the machine. Fig. 10-100 shows an electrical block diagram for this type of sensor. From the diagram you can see that the sensor has a dc power supply voltage of 15–24 volts dc and a two-wire output that supplies 4–20 mA.

This type of sensor is easy to troubleshoot in that you need to measure the power supply voltage and check for an output voltage. If the supply voltage is between 15–24 volts dc, then you can move the rod to different positions and the 4–20 mA output current should change correspondingly. If you want to measure the output signal as a voltage, you will need to place a 500 Ω resistor across the output terminals to cause a 2–10 volt signal to be produced. If the output signal does not change when the rod is repositioned, the electronic circuitry may be faulty and you can test the sensor by substituting it with a known good one.

Typical sensor lengths are 12 in., 18 in., 24 in., 30 in., 36 in., and 48 in. This type of positional sensor can provide resolution to 0.004 in. Another advantage of this type of sensor is that there are no parts to wear since the rod does not touch the doughnut magnet or any other part of the sensor as it travels in and out of the body. Other applications for this type of sensor involve incorporating the magnetostrictive sensor directly into hydraulic or pneumatic cylinders so that the position of the cylinder's rod may be measured.

10.7.5 Proximity Switches

The proximity switch uses a principle called *eddy current killed oscillator,* so it is sometimes called an ECKO switch. The proximity switch is a *noncontact limit switch.*

FIGURE 10-101 Examples of proximity switches. The size of the barrel of the larger switch in this picture is about the size of a 50¢ piece, and the smallest one is about the thickness of a pencil. (Courtesy of Honeywell Micro Switch Division.)

Fig. 10-101 shows several examples of proximity switches. The inductive proximity switch detects the presence or absence of a target, which means that it can be used to tell when the switch comes into close position (proximity) of metal. The switch is an on/off sensor since it indicates the presence or absence of a target. Typical sensing distance for a proximity switch is 0.2 mm to 10 mm. The distance of 0.2 mm is about the thickness of a business card. The capacitive proximity switch can detect the presence of material such as plastic, wood, cardboard, paper, and nonferrous metals such as aluminum.

Fig. 10-102 shows a block diagram of the internal circuit of the proximity switch. From this diagram you can see that a power supply energizes the oscillator portion of the circuit. The oscillator produces magnetic flux lines that emanate out the end of the sensor. When the switch is not sensing anything in the field, the strength of the oscillator circuit is at its maximum. The integrator op amp detects the difference in the oscillator circuit and sends an output to the trigger, which activates the output part of the proximity switch. It does not matter if the metal is moved close to the switch or if the switch is moved close to the metal.

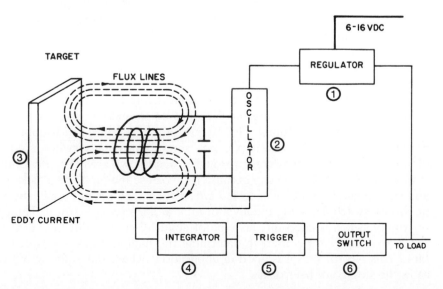

FIGURE 10-102 Block diagram of an eddy current killed oscillator (ECKO) proximity switch. (Courtesy of Micro Switch, a Honeywell Division.)

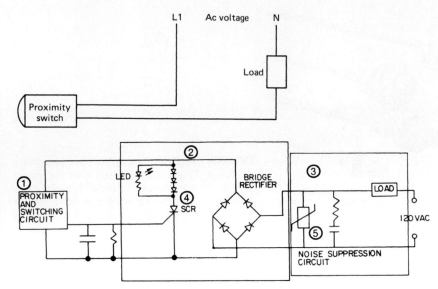

FIGURE 10-103 **(a)** A simple electrical diagram that shows a two-wire proximity switch controlling ac current to the coil of a motor starter. **(b)** An electronic diagram of the proximity switch. Notice that an SCR is used to interrupt current in the dc side of a bridge circuit. (Courtesy of Honeywell Micro Switch Division.)

The proximity switch is available as a switch for ac or dc loads. Fig. 10-103a shows the proximity switch wired to the coil of a motor starter. The proximity switch in this diagram is a two-wire switch, which means it is basically wired in series with the coil and the 110 volt ac power supply. This type of sensor uses the same two wires to get its power and to switch current on and off to the load. The proximity switch will have low impedance (resistance) like a closed switch when no target is present and it will have high impedance like an open switch when a target is sensed. Fig. 10-103b shows the internal electronic circuit for an ac-type switch. From this diagram you can see that the ac voltage passes through a bridge rectifier so that the oscillator portion of the switch can be supplied with dc voltage. An SCR is connected in series with the dc side of the bridge circuit. The gate of the SCR is connected to the oscillator, and when the proximity switch is not near metal, the oscillator will provide enough current to trigger the SCR. When current flows through the SCR, it allows ac current to flow through the ac side of the bridge circuit. When the SCR is not turned on, it will block dc current flow in the bridge, which will also stop ac current flow in the bridge. This causes the proximity circuit to look like an open switch.

The proximity switch is also available to sense nonmetallic targets. These types of switches use a variable capacitance circuit and anything in the target range that changes the capacitance will cause the switch to trip. The capacitive proximity switch can be used to indicate the presence or absence of plastic caps on bottles or other similar applications. Fig. 10-104 shows several typical applications for proximity switches. The first application shows a proximity switch used to detect the caps on bottles as they come through a wash-down area. The proximity switch is very useful to detect the cap because it can sense the metal or plastic in the cap, and the proximity switch is a sealed-type switch. The second application shows a proximity switch detecting the level of milk in a cardboard carton. The capacitive proximity switch is good for this type of application. The third application shows two proximity switches used to determine the high and low levels of liquid in a tank. This type of sensor is good in applications where the liquid may foam, such as in the storage of beer or ale.

Checking for bottle cap in high humidity environment

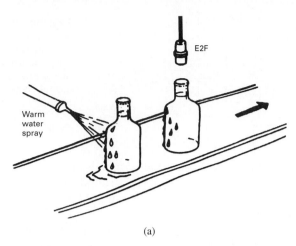

(a)

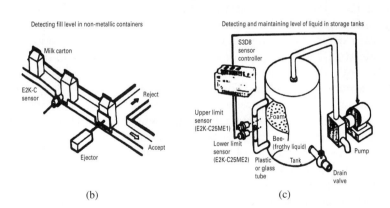

Detecting fill level in non-metallic containers

Detecting and maintaining level of liquid in storage tanks

(b)

(c)

FIGURE 10-104 **(a)** Application of a proximity switch that is used to check for bottle caps in a high-humidity location. **(b)** Example of a proximity switch used to detect the level of milk in a carton-filling application. **(c)** Example of a proximity switch used to determine the level of liquid in a tank. (Courtesy of Omron Electronics.)

10.7.6 Hall-Effect Sensors

Another type of positioning sensor is the *Hall-effect sensor*. This type of sensor works on a simple electromagnetic principle. In 1879 Edwin Hall observed that when a thin sheet of conducting material such as gold foil has current flowing through it, a separate small voltage could be induced in the foil by passing a magnetic field at right angles to the foil. This voltage is called the Hall-effect voltage. From the diagram in Fig. 10-105 you can see that a current will flow through the thin foil

FIGURE 10-105 Diagram of the Hall effect. A current is conducted through a thin piece of foil from terminal I and ground. When a magnetic field is brought perpendicular to the foil, a small voltage called the Hall-effect voltage is produced at the terminals attached to the opposite sides of the foil. (Courtesy of Honeywell Micro Switch Division.)

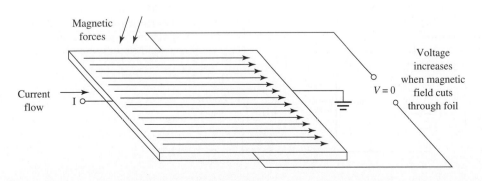

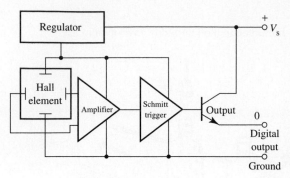

FIGURE 10-106 Electrical block diagram for a Hall-effect sensor used in a current-sourcing circuit. An op amp, Schmitt trigger, and NPN transistor are used to provide an output signal when a magnetic field is brought close to the Hall-effect sensor. (Courtesy of Honeywell Micro Switch Division.)

from the power supply current at the terminals attached to the left and right sides. These terminals are identified as I and ground. When a perpendicular magnetic field is brought close to the foil, a very small Hall-effect voltage will be available at the top and bottom terminals.

When Edwin Hall did his experiment at Johns Hopkins University, the amount of voltage produced by the Hall effect was in the range of 20–30 μV. This very small voltage was not useful until amplifiers such as the op amp were designed to increase the signal to a useful range. Fig. 10-106 shows an electrical block diagram of a current-sourcing circuit for a Hall-effect sensor, and Fig. 10-107 shows an electrical block diagram of a current-sinking, Hall-effect sensor. The basic concepts of both diagrams are very similar. From the diagrams you can see that both sensors are three-wire sensors. This means that two wires, the $+V_s$ and the ground provide dc voltage for the power supply portion of the sensor. Terminal O and ground are used as the output terminals for the sensor. Since this is a three-wire sensor, the ground terminal is part of the power supply and part of the output circuit. The power supply uses a voltage regulator to provide the initial current for the Hall-effect element and voltage for the op amp. The small sensor terminals are connected to the op amp input terminals. When a magnetic field is sensed, a small voltage is sent to the op amp and the output of the op amp is sent to a Schmitt trigger and then to the base of an NPN transistor. When the base of the transistor is biased, it will go into saturation and current will flow through its emitter–collector circuit to provide a digital (on/off) output signal. You should notice in the current-sinking circuit that the transistor provides a path to ground when the transistor is biased to saturation.

Fig. 10-108 shows examples of several Hall-effect sensors. In some applications where the Hall-effect sensor is used to determine if a door is open or closed, the magnet is built into the Hall effect. When the door is closed, a piece of ferrous

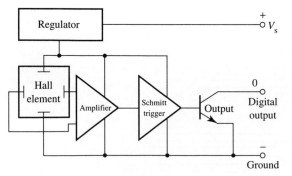

FIGURE 10-107 Electrical diagram of current-sinking output for a Hall-effect sensor. (Courtesy of Honeywell Micro Switch Division.)

FIGURE 10-108 Example of Hall-effect switches. These switches are used for positioning applications such as indicating when a door is open or closed or if a safety guard is in place. (Courtesy of Honeywell Micro Switch Division.)

metal is placed between the magnet and the Hall-effect sensor to interrupt the magnetic field. When the door is open, the piece of metal is removed, and the magnet will cause the sensor to conduct the Hall-effect voltage.

Another application for this type of sensor is in membrane-type keyboards. The membrane-type keyboard is a sealed keyboard used in industrial applications where the keyboard for computers and color graphic operator screens must be sealed so they can be washed down. Each key on the keyboard has a small magnet and Hall-effect sensor. When a key is depressed, the magnet is brought close enough to the sensor to activate the Hall effect.

The Hall-effect sensor is also useful in analog signal applications. This type of circuit uses a semiconductor material as the Hall element. The amount of Hall voltage will be proportional to the strength of the magnetic field. This means that the sensor can produce an analog signal to indicate how close the magnetic field is to the Hall-effect sensor. The distance can easily be converted into a signal that indicates position or distance traveled.

10.7.7 Linear and Rotary Encoders

An encoder is an electrical mechanical device that can monitor motion or position. A typical encoder uses optical sensors to provide a series of pulses that can be translated into motion, position, or direction. Fig. 10-109 shows a diagram and picture of rotary encoders. The diagram in Fig. 10-109b shows that the disk is very thin, and a stationary light-emitting diode (LED) is mounted so that its light will continually be focused through the glass disk. A light-activated transistor is mounted on the other side of the disk so that it can detect the light from the LED. The disk is mounted to the shaft of a motor or other device whose position is being sensed, so that when the shaft turns, the disk turns. When the disk lines up so the light from the LED is focused on the phototransistor, the phototransistor will go into saturation and an electrical square wave pulse will be produced. This figure shows an example of the square wave pulses that are produced by the rotary encoder. This type of disk was used in early applications but the size of the holes in the metal disk limited the amount of accuracy that could be obtained. As more holes were cut in the disk, it became too fragile for industrial use.

10.7.7.1 Incremental Rotary Encoder ■ An encoder with one set of pulses would not be useful since it could not indicate the direction of rotation. Most incremental

Rotary Incremental Optical Encoder

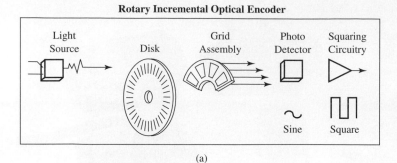

(a)

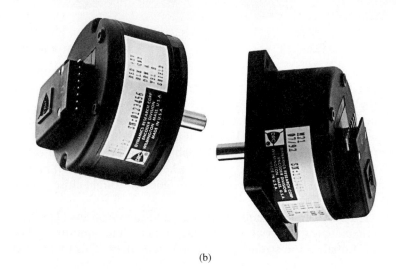

FIGURE 10-109 **(a)** Internal operation of incremental optical encoder. **(b)** Examples of optical rotary encoders. (Courtesy of Dynamics Research Corp.)

(b)

encoders have a second set of pulses that is offset (out of phase) from the first set of pulses, and a single pulse that indicates each time the encoder wheel has made one complete revolution. Fig. 10-110 shows an example of the two sets of pulses that are offset. Since the two sets of pulses are out of phase from each other, it is possible to determine which direction the shaft is rotating by the amount of phase shift between the first set and second set of pulses. The first set of pulses are called

FIGURE 10-110 Examples of the A pulse, B pulse, and the command pulse. If the A pulse occurs before the B pulse, the shaft is turning clockwise, and if the B pulse occurs before the A pulse, the shaft is turning counterclockwise. The C pulse occurs once per revolution.

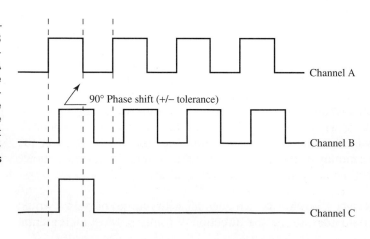

the A pulses, and the second set of pulses are called the B pulses. A third light source is used to detect a single pulse that appears once per revolution. This pulse is called the *command pulse,* which is used to count revolutions of the shaft where the encoder is connected.

Since the incremental encoder only provides a string of pulses, a home switch must be used with this type of encoder to ensure that the encoder is calibrated to the actual location of the home reference point. The early encoder wheels that were made from metal were not too useful as more resolution was needed. Today encoder wheels are made from clear glass that has opaque segments etched in them like bars. As the encoder wheel spins, the opaque segments block the light and where the glass is clear, light is allowed to pass. This provides a pulse train similar to the encoder wheel that has holes drilled in it. Typical glass encoders have from 100–6000 segments. This means that these encoders can provide 3.6° of resolution for the encoder with 100 segments, and 0.06° of resolution for the encoder with 6000 segments. If the shaft of the encoder is connected to a drive shaft for a motor that is connected to a ball screw or a reduction gear, the number of degrees of resolution can be converted into linear position.

It would be impossible to drill hundreds of holes in the encoder wheel to get the higher amounts of resolution because the wheel would not have enough material remaining to give the wheel strength. For this reason modern encoder wheels with high resolution use etched glass wheels. The glass is etched with chemicals to produce alternating opaque segments.

The second pulse train is developed in this type of encoder by placing a second light source and second light receiver at a different angle from the first set. Since the location of the second light source is different from the first, the second pulse train will be shifted from the first just as if two separate sets of holes were drilled. This arrangement allows the encoder wheel to provide both incremental and direction of rotation information with only one set of opaque bars etched in the glass. The second pulse train is used to determine the direction of rotation for the encoder wheel.

Fig. 10-111 shows an example of the etched glass encoder and a diagram of the light source and receiver. From this figure you can see that the glass encoder looks as if it has very thin black lines drawn on it. The black lines are the opaque segments that block light. The diagram from this figure shows only one light source and receiver. A second identical light source and receiver is mounted on the encoder in such a way that it produces the offset pulse train.

10.7.7.2 Absolute Encoders ■ One of the major drawbacks of the incremental encoder is that the number of pulses that are counted are stored in a buffer or external counter. If power loss occurs, the count will be lost. This means that if a machine with an encoder has its electricity turned off each night or for maintenance, the encoder will not know its exact position when power is restored. The encoder must use a home-detection switch to indicate the correct machine position. The incremental encoder uses a *homing routine* that forces the motor to move until a home limit switch is activated. When the home limit switch is activated, the buffer or counter is zeroed and the system knows where it is relative to fixed positional points.

The absolute encoder is designed to correct this problem. It is designed in such a way that the machine will always know its location. Fig. 10-112 shows an example of an absolute encoder. From this figure you can see that this type of encoder has alternating opaque and transparent segments like the incremental encoder, but the absolute encoder uses multiple groups of segments that form concentric circles on the encoder wheel like a "bull's eye" on a target or dart board.

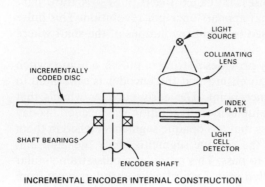

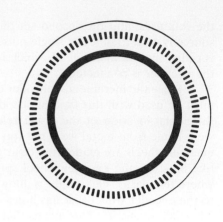

INCREMENTAL ENCODER INTERNAL CONSTRUCTION

Incremental encoder

FIGURE 10-111 Example of an etched glass incremental encoder wheel. (Courtesy of Analog Devices.)

Pulse train for incremental encoder

The concentric circles start in the middle of the encoder wheel and as the rings go out toward the outside of the ring they each have double the number of segments than the previous inner ring. The first ring, which is the innermost ring, has one transparent and one opaque segment. The second ring out from the middle has two transparent and two opaque segments, and the third ring has four of each segment. If the encoder has ten rings, its outermost ring will have 512 segments, and if it has 16 rings it will have 32,767 segments.

Since each ring of the absolute encoder has double the number of segments of the prior ring, the values form numbers for a binary counting system. In this type of encoder there will be a light source and receiver for every ring on the encoder wheel. This means that the encoder with 10 rings has 10 sets of light sources and receivers, and the encoder with 16 rings has 16 light sources and receivers.

The advantage of the absolute encoder is that it can be geared down so that the encoder wheel makes one revolution during the full length of machine travel. If the length of machine travel is 10 in., and its encoder has 16-bit resolution, the resolution of the machine will be 10/65,536 which is 0.00015 in. If the travel for the machine is longer, such as 6 ft, a *coarse resolver* can keep track of each foot of travel, and a second resolver called the *fine resolver* can keep track of the position within 1 ft. This means the coarse encoder can be geared so that it makes one revolution over the entire 6 ft distance, while the fine encoder is geared so that its entire resolution is spread across 1 ft (12 in.).

Since the absolute encoder produces only one distinct number or *bit pattern* for each position within its range, it knows where it is at every point between the two ends of its travel, and it does not need to be homed to the machine each time its power is turned off and on.

10.7.7.3 Linear Encoders ■ Linear encoders have been refined so that they are very useful. This type of encoder operates similarly to the rotary absolute encoder.

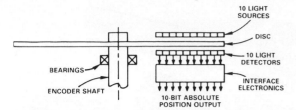

ABSOLUTE OPTICAL ENCODER INTERNAL CONSTRUCTION

Absolute encoder

FIGURE 10-112 Example of an absolute encoder wheel that shows the pattern of concentric circles. This diagram also shows the location of 16 light souces and 16 light receivers that decode the pattern of light as it passes through the 16 concentric circle patterns. (Courtesy of Analog Devices.)

The linear encoder has two identical rectangular pieces of glass that are both etched with opaque and transparent segments. One of the pieces of glass is fixed, and the other moves by a sliding arm that is attached to the movable part of a machine or robot. When the machine or robot moves, the arm moves the sliding piece of glass past the fixed piece of glass. At each point along its movement, the opaque and transparent segments of glass will create unique light patterns (on and off segments) that are decoded into a binary number that indicates a position. The main advantage of this type of encoder is that the size of the glass plates will be the same size as the total distance of the machine travel. This ensures that the machine will know exactly where it is at every point along the travel distance, even if the machine is powered down and power is returned. Fig. 10-113 shows examples of linear encoders. This type of encoder has resolution down to 0.0000001 in. or 0.1 microns.

From the figure you can see that this type of encoder is specifically designed where the travel is linear rather than rotary. These types of encoders are necessary for precision machining, welding, or applications that use lasers. The latest technology requires highly polished or finished surfaces that are held to very tight tolerances and older technology cannot provide the amount of resolution required.

FIGURE 10-113 Examples of a linear encoder, which have a fixed piece of rectangular glass and a movable piece of glass. The two pieces of glass line up at every position to provide a unique pattern of on and off bits that are converted into position. (Courtesy of Dynamics Research Corporation.)

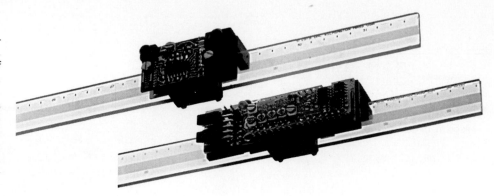

10.7.7.4 Resolvers ■ A resolver is a transducer that uses a stator winding and rotor winding to produce waveforms for shaft angle measurement for positioning. The generic term for all of these types of transducers is *synchro*. Fig. 10-114a shows an example of a resolver with the rotating coil removed from the stationary (stator) coil. This figure also shows two different types of winding relationships between the stator and rotor coils. In Fig. 10-114b you can see that the stator uses three coils that are Y (wye) connected, and the rotor uses a single coil. In Fig. 10-114c you can see the stator has two separate coils that are mounted at 90° to each other. If the transducer uses three coils connected in a Y (wye), it is generally called a *synchro,* and if the stator has two windings it is generally called a *resolver.*

FIGURE 10-114   **(a)** A resolver with its rotor removed from its stator. **(b)** The stator windings connected in a Y (wye) configuration. When the windings are connected in this configuration, the transducer is generally called a synchro. **(c)** Two stator windings are mounted at 90° to each other. When the windings are connected this way, the transducer is generally called a resolver. (Courtesy of Analog Devices.)

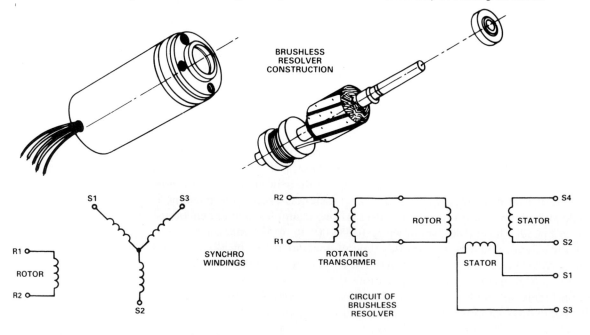

In the operation of the resolver, the rotor is excited with ac voltage. The stator and rotor will act like a generator and a voltage will be produced in the S_1 to S_3 winding and a voltage that is out of phase by 90° will be produced in the S_2 to S_4 winding. Fig. 10-115 shows the waveforms produced by each stator. From the diagram you can see that the phase difference between waveforms of the two windings is 90° at all times. Each waveform represents one full revolution of the rotor shaft (360°).

One of the earliest applications of resolvers was to indicate the position of large guns on U.S. Navy destroyers. In these early resolvers, the sum of the two voltages was sent to an amplifier whose output was connected directly to a motor. In this configuration, a setpoint voltage was also sent to the amplifier, and the motor would begin to turn the guns. This would also turn the shaft of the resolver until the output voltage from the resolver would provide an amount of voltage equal to the setpoint. Since the setpoint voltage and the resolver voltage are out of phase, the two voltages would cancel each other when they were equal, which would mean the sum of the voltage sent to the motor would be zero and the motor would stop moving. It was also important in this early application that the gun mount could only make one revolution. These early resolvers were not useful without modification in applications where the motor shaft turned multiple rotations.

The resolver can determine the location of the rotor within 1° anywhere in one revolution. Since the resolver is attached to the motor shaft to determine the location of the motor shaft, it can be directly connected or it can be connected through a set of gears. If gears are used, a coarse resolver and a fine resolver can be used to determine the location of the shaft anywhere along the entire movement. The coarse resolver is geared so that it makes only one revolution over the entire range of travel, and the fine resolver is geared so that it makes one revolution every 12 in. Resolvers are typically used in robot applications and machine tool application where the location of a robot axis or machine position must be determined continually.

Another advancement in resolver technology occurred when op amps became refined. The op amp has the ability to compare the voltage between the two stator waveforms and determine the exact location of shaft rotation within 0.001 degrees. The op amp can also be used to detect which waveform is leading or lagging the other. This indicates if the rotation of the shaft is clockwise or counterclockwise.

Troubleshooting a resolver is simple because it acts like a generator and its windings act like a transformer. The simplest test for the resolver is to test for ac exciter voltage at the R_2 and R_4 terminals of the rotor. If the exciter voltage is present, a voltage should be present at the S_1 and S_3 stator terminals and the S_2 and S_4 stator terminals because the relationship between the rotor and stator windings is

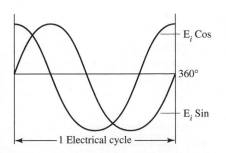

FIGURE 10-115   Waveforms from the two stator windings of a resolver. Notice that the two waveforms are always 90° out of phase.

essentially the same as the relationship between the primary and secondary windings of a transformer. This relationship will be present whether or not the rotor shaft is turning. When the resolver is rotating, the waveform of the stator voltages will be a sine wave like an ac alternator, which can be displayed on an oscilloscope. If an exciter voltage is present, but one or both of the stator voltages are not present, either the rotor or the stator windings have an open circuit. You can disconnect the resolver and test both the rotor and the stator windings for continuity. If any of the windings have an open, the resolver must be replaced.

The second type of problem that occurs with the resolver is the wires that connect the stator and rotor windings to the resolver control circuit may develop an open. Since the resolver must be mounted near the motor shaft, and the detection circuit is mounted near the controls, the amount of wire between these two may be significant, and it may have one or more terminal connections between them. The wires can become loose at any of these terminal connections or they can develop an open anywhere in between. You can determine if the wiring has a problem by testing for exciter voltage at the source (resolver control circuit) and then at the resolver. If voltage is present at the control end of the circuit but not at the resolver end, one of the two wires has an open. The stator circuit can be tested in a similar manner, except the voltage is developed at the stator and it uses the wires to get to the controller. This means that voltage should be tested at the stator and then at the controller. If voltage is present at the stator but not at the controller, an open has developed in the wiring.

10.8 MOTION SENSORS

A wide variety of motion sensors is available to measure acceleration, velocity, and speed. The measurement of velocity and speed of both rotary and linear applications is important. The instruments used to measure these values typically do so with indirect methods, such as using piezoelectric elements to measure stress and calculate the amount of motion that results. This section will discuss several of these types of sensors.

10.8.1 Accelerometers

The definition of acceleration is the rate of change of velocity, and velocity is defined as the rate of change of position. Velocity is commonly called speed, so the definition of acceleration could also be the rate of change of speed. Accelerometers are sensors that convert the motion that represents the aspect of acceleration into an electrical signal. These devices are typically used to measure vibration on machines and structures and other related acceleration detectors and motion detectors for mines, highways, and bridges that are located in areas subject to earthquakes. The most recent application for accelerometers is to measure vibration on large machines with pumps or motors with bearings that tend to wear out. When these devices wear out unexpectedly, they cause downtime that is very expensive. If the vibration is monitored, the amount of wear on bearings and gears can be calculated and the parts can be changed during scheduled maintenance so that failure will not cause downtime.

Acceleration is commonly measured in dv/dt (delta velocity over delta time) or by Newton's law of motion that states $F = ma$. Since force is equal to mass

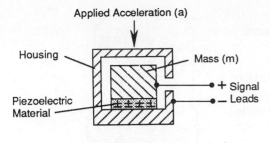

Applied Acceleration (a)

Housing — Mass (m)

Piezoelectric Material — + Signal, − Leads

Basic Accelerometer

FIGURE 10-116 A cut-away diagram of a basic accelerometer. (Courtesy of PCB Piezotronics, Depew, N.Y.)

times acceleration, it is possible to determine acceleration by measuring the force and dividing it by the mass. Most accelerometers use quartz or ceramic crystals to generate a piezoelectric effect that is converted into an electrical output.

When force is exerted on a crystal, it is stressed and through the piezoelectric effect, the crystal generates an electrostatic charge output which is expressed in terms of "PicoCoulombs." A signal conditioner, either internal to the sensor or external, converts the high impedance charge signal from the crystal to a high voltage low impedance output which is compatible with meter, data collectors, and other readout instruments.

Salient characteristics of piezoelectric sensors include high output, high frequency response, wide dynamic operating range, ruggedness, and durability.

Fig. 10-116 shows a diagram of a basic accelerometer consisting of a crystal, seismic mass, and housing. From this diagram you can see that the sensor has a specified amount of mass, called the "seismic" mass. Vibratory forces applied to the base of the accelerometer cause the seismic mass to stress the crystal which generates an electrostatic charge output proportional to acceleration.

Piezoelectric crystals have both a positive and negative polarity output. In figure 10-116 the positive electrode (Output) is connected to the mass and the negative electrode (Ground) is connected to the housing. A signal conditioner converts the high impedance charge output into a usable voltage signal.

There are several different piezo accelerometer design configurations including shear, upright compression, inverted compression, and flexural. Although each design has advantages and disadvantages, the shear design is considered the most accurate since it is least sensitive to temperature and base strain inputs. Shear is also the most widely used design.

For industrial machine vibration monitoring applications, shear structured accelerometers with integral electronics are packaged in robust hermetic sealed housings with durable electrical connectors to withstand tough factory environments.

Piezoelectric sensors are further classified by the way they condition the signal. If the sensor contains internal signal conditioning, it is called the voltage mode or Integrated Circuit Piezoelectric (ICP®) sensor, and if the signal is conditioned externally by a charge amplifier, it is called a charge mode sensor.

Fig. 10-117 shows a picture and diagram of an ICP shear structured accelerometer. From this figure you can see that this type of vibration sensor couples securely to a machine or structure by means of a mounting stud. Vibration sensors can also be magnetically or adhesively mounted. Vibratory forces transmitted into the base of the sensor cause the seismic mass to stress the crystal which generates an output proportional to acceleration.

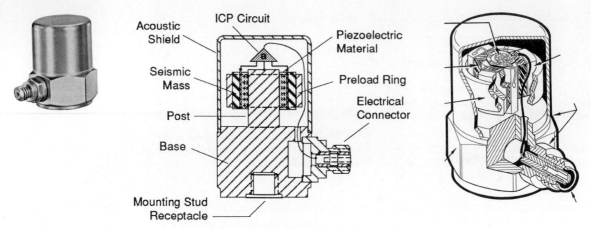

FIGURE 10-117   **(a)** A picture of a shear mode accelerometer. **(b)** A diagram of a shear mode accelerometer. **(c)** A cut-away diagram of a shear mode accelerometer. (Courtesy of PCB Piezotronics, Pepew, N.Y.)

10.8.2 Tachometers

Tachometers measure the angular speed of a rotating shaft. The speed of a shaft is measured in revolutions per minute (rpm). A tachometer could be as simple as a dc or ac generator that can determine the speed of shaft rotation by the amount of voltage the generator produces or the frequency of the output signal. The magnitude of the generator voltage and the frequency of the generated voltage will increase proportionally with speed. Frequency can also be measured by a photocell tachometer. The number of pulses produced by the photocell will increase as the speed of the shaft rotation increases.

Tachometers are used to determine the speed of a motor shaft for motor drives on conveyors, and to determine the speed of rotation of the screw shaft on a plastic injection molding machine. The speed of rotation of the screw on a plastic injection molding machine is important to control because the screw shaft is used to meter the amount of plastic that is drawn into the barrel of the machine for the next injection shot. If the speed is not controlled, the screw will turn at different speeds and more or less plastic will be drawn into the barrel and the amount of plastic being used for each part will be inconsistent.

The speed of the rollers in large rolling mills is also important to measure. In this type of application the rpm of each motor is measured and compared to setpoints. A servo system will increase or decrease the motor speed to keep the rollers at the correct rpm. In packaging applications several motors must be synchronized as the production line changes speeds to ensure the machine will operate correctly. Tachometers are used to measure the speed of each motor and a controller adjusts each motor speed to match the speed of the production line.

There are two basic types of tachometers that use the change in voltage to determine speed. These are the dc generator tachometer and the drag cup tachometer. These types of tachometers operate similarly to an unregulated generator. The faster they turn, the more voltage they produce. The dc generator produces a dc voltage, and the drag cup tachometer produces an ac voltage. These tachometers can provide the direction of rotation information as well as speed, which is useful because most of these types of sensors are used to provide feedback signals. Since these signals are basically a voltage, a simple voltmeter can be used as an indicator.

The *frequency-type tachometer* counts pulses produced by a rotating field tachometer, toothed rotor tachometer, or the photocell tachometer. These types of tachometers produce sine waves or pulses that can be counted. These types of tachometers need a more sophisticated digital circuit to complete the process of *count, store, calculate, display,* and *reset* to get a value displayed that represents rpm. The rotating field and the toothed rotor tachometers produce a waveform and the photocell uses a rotating disk that has a number of windows in it. A light source is positioned so that it will shine light through each window in the disk to a photocell detector as the disk spins. The disk is connected to the tachometer shaft, so when it turns the windows line up with the photocell and the photocell produces a pulse when it is struck by light. In each of these types of tachometers a pulse stream is produced and it is proportional to the speed of the tachometer shaft.

10.9 PH SENSORS

The pH scale indicates how strong an acid or alkaline solution is and is based on the concentration of hydrogen ions (H^+ and OH^-) in a solution. The pH measurement is important for water-quality standards and treatment. The pH scale is a number scale that runs from 0–14. The numbers 0–14 refer to the exponential value of the ion concentration. For example, the log of the concentration of hydrogen ions in water is 10^{-7} (when the sign of the exponent is changed, it becomes the number for the pH scale). The middle of the scale is 7, which indicates that the sample being measured is neutral and it is not an acid or base. Numbers below 7 indicate the sample is an acid with 0 being the strongest acid, and numbers above 7 indicate the sample is a base, with 14 being the strongest base. If process water from a factory is to be returned to a river, it must be neutral, so it will be tested to determine if it is acidic. If it is acidic, base material such as soda ash can be added to it to return it to a neutral substance. Fig. 10-118 shows a table with examples of common materials as they apply to the pH scale. A wide variety of sensors is available to measure pH in industrial applications. These sensors consist of a probe and a signal conditioner. The probe is placed in a location where it can sample the liquid being measured.

◆ 14—Lye 1 molar solution
◆ 13—dilute sodium hydroxide
◆ 12—lime water
◆ 11—household ammonia
◆ 10—milk of magnesia
◆ 9—baking soda
◆ 8—albumin
◆ 7—pure water
◆ 6—tap water (from a well)
◆ 5—black tea
◆ 4—
◆ 3—vinegar
◆ 2—lemon juice
◆ 1—
◆ 0—dilute sufuric acid (battery acid)

FIGURE 10-118 Table showing the pH value for some of the most common household products.

10.10 HUMIDITY SENSORS

Relative humidity is defined as the amount of moisture in the air compared to the total amount of moisture that the air could hold when it is 100% saturated. It is important to understand that as air is heated its ability to hold moisture increases. The amount of moisture in air is important in industrial applications like printing and environmental control of computer rooms or clean rooms where solid-state integrated circuits are manufactured. Other industrial applications where humidity is controlled include painting applications and plastic injection molding and extruding. The moisture content of the raw material in plastic injection molding and extruding must be controlled because the moisture turns to steam at the high temperatures during the injection process and causes unwanted defects in the surface of the parts. In printing applications the humidity in the press room must be strictly controlled to maintain approximately 40% humidity. If the humidity is too high, static electricity builds up as the paper moves across the platens and press rollers and the static charges will use the moisture in the air to move from the paper to the metal of the presses. These static charges are dangerous and can cause fires or explosions when they jump, or they can injure personnel who are near.

Since relative humidity is dependent on temperature, most relative humidity sensors have temperature measurement devices built in or added to them. After these sensors have determined the relative humidity, their signals are generally used in feedback systems for dehumidifiers or humidifiers. If the relative humidity is too high, a dehumidifier is used to remove additional moisture from the air. This is accomplished by passing the air over a cold surface such as an air-conditioning coil. If the temperature is below the dewpoint, the moisture in the air will condense and form moisture droplets that can easily be collected as gravity pulls them to the lowest part of the coil where they go down the drain.

10.11 GAS DETECTORS

In modern industrial applications it is important to measure different types of gases. This may be done to measure the amount of gas such as oxygen and nitrogen that is used as part of a process, or it may be to detect the small amounts of different gases that are determined to be pollutants or dangerous to personnel. In some applications it is possible for workers to become exposed to dangerous gases such as ammonia, chlorine, or hydrogen cyanide, or the amount of oxygen available for breathing may be limited. Personal detectors can be worn by each employee to determine if unwanted gases are present, or they can monitor the amount of oxygen available for breathing. These types of sensors can also be used to detect the amount of gas that remains in an empty tank that must be entered for maintenance.

When gases must be measured, generally mass flow meter devices are used. The amount of gas must be detected and measured to ensure the proper amount is being added to a process, or the amount must be determined for billing purposes. Fig. 10-119 shows an example of a gas detector. You should notice that this type of sensor uses a sealed sensing head and a signal conditioner to provide a signal that can be used as an indicator or alarm.

Loop-Powered Transmitter
No amplifiers or signal conditioners required, just a two-wire connection to any suitable 4-20mA receiving device. Analog output is linearly proportional to gas concentration.

RESDEL™ Shell
Manufactured from a spun epoxy material, the shell is lightweight but durable and highly chemically resistant. In addition, an impregnated inner lining protects the circuitry from external RFI/EMI sources.

Simple Span Adjustment
Touching a magnet to the designated point on the case performs a quick and simple span adjustment.

Sealed Electronics
To enhance its mechanical strength and achieve complete environmental protection, the entire circuitry is encapsulated.

Microprocessor Control
Automatic sensor driver, signal conditioner, temperature compensator and zero and span adjustment — and no controls to adjust.

Simple Zero Adjustment
Touching a magnet to the designated point on the case performs a quick and easy zero adjustment.

Elastomeric Connector
When placed under pressure by tightening the sensor end cap, a network of fine wires is exposed, making contact with the concentric gold rings on the base of the sensor and the transmitter. Thus, connecting pins or sockets are not required allowing simple "pop-in" sensor replacement.

EIT's Electrochemical Sensor
The "heart" of the SENSOR STIK, EIT's amperometric membraned sensor is a sealed, maintenance free device. Considered to be the most advanced of its kind available, each sensor exhibits rapid response time, excellent repeatability and little or no response to interfering gases. Typical calibration frequency is three months with an expected life of two years in most applications. Each sensor carries a full 12 month warranty.

Sensor End Cap
Secures the sensor tightly in place using an o-ring seal. This prevents ingress of water and allows efficient operation of the elastomeric connector. Can be temporarily or permanently replaced with a combination rain shield/calibration adaptor or flowcell.

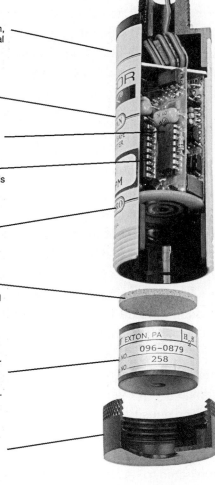

FIGURE 10-119 Cutaway picture of a gas detector. (Courtesy of EIT.)

SOLUTION TO JOB ASSIGNMENT

Each of the tanks requires a level sensor. The tanks that hold dry material should use a level sensor like the paddlewheel sensor, vibrating tines, or the beam breaker if you are trying to measure a single point where the tank is considered full or empty. If a continuous reading is required, a weighing system or a sounding-type sensor that uses a weight and line could be used. A sonic-type sensor could also be used.

The tank that holds the liquid material could use a float, displacer, or conductance sensor to indicate a single point, or it could use a capacitance-type, sonic, or ΔP-type sensor to indicate a continuous level.

The temperature detection system could use thermocouples, RTDs, thermistors, or IC temperature sensors. You would need to match the temperature range of the liquid with each type of sensor (see Fig. 10-15 for thermocouples and Fig. 10-21 for RTDs). Since this is a liquid, you will need to include a thermowell so you can install and remove the sensor for calibration without allowing liquid to leak from the tank.

The temperature of the steam jacket could be determined by a pressure gauge. Since a steam pressure gauge is specified, it could be used to indicate the amount of steam pressure, and the temperature of the steam jacket could be determined from a conversion chart. The type of gauge would normally be a Bourdon tube gauge.

The flow meter to measure the amount of liquid going into the tank should be a fairly accurate type of flow meter similar to the paddlewheel or turbine flow meter. If more accuracy is required, a positive displacement flow meter like the gear type should be used.

The positioning sensors to ensure the tanks are in the correct position for loading could be as simple as a proximity switch. If more accuracy is required, a ball screw mechanism could be used for motion, and an encoder or resolver could be used to determine the correct position. A linear potentiometer, LVDT, or magnetostrictive sensor could be used if the amount of distance is not over 6 ft.

The density sensor should be similar to the one shown in Fig. 10-68 and Fig. 10-69. It could take a sample of the mixture to determine its density. If the density sensor is connected to a closed-loop controller, it could allow more liquid to be added if the mixture is too dry, or more powder to be added if the mixture is too wet.

The formulas for flow are found in Problems 10-3, 10-4, and 10-5. The formulas for temperature conversion for Fahrenheit and Celsius and a table are shown in Fig. 10-11.

QUESTIONS

1. Illustrate the difference between an on-off signal and an analog signal. Provide an example of where to use each type of signal.
2. Explain why it is so important to understand the theory of operation for each of the types of sensors.
3. Provide an application where you would use a Bourdon tube pressure sensor, a strain gauge pressure sensor, and a piezoelectric pressure sensor.
4. Explain the operation of a strain gauge and indicate the type of signal you would expect from it.

5. Explain the operation of a piezoelectric pressure sensor and indicate the type of signal you would get from it.

6. Identify the two broad categories of load cells.

7. Explain how you would troubleshoot a load cell.

8. Compare the terms *laminar flow* and *turbulent flow*.

9. Explain the operation of a sonic type flow meter and explain why you would use this type of flow meter instead of another.

10. Explain the operation of a Coriolis mass flow meter.

11. What is a Reynolds number and what is it used for?

12. Explain the operation of a thermal mass flow meter.

13 Explain why density measurements would be used in industrial applications.

14. Describe the operation of an RF admittance (capacitance) level sensor.

15. Explain how a sounding-level sensor operates.

16. Explain how a sonic-level sensor operates.

17. Provide an application where a linear potentiometer and a rotary potentiometer can be used.

18. Explain the operation of a magnetostrictive position sensor.

19. Describe the basic operation of an accelerometer. Be sure to include the function of the piezoelectric element.

20. Explain the pH scale and give an example of a substance that is an acid, a base, and neutral. Be sure to give the pH strength of the examples that you provide.

TRUE OR FALSE

1. _____ One advantage of a 4–20 mA signal over a 0–10 volt dc signal is that the 4–20 mA signal uses a live zero, which makes it easier to detect a broken wire in a sensor or transducer.

2. _____ A strain gauge is a larger version of a load cell.

3. _____ The positive displacement flow meter provides a more accurate measure of flow than a ΔP flow meter.

4. _____ A turbine flow meter uses a turbine wheel that is turned when fluid flows past it. The number of revolutions of the turbine shaft is then converted to amount of flow.

5. _____ A vortex-type flow meter uses LEDs to measure fluid flow.

6. _____ Explain the operation of a pressure differential (ΔP) type flow sensor that uses an orifice plate to create a pressure drop that is used to calculate flow.

7. _____ Pressure can be used to measure temperature, flow, and level because there is a relationship between each that can be used in calculations.

8. _____ A positioning system that uses an absolute encoder does not need a home switch.

9. _____ An incremental encoder produces two waveforms (A pulse and B pulse) that are out of phase from each other to determine the direction of rotation.

10. _____ An acid has a higher pH number than a base.

MULTIPLE CHOICE

1. Zero and span are provided on most sensors _____.

   a. to make the sensor easier to troubleshoot.

   b. to allow the sensor to be adjusted in the field to match the actual minimum and maximum conditions that are being sensed.

   c. to make the sensor easier to be removed if it is broken.

2. An eddy current killed oscillator (ECKO) proximity switch _____.

    a. can detect ferrous metals.

    b. can detect any metal or nonmetal part since it bounces an echo off the part and measures how long it takes for the signal to return to the sensor.

    c. can detect anything that breaks the beam of light it sends out.

3. Vibrating-tine sensor _____.

    a. detects the level of granular material in a bin when the material covers the vibrating tines.

    b. detects the flow of liquids as they pass the vibrating tines.

    c. detects the level of gases and vapors when they make the tines vibrate more slowly.

4. Light can be used to indicate the level of liquid and solids _____.

    a. when the beam of light changes its frequency as it reflects off solids and liquids.

    b. when the beam of light is broken by the presence of material indicating the material has reached the level at which the sensor is mounted.

    c. when the impedance of the return light beam is 20% higher than the impedance of the light that is sent to the target.

5. You can detect changes in temperature with a(n) _____.

    a. thermocouple.

    b. RTD.

    c. thermistor.

    d. IC temperature sensor.

    e. all the above.

    f. only a and b.

6. The difference between an absolute and an incremental encoder is _____.

    a. the absolute encoder produces a set of pulses that requires a home switch to determine its exact starting location.

    b. the incremental encoder automatically knows its position as soon as power is turned on.

    c. the absolute encoder automatically knows its position when power is turned on.

    d. all the above.

7. A venturi flow meter _____.

    a. uses a pressure differential across a venturi that is converted through a calculation to the flow value.

    b. measures flow directly from the pressure differential across a venturi.

    c. counts the number of revolutions the venturi wheel makes as fluid flows past it.

8. A paddlewheel can be used to measure level _____.

    a. by causing the paddlewheel to turn slowly when material is below its location and determining when it stops as material covers it.

    b. by counting the number of rotations the paddlewheel turns as fluid moves past it and converting this number through a calculation.

    c. by determining the pressure differential across the paddles on the paddlewheel as fluid moves past it.

PROBLEMS

1. Convert the following Fahrenheit temperatures to Celsius: 0°F, 32°F, 72°F, 100°F, 150°F, 200°F, 212°F, 400°F.

2. Convert the following Celsius temperatures to Fahrenheit: 0°C, 50°C, 100°C, 150°C, 200°C, 300°C, 400°C.

3. Compare the following pressures and list them in order of the most pressure to the least pressure: 12 psi, 15 psig, 10 psia.

4. Calculate the amount of resistance a strain gauge would need to produce to balance the bridge circuit shown in Fig. 10-45 if R_1 and R_3 are equal to 250 Ω, and R_2 is set for 400 Ω.

5. Calculate the flow through a pipe that has a 3in. inside diameter when the velocity of the flow is 4 ft per second.

6. Calculate the Reynolds number for a liquid flowing through a pipe with a 1.5in. inside diameter, whose specific gravity is 0.60, flow rate is 10 gpm, and whose viscosity is 1.15 ft^2/sec.

7. Calculate the flow through an orifice plate when the P_1 pressure is 4.3 psi, and P_2 pressure is 3.8 psi, and the k value for the orifice plate is 10.2.

8. If a load cell is rated for 2 mV/volt and its power supply is 9 volts, how much voltage would you expect to see if the load cell was loaded to its maximum weight?

9. Calculate the depth of a tank of water if a pressure differential sensor shows the pressure at the bottom of the tank is 27.5 psi.

10. Calculate the resolution of a ball screw mechanism that has a pitch of 20 threads per inch.

DRAW AND EXPLAIN

1. Draw a diagram of a potentiometer connected to a power supply to provide a change of voltage as the potentiometer wiper moves. Explain how this circuit operates.

2. Draw the electrical diagram for a linear variable differential transformer (LVDT) and explain its operation.

3. Draw a linear encoder and explain how it creates positional data.

4. Draw the electrical diagram for a resolver and explain its operation.

5. Draw a diagram that shows the operation of a Hall-effect sensor and explain how the Hall effect works.

6. Draw the output signals from an incremental encoder and explain what each waveform is used for.

7. Draw a diagram of a circuit that could use the signal from a strain gauge to provide a reading to a meter to indicate pressure.

8. Draw a rack and pinion mechanism and explain how linear motion can be converted to rotary motion and how rotary motion can be turned into linear motion.

9. Draw a diagram of a circuit that a load cell would use to produce a useful signal. Be sure to identify the exciter voltage and the signal voltage.

10. Draw a pitot tube flow meter and explain its operation.

11. Draw a float-level sensor and explain its operation.

12. Draw a displacer-level sensor and explain how it differs from a float-level sensor.

13. Draw a diagram that indicates how a system can use a single float to indicate several different levels of liquid.

14. Draw a Bourdon tube pressure sensor and explain its operation.

REFERENCES

1. Allocca, John A., and Allen Stuart, *Transducers Theory & Applications.* Reston, VA: Reston Publishing, Inc., a Prentice Hall Company, 1984.

2. Jacob, J. Michael, *Industrial Control Electronics.* Englewood Cliffs, NJ: Prentice Hall Publishing, 1988.

3. Johnson, Curtis, *Process Control Instrumentation Technology,* 4th ed. Englewood Cliffs, NJ: Regents Prentice Hall Publishing, 1993.

4. Khazan, Alexander D., *Transducers and Their Elements.* Englewood Cliffs, NJ: Prentice Hall Publishing, 1994.

5. Mahoney, Timothy J., *Industrial Solid-State Electronics*, 2nd ed. Englewood Cliffs, NJ: Prentice Hall Publishing, 1986.

6. *Hall Effect Transducers,* Micro Switch, a Honeywell Division, Freeport, IL, 1982.

11

OUTPUT DEVICES:
AMPLIFIERS, VALVES, RELAYS, VARIABLE-FREQUENCY DRIVES, STEPPER MOTORS, AND SERVOMOTOR DRIVES

OBJECTIVES

After reading this chapter, you will be able to:

1. Identify all the parts of a two-way solenoid valve and explain their operation.
2. Use a table to determine the correct type and size of solenoids to use for an industrial application.
3. Identify all the main parts of a pneumatic-assisted valve and explain their operation.
4. Explain the operation of a hydraulic solenoid valve and identify the function of each port.
5. Identify the main parts of a relay and explain the operation of the coil and contacts.
6. Explain the operation of a full-step stepper motor.
7. Explain the operation of the switching controller for a stepper motor and provide an example of how the step sequence works.

8. Explain the operation of a stepper motor in full-step, half-step, and micro-step modes.
9. Identify the parts of a permanent magnet motor and explain their operation.
10. Calculate the amount of torque a servomotor can provide at its shaft.
11. Calculate the size of the servomotor required for an industrial application.
12. Use a schematic diagram and explain the operation of a DC servo amplifier.
13. Use a schematic diagram and explain the operation of an AC servo amplifier.
14. Use tables and calculations to determine the size of a servomotor and its drive amplifier.
15. Identify the parts of an AC variable-frequency drive and explain their operation.
16. Explain the function of the main parameters of a variable-frequency drive.

You are working in an area of the plant that has received a large machine with a robot loader that puts parts in the machine's fixture, and unloads the parts when it is finished. The machine and its loader have the following motors and controls: The machine has a 20hp AC servomotor that drives its cutting head. The machine also has a DC servomotor. The robot loader uses two brushless DC motors and AC motors to control its motion. The machine has a three-phase AC coolant pump that is turned on and off by a size 1 motor starter. The coolant flow is controlled by a two-way solenoid valve. The fixture to hold the part in the mill is opened and closed by a three-position, four-way, five-port hydraulic solenoid

valve. The conveyor that supplies parts to the parts loader and takes away finished parts is controlled by an AC variable-frequency drive.

Your job is to locate material in this chapter that can be used to determine the correct size of the components, and troubleshoot and verify that all parts of this machine are operating correctly. Write a short list that explains what you would consider when you are determining the size of each component and include a written procedure that indicates what you should check for each of the parts of the machine to determine if it is operating correctly. Be sure to include information about the controller for the brushless DC motor, and the servo amplifiers. The parts that you should include in your report are: AC servomotor for the cutter head, DC and AC servomotors for the axis motion, for the robot parts loader, motor starter that controls the coolant pump, AC variable-frequency drive for the conveyor, solenoid valve that controls the coolant, and hydraulic valve that controls the parts fixture.

11.1 OVERVIEW OF OUTPUT DEVICES

All industrial electronic circuits must have some type of output device to provide final control of the system. The previous chapter discussed in detail input devices and sensors that can be used to provide information to the controller. This chapter explains the various types of output components you will encounter in modern industrial circuits and applications. For example, the output of the system may be a simple amplifier for a valve or motor. The amplifier is required because the output signal from the controller is too small to operate a motor or valve.

Some valves and motor drives have amplifiers that are built into their circuits. The signal from the controller may be digital (on-off) or it can be an analog signal (continuous minimum to maximum such as 0–10 V DC). In some systems the amplifier is a separate component and it will take an input signal and produce a voltage and current to operate a motor or valve. You will need to understand how the individual components that you studied in Chapter 5 are combined into circuits that you studied in Chapter 8. In Chapter 9 you learned the basic concepts of open-loop and closed-loop systems and you learned that the final control element is called the *output*. In this chapter you will be introduced to the wide variety of components such as valves, AC motors, and DC motors that act as the final control element (output) in industrial electronic systems. You will learn the basic theory of operation of each of these elements, and you will learn to determine the criteria for selecting the correct size and how to troubleshoot them when they do not operate correctly.

11.2 SOLENOID VALVES

One of the simplest types of final control devices in industrial circuits is the solenoid valve. The solenoid valve consists of a magnet coil and a movable armature. When the solenoid's coil is energized, the armature is pulled up into the coil and the path for fluid to flow through the valve is open. When the coil is de-energized, spring tension forces the armature down so that the valve is closed off. Fig. 11-1a shows an example of a solenoid with the coil de-energized and the valve closed so no

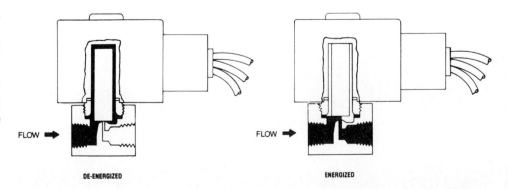

FIGURE 11-1 **(a)** A solenoid valve that is in the de-energized position so that no fluid will flow through it. **(b)** A solenoid valve that is in the energized position so that fluid will flow through it. (Courtesy of Automatic Switch Co.)

FLOW ➡

DE-ENERGIZED

FLOW ➡

ENERGIZED

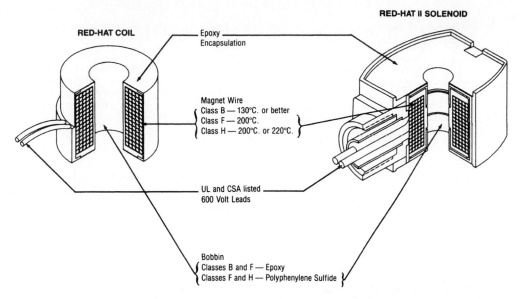

FIGURE 11-2 Cut-away view of coils for solenoid valves. (Courtesy of Automatic Switch Co.)

fluid can flow through it. Fig. 11-1b shows the solenoid with its coil in the energized position and fluid flowing through it. The fluid can be anything from water to refrigerants and chemicals. These types of valves are rather small and the piping that is connected to them is usually less than 1 in. in diameter.

The coil of the solenoid valve is made of multiple turns of wire molded in the shape of a hollow core so that it fits around the armature. Fig. 11-2 shows a cut-away diagram of two solenoid coils so you can see the relationship of the coil and the armature. The coils of wire are encapsulated with epoxy for protection against moisture and heat, and two lead wires are provided for field connections. When the coil is energized, a strong magnetic field is developed and it pulls the armature into the middle of the coil.

When power is first applied to the coil, the amount of current drawn by the coil to begin building the magnetic field will be approximately three times the amount of current that is used after the armature moves to the middle of the coil. Fig. 11-3 shows a diagram of the current flow pattern when a solenoid coil is energized. The initial current flow is called *inrush current*. The current that holds the armature in the energized position is called *holding current*.

If the coil is powered with DC voltage, an inductive voltage is created anytime power to the coil is de-energized. The inductive voltage is called an *inductive kick* and it is up to ten times the applied voltage and is in reverse polarity to the applied

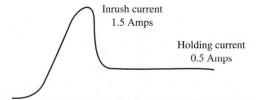

FIGURE 11-3 A diagram of inrush and holding current for a solenoid coil.

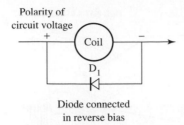

Polarity of
circuit voltage

Coil

D₁

Diode connected
in reverse bias

FIGURE 11-4 A diode is connected in reverse bias across a solenoid coil that is powered by dc voltage. The diode protects other electronic components in the circuit from inductive voltage that occurs when dc voltage to the coil is de-energized.

voltage. A diode or other type of suppression device must be connected across the coil of the solenoid to protect any other electronic components in the circuit that may be damaged by this voltage. The diode is connected in reverse bias across the DC solenoid coil so that when voltage is applied in normal polarity, the diode does not provide a path for current. When the solenoid coil is de-energized, the inductive voltage is the opposite polarity to the power supply, so it will flow through the diode and back into the coil. Since the coil is made of a large length of wire, the energy of the inductive voltage will be dissipated as it moves through the wire. This will render the excessive inductive voltage harmless. The fact that the inductive voltage will travel through the diode in the forward bias direction means the 0.7–1 volt drop across the diode junction will also limit the $V = \angle \dfrac{di}{dt}$ surge. Fig. 11-4 shows an example of the diode connected across the coil of a solenoid that is powered with DC voltage.

Since the armature physically moves inside the coil when power is applied to the coil, it will change the magnetic field, which will cause the current to drop to the holding-current level. If the armature did not move, the current would remain at the inrush level, which exceeds the current rating for the wire. The coil would quickly overheat and the wire would be burned until it caused an open. The level of the inrush current is determined by dividing the applied voltage by the resistance of the wire in the coil. This means that you should never connect voltage to a coil if it is not mounted on an armature. Some technicians test the coil by trying to pull the coil off the armature to see if it is magnetized. A problem occurs if the coil is pulled too hard and actually comes off the armature because the current will increase to the inrush level and cause the coil that is being tested to burn out. If the coil is tested with voltage, it is important that the coil is mounted over the armature.

11.2.1 Typical Voltages and Wattages for Solenoid Valves

Solenoid valves are available in a variety of DC and AC voltages. The common DC voltages are 6, 12, 24, 120, and 240 volts. The common AC voltages at 60 Hz are 24, 120, 240, and 480 volts. The common AC voltages for 50 Hz are 110 and 220 volts.

The power consumption for each valve will be slightly different due to the size of the valve stem that is being moved open and closed and the size of the coil needed to build the magnetic field. Fig. 11-5 shows a data sheet for a typical solenoid valve. This data sheet shows that the wattage for a normally closed solenoid with ⅜-inch openings will range from 6.1–17.1 W for a coil that is powered with AC voltage, and from 10.6–11.6 W for a coil that is powered with DC voltage. This

SPECIFICATIONS

Pipe Size (ins.)	Orifice Size (ins.)	Cv Flow Factor	Operating Pressure Differential (psi)						Max. Fluid Temp. °F.		Standard Solenoid Enclosures Red-Hat II–Types 1, 2, 3, 3S, 4 and 4X						Watt Rating/ Class of Coil Insulation ②	
			Max. AC			Max. DC					Brass Body			S. S. Body				
			Air-Inert Gas	Water	Lt. Oil @ 300 SSU	Air-Inert Gas	Water	Lt. Oil @ 300 SSU	AC	DC	Catalog Number	Constr. Ref. No.	UL Listing	Catalog Number	Constr. Ref. No.	UL Listing	AC	DC
NORMALLY CLOSED (Closed when de-energized), Buna "N" Disc																		
1/8	3/64	.06	750	750	530	650	640	550	180	120	8262G1	1	o	8262G12	1	o	6.1/F	10.6/F
1/8	3/32	.20	275	290	130	150	140	145	180	120	8262G14	1	o	8262G15	1	o	6.1/F	10.6/F
1/8	1/8	.34	155	180	140	80	80	80	180	120	8262G2	1	o	8262G6	1	o	6.1/F	10.6/F
1/4	3/64	.06	750	750	500	500	500	500	180	120	8262G19	16	o	8262G80	11	o	6.1/F	10.6/F
1/4	3/32	.17	360	340	160	150	125	125	180	120	8262G20	16	o	8262G86	11	o	6.1/F	10.6/F
1/4	1/8	.35	140	165	90	65	60	60	180	120	8262G22	16	o	8262G7	11	o	6.1/F	10.6/F
1/4	1/8	.35	300	300	200	75	70	70	180	150	8262G232	17	o	—	—	—	10.1/F	11.6/F
1/4	5/32	.50	180	200	145	40	40	45	180	150	8262G202	4	o	8262G220	12	o	10.1/F	11.6/F
1/4	7/32	.72	90	100	100	25	25	25	180	150	8262G208	4	o	8262G226	12	o	10.1/F	11.6/F
1/4	7/32	.85	40	50	40	17	20	21	180	120	8262G13	2	o	8262G36	11	o	6.1/F	10.6/F
1/4	9/32	.88	60	75	60	18	15	18	180	150	8262G210	4	o	—	—	—	10.1/F	11.6/F
1/4	9/32	.88	90	100	90	25	20	22	180	150	8262G212	6	o	8262G230	13	o	17.1/F	22.6/F
1/4	9/32	.96	27	36	28	15	16	16	180	120	8262G90	2	o	8262G38	11	o	6.1/F	10.6/F
3/8	1/8	.35	160	150	90	65	60	60	180	120	8263G2	3	o	8263G330	3	o	6.1/F	10.6/F
3/8	5/32	.52	100	100	100	35	35	35	180	150	8263G200	5	o	8263G331	5	o	10.1/F	11.6/F
3/8	7/32	.72	100	100	100	25	25	25	180	150	8263G206	5	o	8263G332	5	o	17.1/F	11.6/F
3/8	9/32	.85	100	100	70	—	—	—	180	—	8263G210	7	o	8263G333	7	o	17.1/F	—
NORMALLY OPEN (Open when de-energized), Buna "N" Disc (except where noted)																		
1/8	1/16	.09	500	300	225	400	250	150	180	120	8262G91	8	•	8262G92	8	•	6.1/F	10.6/F
1/8	3/32	.15	275	200	150	190	110	110	180	120	8262G93	8	•	8262G94	8	•	6.1/F	10.6/F
1/8	1/8	.21	125	100	85	80	60	50	180	120	8262G31	8	•	8262G35	8	•	6.1/F	10.6/F
1/4	3/64	.06	750	700	700	500	500	500	140	140	8262G260①	9	•	8262G130①	14	•	10.1/F	11.6/F
1/4	3/32	.17	300	250	230	200	150	125	140	140	8262G261①	9	•	8262G134①	14	•	10.1/F	11.6/F
1/4	1/8	.35	130	110	100	80	60	60	180	150	8262G262	9	•	8262G138	14	•	10.1/F	11.6/F
1/4	5/32	.49	85	75	60	45	30	30	180	150	8262G263	4	•	8262G142	14	•	10.1/F	11.6/F
1/4	7/32	.83	45	45	40	25	20	20	180	150	8262G264	4	•	8262G148	14	•	10.1/F	11.6/F
1/4	9/32	.96	30	25	20	15	15	15	180	150	8262G265	4	•	8262G152	14	•	10.1/F	11.6/F

Notes: ① Cast Urethane disc supplied as standard.   ② On 50 Hertz service, the watt rating for the 6.1/F solenoid is 8.1 watts.

FIGURE 11-5 Technical data sheet for a typical solenoid valve. (Courtesy of Automatic Switch Co.)

means that if the coil is powered with 120 volts and it consumes 17.1 W it would draw 0.14 A for holding current. If you multiply this value by 3, the inrush current is approximately 0.42 A.

Some data sheets provide the rating for solenoid valves as the amount of voltamperes (VA) that the coil consumes. The *apparent power* is determined by measuring the current and the voltage and multiplying them to provide the VA value. *True power*, which is the actual wattage the circuit consumes, is calculated by the formula $P = I^2R$. Wattage is determined by measurement with a wattmeter. The true power (W) and the apparent power (VA) will only be equal values if the circuit load is purely resistive. If the load is magnetic (inductive) or if it has capacitance, the apparent power value will always be larger. The ratio of the true power divided by the apparent power is called the *power factor*. The data sheet shown in Fig. 11-5 also shows typical ratings for the solenoid valve. Notice that these ratings include the pipe size, flow factor, pressure differential, and maximum temperature. The wattage rating for each coil is also provided in the column at the far right.

When you are providing fuse protection for a coil in a circuit, you can size the fuse at the amount of holding current that you measure with an ammeter. The fuse will need to be a dual-element time-delay fuse so that the inrush current can occur without blowing the fuse, while still providing protection at the holding-current level.

- -

PROBLEM 11-1

Select a solenoid valve from the data sheet in Fig. 11-5 that will be used to replace the normally closed solenoid valve that controls the coolant flow to the milling machine that was delivered to your factory. The coolant for cutting is a mixture of light oils. This valve needs to have a $\frac{3}{8}$-inch opening and a flow-factor rating of .72. The minimum pressure for this system is 0 psi and the maximum pressure is 100 psi. The control voltage for this machine is 110 V AC.

SOLUTION

Since this valve is used to control the flow of light oil that is used as a coolant and it needs an opening of $\frac{3}{8}$ in., you should start your search in the last four rows of normally closed part of the table. Since the control voltage for the system is 110 V AC, you can ignore the DC-powered coils. The pressure for the system ranges from 0–100 psi and the fluid is light oil, so you will have to select a **8263G206 Brass Body Valve** listed in the next to the last row at the bottom of the table for normally closed solenoid.

- -

11.2.2 Types of Solenoids

Solenoid valves are manufactured to control the flow of air, water, inert gases, light oils, refrigerants, and other fluids. A large variety of valves is available to provide three-way or four-way control. Fig. 11-6 shows a diagram of a simple on/off pneumatic valve. This type of valve is called a three-way, two-position solenoid valve. The diagram for this type of valve is used in pneumatic or hydraulic diagrams. The symbol for the main valve body is a rectangle that is made by placing two boxes side by side. The ports for the valve are identified in this part of the symbol.

The small rectangle at the far left end with a slash line through it is the symbol for the solenoid coil.

The line at the far right end that looks like the symbol for a resistor is the symbol for a spring.

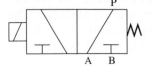

FIGURE 11-6   Three-way, two-position solenoid valve.

This symbol indicates that this valve uses a spring to return the valve to the de-energized position. P identifies the pressure port, and A and B identify the outlet ports.

The reason two boxes are used in the symbol for the valve is that the box that has the symbol of the coil attached indicates the position of the armature and the ports that are connected when the valve is energized. The box that has the spring symbol attached shows the position of the armature and the ports that are connected when the valve is de-energized. For example, when the valve is de-energized, you would read the diagram in the box on the side toward the spring symbol, which shows port A is connected to pressure and port B is blocked. When the valve is energized, you would use the diagram of the ports that is in the box on the left showing that port A is blocked and port B is connected to pressure. You should notice that the ports are only identified once in the diagram since this symbol represents the two positions of the ports of a single valve. Sometimes the symbol will have the ports identified in each part of the diagram.

11.2.3 Applications for Solenoids

Solenoids are used in a wide variety of applications. Simple applications include energizing a valve to allow fluid such as water or solvents to flow. A more widely used application includes control of air for use in pneumatic cylinders or control of oil for hydraulic cylinders. For example, a pneumatic cylinder can be used to move machine tools into place for cutting and drilling applications, or a pneumatic cylinder can be used for one or more axes on a robot. Fig. 11-7 shows an example of a solenoid valve that is used to control the extension and retraction of a pneumatic cylinder for a robot. The cylinder controls the up-and-down action of the robot for pick and place operation. When the cylinder is extended, the vacuum cup on the end of the rod is down on a conveyor to pick up a part. When the cylinder is retracted, it picks the part up and the robot moves from the conveyor to a machining fixture where the cylinder is extended again to drop the part into the fixture.

The solenoid valve has four ports that are labeled P (pressure), E (exhaust), A, and B. You should notice the diagram symbol for the valve shows two boxes with the ports identified in one. It is important to understand the physical valve has four ports identified as P, E, A, and B. The symbol for the valve shows the

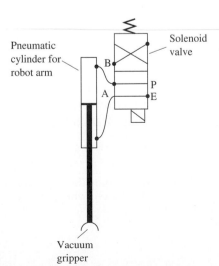

Pneumatic cylinder for robot arm

Solenoid valve

B

A

P

E

Vacuum gripper

FIGURE 11-7 An air cylinder connected to a solenoid valve. The air cylinder is part of a robot arm that uses a vacuum cup to pick up parts.

valve body two times to indicate the position of the valve passages when the coil for the valve is energized and when it is de-energized.

In the diagram of the valve, the ports shown in the box near the coil indicate the position of the valve when the coil is energized. In this condition, the pressure port (P) is connected to the B port, which allows air to fill the cylinder and causes the rod to extend. It is important to understand that the front port on the cylinder is connected to the A port, which is connected to the exhaust port (E) when the valve is energized. This allows a passage for the air in the front part of the cylinder to escape so the valve rod can move forward.

The diagram on the side of the valve near the spring symbol shows the position of the valve passages when the valve is de-energized. When the coil is de-energized, the spring moves the valve spool to a position where the pressure port (P) is connected to the A port and the exhaust port E is connected to the B port. This allows air pressure to enter the front of the cylinder and causes the rod to retract. The rear port of the cylinder is connected to the B port on the valve, which allows air to exhaust and the rod to retract.

11.2.4 Three-Position Solenoid Valves

Another type of solenoid valve used in robotic and other machine motion applications is a three-position valve. This valve is similar to the two-position valve where the solenoid is energized or de-energized. From the diagram in Fig. 11-8 you can see that the valve has two solenoid coils, one on each end and a spring on each end. The actual coil symbol for this valve is more specific in that it shows a pilot valve is used to move the main spool. This means that the solenoid will energize a small valve, and when its spool moves, it will move fluid that will cause the main valve spool to move. The pilot valve in this application is like an amplifier.

It is easier to understand the operation of the valve if you disregard the pilot operation and explain the main spool movement. When the solenoid on the left side of the valve is energized, the spool (armature) in the valve will shift to the position shown in the diagram in the left box for the valve symbol. When the solenoid on the right side of the valve is energized, the valve will shift to the position shown in the diagram in the right box for the valve. When both solenoids are de-energized, spring pressure from the springs on each end will cause the valve spool to shift to a center position. In the diagram of this valve, you can see that the center position causes the spool to block all ports. This is a safety feature used in robots and other motion control systems that allows the valve to move to the center position during an emergency stop condition and blocks the air to all ports of the robot's cylinder, causing all motion to stop.

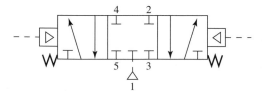

FIGURE 11-8 A three-position, four-way valve with all ports blocked when both solenoids are de-energized.

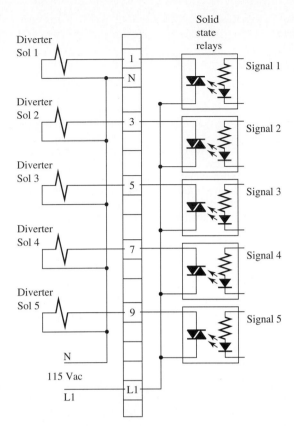

FIGURE 11-9 AC powered solenoid coils controlled by a triac in an opto-coupler. The solenoids are on the left side of the terminal strip in this diagram.

11.2.5 Diagrams of Solenoids as Outputs

The solenoids discussed so far are all used as outputs for both electronic and electromechanical circuits. In the first few chapters of this text you learned about electronic circuits that used a triac in an optocoupler to energize solenoid-type output. Other electronic circuits may use amplifiers or transistors to energize solenoids. Fig. 11-9 shows an example of a circuit that uses a triac in an optocoupler. In this application of a high-speed weighing system solenoids are used to energize a pneumatic ejector system. The solenoids are connected to a terminal board to make troubleshooting simpler.

The ejectors use a small pneumatic cylinder that can extend and retract extremely fast. When an overweight product is detected, the solenoid is energized and the cylinder is extended to push the product off the conveyor and retract to be ready to repeat the operation in less than 0.1 second. This allows the system to weigh 600 products per minute. Since the weighing system is all electronic, electronic circuits may use triacs or transistors to energize and de-energize the solenoid valve. If transistors are used, the solenoid coils will be powered by DC voltage, and a diode must be connected in reverse bias across the coil terminals to protect the circuit against the reverse polarity of inductive voltage that occurs when the solenoid is de-energized.

Fig. 11-10 shows a solenoid coil controlled by a motor control circuit that uses a selector switch and timer. In this type of application a small carrier is used to hold an automobile dashboard while a plastic coating is applied. The plastic coating is a sheet of color film that is draped over the dashboard. When the film is in place,

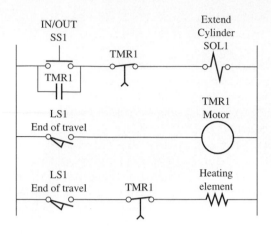

FIGURE 11-10 Electric diagram of selector switch and timer used to control a solenoid valve. The solenoid valve energizes a pneumatic cyclinder that extends an automotive dashboard into a curing oven for a shrinking and curing process.

the operator switches the momentary-type selector switch (SS1) to the in position, which energizes the solenoid (SOL1). A set of normally open (NO) contacts from the timer is used to seal in the momentary contacts of the selector switch. When the solenoid energizes, a spool in the solenoid valve moves so air is directed to a pneumatic cylinder that extends and moves the carrier into a baking oven. When the carrier is moved into the oven and the cylinder is fully extended, a limit switch (LS1) is activated, energizing a timer (TMR1). The second set of contacts in limit switch (LS1) and a second set of contacts in the timer (TMR1) also activate the heating element to start the shrinking and curing process. When the timer times out, its normally closed (NC) contacts open and de-energize the solenoid coil. The second set of timer contacts also opens and de-energizes the heating element. The seal in the contacts of the timer (TMR1) also opens so the solenoid coil remains de-energized.

When the solenoid coil is de-energized, the spring in the valve causes the spool to return. The cylinder is retracted, causing the carrier to return from the oven so the operator can remove the finished dashboard and place another one on the carrier for the process to repeat.

11.3 PROPORTIONAL VALVES

The solenoid valves discussed so far are all on/off type valves, which are either completely open or completely closed. There are many applications that require the valve to be *proportional*, which means that the valve can be open to any value from 0–100%. Proportional valves are typically used in hydraulic applications where larger loads are controlled. For example, a proportional hydraulic valve can be used to control the pressure and flow of plastic in an injection molding machine. Larger valves can be used to control the amount of fluid flowing to a hydraulic motor. Hydraulic motors were introduced in industrial applications in the 1960s and 1970s prior to the advent of solid-state circuits for AC and DC motor speed control. These motors were typically over 50 hp and their speed could be controlled easily by adjusting the amount of fluid that flowed past their vanes. This type of application was very popular when electronic components could only control small amounts of voltage and current. The small amount of current could be controlled at the proportional valve, and a large amount of horsepower could be controlled at the hydraulic motor.

A proportional valve is similar to the pneumatic valves discussed earlier in that it has a spool that can allow ports to change as well as open or close from 0–100%. Frequently proportional valves are used in feedback systems so they are called servo valves. Two types of feedback are available with these valves. One type of feedback is external positioning sensors such as encoders and resolvers. A second type of feedback is located directly inside the valve to indicate the position of the spool. Internal feedback is used where the valve response must be so fast that it cannot wait for an external feedback signal to be returned to the valve amplifier and then to the valve. Applications that require faster response include hydraulic robots and velocity control valves on plastic injection molding machines that control the speed that plastic is injected into the mold.

11.3.1 Types of Control from a Proportional Control Valve

There are three basic types of control valves that are available as proportional control valves. The first type is called the *directional control valve.* This type of valve has *travel functions* and *flow functions* available for control. The travel function includes directional control for a valve. When a valve is connected to a cylinder, this type of control allows the cylinder to extend or retract. The flow function controls the amount of hydraulic fluid flow through the valve. In most applications, fluid flow will be similar to current flow in electricity. This means the greater the fluid flow, the more power the valve will control. When fluid flow is used to control a hydraulic motor, the greater the flow, the faster the motor can turn and the larger the load it can move.

Fig. 11-11 shows an example of a simple proportional valve. From this diagram you can see that the valve is controlled by a control amplifier (op amp) and the op amp is controlled by a potentiometer. You should remember from Chapter 8 that the output of the op amp will be 0–100% depending on the input signal.

The amount of voltage that the op amp sends to the valve will determine the percentage opening for the valve. If the 100% voltage signal is sent to the valve, the valve will be open 100%. If the voltage is 50%, the valve will be open 50%.

The second type of valve, called a *flow control valve,* uses internal positional

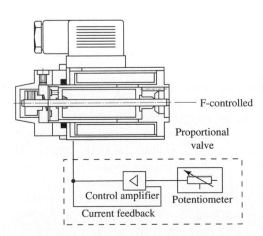

FIGURE 11-11 A control amplifier connected to a proportional valve. The amplifier is used to control the amount of voltage sent to a proportional valve. A potentiometer is used to control the input voltage to the amplifier. (Courtesy of Rexroth Corporation.)

feedback to indicate the position of the valve spool when a signal is provided. The feedback helps the amplifier determine the position of the spool and allows the amplifier to provide more or less voltage if the spool has not moved to the proper position inside the valve. The position of the valve spool is very important to controlling fluid flow. *Fluid flow* is similar to *velocity control* and these two terms may be used interchangeably.

The third type of valve is called a *pressure control valve*. In a hydraulic system, the pump control can be designed to allow the pump to run 100% of the time at 100% rpm, or the system can be designed so a variable-type pump is used and its pumping capacity is adjusted 0–100%. If the pump is running at 100% rpm it will produce maximum pressure anytime the valves in the system are closed. At first this may sound like an unwanted condition, but it can be used as a precise method of control.

The valves in this type of system are connected between the main flow lines and the reservoir tank. When the valve is open, the flow is allowed to return directly back to the reservoir tank rather than be used to move cylinders or hydraulic motors. This type of valve is called a *pressure relief valve*. The valve controls the amount of fluid that is sent to a cylinder or hydraulic motor by the amount of fluid that is not bypassed. This means that if the pressure relief valve is set at 100% open, the cylinder or hydraulic motor will not move because all of the flow from the pump is being sent directly back to the reservoir. If the amplifier sets the valve at 0%, the total flow from the pump is sent to the hydraulic motor or cylinder and this will make them move at maximum speed with maximum power. Fig. 11-12 shows examples of the directional control valve, the flow control valve, and the pressure relief valve.

11.3.2 Proportional Valve Amplifiers

The amplifiers for these valves use op amp circuits to provide variable-voltage or variable-current outputs. The circuits also provide ramp-up and ramp-down capabilities by adding a capacitor and a potentiometer to the circuit. The potentiometer is adjusted to increase or decrease the time constant with the capacitor, which increases or decreases the ramp time. You can review this information in greater detail in Chapter 8 where you first learned about the types of op amps. The arrangement of the capacitor and potentiometer makes this op amp an integrator.

Fig. 11-13 shows the capacitor and potentiometer connected to the op amp integrator to provide this ramp-type signal. The ramp-up signal is needed to ensure the valve is opened smoothly when voltage is first applied and that the machine motion caused by the valve is also smooth. As a technician, you will be expected to locate the potentiometer on the amplifier board and adjust it to match the ramp-up and ramp-down signals so that the machine motion is smooth as the valve opens and closes.

Fig. 11-14 shows an example of a ramp-up signal and a ramp-down signal that are compared to the on–off signal received by the amplifier. When the on-signal is energized to the valve amplifier, it is shown as instantly going from off to on. The amplifier produces the ramp signal and the slope of this signal is determined by the value of the potentiometer and capacitor which cause the time constant. It is important to remember that you can only adjust the potentiometer, since the capacitor is a fixed capacitor. The opening of the valve will be gradual, which will make the machine operation very smooth. When the signal to the valve amplifier is de-energized, the ramp-down signal will gradually close the valve, which will

(a)

(b)

(c)

FIGURE 11-12 **(a)** A directional control valve and amplifier. **(b)** A pressure relief valve and amplifier. **(c)** A flow control valve and amplifier. (Courtesy of Rexroth Corporation.)

again make the machine operation very smooth. If the machine motion is too jerky or too rough, you will need to adjust the slope until the machine operates smoothly. You may have to readjust the slope periodically as the machine wears.

Fig. 11-15 shows a picture of a typical valve amplifier, and Fig. 11-16 shows the diagram of this amplifier. The supply voltage for the amplifier is 24 V dc and it is connected to terminal 24ac and 18ac. Terminal 24ac is the positive terminal.

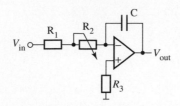

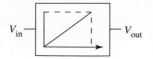

FIGURE 11-13 Typical op amp circuit for a proportional valve. The potentiometer and capacitor provide a time constant that controls the slope of the ramp-up and ramp-down signal. (Courtesy of Rexroth Corporation.)

The circuit in box 1 is a rectifier, filter, and regulator that ensures the dc voltage is pure and without ripple. The output of this circuit is $+/-9$ V dc, and it can be adjusted to $+/- 0.1$ volt with the potentiometer that is connected to terminals 10ac and 14ac. The $+9$ V dc is measured between terminals 10ac and 14ac, and the -9 V dc is measured between terminals 16ac and 14ac. Terminal 14ac is considered the 0 V terminal.

An external analog signal is connected to the amplifier card at terminals 28c and 30ac. The 28c terminal is the plus terminal and the 30ac is the negative terminal. The analog signal voltage should be 0–10 V dc. This signal would typically come from a controller such as a programmable controller or other microprocessor controller that has machine control logic.

The proportional valve is connected to the output terminals of the amplifier on the right side of the diagram. These terminals are marked 22ac and 20ac. The output will be a variable voltage (0–6 volts) with a corresponding variable current (0–800 mA). A voltmeter can be connected to terminals 32c and 14c to provide a panel meter that indicates the status of the valve signal. This small voltmeter is usually mounted near the operator panel or near the valve to give a visual indication of the voltage signal that is being sent to the valve.

The circuit in box 2 is the ramp-up and ramp-down circuit. A switch is connected across terminals 2a and 4ac to activate the ramp-up option, and a switch is

FIGURE 11-14 Typical ramp-up and ramp-down signals. The slope of the ramp is controlled by the capacitor and potentiometer. The ramp provides smooth operation of a machine when hydraulic power is applied and turned off. (Courtesy of Rexroth Corporation.)

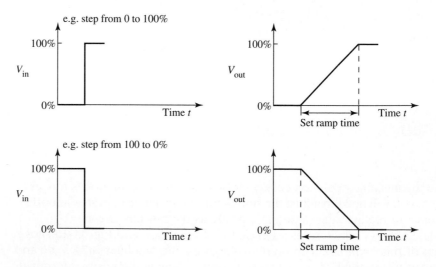

FIGURE 11-15 Amplifier card for proportional valve. This card slides into a card cage and receives an input signal (0–9 volts dc) from a PLC (programmable logic controller) and sends an output signal (0–6 volts dc) to a proportional valve. (Courtesy of Rexroth Corporation.)

connected across terminals 2c and 4ac to activate the ramp-down option. If these switches are not closed, the ramps are not activated.

The circuit in box 3 is the output stage of the amplifier. This circuit utilizes a pulse-width modulation (PWM) circuit. The current is pulsed to conserve energy and minimize the thermal load to the amplifier.

Four potentiometers are identified to provide adjustment for pilot current (R1), maximum current (R2), ramp-up time (R3), and ramp-down time (R4). These potentiometers allow the technician to adjust the amplifier to customize the valve

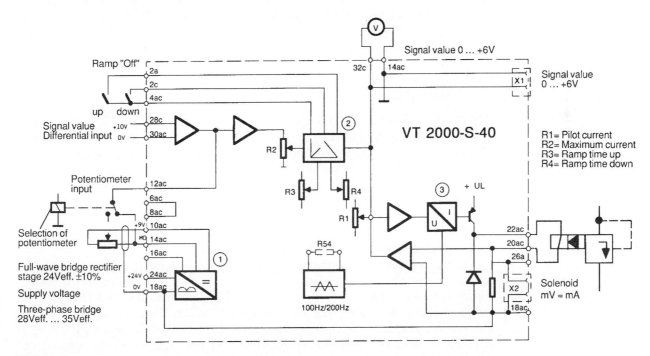

FIGURE 11-16 Electronic diagram of a typical amplifier circuit. The amplifier receives its power supply and input signal on the left side of the circuit, and the proportional valve is connected to the right side of the circuit. (Courtesy of Rexroth Corporation.)

action to the machine's action. At times two similar machines will react slightly differently when the same amount of hydraulic pressure is used to cause machine movement. These differences can be adjusted from machine to machine through the potentiometer so that the machine operation will be as smooth as possible.

11.3.3 Constant Voltage Proportional Valves

Some proportional valves are designed to use constant voltage with variable current. This concept is difficult for a technician to understand when first encountered. At first it would seem that this scenario is impossible because of Ohm's law. You learned in basic circuits class that current would increase as voltage is increased if resistance is constant. The way that these types of circuits operate is that they use a constant-voltage amplifier circuit. The amount of current is adjusted from 0–800 mA when the input signal changes from 0–10 volts. This type of circuit is easy to test and troubleshoot if you are aware of it. For example, if you used a voltmeter to test the constant-voltage valve and amplifier, it would show maximum voltage (6 V dc) even when the valve is at rest and the input signal is at 0 volts. This would tend to confuse you, since the maximum voltage signal is generally associated with the maximum input signal. An ammeter would be needed to test this type of valve because you would see that even though the voltage is constant, the current would change proportionately with the input voltage signal. When the current is at minimum, this type of valve has full voltage (6 volts). You should understand that even though the valve has 6 volts applied, it would not be activated because the voltage is only a potential at this point and there is no current. When the current is changed, the valve will change.

It is important that you do *not* use visual indicators, such as LED indicators or small lamps, to indicate when voltage is present because they will be illuminated all the time since the constant-voltage amplifier has voltage applied at all times and merely varies the current when the valve is adjusted.

11.3.4 Troubleshooting Proportional Amplifiers and Valves

Proportional amplifiers and valves are easy to troubleshoot. If you suspect the valve is bad, you can apply full voltage (6 V dc) or variable voltage (0–6 V dc) from a power supply and bypass the amplifier all together. The only problem that this presents is that it may cause the valve to open 100%, which may cause unwanted machine motion. You could turn the source of hydraulic pressure off prior to this test to render the machine safe. If the valve operates satisfactorily when it is connected directly to a power supply, it is good. If it will not operate when it is connected directly to a power supply, it will not work correctly when it is connected to the valve amplifier.

You can troubleshoot the amplifier by testing the power supply terminals to determine that the correct amount and polarity of voltage is applied. If the correct supply voltage is applied, you can provide an input voltage from 0–9 volts dc. The output signal to the valve should adjust proportionally from 0–800 mA and 0–6 volts. Some amplifiers may not operate if the load (valve) connected to it has an open circuit in the wiring or the solenoid coil. You can determine the condition of the wiring and the coil by taking a current measurement. The valve should draw a small amount of current (1–5 mA) even when the valve is at rest. If voltage is present at the output terminals of the valve amplifier and no current is present, you should suspect an open in the wiring or the solenoid coil. If voltage is present at the valve amplifier terminals and proportional valve terminals and no current is present, you should suspect an open solenoid coil.

....

PROBLEM 11-2

You are asked to troubleshoot a proportional valve and amplifier that are used to control the injection pressure on a plastic injection molding machine. The problem has been listed on the request-for-service tag as: "Valve will not open to build injection pressure." Identify the test points, the amount of voltage, and the type of signal that you would expect to see at each point. Use the amplifier diagram shown in Fig. 11-16 to help you determine your solution.

SOLUTION

Since the service tag states that the valve will not open to build injection pressure, you should suspect that it is not operating at all. You can start at the input or the output section of the amplifier to solve this problem. For this example we will make our first test across terminals 22ac and 20ac where the valve is connected on the lower right side of the diagram. Place a voltmeter across these terminals, or you can place an ammeter in series with either of the terminals. You will need the machine operator to cycle the machine to the step where injection pressure is built up and set the pressure limit to full pressure. Keep the machine on this step while you test for the voltage or current. The voltage should be approximately 10 volts since the control is set for full voltage. If you are measuring current, the amount of current should be 800 mA. If the voltage is 10 volts and the valve does not move, the electrical system is operating correctly and the problem will be in the hydraulic side of the system. If the voltage is less than 1 volt, the problem will be in the amplifier or the signal being sent to the amplifier from the controller.

If the output voltage from the amplifier is too low, your next check will be at the supply voltage terminals (24ac and 18ac) on the bottom left side of the diagram. The voltage at this point should be 24 V dc with the positive voltage at terminal 24ac. If you do not have supply voltage or it is less than 24 volts, you must fix the power supply.

If the supply voltage is 24 volts and the polarity is correct, you can test the voltage as it passes through the amplifier. The first voltage test inside the amplifier can be made across terminals 10ac and 14ac where 9 V dc is available. If you do not have voltage at this point, you would change the amplifier card. If voltage is present, proceed to the intermediate test points in the amplifier.

The next place you can test for voltage is at the signal voltage test points at the top right side of amplifier X1. The signal should be approximately 6 V dc at this point. If you do not have voltage at this point, you should check the ramp-up switch and set it to no ramp for this test. This will ensure that the maximum amount of voltage is provided to the output of the amplifier. If you do not have voltage even after the ramp-up switch is turned off, you have a problem in the amplifier and it must be changed.

You can see that troubleshooting an amplifier requires testing at several terminal points as the voltage moves through the amplifier. You could also remove and replace the amplifier or try a separate power supply of 0–10 volts and test the valve directly without an amplifier. If the valve operates with the external power supply but it will not work correctly with the amplifier, you can replace the amplifier.

If the valve will not work with the external power supply, you can focus your tests on the amplifier and hydraulic system.

• •

Note!
Some amplifiers or controllers have a feature that provides a minimum electrical signal when the controller is requesting full pressure. This is called a *reverse-acting valve* **feature. If the reverse-acting valve feature is selected, the amplifier will provide a minimum electrical signal when the controller requests maximum pressure, and it will produce a maximum electrical signal when the controller requests minimum pressure.**

11.4 PNEUMATIC-ASSISTED CONTROL VALVES

Pneumatic (air) assisted control valves are used to control the flow of water and liquid food products, petroleum products, and other chemicals through large piping. These types of valves are opened and closed by an electrical signal that is converted to air pressure. Pneumatic-operated valves became popular prior to the advent of electronic and solid-state components being used in industrial electronics. Pneumatic systems could provide analog control to open and close very large valves smoothly. The pneumatic controlled valves allowed the valve opening to be anywhere between full closed and full open. The signals for these systems were transmitted over long distances throughout a factory or process plant by transmitting air through plastic tubing or copper tubing. The pneumatic technology has proven so dependable it has been incorporated with modern electronic systems and controls. This means that new technology today can use an electronic or microprocessor controller and use an electronic-to-pneumatic converter that changes a milliamp signal to a pneumatic signal. The electrical signal is proportional 4–20 mA and the air pressure is proportional 3–15 psi.

These types of valves are commonly used as loads in the process industries. They are also used in applications where explosive atmospheres exist such as in spray painting and chemical processing.

11.4.1 Simplified Pneumatic-Assisted Valve

The pneumatic-assisted valve operates on a simple principal of counterbalance. Fig. 11-17 shows a diagram of the basic parts of this type of valve actuator and electrical-to-pressure converter that is called an *I/P* transmitter because it converts a milliamp signal to variable pressure. The top part of the diagram shows the converter, and the bottom left part of the diagram shows the pneumatic-operated valve.

The main part of the converter is a *balance beam*. The balance beam is balanced on the beam pivot. Air from a nozzle is directed on the left end of the beam, which tends to push this end of the beam down. A coil that produces an electromagnetic field and a permanent magnet are connected near the middle of the beam. When the coil is energized and it produces a strong magnetic field, it will tend to pull the beam up. This will cause a force that opposes the air flow directly on the end of the beam. The stronger the current in the coil is, the stronger

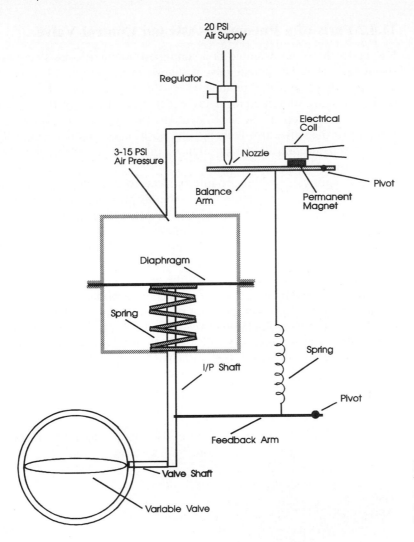

FIGURE 11-17 Simplified diagram of a pneumatic-assisted valve with an electric-to-pneumatic signal converter.

the magnetic field becomes and the more the beam is pulled tighter against the air nozzle. This will cause the beam to restrict the flow of air coming from the nozzle.

A tube is connected at a T connection near the tip of the air nozzle. This tube provides air to the diaphragm of the pneumatic-actuated valve. The amount of air pressure on the diaphragm of the valve will counter the spring pressure inside the valve that pushes the diaphragm up. The difference of the air pressure and spring pressure inside the valve will determine the position of the diaphragm, which will in turn determine the position of the valve stem somewhere between full open and full closed.

The valve mechanism for this valve is not shown. It would be connected to the end of the valve stem at the bottom of this diagram, and it would move up and down into a valve seat. The position of the valve mechanism can be anywhere between completely in the seat, which would shut off all flow through the valve, to completely up out of the seat, which would allow complete flow. A rod is connected to the valve stem that provides a feedback signal. This rod is normally connected to some type of positional sensor like a linear potentiometer or LVDT to convert the positional data to an electrical signal.

11.4.2 Parts of a Pneumatic-Assisted Control Valve

Fig. 11-17 shows an example of a simplified pneumatic-assisted control valve and signal converter. The actual valve is slightly more complex. Fig. 11-18 shows a cut-away picture of an actual pneumatic-assisted control valve. Each of the basic parts of the valve are identified in Figure 11.19. The legend for the numbers is listed in Fig. 11-19. The top of the valve contains the pneumatic operator. When air fills this part of the valve, it will change the position of the diaphragm. The position of the diaphragm controls the position of the valve actuator stem, which pulls the valve out of its seat.

The springs provide a counterbalance to the diaphragm and help to return the valve to its normal position when the air pressure returns to minimum (3 psi). This valve is a normally closed (NC) valve, which means it will be in the closed position when air pressure is at minimum. These types of valves are available as normally open (NO) valves also. The *fail-safe condition* of the system will determine whether NO or NC valves are used. For example, if the system is supplying cooling water for a process, you would want the valve to be an NO valve so that if the air pressure or valve failed, the springs in the valve would always cause the valve stem to cause the valve to be in the open position. If the valve controlled the supply of liquid to a tank, you would want the valve to be an NC type so that if the control part of the system failed, the valve would close and not allow liquid to enter the tank. If an NO valve were used in this type of application, the valve would return to its full open position when a failure to control air pressure or a valve failure occurred, which could allow the tank to overfill.

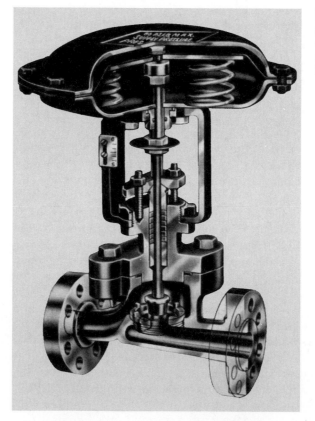

FIGURE 11-18 Cut-away diagram of a pneumatic-assisted control valve. (Courtesy of Masoneilan, Dresser Valve and Controls Division.)

Ref. No.	Description
1	Plug/Stem Assembly
2	Packing Flange Stud
3	Packing Flange Nut
4	Packing Flange
5	Packing Follower
6	Drive Nut
7	Packing
8	Bonnet
13	Body
9	Hex Head Cap Screws
10	Body Gasket
11	Guide Bushing
12	Seat Ring
14	Indicator
15	Hex Jam Nuts
16	Yoke
17	Indicator Plate
18	Machine Screw
19	Actuator Stem
20	Bushing
21	Hex Nut
22	Stem Spacer
23	Lower Case S/A
24	Hex Nut
25	Hex Head Cap Screw
26	Diaphragm
27	Upper Case
28	Diaphragm Plate
29	Spring
30	Spacer
31	Spacer
32	Hex Head Cap Screw
33	O-Ring
34	O-Ring
35	Actuator Vent Plug
36	Information Plate
Ref. No.	

FIGURE 11-19 List to identify parts of the pneumatic-assisted control valve in 11.18. (Courtesy of Masoneilan, Dresser Valve and Controls Division.)

11.4.3 Butterfly Pneumatic-Assisted Control Valves

Another type of pneumatic-assisted control valve is a *butterfly valve*. From the diagram in Fig. 11-20 you can see that the valve mechanism for this type of valve has a valve plate with a rod mounted through its axis that becomes the valve stem.

FIGURE 11-20 Butterfly pneumatic-assisted control valve. (Courtesy of The Foxboro Company.)

The valve's pneumatic control mechanism is linked to the valve stem so that it can be rotated to cause the valve plate to move to a position that is horizontal with the flow or vertical to the flow, which would cause the flow to stop.

The pneumatic control mechanism for this type of valve is similar to the one shown in the previous figure. When air is sent to the pneumatic chamber of the valve, it will press on a diaphragm and cause the valve linkage to move up and down. In this type of valve, the up-and-down movement causes the valve stem for the butterfly valve to rotate to the open or closed position. Springs provide the counterbalance to the diaphragm, which allows the valve to be proportional.

11.4.4 Current-to-Pressure (I/P) Converters

The pneumatic-assisted control valves require a converter to change the proportional electrical signal to a proportional pneumatic signal. In Fig. 11-17 the converter is shown as part of the valve. In some cases the converter is a separate part mounted near the valve. The converter is generally mounted in a location where it can be serviced easily and the valve is generally mounted in the piping.

The proportional electrical signal is generally a 4–20 mA current signal, and the air pressure signal is generally set for 3–15 psi. This type of signal converter is called an *I/P* converter because it changes a current signal (*I*) to a pressure signal (*P*). Fig. 11-21 shows an example of an *I/P* converter. The operation of this type of *I/P* converter is similar to the one shown in Fig. 11-17. It uses a magnetic coil to change the position of a balance beam that controls a small amount of pilot air pressure. The pilot air pressure controls the main air pressure that is regulated at 3–15 psi. The air supply for the *I/P* converter must be approximately 20 psi so that the converter can control the pressure between 3–15 psi.

Fig. 11-22 provides a graph that shows the ratio of current to air pressure. This graph allows you to select a milliamp signal value and determine the amount of air pressure that the *I/P* converter should produce. For example, when the electrical signal is at its minimum (4 mA), the air pressure signal will also be at its minimum (3 psi). When the electrical signal is at its maximum (20 mA), the air

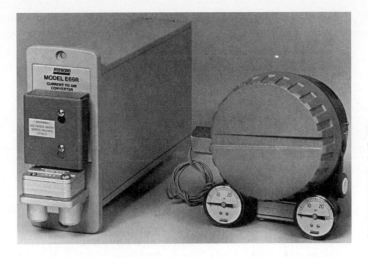

FIGURE 11-21 Current-to-pressure (*I/P*) converter. This device has an air pressure gauge to indicate the supply air, and another gauge to indicate the regulated pressure. (Courtesy of The Foxboro Company.)

pressure signal will be at its maximum (15 psi). The midpoint value for the electrical signal is 12 mA, which provides 9 psi.

The electrical signal for the *I/P* converter originates from an amplifier. The amplifier receives an input signal from a programmable controller or other type of electronic controller. The signal is generally analog, but it could be a digital signal that is sent through a digital-to-analog (D/A) converter prior to being sent to the amplifier. Many newer microprocessor controlled systems are also capable of providing the analog milliamp signal.

11.4.4.1 Applications That Use 3–15 psi ■ You may be familiar with the temperature control systems in large buildings such as the classrooms in your school. Many of these control systems use pneumatic control (3–15 psi) to control the opening and closing of water valves and dampers, which control the temperature and air flow to each room. If the building was built in the 1960s through the 1980s it will more than likely have this type of control. Newer control systems may use a

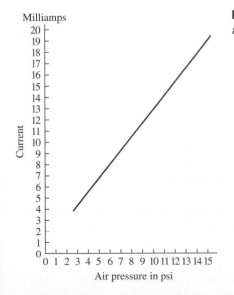

FIGURE 11-22 A graph of the milliamp signal and air pressure signal for an *I/P* converter.

combination of the electrical (4–20 mA) with the pneumatic (3–15 psi). If the system uses a combination of controls, it will use pressure-to-current (*P/I*) converters or transmitters, or current-to-pressure (*I/P*) converters. These types of control systems are also widely used in process control systems and applications where large valves and actuators are used.

11.4.4.2 Calculating mA or psi for *I/P* Transmitters ■ The amount of psi can also be calculated from the amount of the milliamp signal, or you can calculate the amount of the milliamp signal when the amount of psi is known. Each of these signals has an offset and a span that must be taken into account for the calculation. The offset for the milliamp signal is 4 mA, and the span is 16 mA. The span is determined by subtracting the lower value from the higher value (20 mA − 4 mA). The offset for the psi signal is 3 psi, and the span is 12 psi (15 psi − 3 psi). The formula for converting between these two systems is shown in the following equation. The milliamp signal is identified as A_1 and the pressure signal is identified as P_1. When you are converting from one to the other, either P_1 or A_1 will become a known value in the formula. The value 4 in the left side of the equation represents the 4 mA offset, and the value 16 represents the 16 mA span. The value 3 in the right side of the formula represents the pressure offset, and the value 12 represents the 12 psi span.

$$\frac{A_1 - 4}{16} = \frac{P_1 - 3}{12}$$

The formula can be further simplified in terms of an unknown pressure or an unknown milliamp signal. The following equations show these two simplifications.

$$P_1 = \frac{(A_1 - 4)}{16}(12) + 3 \quad \text{or} \quad A_1 = \frac{(P_1 - 3)}{12}(16) + 4$$

PROBLEM 11-3

Use the graph in Fig. 11-21 or the formula to determine the amount of air pressure that you would expect to find at the valve for the following electrical signals: 4 mA, 8 mA, 12 mA, 17.5 mA, and 20 mA.

You would need to know these values if you were troubleshooting the valve and amplifier, or if you were trying to complete a field calibration to ensure the valve moved its full range 0–100% as it received the electrical signal.

SOLUTION

The graph shows that the air pressure ranges from 3–15 psi and the electrical signal ranges from 4–20 mA. Some of the values that line up on the graph will allow you to use the graph to get accurate conversion. For example, the amount of air pressure for a 4 mA signal is 3 psi. The amount of air pressure for 8 mA is 6 psi. The amount of air pressure for 12 mA signal is 9 psi, and the amount of air pressure for the 20 mA signal is 15 psi.

The amount of air pressure for 17.5 mA will not show up accurately on the graph, so you can use the formula to calculate the amount of pressure.

$$P_1 = \frac{(A_1 - 4)}{16}(12) + 3 \qquad P_1 = \frac{(17.5 \text{ mA} - 4)}{16}(12) + 3 \qquad P_1 = 13.125 \text{ psi}$$

11.5 MOTOR-DRIVEN VALVES

Another type of control valve uses electric motors to open or close it. The electric motor is geared down so that its shaft moves slowly and it is limited to one revolution. The time for one revolution will be 10–20 seconds. This means that the valve will require 10–20 seconds to move from open to closed or from the closed position to the open position. Fig. 11-23 shows an example of this type of motor and a set of dampers that the motor can move.

A typical application for these types of motors is to set the position for the dampers in heating or air-conditioning systems. The motor is proportional, which means that it can position the dampers in any position between full open and full closed. The motor can be operated on any AC or DC voltage but 24 volts ac is a a common control voltage for air-conditioning and heating systems. The basic principle of operation for this type of motor is that its shaft will turn as long as power is applied to the motor, and the shaft will remain in position anytime power is disconnected. Limit switches are also used to detect the maximum travel in either direction. When the motor shaft rotates one complete revolution, it will move against the limit switch. This will open and stop the motor's travel in the clockwise direction and set the limit switch contacts so that the motor will be energized in the counterclockwise direction when power is applied again. When the motor shaft moves full travel in the counterclockwise direction, it will hit that limit switch. This will de-energize the motor in the counterclockwise direction and set it so it will start in the clockwise direction when power is applied again.

Potentiometers or other position sensors can be used to provide feedback to tell the exact position of the damper. In many cases, the feedback mechanism can be a temperature sensor. If the dampers are used to control the temperature in an air-conditioning system, the damper motor would open the dampers farther if the temperature is not cold enough. If the temperature is getting too cold, the dampers could be closed off more to limit the amount of cold air entering the conditioned space.

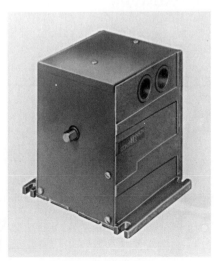

FIGURE 11-23 Electric motor and dampers. (Courtesy of Johnson Controls Inc.)

Dampers are used in many air supply systems for industrial processes. Any type of sensor can be used to provide feedback to the system, such as an air flow sensor. A pressure sensor could also be used to determine the back pressure in the system. This would indicate how far closed the dampers are and how much back pressure or pressure differential is built up across the dampers.

In some applications, limit switches are used as end switches. The limit switches have both NC and NO contacts that are positioned so that the NC set will be opened when the damper travels to its maximum. The NO set will close at this same point to ready the motor for reverse travel when the controller provides the reverse voltage.

This type of motor is also used to move the actuator or valve stem in the valves shown in Figs. 11-18 and 11-20. The valves shown in these figures use air to move the valve actuator. The electric motor can be used to move the valve actuator the same way that the pneumatic diaphragm does. The main difference is that the motor-driven valves are slow to respond, which means that the process they are controlling will have a lot of *dead time*. You should remember from Chapter 9 that dead time will cause the amount of integral action and derivative action in the controller to change. When you are tuning a control system that has dead time, you will need to increase the integral value (time) to make the system control correctly and respond smoothly.

In some applications such as filling a very large tank with liquid, the amount of dead time will not be a problem and the longer actuation time will be acceptable. The electric motor-driven valve is often used when a source of control air is not readily available.

11.5.1 Using a Proportional Amplifier with a Motor-Driven Valve

A proportional amplifier that provides a 4–20 mA signal or an amplifier that provides a 0–800 mA signal can also be used to control a motor-driven valve. When you are asked to troubleshoot a motor-driven valve, you must be careful to identify the type of valve and amplifier being used. If the proportional amplifier is used, it is important to determine its span. You may need to refer to the technical information that is provided with the valve. If the valve and amplifier are proportional, you will need to try a variety of voltages or currents through its span to ensure that it will operate correctly.

The proportional amplifier will have a range for the input signal, usually 4–20 mA or 0–10 volts, and it will also have an output range. Another way to test the drive motor is to provide the maximum output signal that the amplifier normally provides from a different source such as a bench-type power supply. If the motor operates from the test power supply, you can assume the motor is operational. If the motor does not operate from the test power supply, you should be sure to check to see if it has end switches incorporated that may cause an open circuit.

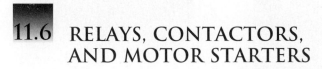

11.6 RELAYS, CONTACTORS, AND MOTOR STARTERS

Relays, contactors, and motor starters are widely used as output devices for a variety of control systems. In some cases the control system consists of switches, such as limit switches, proximity switches, and photoelectric switches, which are wired together to provide the control portion of the circuit. The relay, contactor, and motor starter are all similar to each other in that they each have sets of NO or NC contacts that are controlled by a coil. The relay is typically the smallest of these devices and the current rating for its contacts will be smaller than that of a contactor and motor starter. The basic rule of thumb is that a relay is designated as having contacts that are rated for less than 15 A. If the device has contacts that are rated for more than 15 A, it is called a contactor. This leaves room for disagreement, so one manufacture may call a device with a contact rating of 15 A a relay, while another may call the same device a contactor. The motor starter is similar to the contactor except it has heaters and overload devices to protect the motor that is connected to it against overcurrent.

11.6.1 Types of Relays

Many types of relays are used in industrial applications today. Fig. 11-24 shows a picture of several of the more common types of relays. These relays may be soldered on a printed circuit board or mounted in an electrical panel. Some of the relays use a socket-type mounting arrangement so that they can quickly be removed and replaced during troubleshooting. Some of the relays have a clear plastic case as a cover so that the technician can see whether the contacts are open or closed.

FIGURE 11-24 Types of industrial relays that include plug-in type relays, relays mounted in industrial panels and relays that are mounted in printed circuit boards. (Courtesy of Omron Electronics.)

11.6.2 Methods of Operation for Relays

All relays have the same basic parts including stationary contacts, movable contacts, armature, and magnet. All relays can also be broken into four basic groups that reflect their method of operation. (Fig. 1-2 in Chapter 1 shows these four basic types of mechanisms that use the magnetic field in the coil to move the contacts open or closed.) The four basic types of action are *horizontal action*, *bell-crank action*, *clapper action*, and *vertical action*. It is important to understand that all relays today use one of these four methods to open and close their contacts. The original designs have been slightly modified through the years, but the basic principles still remain.

11.6.3 Sets of NO and NC Contacts

Each relay will have some number of NO, NC, or a combination of both NO and NC contacts. These sets of contacts have become standards and they each have names to indicate the types and arrangement of contacts. Fig. 11-25 shows examples of four arrangements of contacts. The contacts in Fig. 11-25a show a relay with two sets of NO and two sets of NC contacts. Each of these sets of contacts is isolated from the others so that you can control four separate circuits if necessary. When

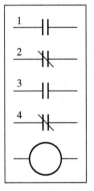

(a) Four-pole control relay

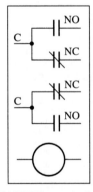

(b) DPDT control relay

FIGURE 11-25 **(a)** Electrical diagram of a relay with two sets of NO and two sets of NC contacts. **(b)** Electrical diagram of a relay with two sets of NO and NC contacts. Each set of NO and NC contacts is connected to a common terminal. This configuration is called double-pole, double-throw (DPDT). **(c)** Electrical diagram of a relay with five sets of NO contacts. **(d)** Electrical diagram of a relay with five sets of NC contacts.

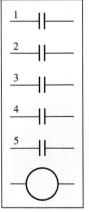

(c) Five-pole control relay, all NO contacts

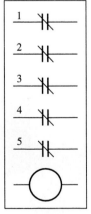

(d) Five-pole control relay, all NC contacts

the coil is energized, the two open sets move to the closed position, and the two closed sets move to the open position.

The relay in Fig. 11-25b shows a double-pole, double-throw (DPDT) set of contacts. The term *double pole* means that each set of contacts has a common terminal connected to both an open and a closed set of contacts. When the coil is energized, the circuit between common and NO is closed, and when the coil is de-energized the circuit between common and NC is closed.

The relay in Fig. 11-25c shows five sets of NO contacts. Each of these sets of contacts is isolated so that it can be used in five separate circuits. The relay in Fig. 11-25d shows five sets of NC contacts. Each of these sets of contacts is also isolated so that it can be used in five separate circuits.

11.6.4 Changing NO Contacts to NC Contacts in the Field

Most relays used in industrial circuits have the ability to have their contacts changed from NO to NC in the field. This means that a technician can remove several screws and mount the contacts upside down and they will change from NO to NC, or vice versa. Fig. 11-26 shows an example of this type of contact. This feature makes adding NO or NC contacts very simple. It also makes it simple to change a set of NO contacts to NC at the time it is needed. Since these contacts are actually cartridges, additional sets of contacts can be added to a relay at anytime as long

ADDING or CONVERTING CONTACT CARTRIDGES HAVING "SWINGAROUND" TERMINALS

General Instructions (Specific cases below.)

1.1 **Adding a contact cartridge:**

As received, accessory cartridges are in the normally open mode with terminal screws adjacent to N.O. symbols. If normally closed mode is desired, convert contact as indicated in Step 1.2 below. When cartridges are inserted, the terminal screws must face the front. The clear cover may face either side. **Do not install more than 8 N.C. contacts per relay.** When installing one cartridge, locate it at an inner pole position. When installing 2 cartridges, locate both in inner or outer (balanced) positions.

1.2 **Converting a contact to its alternate mode (N.O. ⇌ N.C.):**

Withdraw an assembled cartridge for replacement or conversion by inserting the blade of a suitably-sized screwdriver under a terminal screw pressure plate. Slide cartridge out. See Figure 2. Back the terminal screws out of the cylindrical nuts a sufficient amount (approximately 2 turns for a fully-tightened screw) to permit rotation of each screw and nut assembly to its alternate position. See Figure 3.

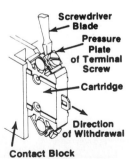

Screwdriver Blade

Pressure Plate of Terminal Screw

Cartridge

Direction of Withdrawal

Contact Block

FIGURE 2

FIGURE 11-26 Example of contacts that can be changed from NO to NC in the field. (Courtesy of Rockwell Automation's Allen Bradley Business.)

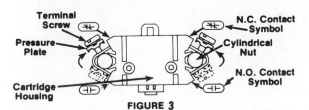

Terminal Screw

Pressure Plate

Cartridge Housing

N.C. Contact Symbol

Cylindrical Nut

N.O. Contact Symbol

FIGURE 3

as there is room. In some cases a *second deck* is added to the relay so more contacts can be added.

11.6.5 Pull-In and Hold-In Current

The coils for each relay can be connected to an electronic circuit board or they can be connected with switches to make a motor control circuit. The main point to remember is that the voltage used to energize the coil should match the coil rating. Just as in the coil of a solenoid, the amount of current that the coil uses when it is first energized and pulls in will be larger than the hold-in current. The *pull-in current* may also be called the *inrush current* and it is a rating that is normally listed in the relay specifications. The inrush current is generally three to five times larger than the *hold-in current*, which may be called the *seal-in* or *sealed current*, and it will also be listed in the specification for the coil. The diagram for pull-in and hold-in current for a relay coil is similar to the example of pull-in and hold-in current for solenoids shown previously in Fig. 11-3.

Some specifications list the coil current ratings in voltamperes (VA). VA is calculated by multiplying the voltage by the current, and you should remember that this rating is the apparent power for the circuit. The wattage for a relay coil will be slightly lower and it can only be measured with a wattmeter.

11.6.6 Types of Relays

A large variety of relays is available for different applications. Fig. 11-27 shows seven different types of industrial relays mounted on a rail. In most installations, the rail is mounted in an electrical panel, and each relay is mounted to the rail. The rail makes it easy to remove and replace the relay when it is faulty. In some cases the basic relay is the same, and a different add-on module is mounted to it. Fig. 11-27a shows a basic relay with an add-on solid-state timer module mounted on it. The add-on timer module allows the timer to convert a basic relay into a time-delay relay. The add-on timer module has one set of NO and one set of NC contacts that are controlled by the delay action.

Fig. 11-27b shows a similar relay with a pneumatic add-on timer. The main difference between the solid-state and pneumatic timer is the method that is used

FIGURE 11-27 Various types of relays available for industrial applications. **(a)** Add-on solid-state time-delay relay. **(b)** Add-on pneumatic time-delay relay. **(c)** AC latching relay. **(d)** AC standard relay. **(e)** AC standard relay. **(f)** Electrically held relay. **(g)** Magnetic latching relay with sealed contacts. (Courtesy of Rockwell Automation's Allen Bradley Business.)

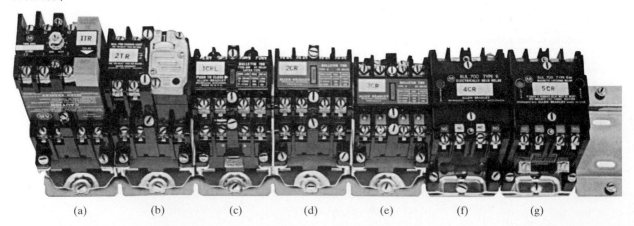

(a)   (b)   (c)   (d)   (e)   (f)   (g)

to create the time delay. Both relays have an adjustment for setting the amount of time delay.

Fig. 11-27c shows a latching relay, which is a special type of relay that has a latch coil and an unlatch coil. This type of relay has a mechanism that will cause the contacts to remain in the energized state even after the latch coil is de-energized. The unlatch coil must be energized to return the relay contacts to their de-energized state. This means that this type of relay can be operated with a pulse of AC voltage instead of continuous voltage.

Fig. 11-27d and Fig. 11-27e show two types of standard relays. These relays are available with four sets of contacts that are convertible so they can be any arrangement of NO and NC contacts.

Fig. 11-27f and Fig. 11-27g show two types of seal-in relays. The contacts in these relays are held in the energized condition like the latch relay. A second coil is required to return the contacts to their normal de-energized condition. Each of the seven sets of contacts can have its coil powered by a variety of AC and DC voltages. The coil can be easily changed to accommodate a different supply voltage, or if the coil is bad.

11.6.7 Contactors

Contactors are similar to relays except they are designed to carry larger currents. Fig. 11-28 shows examples of several sizes of available contactors for industrial applications. The smallest contactor in this figure is about 4 in. tall, and the largest is about 4 ft tall.

Fig. 11-29 shows a table for voltage, current, and horsepower ratings for contactors and motor starters. This table provides NEMA ratings for contactors and motor starters that range from size 00 through size 9. From this table you can see that the current rating for a size 00 is 9 A, and the rating for a size 9 contactor is 2250 A. Additional ratings for total horsepower are also listed in this table. When you are selecting a contactor or motor starter, you must ensure that the contact ratings exceed the load you are going to control with them. Each of these sizes of contactors and motor starters is available with a variety of coil voltages that range from 120–600 V.

· ·

PROBLEM 11-4

Use the table in Fig. 11-29 to determine the proper size of contactor to use for controlling a large electric heating element that pulls 75 A as a continuous load. The voltage for the heating element is 480 volts.

SOLUTION

Since the heating element draws 75 A, you will notice that the size 2 contactor can safely control 45 A and the size 3 contactor can handle 90 A. Since the 75 A load falls between these two values, the larger size 3 starter must be used. You should also notice that the starter size can be used for all voltages up to 600 volts.

· ·

11.6.8 Basic Parts of a Contactor

An example of a contactor with an exploded-view picture is shown Fig. 11-30. From this figure you can see that the contactor has six basic parts. The main section of

FIGURE 11-28 Examples of NEMA rated contactors. The size 00 is approximately 4 in. tall and the size 9 is approximately 4 ft tall. (Courtesy of Rockwell Automation's Allen Bradley Business.)

the contactor consists of mounting plate, base, stationary contacts, and arc hood. The mounting plate provides a means of mounting the contactor in an electrical panel. The base is made of plastic and provides mounting points for the stationary contacts and the remainder of the contactor's moving parts. The arc hood provides a cover for the contacts so that any electrical arc that may occur is contained.

The second section of the contactor includes the contact carrier, movable contacts, armature, and retainer spring. These parts provide the apparatus that allows the armature to move the movable set of contacts against the stationary contacts when the coil is energized. The armature is the part of the contactor that actually moves when the magnetic field in the coil is energized, and the retainer springs keep these parts in place.

The third section of the contactor is the head cover and arc quencher. These parts contain the arc that occurs when the contacts open or close so it cannot cause damage to the contacts or the electrical components mounted near the contactor.

The fourth part of the contactor is the coil. The coil is manufactured in a rectangular shape with two square holes in it. These holes provide a space for the

NEMA Size	Contin-uous Amp. Rating	600 VOLTS MAXIMUM				
		Maximum Horsepower Rating [2] (Full load current must not exceed the "Continuous Ampere Rating")			Maximum Horsepower Rating For Plugging Service [1]	
		Volts	Single Phase	3 or 2 Phase	Single Phase	Three Phase
00	9	120	⅓	¾	—	—
		208	—	1½	—	—
		240	1	1½	—	—
		480	—	2	—	—
		600	—	2	—	—
0	18	120	1	2	½	1
		208	—	3	—	1½
		240	2	3	1	1½
		480	—	5	—	2
		600	—	5	—	2
1	27	120	2	3	1	2
		208	—	7½	—	3
		240	3	7½	2	3
		480	—	10	—	5
		600	—	10	—	5
2	45	120	3	—	2	—
		208	—	15	—	10
		240	7½	15	5	10
		480	—	25	—	15
		600	—	25	—	15
3	90	120	—	—	—	—
		208	—	30	—	20
		240	—	30	—	20
		480	—	50	—	30
		600	—	50	—	30
4	135	120	—	—	—	—
		208	—	50	—	30
		240	—	50	—	30
		480	—	100	—	60
		600	—	100	—	60
5	270	120	—	—	—	—
		208	—	100	—	75
		240	—	100	—	75
		480	—	200	—	150
		600	—	200	—	150
6	540	208	—	200	—	150
		240	—	200	—	150
		480	—	400	—	300
		600	—	400	—	300
7	810	208	—	300	—	—
		240	—	300	—	—
		480	—	600	—	—
		600	—	600	—	—
8	1215	208	—	450	—	—
		240	—	450	—	—
		480	—	900	—	—
		600	—	900	—	—
9	2250	208	—	800	—	—
		240	—	800	—	—
		480	—	1600	—	—
		600	—	1600	—	—

[1] An example is plug-stop or jogging (inching duty) which requires continuous operation with more than five openings per minute.

[2] Non-Motor Loads — When contactors are required to switch non-motor loads such as lighting circuits, ovens, transformer primaries, etc., use the Bulletin 702L contactor.

FIGURE 11-29 National Electrical Manufacturers Association (NEMA) amperage and horsepower table for contactor and motor starter sizes. (Courtesy of Rockwell Automation's Allen Bradley Business.)

magnet yoke to protrude through so that the two feet of the yoke can make contact with the armature when the armature moves to the closed position. You should notice that the yoke is made of laminated steel so that it does not retain the residual magnetism when the coil is de-energized.

The fifth section of the contactor includes the magnet yoke and the yoke retainer. The yoke is mounted so that it protrudes through the middle of the coil. The retainer keeps the yoke in position, and it is removable so that the yoke can

FIGURE 11-30 Exploded view of a contactor showing all of the basic parts. A picture of a completely assembled contactor is also shown for comparison. (Courtesy of Rockwell Automation's Allen Bradley Business.)

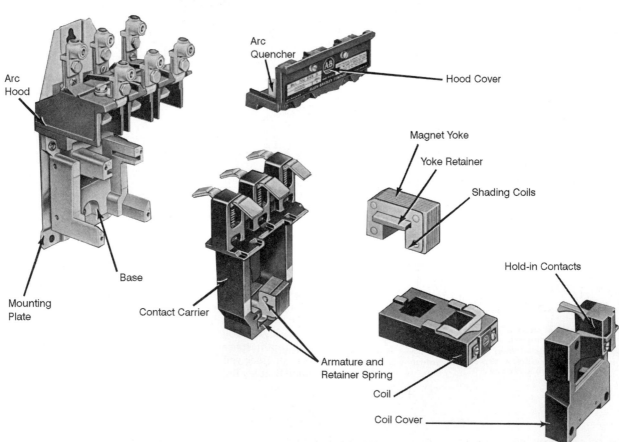

Arc Hood

Arc Quencher

Hood Cover

Magnet Yoke

Yoke Retainer

Shading Coils

Base

Hold-in Contacts

Mounting Plate

Contact Carrier

Armature and Retainer Spring

Coil

Coil Cover

be removed, which allows the coil to be removed. The yoke actually helps to keep the coil in its proper position.

The sixth part of the contactor is the coil cover. The coil cover also provides a place to mount any additional auxiliary contacts. One set of NO auxiliary contacts is usually provided to serve as *hold-in* or *seal-in contacts* for the contactor. The hold-in contacts are connected in parallel with the start push button so that when the start button is momentarily depressed and released, the hold-in contacts will close when the coil is energized to provide an alternate path around the start push button contacts. The current rating for the auxiliary contacts is usually very small because they only need to carry current that is used to energize the coil.

11.6.9 Ratings for Coils

The specification ratings for contactor coils provide information regarding inrush current, sealed current, and operating time. From Fig. 11-31 you can see that the inrush current for a size 00 contactor with a 208 volt coil is 0.29 A. The seal-in current for the same coil is 0.07 A. The inrush current for a size 2 contactor with a 120 volt coil is 1.8 A and the seal-in current for the same coil is 0.25 A.

The operating time is provided so that calculations regarding pick-up time and drop-out time can be made. Pick-up time and drop-out time are important in some applications where the contactors are controlled by microprocessor systems or programmable logic controllers where the sequence of operation is crucial. All of the times in this table are listed in milliseconds.

· ·

PROBLEM 11-5

Determine the amount of inrush current and sealed (holding) current for the coil of the number 3 starter that you selected in the previous problem for use in

FIGURE 11-31 Inrush current, sealed currents and operating times for coils of NEMA rated contactors. (Courtesy of Rockwell Automation's Allen Bradley Business.)

COIL CURRENTS

NEMA Size	No. of Poles	Inrush Current (Amps.) 60 Cycles					Sealed Current (Amps.) 60 Cycles				
		120V	208V	240V	480V	600V	120V	208V	240V	480V	600V
00	1-2-3	0.50	0.29	0.25	0.12	0.07	0.12	0.07	0.06	0.03	0.02
0	1-2-3-4	0.88	0.50	0.44	0.22	0.17	0.14	0.08	0.07	0.04	0.03
1	1-2-3-4	1.54	0.89	0.77	0.39	0.31	0.18	0.10	0.09	0.04	0.04
2	2-3-4	1.80	1.04	0.90	0.45	0.36	0.25	0.14	0.13	0.06	0.05
3	2-3	4.82	2.78	2.41	1.21	0.97	0.36	0.21	0.18	0.09	0.07
3	4	5.34	3.08	2.67	1.33	1.07	0.39	0.23	0.20	0.10	0.08
4	2-3	8.30	4.80	4.15	2.08	1.66	0.54	0.31	0.27	0.14	0.11
4	4	9.90	5.71	4.95	2.47	1.98	0.61	0.35	0.31	0.15	0.12
5	2-3	16.23	9.36	8.11	4.06	3.25	0.81	0.47	0.41	0.20	0.16
6	2-3	0.62	Current Shown is The Very Small AC Current Passing Through Coil of Control Relay. Any Standard Duty 120V Control Station May Be Used In This Circuit.				0.082	Current Shown is The Very Small AC Current Passing Through Coil of Control Relay. Any Standard Duty 120V Control Station May Be Used In This Circuit.			
7	2-3	0.62					0.082				
8	2-3	0.62					0.082				
9	2-3	1.2					0.16				

OPERATING TIME

Size	Approximate Operating Time in Milliseconds 3 Pole Contactors	
	Pick-up	Drop-out
00	28	13
0	29	14
1	26	17
2	32	14
3	35	18
4	41	18
5	43	18
6	88	40
7	88	45
8	118	94
9	118	84

controlling the 75 A electric heating coil. You should remember that the heating element uses 480 volts, three-phase AC voltage, and the coil for the contactor is 208 volts. Use the table shown in Fig. 11-31 to determine the inrush and sealed currents. Use the second table shown in this figure to determine the operating time for pick-up and drop-out times for the size 3 contactor.

SOLUTION

Since the contactor in this application is used to control three-phase voltage, the coil has to move three contacts when it opens and closes. The table in Fig. 11-31 shows that the size 3 starter with three poles and 208 volts coil uses 2.78 A to pull in the contacts. The sealed current for this contactor is 0.21 A. The time it takes to pick up the contacts is 35 msec, and the time for drop-out is 18 msec.

11.6.10 Motor Starters

A motor starter is basically a contactor with an overload block added to it. The overload block consists of a heater and overload contacts. The overload block provides protection against overcurrent for the motor that is connected to the motor starter. This means that any of the contactors shown in previous figures could be used as motor starters by adding an overload block to them. Electrical parts dealers may stock the contactor as an individual unit for use in circuits where the load is noninductive. The overload block is stocked as a separate part and it can be added to the contactor in the field to allow the contactor to become a motor starter. The parts dealers also stock motor starters that have the overload block added during the manufacturing process so that the motor starter can be used without modification. Fig. 11-32 shows a typical motor starter, Fig. 11-33 shows a cut-away diagram of a motor starter, and Fig. 11-34 shows an electrical diagram of a motor starter.

Figs. 11-32 and 11-33 will help you identify all of the main parts of the motor starter. You should notice that all of the parts except the overload block look exactly like the parts of a contactor because the motor starter is actually a contactor with the overload block added at the factory or in the field. You should also notice the parts in the cut-away diagram as they appear in the electrical diagram. The terminals for incoming power are identified as L1, L2, and L3 in the electrical diagram, and they are identified at the top of the motor starter in the picture as *line-side power terminals*. The terminals where the motor is connected are identified as terminals T1, T2, and T3 in the electrical diagram, and they are identified as *load-side terminals* at the bottom of the motor starter. The location of the auxiliary contacts is shown in the picture on the left side of the motor starter just to the left of the terminals for the coil. The terminals for the overload contacts are shown at the bottom left side of the motor starter.

The electrical diagram shows the three-phase voltage is connected to the motor starter at the terminals marked L1, L2, and L3. These terminals are connected in series with the main contacts. The overload block consists of two basic parts: the *heaters* and the *overload contacts*. The heater assembly is connected in series with each set of main contacts. The motor is connected to the motor starter at terminals T1, T2, and T3. The coil is connected in series with the overload contacts. If the motor draws excessive current, it will pass through the heater assembly and build up heat. When the heat reaches a critical point, it will cause the overload contacts to open and interrupt current flow to the coil, which will cause the main

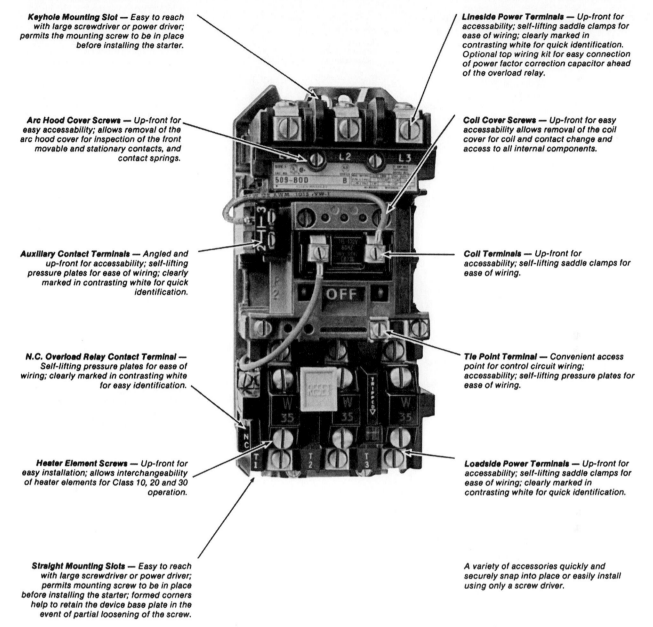

Keyhole Mounting Slot — *Easy to reach with large screwdriver or power driver; permits the mounting screw to be in place before installing the starter.*

Arc Hood Cover Screws — *Up-front for easy accessibility; allows removal of the arc hood cover for inspection of the front movable and stationary contacts, and contact springs.*

Auxiliary Contact Terminals — *Angled and up-front for accessibility; self-lifting pressure plates for ease of wiring; clearly marked in contrasting white for quick identification.*

N.C. Overload Relay Contact Terminal — *Self-lifting pressure plates for ease of wiring; clearly marked in contrasting white for easy identification.*

Heater Element Screws — *Up-front for easy installation; allows interchangeability of heater elements for Class 10, 20 and 30 operation.*

Straight Mounting Slots — *Easy to reach with large screwdriver or power driver; permits mounting screw to be in place before installing the starter; formed corners help to retain the device base plate in the event of partial loosening of the screw.*

Lineside Power Terminals — *Up-front for accessibility; self-lifting saddle clamps for ease of wiring; clearly marked in contrasting white for quick identification. Optional top wiring kit for easy connection of power factor correction capacitor ahead of the overload relay.*

Coil Cover Screws — *Up-front for easy accessibility allows removal of the coil cover for coil and contact change and access to all internal components.*

Coil Terminals — *Up-front for accessibility; self-lifting saddle clamps for ease of wiring.*

Tie Point Terminal — *Convenient access point for control circuit wiring; accessability; self-lifting pressure plates for ease of wiring.*

Loadside Power Terminals — *Up-front for accessibility; self-lifting saddle clamps for ease of wiring; clearly marked in contrasting white for quick identification.*

A variety of accessories quickly and securely snap into place or easily install using only a screw driver.

FIGURE 11-32 A typical motor starter. The motor starters are rated in size from 00–9 by NEMA. (Courtesy of Rockwell Automation's Allen Bradley Business.)

contacts to open and stop current flow to the motor. This provides protection for the motor against overcurrent.

Fig. 11-34 shows an electrical diagram of the motor starter coil connected to a stop-start circuit. This circuit shows how the auxiliary contacts are connected in parallel with the start push button so that it will seal in the circuit when the push button is depressed momentarily. In Fig. 11-35 the auxiliary contacts appear to be connected in series with the coil, but the auxiliary contacts only use terminal 3 to

CUT-AWAY STARTER SAMPLE

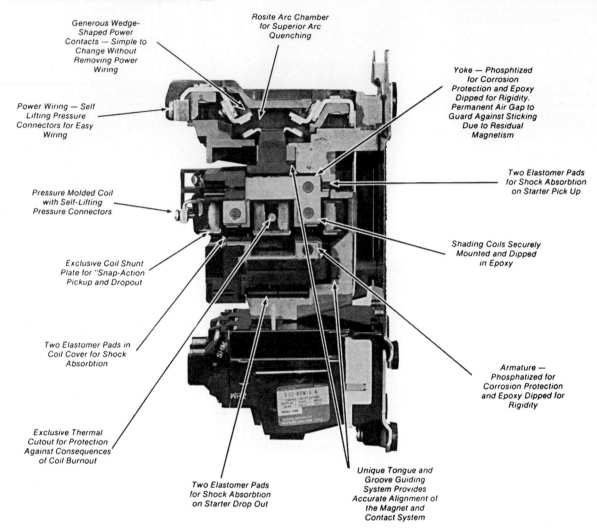

Generous Wedge-Shaped Power Contacts — Simple to Change Without Removing Power Wiring

Rosite Arc Chamber for Superior Arc Quenching

Yoke — Phosphtized for Corrosion Protection and Epoxy Dipped for Rigidity. Permanent Air Gap to Guard Against Sticking Due to Residual Magnetism

Power Wiring — Self Lifting Pressure Connectors for Easy Wiring

Two Elastomer Pads for Shock Absorbtion on Starter Pick Up

Pressure Molded Coil with Self-Lifting Pressure Connectors

Shading Coils Securely Mounted and Dipped in Epoxy

Exclusive Coil Shunt Plate for "Snap-Action Pickup and Dropout

Two Elastomer Pads in Coil Cover for Shock Absorbtion

Armature — Phosphatized for Corrosion Protection and Epoxy Dipped for Rigidity

Exclusive Thermal Cutout for Protection Against Consequences of Coil Burnout

Two Elastomer Pads for Shock Absorbtion on Starter Drop Out

Unique Tongue and Groove Guiding System Provides Accurate Alignment of the Magnet and Contact System

FIGURE 11-33 Cut-away picture of a motor starter. (Courtesy of Rockwell Automation's Allen Bradley Business.)

make the parallel connection to the coil. This can be seen more easily in the diagram of the start–stop circuit.

11.6.11 Operation of the Overloads

Modern motor starters use one of three basic methods of operation for overload block. Fig. 11-36 shows two typical overload blocks and a Type W and Type J heating element. The heating element is wired in series with the motor so that all of the current that the motor draws goes through the heating element and then is transferred to the overload assembly. Fig. 11-37 shows diagrams of the three basic types of heater and overload assemblies. The operation of the different types of overload assemblies will be explained so you can see why several different types are used.

The first diagram in Fig. 11-37 shows the operation of a *eutectic alloy overload*.

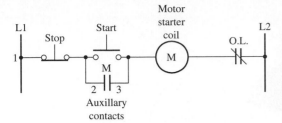

FIGURE 11-34 Stop push button and start push button connected to motor starter coil. Notice that the overload contacts are connected to series with the coil, and the auxiliary contacts are connected in parallel with the start push button.

This type of overload assembly consists of a ratchet, pawl, pivot, and heater. The pawl acts like a trigger mechanism that is spring loaded against the ratchet. The spring provides pressure that tries to turn the pawl. If the pawl is allowed to turn, it will strike the contact actuator called the *pivot*. The pivot will cause the NC overload contacts to open. The shaft of the ratchet is held in place by the eutectic alloy, which is similar to solder and used to make connections on a terminal board.

When the alloy is cold, it will hold the shaft of the ratchet in place, and when excessive current flows through the heater, it will provide excess heat to the alloy. Since the alloy has a low melting point, it will melt and allow the ratchet to move freely. When the ratchet moves, the pawl is also allowed to move, and its movement will cause the pivot to move and open the overload contacts. When the overload contacts open, current to the coil is interrupted, and the main contacts to the motor will open, which stops current flow to the motor. When this occurs, the current flow to the motor stops providing heat to the heater, which allows the alloy to cool down. After several seconds, the alloy is cool enough to allow the overload assembly to be reset.

The overload assembly is reset by depressing the reset button on the front of the motor starter. The reset button moves the ratchet and pawl back to the cocked condition so that it is ready for the next overcurrent condition. When the ratchet and pawl return to the cocked (reset) condition, the overload contacts return to the closed position. When the start push button is depressed, current can again flow to the motor starter coil, which will cause the main contacts to close and provide current to the motor again. If the condition that caused the motor to draw excessive current is not fixed, the motor will again draw excessive current and the

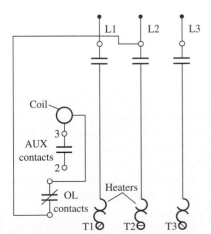

FIGURE 11-35 Electrical diagram of a motor starter. The coil and overload contacts are also shown. Incoming power is connected to terminals L1, L2, and L3, and the motor is connected to terminals T1, T2, and T3.

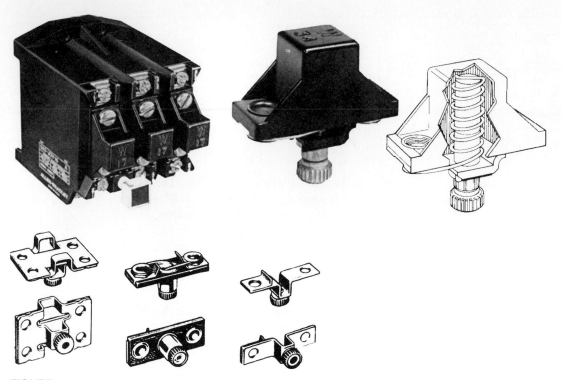

FIGURE 11-36 Example of a Type W and Type J heating element. (Courtesy of Rockwell Automation's Allen Bradley Business.)

overload process will recur. It is important to remember that the motor current flows through the heater element, and as long as the motor current is within specifications, the amount of heat from the heating element will not be sufficient to trip the overload mechanism. When the motor draws excessive current, the additional heat from the heater element will cause the overload to trip.

The second type of overload relay is called a *bimetal overload*. The bimetal overload assembly uses two dissimilar metals that are bonded together to provide the movement to open the overload contacts. The bimetal strip receives excess heat from the heater when the motor draws excessive current. When this occurs, the bimetal strip begins to heat up. The heat in the bimetal strip will cause the two dissimilar metals to expand at different rates. This action causes the metal to move at one end and trip the *trip bar latch mechanism*. When the latch mechanism is tripped, it will open the overload contacts and interrupt current flow to the motor starter coil. When the overloads are allowed to cool down, the reset button can be depressed and reset the latch mechanism.

The third type of overload mechanism is called the *phase-loss sensitivity overload relay*. This type of overload assembly is similar to the bimetal overload except it uses the difference of current in any of the phases to trip the overload. The difference in current can be a loss of phase or an overload in any or all of the phases. This type of overload also has two trip bars. The bottom trip bar, called the moving trip bar, will be actuated when the unbalanced overload condition occurs. When the moving trip bar moves to the right, it will open the overload contacts, which interrupts current to the motor starter coil.

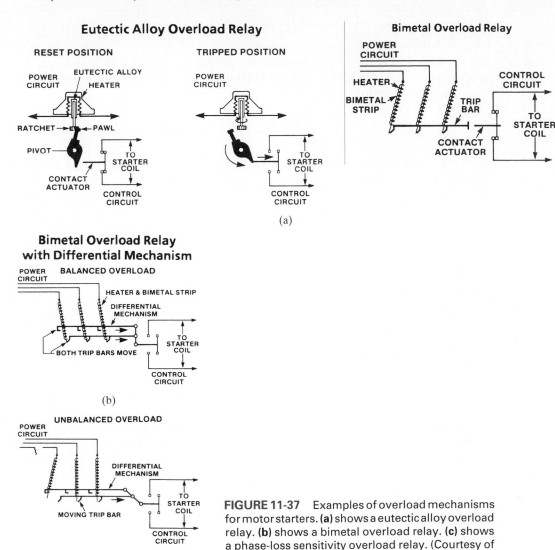

FIGURE 11-37 Examples of overload mechanisms for motor starters. **(a)** shows a eutectic alloy overload relay. **(b)** shows a bimetal overload relay. **(c)** shows a phase-loss sensitivity overload relay. (Courtesy of Rockwell Automation's Allen Bradley Business.)

11.6.12 Sizing Motor Starters

Motor starters use the same table as contactors for determining the correct size for controlling a motor. The table in Fig. 11-29 includes ratings for motor starters and the loads are listed as horsepower ratings. The table also provides maximum ratings for each motor starter when it is used in plugging applications. *Plugging* means that two legs of the voltage supplied to a three-phase motor are switched to cause the motor's shaft to rotate in the opposite direction. The plugging (reverse) voltage is only applied for a short period of time, usually less than 1 second, so that the motor shaft stops quickly. The voltage is disconnected before the motor's shaft can actually start to turn in the opposite direction.

Plugging causes the motor to draw very large currents for a short period of time so the size of the motor starter must be downsized to account for the additional heat that tends to build up when the larger currents are pulled.

· ·

PROBLEM 11-6

Use the table in Fig. 11-29 to determine the proper size motor starter to select for an application that uses a 20hp motor powered by 480 volts three phase. This motor is not used for plugging.

SOLUTION

Since the motor is a three-phase motor, and it uses 480 volts, you will need to move through the table to locate these two conditions. The table lists 15hp and 25hp motors. Since the 20hp motor is not listed, you should select the next higher motor size that is listed. In this application the size 2 motor starter would be selected since it can be used for motors up to 25hp at 480 volts three phase.

· ·

11.6.13 Sizing Heater Elements for Motor Starters

Another important step in installing or replacing a motor starter is to select the correct size of heating element so that it matches the current that the motor draws. The correct size of heating element will ensure that the motor starter senses the correct amount of overload current for a specific motor. From the previous tables you have learned that 11 different sizes of motor starters are available (size 00–size 9). The motor starter size refers to the size of its contacts, which determines the amount of full-load motor current the contacts can safely switch on and off. For example, the size 1 motor starter is used to energize three-phase motors from 3hp through 10hp. The 3hp motor typically draws 4.0 A and the 10hp motor draws 12.4 A. The motor starter's contacts are large enough to handle up to 27 A. When a motor starter is manufactured, it is not known what size motor will be connected to it, so it may be shipped without heating elements. If the size 1 motor starter is to be used for an application that has a 3hp motor, the heating elements will need to be selected to set the overcurrent limit at 4 A to ensure that the overloads will trip at 4 A and the 3hp motor will not receive overcurrent. If the application used a 10hp motor, the heating element would be selected for 12.4 A so that the motor starter would trip its overload at 12.4 A. You can see that the selection of the heating element allows the motor starter to be customized to protect the motor that is connected to it for the exact amount of overcurrent the motor draws.

As a technician you will be responsible for looking up the full-load amperage (FLA) rating for the motor you are working with and selecting the correct heating element size to protect the motor. Fig. 11-38 shows one table for selecting heating elements for size 0 through size 4 starters. Other tables provide selection data for all of the motors and motor starters that you will encounter. If you are selecting the heating element for the 3hp power motor that is connected to a size 1 motor starter, you need to look at the motor name plate to determine the exact amount of FLA for the motor. For this example, the motor name plate indicates the motor draws 4 A. From this table you can see that size W43 is the closest heater size since it is rated for 4.11 A when used in a size 1 motor starter. If you were selecting the heating element for the 10hp motor whose FLA is 12.4 A, you would select a size W55 heating element since its rating is 12.8 A. The heating element size is very seldom the exact size for the FLA of the motor, so you will need to select the next larger size heater to protect the motor.

TABLE 144

Heater Type Number	Full Load Amperes				
	Size 0	Size 1	Size 2	Size 3	Size 4
W10	0.18	0.18	. . . .	. . . .	. . . .
W11	0.20	0.20	. . . .	. . . .	. . . .
W12	0.22	0.22	. . . .	. . . .	. . . .
W13	0.24	0.24	. . . .	. . . .	. . . .
W14	0.26	0.26	. . . .	. . . .	. . . .
W15	0.29	0.29	. . . .	. . . .	. . . .
W16	0.32	0.32	. . . .	. . . .	. . . .
W17	0.35	0.35	. . . .	. . . .	. . . .
W18	0.38	0.38	. . . .	. . . .	. . . .
W19	0.42	0.42	. . . .	. . . .	. . . .
W20	0.46	0.46	. . . .	. . . .	. . . .
W21	0.51	0.51	. . . .	. . . .	. . . .
W22	0.56	0.56	. . . .	. . . .	. . . .
W23	0.62	0.62	. . . .	. . . .	. . . .
W24	0.68	0.68	. . . .	. . . .	. . . .
W25	0.75	0.75			. . . .
W26	0.82	0.82			. . . .
W27	0.90	0.90			. . . .
W28	0.99	0.99			. . . .
W29	1.09	1.09			. . . .
W30	1.20	1.20	. . . .	. . . .	. . . .
W31	1.32	1.32	. . . .	. . . .	. . . .
W32	1.45	1.45	. . . .	. . . .	. . . .
W33	1.59	1.59	. . . .	. . . .	. . . .
W34	1.75	1.75	. . . .	. . . .	. . . .
W35	1.93	1.93	. . . .	. . . .	. . . .
W36	2.12	2.12	. . . .	. . . .	. . . .
W37	2.33	2.33	. . . .	. . . .	. . . .
W38	2.56	2.56	. . . .	. . . .	. . . .
W39	2.81	2.81	. . . .	. . . .	. . . .
W40	3.09	3.09	. . . .	. . . .	. . . .
W41	3.40	3.40	. . . .	. . . .	. . . .
W42	3.74	3.74	. . . .	. . . .	. . . .
W43	4.11	4.11	. . . .	. . . .	. . . .
W44	4.52	4.52	. . . .	. . . .	. . . .
W45	4.97	4.97	. . . .	. . . .	. . . .
W46	5.46	5.46	5.60	. . . .	. . . .
W47	6.01	6.01	6.15	. . . .	. . . .
W48	6.60	6.60	6.76	. . . .	. . . .
W49	7.26	7.26	7.43	. . . .	. . . .
W50	7.98	7.98	8.17	. . . .	. . . .
W51	8.78	8.78	8.98	. . . .	. . . .
W52	9.65	9.65	9.87	. . . .	. . . .
W53	10.6	10.6	10.8	. . . .	. . . .
W54	11.7	11.7	11.9	. . . .	. . . .
W55	12.8	12.8	13.1	. . . .	. . . .
W56	14.1	14.1	14.4	. . . .	. . . .
W57	15.4	15.4	15.7	. . . .	. . . .
W58	16.8	16.8	17.1	. . . .	. . . .
W59	18.3	18.3	18.6	. . . .	. . . .
W60	. . . .	19.8	20.1	. . . .	. . . .
W61	. . . .	21.3	21.7	25.5	. . . .
W62	. . . .	22.7	23.1	28.1	. . . .
W63	. . . .	24.4	24.8	31.0	32.0
W64	. . . .	26.2	28.6	34.0	35.0
W65	. . . .	28.2	30.5	37.0	38.5
W66	. . . .	. . . .	33.0	40.0	42.5
W67	. . . .	. . . .	35.5	43.5	46.5
W68	. . . .	. . . .	38.0	47.0	51
W69	. . . .	. . . .	40.5	51	55
W70	. . . .	. . . .	43.5	55	59
W71	. . . .	. . . .	47.0	59	64
W72	. . . .	. . . .	. . . .	63	69
W73	. . . .	. . . .	. . . .	67	74
W74	. . . .	. . . .	. . . .	71	79
W75	. . . .	. . . .	. . . .	76	84
W76	. . . .	. . . .	. . . .	80	90
W77	. . . .	. . . .	. . . .	85	96
W78	. . . .	. . . .	. . . .	90	102
W79	. . . .	. . . .	. . . .	. . . .	107
W80	. . . .	. . . .	. . . .	. . . .	113
W81	. . . .	. . . .	. . . .	. . . .	118
W82	. . . .	. . . .	. . . .	. . . .	124
W83	. . . .	. . . .	. . . .	. . . .	130
W84	. . . .	. . . .	. . . .	. . . .	135
W85	. . . .	. . . .	. . . .	. . . .	. . . .

FIGURE 11-38 Table for heating element selection for Allen-Bradley Type W heaters. (Courtesy of Rockwell Automation's Allen Bradley Business.)

11.6.13.1 Classifications of Overload and Heater Elements for Motor Starters ■
Overloads and heater elements for motor starters are classified by the amount of time the overcurrent condition can occur. A *Class 10* overload is rated to trip after experiencing an overcurrent condition for 10 seconds. A *Class 20* overload is rated to trip when the overcurrent exists for 20 seconds and the *Class 30* is rated to trip when the overcurrent exists for 30 seconds. The amount of time the overcurrent is allowed to occur will change with the type of application for which the motor is used and the type of motor that is being used. In some cases the load will seldom cause an overcurrent and it will be of short duration, so the Class 10 is recommended. In other applications such as extruders for plastic presses and in hydraulic pump applications, it is not uncommon for an overload condition to occur for 10–20 seconds and then the motor current will return to normal. Since this overcurrent exists for a short period of time, it will not damage the motor, so the Class 2 or Class 3 overload will be used to provide the motor time to get back to normal overload current levels without tripping the overloads. If a serious problem occurs and the overload exists for more than the rated time for the overload, the heating element will send enough heat to the overload assembly to cause it to trip.

••

PROBLEM 11-7

You are connecting a 7.5hp motor to a new size 1 motor starter and you must select the proper size heater element. The FLA rating for this motor on its data plate is 8.5 A. Use Fig. 11-38 to help you determine the proper selection.

SOLUTION

In Fig. 11-38 you should use the column for the size 1 motor starter and move down the table until you reach the rating for 8.5 A or the next highest rating. Since 8.5 A is not listed, the next larger rating is for 8.78 A. The heater element with this rating is a size W51.
••

11.6.14 Solid-State Motor Starters

Solid-state motor starters combine solid-state circuitry to provide the function of a traditional motor starter. Fig. 11-39 shows several solid-state motor starters. From this figure you can see that the motor starter has six potentiometers mounted on the face of the starter. These potentiometers provide a means to set the ramp-up and ramp-down functions for the motor. This feature is called *soft starting.* The traditional motor starter is called an *across-the-line starter,* which means that the main contacts close and provide full voltage to the starter the instant the coil is energized. The solid-state starter provides voltage that increases slowly (ramps) to allow the motor to start turning its shaft slowly and increasing to full rpm. The solid-state motor starter also provides adjustable overloads. Additional potentiometers on the face of the starter allow you to adjust the percentage of overload from 0–400% and the amount of time the overload current is allowed to occur.

The solid-state motor starter uses triacs and transistors for power control in AC motor starters, and SCRs and transistors for DC motor starters. These controls have provided the function of a traditional motor starter, yet provide additional protection for the motor. This is possible by using solid-state devices on the inexpensive models, and by using microprocessors on the more expensive models.

FIGURE 11-39 Solid-state motor starters that provides soft-starting functions for motors. (Courtesy of Rockwell Automation's Allen Bradley Business.)

11.7 VARIABLE-FREQUENCY DRIVES

Variable-frequency drives for AC motors have been the innovation that has brought the use of AC motors back into prominence. The AC induction motor can have its speed changed by changing the frequency of the voltage used to power it. This means that if the voltage supplied to an AC motor is 60 Hz, the motor will run at its rated speed. If the frequency is increased above 60 Hz, the motor will run faster than its rated speed, and if the frequency of the supply voltage is less than 60 Hz, the motor will run less than its rated speed. The variable-frequency drive is the electronic controller specifically designed to change the frequency of voltage supplied to the motor.

In the 1960s, frequency drives had rather small solid-state components that limited the amount of current the drive could supply to the motor. This tended to limit the size of motor that could be controlled by a frequency drive and they were not commonly used. When larger transistors became available in the 1980s, the variable-frequency drives allowed the largest motors to have their speed controlled. These earliest drives utilized linear amplifiers to control all aspects of the drive. Jumpers and dip switches were used to provide ramp-up (acceleration) and ramp-down (deceleration) features by switching larger or smaller resistors into circuits with capacitors to create different slopes.

The advent of the microprocessor has allowed the variable-frequency drive to become a very versatile controller that not only controls the speed of the motor, but protects the motor against overcurrent during ramp up and ramp down conditions. Newer drives also provide methods of braking, power boost during ramp up, and a variety of control during ramp down. The biggest savings that the variable frequency drive provides is that it can ensure that the motor does not pull excessive

FIGURE 11-40 Picture of a variable-frequency drive. (Courtesy of Rockwell Automation's Allen Bradley Business.)

current when it starts, so the overall demand factor for the entire factory can be controlled to keep the utility bill as low as possible. This feature alone can provide payback in excess of the price of a drive in less than one year. It is important to remember that when motors are started with a traditional motor starter they will draw locked rotor amperage (LRA) while they are starting. When the locked rotor amperage occurs across many motors in a factory, it pushes the electrical demand factor too high which results in the factory paying a penalty for all of the electricity consumed during the billing period. Since the penalty can be as much as 15% to 25%, the savings on a $25,000 electric bill can be used to purchase drives for virtually every motor in the factory even though the application does not require variable speed.

Today the variable frequency drive is perhaps the most common type of output or load for a control system. As applications become more complex the frequency drive has the ability to control the speed of a motor, the direction the motor shaft is moving, the torque the motor provides to the load and any other motor parameter that can be sensed. Newer drives have a variety of parameters that can be controlled by numbers that are programmed into it or downloaded from another microprocessor controlled system such as a programmable controller. These drives are also available in smaller sizes that are cost-efficient and take up less space. Fig. 11-40 shows a picture of a variable-frequency drive.

11.7.1 Block Diagram of a Variable-Frequency Drive

The block diagram of the variable-frequency drive can be divided into three major sections: the power conversion section, the microprocessor control section (CPU), and the control section that includes the external switches and signals to control

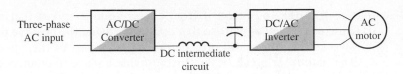

the drive operations, and the power section where AC voltage is converted to DC and then DC voltage is inverted back to three-phase AC voltage.

Fig. 11-41 shows a block diagram of the power section of the variable-frequency drive as three separate sections to indicate the three main functions: the rectifier, the filter, and the switching section that uses regular transistors, darlington pair transistors, or insulated gate bipolar transistors (IGBT) to *invert* the DC voltage back to AC voltage with the proper frequency. The individual electronic components in these sections have been explained in detail in Chapter 5. We will review the major components at this point and explain how they work together to provide the major drive functions.

11.6.2 A Typical Drive Installation

Figure 11-44 shows a typical variable-frequency drive installation. This diagram shows the wires that supply power to the drive, the wires that provide voltage from the drive to the motor, and all of the necessary input and output signals that the drive needs for operation. From the diagram you can see that the power source for the variable-frequency drive is provided at terminals R, S, and T by three-phase AC voltage. The value of this voltage can be 208, 240, or 480 volts. The three phase voltage is converted to DC voltage in the rectifier section of the drive where six diodes are connected as a 3-phase full wave bridge rectifier. On larger drives the diodes can be replaced with silicon controlled rectifiers (SCRs).

The next components in this circuit are the choke (L1) and the capacitors that make up the filter section for the drive. The capacitors and choke provide a filter that remove all of the ripple and any trace of the original frequency. The voltage at this point in the drive is pure DC voltage and it will be approximately 670 volts.

The output section of this drive contains three pairs of insulated gate bipolar transistors (IGBTs). These transistors are turned on by a pulse width modulation (PWM) control circuit that times the conduction of each IGBT so that a PWM wave is produced that looks like a sine wave output. The transistors are turned on and off approximately 12 times for each half cycle. Each time the transistor is turned on, its amplitude will be adjusted so that the overall shape of the waveform looks like a sine wave. The time each transistor is turned on is adjusted as the frequency for the output signal is adjusted. The overall frequency for the drive output signal to the motor will be determined by the frequency of the PWM sine wave. The frequency can be adjusted from 0–400 Hz on some drives and typically it can be adjusted from 0–120 Hz. The amplitude of the signal will change to change the voltage of the signal. The voltage and current for the output signal will be adjusted to provide the correct amount of torque to the motor load. The drive will maintain a volts-per-hertz ratio (V/Hz ratio) to ensure that the motor has sufficient power to provide torque to respond to changes in the load. The V/Hz ratio can be adjusted slightly to provide more voltage at lower frequencies if the motor is used in these applications where larger loads must be moved accurately at lower speeds. In the block diagram you should also see that a diode is connected in reverse bias across each IGBT to protect it from excess voltage spikes that may occur. The IGBTs are controlled as pairs so that one will provide the positive part of the PWM sine wave and the other will provide the negative part of the wave.

The output terminals of the drive provide a place to connect the three motor leads. These terminals are identified as U, V, and W. The markings R, S, and T for the input voltage and U, V, W for the output terminals are worldwide standards. Some drives manufactured in the United States prior to 1990 may still be identified as L1, L2, and L3 for input terminals and T1, T2, and T3 for output terminals where the motor is connected.

11.7.3 External Control Switches and Contacts for the Drive

In the previous application the drive is enabled and controlled by external switches or contacts. The National Electric Code and local codes will specify the exact number of external controls required for each drive. The diagram in Fig. 11-42 shows typical control switches. You can see at the upper left side of the drive that an NO start push button is connected to terminals TB2-7 and TB2-6, and an NC stop push button is connected between terminals TB2-7 and TB2-8. The voltage for these circuits is provided internally from TB2-7. Power must be received continually through the stop push button, and when the start push button is depressed, a pulse signal will be received and the drive will begin the start sequence. The start sequence will provide a ramp that is used to start the motor by slowly increasing the frequency and voltage. Anytime the stop button is opened, the drive will be stopped and if a ramp-down sequence is programmed, the drive will come to a stop gradually. If a ramp-down sequence is not programmed, the motor can be stopped with braking, or it can be allowed to coast to a stop.

The second set of switches provides a set of terminals where an external set of contacts from a control relay can be connected to provide a *drive-enable function*. When the contacts are connected to terminals TB2-11 and TB2-12, power will flow from terminal TB2-12 through the closed contacts to terminal TB2-11 to enable the drive. When the drive is in the enable condition, power is allowed to flow through the drive to the motor. Anytime these contacts are opened, the drive will become disabled, and the drive will not send voltage to the motor.

The drive can be switched from forward to reverse with a remote switch. The external switch is connected between terminals TB2-12 and TB2-13 to provide the reverse signal to the drive. Anytime the switch is closed and this circuit is made, the drive will reverse the phase sequence to the motor, which provides the same effect as physically swapping leads U and W. You should remember that whenever any two leads of a three-phase power supply to a motor are switched, the phase relationship between these two leads is changed and the motor will run in the opposite direction. This is an important feature of variable-frequency drives in that they can reverse the direction of a motor without using expensive reversing motor starters.

A jog push button can be connected across terminals TB2-14 and TB2-12 to provide the *jog function*. When this switch is closed, the drive will check the jog parameter and produce the programmed frequency for this parameter. For example, if you want the motor to turn at 20% rpm at jog speed, you would enter 20% into the program as the jog parameter and the drive will produce a frequency that causes the motor to rotate 20% anytime the jog push button is depressed.

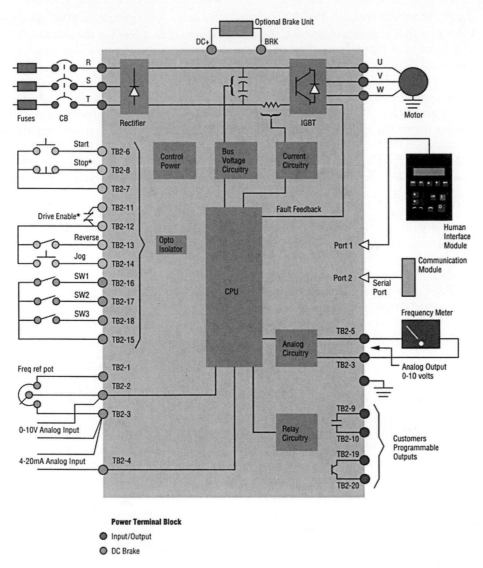

FIGURE 11-42 Block diagram of variable-frequency drive that also shows the components that are connected to the drive to provide additional control. (Courtesy of Rockwell Automation's Allen Bradley Business.)

The external switches identified as SW1, SW2, and SW3 are connected between terminal TB2-15 and terminals TB2-16, TB2-17, and TB2-18. If switch 1 is closed, the drive will provide the frequency to operate the drive at the frequency that is entered for program 1. This allows an operator to change fixed speeds for a drive if it is used as a stand-alone control. For example, you may have an application where you have a frequency drive controlling the speed of a conveyor. An operator can close switch 1, 2, or 3 to manually set the conveyor speed. If switch SW1 is closed, the drive will operate at 60% rpm, which is the value that is entered into the program for speed 1. If SW2 is closed, the drive will operate at 80%, which is the value that is entered into the program for speed 2. If SW3 is closed, the drive will operate at 100% rpm, which is the speed entered into the program parameter for speed 3.

If you wanted to provide the operator with the ability to set the speed at any value from 0–100% rpm, you could connect a remote potentiometer 10 kΩ to terminals TB2-1, TB2-2, and TB2-3. When the drive is set for external signal, the voltage from the potentiometer will provide the reference signal for the drive. Terminals TB2-2 and TB2-3 are used for an external voltage signal (0–10 volts DC), which could come from some other type of microprocessor controller such as a PLC. If the external control signal is going to be a milliamp signal, terminals TB2-3 and TB2-4 would be used.

Two ports are provided on the right side of the drive to accept serial inputs. A hand-held programmer is connected to port 1 and it can be used for entering programming parameters or to check parameters for troubleshooting. The second serial port can be used to establish serial communications from a portable computer, which allows the parameters to be loaded from a disk or stored to a disk. This provides a means of reloading the parameters in case of a programming problem or if a drive is changed.

11.7.4 External Outputs for the Drive

The drive has two types of on/off outputs available and they are shown on the right side of the diagram in Fig. 11-42 for the drive. A set of NO contacts is provided at terminals TB2-9 and TB2-10. These contacts are controlled by the drive and their function can be selected in the drive parameters. For example, these contacts can be selected to close when the drive is enabled, or they can be used to indicate the drive is in a fault condition. A solid-state output is available at terminals TB2-19 and TB2-20. This circuit is basically the emitter and collector of a transistor. When the drive controls this circuit, it sends a signal to the transistor base, which in turn causes the collector–emitter circuit to go to saturation.

Another output that is available from the drive is a frequency signal that can be sent to a frequency meter. The frequency meter can be located on the operator panel where it will be used to indicate the speed signal that the drive is sending to the motor.

11.7.5 Solid-State Circuits for Variable-Frequency Drives

As you know, the solid-state circuitry of a variable-frequency drive can be described as having three sections. The first section of the drive is called the rectifier section or converter. This section consists of a three-phase bridge rectifier. The second section of the drive is called the DC intermediate section and it contains the filter components. The third block of the drive is called the inverter section because this is where the DC voltage is turned back into three-phase AC voltage.

Fig. 11-43 shows an electronic diagram of an Allen-Bradley 1336 drive. Each section of the drive is now shown with the actual components connected as you would find them when you opened the drive to troubleshoot it. The rectifier section of this drive utilizes a three-phase bridge rectifier, which is actually a module. This means it can be removed and replaced as a unit rather quickly if it fails. Chapter 4 shows a picture of this type of module and explains its internal operations. The

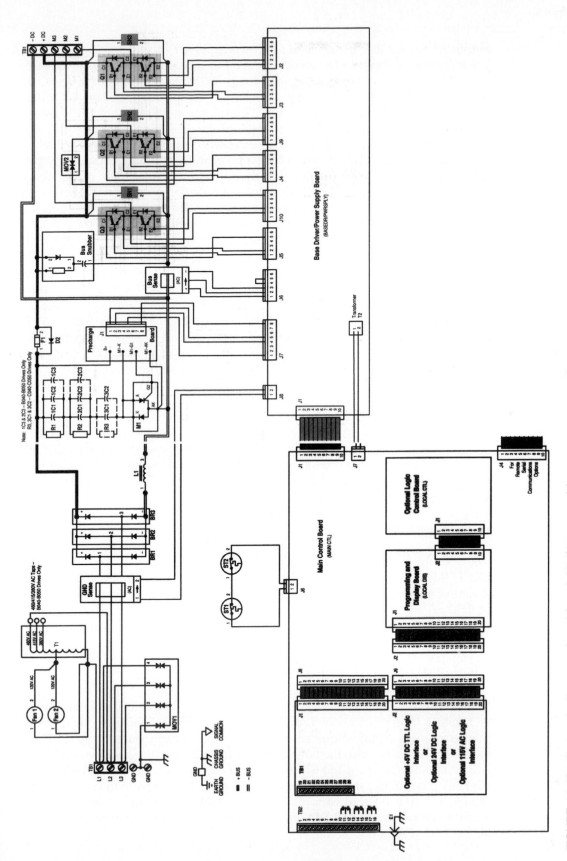

FIGURE 11-43 Electronic diagram of an AC variable-frequency drive. The major parts of the drive include the rectifier, the filter, and the output transistors. (Courtesy of Rockwell Automation's Allen Bradley Business.)

output of the rectifier section is six half-waves. A set of metal-oxide varistors is connected to the input of the rectifier section to protect against voltage surges.

When the input voltage to the drive is 480 volts ac, the output dc voltage from the rectifier section will be approximately 670 volts dc. The pulsing DC voltage is applied to the DC bus on this system. The DC bus is identified by the +DC and the −DC wires that run through the length of the drive circuit.

The filter section of the drive uses capacitors and an inductor to filter the voltage and current. The capacitors have a precharge circuit that allows the capacitors to reach full charge slowly so that they are not damaged. The capacitors are connected in parallel with the DC bus, and the inductor is connected in series with the negative DC bus wire. A set of resistors is provided to discharge the capacitors anytime power is removed. You should always allow sufficient time for the capacitors to discharge before you try to work on the solid-state components in the drive. The filter allows the pulsing DC voltage to be changed to pure DC.

The output section of the drive converts the DC voltage back to three-phase voltage. This section uses pulse-width modulation (PWM) techniques to switch three pairs of transistors on and off up to 12 times during each half-wave to produce a three-phase output. The amplitude of the signal determines the amount of voltage for the AC voltage, and the frequency of the signal will determine the frequency of the output of the drive. In most cases the output frequency can be any value between 0–120 Hz. Some drives allow the upper frequency to reach 400 Hz.

The output transistors are connected across the DC bus. One transistor is connected to the positive DC bus wire, and when it is switched on and off, it will provide the positive half-cycle for one phase of the AC signal. The second transistor is connected to the negative DC bus wire, and when it is switched on and off, it will provide the negative half-cycle for one phase of the AC signal. The base drive board provides the PWM signals for all of the transistors. The drive has a microprocessor that accepts the command signal and determines the correct frequency and voltage for the output transistors.

11.7.6 Pulse-Width Modulation Waveforms for Variable-Frequency Drives

Fig. 11-44 shows a typical waveform for a pulse-width modulation (PWM) circuit in the AC variable-frequency drive. The transistors in the PWM circuit are switched on and off approximately 12 times each half-cycle. The on and off cycles create the overall frequency waveform for the output signal. The output waveform of the PWM section looks like a traditional three-phase signal to the motor. If you place an isolated-case oscilloscope across any two of the output leads of the drive, you will see a signal that looks similar to the one in the diagram.

If you are using a digital voltmeter to measure the output voltage on a variable-frequency drive, you must be aware that some digital voltmeters will not read the AC voltage from this section accurately because of the switching frequency of the transistors. The digital voltmeters tend to read the drive's output voltage higher than it actually is because the voltmeter may be fast enough to sample some of the individual waveforms created when the transistors are switched on and off rapidly. An analog meter may show the voltage more accurately because the needle cannot change as fast as the transistor is switched on and off. For this reason some drive manufacturers provide an LED display to show an accurate voltage reading right on the face of the drive.

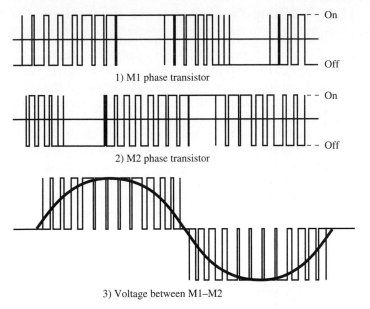

1) M1 phase transistor

2) M2 phase transistor

3) Voltage between M1–M2

FIGURE 11-44 Output waveform of the PWM section of the variable-frequency drive. Notice all of the points where the transistor is switched on and off inside each half-wave. (Courtesy of Rockwell Automation's Allen Bradley Business.)

11.7.7 Insulated Gate Bipolar Transistor Drives

In newer drives, the transistors may be replaced by insulated gate bipolar transistors (IGBTs). The IGBTs are used because they can be switched on and off at much higher frequencies that do not conflict with other signals. The higher frequencies are also used outside of the audible range for humans, so the drive will not emit a hum that humans can hear. The high frequency is divided so that the output signal for the drive is within the 0–120 Hz range.

Fig. 11-45 shows two pairs of IGBTs used to produce two phases of a typical output section for a variable-frequency drive using IGBTs instead of bipolar transistors. From this section you can see that one IGBT of each pair is connected to the positive bus and a second is connected to the negative bus. The IGBTs operate similarly to the transistors in that they are cycled on and off at high frequencies within the overall waveform of a sine wave.

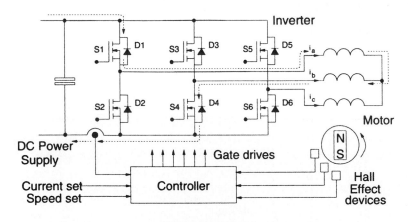

Fig.2 BDCM system

FIGURE 11-45 Insulated gate bipolar transistors (IGBTs) connected to the output stage of the variable-frequency drive. The IGBTs are used instead of traditional bipolar transistors in newer drives. (Courtesy of Philips Semiconductors.)

11.7.8 Variable-Frequency Drive Parameters

Variable-frequency drives have had the ability to provide features such as ramp-up speed, ramp-down speed, boost voltage, and braking functions since they were first designed. Prior to having a microprocessor, these features and functions were designed into the op amp circuits that controlled the drive and were enabled by the placement of jumpers or the setting of dip switches. After microprocessor chips were integrated into the drive, these features have become programmable and in many cases they have become proportional so that you can ask for braking, but you can limit the braking to 60% for 3 seconds. If you select ramp up, you can select more than one ramp speed. Then you can integrate the selection with an external switch so that the ramp speed can be selected by external conditions. The numbers that are programmed into the drive to select these features are called *parameters*. The type and number of parameters will vary from drive to drive. The Allen-Bradley 1336 drive has 89 parameters while the Allen-Bradley 1305 drive

FIGURE 11-46 List of parameters for a variable-frequency drive. (Courtesy of Rockwell Automation's Allen Bradley Business.)

Parameter Number	Description	Units	Min/Max Values	Factory Selling
0	Parameter Mode	None	0/0	—
1	Output Volts	Volts	0/575	0
2	Output Current	% Rated	0/200	0
3	Output Power	% Rated	0/200	0
4	Last Fault	Code	0/37	0
5	Frequency Select 1	Code	0/5	0
6	Frequency Select 2	Code	0/5	0
7	Accel Time 1	Seconds	0.0/500	5.0
8	Decal Time 1	Seconds	0.0/600	6.0
9	DC Boost Select	Code	0/12	2
10	Stop Select	Code	0/2	0
11	Decal Frequency Hold	Off/On	0/1	0
12	DC Hold Time	Seconds	0/15	0
13	DC Hold Volts	Volts	0/115	0
14	Auto Restart	Off/On	0/1	1
15	Factory Set—Do Not Change	None	0/0	0
16	Minimum Frequency	Hertz	0/120	0
17	Base Frequency	Hertz	40/120	60
18	Base Volts	Volts	115/575	460/575 ○
19	Maximum Frequency	Hertz	40/250	60
20	Maximum Volts	Volts	115/575	460/575 ●
21	Local Run	Off/On	0/1	1
22	Local Reverse	Off/On	0/1	1
23	Local Jog	Off/On	0/1	1
24	Jog Frequency	Hertz	.0/120	0
25	Analog Output	Code	0/1	0

Parameter Number	Description	Units	Min/Max Values	Factory Selling
26	Preset/2nd Accel	Code	0/1	0
27	Preset Frequency 1	Hertz	0.0/250	0.0
28	Preset Frequency 2	Hertz	0.0/250	0.0
29	Preset Frequency 3	Hertz	0.0/250	0.0
30	Accel Time 2	Seconds	0.0/600	5.0
31	Decel Time 2	Seconds	0.0/600	5.0
32	Skip Frequency 1	Hertz	0/250	250
33	Skip Frequency 2	Hertz	0/250	250
34	Skip Frequency 3	Hertz	0/250	250
35	Skip Frequency Band	Hertz	0/15	0
36	MOPC	% Rated	50/150	150
37	Serial Baud Rate	Code	0/1	1
38	Overload Current	% Rated	50/115	100
39	Fault Clear	Off/On	0/1	1
40	Power Fault	On/Off	0/1	0
41	Motor Type	Code	0/2	0
42	Slip Compensation	Hertz	0.0/5.0	0.0
43	Dwell Frequency	Hertz	0/120	0
44	Dwell Time	Seconds	0/10	0
45	PWM Frequency	WM	0.4/2.0	0.4
46	Pulse Scale Factor	Ratio	1/255	64
47	Language	Code	0/5	0
48	Start Boost	Volts	0/115	0
49	Break Frequency	Hertz	0/120	0
50	Break Volts	Volts	0/230	0
	Parameters 51-69 Can Only Be Accessed Thru the Serial Port			
70	Base Driver Board Version	Code	—	—
71	Control Board Version	Code	—	—
72	Activate Parameters 73-76	Off/On	0/1	0
73	Preset Frequency 4	Hertz	0.0/250	0.0
74	Preset Frequency 5	Hertz	0.0/250	0.0
75	Preset Frequency 6	Hertz	0.0/250	0.0
76	Preset Frequency 7	Hertz	0.0/250	0.0
77	Above Frequency Contact	Hertz	0/250	0
78	Traverse Period	Seconds	0.0/30	0.0
79	Maximum Traverse	Hertz	0.0/100	0.0
80	Inertia Compensation	Hertz	0.0/20	0.0
81	Soft Start/Stop Enable	Off/On	0/1	0
82	Amp Limit Fault Enable	Off/On	0/1	0
83	Run Boost	Volts	0/115	0
84	Analog Inverse	Off/On	0/1	0
85	Restart Tries	Code	0/9	0
86	Fault Buffer 0	Code	0/37	0
87	Fault Buffer 1	Code	0/37	0

has 136 parameters. Fig. 11-46 shows a typical list of parameters for a variable-frequency drive.

The drive has a set of standard values to be used for the parameters. These parameters are called the *default settings* or *factory settings*. It is important to record the actual parameter settings if they are different from the factory settings, so that the proper settings can be put into the drive if it is ever removed and replaced with a new one. The parameters provide a means to customize the drive to any specific application. Some manufacturers provide software that can be used on a portable computer to save and load the parameters from a drive. The parameters can also be saved and loaded from the PLC program, which means they can also be changed while the system is in operation to provide custom parameters for multiple recipes.

The parameters can be described in groups by their function. Examples of these groups include metering, setup, advanced setup, frequency settings, diagnostics, faults, and process displays. Examples of the setup group include minimum frequency and maximum frequency. This allows the minimum and maximum frequency to be fixed so that the motor does not run less than the minimum value and more than the maximum value.

Another setup parameter is acceleration time, which determines the acceleration ramp. For example, if you select the ramp time as 10 seconds, the drive will increase the frequency proportionally from the programmed minimum frequency to the programmed maximum frequency in 10 seconds. If you wanted the motor to ramp up to speed more quickly, you would shorten the ramp time. If you wanted the motor to ramp up more slowly, you would increase the ramp time. Most drives have more than one acceleration ramp parameter and each ramp is enabled to the input switches that were previously discussed. This means that you could have up to three acceleration ramps that would be selected by switch 1, 2, or 3. The deceleration time is also programmable. Three deceleration ramps are available for this drive.

Another setup parameter available for the drive is overload current limit. This parameter will act like a programmable fuse. You can select any value from 100–115%. The other variable that operates with the overload current is the amount of time this current can exist. Since the drive is controlled by a microprocessor, it monitors the current and voltage and turns off the drive if these values become excessive.

The drive also provides metering parameters. These can be integrated with protection, display, and fault functions. Most drives have displays that are built into the face of the drive. The technician can use the display on the face of the drive to observe the amount of input voltage, the amount of current, the frequency, the temperature, the output voltage, and the last fault that was recorded.

Advanced setup parameters include the type of braking the drive will use. The choice for this parameter is no braking (coasting to a stop) or braking, and the amount of braking voltage and the amount of time the braking voltage should be applied are determined. Another advanced setup parameter is called *DC boost voltage*, which is DC voltage that can be applied with AC frequency during starting or at times when the motor needs more torque. The DC boost voltage makes the magnetic field in the motor stronger to reduce the amount of slip the motor has. This will allow the motor to provide additional power in applications where more starting torque is needed or when more torque is needed during specific loading conditions.

The drive has the ability to test hundreds of points in its circuit boards for changes in voltage, current, frequency, and temperature. The present value of each of these variables at the input and output stages of the drive can be compared against the value set into the parameter. If the value is exceeded, the drive can indicate each occurrence with a fault code. The drive also changes the state of contacts that can be used to enable a fault indicator lamp or horn. If the fault has occurred, it will be stored in the drive where it can be brought to the display by pressing a series of keys on the front panel of the drive. The serial port connection provides a method to send the fault codes to an external controller such as a PLC where they can be logged with the date and time they occurred and they can also be printed as they occur. Some drives have the capacity to store multiple faults so that the technician can review the last five faults. This provides a means to detect and store multiple faults if more than one problem occurs during the fault condition. Typical fault conditions are low voltage, high voltage, high current, and overtemperature.

11.7.9 Typical Applications for Variable-Frequency Drives

Variable-frequency drives are widely used to control the speed of conveyor systems, blower speeds, pump speeds, machine tool speeds, and other applications that require variable speed with variable torque. In some applications such as speed control for a conveyor, the drive is installed with a remote potentiometer that personnel can adjust manually to set the speed for the conveyor. In this type of application, the personnel who use the conveyor can manually set the motor speed with the minimum and maximum frequency that is programmed into the parameters.

In other applications such as blower speed control or pump speed control, the variable-frequency drive can be controlled by a 4–20 mA or 0–10 volt input signal that comes from a microprocessor controller. When the input signal for the drive is provided from a controller, the system is considered to be a closed loop. For example, in a system where the drive is controlling the speed of a blower, a temperature sensor can be used to determine the temperature of a room. If it is too cold and the blower is moving warm air, the speed of the blower can be increased. If the temperature becomes too hot, the speed of the blower can be slowed until the temperature returns to the correct setpoint. If the variable-frequency drive is used to control the level of product in a tank by varying the speed of a pump, a level sensor can be connected to a controller. If the level of the tank is becoming low, the drive can increase the speed of the motor and pump, and if the level is too high, the speed can be reduced.

Fig. 11-47 shows an example of the field wiring connections at the terminal boards of a drive for a typical application. In this case the drive diagram is for an Allen-Bradley 1336 drive. The top diagram shows the external control signals that are connected to terminal board 2 (TB2) of the drive. You can see that the command signal can be an external potentiometer connected to terminals 1, 2, and 3, or it can be a 0–10 volt signal connected to terminals 4 and 5, or it can be a 4–20mA signal connected to terminals 4–6. This terminal board also provides NO and NC contacts to indicate the drive is at speed, running, faulted, or has a drive alarm. These contacts are used as part of fault or safety circuits for the system.

Terminal board 3 (TB3) in this figure shows the start and stop switches that must be connected to the drive to provide the signals to cause it to start and become enabled. The signal voltage for this board can be selected as 5 volts TTL, 28 volts DC, or 115 volts AC supply. The diagram shown uses 115 volts AC. Start, stop, and jog buttons are connected to terminals 19, 20, and 22. A 115 volt signal must also be provided at terminal 30 to enable the drive. Terminals 24 and 26 provide the inputs for switch SW1 and SW2, which will be used to make a two-bit binary code to indicate the acceleration and deceleration parameters. Terminal 23 is used to set the motor in reverse from a remote switch. Terminals 21, 25, and 29 are the common for this board and since they are connected inside the board, the common from the transformer need only be connected to one of these terminals.

After the 115 volt AC signals that are used as inputs for this board are received, they must be isolated and converted before they are sent to the microprocessor. The isolation and rectification portion of this board is shown at the top part of the diagram in Fig. 11-47. From this diagram you can see that after each 115 volt AC

FIGURE 11-47 **(a)** Diagram of TB2 for a 1336 Allen-Bradley drive. **(b)** Diagram of TB3 for a 1336 Allen-Bradley drive. **(c)** Diagram of TB1 that shows the power connections to the drive and the terminal connections for the motor. (Courtesy of Rockwell Automation's Allen Bradley Business.)

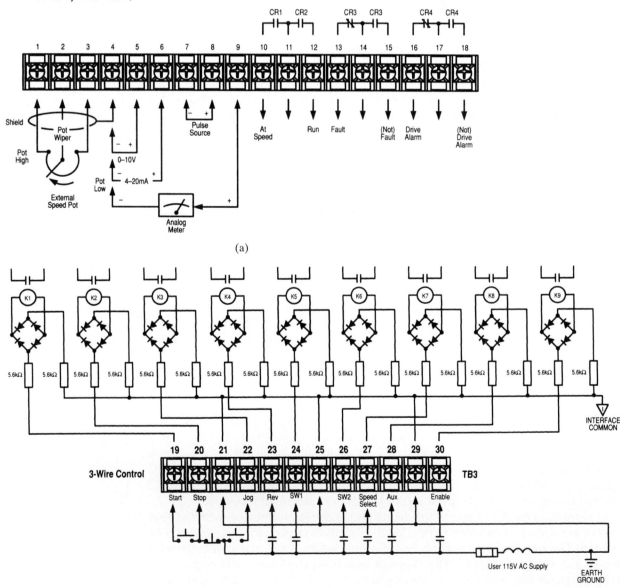

(a)

(b)

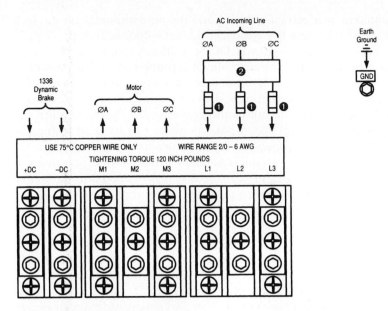

❶ User supplied drive input fuses.

❷ Motor disconnecting means including branch circuit, short circuit, and ground fault protection.

(c)

FIGURE 11-47 (Continued)

signal is received, it is rectified and directed through a DC-powered relay to provide isolation before it is sent to the processor. The diagram also shows each signal connected to a set of resistors to drop the voltage and to a bridge rectifier to change the AC voltage to DC. The output of the bridge rectifier is connected to the coil of a DC relay. The contacts of the relay are connected to the processor inputs. Since this signal comes through a relay, there is total isolation between the 115 volt AC used in the field switches connected to this board and the contacts that are connected directly to the processor section of the drive.

The high-voltage terminal board for the variable-frequency drive is designated as TB1. This terminal board provides connections at the far right side for L1, L2, and L3 where the three-phase 480 volt AC is connected. Terminals M1, M2, and M3 are connections for the output of the variable-frequency drive to the three-phase motor. Terminals −dc and +dc are located at the far left side of the terminal block and they provide DC voltage directly from the DC bus. This voltage is provided for a dynamic brake that can be added to the three-phase motor. The dynamic brake provides a means to stop the motor shaft quickly. The operation of the dynamic brake will be covered in Chapter 13.

11.7.10 Operating the Variable-Frequency Drive

The variable-frequency drive can be operated as a stand-alone unit or it can be controlled by a PLC or other electronic control system that sends a 0–10 volt or a 4–20 mA signal. When the drive is operated as a stand-alone system, a potentiometer may be provided directly on the face of the drive. The operator of the machinery can set the speed for the drive by adjusting the potentiometer from 0–100%. In some applications, the potentiometer is mounted on the machine's front panel instead of on the face of the drive. The reason for this is that in some cases the drive must be mounted in an electrical cabinet, and the potentiometer on the face of the drive is not accessible when the electrical cabinet's door is closed.

If the drive is controlled by a 0–10 volt DC or 4–20 mA signal, the command signal will be changed by the logic of the controller it is connected to. The controller generally will have some type of sensor that indicates speed or position if the system is a closed-loop system. Or the command signal can be generated from a predefined table in the controller that sets the motor to the proper speed for the application. In this type of application, the local speed potentiometer can be used to jog the motor or when the motor must be operated for maintenance.

Specific applications that use variable-frequency drives will be presented in Chapter 12. You will get a better idea of how the drive is connected to the motor circuit and the function that variable speed provides in the application.

11.7.11 Scalar Drives and Vector Drives

Some companies provide additional circuits inside the drive to bring in the closed-loop feedback signals from encoders or resolvers. These circuits will decode the signals from the sensors and scale them to the level required. This feature makes the drive operate more like a package system. This also makes it easier for design engineers to put a complete system together where all of the components match. Allen-Bradley calls its drive with closed-loop capability a *vector drive*. Its *scalar drive* can be operated as either an open-loop or closed-loop drive. Other manufactures provide similar packages. It is important to remember that the vector and scalar drives are more expensive and more complex than a simple variable-frequency drive. A fractional horsepower variable-frequency drives cost less than $500 today.

11.7.12 Variable-Voltage Input Drives

The variable-voltage input (VVI) drive is the technology that was used in some of the earliest AC variable-frequency drives. Since these earlier drives did not have microprocessor chips to establish the transistor driver signals, they used existing technology such as oscillators. Fig. 11-48 shows a block diagram of this type of drive and a typical waveform of the output.

From the block diagram you can see that the basic parts of the drive such as the rectifier section, filter, and inverter sections are much the same as modern drives. The major difference is that SCRs are used for the rectifier section instead of diodes. The reason for this is that diodes were not manufactured as large as SCRs at this time, so the SCRs were used because they were more durable. Also it is possible to adjust the SCR's timing with a signal from the regulator section of the drive to change the amount of voltage and current delivered to the drive. The amount provided sufficient change to meet most of the varying demands of torque.

From the waveform diagram you can see that the voltage waveform was a six-step signal. The switching device in the inverter section would be switched on and off at specific points to provide the six-step signal. The current waveform looks more like a sine wave because the inductance of the motor helps to cause the phase shift needed to smooth out the square wave shape of the six-step signal. You will encounter a few drives with this early technology because they have paid for themselves over and over. In some cases the drives are changed and a newer microprocessor programmable drive is used whenever there is a problem with the original drive. It is also more practical to change to a modern drive that can provide a wider range of torque for all applications the motor may encounter.

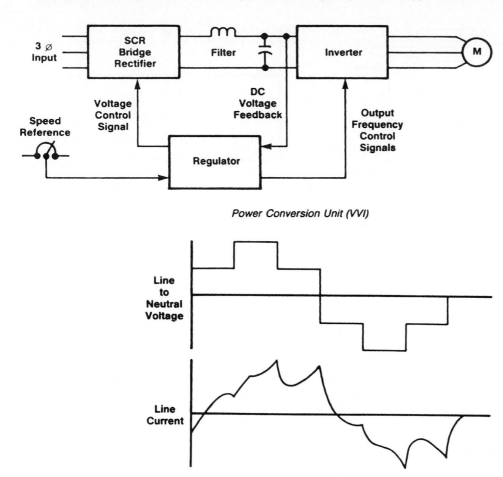

Power Conversion Unit (VVI)

VVI Output Waveforms

FIGURE 11-48 Block diagram and waveform for a variable-voltage input drive. This is one of the earliest types of variable-frequency drives. (Courtesy of Rockwell Automation's Allen Bradley Business.)

11.7.13 Current-Source Input Drives

Another type of drive technology used in early variable-frequency drives is called current-source input (CSI). Fig. 11-49 shows a block diagram and the voltage and current waveforms for this drive. From the block diagram you can see that this drive also did not have a microprocessor. This means that all of the drive for the waveforms was established from oscillators and other circuits that combined with op amps to make the closed-loop portion of the drive. The input-speed potentiometer set a reference voltage on the op amps in the *speed* or *voltage control block*. The feedback signal from the output lines to the motor is compared to the setpoint in this block. The output of the speed control block is sent to the *current regulator block* that controls the firing angle on the SCRs in the rectifier section. The voltage and current are changed to meet the changing demands of the torque needed to move the load.

A companion signal is sent from the speed control block to the *frequency control block*, which adjusts the speed of the motor by changing the frequency that is sent to it. The speed potentiometer sets the percentage of speed that is needed and the frequency control adjusts the frequency of the inverter section. The inverter section of the drive uses large SCRs to switch the voltage and current.

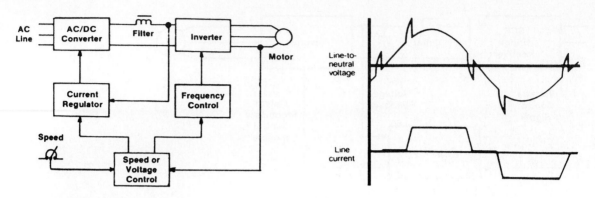

Block Diagram for a Typical CSI Drive

Typical CSI Voltage and Current Waveforms

FIGURE 11-49 Block diagram with waveform of a current-source input drive. (Courtesy of Rockwell Automation's Allen Bradley Business.)

The waveform for this drive is also shown in the figure. From this diagram you can see that the current is adjusted similarly to the six-step voltage waveform of the VVI drive. The voltage signal for the CSI drive is smoother and looks more like a traditional sine wave. As a technician you may find a few of these drives still operational. As a rule, however, they are generally replaced as they encounter problems.

11.8 DC DRIVES

Variable DC drives have been used to control DC motors longer than variable-frequency drives have been used to control AC motors. The first motor speed control used DC motors because of the simplicity of controlling the voltage to the armature and field of a DC motor. The main obstacle in using DC motors is the increased amount of maintenance involved because the DC motor has brushes and a commutator. Early speed control for DC motors consisted of large resistors that were switched in the motor circuit to reduce the amount of voltage supplied to it. The resistors created problems because of the heat buildup.

DC drives designed in the 1970s and 1980s combined op amp circuits to provide ramping capability with SCR firing circuits to control large voltages. Today modern DC drives utilize the latest solid-state power-switching technology combined with microprocessors to provide programmable features. When you analyze the diagrams for the DC drive you will notice that much of the circuitry looks similar to the AC drive. The main difference is that the rectifier stage and output stage of the DC drive are combined because the DC drive simply adjusts the DC voltage and current rather than invert it back to AC. Since the output voltage for the drive is DC, SCRs will be used in rectifier circuits. The newest drives have programmable parameters similar to AC drives in that they set the maximum voltage, current, and speed, as well as provide protection against overcurrent, overtemperature, phase loss of incoming power, and field loss.

Fig. 11-50 provides a picture of a typical DC drive as a stand-alone product,

FIGURE 11-50 Typical DC drive shown as a stand-alone drive and mounted in a panel. (Courtesy of Rockwell Automation's Allen Bradley Business.)

and as you would find it when mounted in a panel. From this picture you can see that it is difficult to distinguish a DC drive from an AC drive by its physical features. Fig. 11-51 shows a diagram of the simple DC drive and you can see that the electronic circuits of the drive are slightly different from the AC drive. In the diagram of the DC drive, you can see that three-phase supply voltage is provided at the top of the diagram. This incoming voltage is sent directly to an isolation transformer and then to a three-phase bridge rectifier in the *armature power converter circuit*. The three-phase rectification in the armature power converter circuit is similar to the rectifier section of the AC drive except large SCRs are used instead of diodes.

The SCRs are used for the rectification section because they can provide voltage control as well as rectification. This simplifies the drive somewhat since rectification and voltage control are combined in one circuit. Op amps provide speed ramps and current ramps for the SCR firing (control) circuit. In the older drives, the op amps were used as stand-alone ramping circuits. In modern drives that have microprocessors, the firing circuits are controlled by digital-to-analog (D/A) circuits that integrate linear circuits with the processor. The voltage from the armature power converter circuit is sent directly to the armature. The DC motor is shown in this diagram as a shunt field and armature. A tachometer is shown connected to the armature as a dotted line, which means it is physically connected to the motor shaft.

The rectifier section may use six SCRs as a bridge similar to the diode bridge rectifiers in AC drives. Or larger drives may connect two SCRs in parallel for each of the six sections of the bridge to provide a 12 SCR full-wave rectifier circuit. When SCRs are connected in parallel, the current rating of the rectifier is nearly doubled.

The firing circuit for the SCRs is synchronized with the three-phase incoming voltage. The firing circuit also receives an input signal called a *reference signal* or *command signal* from the speed amp and the current amp. The speed amp receives

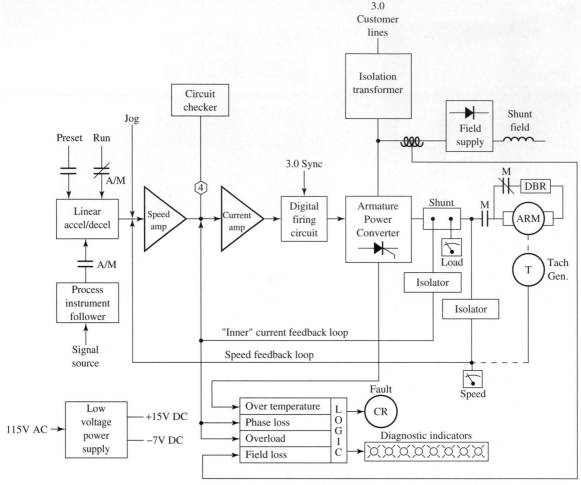

FIGURE 11-51 Block diagram of a DC drive. Three-phase voltage is supplied at the top of the diagram, and the DC motor is shown as a shunt field and armature. (Courtesy of Rockwell Automation's Allen Bradley Business.)

a feedback signal from a tachometer, and the current amp receives a signal from a current transducer (shunt) that is connected in series with the armature. As the current in the wire to the armature increases or decreases, the voltage across the shunt will increase or decrease and provide a feedback signal to the current amplifier.

In the diagram you can also see that DC field voltage is provided by a smaller diode bridge. The AC voltage supply for this bridge rectifier is tapped off of the output of the isolation transformer prior to the main rectifier in the armature power converter. Since this voltage comes from a diode bridge rectifier, it will be constant. Speed control for the DC motor is provided by keeping the shunt field voltage constant and by varying the armature voltage and current.

Fault circuits are provided in the drive to test for overtemperature, phase loss, overload conditions, and the loss of field current in the motor. Indicator lamps are provided on the front of the drive to show when a fault has occurred. A speed indicator is also provided on the face of the drive to show the actual speed of the motor. The speed indicator receives its signal from the tachometer that is connected to the shaft of the DC motor.

11.8.1 Modern Programmable DC Drives

Modern DC drives combine a microprocessor with the drive circuitry shown in the previous diagram. The major difference that the microprocessor provides is that it allows programmable parameters to be used with the drive to set maximum voltages and currents, and provide a variety of ramp-up and ramp-down signals. These parameters are similar to the ones listed for the AC drive presented in the previous section. Another feature the microprocessor provides is closed-loop control such as PID (proportional, integral, and derivative) control. In Chapter 8 you learned about closed-loop PID control. Fig. 11-52 shows a block diagram of a modern DC drive that incorporates the closed-loop PID control that is used for speed (velocity) control. A separate control loop for current is also used. In this diagram you can see that three-phase voltage is supplied at the bottom of the drive and is converted to variable DC voltage by the three-phase SCR rectifier section. An encoder is used as the feedback sensor to provide the velocity feedback signal to the velocity processor. Current sensors can also be used to provide a second loop for current control.

Another feature of the microprocessor controlled drive is that it can be programmed from a programming terminal or programming software that is in a laptop computer. The programming parameters can also be loaded or *changed on the fly* from a PLC. This provides equal functions to the AC drive.

FIGURE 11-52 Block diagram of microprocessor controlled DC drive. This drive provides PID control of the DC motor speed. (Courtesy of Rockwell Automation's Allen Bradley Business.)

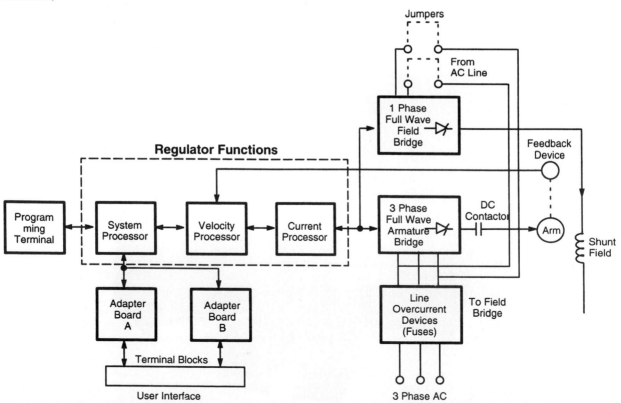

11.9 STEPPER MOTORS

Stepper motors provide a means for precise positioning and speed control without the use of feedback sensors. The basic operation of a stepper motor allows the shaft to move a precise number of degrees each time a pulse of electricity is sent to the motor. Since the shaft of the motor moves only the number of degrees that it was designed for when each pulse is delivered, you can control the pulses that are sent and control the positioning and speed. The rotor of the motor produces torque from the interaction between the magnetic field in the stator and rotor. The strength of the magnetic fields is proportional to the amount of current sent to the stator and the number of turns in the windings.

Fig. 11-53 shows a picture of typical stepper motors and their controllers. These motors are available for a variety of motion control applications such as single- and multiple-axis control for robots, or machine tools such as mills and lathes. Smaller motors are used in printers and in some computer disk drives.

11.9.1 Types of Stepper Motors

Three basic types of stepper motors include the *permanent magnet motor,* the *variable reluctance motor,* and the *hybrid motor,* which is a combination of the previous two. Fig. 11-54 shows a cut-away diagram of a typical permanent magnet stepper motor. The rotor for the permanent magnet motor is called a *canstack rotor* and a diagram of it is shown in Fig. 11-55. The canstack rotor shows that the permanent magnet motor can have multiple rotor windings, which means that the shaft for this type of stepper motor will turn fewer degrees as each pulse of current is received at the stator. For example, if the rotor has 50 teeth and the stator has

FIGURE 11-53 Two sizes of typical stepper motors with their controllers. The larger motor is approximately 8 in. in diameter, and the smaller motor is approximately 3 in. in diameter. (Courtesy of Parker Compumotor Division.)

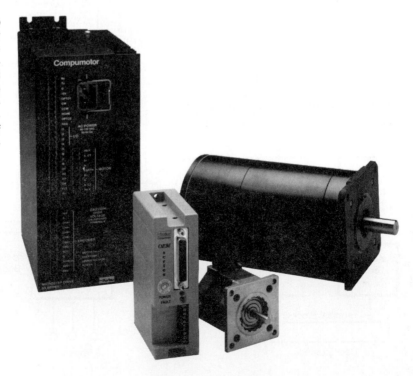

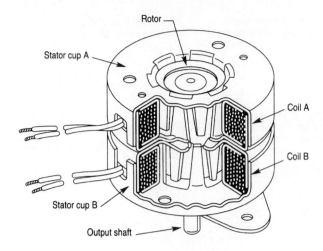

FIGURE 11-54 Cut-away diagram of a permanent magnet stepper motor. (Courtesy of Parker Compumotor Division.)

8 poles with 5 teeth each (total of 40 teeth), the stepper motor is able to move 200 distinct steps to make one complete revolution. This means that shaft of the motor will turn 1.8° per step. The main feature of the permanent magnet motor is that a permanent magnet is used for the rotor, which means that no brushes are required. The drawback of this type of motor is that it has relatively low torque and must be used for low-speed applications.

The *variable reluctance motor* does not use permanent magnets, so the field strength can be varied. The amount of torque for this type of motor is still small so it is generally used for small positioning tables and other small positioning loads. Since this type of motor does not have permanent magnets, it cannot use the same type of stepper controller as other types of stepper motors.

The *hybrid stepper motor* is the most widely used and combines the principles of the permanent magnet and the variable reluctance motors. Fig. 11-56 shows an example of a hybrid stepper motor. Most hybrid stepper motors have two phases and operate on the principle used to explain the 12-step motor previously.

11.9.2 Stepper Motor Theory of Operation

The stepper motor uses the theory of operation for magnets to make the motor shaft turn a precise distance when a pulse of electricity is provided. You learned

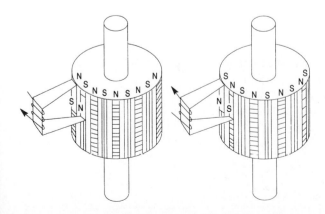

FIGURE 11-55 Canstack rotor that is used in permanent magnet stepper motors. (Courtesy of Parker Compumotor Division.)

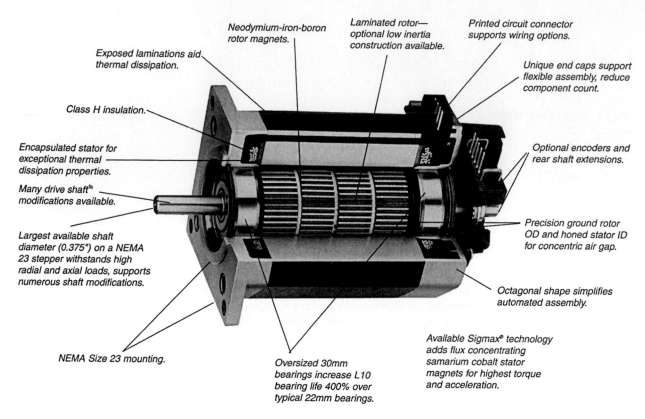

Exposed laminations aid thermal dissipation.

Neodymium-iron-boron rotor magnets.

Laminated rotor— optional low inertia construction available.

Printed circuit connector supports wiring options.

Unique end caps support flexible assembly, reduce component count.

Class H insulation.

Encapsulated stator for exceptional thermal dissipation properties.

Many drive shaft modifications available.

Largest available shaft diameter (0.375") on a NEMA 23 stepper withstands high radial and axial loads, supports numerous shaft modifications.

Optional encoders and rear shaft extensions.

Precision ground rotor OD and honed stator ID for concentric air gap.

Octagonal shape simplifies automated assembly.

NEMA Size 23 mounting.

Oversized 30mm bearings increase L10 bearing life 400% over typical 22mm bearings.

Available Sigmax® technology adds flux concentrating samarium cobalt stator magnets for highest torque and acceleration.

FIGURE 11-56 Hybrid stepper motor combines features of the permanent magnet stepper and the variable reluctance stepper motors. (Courtesy of Pacific Scientific.)

previously that like poles of a magnet repel and unlike poles attract. Fig. 11-57 shows a typical cross-sectional view of the rotor and stator of a stepper motor. From this diagram you can see that the stator (stationary winding) has four poles, and the rotor has six poles (three complete magnets). The rotor will require 12 pulses of electricity to move the 12 steps to make one complete revolution. Another way to say this is that the rotor will move precisely 30° for each pulse of electricity that the motor receives. The number of degrees the rotor will turn when a pulse of electricity is delivered to the motor can be calculated by dividing the number

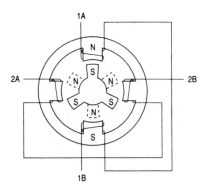

FIGURE 11-57 Diagram that shows the position of the six-pole rotor and four-pole stator of a typical stepper motor. (Courtesy of Parker Compumotor Division.)

of degrees in one revolution of the shaft (360°) by the number of poles (north and south) in the rotor. In this stepper motor 360° is divided by 12 to get 30°.

When no power is applied to the motor, the residual magnetism in the rotor magnets will cause the rotor to *detent* or align one set of its magnetic poles with the magnetic poles of one of the stator magnets. This means that the rotor will have 12 possible detent positions. When the rotor is in a detent position, it will have enough magnetic force to keep the shaft from moving to the next position. This is what makes the rotor feel like it is *clicking* from one position to the next as you rotate the rotor by hand with no power applied.

When power is applied, it is directed to only one of the stator pairs of windings, which will cause that winding pair to become a magnet. One of the coils for the pair will become the north pole, and the other will become the south pole. When this occurs, the stator coil that is the north pole will attract the closest rotor tooth that has the opposite polarity, and the stator coil that is the south pole will attract the closest rotor tooth that has the opposite polarity. When current is flowing through these poles, the rotor will now have a much stronger attraction to the stator winding, and the increased torque is called *holding torque.*

By changing the current flow to the next stator winding, the magnetic field will be changed 90°. The rotor will only move 30° before its magnetic fields will again align with the change in the stator field. The magnetic field in the stator is continually changed as the rotor moves through the 12 steps to move a total of 360°. Fig. 11-58 shows the position of the rotor changing as the current supplied to the stator changes.

In Fig. 11-58a you can see that when current is applied to the top and bottom stator windings, they will become a magnet with the top part of the winding being the north pole, and the bottom part of the winding being the south pole. You should notice that this will cause the rotor to move a small amount so that one of its south poles is aligned with the north stator pole (at the top), and the opposite end of the rotor pole, which is the north pole, will align with the south pole of the stator (at the bottom). A line is placed on the south-pole piece that is located at the 12 o'clock position in Fig. 11-58a so that you can follow its movement as current

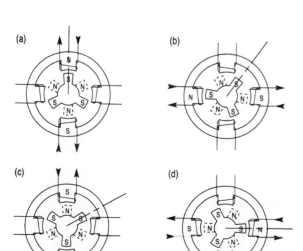

FIGURE 11-58 Movement of the stepper motor rotor as current is pulsed to the stator. (Courtesy of Parker Compumotor Division.)

is moved from one stator winding to the next. In Fig. 11-58b current has been turned off to the top and bottom windings, and current is now applied to the stator windings shown at the right and left sides of the motor. When this occurs, the stator winding at the 3 o'clock position will have the polarity for the south pole of the stator magnet, and the winding at the 9 o'clock position will have the north-pole polarity. In this condition, the next rotor pole that will be able to align with the stator magnets is the next pole in the clockwise position to the previous pole. This means that the rotor will only need to rotate 30° in the clockwise position for this set of poles to align itself so that it attracts the stator poles.

In Fig. 11-58c you can see that the top and bottom stator windings are again energized, but this time the top winding is the south pole of the magnetic field and the bottom winding is the north pole. This change in magnetic field will cause the rotor to again move 30° in the clockwise position until its poles will align with the top and bottom stator poles. You should notice that the original rotor pole that was at the 12 o'clock position when the motor first started has now moved three steps in the clockwise position.

In Fig. 11-58d you can see that the two side stator windings are again energized, but this time the winding at the 3 o'clock position is the north pole. This change in polarity will cause the rotor to move another 30° in the clockwise direction. You should notice that the rotor has moved four steps of 30° each, which means the rotor has moved a total of 120° from its original position. This can be verified by the position of the rotor pole that has the line on it, which is now pointing at the stator winding that is located in the 3 o'clock position.

11.9.3 Switching Sequence for Full- and Half-Step Motors

The stepper motor described in the previous section uses a four-step switching sequence, which is called a full-step switching sequence. Fig. 11-59 shows a switching diagram and a table that indicates the sequence for the four switches used to control the stepper motor. The diagram shows four switches with four separate amplifiers. The diagram for the motor shows the same four windings that were discussed in the theory of operation the previous section. Each of the windings is tapped at one end and they are connected through a resistor to the negative terminal of the power supply.

The table shows the sequence for energizing the coils. During the first step of the sequence, switches SW1 and SW3 are on and the other two are off. During the second step of the sequence, switches SW1 and SW4 are on and the other two are off. During the third step of the sequence, SW2 and SW4 are on and the other two are off. During the fourth step of the sequence, SW2 and SW3 are on and the other two are off. This sequence continues through four steps, and then the same four steps are repeated again. These steps cause the motor to rotate one *step* or *tooth* on the rotor when a pulse is applied by closing two of the switches. Fig. 11-60 shows the position of the poles during each step when the motor is in full-step mode.

11.9.3.1 Half-Step Switching Sequence ■ Another switching sequence for the stepper motor is called an *eight-step* or *half-step sequence*. The switching diagram for the half-step sequence is shown in Fig. 11-61. The main feature of this switching sequence is that you can double the resolution of the stepper motor by causing the rotor to move half the distance it does when the full-step switching sequence is used. This means that a 200-step motor, which has a resolution of 1.8°, will have a

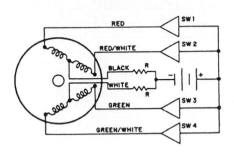

	FOUR STEP INPUT SEQUENCE (FULL-STEP MODE)*			
STEP	SW1	SW2	SW3	SW4
1	ON	OFF	ON	OFF
2	ON	OFF	OFF	ON
3	OFF	ON	OFF	ON
4	OFF	ON	ON	OFF
1	ON	OFF	ON	OFF

FIGURE 11-59 **(a)** Diagram of switching circuits for stepper motor. **(b)** The switching sequence for a four-step (full-step) switching mode. (Courtesy of Superior Electric, Warner Electric.)

resolution of 400 steps and 0.9°. The half-step switching sequence requires a special stepper motor controller, but it can be used with a standard hybrid motor. The way the controller gets the motor to reach the half-step is to energize both phases at the same time with equal current.

In this sequence the first step has SW1 and SW3 on, and SW2 and SW4 are off. The sequence for the first step is the same as the full-step sequence. The second step has SW1 on and all of the remaining switches are off. This configuration of switches causes the rotor to move an additional half-step. The third step has SW1 and SW4 on, and SW2 and SW3 are off, which is the same as step 2 of the full-step sequence. The sequence continues for eight steps and then repeats. The main difference between this sequence and the full-step sequence is that steps 2, 4, 6, and 8 are added to the full-step sequence to create the half-step moves.

11.9.4 Microstep Mode

The full-step and half-step motors tend to be slightly jerky in their operation as the motor moves from step to step. The amount of resolution is also limited by the number of physical poles that the rotor can have. The amount of resolution (number of steps) can be increased by manipulating the current that the controller sends to the motor during each step. The current can be adjusted so that it looks similar to a sine wave.

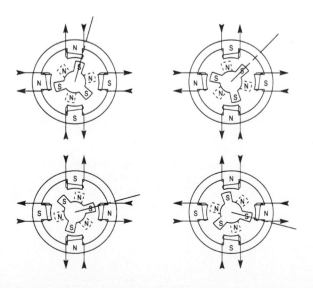

FIGURE 11-60 The diagrams that show the position of each pole while the motor is in full-step mode. The diagrams a, b, c, and d show the movement of the rotor in sequence. (Courtesy of Parker Compumotor Division.)

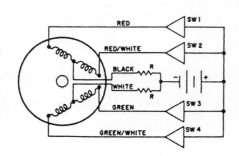

EIGHT STEP INPUT SEQUENCE HALF-STEP MODE*			
STEP	SW1	SW2	SW3	SW4
1	ON	OFF	ON	OFF
2	ON	OFF	OFF	OFF
3	ON	OFF	OFF	ON
4	OFF	OFF	OFF	ON
5	OFF	ON	OFF	ON
6	OFF	ON	OFF	OFF
7	OFF	ON	ON	OFF
8	OFF	OFF	ON	OFF
1	ON	OFF	ON	OFF

FIGURE 11-61 **(a)** The stepper motor with its switches. **(b)** The switching sequence for the eight-step input (half-step mode). (Courtesy of Superior Electric, Warner Electric.)

Fig. 11-62 shows the waveform for the current to each phase. From this diagram you can see that the current sent to each of the two sets of windings is timed so that it is always *out of phase* with each other. The fact that the current to each individual phase increases and decreases like a sine wave and that is always out of time with the other phase will allow the rotor to reach hundreds of intermediate steps. In fact it is possible for the controller to reach as many as 500 microsteps for a full-step sequence, which will provide 100,000 steps for each revolution.

The voltage sent to the motor is now a sine wave. The motor for this type of application is generally a permanent magnet brushless DC motor. When the sine wave is sent to the motor at 60 Hz it will cause the motor shaft to rotate at 72 rpm. The motor windings will require a capacitor to be wired in series for this type of application.

11.9.5 Stepper Motor Amplifier Circuits

The stepper motor system consists of a translator circuit that receives a signal that includes the number of steps and the direction. The translator circuit sends four individual control signals to the switch set circuit and the switch set circuit sends power signals to each of the two phases (windings) in the stepper motor. Fig. 11-63 shows a diagram of the typical stepper motor system. You should notice that the signal for the number of steps is included with this diagram and it is a series of square wave pulses (one for each step the rotor should move). The signal for the direction is a constant-voltage signal that is either positive or negative.

The translator typically receives its signal from a programmable logic controller (PLC) or other type of microprocessor controller. In some systems the controller is specifically designed to provide motion control or sequential control. The controller sends a command signal that consists of the number of steps the rotor should turn, and the direction signal indicates the direction. The step signals can be detected with LEDs or with an oscilloscope as they are sent to the translator. This means

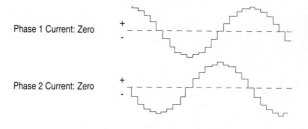

Phase 1 Current: Zero

Phase 2 Current: Zero

FIGURE 11-62 Phase current diagram for a stepper motor controller in microstep mode. (Courtesy of Parker Compumotor Division.)

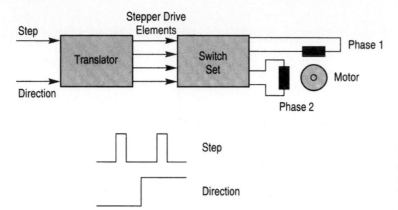

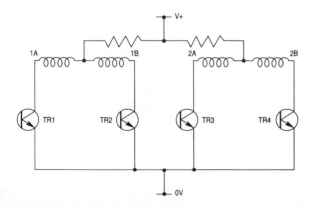

FIGURE 11-63 A diagram of a translator and stepper motor. (Courtesy of Parker Compumotor Division.)

that if you are troubleshooting the translator and you want to know if it is receiving the pulses that represent the steps, you can use an LED indicator or scope to see these pulses. The direction of rotation signal can be detected with a voltmeter to determine if the signal is a positive voltage or negative voltage.

11.9.5.1 Unipolar Drive Amplifier ■ The simplest type of drive amplifier for a stepper motor is the *unipolar drive*. From the electrical diagram in Fig. 11-64 you can see the circuit consists of two transistors that are connected to each winding set (phase). You should also notice that the two windings that make up each set have a terminal connection at the point where the two windings are connected. The positive voltage supply is connected at this point. Each winding has a transistor emitter–collector circuit connected in series with it, and the negative terminal of the voltage supply is connected to the emitter of each transistor.

This type of amplifier is called a unipolar drive because current can only flow in one direction at any one time. The motor must be a *bipolar type* so that current can be reversed in the second segment of the winding to get the motor to run in the reverse direction.

11.9.5.2 Chopper Drive Amplifier ■ The chopper drive amplifier provides a means to control the current in the stepper windings to provide better torque control. Fig. 11-65 shows an electrical diagram of this type of drive amplifier. You can see that each winding segment has four transistors connected to it. The transistors are

FIGURE 11-64 A unipolar drive amplifier circuit. The transistors act as the switches to provide the power drive current waveforms for the motor. (Courtesy of Parker Compumotor Division.)

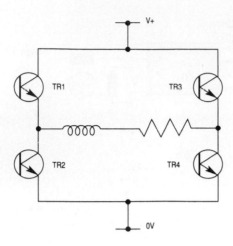

FIGURE 11-65 A chopper amplifier circuit for a stepper motor. This type of circuit is also called a recirculating chopper amplifier. (Courtesy of Parker Compumotor Division.)

identified as TR1 through TR4. The circuit also has two diodes and small resistor (approximately 0.1 Ω). Current is supplied to the winding by energizing TR1 and TR4, or by energizing TR2 and TR3. The direction of current flow will determine the polarity of the stator pole.

In the chopper circuit the combination of transistors allows the current to be recirculated through the winding if the current requirement to provide the motor torque at any instant is reached. This means that no current is wasted, and the drive amplifier and motor are more efficient. In a typical resistance-limited (*R-L*) drive the motor draws maximum current when the rotor is not turning which results in wasting up to 90% of current to the motor. In the chopper amplifier the chopper circuit allows the transistors to *chop* the current so that it can control the recirculation through the windings so that little current is wasted.

11.9.5.3 Power Dumping Control for a Chopper Drive Amplifier ■ One of the problems with permanent magnet motors is that they will generate a current anytime the rotor is turning. This generated current may cause a problem when the motor starts to decelerate and the current will become stronger than the supplied current. If this occurs, the excess current will damage the switching circuit components.

One method to control this problem is to include a power dumping circuit with the switching transistors. Fig. 11-66 shows an example of the location of the power dump circuit. In this diagram the power dump circuit is represented as a block.

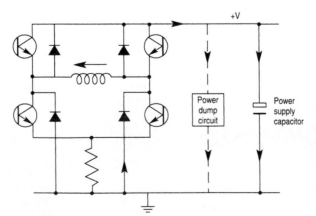

FIGURE 11-66 The dump circuit is shown connected across the $+V$ and ground in parallel with the power supply capacitor. This circuit recirculates the excess current that is generated when the motor decelerates. The regeneration circuit makes the stepper motor more efficient. (Courtesy of Parker Compumotor Division.)

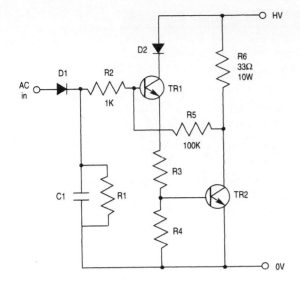

FIGURE 11-67 The detailed electrical diagram of the dumping circuit for the stepper motor chopper amplifier shown in the previous figure. (Courtesy of Parker Compumotor Division.)

Fig. 11-67 shows the detailed electronic circuit for the power dump circuit. The power dumping circuit has a detector to check the current threshold and turn off all transistors if the threshold is exceeded. When all of the transistors are turned off, the current will be isolated from the circuit components, and a capacitor will allow a path for the current to circulate through the winding until it dissipates below the threshold and becomes harmless. This type of circuit also provides a means to increase the efficiency of the drive slightly because the excess current that is built up during deceleration is regenerated instead of wasted.

The components in the power dump circuit use the rectifier and capacitors to establish a reference voltage from the AC applied voltage. The AC applied voltage is separate from the dc voltage the circuit will see from the motor windings. When the motor decelerates, the voltage that is generated by the stepper motor rotor will exceed the reference voltage, which will turn on transistor TR2. When TR2 is turned on, it will provide a path between the two potentials (HV and 0V) through the 33 Ω resistor. This path allows the excess current to be regenerated, which will cause it to dissipate. When the current has dissipated below the threshold, the transistor is turned off again and the circuit is waiting for the threshold to be exceeded again.

11.9.6 Stepper Motor Applications

Stepper motors are used in a wide variety of applications in industry including computer peripherals, business machines, motion control, and robotics, which are included in process control and machine tool applications. A complete list of applications is shown in Fig. 11-68.

11.9.7 Selecting the Proper Size Stepper Motor

When a stepper motor is selected, eight different things must be considered:

1. operating speed in steps/second
2. torque in ounce-inches

Application	Use
Computer Peripherals	
Floppy Disc	position magnetic pickup
Printer	carriage drive
Printer	rotate character wheel
Printer	paper feed
Printer	ribbon wind/rewind
Printer	position matrix print head
Tape Reader	index tape
Plotter	X-Y-Z positioning
Plotter	paper feed
Business Machines	
Card Reader	position cards
Copy Machine	paper feed
Banking Systems	credit card positioning
Banking Systems	paper feed
Typewriters (automatic)	head positioning
Typewriters (automatic)	paper feed
Copy Machine	lens positioning
Card Sorter	route card flow
Process Control	
Carburetor Adjusting	air-fuel mixture adjust
Valve Control	fluid gas metering
Conveyor	main drive
In-Process Gaging	parts positioning
Assembly Lines	parts positioning
Silicon Processing	I. C. wafer slicing
I. C. Bonding	chip positioning
Laser Trimming	X-Y positioning
Liquid Gasket Dispensing	valve cover positioning
Mail Handling Systems	feeding and positioning letters
Machine Tool	
Milling Machines	X-Y-Z table positioning
Drilling Machines	X-Y table positioning
Grinding Machines	downfeed grinding wheel
Grinding Machines	automatic wheel dressing
Electron Beam Welder	X-Y-Z positioning
Laser Cutting	X-Y-Z positioning
Lathes	X-Y positioning
Sewing	X-Y table positioning

FIGURE 11-68 Applications that use stepper motors. (Courtesy of Parker Compumotor Division.)

3. load inertia in lb-in.^{2}

4. required step angle

5. time to accelerate in milliseconds

6. time to decelerate in milliseconds

7. type of drive to be used

8. size and weight considerations

Some of this information will be provided from application specifications, such as the size and weight considerations, step angle, and the operating speed. Other information must be calculated. Several formulas are provided to help you with these calculations.

Torque (ounce-inches)

$$T = Fr$$

where F = force in ounces

r = radius in inches

Load inertia (I = Moment of inertia (lb-in.^{2})

$$I \text{ (lb-in.}^2) = \frac{Wr^2}{2} \quad \text{for a disk}$$

$$I \text{ (lb-in.}^2) = \frac{Wr^2}{2}(r_1^2 + r_2^2) \quad \text{for a hollow cylinder}$$

where W = weight in pounds

r = radius in inches of solid cylinder or disk

r_1 = inner radius of hollow cylinder

r_2 = outer radius of hollow cylinder

The formula for equivalent inertia to overcome friction in the system and enough torque to start or stop all inertia loads is:

$$T = I\alpha/24$$

where T is torque in ounce-inches

I is the moment of inertia in lb-in.^{2}

α is angular acceleration in radians per second^2

1/24 is the conversion factor for converting gravitational units (in.-sec^2) to units of mass (lb-in.^{2})

The formula for calculating the torque required to rotationally accelerate an inertia load is:

$$T = 2 \times I_0 \frac{\omega'}{t} \times \frac{\pi\phi}{180} \times \frac{1}{24}$$

where T = torque in ounce-inches

I_0 = inertial load in lb-in.^{2}

π = 3.1416

ϕ = step angle in degrees

$\omega' =$ step rate in steps per second

$t =$ time in seconds

....

PROBLEM 11-8

Calculate the torque required to accelerate a load that has inertia of 9.2 lb-in.^{2}, a step angle of 1.8°, and an acceleration from 0–1000 steps per second in 0.5 second.

SOLUTION

$$T = 2 \times 9.2 \times \frac{1000}{0.5} \times \frac{\pi \times 1.8}{180} \times \frac{1}{24}$$

Torque = 48.2 ounce-inches

....

11.10 LINEAR STEPPER MOTORS

The linear stepper motor has been made flat instead of round so its motion will be along a straight line instead of rotary. A picture of a linear motor and its amplifier is shown in Fig. 11-69, and the basic parts of the linear motor are shown in Fig. 11-70. In this diagram you can see the motor consists of a *platen* and a *forcer*. The platen is the fixed part of the motor and its length will determine the distance the motor will travel. It has a number of teeth that are like the rotor in a traditional

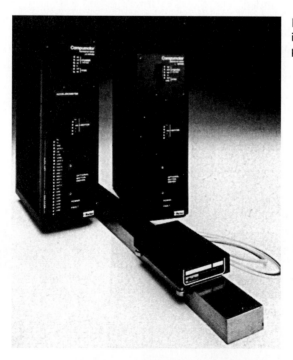

FIGURE 11-69 A linear motor and its amplifier. (Courtesy of Parker Compumotor Division.)

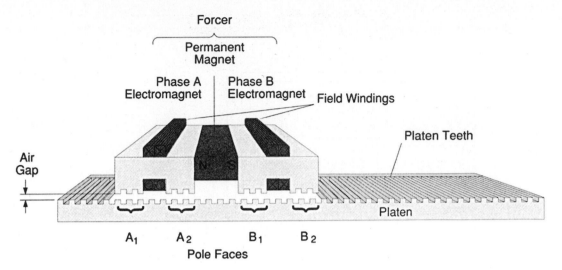

FIGURE 11-70   The forcer is shown on top of the platen of a linear motor. The electromagnets are identified on the forcer. (Courtesy of Parker Compumotor Division.)

stepper motor except it is passive and is not a permanent magnet. The forcer consists of four pole pieces that each have three teeth. The pitch of each tooth is staggered with respect to the teeth of the platen. It uses mechanical roller bearings or air bearings to ride above the platen on an air gap so that the two never physically come into contact with each other. The magnetic field in the forcer is changed by passing current through its coils. This action causes the next set of teeth to align with the teeth on the platen and causes the forcer to move from tooth to tooth over the platen in linear travel. When the current pattern is reversed, the forcer will reverse its direction of travel. A complete switching cycle consists of four full steps, which moves the forcer the distance of one tooth pitch over the platen. The typical resolution of a linear motor is 12,500 steps per inch, which provides a high degree of resolution. The typical load for a linear motor is low mass that requires high-speed movements.

11.10.1 Operation of the Linear Stepper Motor

The forcer consists of two electromagnets that are identified in Fig. 11-70 as magnet A and magnet B and one permanent magnet. The permanent magnet is a strong rare-earth permanent magnets. The electromagnets are formed in the shape of teeth so that their magnetic flux can be concentrated. In the diagram you can see that the forcer has four sets of teeth and these teeth are spaced in quadrature so that only one set of teeth is aligned with the teeth on the platen at any time.

When current is applied to the coil (field winding) of the electromagnets, their magnetic flux passes through the air gap between the forcer and the platen, causing a strong attraction between the two. The magnetic flux from the electromagnets also tends to reinforce the flux lines of one of the permanent magnets and cancels the flux lines of the other permanent magnet. The attraction of the forces at the time when peak current is flowing is up to ten times the holding force.

When a pattern of energizing one coil and then another is established, the resulting magnetic field will pull the motor in one direction from one tooth to the

LX System Diagram

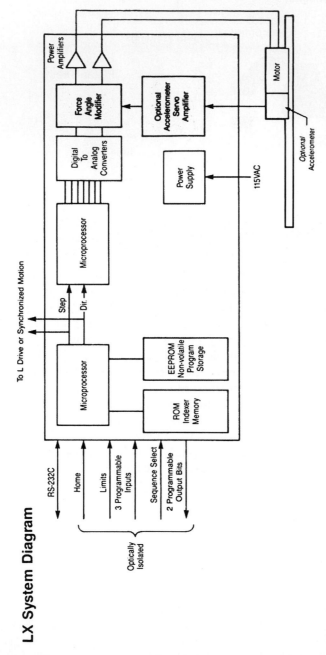

FIGURE 11-71 A block diagram of a linear motor controller. (Courtesy of Parker Compumotor Division.)

next. When current flow to the coil is stopped, the forcer will align itself to the appropriate tooth set and create a holding force that tends to keep the forcer from moving left or right to another tooth. The linear stepper motor controller sets the pattern for energizing and de-energizing the field coils so that the motor moves smoothly in either direction. By reversing the pattern, the direction the motor travels is reversed.

Fig. 11-71 shows a block diagram of the linear stepper motor controller. From this diagram you can see that it has a microprocessor that interfaces with a digital-to-analog converter, a force angle modifier, and a power amplifier. It also has a power supply for the amplifiers and it may have an accelerometer amplifier as an option. The microprocessor has ROM and EPROM memory to store programs.

11.10.2 Linear Motor Applications

The applications for a linear motor tend to be straight-line motion. These types of applications are slightly different from traditional stepper motor applications where the rotary motion is converted to linear motion with a ball and screw, rack and pinion, or other method. Fig. 11-72 shows the linear motor used in a coil winding positioner application. The linear motor in this application is teamed with a servomotor that controls the speed of the coil winding mechanism. The linear motor determines the exact location of the next coil that is added to the spool. The speed of the linear motor can be increased or decreased when the machine is spooling larger-

Coil Winder

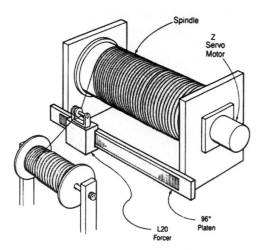

FIGURE 11-72   A linear stepper motor used in a coil winding application. The linear motor is used to control the position of the coil winder. (Courtesy of Parker Compumotor Division.)

A Compumotor L-L20-P96 system acts as the traverse element to guide the wire, while a Z Series servo motor rotates the spindle. Both axes are coordinated by a Compumotor 4000 indexer preprogrammed to produce a number of different coil types. Precise position control and mechanical simplicity over a long length of travel are provided by the linear motor.

Semiconductor wafer transport

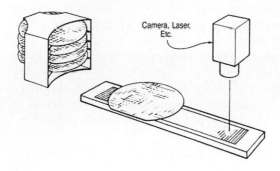

In this application, the linear motor acts as a transport for semiconductor wafers. The L20 linear motor system offers increased throughput and gentle handling of the wafer.

FIGURE 11-73 A linear stepper motor used to transport a silicon semiconductor wafer through a laser inspection station. (Courtesy of Parker Compumotor Division.)

diameter or smaller-diameter wire. The ability of the linear motor to provide small incremental steps makes it a good match for this application.

Fig. 11-73 shows a second application where the linear motor is used to transport a semiconductor wafer through a precision laser inspection station. The linear motor provides excellent locating ability for this application.

11.11 SERVOMOTORS

Servomotors are used in a variety of applications in industrial electronics that include precision positioning as well as motion control of larger motors. The main limitation of a stepper motor is that it cannot move large loads. The servomotor can be larger than 50 hp if necessary. Basically any motor can be used in a servo system, but the permanent magnet motor is typically used for smaller loads that are powered by DC voltage. Servo systems that use AC voltage can use large three-phase induction motors. The main components of a servo system are shown in Fig. 11-74. In this diagram you can see that the system includes a controller, amplifier, motor, and feedback sensor such as an encoder or resolver for position information, and a tachometer (generator) for velocity information.

The diagram in Fig. 11-74 shows an analog servo system. In this type of system, the amplifier uses analog amplifiers to compare the reference signal to the feedback signal and produce a power signal that is sent to the motor. You may need to

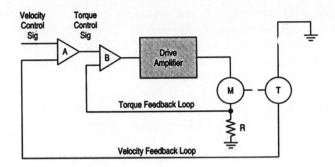

FIGURE 11-74 Block diagram of a typical servo control system. (Courtesy of Parker Compumotor Division.)

review Chapter 9 that explains closed-loop systems. In this diagram the motor is identified by the letter M, and the tachometer is identified by the letter T. This type of servo system is only checking for velocity and torque. This application is used on the main spindle motors for machine tool equipment like mills and lathes.

Fig. 11-75 shows a diagram of a servo system that has both digital and analog components. This servo diagram is slightly different from the first in that it has a positioning loop and a velocity loop. From this diagram you can see that the *axis feedrate command signal* is sent to the first section of this control loop at the far left point of the circuit. The box represented by a ∫ indicates that this section receives a frequency-type signal (pulse) and converts it to a filtered digital signal. The output of this section is called the *positioning command signal* and it is sent to the circle that represents the summing junction. The positioning feedback signal from an encoder is also sent to the summing junction as a digital value. The summing junction compares the command signal to the feedback signal and the result is called the *error*.

The error signal is sent to the gain calculation block where the signal is multiplied by the gain factor. The output of the gain calculation block is sent to a

FIGURE 11-75 A servo system with velocity and position loops. This system uses analog and digital components. (Courtesy of Rockwell Automation's Allen Bradley Business.)

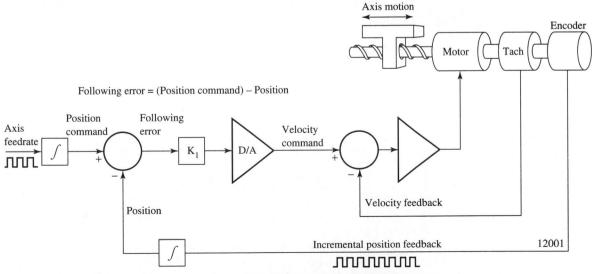

D/A amplifier. The D/A amplifier takes the digital value and turns it into an analog voltage that is sent to the next section of the controller.

The next section of the controller is the velocity loop. This loop uses the analog voltage from a tach-generator as a feedback signal. The circle represents another summing junction, and if the velocity of the motor is too slow, voltage is added to the original command signal. If the velocity is too fast, voltage is subtracted from the command signal. The signal from this section is sent to a final amplifier where the voltages and currents are increased to values large enough to run the motor.

The motor shaft is connected to a ball-screw mechanism that operates a machine tool. The axis motion is only two directions (back and forth). The tachometer and encoder are connected to the motor shaft, and they are generally built into the servomotor chassis.

11.11.1 Simple Servo Amplifiers

The output stage of all servo amplifiers is an analog circuit. The analog circuit provides a means to allow the voltage and current for the motor to be adjusted to control position, velocity, and torque. The feedback and comparator stages can be any mixture of digital and analog devices. For example, if the feedback section uses a resolver, the output of this device is analog, so the section it works with is generally also analog. If the feedback device is an encoder, its output is digital, and the digital signal can be converted through a frequency-to-voltage converter so that the signal is usable in an analog circuit. Or it can be filtered and can use a digital value. The advent of microprocessors has allowed the digital values to be used through every part of the servo controller except the final output stage.

Fig. 11-76 shows a diagram of the components in a typical servo linear amplifier. The circuit shows the motor winding connected to a set of transistors (Q_1 and Q_2). The transistors can control positive ($+V$) or negative ($-V$) voltage to make the motor turn in the clockwise or counterclockwise direction. The transistors can be pulsed on and off as in a pulse-width modulation (PWM) circuit, or they can be ramped up and down as a simple linear circuit. The base of the transistors can be controlled by a controller section of an amplifier that is completely linear. Or the controller can be digital with a D/A converter to provide the analog control signal to the base of Q_1 and Q_2, or the controller can be completely digital and the base

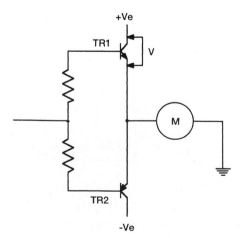

FIGURE 11-76 A typical linear amplifier for a servo system. (Courtesy of Parker Compumotor Division.)

of transistors Q_1 and Q_2 can be pulsed directly by the digital controller. Typically IGBTs (insulated gate bipolar transistors) are used in modern servo amplifiers where PWM or other switching circuits are used. The IGBTs allow the transistors to be switched on and off at frequencies that limit the *harmonic hum* in a motor or amplifier. The high-pitched hum represents both audio and electronic noise that must be eliminated or controlled.

The amplifier circuits for DC servomotors are similar to the AC circuits used for pulse-width modulation or other switching systems. In fact the complete amplifier for AC servomotors will be similar to the variable-frequency drive amplifiers shown earlier in this chapter.

11.11.1.1 Early Amplifiers (Push-Pull Amps) ■ The design of amplifiers has changed rapidly over the last 15 years because transistors, triacs, and SCRs have become able to handle larger voltages and currents without damaging themselves. It is easy to see the advantages that the changes in these devices have brought to motor drive amplifiers, but you must keep in mind that the early amplifiers were built so well that you will run into them even today when you are asked to troubleshoot a drive system. For this reason it is prudent to learn their basic parts and functions so you will be able to troubleshoot and analyze them. It is also a good idea to understand their basic operation because this is what has been modified to make the newer drives more efficient and more powerful.

One of the earliest types of linear amplifiers is called a *push-pull amplifier*, which was designed so that two transistors switched on and off to share the current load for the motor. Fig. 11-77 shows an example of this type of amplifier, and you can see that Q_2 and Q_3 are the power transistors. They are connected to the primary winding of transformer T_2. The servomotor winding is connected to the secondary winding of transformer T_2.

The operation of the push-pull amplifier begins with a sine wave signal that enters the input of the push-pull amplifier through capacitor C_1. Capacitor C_1 makes sure that the input signal is a pure sine wave with no DC bias. The base circuit of transistor Q_1 has a DC bias on it of approximately 13.5 volts. The base–emitter junction needs only 0.7 volt to turn it on, the rest of the DC bias voltage providing DC current through R_2 and R_3. This causes the sine wave input signal to practically

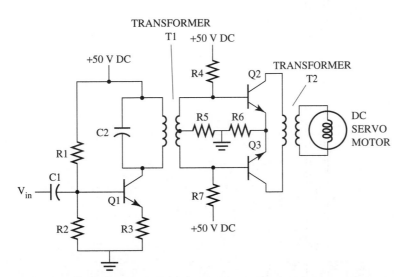

FIGURE 11-77 Push-pull amplifier for an AC servomotor. This diagram shows the power stages of the amplifier.

turn transistor Q_1 off at its minimum and cause Q_1 to be driven almost into saturation at the sine wave's maximum. This causes current to flow through the primary winding of transformer T_1. Notice the secondary of T_1 is center tapped to ground. When the positive half of the sine wave appears on the secondary, it appears across the entire secondary. Because of the center tap, only the upper portion of the secondary sees a positive voltage, and this forward bias transistor Q_2 allows it to conduct. Transistor Q_3 is turned off because it sees a negative voltage at its base. With Q_2 conducting, current flows up through the primary of T_2, providing a positive pulse to the secondary of T_2. When Q_3 conducts, a negative pulse is provided to the primary of T_2.

11.11.1.2 Chopper Amplifiers ■ Another type of early amplifier for a servomotor is called the *chopper amplifier* (see Fig. 11-78). In this type of amplifier the positive rectangular DC pulses arrive at the input of the amplifier circuit at capacitor C_1. These pulses arrive at the base of Q_1 as narrow spikes, which momentarily turn Q_1 on. This in turn momentarily turns Q_2 on, which allows current to flow through the primary of transformer T_1. Now the primary of transformer T_1 is really an *L-C* tank circuit. (Remember that the primary winding of the transformer is actually a big inductor.) When this tank circuit is hit by a pulse, it will produce a cycle or two of pure sine wave. When hit, in other words, the tank circuit will ring like a bell. The amplifier circuit is the clapper that rings the bell. Notice the secondary of T_1 is center tapped to $-60\ V_P$. The secondary of T_1 sees a pure AC sine wave, and to this AC signal, the $-60\ V_P$ appears as a ground. This means that for the positive half-cycle of the sine wave, Q_3 would see a positive pulse, and Q_4 would see a negative pulse. Both power transistors are NPN transistors, so a positive bias is needed at the base to cause them to conduct. As both bases are grounded, Q_4 would go into conduction because its emitter is lower than its base, giving it a forward base–emitter bias. The output of the tapped control winding would then be a sine wave. It should be noticed that the tapped control winding has $+60\ V_P$ on it, and the secondary of T_1 has $-60\ V_P$ on it. This means that the output of the tapped control winding is going to be a $120V_P$ sine wave.

FIGURE 11-78 Output stages of a chopper amplifier for an AC servomotor.

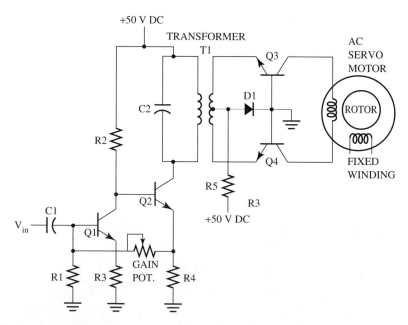

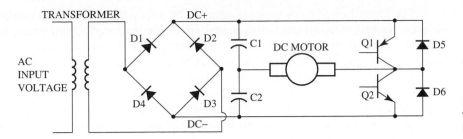

FIGURE 11-79 Two-transistor amplifier for a DC servomotor.

11.11.2 Amplifiers for DC Servomotors

The amplifiers for DC servomotors are slightly different from the push-pull amplifier and the chopper amplifier in that the power transistors can have a constant bias on their base rather than a pulsed signal. Fig. 11-79 shows an example of a two-transistor amplifier for a DC servomotor. The power supply for this amplifier is AC voltage. The first part of this circuit is the bridge rectifier that provides a DC voltage at the DC bus. The output stage of this amplifier uses two transistors and two capacitors that are connected across the DC motor armature.

The base of each of the power transistors is controlled by a switching circuit. This circuit can be controlled by an analog circuit or from a microprocessor. When the direction signal indicates the motor should run in the forward direction, the top transistor is biased on so that positive voltage is provided to the right-side terminal of the armature. The amount of bias voltage to the transistor base will increase or decrease to change the speed of the motor. When the direction signal indicates the motor should run in the opposite direction, the bottom transistor will be biased on and negative voltage is applied to the right side of the motor armature. A diode is connected in reverse bias across the emitter–collector terminals of each power transistor to limit the effects of voltage transients on the transistors. When a transient occurs, the diode provides a path to route the excess voltage and current back into the motor winding where it will be dissipated harmlessly.

11.11.2.1 Four-Transistor Amplifier for DC Servomotor ■ One of the drawbacks of a two-transistor amplifier is that the transistors must handle large amounts of current. Fig. 11-80 shows an example of a four-transistor amplifier for a DC servomotor. The four-transistor amplifier is commonly called a *bridge driver*. In this diagram you can see that the bridge rectifier is drawn as a rectangle but its operation is identical to the one shown in the two-transistor amplifier circuit. You should remember that it is easier to see the operation of a bridge rectifier in this configuration when three-phase power supply is used.

FIGURE 11-80 Four-transistor amplifier for a DC servomotor.

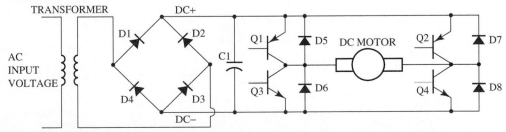

The base of each transistor is controlled by a switching circuit. Again the bias of each transistor is a continuous signal that can be varied from minimum to maximum. When the amplifier is set to run the motor in the clockwise direction, transistors Q2 and Q3 are biased on so that positive voltage is applied to the right side of the motor armature. When the motor is set to run in the counterclockwise direction, transistors Q1 and Q4 are biased on so that positive voltage is directed to the left side of the armature. The amount of bias voltage will determine the amount of voltage each transistor passes to the armature, which will in turn change the speed of the motor.

11.11.3 AC Servo Amplifiers

The amplifier for AC three-phase motors includes a pulse-width modulation circuit for voltage, current, and frequency control. Fig. 11-81 shows an example of this type of amplifier. From the diagram you can see that this circuit is designed specifically for a three-phase trapezoidal motor. The transistors in the amplifier are connected in an *H-bridge* configuration. The motor windings are connected as a three-phase wye with no external wires connected to the wye point. This type of motor is also called a *star connection* when it is used with brushless AC servomotors.

The drive logic and PWM switching controller is shown in the diagram as a block that is identified as a logic and PWM circuit. This block shows six arrows pointing away from it and pointing to the transistors. These arrows represent the six circuits for the base of each of the six transistors. The block below the PWM circuit represents the current-sensing part of the amplifier. This section of the amplifier uses a recirculating chopper system to control the current in a manner that is similar to the chopper circuit in the DC amplifier. The signals for this section of the amplifier come from the voltage that is developed across the series resistors connected between the transistor section and the motors. As you know, the amount of current flowing to the motor will determine the amount of voltage drop across these resistors.

This amplifier has a velocity amplifier that receives the original command

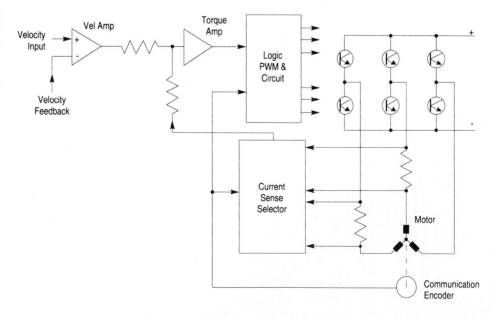

FIGURE 11-81 An AC servo drive amplifier specifically designed to operate with an AC trapezoidal brushless servomotor. (Courtesy of Parker Compumotor Division.)

signal for the amplifier and the velocity feedback. The op amp provides an output that represents the difference (error) between the command signal and the feedback signal. The output of the velocity amp is sent to the torque amp, where it is combined with the feedback from the current-sensing block. The output from this op amp is sent to the logic and PWM circuit block where it acts as the command signal. The position encoder provides the feedback signal for this block. This means that the velocity and position amplifiers are actually a closed-loop system within a closed-loop system. The gain for each of these amplifiers must be tuned so that the system has the best torque response and smooth acceleration and deceleration.

The feedback mechanism is generally a brushless DC tach generator, or an AC generator. Each of these feedback mechanisms provides smooth feedback voltages. If an encoder is used, its binary (digital) signal must be converted to an analog signal through a D/A converter or a frequency-to-analog F/A type converter if the signal is produced as a frequency.

11.11.4 Digital Servo Drives

The advent of the microprocessor chip has allowed a new family of servo drives for both AC and DC servomotors. Fig. 11-82 shows an example of a typical microprocessor controlled servo amplifier for an AC single-phase motor. The diagram for a three-phase motor would have one additional set of blocks for the D/A converter, PWM controller, and the set of H-bridge transistors. The H-bridge transistors will operate exactly like the H-bridge shown earlier in the example of DC servomotor drives.

In the diagram for the single-phase AC motor drive, you can see that command signals that indicate the amount of distance to travel (step) and the direction are provided to the microprocessor on the left side. Since these are both digital signals, they can come from a program or from a keypad that is manipulated by a machine operator. The command signals are compared with the digital feedback signals. In a digital controller, this comparison is in the form of a mathematical problem, which determines the *difference* (subtracts one from the other), and the result is called the *error signal.* The error signal can be positive or negative as the feedback signal changes, indicating that the drive amplifier is causing the motor shaft to move.

The error signal is a digital value and it is sent to the digital-to-analog converter (D/A) block of the drive where it is converted to an analog voltage or current. The analog voltage or current signal is used to control the pulse-width modulation (PWM) block, which provides a bias signal for the transistors that are used in the

FIGURE 11-82 Microprocessor controlled servo drive amplifier. This diagram is for a two-phase AC brushless servomotor. (Courtesy of Park Compumotor Division.)

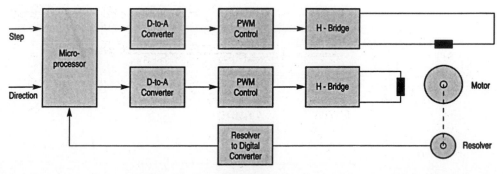

H-bridge. The PWM choppers control the base of each of the transistors in the H-bridge. In the diagram the drive has two H-bridge sections because the single-phase motor has only two stator windings. A three-phase motor would require the additional circuit to control the third stator winding.

The feedback signal for this system can be either digital (encoder) or analog (resolver) since the signal will be converted to a pure digital signal before it is sent to the microprocessor. If the signal is a digital value from an encoder, it may be manipulated inside the microprocessor to make sure its resolution and accuracy are calibrated to the machine. This is important if the machine uses any gears or transmissions between the motor and the moving parts of the machine.

11.11.5 Servomotors

Servomotors are available as AC or DC motors. Early servomotors were generally DC motors because the only type of control for large currents was through SCRs for many years. As transistors became capable of controlling larger currents and switching the large currents at higher frequencies, the AC servomotor became used more often. Early servomotors were specifically designed for servo amplifiers. Today a class of motors is designed for applications that may use a servo amplifier or a variable-frequency controller, which means that a motor may be used in a servo system in one application, and used in a variable-frequency drive in another application. Some companies also call any closed-loop system that does not use a stepper motor a servo system, so it is possible for a simple AC induction motor that is connected to a velocity controller to be called a servomotor. (See Figs. 11-83 and 11-84.)

Some characteristics of all servomotors are:

- Ability to produce high torque at any speed.
- Must not overheat at low speeds or at standstill.
- Ability to change direction and accelerate quickly.
- Ramp up and down smoothly to reach position accurately.
- Limit drift during repeated operations.
- Be able to accelerate and decelerate smoothly under any load within range.

FIGURE 11-83   Typical PM servomotors. (Courtesy of Pacific Scientific.)

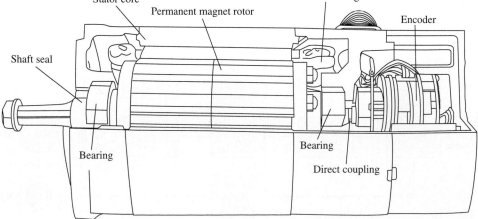

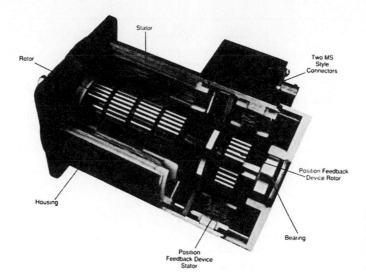

Stator

Rotor

Two MS Style Connectors

Position Feedback Device Rotor

Housing

Bearing

Position Feedback Device Stator

FIGURE 11-84 Cutaway picture of a permanent magnet servomotor. (Courtesy of Pacific Scientific.)

Some changes that must be made to any motor that is designed as a servomotor includes the ability to operate at a range of speeds without overheating, the ability to operate at zero speed and retain sufficient torque to hold a load in position, and the ability to operate at very low speeds for long periods of time without overheating. Older type motors have cooling fans that are connected directly to the motor shaft. When the motor runs at slow speed, the fan does not move enough air to cool the motor. Newer motors have an separate fan mounted so it will provide optimum cooling air. This fan is powered by a constant voltage source so that it will turn at maximum RPM at all times regardless of the speed of the servomotor. One of the most usable types of motors in servo systems is the permanent magnet (PM) type motor. The voltage for the field winding of the permanent magnet type motor can be AC voltage or DC voltage. The permanent magnet type motor is similar to other PM type motors presented previously. Fig. 11-83 shows a cut-away picture of a PM motor and Fig. 11-84 shows a cut-away diagram of a PM motor. From the picture and diagram you can see the housing, rotor and stator all look very similar to the previous type PM motors. The major difference with this type of motor is that it may have gear reduction to be able to move larger loads quickly from a stand still position. This type of PM motor also has an encoder or resolver built into the motor housing. This ensures that the device will accurately indicate the position or velocity of the motor shaft.

11.11.5.1 Brushless Servomotors ■ The brushless servomotor is designed to operate without brushes. This means that the commutation that the brushes provided must now be provided electronically. Electronic commutation is provided by switching transistors on and off at appropriate times. Fig. 11-85 shows three examples of the voltage and current waveforms that are sent to the brushless servomotor. Fig. 11-86 shows an example of the three windings of the brushless servomotor. The main point about the brushless servomotor is that it can be powered by either ac voltage or dc voltage.

Fig. 11-85 shows three types of voltage waveforms that can be used to power the brushless servomotor. Fig. 11-85a shows a trapezoidal EMF (voltage) input and

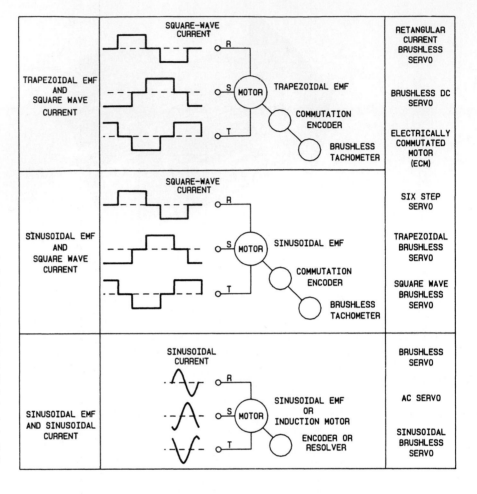

FIGURE 11-85 **(a)** Trapezoidal input voltage and square wave current waveforms. **(b)** Sinusoidal input voltage and sinusoidal voltage and square wave output voltage waveforms. **(c)** Sinusoidal input voltage and sinusoidal current waveforms. This has become the most popular type of brushless servomotor control. (Courtesy of Electro-Craft, A Rockwell Automation Business.)

a square wave current input. Fig. 11-85b shows a sinusoidal waveform for the input voltage and a square wave current waveform. Fig. 11-85c shows a sinusoidal input waveform and a sinusoidal current waveform. The sinusoidal input and sinusoidal current waveform are the most popular voltage supplies for the brushless servomotor.

Fig. 11-86 shows three sets of transistors that are similar to the transistors in the output stage of the variable-frequency drive. In Fig. 11-86a the transistors are connected to the three windings of the motor in a similar manner as in the variable-frequency drive. In Fig. 11-86b the diagram of the waveforms for the output of the transistors is shown as three separate sinusoidal waves. The waveforms for the control circuit for the base of each transistor are shown in Fig. 11-86c. Fig. 11-86d shows the back EMF for the drive waveforms.

11.11.6 Servomotor Controllers

Servomotor controllers have become more than just amplifiers for a servomotor. Today servomotor controllers must be able to make a number of decisions and provide a means to receive signals from external sensors and controls in the system, and send signals to host controllers and PLCs that may interface with the servo system. Fig. 11-87 shows a picture of several servomotors and their amplifiers.

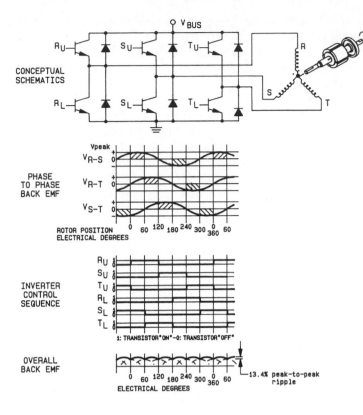

FIGURE 11-86 **(a)** Transistors connected to the three windings of the brushless servomotor. **(b)** Waveforms of the three separate voltages that are used to power the three motor windings. **(c)** Waveforms of the signals used to control the transistor sequence that provides the waveforms for the previous diagram. **(d)** Waveform of the overall back EMF. (Courtesy of Electro-Craft, A Rockwell Automation Business.)

FIGURE 11-87 Example servomotors and amplifiers. (Courtesy of Pacific Scientific.)

The components in this picture look similar to a variety of other types of motors and controllers.

Fig. 11-88 shows a diagram of the servomotor controller so that you can see some of the differences from other types of motor controllers. The controller in this diagram is for a DC servomotor. The controller has three ports that bring signals in or send signals out of the controller. The power supply, servomotor, and tachometer are connected to port P3 at the bottom of the controller. You can see that the supply voltage is 115 volt AC single phase. A main disconnect is connected in series with the L1 wire. The L1 and N lines supply power to an isolation step-down transformer. The secondary voltage of the transformer can be any voltage between 20–85 volts. The controller is grounded at terminal 8. You should remember that the ground at this point is only used to provide protection against short circuits for all metal parts in the system.

The servomotor is connected to the controller at terminals 4 and 5. Terminal 5 is + and terminal 4 is −. Terminal 3 provides a ground for the shield of the wires that connect the motor and the controller. The tachometer is connected to terminals 1 and 2. Terminal 2 is + and terminal 1 is −. The shield for this cable is grounded

FIGURE 11-88 Diagram of a servo controller. This diagram shows the digital (on-off) signals and the analog signals that are sent to the controller, and the signals the controller sends back to the host controller or PLC. (Courtesy of Electro-Craft, A Rockwell Automation Business.)

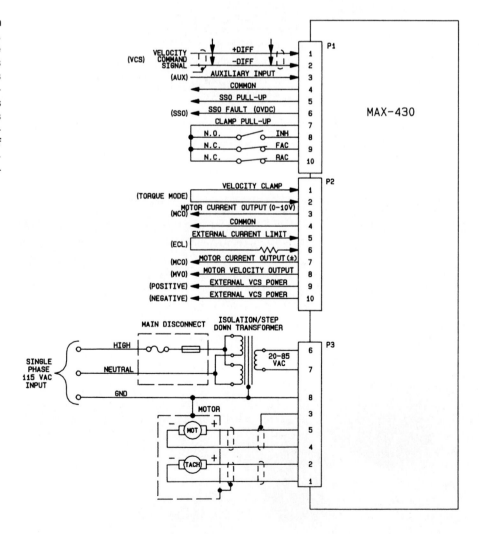

to the motor case. The wires connected to this port will be larger than wires connected to the other ports, since they must be capable of carrying the larger motor current. If the motor uses an external cooling fan, it will be connected through this port. In most cases the cooling fan will be powered by single-phase or three-phase AC voltage that remains at a constant level such as 110 volts AC or 240 volts AC.

The command signal is sent to the controller through port P1. The terminals for the command signal are 1 and 2. Terminal 1 is + and terminal 2 is −. This signal is a type signal, which means that it is not grounded or does not share a ground potential with any other part of the circuit. Several additional auxiliary signals are also connected through port 1. These signals include inhibit (INH), which is used to disable the drive from an external controller, and forward and reverse commands (FAC and RAC), which tell the controller to send the voltage to the motor so that it will rotate in the forward or reverse direction. In some applications, the forward maximum travel limit switch and reverse maximum travel limit switch are connected so that if the machine travel moves to the extreme position so that it touches the overtravel limit switch, it will automatically energize the drive to begin travel in the opposite direction.

Port P1 also provides several digital output signals that can be used to send fault signals or other information such as "drive running" back to a host controller or PLC. Port P1 basically is the interface for all digital (on-off) signals.

Port P2 is the interface for analog (0–max) signals. Typical signals on this bus include motor current and motor velocity signals that are sent from the servo controller back to the host or PLC where they can be used in verification logic to ensure the controller is sending the correct information to the motor. Input signals from the host or PLC can also be sent to the controller to set maximum current and velocity for the drive. In newer digital drives, these values are controlled by *drive parameters* that are programmed into the drive.

11.11.7 PWM Servo Amplifier

The PWM servo amplifier is used on small-size servo applications that use DC brush-type servomotors. Fig. 11-89 shows a diagram for this type of amplifier. From the diagram you can see that single-phase AC power is provided to the amplifier as the supply at the lower left part of the diagram. The AC voltage is rectified and sent to the output section of the drive that is shown in the top right corner of the diagram. The output section of the drive uses four IGBTs to create the pulse-width modulation waveform. The IGBTs are connected so that they provide 30–120 volts DC and up to 30 A to the brush-type DC servomotor. The polarity of the motor is indicated in the diagram.

The remaining circuits show a variety of fault circuits in the middle of the diagram that originate from the *fault logic board* and provide an output signal at the bottom of the diagram. You should notice that the fault output signals include overvoltage, overtemperature, and overcurrent. A fourth signal is identified as SSO (system status output), which indicates the status of the system as faulted anytime a fault has occurred. A jumper is used to set the SSO signal as an open collector output with a logic level "1" indicating the drive is ready, or as a normally closed relay indicating the drive is ready.

The input terminals at the bottom right part of the diagram are used to *enable* or *inhibit* the drive, and to select forward amplifier clamp (FAC) or reverse amplifier

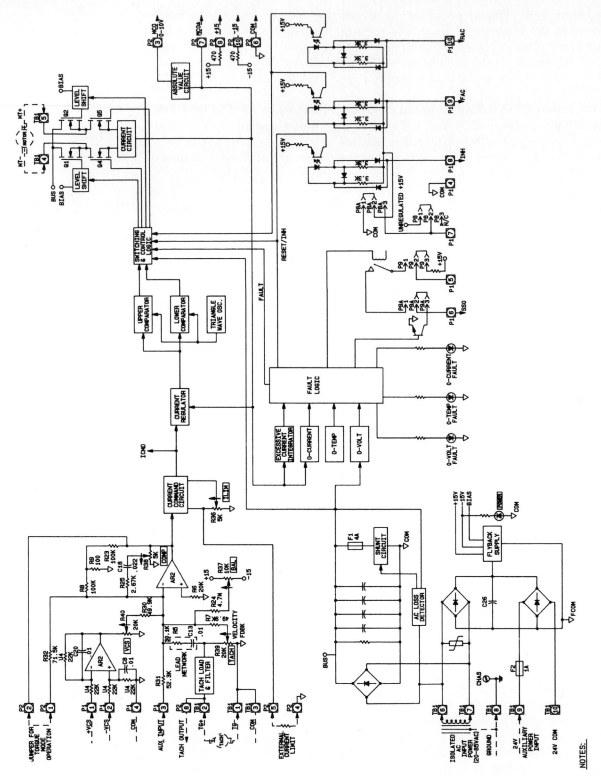

FIGURE 11-89 Diagram of a pulse-width modulation (PWM) amplifier with a brush-type DC servomotor. (Courtesy of Electro-Craft, A Rockwell Automation Business.)

554

clamp (RAC). The inhibit signal is used as a control signal, since it inhibits the output stage of the amplifier if it is high. The FAC and RAC signals limit the current in the opposite direction to 5%.

The input signals are shown in the diagram at the upper left side. The VCS (velocity command signal) requires a +VCS and a −VCS signal to provide the differential signal.

11.11.8 Applications for Servo Amplifiers and Motors

You will get a better idea of how servomotors and amplifiers operate if you see some typical applications. Fig. 11-90 shows an example of a servomotor used to control a press feed. In this application sheet material is fed into a press where it is cut off to length with a knife blade or sheer. The sheet material may have a logo or other advertisement that must line up registration marks with the cut-off point. In this application the speed and position of the sheet material must be synchronized with the correct cut-off point. The feedback sensor could be an encoder or resolver that is coupled with a photoelectric sensor to determine the location of the registration mark. An operator panel is provided so that the operator can jog the system for maintenance to the blades, or when loading a new roll of material. The operator panel could also be used to call up parameters for the drive that correspond to each type of material that is used. The system could also be integrated with a programmable controller or other type of controller and the operator panel could be used to select the correct cut-off points for each type of material or product that is run.

11.11.8.1 An Example of a Servo Controlled In-Line Bottle-Filling Application ■
A second application is shown in Fig. 11-91. In this application multiple filling heads line up with bottles as they move along a continuous line. Each of the filling heads

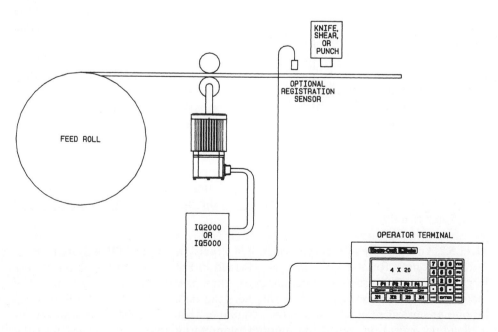

PRESS FEED APPLICATION

FIGURE 11-90 Application of a servomotor controlling the speed of material as it enters a press for cutting pieces to size. (Courtesy of Electro-Craft, A Rockwell Automation Business.)

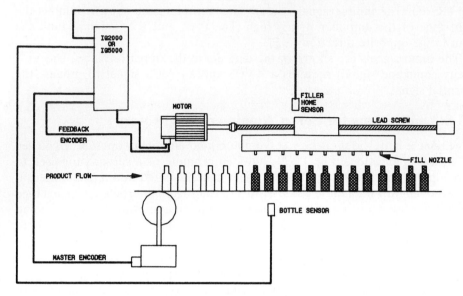

FIGURE 11-91 Application of a beverage-filling station controlled by a servomotor. (Courtesy of Electro-Craft, A Rockwell Automation Business.)

must match up with a bottle and track the bottle while it is moving. Product is dispensed as the nozzles move with the bottles. In this application ten nozzles are mounted on a carriage that is driven by a ball-screw mechanism. The ball-screw mechanism is also called a lead screw. When the motor turns the shaft of the ball screw, the carriage will move horizontally along the length of the ball-screw shaft. This movement will be smooth so that each of the nozzles can dispense product into the bottles with little spillage.

The servo drive system utilizes a positioning drive controller with software that allows the position and velocity to be tracked as the conveyor line moves the bottles. A master encoder tracks the bottles as they move along the conveyor line. An auger feed system is also used just prior to the point where the bottles enter the filling station. The auger causes a specific amount of space to be set between each bottle as it enters the filling station. The bottles may be packed tightly as they approach the auger, but as they pass through the auger their space is set exactly so that the necks of the bottles will match the spacing of the filling nozzles. A detector is also in conjunction with the dispensing system to ensure that no product is dispensed from a nozzle if a bottle is missing or large spaces appear between bottles.

The servo drive system compares the position of the bottles from the master encoder to the feedback signal that indicates the position of the filling carriage that is mounted to the ball screw. The servo drive amplifier will increase or decrease the speed of the ball-screw mechanism so that the nozzles will match the speed of the bottles exactly.

11.11.8.2 An Example of a Servo Controlled Precision Auger Filling System ■ A third application for a servo system is provided in Fig. 11-92. In this application a large filling tank is used to fill containers as they pass along a conveyor line. The material that is dispensed into the containers can be a single material fill or it can be one of several materials added to a container that is dumped into a mixer for a blending operation. Since the amount of material that is dispensed into the container must be accurately weighed and metered into the box, an auger that is

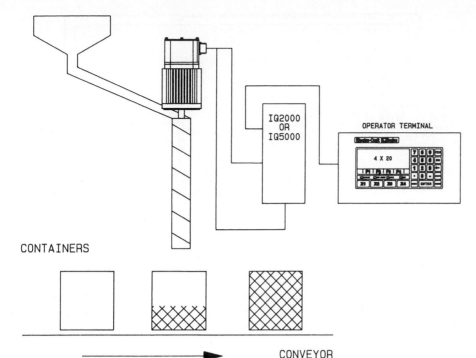

FIGURE 11-92 Application of a precision auger filling station controlled by a servomotor. (Courtesy of Electro-Craft, A Rockwell Automation Business.)

controlled by a servo system is used. The feedback sensor for this system can be a weighing system such as the load cell discussed in earlier chapters. The command signal can come from a programmable controller or the operator can enter it manually by selecting a recipe from the operator's terminal. The amount of material can be different from recipe to recipe.

The speed of the auger can be adjusted so that it runs at high speed when the container is first being filled, and the speed can be slowed to a point where the final grams of material can be metered precisely as the container is filled to the proper point. As the price of material increases, precision filling equipment can provide savings as well as quality in the amount of product used in the recipe.

11.11.8.3 An Example of a Label Application Using Servomotors ■ The fourth application has a servomotor controlling the speed of a label-feed mechanism that pulls preprinted labels from a roll and applies them to packages as they move on a continuous conveyor system past the labeling mechanism. The feedback signals are provided by an encoder that indicates the location of the conveyor, tach generator that indicates the speed of the conveyor, and a sensor that indicates the registration mark on each label. The servo positioning system is controlled by a microprocessor that sets the error signal, and the servo amplifier that provides power signals to the servomotor. This application is shown in Fig. 11-93.

11.11.8.4 An Example of a Random Timing Infeed System Controlled by a Servomotor ■ The fifth application is presented in Fig. 11-94 and it shows a series of packaging equipment that operates as three separate machines. The timing cycle of each station of the packaging system is independent from the others. The packaging system consists of an *infeed conveyor*, a *positioning conveyor*, and a *wrapping station*.

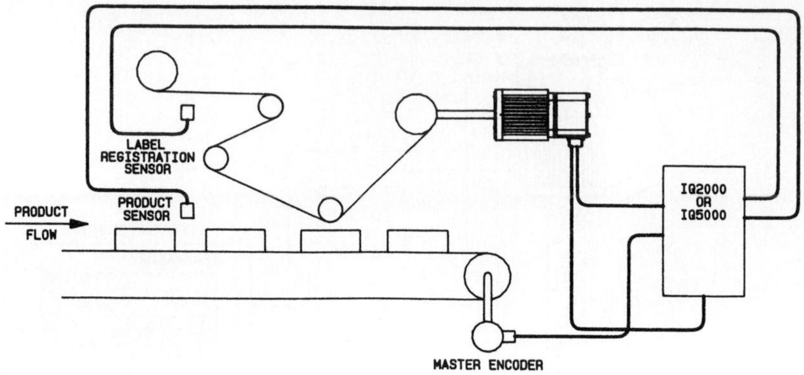

FIGURE 11-93 Example of a labeling application controlled by a servomotor. (Courtesy of Electro-Craft, A Rockwell Automation Business.)

The infeed conveyor and the wrapping station are mechanically connected so that they run at the same speed. The position of the packages on the wrapping station must be strictly controlled so that the packages do not become too close to each other. A piece of metal called a *flight* is connected to the wrapping station conveyor at specific points to ensure each package stays in position. A sensor is mounted at the beginning of the positioning conveyor to determine the front edge of the package when it starts to move onto the positioning conveyor. A second sensor is positioned at the bottom of the packaging conveyor to detect the flights. Both of these signals from the sensors are sent to the servomotor to provide information so the servo can adjust the speed of the positioning conveyor so that each package aligns with one of the flights as it moves onto the packaging conveyor. This application shows that the servo positioning controller can handle a variety of different signals from more than one sensor because the controller uses a microprocessor.

11.11.9 Selecting a Servomotor for an Application

When you must select a servomotor for initial installation or for a replacement, you will need to consider the following data which are shown in the table in Fig.

FIGURE 11-94 Example of a packaging system with random timing functions controlled by a servomotor. (Courtesy of Electro-Craft, A Rockwell Automation Business.)

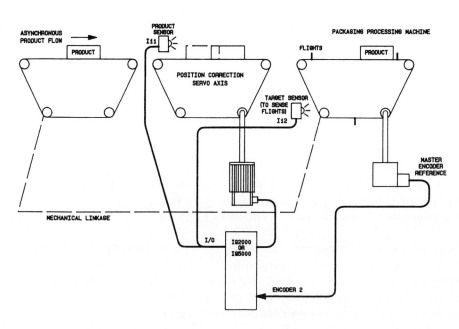

Technical Data		Z-605 ZX-605 ZXF-605	Z-606 ZX-606 ZXF-606	Z-610 ZX-610 ZXF-610	Z-620 ZX-620 ZXF-620	Z-630 ZX-630 ZXF-630	Z-635 ZX-635 ZXF-635	Z-640 ZX-640 ZXF-640	Z-910 ZX-910 ZXF-910	Z-920 ZX-920 ZXF-920	Z-930 ZX-930 ZXF-930	Z-940 ZX-940 ZXF-940
Continuous stall torque												
	oz-in	346	633	867	1,743	2,475	2,475	4,114	2,407	4,263	5,990	9,021
	lb-in	22	40	54	109	155	155	257	150	266	374	564
	Nm	2.44	4.47	6.12	12.31	17.48	17.48	29.05	17.0	30.1	42.3	63.7
Peak torque	oz-in	1,083	1,954	1,733	3,486	4,951	4,951	8,228	5,205	8,525	11,980	18,041
	lb-in	68	122	108	218	309	309	514	325	533	749	1,128
	Nm	7.65	13.80	12.24	24.62	34.96	34.96	58.10	35.4	61.5	84.6	127.5
Rated power	hp	2	2.1	4.2	5.6	5.4	6.1	5.9	9.6	10.4	11.0	11.1
	k Watts	1.49	1.57	3.13	4.18	4.03	4.5	4.40	7.2	7.8	8.2	8.3
Rated speed	rpm	6,200	3,600	7,000	3,700	2,500	3,000	1,600	5,000	3,150	2,300	1,500
	rps	103	60	117	62	42	50	27	83.3	52.5	38.3	25.0
Rated current (line)	A (rms)	5	5.3	14.1	14.1	14.1	14.1	14.1	27.2	27.7	28.3	28.3
Peak current (3.3 sec max)	A (rms)	16.6	17.2	28.2	28.2	28.2	28.2	28.2	56.6	56.6	56.6	56.6
Max cont AC input power (3 phase 240 VAC)	A (rms)	6	6	15	15	15	15	15	30	30	30	30
Rotor inertia	oz-in^2 (mass)	5.45	9.45	13.73	35.87	50.79	56.21	111.21	50.79	111.21	166.21	459.48
	oz-in-sec^2	0.01	0.02	0.04	0.09	0.13	0.15	0.29	0.132	0.288	0.431	1.190
	kg m^2 x 10^{-6}	99.6	172.9	251.2	656	929	1,028	2,034	929	2,034	3,040	8,404
Motor weight	lbs	10.0	14.0	17.0	29.0	32.0	37	51.0	32.0	57.0	65.0	112.0
	kg	4.5	6.4	7.7	13.2	14.5	16.8	23.2	15.0	26.0	29.0	51.0
Shipping weight	lbs	52.0	55.0	58.0	71.0	74.0	79.0	93.0	89.0	114.0	122.0	169.0
	kg	23.6	25.0	26.4	32.3	33.6	35.9	42.3	40.0	52.0	55.0	77.0

FIGURE 11-95 Technical data table for selecting servomotors. The motors are listed across the top of the table and the data are listed down the side. (Courtesy of Parker Compumotor Division.)

11-95. The first point to consider is whether the motor will be used for continuous duty or intermittent duty. The table shows the amount of continuous stall torque each motor can produce. The second rating from the table is the amount of peak torque. Both the continuous torque and the peak torque are listed in oz-in. (ounce-inches), lb-in. (pound-inches) or in-lb. (inch-pound), and in Newton-meters (Nm).

The conversion of ounce-inches, pound-inches, and Newton-meters follows:

$$1 \text{ inch-pound} = 16 \text{ ounce-inches}$$

$$1 \text{ inch-pound} = 0.112085 \text{ Newton-meter}$$

The rated power for each motor is listed in horsepower (hp) and kilowatts (kW). The calculation for determining horsepower and the conversion from horsepower to watts follows.

Calculate Horsepower

$$\text{Horsepower} = \frac{\text{Torque} \times \text{Speed}}{16,800}$$

$$\text{Torque} = \text{oz-in}$$
$$\text{Speed} = \text{revolutions per second}$$

The horsepower calculation uses the torque available at the specified speed.

$$1 \text{ Horsepower} = 746 \text{ watts}$$

The rated speed is listed on the data table as rpm (revolutions per minute). The rated current is listed in A rms. The peak current and the maximum continuous AC input power are listed in A rms. The rotor inertia is rated in oz-in.[2] (mass). The model numbers for each motor are listed across the top of this table.

PROBLEM 11-9

Select the proper motor from the table shown in Fig. 11-95. The load for the motor is 6 hp, the rated speed is 3000 rpm, and the rated torque required for the load is at least 4,500 oz-in.

SOLUTION

Since the load requires a minimum of 6 hp and requires at least 4,500 oz-in. of torque, the Z-635, ZX-635, or ZFX 635 should be the smallest motor selected. The motor could be the next larger size, but it is not advisable to select too large a motor.

SOLUTION TO JOB ASSIGNMENT

The solution to your job assignment should refer to diagrams and tables in this chapter. To determine the size of the AC servomotor for the cutter head, you should use the calculations and table found in Section 11.10. The diagram in Fig. 11-81 should be used to troubleshoot or check out the drive to ensure that is operating correctly.

To determine the size of the DC servomotor for the parts loader axis motors, you should use the calculation and tables found in Section 11.11. The diagrams in Figs. 11-76, 11-77, 11-78, 11-79, 11-80, 11-81, 11-85, 11-86, and 11-89 should be used to troubleshoot or check out the servo drive to ensure it is operating correctly.

To determine the size of the stepper motor for the parts loader, you should use the calculation and tables found in Sections 11.9 and 11.10. The diagrams in Figs. 11-58, 11-59, 11-60, 11-64, 11-65, 11-66, and 11-67 should be used to troubleshoot or check out the drive to ensure it is operating correctly.

To determine the size of the AC variable-frequency drive motor for the conveyor, you should use the calculation and tables found in Section 11.7. The diagrams in Figs. 11-43, 11-44, and 11- 47 should be used to troubleshoot or check out the drive to ensure it is operating correctly.

To determine the size of the motor starter and motor for the coolant pump, you should use the calculation and tables (Figs. 11-29 and 11-38) found in Section 11.6.10. The diagrams in Figs. 11-25, 11-26, 11-27, 11-34, 11-35, and 11-37 should be used to troubleshoot or check out the starter and motor to ensure they are operating correctly.

To determine the size of the solenoid valve for the coolant, you should use the calculation and tables found in Section 11.2. The diagrams in Figs. 11-1, 11-6, 11-9, and 11-10 should be used to troubleshoot or check out the solenoid valve to ensure it is operating correctly.

To determine the size of the hydraulic valve for the parts fixture, you should use the calculation and tables found in Section 11.3. The diagrams in Figs. 11-8 and 11-11 should be used to troubleshoot or check out the hydraulic valve to ensure it is operating correctly.

QUESTIONS

1. Identify all the parts of a two-way solenoid and explain their operation.
2. Why is the symbol for the internal operation of a two-way valve drawn twice in the diagram for a pneumatic valve, and three times for a three-way pneumatic valve?
3. Explain the operation of a three-position, four-way hydraulic solenoid valve and the function of all of the ports.
4. List three applications where proportional valves could be used.
5. Explain the operation of a pneumatic-assisted valve and how it is controlled by a 4–20 mA signal.
6. Explain how a 4–20 mA signal is converted to 3–15 psi to control the pneumatic-assisted valve.
7. List three applications where pneumatic-assisted valves could be used.
8. List the basic sections of a variable-frequency drive and explain the operation of each.
9. Explain how a variable-frequency drive changes the speed of an AC motor.
10. List four common parameters used in a variable-frequency drive and explain what function parameters provide for the drive.
11. List the basic parts of a stepper motor and explain the function of each.
12. Explain what a hybrid stepper motor is and why it is the most widely used stepper motor.
13. List ten applications where stepper motors are used.
14. How is a linear motor similar to a stepper motor?
15. Explain the operation of a push-pull amplifier and chopper amplifier used in servo controllers.
16. List two applications for a linear motor and explain why the linear motor is a better choice than a stepper motor for these applications.
17. Use the diagrams in this chapter to explain the difference between the brush-type and brushless-type servomotors.
18. List six characteristics of all servomotors.
19. If transistor Q_2 in the chopper amplifier diagram of Fig. 11-78 opened, what would be the effect to the amplifier and servomotor?
20. If transistor Q_1 in the chopper amplifier diagram of Fig. 11-78 shorted, what would be the effect to the amplifier and servomotor?

TRUE OR FALSE

1. _____ Inrush current is larger than holding current for a solenoid.

2. _____ The diode that is connected in reverse bias across a solenoid coil or other inductive load that is powered by DC voltage is there for the purpose of rectification to ensure that the voltage that reaches the coil is always DC.

3. _____ A proportional valve is more efficient than an on/off valve because it can turn on and off completely with only a small proportion of the amount of energy the on/off valve needs.

4. _____ The function of the heaters and overload mechanism in a motor starter is to monitor the amount of current the motor is using, and trip the overload contacts open if the amount of current becomes too large.

5. _____ The variable-frequency drive creates the three-phase output waveform by switching transistors on and off at the appropriate time.

6. _____ IGBTs are used in modern variable-frequency drives instead of traditional transistors because they can withstand more voltage and current and because they turn on and off at a higher frequency, making the drive quieter.

7. _____ Stepper motors are generally used in open-loop applications where a position sensor is not needed because the stepper moves a precise amount of distance each time a pulse is provided to the motor.

8. _____ A linear motor is similar to a stepper motor in that it is approximately the same size and shape as a stepper motor and, looking at them, it would be difficult to tell them apart.

9. _____ The DC voltage on the DC bus of a variable-frequency drive that is powered with 480 volts AC can be higher than 600 volts DC because several step-up transformers are used inside the drive to increase the DC voltage.

10. _____ The push-pull amplifier uses a single transistor that is pulsed on and off to control the voltage to a servomotor.

MULTIPLE CHOICE

1. The main parts of a motor starter include _____.
   a. the coil, NO main contacts, heater, and solenoid.
   b. the coil, NO main contacts, NO auxiliary contacts, heater, and NC overload contacts.
   c. the coil, NO main contacts, NC main contacts, NO auxiliary contacts, heater contacts, and overloads.

2. The operation of a solid-state motor starter is different from an electromechanical motor starter _____.
   a. because it uses a bridge rectifier to change AC to DC voltage.
   b. because it uses MOVs to protect the coil from overvoltage conditions.
   c. because the solid-state motor starter can turn on voltage gradually to the motor to limit inrush current.

3. The operation of a stepper motor and a servomotor _____.
   a. is different because the voltage to the stepper motor must be pulses that cause the motor to rotate a step at a time.
   b. is different because the voltage to the servomotor can be only regular DC voltage and the voltage to the stepper motor is pulsed.
   c. is nearly the same because the stepper and servo are actually two different names for the same type of motor.

4. The main parts of a relay include _____.
   a. coil and one or more sets of NO or NC contacts.
   b. coil, NO contacts, heater, and NO overload contacts.
   c. coil, NO contacts, heater, and NC overload contacts.

5. The operation of a linear motor _____.

    a. is similar to the stepper motor except its travel is linear rather than rotary.

    b. is similar to the servomotor except its travel is linear rather than rotary.

    c. is actually a rotary-type stepper motor that uses a ball-screw mechanism to create linear travel.

6. The brushless servomotors generally are powered with sinusoidal waveforms _____.

    a. because sinusoidal waveforms are much more efficient than a pure DC waveform and this prevents the motor from overheating.

    b. because the sinusoidal waveforms provide electronic commutation.

    c. because the sinusoidal waveforms provide more torque than pure DC waveforms.

PROBLEMS

1. Use the tables provided in this chapter to determine the catalog number for the correct size of a normally opened solenoid for a water system that has an orifice size of $\frac{1}{4}$ in. and a flow factor rating of .83. The maximum operating pressure differential for this valve should be 45 psi.

2. Calculate the amount of current and air pressure required to open the pneumatic-assisted valve 34%.

3. Draw the symbol for a three-way, two-position solenoid valve.

4. Use the tables provided in this chapter and select the proper size motor starter for a 60hp three-phase motor that operates on 480 volts AC.

5. Determine the inrush current and seal-in current for a size 3 starter that has three poles. What would the pick-up and drop-out times be for this coil?

6. Use the table provided in Fig. 11-38 to select the proper size heater elements for a size 2 motor starter when the motor draws 15 A.

7. Draw the diagram of a unipolar drive amplifier used in stepper motor controllers. Explain the operation of this circuit.

8. Draw the diagram of a chopper drive amplifier used in stepper motor controllers. Explain the operation of this circuit.

9. Calculate the torque required to accelerate a load that has inertia of 8 lb-in.^{2}, a step angle of 3.6°, and an acceleration of 0–1000 steps in 1 second.

10. Draw a block diagram of a servo system listing all the parts of the system and explaining the function of each.

11. Draw a diagram of an H-bridge circuit that is used to control a DC servomotor and explain its operation.

REFERENCES

1. *Air Control Valves and Air Preparation Units Catalog #0600*. Parker Fluidpower, Parker Hannifin Corporation, Pneumatic Division, P.O. Box 901, Richland, Michigan 49083, 1986.

2. *Allen-Bradley Industrial Controls Catalog*. Allen-Bradley Headquarters, 1201 South Second Street, Milwaukee, WI 53204, 1995.

3. *Allen-Bradley 1336 Variable Frequency Drive Maintenance Manual*. Allen-Bradley Headquarters, 1201 South Second Street, Milwaukee, WI 53204, 1995.

4. *Compumotor Digiplan Positioning Control Systems and Drives*. Compumotor Division, 5500 Business Park Drive, Rohnert Park, CA 94928, 1994.

5. *Design Engineers Guide to AC Synchronous Motors*. Superior Electric, 383 Middle Street, Bristol, CT 06010, 1992.

6. *Design Engineers Guide to DC Stepping Motors,* Superior Electric, 383 Middle Street, Bristol, CT 06010, 1992.

7. *Electro-Craft Catalog*. Reliance Motion Control, 6950 Washington Avenue South, Eden Prairie, MN 55344, 1994.

12

AC AND DC MOTORS

OBJECTIVES

After reading this chapter, you will be able to:

1. Identify the basic parts of a DC motor and explain its operation.
2. Explain how to change the speed of a DC motor.
3. Explain how to change the rotation of a DC motor.
4. Explain the difference between series, shunt, and compound DC motors.
5. Identify the basic parts of a three-phase AC motor and explain its operation.

6. Explain how to change the speed of an AC motor.
7. Explain how to change the rotation of an AC motor.
8. Identify the basic parts of a single-phase AC motor and explain its operation.
9. Explain the difference between a split-phase, capacitor-start, and permanent-split capacitor motors.
10. Use motor frame-size data to select new motors.

You are working on a large machine with several different AC and DC motors to make the various parts of the machine operate. You need to set up the machine and make a few modifications for a new product that the machine will make. The machine has an AC three-phase motor to drive the main mechanism and a DC motor that is used to drive an outfeed conveyor. A series DC motor is permanently mounted on this machine to lift large rollers into place for maintenance. The machine also has several small AC three-phase pump motors, and several small fractional horsepower AC single-phase motors.

Your job is to check the large AC motor and ensure that it is wired as a wye-connected motor, and that it is turning in the proper direction. Make changes as necessary. You will also need to locate the brushes and commutator on each of the DC motors and check them to ensure they are in good working order. Also check the rotation of each DC motor and ensure that it is rotating in the proper direction and make changes if necessary.

One of the fractional horsepower single-phase motors is having trouble starting so you will need to check the centrifugal switch and make the necessary repairs or replace the motor. You will need to take data from all of the motors on this machine for your maintenance records and determine what types of spare motors you should keep on hand when a motor needs to be changed out.

12.1 INTRODUCTION

DC motors are commonly used to operate machinery in a variety of applications on the factory floor. DC motors were one of the first types of energy converters used in industry. It is important to remember that the earliest machines required speed control and DC motors could have their speed changed by varying the voltage sent to them. The earliest speed controls for DC motors were nothing more than large resistors.

DC motors required large amounts of DC voltage for operation. This means that a source for the DC voltage is needed at the factory. This creates a problem because DC voltage cannot be generated and distributed over a long distance, so AC voltage is the industry standard. One way to provide the DC voltage is to use generators that are set up at the factory site where large AC motors are used to turn them to produce the amount of required DC voltage. This system uses a large AC motor to drive a DC generator directly at a constant speed. The field current in the generator is regulated to adjust the level of DC voltage from the generator, which in turn is used to vary the speed of any DC motor that the generator powers. This system, called a Ward-Lennard system, was popular until solid-state diodes became available for rectifying large amounts of AC voltage to DC for use in motor-driven circuits. Once solid-state diodes and SCRs became available, DC motors became more usuable in industry.

During the 1950s and 1960s the use of DC motors became more prevalent in machinery control because their speed and torque were easy to control with simple SCR controllers. The SCR could rectify AC voltage to DC, provide current and voltage control at the same time, and were capable of being paralleled for larger loads up to 1000 A. As solid-state controls became more reliable in the late 1960s and the 1970s, a wide variety of low-cost AC motor speed controls became available.

During these years transistors could handle larger loads, and microprocessors became relatively inexpensive so that they could be used to make variable-frequency AC motor controls. At this time, you had a choice of using good-quality AC or DC motors for all types of special speed and torque applications.

As you read this chapter, you will learn about the concepts of controlling a DC motor's speed, its torque, and being able to reverse the direction of its rotation. It is also important to be able to recognize the features that make the series, shunt, and compound DC motors different from each other. You should also have a good understanding of the basic parts of the DC motor so that when you must troubleshoot a DC motor circuit, you will be able to recognize a malfunctioning component and make repairs or replace parts as quickly as possible.

It is also important to understand the methods of controlling the speed of AC motors, their direction of rotation, and the amount of torque they can develop, since these are the principles that motor drives use to control motors. Operation of special drive controllers is easy to understand if you know what electrical principle they are trying to alter to provide control for the motor. If you do not understand the motor principle, it is doubtful that you will fully understand the motor control device, and this will make the system nearly impossible to troubleshoot. If you understand these concepts, you will easily be able to understand the next generation of controls that will be produced during the next ten years.

Magnetic theories are provided with a discussion of basic DC motor components. Additional information will explain the difference between series, shunt, and compound motors. Diagrams are provided to explain the methods of controlling speed, rotation, and torque. These diagrams are useful for making field wiring connections and for testing motors during troubleshooting procedures.

12.2 MAGNETIC THEORY

DC motors operate on the principles of basic magnetism. You should remember that a coil of wire can be magnetized when current is passed through it. When this principle was used in relay coils, the polarity of the current was not important. When the current is passed through a coil of wire to make a field coil for a motor, the polarity of the current will determine the direction of rotation for the motor.

The polarity of the current flowing through the coil of wire will determine the location of the north and south magnetic poles in the coil of wire. Another important principle involves the amount of current that is flowing through the coil. The amount of current was not important as long as enough current was present to move the armature of the relay or solenoid. In a DC motor, the amount of current in the windings will determine the speed (rpm) of the motor shaft and the amount of torque that it can produce.

You should remember from basic magnetic theory that the left-hand rule of current flow through a coil of wire helps you understand that the direction of current flow will determine the magnetic polarity of the coil. The left-hand rule is used to show you a principle from which several facts can be determined.

The first fact that you should understand is that the direction of current flow will determine which end of a coil of wire is negative or positive. This will determine which end of the coil will be the north pole of the magnet and which end will be the south pole. It is also easy to see that by changing the direction of the current flow in the coil of wire, the magnetic poles will be reversed in the coil. This is important to understand because the direction of the motor's rotation is determined by the changing magnetic field.

Another basic concept about magnets that you should remember is the relationship between two like poles and two unlike poles. When the north poles of two different magnets are placed close to each other, they will repel each other. When the north pole of one magnet is placed near the south pole of another magnet, the two poles will attract each other very strongly.

The third principle that is important to understand with the coil of wire is that the strength of the magnetic field can be varied by changing the amount of current flowing through the wire in the coil. If a small amount of current is flowing, a small number of flux lines will be created and the magnetic field will be relatively weak. If the amount of current is increased, the magnetic field will become stronger. The strength of the magnetic field can be increased to the point of saturation. A magnetic coil is said to be saturated when its magnetic strength cannot be increased by adding more current.

Saturation is similar to filling a drinking glass with water. You cannot get the level of the glass any higher than full. Any additional water that is put into the glass when it is full will not increase the amount of water in the glass. The additional water will run over the side of the glass and be wasted. The same principle can be applied to a magnetic coil. When the strength of the magnetic field is at its strongest point, additional electric current will not cause the field to become any stronger.

FIGURE 12-1 A typical DC motor.

12.3 DC MOTOR THEORY

The DC motor has two basic parts: the rotating part that is called the *armature*, and the stationary part that includes coils of wire called the *field coils*. The stationary part is also called the *stator*. Fig. 12-1 shows a picture of a typical DC motor, Fig. 12-2 shows a picture of a DC armature, and Fig. 12-3 shows a picture of a typical stator. From the picture in Fig. 12-2 you can see the armature is made of coils of wire wrapped around the core, and the core has an extended shaft that rotates on bearings. You should also notice that the ends of each coil of wire on the armature are terminated at one end of the armature. The termination points are called the *commutator*, and this is where the brushes make electrical contact to bring electrical current from the stationary part to the rotating part of the machine.

The picture in Fig. 12-3 shows the location of the coils that are mounted inside the stator. These coils will be referred to as field coils in future discussions and they may be connected in series or parallel with each other to create changes of torque in the motor. You will find the size of wire in these coils and the number of turns of wire in the coil will depend on the effect that is trying to be achieved.

It will be easier to understand the operation of the DC motor from a basic diagram that shows the magnetic interaction between the rotating armature and the stationary field coils. Fig. 12-4 shows three diagrams that explain the DC motor's operation in terms of the magnetic interaction. In Fig. 12-4a you can see that a bar magnet has been mounted on a shaft so that it can spin. The field winding is one long coil of wire that has been separated into two sections. The top section is

FIGURE 12-2 The armature (rotor) of a DC motor has coils of wire wrapped around its core. The ends of each coil are terminated at commutator segments located on the left end of the shaft. The brushes make contact on the commutator to provide current for the armature.

FIGURE 12-3 The stationary part of a DC motor has the field coils mounted in it.

connected to the positive pole of the battery and the bottom section is connected to the negative pole of the battery. It is important to understand that the battery represents a source of voltage for this winding. In the actual industrial-type motor this voltage will come from the DC voltage source for the motor. The current flow in this direction makes the top coil the north pole of the magnet and the bottom coil the south pole of the magnet.

The bar magnet represents the *armature* and the coil of wire represents the *field*. The arrow shows the direction of the armature's rotation. Notice that the arrow shows the armature starting to rotate in the clockwise direction. The north pole of the field coil is repelling the north pole of the armature, and the south pole of the field coil is repelling the south pole of the armature.

As the armature begins to move, the north pole of the armature comes closer

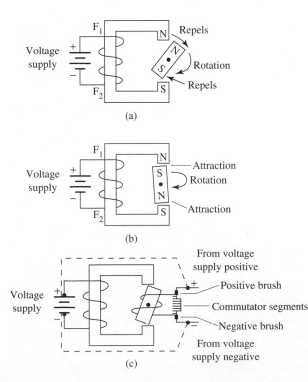

FIGURE 12-4 **(a)** Magnetic diagram that explains the operation of a DC motor. The rotating magnet moves clockwise because like poles repel. **(b)** The rotating magnet is being attracted because the poles are unlike. **(c)** The rotating magnet is now shown as the armature coil, and its polarity is determined by the brushes and commutator segments.

to the south pole of the field, and the south pole of the armature is coming closer to the north pole of the field. As the two unlike poles near each other, they begin to attract. This attraction becomes stronger until the armature's north pole moves directly in line with the field's south pole, and its south pole moves directly in line with the field's north pole (Fig. 12-4b).

When the opposite poles are at their strongest attraction, the armature will be "locked up" and will resist further attempts to continue spinning. For the armature to continue its rotation, the armature's polarity must be switched. Since the armature in this diagram is a permanent magnet, you can see that it would lock up during the first rotation and not work. If the armature is an electromagnet, its polarity can be changed by changing the direction of current flow through it. For this reason the armature must be changed to a coil (electromagnet) and a set of commutator segments must be added to provide a means of making contact between the rotating member and the stationary member. One commutator segment is provided for each terminal of the magnetic coil. Since this armature has only one coil, it will have only two terminals, so the commutator has two segments.

Since the armature is now a coil of wire, it will need DC current flowing through it to become magnetized. This presents another problem; since the armature will be rotating, the DC voltage wires cannot be connected directly to the armature coil. A stationary set of carbon brushes is used to make contact to the rotating armature. The brushes ride on the commutator segments to make contact so that current will flow through the armature coil.

In Fig. 12-4c you can see that the DC voltage is applied to the field and to the brushes. Since negative DC voltage is connected to one of the brushes, the commutator segment the negative brush rides on will also be negative. The armature's magnetic field causes the armature to begin to rotate. This time when the armature gets to the point where it becomes locked up with the magnetic field, the negative brush begins to touch the end of the armature coil that was previously positive and the positive brush begins to touch the end of the armature coil that was negative. This action switches the direction of current flow through the armature, which also switches the polarity of the armature coil's magnetic field at just the right time so that the repelling and attracting continues. The armature continues to switch its magnetic polarity twice during each rotation, which causes it to continually be attracted and repelled with the field poles.

This is a simple two-pole motor that is used primarily for instructional purposes. Since the motor has only two poles, the motor will operate rather roughly and not provide too much torque. Additional field poles and armature poles must be added to the motor for it to become useful for industry.

12.4 DC MOTOR COMPONENTS

The armature and field in a DC motor can be wired three different ways to provide varying amounts of torque or different types of speed control. The armature and field windings are designed slightly differently for different types of DC motors. The three basic types of DC motors are the *series motor*, the *shunt motor*, and the *compound motor*. The series motor is designed to move large loads with high starting torque in applications such as a crane motor or lift hoist. The shunt motor

FIGURE 12-5 A cutaway picture of a DC motor.

is designed slightly differently, since it is made for applications such as pumping fluids, where constant-speed characteristics are important. The compound motor is designed with some of the series motor's characteristics and some of the shunt motor's characteristics. This allows the compound motor to be used in applications where high starting torque and controlled operating speed are both required.

It is important that you understand the function and operation of the basic components of the DC motor, since motor controls will take advantage of these design characteristics to provide speed, torque, and direction of rotation control. Fig. 12-5 shows a cut-away picture of a DC motor and Fig. 12-6 shows an exploded-view diagram of a DC motor. In these figures you can see that the basic components include the armature assembly, which includes all rotating parts; the frame assembly, which houses the stationary field coils; and the end plates, which provide bearings

FIGURE 12-6   An exploded view of a DC motor. This diagram shows the relationship of all of the components.

for the motor shaft and a mounting point for the brush rigging. Each of these assemblies is explained in depth so that you will understand the design concepts used for motor control.

12.4.1 Armature

The armature is the part of a DC motor that rotates and provides energy at the end of the shaft. It is basically an electromagnet, since it is a coil of wire that has to be specially designed to fit around core material on the shaft. The core of the armature is made of laminated steel and provides slots for the coils of wire to be pressed onto. Fig. 12-7a shows a sketch of a typical DC motor armature. Fig. 12-7b shows the laminated steel core of the armature without any coils of wire on it. This gives you a better look at the core.

The armature core is made of laminated steel to prevent the circulation of eddy currents. If the core were solid, magnetic currents would be produced that would circulate in the core material near the surface and cause the core metal to heat up. These magnetic currents are called *eddy currents*. When laminated steel sections are pressed together to make the core, the eddy currents cannot flow from one laminated segment to another, so they are effectively canceled out. The laminated core also prevents other magnetic losses called *flux losses*. These losses

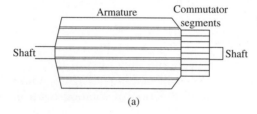

(a)

FIGURE 12-7 **(a)** Armature and commutator segments. **(b)** Armature prior to the coil's wire being installed. **(c)** Coil of wire prior to being pressed into the armature. **(d)** A coil pressed into the armature. The end of each coil is attached to a commutator segment.

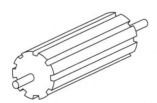

(b) Armature with coils missing

(c) Copper wire roll ready to be pressed onto armature

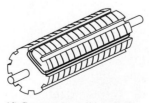

(d) Copper wire coil installed in armature set

tend to make the magnetic field weaker so that more core material is required to obtain the same magnetic field strengths. The flux losses and eddy current losses are grouped together by designers and called *core losses*. The laminated core is designed to allow the armature's magnetic field to be as strong as possible since the laminations prevent core losses.

Notice that one end of the core has commutator segments. There is one commutator segment for each end of each coil. This means that a armature with four coils will have eight commutator segments. The commutator segments are used as a contact point between the stationary brushes and the rotating armature. When each coil of wire is pressed onto the armature, the end of the coil is soldered to a specific commutator segment. This makes an electrical terminal point for the current that will flow from the brushes onto the commutator segment and finally through the coil of wire. Fig. 12-7c shows the coil of wire before it is mounted in the armature slot, and Fig. 12-7d shows the coil mounted in the armature slot and soldered to the commutator segment.

The shaft is designed so that the laminated armature segments can be pressed onto it easily. It is also machined to provide a surface for a main bearing to be pressed on at each end. The bearing will ride in the end plates and support the armature when it begins to rotate. One end of the shaft is also longer than the other, since it will provide the mounting shaft for the motor's load to be attached. Some shafts have a key way or flat spot machined into them so that the load that is mounted on it can be secured. You must be careful when handling a motor that you do not damage the shaft, since it must be smooth to accept the coupling mechanism. It is also possible to bend the shaft or cause damage to the bearings so that the motor will vibrate when it is operating at high speed. The commutator is made of copper. A thin section of insulation is placed between each commutator segment. This effectively isolates each commutator segment from all others.

12.4.2 Motor Frame

The armature is placed inside the frame of the motor where the field coils are mounted. When the field coils and the armature coils become magnetized, the armature will begin to rotate. The field winding is made by coiling up a long piece of wire. The wire is mounted on laminated pole pieces called field poles. Similar to an armature, these poles are made of laminated steel or cast iron to prevent eddy current and other flux losses. Fig. 12-8 shows the location of the pole pieces inside a DC motor frame.

The amount of wire that is used to make the field winding will depend on the type of motor that is being manufactured. A series motor uses heavy-gauge wire for its field winding so that it can handle very large field currents. Since the wire is a large gauge, the number of turns of wire in the coil will be limited. If the field winding is designed for a shunt motor, it will be made of small-gauge wire and many turns can be used.

After the coils are wound, they are coated for protection against moisture and other environmental elements. After they have been pressed onto the field poles, they must be secured with shims or bolts so that they are held rigidly in place. Remember: When current is passed through the coil, it will become strongly magnetized and attract or repel the armature magnetic poles. If the field poles are not rigidly secured, they will be pulled loose when they are attracted to the armature's magnetic field and then pressed back into place when they become repelled. This action will cause the field to vibrate and damage the outer protective insulation

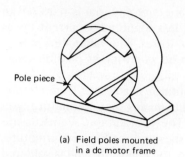

(a) Field poles mounted
in a dc motor frame

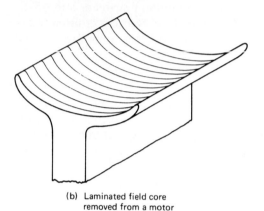

(b) Laminated field core
removed from a motor

FIGURE 12-8 **(a)** This diagram shows the location of the pole pieces in the frame of a DC motor. **(b)** This diagram shows an individual pole piece. You can see that it is made of laminated sections. The field coils are wound around the pole pieces.

and cause a short circuit or a ground condition between the winding and the frame of the motor.

The ends of the frame are machined so that the end plates will mount firmly into place. An access hole is also provided in the side of the frame or in the end plates so that the field wires can be brought to the outside of the motor, where DC voltage can be connected.

The bottom of the frame has the mounting bracket attached. The bracket has a set of holes or slots provided so that the motor can be bolted down and securely mounted on the machine it is driving. The mounting holes will be designed to specifications by frame size. The dimensions for the frame sizes are provided in tables printed by motor manufacturers. Since these holes and slots are designed to a standard, you can predrill the mounting holes in the machinery before the motor is put in place. The slots are used to provide minor adjustments to the mounting alignment when the motor is used in belt-driven or chain-driven applications. It is also important to have a small amount of mounting adjustment when the motor is used in direct-drive applications. It is very important that the motor be mounted so that the armature shaft can turn freely and not bind with the load.

12.4.3 End Plates

The end plates of the motor are mounted on the ends of the motor frame. Figs. 12-5 and 12-6 show the location of the end plates in relation to the motor frame. The end plates are held in place by four bolts that pass through the motor frame. The bolts can be removed from the frame completely so that the end plates can be removed easily for maintenance. The end plates also house the bearings for the armature shaft. These bearings can be either sleeve or ball type. If the bearing is a ball-bearing type, it is normally permanently lubricated. If it is a sleeve type, it will require a light film of oil to operate properly. The end plates that house a sleeve-type bearing will have a lubrication tube and wicking material. Several drops of lubricating oil are poured down the lubrication tube, where they will saturate the wicking material. The wicking is located in the bearing sleeve so that it can make contact with the armature shaft and transfer a light film of oil to it. Other types of sleeve bearings are made of porous metal so that it can absorb oil to be used to create a film between the bearing and the shaft.

It is important that the end plate for a sleeve bearing be mounted on the motor frame so that the lubricating tube is pointing up. This position will ensure that gravity will pull the oil to the wicking material. If the end plates are mounted so that the lubricating tube is pointing down, the oil will flow away from the wicking and it will become dry. When the wicking dries out, the armature shaft will rub directly on the metal in the sleeve bearing, which will cause it to quickly heat up, and the shaft will seize to the bearing. For this reason it is also important to follow lubrication instructions and oil the motor on a regular basis.

12.4.4 Brushes and Brush Rigging

The brush rigging is an assembly that securely holds the brushes in place so that they will be able to ride on the commutator. It is mounted on the rear end plate so that the brushes will be accessible by removing the end plate. An access hole is also provided in the motor frame so that the brushes can be adjusted slightly when the motor is initially set up. The brush rigging uses a spring to provide the proper amount of tension on the brushes so that they make proper contact with the commutator. If the tension is too light, the brushes will bounce and arc, and if the tension is too heavy, the brushes will wear down prematurely.

The brush rigging is shown in Fig. 12-5 and Fig. 12-6. Notice that it is mounted on the rear end plate. Since the rigging is made of metal, it must be insulated electrically when it is mounted on the end plate. The DC voltage that is used to energize the armature will pass through the brushes to the commutator segments and into the armature coils. Each brush has a wire connected to it. The wires will be connected to either the positive or negative terminal of the DC power supply. The motor will always have an even number of brushes. Half of the brushes will be connected to positive voltage and half will be connected to negative voltage. In most motors the number of brush sets will be equal to the number of field poles. It is important to remember that the voltage polarity will remain constant on each brush. This means that for each pair, one of the brushes will be connected to the positive power terminal, and the other will be connected permanently to the negative terminal.

The brushes will cause the polarity of each armature segment to alternate from positive to negative. When the armature is spinning, each commutator segment will come in contact with a positive brush for an instant and will be positive during that time. As the armature rotates slightly, that commutator segment will come in contact with a brush that is connected to the negative voltage supply and it will become negative during that time. As the armature continues to spin, each commutator segment will be alternately powered by positive and then negative voltage.

The brushes are made of carbon-composite material. Usually the brushes have copper added to aid in conduction. Other material is also added to make them wear longer. The end of the brush that rides on the commutator is contoured to fit the commutator exactly so that current will transfer easily. The process of contouring the brush to the commutator is called *seating*. Whenever a set of new brushes is installed, the brushes should be seated to fit the commutator. The brushes are the main part of the DC motor that will wear out. It is important that their wear be monitored closely so that they do not damage the commutator segments when they begin to wear out. Most brushes have a small mark on them called a wear mark or wear bar. When a brush wears down to the mark, it should be replaced. If the brushes begin to wear excessively or do not fit properly on the commutator, they will heat up and damage the brush rigging and spring mechanism. If the brushes have been overheated, they can cause burn marks or pitting on the commutator segments and also warp the spring mechanism so that it will no longer hold the brushes with the proper amount of tension. Fig. 12-5 and Fig. 12-6 show the location of the brushes riding on the commutator.

If the spring mechanism has been overheated, it should be replaced and the brushes should be checked for proper operation. If the commutator is pitted, it can be turned down on a lathe. After the commutator has been turned down, the brushes will need to be reseated.

After you have an understanding of the function of each of the parts or assemblies of the motor, you will be able to understand better the operation of a basic DC motor. Operation of the motor involves the interaction of all the motor parts. Some of the parts will be altered slightly for specific motor applications. These changes will become evident when the motor's basic operation is explained.

12.5 DC MOTOR OPERATION

The DC motor you will find in modern industrial applications operates very similarly to the simple DC motor described earlier in this chapter. Fig. 12-9 shows an electrical diagram of a simple DC motor. Notice that the DC voltage is applied directly to the field winding and the brushes. The armature and the field are both shown as a coil of wire. In later diagrams, a field resistor will be added in series with the field to control the motor speed.

When voltage is applied to the motor, current begins to flow through the field coil from the negative terminal to the positive terminal. This sets up a strong magnetic field in the field winding. Current also begins to flow through the brushes into a commutator segment and then through an armature coil. The current continues to flow through the coil back to the brush that is attached to other end of the coil and returns to the DC power source. The current flowing in the armature coil sets up a strong magnetic field in the armature.

The magnetic field in the armature and field coil causes the armature to begin to rotate. This occurs by the unlike magnetic poles attracting each other and the

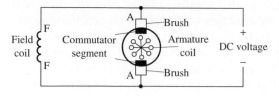

FIGURE 12-9 Simple electrical diagram of DC shunt motor. This diagram shows the electrical relationship between the field coil and armature.

like magnetic poles repelling each other. As the armature begins to rotate, the commutator segments will also begin to move under the brushes. As an individual commutator segment moves under the brush connected to positive voltage, it will become positive, and when it moves under a brush connected to negative voltage it will become negative. In this way, the commutator segments continually change polarity from positive to negative. Since the commutator segments are connected to the ends of the wires that make up the field winding in the armature, it causes the magnetic field in the armature to change polarity continually from north pole to south pole. The commutator segments and brushes are aligned in such a way that the switch in polarity of the armature coincides with the location of the armature's magnetic field and the field winding's magnetic field. The switching action is timed so that the armature will not lock up magnetically with the field. Instead the magnetic fields tend to build on each other and provide additional torque to keep the motor shaft rotating.

When the voltage is de-energized to the motor, the magnetic fields in the armature and the field winding will quickly diminish and the armature shaft's speed will begin to drop to zero. If voltage is applied to the motor again, the magnetic fields will strengthen and the armature will begin to rotate again.

12.6 TYPES OF DC MOTORS

Three basic types of DC motors are used in industry today: the series motor, the shunt motor, and the compound motor. The series motor is capable of starting with a very large load attached, such as lifting applications. The shunt motor is able to operate with rpm control while it is at high speed. The compound motor, a combination of the series motor and the shunt motor, is able to start with fairly large loads and have some rpm control at higher speeds. In the remaining sections of this chapter we show a diagram for each of these motors and discuss their operational characteristics. As a technician you should understand methods of controlling their speed and ways to change the direction of rotation because these are the two parameters of a DC motor you will be asked to change as applications change on the factory floor. It is also important to understand the basic theory of operation of these motors because you will be controlling them with solid-state electronic circuits. You will need to know if problems that arise are the fault of the motor or the solid-state circuit.

12.6.1 DC Series Motors

The series motor provides high starting torque and is able to move very large shaft loads when it is first energized. Fig. 12-10 shows the wiring diagram of a series motor. From the diagram you can see that the field winding in this motor is wired in series with the armature winding. This is the attribute that gives the series motor its name.

Since the series field winding is connected in series with the armature, it will carry the same amount of current that passes through the armature. For this reason the field is made from heavy-gauge wire that is large enough to carry the load. Since the wire gauge is so large, the winding will have only a few turns of wire. In some larger DC motors, the field winding is made from copper bar stock rather

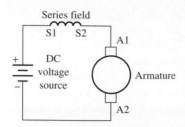

FIGURE 12-10 Electrical diagram of series motor. Notice that the series field is identified as S1 and S2.

than the conventional round wire used for power distribution. The square or rectangular shape of the copper bar stock makes it fit more easily around the field pole pieces. It can also radiate more easily the heat that has built up in the winding due to the large amount of current being carried.

The amount of current that passes through the winding determines the amount of torque the motor shaft can produce. Since the series field is made of large conductors, it can carry large amounts of current and produce large torques. For example, the starter motor that is used to start an automobile's engine is a series motor and it may draw up to 500 A when it is turning the engine's crankshaft on a cold morning. Series motors used to power hoists or cranes may draw currents of thousands of amperes during operation.

The series motor can safely handle large currents since the motor does not operate for an extended period. In most applications the motor will operate for only a few seconds while this large current is present. Think about how long the starter motor on the automobile must operate to get the engine to start. This period is similar to that of industrial series motors.

12.6.2 Series Motor Operation

Operation of the series motor is easy to understand. In Fig. 12-10 you can see that the field winding is connected in series with the armature winding. This means that power will be applied to one end of the series field winding and to one end of the armature winding (connected at the brush).

When voltage is applied, current begins to flow from negative power supply terminals through the series winding and armature winding. The armature is not rotating when voltage is first applied, and the only resistance in this circuit will be provided by the large conductors used in the armature and field windings. Since these conductors are so large, they will have a small amount of resistance. This causes the motor to draw a large amount of current from the power supply. When the large current begins to flow through the field and armature windings, it causes a strong magnetic field to be built. Since the current is so large, it will cause the coils to reach saturation, which will produce the strongest magnetic field possible.

12.6.3 Producing Back EMF

The strength of these magnetic fields provides the armature shafts with the greatest amount of torque possible. The large torque causes the armature to begin to spin with the maximum amount of power. When the armature begins to rotate, it begins to produce voltage. This concept is difficult for some students to understand since the armature is part of the motor at this time.

You should remember from the basic theories of magnetism that anytime a magnetic field passes a coil of wire, a current will be produced. The stronger the magnetic field is or the faster the coil passes the flux lines, the more current will be generated. When the armature begins to rotate, it will produce a voltage that is of opposite polarity to that of the power supply. This voltage is called *back voltage*, *back EMF* (electromotive force), or *counter EMF*. The overall effect of this voltage is that it will be subtracted from the supply voltage so that the motor windings will see a smaller voltage potential.

When Ohm's law is applied to this circuit, you will see that when the voltage is slightly reduced, the current will also be reduced slightly. This means that the series motor will see less current as its speed is increased. The reduced current will mean that the motor will continue to lose torque as the motor speed increases. Since the load is moving when the armature begins to pick up speed, the application will require less torque to keep the load moving. This works to the motor's advantage by automatically reducing the motor current as soon as the load begins to move. It also allows the motor to operate with less heat buildup.

This condition can cause problems if the series motor ever loses its load. The load could be lost when a shaft breaks or if a drive pin is sheared. When this occurs, the load current is allowed to fall to a minimum, which reduces the amount of back EMF that the armature is producing. Since the armature is not producing a sufficient amount of back EMF and the load is no longer causing a drag on the shaft, the armature will begin to rotate faster and faster. It will continue to increase rotational speed until it is operating at a very high speed. When the armature is operating at high speed, the heavy armature windings will be pulled out of their slots by centrifugal force. When the windings are pulled loose, they will catch on a field winding pole piece and the motor will be severely damaged. This condition is called *runaway* and you can see why a DC series motor must have some type of runaway protection. A centrifugal switch can be connected to the motor to de-energize the motor starter coil if the rpm exceeds the set amount. Other sensors can be used to de-energize the circuit if the motor's current drops while full voltage is applied to the motor. The most important part to remember about a series motor is that it is difficult to control its speed by external means because its rpm is determined by the size of its load. (In some smaller series motors, the speed can be controlled by placing a rheostat in series with the supply voltage to provide some amount of change in resistance to control the voltage to the motor.)

Fig. 12-11 shows the relationship between series motor speed and armature current. From this curve you can see that when current is low (at the top left), the motor speed is maximum, and when current increases, the motor speed slows down (bottom right). You can also see from this curve that a DC motor will run away if the load current is reduced to zero. (It should be noted that in larger series machines used in industry, the amount of friction losses will limit the highest speed somewhat.)

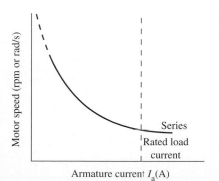

FIGURE 12-11 The relationship between series motor speed and the armature current.

12.6.4 Reversing the Rotation of the Motor

The direction of rotation of a series motor can be changed by changing the polarity of either the armature or field winding. It is important to remember that if you simply changed the polarity of the applied voltage, you would be changing the polarity of both field and armature windings and the motor's rotation would remain the same.

Since only one of the windings needs to be reversed, the armature winding is typically used because its terminals are readily accessible at the brush rigging. Remember that the armature receives its current through the brushes, so that if their polarity is changed, the armature's polarity will also be changed. A reversing motor starter is used to change wiring to cause the direction of the motor's rotation to change by changing the polarity of the armature windings. Fig. 12-12 shows a DC series motor that is connected to a reversing motor starter. In this diagram the armature's terminals are marked A1 and A2 and the field terminals are marked S1 and S2.

When the forward motor starter is energized, the top contact identified as F closes so the A1 terminal is connected to the positive terminal of the power supply and the bottom F contact closes and connects terminals A2 and S1. Terminal S2 is connected to the negative terminal of the power supply. When the reverse motor starter is energized, terminals A1 and A2 are reversed. A2 is now connected to the positive terminal. Notice that S2 remains connected to the negative terminal of the power supply terminal. This ensures that only the armature's polarity has been changed and the motor will begin to rotate in the opposite direction.

You will also notice the normally closed (NC) set of R contacts connected in series with the forward push button, and the NC set of F contacts connected in series with the reverse push button. These contacts provide an *interlock* that prevents the motor from being changed from forward to reverse direction without stopping

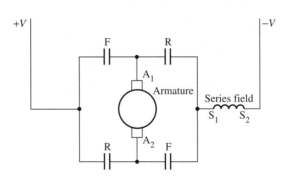

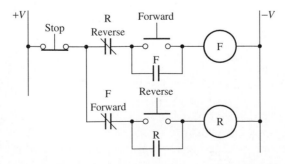

FIGURE 12-12 DC series motor connected to forward and reverse motor starter.

the motor. The circuit can be explained as follows: when the forward push button is depressed, current will flow from the stop push button through the NC R interlock contacts, and through the forward push button to the forward motor starter (FMS) coil. When the FMS coil is energized, it will open its NC contacts that are connected in series with the reverse push button. This means that if someone depresses the reverse push button, current could not flow to the reverse motor starter (RMS) coil. If the person depressing the push buttons wants to reverse the direction of the rotation of the motor, he or she will need to depress the stop push button first to deenergize the FMS coil, which will allow the NC F contacts to return to their NC position. You can see that when the RMS coil is energized, its NC R contacts that are connected in series with the forward push button will open and prevent the current flow to the FMS coil if the forward push button is depressed. You will see a number of other ways to control the FMS and RMS starter in later discussions and in the chapter on motor controls.

12.6.5 Installing and Troubleshooting the Series Motor

Since the series motor has only two leads brought out of the motor for installation wiring, this wiring can be accomplished rather easily. If the motor is wired to operate in only one direction, the motor terminals can be connected to a manual or magnetic starter. If the motor's rotation is required to be reversed periodically, it should be connected to a reversing starter.

Most DC series motors are used in direct-drive applications. This means that the load is connected directly to the armature's shaft. This type of load is generally used to get the most torque converted. Belt-drive applications are not recommended since a broken belt would allow the motor to run away. After the motor has been installed, a test run should be used to check it out. If any problems occur, the troubleshooting procedures should be used.

The most likely problem that will occur with the series motor is that it will develop an open in one of its windings or between the brushes and the commutator. Since the coils in a series motor are connected in series, each coil must be functioning properly or the motor will not draw any current. When this occurs, the motor cannot build a magnetic field and the armature will not turn. Another problem that is likely to occur with the motor circuit is that circuit voltage will be lost due to a blown fuse or circuit breaker. The motor will respond similarly in both of these conditions.

The best way to test a series motor is with a voltmeter. The first test should be for applied voltage at the motor terminals. Since the motor terminals are usually connected to a motor starter, the test leads can be placed on these terminals. If the meter shows that full voltage is applied, the problem will be in the motor. If it shows that no voltage is present, you should test the supply voltage and the control circuit to ensure that the motor starter is closed. If the motor starter has a visual indicator, be sure to check to see that the starter's contacts are closed. If the overloads have tripped, you can assume that they have sensed a problem with the motor or its load. When you reset the overloads, the motor will probably start again but remember to test the motor thoroughly for problems that would cause an overcurrent situation.

If the voltage test indicates that the motor has full applied voltage to its terminals but the motor is not operating, you can assume that you have an open in one of the windings or between the brushes and the armature and continue testing. Each of these sections should be disconnected from each other and voltage should be removed so that they can be tested with an ohmmeter for an open. The series field coils can be tested by putting the ohmmeter leads on terminals S1 and S2. If the meter indicates that an open exists, the motor will need to be removed and sent to be rewound or replaced. If the meter indicates that the field coil has continuity, you should continue the procedure by testing the armature.

The armature can also be tested with an ohmmeter by placing the leads on the terminals marked A1 and A2. If the meter shows continuity, rotate the armature shaft slightly to look for bad spots where the commutator may have an open or the brushes may not be seated properly. If the armature test indicates that an open exists, you should continue the test by visually inspecting the brushes and commutator. You may also have an open in the armature coils. The armature must be removed from the motor frame to be tested further. When you have located the problem, you should remember that the commutator can be removed from the motor while the motor remains in place and it can be turned down on a lathe. When the commutator is replaced in the motor, new brushes can be installed and the motor will be ready for use.

It is possible that the motor will develop a problem but still run. This type of problem usually involves the motor overheating or not being able to pull its rated load. This type of problem is different from an open circuit because the motor is drawing current and trying to run. Since the motor is drawing current, you must assume that there is not an open circuit. It is still possible to have brush problems that would require the brushes to be reseated or replaced. Other conditions that will cause the motor to overheat include loose or damaged field and armature coils. The motor will also overheat if the armature shaft bearing is in need of lubrication or is damaged. The bearing will seize on the shaft and cause the motor to build up friction and overheat.

If either of these conditions occurs, the motor may be fixed on site or be removed for extensive repairs. When the motor is restarted after repairs have been made, it is important to monitor the current usage and heat buildup. Remember that the motor will draw DC current so that an AC clamp-on ammeter will not be useful for measuring the DC current. You will need to use an ammeter that is specially designed for very large DC currents. It is also important to remember that the motor can draw very high locked-rotor current when it is starting, so the ammeter should be capable of measuring currents up to 1000 A. After the motor has completed its test run successfully, it can be put back into operation for normal duty. Anytime the motor is suspected of faulty operation, the troubleshooting procedure should be rechecked.

12.6.6 DC Series Motor Used as a Universal Motor

The series motor is used in a wide variety of power tools such as electric hand drills, saws, and power screwdrivers. In most of these cases, the power source for the motor is AC voltage. The DC series motor will operate on AC voltage. If the motor is used in a hand drill that needs variable-speed control, a field rheostat or other type of current control is used to control the speed of the motor. In some newer tools, the current control uses solid-state components to control the speed of the motor. You will notice that the motors used for these types of power tools have brushes and a commutator, and these are the main parts of the motor to wear out. You can use the same theory of operation provided for the DC motor to troubleshoot these types of motors.

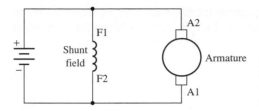

FIGURE 12-13 Diagram of DC shunt motor. Notice the shunt coil is identified as a coil of fine wire with many turns that is connected in parallel (shunt) with the armature.

12.7 DC SHUNT MOTORS

The shunt motor is different from the series motor in that the field winding is connected in parallel with the armature instead of in series. You should remember from basic electrical theory that a parallel circuit is often referred to as a shunt. Since the field winding is placed in parallel with the armature, it is called a shunt winding and the motor is called a shunt motor. Fig. 12-13 shows a diagram of a shunt motor. Notice that the field terminals are marked F1 and F2, and the armature terminals are marked A1 and A2. You should notice in this diagram that the shunt field is represented with multiple turns using a thin line.

The shunt winding is made of small-gauge wire with many turns on the coil. Since the wire is so small, the coil can have thousands of turns and still fit in the slots. The small-gauge wire cannot handle as much current as the heavy-gauge wire in the series field, but since this coil has many more turns of wire, it can still produce a very strong magnetic field. Fig. 12-14 shows a picture of a DC shunt motor.

12.7.1 Shunt Motor Operation

The shunt motor has slightly different operating characteristics than the series motor. Since the shunt field coil is made of fine wire, it cannot produce the large current for starting like the series field. This means that the shunt motor has very low starting torque, which requires that the shaft load be rather small.

When voltage is applied to the motor, the high resistance of the shunt coil keeps the overall current flow low. The armature for the shunt motor is similar to the series motor and it will draw current to produce a magnetic field strong enough to cause the armature shaft and load to start turning. Like the series motor, when the armature begins to turn, it will produce back EMF. The back EMF will cause the current in the armature to begin to diminish to a very small level. The amount

FIGURE 12-14 Typical DC shunt motor. These motors are available in a variety of sizes. This motor is a 1 hp (approximately 8 in. tall).

of current the armature will draw is directly related to the size of the load when the motor reaches full speed. Since the load is generally small, the armature current will be small. When the motor reaches full rpm, its speed will remain fairly constant.

12.7.2 Controlling the Speed of the Motor

When the shunt motor reaches full rpm, its speed will remain fairly constant. The reason the speed remains constant is due to the load characteristics of the armature and shunt coil. You should remember that the speed of a series motor could not be controlled since it was totally dependent on the size of the load in comparison to the size of the motor. If the load was very large for the motor size, the speed of the armature would be very slow. If the load was light compared to the motor, the armature shaft speed would be much faster, and if no load was present on the shaft, the motor could run away.

The shunt motor's speed can be controlled. The ability of the motor to maintain a set rpm at high speed when the load changes is due to the characteristic of the shunt field and armature. Since the armature begins to produce back EMF as soon as it starts to rotate, it will use the back EMF to maintain its rpm at high speed. If the load increases slightly and causes the armature shaft to slow down, less back EMF will be produced. This will allow the difference between the back EMF and applied voltage to become larger, which will cause more current to flow. The extra current provides the motor with the extra torque required to regain its rpm when this load is increased slightly.

The shunt motor's speed can be varied in two different ways. These include varying the amount of current supplied to the shunt field and controlling the amount of current supplied to the armature. Controlling the current to the shunt field allows the rpm to be changed 10–20% when the motor is at full rpm.

This type of speed control regulation is accomplished by slightly increasing or decreasing the voltage applied to the field. The armature continues to have full voltage applied to it while the current to the shunt field is regulated by a rheostat that is connected in series with the shunt field. When the shunt field's current is decreased, the motor's rpm will increase slightly. When the shunt field's current is reduced, the armature must rotate faster to produce the same amount of back EMF to keep the load turning. If the shunt field current is increased slightly, the armature can rotate at a slower rpm and maintain the amount of back EMF to produce the armature current to drive the load. The field current can be adjusted with a field rheostat or an SCR current control.

The shunt motor's rpm can also be controlled by regulating the voltage that is applied to the motor armature. This means that if the motor is operated on less voltage than is shown on its data plate rating, it will run at less than full rpm. You must remember that the shunt motor's efficiency will drop off drastically when it is operated below its rated voltage. The motor will tend to overheat when it is operated below full voltage, so motor ventilation must be provided. You should also be aware that the motor's torque is reduced when it is operated below the full voltage level.

Since the armature draws more current than the shunt field, the control resistors were much larger than those used for the field rheostat. During the 1950s and 1960s SCRs were used for this type of current control. The SCR was able to control the armature current since it was capable of controlling several hundred amperes. In Chapter 11 we provided an in-depth explanation of the DC motor drive.

12.7.3 Torque Characteristics

The armature's torque increases as the motor gains speed due to the fact that the shunt motor's torque is directly proportional to the armature current. When the motor is starting and speed is very low, the motor has very little torque. After the motor reaches full rpm, its torque is at its fullest potential. In fact, if the shunt field current is reduced slightly when the motor is at full rpm, the rpm will increase slightly and the motor's torque will also increase slightly. This type of automatic control makes the shunt motor a good choice for applications where constant speed is required, even though the torque will vary slightly due to changes in the load. Fig. 12-15 shows the torque/speed curve for the shunt motor. From this diagram you can see that the speed of the shunt motor stays fairly constant throughout its load range and drops slightly when it is drawing the largest current.

12.7.4 Reversing the Rotation of the Motor

The direction of rotation of a DC shunt motor can be reversed by changing the polarity of either the armature coil or the field coil. In this application the armature coil is usually changed, as was the case with the series motor. Fig. 12-16 shows the electrical diagram of a DC shunt motor connected to a forward and reversing motor starter. You should notice that the F1 and F2 terminals of the shunt field are connected directly to the power supply, and the A1 and A2 terminals of the armature winding are connected to the reversing starter.

When the FMS is energized, its contacts connect the A1 lead to the positive power supply terminal and the A2 lead to the negative power supply terminal. The F1 motor lead is connected directly to the positive terminal of the power supply and the F2 lead is connected to the negative terminal. When the motor is wired in this configuration, it will begin to run in the forward direction.

When the RMS is energized, its contacts reverse the armature wires so that the A1 lead is connected to the negative power supply terminal and the A2 lead is connected to the positive power supply terminal. The field leads are connected directly to the power supply, so their polarity is not changed. Since the field's polarity has remained the same and the armature's polarity has reversed, the motor will begin to rotate in the reverse direction. The control part of the diagram shows that when the FMS coil is energized, the RMS coil is locked out.

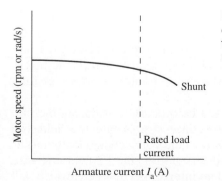

FIGURE 12-15 A curve that shows the armature current versus the armature speed for a shunt motor. Notice that the speed of a shunt motor is nearly constant.

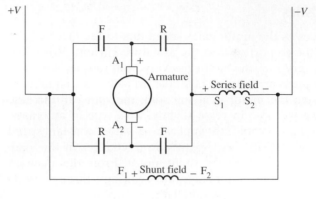

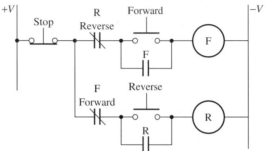

FIGURE 12-16 Diagram of a shunt motor connected to a reversing motor starter. Notice that the shunt field is connected across the armature and it is not reversed when the armature is reversed.

12.7.5 Installing the Shunt Motor

The shunt motor can be installed easily. The motor is generally used in belt-drive applications. This means that the installation procedure should be broken into two sections, which include the mechanical installation of the motor and its load, and the installation of electrical wiring and controls.

When the mechanical part of the installation is completed, the alignment of the motor shaft and the load shaft should be checked. If the alignment is not true, the load will cause an undue stress on the armature bearing and there is the possibility of the load vibrating and causing damage to it and the motor. After the alignment is checked, the tension on the belt should also be tested. As a rule of thumb, you should have about $\frac{1}{2}$ to $\frac{1}{4}$ inch of play in the belt when it is properly tensioned.

Several tension measurement devices are available to determine when a belt is tensioned properly. The belt tension can also be compared to the amount of current the motor draws. The motor must have its electrical installation completed to use this method.

The motor should be started, and if it is drawing too much current, the belt should be loosened slightly but not enough to allow the load to slip. If the belt is slipping, it can be tightened to the point where the motor is able to start successfully and not draw current over its rating.

The electrical installation can be completed before, after, or during the mechanical installation. The first step in this procedure is to locate the field and armature leads in the motor and prepare them for field connections. If the motor is connected to magnetic or manual across the line starter, the F1 field coil wire can be connected to the A1 armature lead and an interconnecting wire, which will be used to connect these leads to the L1 terminal on the motor starter. The F2 lead can be connected to the A2 lead and a second wire, which will connect these leads to the L2 motor starter terminal.

When these connections are completed, field and armature leads should be replaced back into the motor and the field wiring cover or motor access plate should be replaced. Next the DC power supply's positive and negative leads should be connected to the motor starter's T1 and T2 terminals, respectively.

After all of the load wires are connected, any pilot devices or control circuitry should be installed and connected. The control circuit should be tested with the load voltage disconnected from the motor. If the control circuit uses the same power source as the motor, the load circuit can be isolated so the motor will not try to start by disconnecting the wire at terminal L2 on the motor starter. Operate the control circuit several times to ensure that it is wired correctly and operating properly. After you have tested the control circuit, the lead can be replaced to the L2 terminal of the motor starter and the motor can be started and tested for proper operation. Be sure to check the motor's voltage and current while it is under load to ensure that it is operating correctly. It is also important to check the motor's temperature periodically until you are satisfied the motor is operating correctly.

If the motor is connected to a reversing starter or reduced-voltage starting circuit, their operation should also be tested. You may need to read the material in Section 15.3.6 to fully understand the operation of these methods of starting the motor using reduced-voltage methods. If the motor is not operating correctly or develops a fault, a troubleshooting procedure should be used to test the motor and locate the problem.

12.7.6 Troubleshooting the Shunt Motor

When the DC shunt motor develops a fault, you must be able to locate the problem quickly and return the motor to service or have it replaced. The most likely problems to occur with the shunt motor include loss of supply voltage or an open in either the shunt winding or the armature winding. Other problems may arise that cause the motor to run abnormally hot even though it continues to drive the load. The motor will show different symptoms for each of these problems, which will make the troubleshooting procedure easier.

When you are called to troubleshoot the shunt motor, it is important to determine if the problem occurs while the motor is running or when it is trying to start. If the motor will not start, you should listen to see if the motor is humming and trying to start. When the supply voltage has been interrupted due to a blown fuse or a de-energized control circuit, the motor will not be able to draw any current and it will be silent when you try to start it. You can also determine that the supply voltage has been lost by measuring it with a voltmeter at the starter's L1 and L2 terminals. If no voltage is present at the load terminals, you should check for voltage at the starter's T1 and T2 terminals. If voltage is present here but not at the load terminals, it indicates that the motor starter is de-energized or defective. If no voltage is present at the T1 and T2 terminals, it indicates that supply voltage has been lost prior to the motor starter. You will need to check the supply fuses and the rest of the supply circuit to locate the fault.

If the motor tries to start and hums loudly, it indicates that the supply voltage is present. The problem in this case is probably due to an open field winding or armature winding. It could also be caused by the supply voltage being too low.

The most likely problem will be an open in the field winding since it is made from small-gauge wire. The open can occur if the field winding draws too much current or develops a short circuit between the insulation in the coils. The best way to test the field is to remove supply voltage to the motor by opening the disconnect or de-energizing the motor starter. Be sure to use a *lockout* when you are working on the motor after the disconnect has been opened. The lockout is a device that is placed on the handle of the disconnect after the handle is placed in the off position, and it allows a padlock to be placed around it so it cannot be removed until the technician has completed the work on the circuit. If lockout has extra holes, additional padlocks can be placed on it by other technicians who are also working on this system. This ensures that the power cannot be returned to the system until all technicians have removed their padlocks. The lockout will be explained in detail in the chapter on motor controls later in this text.

After power has been removed, the field terminals should be isolated from the armature coil. This can be accomplished by disconnecting one set of leads where the field and armature are connected together. Remember that the field and armature are connected in parallel and if they are not isolated, your continuity test will show a completed circuit even if one of the two windings has an open.

When you have the field coil isolated from the armature coil, you can proceed with the continuity test. Be sure to use the R × 1k or R × 10k setting on the ohmmeter because the resistance in the field coil will be very high since the field coil may be wound from several thousand feet of wire. If the field winding test indicates the field winding is good, you should continue the procedure and test the armature winding for continuity.

The armature winding test may show that an open has developed from the coil burning open or from a problem with the brushes. Since the brushes may be part of the fault, they should be visually inspected and replaced if they are worn or not seating properly. If the commutator is also damaged, the armature should be removed, so the commutator can be turned down on a lathe.

If either the field winding or the armature winding has developed an open circuit, the motor will have to be removed and replaced. In some larger motors it will be possible to change the armature by itself rather than remove and replace the entire motor. If the motor operates but draws excessive current or heats up, the motor should be tested for loose or shorting coils. Field coils may tend to come loose and cause the motor to vibrate and overheat, or the armature coils may come loose from their slots and cause problems. If the motor continues to overheat or operate roughly, the motor should be removed and sent to a motor rebuilding shop so that a more in-depth test may be performed to find the problem before the motor is permanently damaged by the heat.

12.8 DC COMPOUND MOTORS

The DC compound motor is a combination of the series motor and the shunt motor. It has a series field winding that is connected in series with the armature and a shunt field that is in parallel with the armature. The combination of series and shunt winding allows the motor to have the torque characteristics of the series

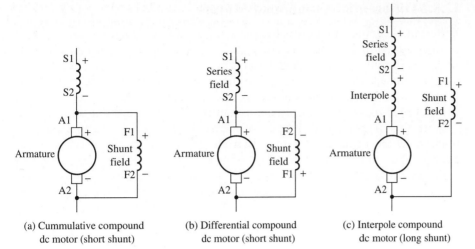

(a) Cummulative compound
dc motor (short shunt)

(b) Differential compound
dc motor (short shunt)

(c) Interpole compound
dc motor (long shunt)

FIGURE 12-17 **(a)** Diagram of a cumulative compound motor. **(b)** Diagram of a differential compound motor. **(c)** Diagram of an interpole compound motor.

motor and the regulated speed characteristics of the shunt motor. Fig. 12-17 shows a diagram of the compound motor. Several versions of the compound motor are also shown in this diagram.

12.8.1 Cumulative Compound Motors

Fig. 12-l7a shows a diagram of the cumulative compound motor. It is so called because the shunt field is connected so that its coils are aiding the magnetic fields of the series field and armature. The shunt winding can be wired as a *long shunt* or as a *short shunt*. Fig. 12-17a and Fig. 12-17b show the motor connected as a short shunt where the shunt field is connected in parallel with only the armature. Fig. 12-17c shows the motor connected as a long shunt where the shunt field is connected in parallel with both the series field, interpoles, and the armature.

Fig. 12-17a also shows the short shunt motor as a cumulative compound motor, which means the polarity of the shunt field matches the polarity of the armature. You can see in this figure that the top of the shunt field is positive polarity and that it is connected to the positive terminal of the armature. In Fig. 12-17b you can see that the shunt field has been reversed so that the negative terminal of the shunt field is now connected to the positive terminal of the armature. This type of motor is called a differential compound because the polarities of the shunt field and the armature are opposite.

The cumulative compound motor is one of the most common DC motors because it provides high starting torque and good speed regulation at high speeds. Since the shunt field is wired with similar polarity in parallel with the magnetic field aiding the series field and armature field, it is called cumulative. When the motor is connected this way, it can start even with a large load and then operate smoothly when the load varies slightly.

You should recall that the shunt motor can provide smooth operation at full speed but it cannot start with a large load attached, and the series motor can start with a heavy load, but its speed cannot be controlled. The cumulative compound motor takes the best characteristics of both the series motor and shunt motor, which makes it acceptable for most applications.

12.8.2 Differential Compound Motors

Differential compound motors use the same motor and windings as the cumulative compound motor but they are connected in a slightly different manner to provide slightly different operating speed and torque characteristics. Fig. 12-17b shows the diagram for a differential compound motor with the shunt field connected so its polarity is reversed to the polarity of the armature. Since the shunt field is still connected in parallel with only the armature, it is considered a short shunt.

In this diagram you should notice that F1 and F2 are connected in reverse polarity to the armature. In the differential compound motor the shunt field is connected so that its magnetic field opposes the magnetic fields in the armature and series field. When the shunt field's polarity is reversed like this, its field will oppose the other fields and the characteristics of the shunt motor are not as pronounced in this motor. This means that the motor will tend to overspeed when the load is reduced just like a series motor. Its speed will also drop more than the cumulative compound motor when the load increases at full rpm. These two characteristics make the differential motor less desirable than the cumulative motor for most applications.

12.8.3 Compound Interpole Motors

The compound interpole motor is built slightly differently from the cumulative and differential compound motors. This motor has interpoles added to the series field (Fig. 12-17c). The interpoles are connected in series between the armature and series winding. It is physically located behind the series coil in the stator. It is made of wire that is the same gauge as the series winding and it is connected so that its polarity is the same as the series winding pole it is mounted behind. Remember that these motors may have any number of poles to make the field stronger.

The interpole prevents the armature and brushes from arcing due to the buildup of magnetic forces. These forces are created from counter EMF called *armature reaction*. They are so effective that normally all DC compound motors that are larger than $\frac{1}{2}$ hp will utilize them. Since the brushes do not arc, they will last longer and the armature will not need to be cut down as often. The interpoles also allow the armature to draw heavier currents and carry larger shaft loads.

When the interpoles are connected, they must be tested carefully to determine their polarity so that it can be matched with the series winding. If the polarity of the interpoles does not match the series winding it is mounted behind, it will cause the motor to overheat and may damage the series winding.

12.8.4 Reversing the Rotation of the DC Compound Motor

Each of the compound motors shown in Fig. 12-17 can be reversed by changing the polarity of the armature winding. If the motor has interpoles, the polarity of the interpole must be changed when the armature's polarity is changed. Since the interpole is connected in series with the armature, it will be reversed when the armature is reversed. The interpoles are not shown in the diagram to keep it simplified. The armature winding is always marked as A1 and A2 and these terminals should be connected to the contacts of the reversing motor starter.

12.8.5 Controlling the Speed of the Motor

The speed of a compound motor can be changed very easily by adjusting the amount of voltage applied to it. In fact, it can be generalized that prior to the late 1970s, any industrial application that required a motor to have a constant speed would be handled by an AC motor, and any application that required the load to be driven at variable speeds would automatically be handled by a DC motor. This statement was true because the speed of a DC motor was easier to change than an AC motor. Since the advent of solid-state components and microprocessor controls, this condition is no longer true. In fact, today a solid-state AC variable-frequency motor drive can vary the speed of an AC motor as easily as that of DC motors. This brings about a condition where you must understand methods of controlling the speed of both AC and DC motors. Information about AC motor speed control is provided in Chapter 11.

Fig. 12-18 shows the characteristic curves of the speed versus armature current for the compound motors. From this diagram you can see that the speed of a differential compound motor increases slightly when the motor is drawing the armature highest current. The increase in speed occurs because the extra current in the differential winding causes the magnetic field in the motor to weaken slightly because the magnetic field in the differential winding opposes the magnetic in series field. As you learned earlier in the speed control of shunt motors, the speed of the motor will increase if the magnetic field is weakened.

Fig. 12-18 also shows the characteristic curve for the cumulative compound motor. This curve shows that the speed of the cumulative compound motor decreases

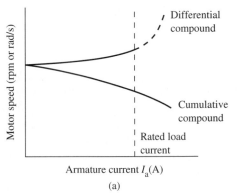

(a)

FIGURE 12-18 **(a)** Characteristic curve of armature current versus speed for the differential compound motor and cumulative compound motor. **(b)** Composite of the characteristic curves for all of the DC motors.

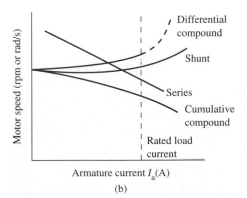

(b)

slightly because the field is increased, which slows the motor because the magnetic field in the shunt winding aids the magnetic field of the series field.

12.9 AC MOTORS

Today AC motors are more widely used in industrial applications than DC motors. They are available to operate on single-phase or three-phase supply voltage systems. This allows the motor control designer to choose the type of motor to fit the application. Most single-phase motors are less than 3 hp; although some larger ones are available, they are not as common. Three-phase motors are available up to several thousand horsepower, although most of the motors that you will be working with will be less than 50 hp.

The AC motor provides several advantages over DC motors. One advantage the AC motor has is its design eliminates the need for brushes and commutators. The second advantage is that its rotating member is made of laminated steel rather than wire that is pressed on a core, which reduces maintentance. The AC motor does not need brushes and commutators since it creates the flux lines in its rotating member by *induction*. The induction process that is used to get the current into the rotating member is similar to the induction that occurs between the primary and secondary windings of a transformer. This is possible in an AC motor because supply voltage is sinusoidal.

The rotating field in the AC motor is called the *rotor*, and the stationary field is called the *stator*. The design of the rotor is different from the rotating armature in the DC motor because it is made completely of laminated steel rather than having copper coils pressed on a laminated steel core.

This allows the AC motor to operate longer than the DC motor with less periodic maintenance, which means that more AC motors are used in industry than DC motors. You need to be aware that the main reason DC motors were used in industry in the 1940s through the 1960s is that their speed could be controlled more easily than controlling the speed of AC motors. With the advent of variable-frequency drives, the speed of all AC motors can be adjusted more easily than DC motors, and the AC motor requires less maintenance since it does not have brushes.

In this section we introduce each of the different types of AC motors and explain their basic parts, theory of operation, methods of controlling their speed and torque, changing the direction of rotation, and procedures for installation and troubleshooting. This basic information will also allow you to understand methods that motor controls use to take advantage of the motor's design to provide control.

We will also introduce the basic parts that are found in all AC motors and explain their operation and function. After the operation of a basic three-phase motor is explained, each type of AC motor is introduced and its special design features and applications for which it is best suited are discussed. You will be able to use this information in recognizing the type of AC motor you are working with; understand the theory of its operation, which will allow you to install and interface it to motor controls; and be able to troubleshoot the motor and quickly determine what faults it has. The three-phase motor is presented first, since some of the parts of a single-phase motor are designed specifically to compensate for the differences between three-phase and single-phase voltage. If you fully understand the characteristics of three-phase voltage and how three-phase motors take advantage of them, you will easily understand single-phase motors.

12.10 CHARACTERISTICS OF THREE-PHASE VOLTAGE

The three-phase AC motor is an induction motor. It is designed specifically to take advantage of the characteristics of the three-phase voltage that it uses for power. Fig. 12-19 shows a diagram of three-phase voltage. From this diagram you should notice that each of the three phases represents a separately generated voltage. You should recall that the three-phase generator has three separate windings that produce the three voltages slightly out of phase with each other. The units of measure for this voltage are electrical degrees. An electrical degree represents one sine wave in 360 degrees. The sine wave may be produced once during each rotation of the generator's shaft or twice during each generator's shaft rotation. If the sine wave is produced by one rotation of the generator's shaft, 360 electrical degrees are equal to 360 mechanical degrees. If the sine wave is produced twice during each shaft's rotation, 360 electrical degrees are equal to 180 mechanical degrees. Since this can tend to be confusing, all electrical diagrams are presented in terms of 360 electrical degrees being equal to 360 mechanical degrees. In this way the degrees will be the same and you will not have to try and figure out if electrical or mechanical degrees are being used in the example.

The first voltage shown in the diagram is called A phase and it is shown starting at 0° and peaking positively at the 90° mark. It passes through 0 volts again at the 180° mark and peaks negatively at the 270° mark. After it peaks negatively, it returns to 0 volts at the 360° mark, which is also the 0° point. The second voltage is called B phase and starts its zero-voltage point 120° later than A phase. B phase

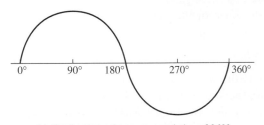

(a) Single-phase sine wave consisting of 360°

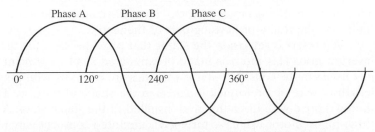

(b) Three-phase sine wave; each sine wave is 120° out of phase with the next

FIGURE 12-19 Diagram of three-phase voltage. Notice the phase shift between each voltage sine wave is 120°.

peaks positively, passes through zero voltage, and passes through negative peak voltage as A phase does, except that it is always 120° later than A phase. This means that B phase is increasing in the positive direction when A phase is passing through its zero voltage at the 180° mark.

The third voltage shown on this diagram is called C phase. It starts at its zero-voltage point 240° after A phase starts at its zero-voltage point. This puts B phase 120° out of phase with A phase and C phase 120° out of phase with B phase.

The ac motor takes advantage of this characteristic to provide a rotating magnetic field in its stator and rotor that is very strong because three separate fields rotate 120° out of phase with each other. Since the magnetic fields are induced from the applied voltage, they will always be 120° out of phase with each other. Do not worry about the induced magnetic field being 180° out of phase with the voltage that induced it. At this time this phase difference is not as important as the 120° phase difference between the rotating magnetic fields.

Since the magnetic fields are 120° out of phase with each other and are rotating, one will always be increasing its strength when one of the other phases is losing its strength by passing through the zero-voltage point on its sine wave. This means that the magnetic field produced by all three phases never fully collapses and its average is much stronger than that of a field produced by single-phase voltage.

 12.11 THREE-PHASE MOTOR COMPONENTS

The AC induction motor has three basic parts: the stator, which is the stationary part of the motor; the rotor, which is the rotating part of the motor; and the end plates, which house the bearings that allow the rotor to rotate freely. This section will provide information about each of these parts of an AC motor. Fig. 12-20 shows a cut-away diagram of a three-phase motor, and Fig. 12-21 shows an exploded-view picture of a three-phase motor. These figures will provide you with information about the location of the basic parts of a motor and how they work together.

12.11.1 Stator

The stator is the stationary part of the motor and is made of several parts. Fig. 12-22 shows the stator of a typical induction motor. The stator is the frame for the motor housing the stationary winding with mounting holes for installation. The mounting holes for the motor are sized according to NEMA standards for the motor's frame type. Some motors will also have a lifting ring in the stator to provide a means for handling larger motors. The lifting ring and mounting holes are actually built into the frame or housing part of the stator.

An insert is set inside the stator that provides slots for the stator coils to be inserted into. This insert is made of laminated steel to prevent eddy current and flux losses in the coils. The stator windings are made by wrapping a predetermined length of wire on preformed brackets in the shape of the coil. These windings are then wrapped with insulation and installed in the stator slots. A typical four-pole, three-phase motor will have three coils mounted consecutively in the slots to form a group. The three coils will be wired so that they each receive power from a

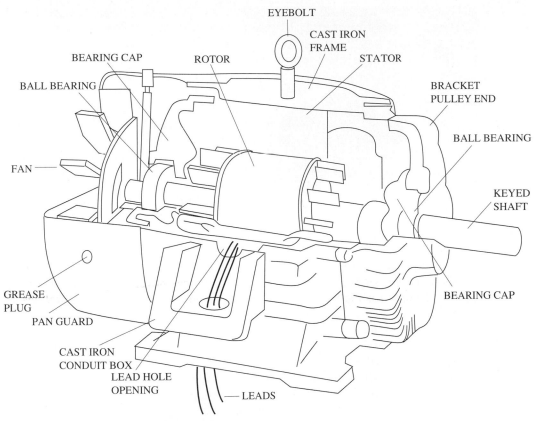

FIGURE 12-20 Cut-away diagram of a three-phase motor. (Courtesy of Century.)

separate phase of three-phase power supply. Three groups are connected together to form one of the four poles of the motor. This grouping is repeated for each of the other three poles so that the motor has a total of 36 coils to form the complete four-pole stator. It is not essential that you understand how to wind the coils or put them into the stator slots; rather, you should understand that these coils are connected in the stator, and 3, 6, 9, 12, or 15 wires from the coil connections will

FIGURE 12-21 Exploded view of a three-phase motor. This picture shows the relative location of all of the parts of the motor.

FIGURE 12-22 Example of a typical stator for an AC motor.

be brought out of the frame as external connections. The external connection wires can be connected on the factory floor to allow the motor to be powered by 208/230 or 480 volts, or they allow the motor to be connected to provide the correct torque response for the load. Other changes can also be made to these connections to allow the motor to start so it uses less locked-rotor current.

After the coils are placed in the stator, their ends (leads) will be identified by a number that will be used to make connections during the installation procedure. The coils are locked into the stator with wedges that keep the coils securely mounted in the slots and allow them to be removed and replaced easily if the coils are damaged or become defective due to overheating.

12.11.2 Rotor

The rotor in an AC motor can be constructed from coils of wire wound on laminated steel or it can be made entirely from laminated steel without any wire coils. A rotor with wire coils is called a *wound rotor* and it is used in a wound-rotor motor. The wound-rotor motor was popular in the 1950s since it could produce more torque than a similar-size induction motor. The main drawback of the wound-rotor motor is that it requires the use of brushes and slip rings to transfer current to it. Since modern AC induction motors can now produce adequate torque, the wound-rotor motor is not often used because the brushes and slip rings require too much maintenance.

Motors that use a laminated steel rotor are called *induction motors* or *squirrel-cage induction motors*. The core of the rotor is made of die-cast aluminum in the shape of a squirrel cage. Laminated sections are pressed onto this core or the core is molded into laminated sections when the squirrel-cage rotor is manufactured. Fig. 12-23 shows a diagram and picture of a squirrel-cage rotor and you can see the skeleton of the squirrel-cage core. The fins or blades are built into the rotor for cooling the motor and it is important that these fan blades are not damaged or broken, since they provide all of the cooling air for the motor and they are balanced so that the rotor will spin evenly without vibrations.

12.11.3 Motor End Plates

The end plates house the bearings for the motor. The end plate and bearing can be seen in the picture of the rotor that is shown in Fig. 12-23b. If the motor is a fractional-horsepower motor, it will generally use sleeve-type bearings and if the

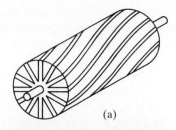

(a)

(b)

FIGURE 12-23 **(a)** Diagram of a squirrel-cage rotor for an AC motor. **(b)** Picture of a squirrel-cage rotor for an AC motor.

motor is one of the larger types, it will use ball bearings. Some ball bearings on smaller motors will be permanently lubricated, while the larger motor bearings will require periodic lubrication. All sleeve bearings will require a few drops of lubricating oil periodically.

The end plates are mounted on the ends of the motor and held in place by long bolts inserted through the stator frame. When nuts are placed on the bolts and tightened, the end plates will be secured in place. If the motor is an open type, the end plates will have louvers to allow cooling air to circulate through the motor. An access plate may also be provided in the rear end plate to allow field wiring if one is not provided in the stator frame.

If the motor is not permanently lubricated, the end plate will provide an oiler tube or grease fitting for lubrication. It is important that the end plates are mounted on the motor so that the oiler tube or grease fitting is above the shaft so that gravity will allow lubrication to reach the shaft. If the end plate is rotated so that the lubrication point is mounted below the shaft, gravity will pull all of the lubrication away from the shaft and the bearings will wear out prematurely. If you need to remove the end plates for any reason, they should be marked so that they will be replaced in the exact position from which where they were removed. This also helps to align the holes in the end plate with the holes in the stator so that the end plates can be reassembled easily.

12.12 OPERATION OF AN AC INDUCTION MOTOR

The basic principle of operation of an inductive motor is based on the fact that the rotor receives its current by induction rather than with brushes and slip rings or brushes and commutators. Current can be induced into the rotor by being in close proximity to the stator. This action is similar to the action of the primary and secondary coils of a transformer. You should remember from your study of transformers and generators that when a coil of wire is allowed to pass across magnetic flux lines, a current will be generated in the coil. This current will be 180° out of phase with the current that produced it.

When the induced current begins to flow in the squirrel-cage conductors embedded in the laminated segments of the squirrel-cage rotor, a magnetic field will be built in the rotor. The magnetic field produced by the induced current will be similar to the magnetic field produced by current that is provided to the rotor with brushes.

Since the current is induced into the rotor, no brushes are required. The strength of the field in the rotor is not as strong as a field that could be produced by passing current through brushes to a coil of wire on the rotor. This means that the magnetic field in the stator must be much stronger to compensate for the weaker field in the rotor to produce sufficient horsepower and torque.

When three-phase voltage is applied to the stator windings, a magnetic field is formed. The natural characteristic of three-phase voltage will cause the magnetic field to move from coil to coil in the stator, which appears as though the field is rotating. Fig. 12-24 shows an experiment your instructor can perform in the laboratory to prove that the magnetic field in the stator will actually rotate. From this figure you can see that the end plates and rotor have been removed from the stator. The three stator windings are connected to three-phase voltage through a switch. A large ball bearing approximately 1 inch in diameter is placed in the stator, and the switch is closed to energize the stator winding. When the magnetic field begins to form in the stator, it will begin to pull on the ball bearing. You may need to give the ball bearing a slight push with a plastic rod to get it to begin to rotate. Do not use a metal object such as a screwdriver to get the ball bearing to move because the tip of the screwdriver will be attracted to the stator by its magnetic field.

FIGURE 12-24 Experiment that shows a ball bearing placed in the stator of a three-phase AC motor. When three-phase voltage is applied to the rotor, a rotating magntic field will be established, and the ball bearing will chase the rotating field around the stator. When the rotor is placed in the stator, it will rotate in step with the rotating magnetic field.

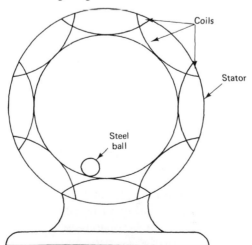

Coils

Stator

Steel
ball

Once you get the ball bearing to rotate, it will continue to rotate at the speed of the rotating magnetic field in the stator until the switch is opened and the stator's magnetic field has collapsed. If you reverse two of the three-phase supply voltage wires, the magnetic field will begin to rotate in the opposite direction, and when you give the ball bearing a push to start it, it will also rotate in the opposite direction.

12.12.1 Induced Current in the Rotor

When the magnetic field in the stator cuts across the poles of the squirrel-cage rotor, a current is induced in the rotor. This current is out of phase with the applied current, but it is strong enough to cause the rotor to start to turn. The speed of the rotor is determined by the number of poles in the stator and the frequency of the incoming AC voltage. A formula is provided to determine the operating speed of the motor:

$$\text{Operating speed of motor} = \frac{F \times 120}{P}$$

where F is the frequency of the applied voltage, 120 is a magnetic constant, and P is the number of poles. It should also be noted at this time that this formula calculates the speed of the rotating field, and the actual speed of the rotor will be slightly less due to slip. The concept of slip will be explained in later sections.

The full rpm is called *synchronous speed*. From this formula we calculate that a two-pole motor will operate at 3600 rpm, a four-pole motor will operate at 1800 rpm, a six-pole motor will operate at 1200 rpm, and an eight-pole motor will operate at 900 rpm. These speeds do not include any slip or losses due to loads. From this example, you can see that the only way an AC induction motor can have its speed changed is to change the number of poles it has, or change the frequency of the voltage supplied to it.

When power is first applied, the stator field will draw very high current since the rotor is not turning. This current is called *locked-rotor amperage* (LRA) and is sometimes referred to as *inrush current*. When LRA moves through the stator, its magnetic field is strong enough to cause the rotor to begin to rotate. As the rotor starts moving, it will begin to induce current into its laminated coils and build up torque. This causes the rotor to spin faster, until it begins to catch up with the rotating magnetic field.

As the rotor turns faster, it will begin to produce voltage of its own. This voltage is called *back EMF* or *counter EMF*. The counter EMF opposes the applied voltage, which has the effect of lowering the difference of potential across the stator coils. The lower potential causes current to become lower when the motor is a full load. The *full-load amperage* is referred to as FLA and will be as much as six to ten times smaller than the inrush current (LRA). The stator will draw just enough current to keep the rotor spinning.

When the load on the rotor increases, it will begin to slow down slightly. This causes the counter EMF to drop slightly, which makes the difference in potential greater and allows more current to flow. The extra current provides the necessary torque to move the increased load and the rotor's speed catches up to its rated level. In this way, the squirrel-cage induction motor is allowed automatically to regulate the amount of current it requires to pull a load under varying conditions. The rotor will develop maximum torque when the rotor has reached 70–80% of synchronous speed. The motor can make adjustments anywhere along its torque range. If the load becomes too large, the motor shaft will slow to the point of stalling and the motor will overheat from excess current draw. In this case the motor must be wired for increased torque, or a larger-horsepower motor should be used.

12.13 CONNECTING MOTORS FOR TORQUE SPEED AND HORSEPOWER CONDITIONS

The squirrel-cage induction motor can be connected in several different ways to produce constant torque, a change of speed, or a change of horsepower ratings. They can also be connected for variable torque. The load conditions or applications will dictate the method you select to connect the motor. These connections can be made after the motor is installed on the factory floor or when the motor is being installed in the machinery it is driving. The next sections will provide examples of each way the motor can be connected to make these changes.

12.13.1 Wye-Connected Motors

The squirrel-cage induction motor may have 6, 9, or 12 terminal leads from the ends of its coils brought out of the motor frame for field wiring connections. Field wiring connections are the connections that you will change as a technician right on the factory floor with the motor mounted in place. Fig. 12-25 shows the diagram of a nine-lead motor. Notice that the coils are positioned in the shape of the letter Y. This motor is called a wye-connected motor or star-connected motor.

The wye motor terminals are numbered in the clockwise direction. The two ends of the first coil are numbered T1 and T4, the second coil is numbered T2 and T5, and the third coil is numbered T3 and T6. The outside coils are isolated from each other while the inside coils are connected together at the wye point. The

FIGURE 12-25 Nine-line, wye-connected motor.

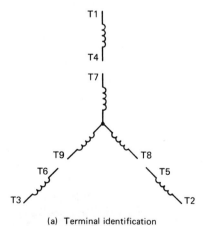

(a) Terminal identification

3φ Applied voltage

L1	L2	L3	Together		
T1	T2	T3	T4 T7	T5 T8	T6 T9

(b) Field wiring connections

second diagram shows the motor connected to three-phase line voltage at T1, T2, and T3. A table is also provided to show which terminals are connected together and which ones are connected to line voltage for the motor to operate correctly. The information provided in this table is similar to data provided on the data plate of the motor.

You will need to make these connections in the field when you are installing the motor. Wire nuts or lugs should be used to make the connections for the coil terminals and supply voltage wires. The diagram is shown for the motor so it can be connected for high voltage. The diagram indicates the high voltage is 480 volts AC. You should be aware that in your area of the country the exact amount of high voltage may vary from 440 volts AC through 480 volts AC. The diagrams will be marked so that 480 volts AC represents any high voltage (440, 460, or 480 volts AC), 240 volts AC represents any low voltage (220, 230, or 240 volts AC) for a delta system, and 208 volts AC represents the low-voltage wye system.

12.13.2 Delta-Connected Motors

Another method of connecting the nine leads of the squirrel-cage induction motor is in a series circuit called a *delta configuration*. Fig. 12-26 shows the nine-lead motor connected in a delta configuration. The term *delta* is used because the formation of the coils in this diagram resembles the Greek capital letter delta (Δ). These connections are also made when the motor is being installed on the factory floor.

A table is also shown in this figure that provides you with the proper terminal connections as they would be shown on the motor's data plate. The motor is also shown connected to three-phase line voltage at terminals T1, T2, and T3. You should note that the terminals are numbered in a clockwise direction starting at the top of the delta. Each tip of the delta is numbered T1, T2, and T3. The other ends of each of the first coils are numbered T4 and T9, and the ends of the second coils are marked T5 and T7, while the ends of the third coil are marked T6 and T8. Terminals T1, T2, and T3 are the midpoints in these coils.

When the motor terminals are wired for operation, terminals T1, T2, and T3 are connected to the power supply at terminals T1, T2, and T3. Terminal T4 is connected to T7, T8 to T5, and T6 to T9 to complete the series circuit. Since these coils are in a series circuit, when three-phase power is supplied, it will come in T1 through the winding and go out terminal T2. Another of the phases will come in

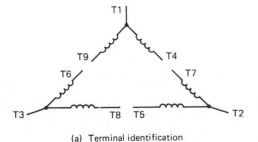

(a) Terminal identification

FIGURE 12-26 A nine-lead, delta-connected motor.

Three-phase voltage

L1	L2	L3	Together		
T1	T2	T3	T4 T7	T5 T8	T6 T9

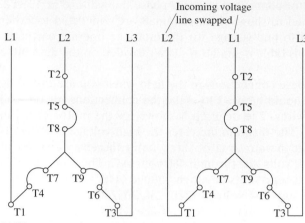

FIGURE 12-27 Diagram of the wye-connected motor for clockwise and counterclockwise operation.

(a) Wye-connected motor for clockwise rotation for 480 V

(b) Wye-connected motor for counterclockwise rotation for 480 V

T3 and go out T2, and the final phase will come in T2 and go out T3. When the sine wave reverses itself, the currents will reverse and come in from the opposite terminals. This means that at any one instant in time, two of the three wires from the power supply will be used to make a complete circuit. Since the AC three-phase voltage is 120° out of phase and the windings in the motor are also 120° out of phase, the three-phase current will energize each coil in such a way as to cause the magnetic field to rotate.

12.14 REVERSING THE ROTATION OF A THREE-PHASE INDUCTION MOTOR

The rotation of a wye- or delta-connected motor can be changed by exchanging any two of the three phases of the incoming voltage. Fig. 12-27 shows diagrams for a wye-connected motor and Fig. 12-28 shows the diagrams for a delta-connected

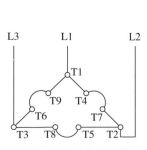

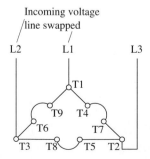

FIGURE 12-28 Diagram of a delta-connected motor for clockwise and counterclockwise operation.

(a) Delta-connected motor for clockwise rotation for 480 V

(b) Delta-connected motor for counterclockwise rotation for 480 V motor for clockwise (forward) and counterclockwise (reverse) rotation. From these diagrams you can see that T1 and T2 supply voltage terminals have been exchanged in the diagram for motor reversal. In industrial applications, terminals T1 and T2 are generally switched by the contacts of the reversing motor starter. These diagrams will be useful for installation connections and troubleshooting.

This wiring configuration is also used when the motor is connected for *plugging* use. When the motor is used for plugging, it is reversed while running at full rpm. When the motor's stator is quickly reconnected for reverse rotation by switching two of the three input voltage lines, it will quickly build up a reverse magnetic field that will begin to rotate in the opposite direction. The rotor will follow this quick change in rotation and begin to rotate in the opposite direction. This will cause the load to decrease its rpm from full in the clockwise direction to zero, and begin to rotate counterclockwise. The moment the rotor begins to rotate in the opposite direction, the power is de-energized and the rotor shaft is stopped from rapid reverse torque.

12.15 CONNECTING MOTORS FOR A CHANGE OF VOLTAGE

Delta- and wye-connected motors may have numerous sets of coils so that they can be wired to operate at two separate voltages. The extra windings are connected in series for the motor to operate at the higher voltage, and they are connected in parallel for the lower voltage. The higher voltage in these applications is usually 440 or 480 volts, while the low voltage is usually 208 or 220 volts. The actual voltage will be specified on the motor's nameplate.

The dual-voltage option provides a larger variety of choices when the motor must be connected to power distribution systems. You should remember from Chapter 3 that it is important to balance the loads on the power distribution system and not overload any of the transformers. When the motors are connected to operate at higher voltages, their current draw is reduced by half. The motor will use the same amount of wattage in both configurations, but smaller-gauge wire and smaller contacts and switch gears can be used throughout the circuit when the motor is connected to the higher voltage.

The lower voltage is also useful when no high voltage is available in the area of the factory where the motor must be installed. The dual-voltage motor can be connected for the lower voltage and a savings can be realized by not having to install extra-long power cables to reach the remote source.

Fig. 12-29 shows a set of diagrams and tables that indicates the proper connections for the motor to operate at both high and low voltage. The diagrams are presented as wiring diagrams with the numbers on the terminal leads. You should remember that the motor can be connected as either a delta motor or a wye motor, which will affect its starting torque and LRA characteristics, so high- and low-voltage diagrams are presented for each of these types of motors.

Fig. 12-29a shows a wye-connected motor wired for high voltage and for low voltage. Notice that the six coils are configured as three sets of parallel coils when the motor is connected for low voltage and as a series circuit when they are connected for high voltage. Since all the coils are used in the configurations for high and low voltage, the motor will have the same amount of torque and horsepower in both cases.

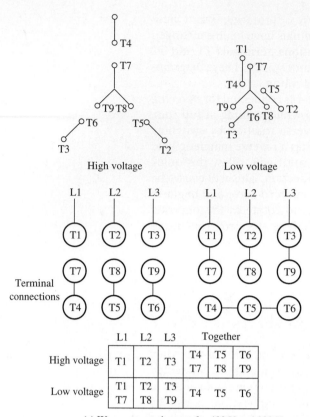

(a) Wye-connected motor for 480 V and 208 V

	L1	L2	L3	Together		
High voltage	T1	T2	T3	T4 T7	T5 T8	T6 T9
Low voltage	T1 T7	T2 T8	T3 T9	T4	T5	T6

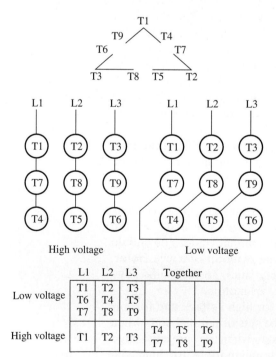

(b) Delta-connected motor for 480 V and 240 V

	L1	L2	L3	Together		
Low voltage	T1 T6 T7	T2 T4 T8	T3 T5 T9			
High voltage	T1	T2	T3	T4 T7	T5 T8	T6 T9

FIGURE 12-29 **(a)** Diagram of a wye-connected motor for high and low voltages. **(b)** Diagram of a delta-connected motor for high and low voltages.

The diagram that shows the coils is presented to give you a picture of how the coils look in series and parallel, but this diagram is sometimes confusing when you must make changes in the field. For this reason a second diagram is presented that shows only terminals. Each of the terminals is numbered as you would find it on a motor, and the heavy line shows the connections that must be made to complete the connections.

A table is provided in this figure that indicates the terminals that are connected to each other, to the power supply, and that are left open. When a terminal is left open, it means that a wire nut should be placed over the terminal end and wrapped with electrical tape to secure it. Do not cut this wire short, as it may be needed when the motor is reconnected later. Remember, machinery is moved around the factory rather frequently. This means that the motor may need to be connected for a different voltage when it is moved in a few years, or a new service entrance and power distribution system may be installed and several motors may be changed to operate on the new voltage.

Fig. 12-29b shows the coils of a delta-connected motor wired for high and low voltage. This diagram shows the coils for the motor connected for low voltage wired in parallel and the coils for a high-voltage connection wired in series. A terminal diagram and table are presented for the delta motor. Remember that these diagrams are presented to show you the variety of methods that manufacturers use to indicate the terminal connections for wiring their motor for high or low voltage.

It should also be pointed out at this time that all squirrel-cage motors that have nine leads brought out of their frame can be connected for high or low voltage. The diagrams presented in this figure are usable on all name brands of motors. This is important because the motor data plate is usually painted over, damaged, or removed when you need it to make field wiring connections. This means that you will be able to use the diagrams found in this figure to make the connections.

12.16 CONNECTING THREE-PHASE MOTORS FOR A CHANGE OF SPEED

Some delta-connected and wye-connected motors can be wired to operate at two different speeds. Unlike the nine-lead motor, which allows all motors to be wired for high or low voltage, not all motors are manufactured to be reconnected for a change of speed. The motor must be specifically manufactured with enough leads brought out of the frame to make the changes required to allow the motor to operate at the different speeds.

Some motors have enough leads brought out to operate at two different speeds, while other motors can be reconnected to operate at up to four speeds. It is important that you understand that a motor's speed is changed by changing the number of poles that are used. You should remember that when the motor uses eight poles it will operate at 900 rpm, with six poles it will operate at 1200 rpm; with four poles it will operate at 1800 rpm, and the two-pole motor will operate at 3600 rpm. This means that the motor would provide less horsepower when poles are removed form the circuit completely to allow the motor to operate at a higher speed.

Some motors provide a means to reconnect the extra poles back into the circuit to keep the overall horsepower rating of the motor constant. Changes in the

THREE PHASE MOTORS

NOTE: THE FOLLOWING DIAGRAMS ARE TYPICAL MOTOR CONNECTION ARRANGEMENTS, CONFORMING TO NEMA STANDARDS. NOT ALL POSSIBLE ARRANGEMENTS ARE SHOWN.

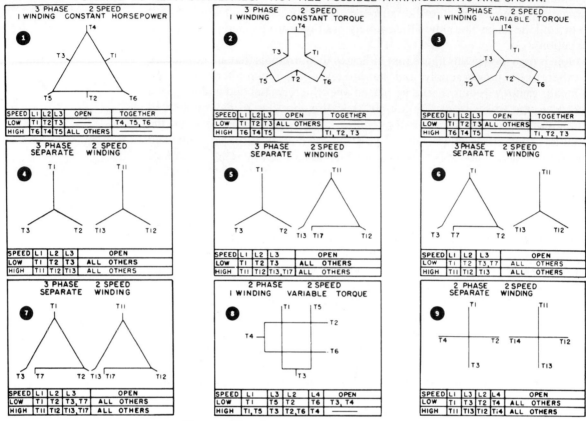

FIGURE 12-30 Two-speed diagrams for six-lead and twelve-lead motors. (Courtesy of Electrical Apparatus Service Association.)

connections can be made to allow the motor to provide constant torque regardless of the speed at which it is operating.

Fig. 12-30 shows a series of diagrams that provides methods of connecting these multilead motors for a change of speed. These diagrams allow you to select the torque and horsepower requirements for your application. From this figure you can see that the diagrams are listed in terms of the number of windings the motor has. The first group of diagrams (1, 2, and 3) shows diagrams for two-speed motors with one winding. These motors will have six leads brought out of the motor for the field connections that are listed at the bottom of each diagram. Diagram 1 is provided to allow the motor to operate at constant horsepower, diagram 2 is provided to allow the motor to operate with a constant torque, and diagram 3 is provided to allow the motor to operate with variable torque.

The second group of diagrams (4, 5, 6, and 7) shows diagrams for two-speed motors with two separate windings. These motors have 12 leads brought out to make two separate windings for field connections. Diagram 4 shows both windings connected as a wye, and diagram 5 shows the first winding connected as a wye, and the second winding connected as a delta. Diagram 6 shows the first winding connected as a delta and the second winding connected as a wye. Diagram 7 shows both windings connected as a delta.

12.16.1 Interpreting the Wiring Diagrams and Tables

The diagrams in Fig. 12-30 show two ways of representing the connections that must be made to the motor terminals. One way shows the connections with an electrical diagram, and the second way is to show a table that indicates which terminal numbers should be connected. When you are ready to make the motor connections, you should notice that the terminals in the motor are marked with numbers which are stamped into the wire material, or identified with a metal tag that has a number on it. The metal tag is crimped on each wire near the terminal end. After you have located all the terminals and have identified them by their numbers, you are ready to make the connections shown in the diagram for your application.

The table lists the terminals that should be connected to the three input-voltage wires. These are identified as L1, L2, and L3, and any terminal number listed in the category under the line number should be connected to that line. Be sure to check the row indicating the speed for which the motor will be wired. The second column lists the wires that are left open. This means that the wires listed in this column should not be connected to anything. They are supposed to remain unconnected and they should have an insulated wire nut or cap placed over the end of the wire securely so that it does not come in contact with any metal parts of the motor or other energized wires. In some tables any lead that is not listed in another column should be left open.

The third column lists the leads or wires that should be connected together. This means that the wires listed in this column should be connected together and no power should be connected to these leads. The leads must be secured together with a wire nut and wrapped with insulating tape because they will be energized. In some diagrams no terminals will be listed in this column, which means that all the leads are used in one of the other columns. Be sure that you account for every lead before you return all the leads back into the motor and replace the field wiring access cover.

These diagrams are extremely useful when the application you are working with requires the motor to be reconnected on the factory floor. Many times these diagrams are not readily available when you need them, so this provides that much needed reference. These motors can also be reversed by exchanging two of the three supply voltage lines. This allows the motors to be used in the widest possible number of applications.

It is important to understand that you will be required to make these changes yourself or direct someone to make them for you. This means that you must understand the concept of changing the connections of motor leads to make the motor fit the application.

12.17 MOTOR DATA PLATES

The motor's data plate lists all the pertinent data concerning the motor's operational characteristics. It is sometimes called the name plate. Fig. 12-31 shows an example of a data plate for a typical AC motor. The data plate contains information about

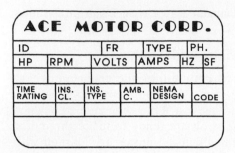

FIGURE **12-31** Data plate for typical AC motor.

the ID (identification number), FR (frame and motor design), motor type, phase, horsepower rating, rpm, volts, amps, frequency, service factor (SF), duty cycle (time), insulation class, ambient temperature rise, and NEMA design and code. Each of these features will be discussed in detail in the next sections.

12.17.1 Identification Number (ID)

The identification number for a motor will basically be a model and serial number. The model number will include information about the type of motor, and the serial number will be a unique number that indicates where and when the motor was manufactured. These numbers will be important when a motor is returned for warranty repairs, or if an exact replacement is specified as a model for model exchange.

12.17.2 Frame Type

Every motor has been manufactured to specifications that are identified as a *frame size*. Fig. 12-32 shows a table with frame size data. These data include the distance between mounting holes in the base of the motor, the height of the shaft, and other

FIGURE **12-32** NEMA frame dimensions table for physical sizes for motors. (Courtesy of Electrical Apparatus Service Association.)

IEC FRAME DIMENSIONS*
FOOT-MOUNTED THREE-PHASE MOTORS

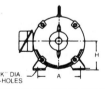

*Letters in parentheses are NEMA equivalent to IEC.

FRAME	H	(D)	A	(2E)	B	(2F)	C	(BA)	K	(H)	D	(U)	E	(N-W)
	mm	IN	mm	IN	mm	IN	mm	IN	mm	IN	mm	IN	mm	IN
56	56	2.2	90	3.54	71	2.8	36	1.42	5.8	.228	9	.354	20	.787
63	63	2.48	100	3.94	80	3.15	40	1.57	7	.276	11	.433	23	.905
71	71	2.8	112	4.41	90	3.54	45	1.77	7	.276	14	.551	30	1.18
80	80	3.15	125	4.92	100	3.94	50	19.7	10	.394	19	.748	40	1.57
90S	90	3.54	140	5.51	100	3.94	56	2.2	10	.394	24	.945	50	1.97
90L	90	3.54	140	5.51	125	4.92	56	2.2	10	.394	24	.945	50	1.97
100L	100	3.94	160	6.3	140	5.5	63	2.48	12	.472	28	1.1	60	2.36
112M	112	4.41	190	7.48	140	5.5	70	2.75	12	.472	28	1.1	60	2.36
132S	132	5.2	216	8.5	140	5.5	89	3.5	12	.472	38	1.5	80	3.15
132M	132	5.2	216	8.5	178	7	89	3.5	12	.472	38	1.5	80	3.15
160M	160	6.3	254	10	210	8.27	108	4.25	15	.591	MANUFACTURERS DO NOT AGREE BEYOND THE 132 FRAME.			
160L	160	6.3	254	10	254	10	108	4.25	15	.591				

critical data about physical dimensions. These data are then given a number such as 56. The frame number indicates that any motor that has the same frame number will have the same dimensions even though it may be made by a different company. This allows users to stock motors from more than one manufacturer or replace a motor with any other with the same frame information with confidence that it will be an exact replacement.

12.17.3 Motor Type

The motor-type category on the data plate refers to the type of ventilation the motor uses. This includes the open type, which provides flow-through ventilation from the fan mounted on the end of the rotor. In some motors that are rated for variable-speed duty (used with variable-frequency drives), the fan will be a separate motor that is built into the end of the rotor. The fan motor will be connected directly across the supply voltage, so it will maintain a constant speed to provide continual cooling regardless of the motor speed.

Another type of motor, the enclosed motor, is not air cooled with a fan. Instead it is manufactured to allow heat to dissipate quickly to and from the inside of the motor outward to the frame. In most cases the frame has fins built into it on the outside to provide more area for cooling air to reach.

12.17.4 Phase

The phase of the motor will be indicated as single phase or three phase. The number may be indicated as 1∅ or 3∅, or the number may be listed by itself.

12.17.5 Horsepower Rating (hp)

The horsepower rating will be indicated as a fractional horsepower (number less than 1.0) or a larger horsepower. Fractional-horsepower numbers may be listed as fractions ($\frac{1}{2}$) or decimals (0.5).

12.17.6 rpm (Speed)

The motor speed will be indicated as rpm. This will be the rated speed for the motor, and it will not account for slip. The actual speed of the motor will be less because of slip. As you know the speed of the motor is determined by the number of poles and the frequency of the AC voltage. Typical speed for a two-pole motor is 3600 rpm, a four-pole motor is 1800, and an eight-pole motor is 1200. The actual speed of the motor rated for 3600 rpm will be approximately 3450, for the 1800 rpm motor it will be approximately 1750, and for the 1200 rpm motor it will be approximately 1150. The actual amount of slip will be indicated by the *motor design letter* (see Section 12.16.14).

12.17.7 Volts

Voltage ratings for single-phase motors will be listed as 115 volts, 208 volts, or 230 volts. Three-phase motors have typical voltage ratings of 208, 240, 440, 460, 480, and 550 volts. Other voltages may be specified for some special-type motors. You should always ensure that the power supply voltage rating matches the voltage rating of the motor. If the rating and the power supply voltage do not match, the motor will overheat and be damaged.

12.17.8 Amps

The amps rating is the amount of full-load current (FLA) the motor should draw when it is under load. This rating will help the designer calculate the proper wire size, fuse size, and heater size in motor starters. The supply wiring for the motor circuit should always be larger than the amps rating of the motor. The NEC (National Electric Code) provides information to help you determine the exact fuse size and heat sizes for each motor application.

12.17.9 Frequency

The frequency of a motor will be listed in hertz (Hz). Typical frequency ratings for motors in the United States is 60 Hz. Motors manufactured for use in some parts of Canada and all of Europe and Asia will be rated for 50 Hz. You must be sure that the frequency of the motor matches the frequency of the power supplied to the motor.

12.17.10 Service Factor (SF)

The *service factor* is a rating that indicates how much a motor can be safely over-loaded. For example, a motor that has an SF of 1.15 can be safely overloaded by 15%. This means that if the motor is rated for 1 hp, it can actually carry 1.15 hp safely. To determine the overload capability of a motor, you multiply the rated horsepower by the service factor. The motor is capable of being overloaded because it is designed with ways of dissipating large amounts of heat.

12.17.11 Duty Cycle

The *duty cycle* of a motor is the amount of time the motor can be operated out of every hour. If the motor's duty cycle is listed as continuous, it means the motor can be run 24 hours a day and does not need to be turned off to cool down. If the duty cycle is rated for 20 minutes, it means the motor can be safely operated for 20 minutes before it must be shut down to be allowed to cool. The motor with this rating should be shut down for 40 minutes of every hour of operation to be allowed to cool.

Another way to specify the duty cycle of a motor is called the *motor rating*. The motor rating on the data plate refers to the type of duty the motor is rated for. The types of duty include continuous duty, intermittent duty, and heavy duty, which includes jogging and plugging duty. Continuous duty includes applications where the motor is started and allowed to operate for hours at a time. The intermittent duty includes operations where the motor is started and stopped frequently. This type of application allows the motor to heat up because it will draw LRA more often than will a motor rated for continuous duty.

Motors that are rated for jogging and plugging are built to withstand very large amounts of heat that will build up when the motor will draw large LRA during starting and stopping. Since the motor can be reversed when it is running in the forward direction for plugging applications, it will build up excessive amounts of heat. Motors with this rating must be able to get rid of heat as much as possible to withstand the heavy-duty applications.

12.17.12 Insulation Class

The insulation class of a motor is a letter rating that indicates the amount of temperature rise the insulation of the motor wire can withstand. The numbers in

Class	Temperature Rise °C
A	105
B	130
F	155
H	180

FIGURE 12-33 Insulation class for motors. This table indicates the amount of temperature for which the wire's insulation is rated.

the insulation class are listed in degrees Celsius (°C). The table in Fig. 12-33 shows typical insulation classes for motors. The insulation class and other temperature-related features of a motor will help determine the temperature rise the motor can withstand.

12.17.13 Ambient Temperature Rise

The ambient temperature rise is also called the Celsius rise. It is the amount of temperature rise the motor can withstand during normal operation. This value is listed in degrees Celsius (°C). A typical open motor can withstand a rise of 40°C (104°F) and an enclosed motor can withstand a 50°C (122°F) rise. This means the motor should not be exposed to environments where the temperature is 104°F above the ambient. If the ambient is considered 72°F, it means the motor is limited to temperatures of 176°F. Another classification that will help determine the amount of temperature a motor will be able to withstand is the type of insulation the motor winding has. The classes of insulation that are used with motors and the amount of temperature these classes can handle are listed in Section 12.16.12.

12.17.14 NEMA Design

The National Electric Manufacturers Association (NEMA) provides motor design ratings. The motor design is listed on the data plate by a letter A, B, C, or D. This designation is determined by the type of wire, insulation, and rotor that are used in the motor and are not effected by the way the motor might be connected in the field.

Type A motors have low rotor circuit resistance and have approximate slip of 5–10% at full load. These motors have low starting torque with a high locked-rotor amperage (LRA). This type of motor tends to reach full speed rather rapidly.

Type B motors have low to medium starting torque and usually have slip of less than 5% at full load. These motors are generally used in fans, blowers, and centrifugal pump applications.

Type C motors have a very high starting torque per ampere rating. This means that they are capable of starting when the full load is applied for applications such as conveyors, crushers, and reciprocating compressors such as air-conditioning and refrigeration compressors. These motors are rated to have slip of less than 5%.

Type D motors have a high starting torque with a low LRA rating. This type of motor has a rotor made of brass rather than copper segments. It is rated for slip of 10% at full load. Normally this type of motor will require a larger frame to produce the same amount of horsepower as a type A, B, or C motor. These motors are generally used for applications with a rapid decrease of shaft acceleration, such as a punch press that has a large flywheel.

These standards are set by NEMA, and a motor must meet all the requirements of the standard to be marked as a type A, B, C, or D. This allows motors made by several manufacturers to be compared on an equal basis according to application.

NEMA Code Letter	Locked-Rotor kVA per hp
A	0–3.15
B	3.15–3.55
C	3.55–4.00
D	4.00–4.50
E	4.50–5.00
F	5.00–5.60
G	5.60–6.30
H	6.30–7.10
J	7.10–8.00
K	8.00–9.00
L	9.00–10.0
M	10.0–11.2
N	11.2–12.5
P	12.5–14.0
R	14.0–16.0
S	16.0–18.0
T	18.0–20.0
U	20.0–22.4
V	22.4 and up

FIGURE 12-34 Table with locked-rotor amperage (LRA) ratings. The ratings are listed as amount of kVa per horsepower. (Courtesy of NEMA National Electrical Manufacturers Association.)

12.17.15 NEMA Code Letters

NEMA code letters use letters of the alphabet to represent the amount of locked-rotor amperage (LRA) in kVA per horsepower a motor will draw when it is started. From the table in Fig. 12-34 you can see that letters in the front of the alphabet indicate low LRA ratings, and letters in the back of the alphabet indicate higher LRA ratings. It is important to remember that the number in the table is not the amount of LRA the motor will draw, but rather it is the number that must be multiplied by the horsepower rating of the motor.

12.18 THREE-PHASE SYNCHRONOUS MOTORS

A synchronous motor is an ac motor designed to run at synchronous speed without any slip. As you know, the induction motor must have slip of approximately 9–10% to operate a maximum torque. Slip is required in an induction motor to allow the rotor to draw enough current to carry its load.

The synchronous motor is designed to operate with no slip by exciting the rotor with dc current once the motor reaches operating rpm. The motor can have dc applied from a dc power source or it can be developed through the use of diodes from a separately generated ac current. This current is produced by a small generator located on the end of the synchronous motor shaft. Prior to the use of modern variable-frequency drives, the synchronous motor was used in industrial applications where the loss of rpm due to slip in an induction motor could not be tolerated and the extra rpm was needed for efficiency. Today the variable-frequency drive can

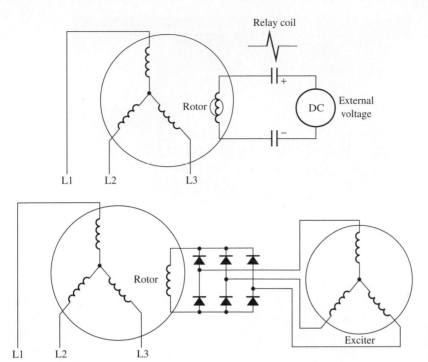

FIGURE 12-35 Electrical diagram of a synchronous motor.

be used to increase the speed of an induction motor to make up the lost rpm so the synchronous motor is not needed as much as it once was.

Fig. 12-35 shows a diagram of a synchronous motor with the dc voltage developed from an outside source and from an internal source. The diagram, which shows diodes being used to rectify the ac from an internal source, is the most common type of synchronous motor used today. This motor also provides an additional function in that it is capable of correcting power factor. The amount of dc current used to excite the rotor also determines the amount of improvement in the power factor.

When it is started, operation of the synchronous motor is similar to that of an induction motor. This means that the motor is started as an induction motor until the motor reaches full rpm. When the motor is at its highest inductive rpm, a switch is closed to provide exciter current to the rotor from the external or internal source.

When the rotor is energized, it will cause its magnetic poles to lock in step with the rotating magnetic field of the motor's stator. Since the speed of the magnetic field is determined by the number of poles in the motor and the frequency of the applied voltage, the speed of the rotor will be locked into this magnetic field's speed and they will rotate in unison. The amount of current in the rotor can be adjusted to provide more torque for the rotor. For this reason, it is important that the motor be started without a heavy load applied. The load can be increased as the motor is synchronized.

12.19 AC SINGLE-PHASE MOTORS

Single-phase motors are used frequently in large and small industries. They are especially useful in applications where motors of less than 1 horsepower are required

FIGURE 12-36 A typical single-phase motor. (Courtesy of GE Motors & Industrial Systems, Fort Wayne, Indiana.)

and in locations where a motor is required and three-phase voltage is not available. In these applications the single-phase motor can be installed on the single-phase voltage source and produce the required horsepower.

The single-phase motor uses a theory of operation that is similar to that of the three-phase motor. There are several minor changes in the single-phase motor design to achieve the same function as a three-phase motor. These changes include the addition of several components and the modification of several others. Fig. 12-36 shows a picture of a typical single-phase motor. It is important to note at this time that a single-phase voltage source may consist of L1 and L2 for 208 and 230 volt systems, or L1 and neutral for a 115 volt system.

12.20 SINGLE-PHASE MOTOR COMPONENTS

Like the three-phase motor, the single-phase motor has three basic components: the *stator, rotor,* and *end plates.* Fig. 12-37 shows a cut-away picture of a single-phase motor. You can see the stator, rotor, and end plates in this figure. Fig. 12-38 also shows a picture of a stator with its rotor removed. This picture gives you a better idea of the relationship between the stator and rotor.

12.20.1 Stator

The stator is the frame of the motor which houses the windings. Since the single-phase motor uses single-phase voltage, it will need a way to produce the starting torque that three-phase voltage produces naturally in a three-phase motor. The single-phase motor has a special starting winding that is used to provide sufficent phase shift to provide starting torque. The motor also has a run winding that is similar to the windings in a three-phase motor. The start winding is made of very fine gauge wire, which has many more turns than the run winding. The run winding is made from wire that is sized to carry the current for the motor at full-load amperage (FLA). This means the run winding wire will be much larger than the start winding and usually in the range of 12- to 16-gauge wire.

FIGURE 12-37 Cutaway picture of a typical single-phase motor. (Courtesy of Century.)

The start winding is also placed in the stator offset from the run winding to give a larger phase difference. This physical phase shift will enhance a shift in the magnetic field produced by two windings. Since the start winding is made of very fine wire that has many turns, it can produce a strong magnetic field for a short period of time. Fig. 12-39a shows the locations of the start and run windings in the stator. You should notice that four run and four start windings are shown inside the stator. The start windings are shown located toward the inside of the stator, where they will be closer to the rotor, and the run winding is shown placed behind the start winding. The start and run windings are connected together in parallel in the motor to provide the magnetic phase shift.

Several electrical diagrams are also presented in this figure to show you methods of representing the single-phase motor. In Fig. 12-39b the windings are shown placed at right angles to each other. This is done to remind you that the windings are physically offset in the stator to produce more of a magnetic phase shift. You should also notice that the run winding is always represented by the larger coil and its terminals are numbered 1 and 4. The start winding is shown as

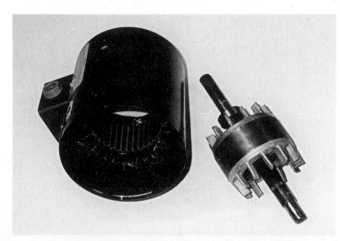

FIGURE 12-38 A stator from a single-phase AC motor with its rotor removed.

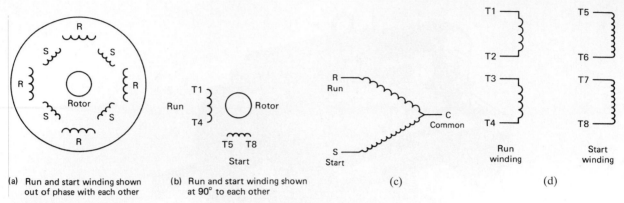

(a) Run and start winding shown out of phase with each other

(b) Run and start winding shown at 90° to each other

(c)

(d)

FIGURE 12-39 **(a)** Diagram of start and run winding locations inside the motor stator. **(b)** Diagram that shows the start winding placed at 90° to the run winding in the stator. **(c)** Diagram that shows one end of the start and run windings connected together at a point called common (C). **(d)** The run winding is shown in two parts and the start winding is shown in two parts.

the smaller coils and its terminals are numbered 5 and 8. The rotor is shown in these diagrams as a circle in the middle of the windings.

In Fig. 12-39c the windings are shown connected in parallel with each other, which is how you would indicate their electrical relationship. You should also notice that in this diagram the run winding is identified with the letter R and the start winding is identified with the letter S. The point where the two windings are connected at the bottom end of the parallel circuit is called the common point and is identified by the letter C. In some motor theory, it is referred to as terminal C, even though it is the point where the two windings are connected together. This type of diagram is used frequently to show the windings of single-phase compressor motors used in air-conditioning systems.

Fig. 12-39d show the windings of a single-phase motor as you would normally see them in diagrams of open-type motors that are used to power small machines in a factory. The windings are shown as two sets. The run winding is shown in two parts. The first part is a winding that is identified by terminal numbers T1–T2, and the second part is identified as T3 and T4. The run winding consists of two parts so that the motor can be connected for high voltage (230 volts) or low voltage (115 volts). If the motor is connected for high voltage, the two parts of the run winding are connected in series, and if the motor is connected for low voltage, the two run windings are connected in parallel. A detailed discussion of high- and low-voltage connections will be presented later in this section.

The start winding is also shown in two parts in Fig. 12-39d. In this diagram you can see that the first segment of the start winding is identified as terminals T5 and T6 and the second set is identified as terminals T7 and T8. If T6 and T7 are permanently connected inside the motor, the two ends of the start winding will be identified as T5 and T8.

12.20.2 Terminal Identification

The terminals on a single-phase motor have a standard identification method. The ends of the start and run windings are numbered to help you identify and locate them when you must install or troubleshoot the motors.

Fig. 12-40 shows the standard numbering method for a split-phase motor. In this figure two diagrams are presented to show an eight-lead and a four-lead motor.

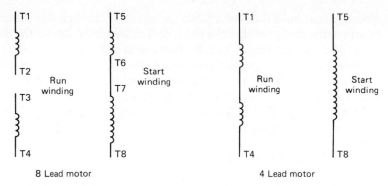

(a) Terminal identification for single-phase motors

T1	Blue	T5	Black
T2	White	T8	Red
T3	Orange	P1	No color assigned
T4	Yellow	P2	Brown

(b) Color codes for terminals

FIGURE 12-40 Terminal identification and color codes for single-phase motors.

In this figure the two windings are shown connected in parallel as they operate electrically. The run winding is shown in two sections. The numbering starts at the top of the diagram, with the terminals of the first section being numbered 1 and 2. The second section's terminals are numbered 3 and 4. The start winding is also shown in two sections, with the terminals of the top section numbered 5 and 6, while the bottom terminals are marked 7 and 8.

The run and start winding terminals can also be identified by the amount of resistance each has. Since the start winding is made of many turns of very fine wire, its resistance will be much higher than that of the run winding. If the terminal identification is missing, you can use a continuity test to group the terminal leads into coils. Then by measuring their resistance, you can compare the readings. The highest readings will belong to the start windings, while the lower readings indicate that the winding is the run winding. This figure also shows a table with the color code for each of the motor terminal conductors for single-phase motors.

The color codes may be used if the terminal identification is not used. You should remember that some manufacturers identify their leads with terminal markers, so the colors of wires that are used in their motors have no meaning. It is also important to understand that some manufacturers do not adhere to the color codes so you must always confirm the terminal markings with an ohm test.

12.20.3 Rotor and End Plates

The rotor in the single-phase motor is similar to the rotor in the three-phase induction motor. The single-phase rotor also has the basic shape of a squirrel cage, so it is also called a squirrel-cage rotor. It has fan blades cast into the aluminum frame to provide cooling air for the motor.

The ends of the rotor provide the shaft for the load and the bearings. The rear part of the shaft is machined to mount inside the shaft bearing, and the front part of the shaft is extended 3 to 4 inches beyond the front bearing to provide a means of mounting pulleys or gears to drive the load.

The end plates are located on each end of the stator. They are secured in place by four bolts that are inserted completely through the stator. They house the bearings for the rotor shaft to ride on. The bearings can be the sleeve type or ball type. The ball bearing is usually lubricated for life and sealed, while the sleeve bearing must be lubricated frequently with several drops of high-grade electric motor oil. You should recall that the sleeve bearing uses felt wicking to hold the excess lubricating oil in contact to the shaft. This means that the end plates must be mounted with the lubricating port pointing upward so that the oil will be pulled to the wicking by gravity.

12.20.4 Centrifugal Switch for the Start Winding

Since the start winding can only stay in the circuit for a short time because its wire is too small and it will heat up rapidly, a switch is provided to disconnect it from the circuit as soon as the motor is started. This switch is a centrifugal switch that mechanically senses the speed of the shaft and opens when the shaft reaches approximately 90–95% of full rpm. Fig. 12-41a shows a picture of the end switch and Fig. 12-41b shows the end switch mounted in the end plate. Fig. 12-41c shows the actuator that moves the centrifugal switch open when the proper speed is reached and allows the switch to close when the motor stops. Fig. 12-41d shows the actuator mounted on the motor shaft.

The centrifugal switch is mounted in the rear end plate. It is commonly referred to as the end switch. The switch has two distinct parts, the switch and actuator. The switch is mounted in the end plate, and the actuator is mounted on the rotor shaft so that it will come in contact with the end switch when the rotor reaches full rpm. The switch is shown removed from the end plate in Fig. 12-41a and it is shown mounted in the end switch in Fig. 12-41b.

The end switch is made of spring steel, which provides tension to keep the switch contacts closed. Whenever the centrifugal actuator is not pressing on the switch, the contacts will remain in the closed position. When the actuator moves along the shaft slightly, it will provide enough force to cause the switch contacts to snap open.

In Fig. 12-41c you can see that the actuator has a weight built into its outer edges. These weights, called flyweights, are hinged on the inside near the rotor and

FIGURE 12-41 A centrifugal switch removed from its end plate. The flyweight mechanism has also been removed from the motor shaft.

allowed to move or swing at the outer edge. Since the outer edge is heavier, the centrifugal force caused by the shaft rotation will cause them to move away from the shaft. Since the actuator is hinged to the inside, this action will cause the actuator to move along the length of the shaft slightly in the direction of the switch. The movement is only $\frac{1}{2}$ to $\frac{3}{4}$ inch, but it is sufficient to actuate the end switch to the open position. Fig. 12-41d shows the actuator mounted on the shaft.

Since the flyweights snap over center to overcome the return spring's tension, you will hear a distinct snap when the motor reaches approximately 95% full speed, which indicates that the end switch has opened, and again after the motor is de-energized and the rotor shaft is coasting to a stop. When you hear the snap as the motor is coasting to a stop, it indicates that the end switch has returned to its closed position.

When the motor is de-energized, the rotor will decelerate to a stop and the centrifugal actuator will return to its original position with the aid of return springs. When the flyweights return to their normal position, the actuator moves back away from the switch and allows it to return to its closed position so that it is ready for the next time the motor is started.

12.20.5 Thermal Overload Protector

Single-phase motors provide a bimetal switch for use as a built-in overload device. This overload is mounted in the rear end plate near the centrifugal switch assembly and terminal board. The thermal overload consists of a heater and contacts. When the motor is starting and running, all its current is pulled through the heating element. If the motor draws excessive current, the heating element will become warm enough to cause the bimetal contacts to snap open and de-energize the motor windings. When the bimetal cools down, the bimetal will cool again and snap closed, which will re-energize the motor. If the same fault still exists, the motor will overheat again and continue to cycle off through its overload.

Some overloads do not reset automatically. Instead, they have a reset button that must be depressed manually to close the overload contacts. This necessitates someone going to the motor when it trips the overload and resetting it manually. At that time the motor and its loads should be inspected to ensure that the motor is operating correctly.

12.21 CHANGING VOLTAGE AND SPEEDS OF SINGLE-PHASE MOTORS

The single-phase motor is available for connection on 110 or 230 V. If the motor is connected for 115 V, it can be reconnected on the factory floor for 230 V. It is also possible to reconnect a motor from 230 V to 115 V. This allows the motor to be used in any voltage application.

Figure 12-42a shows the two coils of the run winding connected in parallel for low voltage. The start winding is then connected in parallel with these coils. The diagram in this figure shows that leads T1, T3, and T5 are connected to L1 and that T2, T4, and T8 are connected to L2. Figure 12-42b shows the connections for high voltage. From this diagram you can see that the two coils for the run

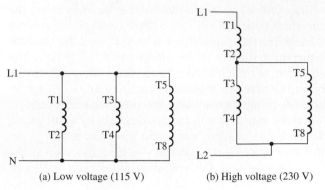

FIGURE 12-42 **(a)** Diagram for single-phase motor wired for 115 volts (low voltage). **(b)** Diagram for single-phase motor wired for 230 volts (high voltage).

(a) Low voltage (115 V)

(b) High voltage (230 V)

winding are connected in series. The start winding is connected in parallel across the lower coil (T3–T4) of the run winding. When 230 V is applied to the two coils of the run winding, the voltage divides equally, 115 V across each coil.

The start winding will also receive 115 V from the terminals to which it is connected. From these connections you can see that each coil section is rated for 115 V. If the motor is connected for low voltage, all the coil sections are connected in parallel so they will each have 115 V applied to them. When the motor is connected to 230 V, the run winding acts as a voltage divider so that each coil still receives 115 V. The single-phase motor is also available for dual speeds. This type of motor must have the number poles reduced to increase the speed of the motor. In most cases the motor will lose some of its horsepower rating when it is operated on the faster speed, since fewer poles are used.

12.22 INCREASING THE STARTING TORQUE OF A SINGLE-PHASE MOTOR

The torque of a single-phase motor can be changed by adding capacitors to the start or run winding of the motor. When the single-phase motor is used without capacitors, as has been shown in the diagrams presented so far in this section, the motor is called a split-phase motor.

When a start capacitor is connected in series with the start winding and centrifugal switch, the motor is called a *capacitor start, induction run (CSIR) motor*. When the motor has a start capacitor in series with the start winding and a run capacitor is connected permanently across the run and start terminal, the motor is called a *two-capacitor motor* or a *capacitor start, capacitor run (CSCR) motor*. If the motor has only a run capacitor connected permanently across the start and run winding, it is called a *permanent split capacitor (PSC) motor*. If the rotor of the single-phase motor is made of copper wire rather than a squirrel-cage rotor, it is called a *wound-rotor motor* or *repulsion start motor*.

The following sections will explain the operation of each of these motors.

Methods of reversing these motors are also presented with their diagrams. At the end of this section methods of troubleshooting each of these types of motors are provided. You should gain an understanding of how these motors operate and how their rotation is reversed. It will also help to understand the methods of reconnecting the motor to operate on dual voltage or dual speeds. This information is important when you must connect motor control devices to them such as reversing starters or dual-voltage starters.

12.23 SPLIT-PHASE MOTORS

The *split-phase motor* is a single-phase motor that does not have any capacitors or other devices in its circuit to alter its torque characteristics. Diagrams of this motor are presented in Fig. 12-43. This motor is also called the split-phase motor or the ISIR (induction start, induction run) motor, since it uses only induction to start and run.

This type of motor has the lowest starting torque of all single-phase motors. It uses the physical displacement of the run and start windings in the stator to provide the phase shift required to start the rotor moving. You should remember that the three-phase motor uses the 120° phase shift that naturally occurs in the three-phase voltage to cause starting torque. Since the single-phase motor does not have a natural phase shift, the split-phase motor uses the difference of the coil size to create a phase difference along with physically locating the start winding out of phase with the run winding to cause a magnetic phase shift that is large enough to cause the rotor to start spinning.

When voltage is first applied to the motor's stator, the rotor is not turning

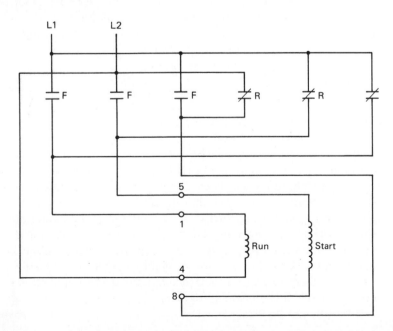

FIGURE 12-43 Diagram of a split-phase motor connected to a forward and reversing motor starter. The run winding (T1–T4) remains connected the same way in both foward and reverse operation. The start winding (T5–T8) gets reversed to make the motor run in the opposite direction.

and the windings will draw maximum current. This current is called *inrush current* or *locked-rotor amperage* (LRA). After the rotor starts to turn, it will induce current from the stator and produce its own magnetic field. This field will cause the rotor to increase speed until it reaches its rated speed. The rated speed is determined by the number of poles the motor uses and the frequency of the applied voltage. This means that a two-pole motor will operate at 3600 rpm, a four-pole motor will operate at 1800 rpm, a six-pole motor will operate at 1200 rpm, and an eight-pole motor will operate at 900 rpm.

12.23.1 Counter EMF

When the rotor increases its rpm, it will begin to generate a back voltage or countervoltage. This voltage is also called *counter EMF (electromotive force)*. The difference between the counter EMF and the applied voltage at full speed may be only 1 to 2 V. This voltage is called the potential difference between the two voltages and it is responsible for causing the current flow required to keep the rotor spinning at its rated speed. The current required to keep the rotor spinning at its rated speed when it is driving a load is called *full-load current* (FLA).

If the load is increased, the rotor's rpm is slowed slightly, which will cause the counter EMF to be reduced slightly. Since the counter EMF has decreased and the applied voltage has remained the same, the difference of potential will increase and cause additional current to flow. The increased current will cause the motor to produce more torque, which will cause the rotor to come back up to speed. This feature makes the motor self-regulate its load. This characteristic of single-phase motors helps them operate at fairly constant speed throughout their load range. If the motor is overloaded too heavily, the rotor will slow down to the point where it will stall. During this time it will continue to draw excessive current and overheat severely.

Methods of reversing the split-phase motor are presented in Fig. 12-43. In this figure you can see that terminals 5 and 8 of the start winding are reversed to get the rotor spinning in the opposite direction. Once the rotor begins spinning in one direction, the magnetic phase shift that is created will cause the motor to continue to rotate in that direction. If you find that the application requires more starting torque, a start capacitor should be added to the split-phase motor to make it a capacitor start, induction run motor.

12.23.2 Applications for Split-Phase Motors

The split-phase motor is used for general-purpose loads. The loads are generally belt driven or small direct-drive loads like small drill presses, shop grinders, air conditioning, and heating belt-driven blowers and small belt-driven conveyors. The main feature of the split-phase motor is that it can be used in areas of the factory where three phase has not been distributed, or on small loads on the factory floor where fractional-horsepower motors can handle the load. The motor does not provide a lot of starting torque, so the load must be rather small or belt driven, where mechanical advantage can be utilized to help the motor start.

The motor is inexpensive and it can be replaced when it wears out rather than trying to rewind it. It is also available in a variety of frame sizes, which allows it to be mounted easily in most machinery. If the application requires too much torque for the split-phase motor, one of the other motors, such as the capacitor start motor, can be used.

12.24 CAPACITOR START, INDUCTION RUN MOTORS

The capacitor start, induction run motor (CSIR) is a split-phase motor with a starting capacitor connected in series with its start winding. This modification to the split-phase motor's design is accomplished when the motor is manufactured, so the capacitor start motor is a separate choice in the manufacturer's catalog when the motor is being specified and selected for the application. Since the CSIR motor can provide more starting torque than the split-phase motor, it would be selected for direct-drive and other applications that require more power during startup. Fig. 12-44 shows a picture of a typical CSIR motor with the start capacitor mounted on top of the motor. The start capacitor is about the size of a small orange juice can. The case for the start capacitor is made of plastic.

12.24.1 Electrical Diagram for a CSIR Motor

A start capacitor is connected in series with the start winding of the CSIR motor to provide a larger phase shift when voltage is first applied to the motor. The increased phase shift causes a stronger magnetic field to pull on the rotor to cause it to begin to spin. The electrical diagram for the CSIR motor connected for forward rotation and for reverse rotation is provided in Fig. 12-45. The start capacitor is usually rated for over 100 microfarads (μF) (up to 1000 μF).

The start capacitor can provide large amounts of capacitance for a short period of time, such as during starting. After the motor is started, the capacitor must be removed from the circuit so that it will not overheat. This is accomplished with the end switch. The electrical diagram in this figure shows the capacitor connected in series with the start winding. When voltage is first applied to the motor, both the start winding and the run winding will be energized. When the rotor reaches nearly

FIGURE 12-44 Picture of a capacitor start, induction run (CSIR) motor. Notice the start capacitor is mounted on the top of the motor. (Courtesy of GE Motors & Industrial Systems, Fort Wayne, Indiana.)

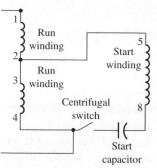

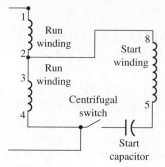

(a) Capacitor start, induction run (CSIR)
motor wired for forward direction

(b) CSIR motor wired for reverse rotation

FIGURE 12-45 **(a)** Electrical diagram of a capacitor start, induction run (CSIR) motor connected for forward rotation. **(b)** Electrical diagram of a CSIR motor connected for reverse rotation.

full rpm, the end switch will open and disconnect the start winding and capacitor from the circuit. This means that no current will flow through either the start winding or the capacitor and they can cool down and be ready for the next time the motor is started. When the motor is de-energized and the rotor slows to a stop, the end switch is closed again and the start winding and capacitor are reconnected to the circuit for the next start.

Since the start capacitor is physically mounted on top of the motor, a metal cover is placed over it to protect it from damage. Since the capacitor's case is made of plastic, it can easily be cracked and damaged if the motor is used in a harsh industrial environment.

12.24.2 Connecting the CSIR Motor for Dual-Voltage or Dual-Speed Applications

The capacitor start motor is similar to the split-phase motor, since its terminals are identified in the same way and it can be connected to operate on either 115 or 230 volt single-phase voltage. Like the split-phase motor, the CSIR motor will generally lose horsepower when it is operated at its higher speed since two poles are deleted from the motor to gain the speed. Some factories prefer to stock dual-voltage, two-speed motors which will cover nearly every application with their equipment. This means that they do not have to stock four separate types of motors for different applications, such as high speed, low voltage; high speed, high voltage; low speed, low voltage; and low speed, high voltage. By stocking the variable-speed, variable-voltage motor, they only need to stock different horsepower sizes, such as $\frac{1}{2}$ and 1 hp; the other changes can be made by reconnecting the motor for the correct voltage or speed at the time the motor is installed.

12.24.3 Connecting the CSIR Motor for a Change of Rotation

All of these motors can easily have their rotation reversed by changing the start winding leads. Diagrams that show the CSIR motor connected for both the forward and reverse operations are provided in Fig. 12-45a and Fig. 12-45b. From these diagrams you can see that the end switch (centrifugal switch) and start capacitor are connected near the end where terminal 8 is located. When the motor is reversed, terminals 5 and 8 are switched.

When the motor is operating in the clockwise direction, T5 is connected to

T3 and T8 is connected to T4. When the motor is connected for counterclockwise rotation, T8 is connected to T3, and T5 is connected with T4. This reverses the current flow in the start winding with respect to the current flow in the run winding, and the rotor will begin to spin in the opposite direction. Remember: Motor rotation is determined by looking at the motor shaft from the end opposite where the load is connected on the shaft. It is also important to remember that the start winding must be connected so that it only receives 115 volts. For this reason the start winding is always connected across the bottom part of run winding.

12.24.4 Applications for CSIR Motors

The CSIR motor can be used for a wider variety of applications where more motor starting torque is required than a split-phase motor can supply. These applications include direct-drive water pumps, air compressors, and larger conveyors. The capacitor start motor is also used to drive hermetic compressors for single-phase air-conditioning systems on rooftops and in window units. The hermetic compressor is the sealed motor that is used to drive the refrigeration pump. Since this is a direct-drive application, the CSIR motor is generally used. The split-phase motor may be used on the smaller hermetic motors and systems that use a capillary tube for their refrigerant metering valve.

Since the compressor is hermetically sealed, it is not practical to place the end switch inside the compressor housing with the motor because it is likely to wear out and require replacement. It is also dangerous to allow the end switch to be mounted inside the compressor housing because it may cause a spark that could ignite fumes from the oil that is used in the compressor for lubrication and cause an explosion. Since an end switch cannot be used, a current relay is used to disconnect the start winding after the motor starts. An electrical diagram of the current relay is provided in Fig. 12-46.

The current relay is generally connected directly on the motor terminals. You can see from the electrical diagram that the current relay consists of a coil and a set of contacts. The coil is connected between terminals L1 and M_R and the NO contacts are connected between terminals L1 and S. You should also notice that the motor is shown as a run winding and a start winding. One end of the run winding and one end of the start winding are connected together at the right side and this point is identified as the common (C).

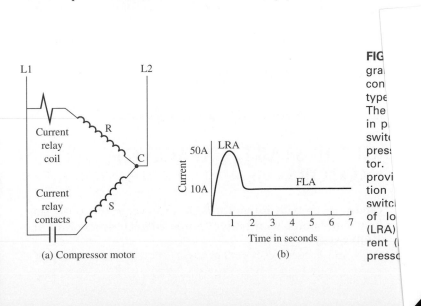

(a) Compressor motor

(b)

The diagram shows the current relay connected to the motor. Notice that the coil of the current relay is connected in series with the run winding and it will remain in this part of the circuit even when the motor is running. The contacts of the current relay are connected in series with the start winding. Since a start capacitor is used with this motor, the contacts (terminal S) of the relay are connected to the start capacitor and the other terminal of the start capacitor is connected to the start winding of the motor.

12.24.5 Current Relay Operation

From Fig. 12-46b you can see that when voltage is applied to the compressor, the run winding will pull locked-rotor amperage (LRA) of up to 40 A. Since the coil of the current relay is in series with the run winding, it will also see this large current, which will be strong enough to pull the current relay's contacts closed. When the contacts close, it will provide a path for voltage to reach the start winding. This voltage will also pass through the start capacitor, which causes a phase shift in the magnetic field in the start winding and the torque will be strong enough to start the compressor motor even though it is under 70–80 psi of pressure.

The current relay has a return spring connected to its contacts that tries to pull them open. After the motor starts and picks up speed, the rotor will produce enough counter EMF to allow the full-load current (FLA) to drop substantially. When the motor's current drops to the FLA level of 3 to 4 A, the spring will pull the contacts open. This drop in current from LRA to FLA occurs in 1 to 2 seconds as the motor reaches full speed. When the contacts open, the start capacitor and start winding are removed from the circuit and their current drops to zero. Since the run winding continues to be energized, it will draw FLA and the motor will continue to run. A graph of the locked rotor current and full-load current is also shown in Fig. 12-45b. You can see that LRA only exists for several seconds while the rotor is coming up to speed. This current is strong enough to pull the current relay's contacts closed for the few seconds the motor requires to start.

Since the current relay is mounted on the outside of the compressor, it can easily be replaced if it becomes faulty. It can also be tested easily by checking voltage from L2 to the S terminal of the motor during the time the motor is starting. If no voltage is present at terminal S during start, the start capacitor should be removed from the circuit and an attempt made to restart the motor. Allow the motor to try to start for only a few seconds during this test because continued LRA current will damage the motor.

If terminal S on the motor receives voltage when the capacitor is removed, you can assume that the capacitor is open. If the terminal still does not receive voltage, the current relay is faulty and should be replaced.

12.25 CAPACITOR START, CAPACITOR RUN MOTORS

The capacitor start, capacitor run (CSCR) motor adds a run capacitor to the start capacitor, which provides the motor with better torque characteristics when the motor is operating at full speed.

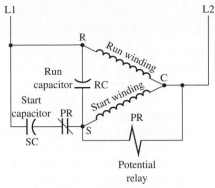

(a) Diagram of potential relay connected to
a capacitor start, capacitor run motor

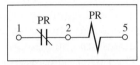

(b) Diagram of potential relay

FIGURE 12-47 **(a)** Capacitor start, capacitor run motor connected to a potential relay. The potential relay is used to energize and de-energize the start winding. **(b)** Diagram of a potential relay.

The run capacitor is usually oval or square shaped and has a metal housing rather than plastic. The metal housing allows the run capacitor to radiate any heat that is built up inside it, since it is connected to remain in the run winding circuit at all times.

From the diagram in Fig. 12-47a you can see that the run capacitor is essentially connected to the capacitor start, induction run motor. This type of motor is used almost exclusively for hermetic compressor motors in air-conditioning systems. You will find air-conditioning systems connected to most modern electronic panels today to provide the additional cooling that is required when computers, motor drives, and other amplifiers are placed in an enclosed cabinet.

For this motor, the run capacitor is connected in parallel with the run winding. It will be in the circuit during starting and remain in the circuit while the motor is running. During the time the motor is starting, the additional capacitor provides a little more phase shift than the start capacitor alone, which gives the motor more starting torque.

After the motor is running, the start capacitor is disconnected from the start winding circuit when the potential relay contacts open, but the run capacitor remains in the circuit because it's wired directly across the R and S terminals. If the load increases slightly because the compressor is trying to pump more refrigerant, the run capacitor will provide a small phase shift to give the rotor more torque and regain the loss of rpm quickly. The larger the load increase is, the more speed the rotor will lose, which will cause additional current to be drawn. The increase in current will pass through the run capacitor and cause it to provide a larger phase shift, which in turn provides the rotor with more torque. The run capacitor allows the speed of the motor to remain fairly constant when the load is constantly varying.

12.25.1 Using a Potential Relay to Start a CSCR Motor

The capacitor start, capacitor run motor is used primarily in starting large single-phase hermetic compressors. Since the hermetic motor cannot use an end switch, a starting relay similar to the current relay must be used. The current relay requires that its coil be connected in series with the run winding of the compressor motor. In the CSCR motor, the run current can be exceedingly large when the motor is starting (LRA). This current is too large for the current relay coil and would tend to burn it out. For this reason a different type of starting relay, which senses the amount of counter EMF for actuation, is used to start the CSCR motor. Since this relay uses the difference of potential between the applied voltage and the counter EMF produced by the rotor, it is called a potential relay.

Fig. 12-47b shows a diagram of a potential relay. Notice that its terminals are identified as 1, 2, and 5. It has an NC set of contacts between terminals 1 and 2 and a high-resistance coil between terminals 2 and 5. The diagram in Figure 12-47a shows the potential relay connected to a motor with a start and run capacitor.

You should notice that the potential relay's contacts are connected in series with the start capacitor and the motor's start winding. The run capacitor is connected between the S and R terminals of the motor, where it will remain even when the motor is running. You should also notice the dot on the curved side of the capacitor symbol. It represents the side of the run capacitor that should be connected to the line side of the power supply. This dot corresponds to a red mark on one of the run capacitor terminals. This terminal is marked because it is the terminal that is closest to the outside of the metal can that the capacitor is mounted in. If the capacitor is shorting out, it will short to this side of the circuit and cause a fuse to blow and the motor would not be damaged. If this capacitor is connected in the circuit so that the identified terminal is connected to the start winding, short-circuit current would be drawn through the motor windings and the motor would be damaged.

The operation of the potential relay is controlled by the amount of counter EMF that the rotor produces. When voltage is applied to the motor through L1 and L2, the run winding is energized directly and the start winding is powered through the start capacitor and NC set of contacts of the potential relay. The applied voltage is not strong enough to activate the potential relay's coil.

When the motor starts and the rotor begins to spin, a counter EMF is produced between motor terminals S and C. The counter EMF will become large enough to energize the potential relay coil when the motor reaches approximately 75% rpm. When the coil is energized, it pulls its NC contacts open, which de-energizes the start capacitor and start winding. This effectively removes the start winding from the source of applied voltage.

The run capacitor is connected to the R and S terminals on the motor so that it is in circuit between the run winding and the start winding and will allow a very small amount of current to flow through from L1 to the the start winding. When the load increases, such as when the compressor must pump more refrigerant, the rotor will begin to slow down and the counter EMF will be reduced slightly. This allows the run capacitor to provide a small phase shift, as it adds more current to the start winding. This increased current with the phase shift provides additional torque and the rotor's speed will again be increased to full rpm. This allows the CSCR motor to operate at a fairly constant speed. The CSCR motor is not used

FIGURE 12-48 Examples of permanent split-capacitor (PSC) motors. Notice the run capacitor mounted on the second motor. (Courtesy of GE Motors & Industrial Systems, Fort Wayne, Indiana.)

in open motor applications, but the run capacitor concept is used frequently in a motor called the permanent split-capacitor motor.

12.26 PERMANENT SPLIT-CAPACITOR MOTORS

The permanent split-capacitor (PSC) motor uses only a run capacitor to provide the phase shift required to start the motor. Fig. 12-48 shows examples of PSC motors, and Fig. 12-49 shows two diagrams of the PSC motor. In the diagrams you

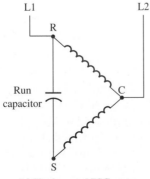

(a) Single-speed PSC motor

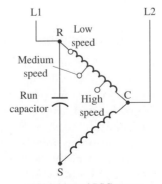

(b) Multispeed PSC motor

FIGURE 12-49 (a) Electrical diagram of a PSC motor. **(b)** Electrical diagram of a multispeed PSC motor.

can see that the run capacitor is connected between the run and start windings and no disconnecting switch or relay is required to de-energize the start winding from the applied voltage when the motor has started. You should notice in the picture of the motors that the run capacitor is oval shaped. The run capacitor has a metal case, which allows it to dissipate extra heat that is built up in the capacitor since it remains in the circuit at all times.

When voltage is applied to the motor, current will flow through the run winding to the common terminal. At this same time current will flow through the run capacitor to the start winding. When the current flows through the run capacitor, it will provide a phase shift that is large enough to start the motor. As the rotor's speed increases, a counter EMF will be produced in the start winding that will limit current through it to less than 1 A when the motor reaches full speed. The small amount of current in the start winding when the motor is operating at full speed is small enough so that it will not cause the start winding to overheat.

When the motor shaft sees an increase in its load, it will slow down slightly. The decrease in the rotor's rpm causes a decrease in the counter EMF, which makes a larger potential difference between it and the applied voltage. The larger potential difference causes an increase in the current in the start winding, which will cause an increase in rotor torque that increases the rotor's rpm.

This characteristic allows the PSC motor to operate with a constant speed under varying load conditions without using any mechanical devices. The PSC motor is generally used for applications such as small hermetic compressors, blade fan loads, and other loads that require constant speed.

12.26.1 Connecting a PSC Motor for a Change of Speed

The PSC motor is available as a variable-speed motor. Since the PSC motor does not require a centrifugal switch, it does not have an end plate or access plate. Instead, all the leads that the motor has provided are brought out of the motor together near the end plate. These leads are generally color coded to identify their speed. Fig. 12-48b shows a diagram for a multispeed PSC motor. The color leads are connected at various points along the run winding coil. If power is applied at the very end of the run winding, all the poles of the run winding are used, and the motor will operate at its lowest speed. If power is connected at the terminal marked high speed, only one of the poles is used, and the motor will operate at its highest speed. You should remember that the motor will lose torque as its speed is increased because fewer windings are used to gain the extra speed.

12.26.2 Changing Voltages or Direction of Rotation

Since the PSC motor is rather specialized and inexpensive, it generally is not made to have its direction of rotation changed or be reconnected to operate on a different voltage. Instead, it is common practice to stock a clockwise and counterclockwise motor for each voltage application that you have in the facility. Since these motors are generally used for blade fan applications, they are commonly used for small air-conditioning units used to cool small offices erected on the factory floor or the main offices of the facility. They are also used in the air-conditioning units to cool electrical control panels on larger equipment that have numerous electronic boards or motor drives mounted directly in the cabinet.

One problem that the PSC motor has is that it may run in the wrong direction when it is used to drive a condenser fan on an outdoor air-conditioning unit. This condition is caused when it is exceedingly windy and the wind blows across the fan blade when the motor is de-energized, causing the motor rotor to spin in the opposite direction than it would normally operate. When voltage is applied to the motor, the rotor will continue to rotate in the direction the wind is blowing when power is applied. If the rotor is stationary when voltage is applied, the rotor will spin in the proper direction. If the wind is blowing the fan blade in the wrong direction, it will continue to spin in that direction when voltage is applied. If the fan is running in the wrong direction, it will cause insufficient air movement across the coils. This will cause the air conditioner to overheat and cause high pressure in the refrigerant coils.

A ratchet mechanism is mounted on the motor shaft of condenser fans when this is a problem. The ratchet allows the motor to spin in the correct direction and prevents the motor shaft from spinning in the wrong direction when the wind is blowing.

12.27 SHADED-POLE MOTORS

Shaded-pole motors are commonly found in applications that require light-duty fans such as small window air conditioners and exhaust fans used in rest rooms. If you are a maintenance electrician or technician, you may be requested to service all of the electrical equipment in the factory, including the office areas. If this is the case, you will run into shaded-pole motors. Shaded-pole motors are also used in the paddlewheel-level sensor described in Chapter 10.

Fig. 12-50 shows a diagram of the shaded-pole motor. From this diagram you can see that the motor has only one winding. It does not have a start winding and a run winding like other single-phase motors. Instead, it has a shading pole that provides the magnetic field phase shift that is required to start the motor. The shaded-pole motor has a copper bar that is inserted around the front of the run winding. The bar is connected at the ends to make a complete circuit called a *pole*.

When voltage is applied to the motor to start it, current will flow through the run winding and build up a magnetic field. A current will be induced in the single winding of the shading pole, and it will cause a phase shift to occur that is large enough to make the rotor start to spin. Once the rotor starts to spin, it will begin to build its own magnetic field and come up to full rpm.

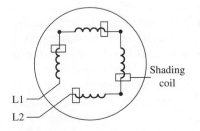

FIGURE 12-50 Diagram of a shaded-pole motor. Notice the shading pool located near each winding.

The shading pole also helps the motor when its load changes at full rpm. If the motor shaft begins to slow down, the phase shift in the shading coil becomes stronger and provides enough torque to bring the rotor back up to full speed. Another unique feature of the shaded-pole motor is that it can withstand LRA for an extended period. Since the motor does not have a start winding, the run winding is large enough to carry locked-rotor current if the rotor becomes stuck. This is important since it provides burnout protection without any additional devices or equipment being added to the motor.

You should remember from Chapter 10 that the shaded-pole motor is used as a level indicator. A paddle is attached to the shaft of the shaded-pole motor and it is turned very slowly. The paddle is mounted in a bin where granular material is stored. When the height of the material increases to the point where it covers the paddle, the paddle will stop turning and stall the motor. Since the motor is a shaded-pole motor, its current will increase, but the extra current will not damage the motor. A sensor is used to detect the change in current, which indicates the level of material is covering the paddle.

Since the shaded-pole motor has these characteristics, it is commonly used for small fan applications. If the fan becomes immovable for any reason, such as lack of lubrication or dirt, the motor will become warm, but it will not overheat and destroy itself like the split-phase or capacitor start motors.

12.28 REPULSION START MOTORS

The repulsion start motor was the most common single-phase motor in use prior to the squirrel-cage motor. After the 1960s very few repulsion start motors were installed because they require brushes and a commutator to operate. The rotor for this type of motor is slightly different from the rotor of the squirrel-cage rotor since it uses copper wire to make its magnetic field. Another feature that makes the rotor different is that it has a wire that connects the commutator segments with a shorting mechanism, which is used in conjunction with the brushes. Since this motor was designed before squirrel-cage motor theory and technology became prevalent, the rotor was patterned after the wound rotor that is used in DC motors. The rotor was made of laminated sections with coils of wire pressed into place and their terminal ends brought out to commutator segments.

When the motor was being started, current was directed to the rotor coils through the brushes. After the rotor was spinning fast enough, the brushes were disconnected from the applied voltage and shorted so that the rotor would act like an inductive rotor. In some motors, the brushes remained connected to the applied voltage, but they were lifted slightly so that they would not make contact with the commutator. At the same time a shorting mechanism would short the commutator segments to complete the circuit on each coil so that it could conduct the induced current like a squirrel-cage rotor in an induction motor.

In both of these types of motors the rotor would start the motor as a repulsion start motor, and after the rotor came up to speed the motor would operate like an induction motor. This would give the motor the maximum amount of starting torque.

Since the rotor required brushes and some kind of lifting or shorting mechanism, it would require an excessive periodic maintenance. This made the motors too expensive to maintain and they were soon replaced with squirrel-cage motors.

12.29 TROUBLESHOOTING THREE-PHASE AND SINGLE-PHASE MOTORS

Three-phase and single-phase motors are similar in their operation. Since each of the motors has some parts that are different, there are a few differences in the tests that should be made for each motor. The troubleshooting procedure should be broken into several sections. These sections are based on the types of symptoms the motor presents when it is not operating correctly. The procedures listed in this section are presented in sequence of the problems that are most likely to occur to problems that are least likely to occur. You should adopt the same type of philosophy when you begin to troubleshoot the motor. You should always begin by looking for the faults that are most likely to occur, and move on to faults that are least likely to occur. You should also perform simple tests first and complex tests later. This type of procedure will allow you to find the majority of problems or faults with the simple tests.

12.29.1 Motor Will Not Turn When Power Is Applied

The most common problem that you will encounter with a motor is when voltage has been applied and the shaft does not turn. This can occur with the three-phase or single-phase motor. The problems that cause this symptom are the loss of voltage to the motor and an open circuit somewhere in the circuit including the voltage supply and the motor windings. You can test for voltage at all terminal points and determine where the loss has occurred.

12.29.2 Motor Hums But Will Not Turn When Power Is Applied

If the motor hums but will not turn when power is applied, it indicates that one phase of a three-phase motor is open, or the start winding of a single-phase motor is open. You can use a clamp-on ammeter to locate the line that is not drawing current in the three-phase motor. After you have determined which line is not drawing current, you can look for a problem such as an open wire or open fuse. It is important to remember that if the motor continues to try to start, additional fuses may also blow and the motor will no longer hum.

    If the motor is a single-phase motor and it hums but will not start, you can remove the voltage source and test the start winding for resistance. If the winding has high current or is open, you will need to change the centrifugal switch, or change the motor. Be sure to test the capacitor also if one is used to start the single-phase motor. This problem may be caused by having too large a load for the motor or a bearing that needs lubrication.

SOLUTION TO JOB ASSIGNMENT

Your solution to the job assignment should include reference to the cut-away diagram of a three-phase AC motor in Fig. 12-20 so that you can review all of the basic parts. You should include the diagram from Fig. 12-25 for the wiring diagram of a wye-connected motor, and the diagram from Fig. 12-27 to show how you would reverse the direction of rotation if needed. Refer to Sections 12.10 through 12.15 for additional information.

Your solution for the DC series motor should include a reference to the cut-away diagrams of the DC motor in Figs. 12-5 and 12-6 to be able to identify the location of the brushes and motor terminals. You should include the diagram in Fig. 12-10 or Fig. 12-12 to show how you would reverse the direction of rotation for the DC motor and refer to the information in Section 12.6.4 to explain how you would change the direction of rotation.

Your solution for checking the single-phase AC motor should include the cut-away diagrams in Figs. 12-37 and 12-39 and you should refer to the diagrams of the centrifugal switch in Fig. 12-41 to explain how you would test the switch to ensure it is operating correctly. Refer to Sections 12.19 through 12.27 for additional information.

Your solution should include a typical data plate shown in Fig. 12-31 and you should refer to Section 12.16 to explain which data make the single-phase AC motors unique. Be sure to explain which motors can be exchanged for other motors. For example, if you stock a split-phase motor, you can add a capacitor to it to replace a capacitor start, induction run motor.

QUESTIONS

1. What is the function of the armature in a DC motor?
2. What is the function of the field in a shunt-type DC motor?
3. Explain how the forward and reverse motor starter circuits shown in Fig. 12-12 provides lockout protection so the motor cannot be switched directly from the forward direction to the reverse direction.
4. What does the term *induction* mean in regard to an induction-type AC motor?
5. Explain how you can use a resistance test to detect the start and run winding of a split-phase AC motor.
6. What is counter EMF in a motor and what function does it provide?
7. Explain how you can change the speed of a DC shunt motor.

TRUE OR FALSE

1. _____ The series field of a DC series motor is made of very fine wire so that its magnetic field will be as strong as possible.

2. _____ The armature is the rotating part of a DC motor.

3. _____ The brushes in a DC motor make contact between the stationary part of the motor and the armature.

4. _____ The three-phase, wye-connected motor has more starting torque than the delta-connected, three-phase motor.

5. _____ The synchronous AC motor uses DC voltage to provide extra field current so the motor does not have slip.

6. _____ An induction-type AC motor needs slip to provide torque for the rotor.

7. _____ Inrush current (LRA) is always larger than full-load current (FLA).

8. _____ The speed of a DC shunt-type motor is adjusted by changing the frequency of the supply voltage for the motor.

9. _____ The direction of rotation for a split phase AC motor can be changed by changing the polarity of the supply voltage.

10. _____ The direction of rotation for a three-phase AC motor can be changed by exchanging L1 and L2 of the supply voltage.

MULTIPLE CHOICE

1. The _____ is connected in series with the start winding for a permanent split-capacitor (PSC) motor.
   a. start capacitor
   b. run capacitor
   c. potential relay

2. The start winding is connected in _____ with the run winding for a split-phase AC single-phase motor.
   a. series
   b. parallel
   c. series parallel

3. The _____ motor has the most torque of all of the single-phase AC motors.
   a. split-phase
   b. capacitor start, induction run
   c. capacitor start, capacitor run

4. The speed of an AC motor can be increased by _____.
   a. decreasing the frequency of the supply voltage to the motor.
   b. increasing the frequency of the supply voltage to the motor.
   c. increasing the amount of voltage to the motor.

5. The single-phase induction motor needs a start winding _____.
   a. to create a phase shift that will provide starting torque.
   b. to increase the speed of the motor.
   c. to create additional running torque.

6. The shaded-pole single-phase motor _____.
   a. uses a relay to remove its start winding from the circuit after the motor starts.
   b. uses a capacitor to remove the start winding from the circuit after the motor starts.
   c. does not have a start winding since it has a shading pole to provide the phase shift for starting.

7. The rated speed of a four-pole AC motor is _____.
   a. 3600 rpm.
   b. 1800 rpm.
   c. 1200 rpm.

PROBLEMS

1. Calculate the rpm of a two-pole AC motor.

2. Draw the diagram of a delta-connected, three-phase motor that is connected for high voltage and low voltage and identify their terminals. Explain how you would reverse the direction of rotation for these motors.

3. Draw the wiring diagram of a wye-connected, three-phase motor that is connected for high voltage and low voltage and identify their terminals. Explain how you would reverse the direction of rotation for these motors.

4. Use the table in Fig. 12-34 to determine the locked-rotor current for a 2hp motor that has a NEMA code letter C.

5. Draw the wiring diagram for a split-phase, capacitor start, induction run, single-phase AC motor and identify its terminals.

6. Draw a sketch of a basic single-phase induction motor and identify the rotor, stator, and centrifugal switch.

7. Draw the wiring diagram for a PSC single-phase AC motor.

8. Draw the wiring diagram of a DC series, DC shunt, and any one of the DC compound motors and identify their terminals.

9. Determine the distance between the bolt holes for a 56 frame motor. (Use the table in Fig. 12-32.)

10. Explain how you would troubleshoot a three-phase AC motor that fails to start and the motor windings do not make any noise or draw any current.

11. Explain how you would troubleshoot a single-phase AC motor that has a centrifugal switch and fails to start.

REFERENCES

1. Kissell, Thomas E., *Modern Industrial Electrical Motor Controls*. Englewood Cliffs, NJ: Prentice Hall Publishing, 1990.

2. Kosow, Irving L., *Electrical Machinery and Transformers*, 2nd ed. Englewood Cliffs, NJ: Prentice Hall Publishing, 1991.

3. Richardson, Donald V, *Rotating Electric Machinery and Transformer Technology*. Reston, VA. Reston Publishing Company, Inc., A Prentice Hall Company, 1978.

4. Rosenberg, Robert, and August Hand, *Electric Motor Repair*, 3rd ed. New York: Holt, Rinehart and Winston, 1986.

13

CASE STUDIES OF FOUR INDUSTRIAL APPLICATIONS

Your company is considering diversifying and is reviewing four small companies that it has a chance to purchase. The president of your company has asked you to review the industrial electronics involved in each of the applications that these companies are using. The applications include a beverage-filling application, an aluminum can stock making operation, a resin dryer for a plastic injection molding application, and a brake drum machining application. You need to write a report on each of these applications that includes what types of industrial electronic controls they are using and the pros and cons of each technology in terms of keeping the systems maintained and calibrated. You also need to include if the technology can be integrated into other electronic controlled systems that your company already uses.

1

NO CAN / NO FILL SENSING FOR BEVERAGE FILLERS

13.1.1 Overview of the System

To fill beverage containers with beverage, the empty containers move on a large conveyor line under a filler system that dispenses the proper amount of beverage in the container. The containers move past the filling mechanism at rates of 650–1000 bottles per minute. Fig. 13-1 shows an example of containers moving down the conveyor line to the filler. The filler system must also place a cap on plastic-type bottles or a lid on aluminum cans. This part of the system is called the *closer*. The containers are positioned by a large screw mechanism called an *auger*, so that they are evenly spaced as they enter the *carousel* for the filler. The carousel consists of six to ten multiple stations that continually rotate to ensure that each container is synchronized with the filler. The filler will have a dispensing mechanism directly above each station on the carousel. When a container enters the carousel, it is precisely located under the dispensing mechanism. The beverage is dispensed at a rate that ensures the container is completely full before it leaves the carousel. At the end of the carousel the cap or lid is placed on the container and it is sealed. Fig. 13-2 shows an example of a six-station carousel that is part of the filler system.

The no can/no fill sensing is used to detect the presence or absence of containers as they move into the filler/closer system. If a container is missing, a proximity switch detects the open space and signals the filler that one of the positions on the carousel will be empty and it should not dispense beverage when the empty position moves through the carousel. The part of the system that puts a cap on the container must also be signaled so that it does not try to place a cap. The sensing system must be able to *sense, decide, and create* an output in less than 100 msec when the line is running at 600 cans/min and in less than 60 msec when the line is running at 1000 can/min. Since three distinct actions must be made, the sensing time must be less than one third of the total reaction time.

When sensors are connected to the dedicated microprocessor controller, it typically takes two read/write cycles to detect the sensor signal and enter it into its data system so it can be used in the logic part of the system. This means the sensing time and microprocessor scan time must be less 16.6 msec for the 600-can/min operation and less than 10 msec for the 1000 can/min operation. You can soon see why this type of sensing and reacting must be controlled by a dedicated microprocessor rather than a PLC.

Another problem that faces the system is that the sensors must be able to cycle 60,000 times per hour, which adds up to 1,440,000 times per day, and 525,600,000 times per year.

13.1.2 Typical Operation

During typical operation the sensors in the filler/closer system must detect each container and synchronize it with the carousel which synchronizes the container with the filler. When the sensors indicate a container is present, the controller emits

639

FIGURE 13-1 Clear plastic beverage containers move at high speed to a filler/closer where soda pop is dispensed into the container and a cap is placed on it. (Courtesy of Hyde Park Electronic Inc.)

FIGURE 13-2 A six-station carousel that is part of the filler system. (Courtesy of Hyde Park Electronic Inc.)

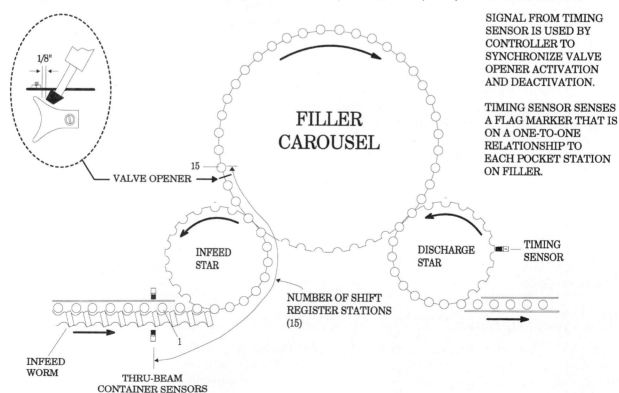

a pulse to activate the trip mechanism on the dispensing mechanism at the proper moment to ensure the container receives the proper amount of beverage. The dispenser mechanism is adjusted so that the proper amount of beverage is dispensed to ensure each container is filled to the same level. The dispenser for the cap or lid is also signaled so that the cap or lid drops onto the container at the proper time.

When a missing container is detected, the signal is placed into the controller so that the dispenser solenoid is not activated during the cycle when the missing container is moving through the carousel. The electronic control of the system is much more complex and it is covered separately in the next section.

13.1.3 Using a Shift Register to Control Signal Timing

The controller uses a shift register to control the signal timing. The length of the shift register is determined by the speed of the line. For example, the length of the shift register may be 2 when the line is moving at slow speeds and it may be 10 when the system speed is 600 can/min.

Fig. 13-3 shows an example of the proximity switch that is used to detect the presence or absence of a container on the line as the containers pass at high speeds. A second proximity switch is used to sense teeth on a gear that drives the system. The second proximity switch is used as the clock input to ensure the system remains synchronized.

The container detect proximity switch is aligned to detect the leading edge of each container. When a container is present at the detect proximity switch when the clock proximity sends a pulse, a binary 1 is loaded into the shift register position as data to indicate a container is present. If no container is detected when the clock pulse is sent, a 0 is loaded into the shift register to indicate a space is present. Fig. 13-4 shows a timing diagram that indicates the condition of the shift register and the valve mechanism when a container is present and when a container is absent. Each step of the sequence is numbered 1–16 and these numbers are indicated in the diagram.

At point (1) in the diagram the proximity sensor starts sensing when the

FIGURE 13-3 Proximity switches are shown detecting containers. (Courtesy of Hyde Park Electronic Inc.)

VOLUMETRIC BEVERAGE FILLER

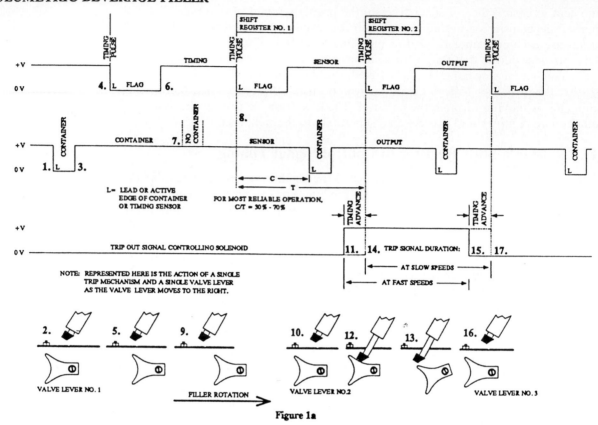

Figure 1a

STEPS IN OPERATION

(Reference to numbers above)

1. Container sensor starts sensing container (lead edge of container signal) - container presence information is loaded into shift register.   2. Valve lever No. 1 and trip mechanism in to-fill position - moving to right.   3. Container sensor stops sensing container.   4. Timing sensor starts sensing flag (lead edge of timing mark.   5. Valve lever No. 1 passes valve opener - in fill position, filler valve open.   6. Timing sensor stops sensing flag.   7. Container sensor did not sense container. Missing container information, including Shift Register number (for this example, set to 2), loaded into shift register at next timing sensor leading edge (8.).   9. Valve lever No. 1 still in fill position, moving to right.   10. Valve lever No. 2 tripped for normal to-fill operation.   11. At high speeds, timing advance receives no-container information at Shift Register No. 2 position, valve opener to be extended at this point...   12. resulting in plunger making contact with valve lever No. 2 and...   13. moving the valve lever to the no-fill position.   14. Same as 11, except at slow speeds, plunger is extended at this point   15. At high speeds, timing advance causes plunger to be retracted at this point...   16. resulting in plunger being returned to normal to-fill position, ready for valve lever No. 3   17. Same as 15, except at slow speeds, plunger is retracted at this point.

NOTE: Shape of trip mechanism and direction of filler valve movement not necessarily relevant to existing filler equipment, but for explanation purposes only.

FIGURE 13-4 Diagram of the signals for the no container/no fill controller for a rotary piston-type filler mechanism. (Courtesy of Hyde Park Electronic Inc.)

leading edge of the container is detected and the binary 1 is loaded into the shift register. At point (2) the filler trip mechanism is in the normal to-fill position, and valve roller #1 is in the no-fill position moving to the left toward the trip mechanism. At point (3) the container sensor stops sensing the container as it passes the proximity switch. At point (4) the timing proximity sensor detects the leading edge of the timing gear for the system and starts the sensing flag (leading edge of timing pulse). At point (5) the valve roller #1 activates the trip mechanism downward to the fill position. At point (6) the timing sensor stops the sensing flag. At point (7) the container sensor did not see a container (note the absence of C in the diagram) between the timing pulses at the C/T (counter timer) position. The "no container" information including the shift register number (for this example, set to a 2) is loaded into the shift register at the next clock signal, which occurs at point (8). At point (9) valve roller #1 has reached the fill position and the fill valve is open to begin dispensing beverage. At point (10) valve roller #2 is in the no-fill position, moving to the left toward the next trip mechanism (not shown). At point (11) the timing advance receives the high-speed "no-container" information into the shift register 2 position, which causes the trip solenoid to be actuated at this point. At point (12) the trip mechanism activates upward to allow valve roller #2 to continue moving past the trip station in the no-fill position. At point (13) the same signals that are shown in points 11 and 12 are shown for lower-speed applications. The main difference between the high-speed filling and low-speed filling is the duration of the signals.

At point (14) a high-speed timing advance is added to account for the signals occurring at higher speeds. The timing advance causes the solenoid to be actuated at a point in time prior to when it needs to open. This *feedforward* action provides sufficient time for the electromechanical action of the valve to allow the valve to fully open even at high speeds. You must remember that it may take a solenoid valve several milliseconds to become magnetized and then fully open, so the signal must be sent to the valve several milliseconds before the container is in position to receive the beverage. The amount of time for the time delay for this signal is fully adjustable so that the solenoid valve operation on the filler can be synchronized with the speed of the containers passing through the carousel under the filler valves to ensure the beverage is dispensed at just the right time.

At point (15) the trip mechanism is tripped back to the downward position to the normal to-fill position (with valve roller #2 moving in the throat of the trip mechanism). The mechanism is ready for valve roller #3 at this point. At point (16) the signals are the same as at point (14) except this part of the diagram represents the signals at lower speeds. The solenoid is actuated at lower speeds between point (14) and point (16).

13.1.4 Electrical Diagram of Sensors and Valves Connected to the Controller

Fig. 13-5 provides a wiring diagram that shows the connection of the container present/absent proximity switch, the timing proximity switch that provides the clock signal, and the two output signals to cylinder A and cylinder B. The selector switch (FILL, TRIP, AUTO) is also shown wired to the controller.

From this diagram you can see that the proximity switches are three wire switches. Two of the wires provide +24 and −24 to the sensor. The third wire is the sensor signal and it is designated *sensor input*. You should also notice that 120 volt AC is supplied to the controller at the top left terminals that are identified as

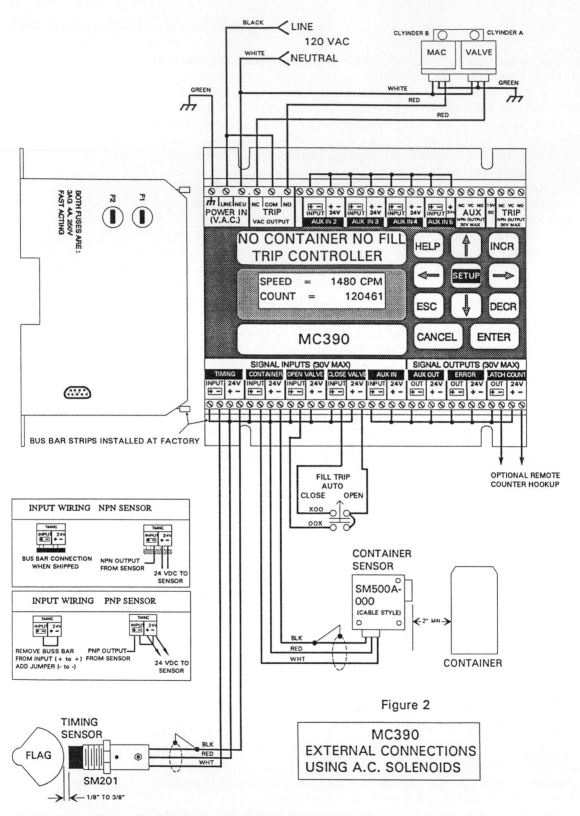

FIGURE 13-5 Wiring diagram for the no container/no fill controller and its inputs and outputs. (Courtesy of Hyde Park Electronic Inc.)

VAC and COM. The output cylinders are connected to the terminals identified as TRIP OUTPUT VAC.

13.1.5 Troubleshooting the System

Since the system is controlled by a dedicated microprocessor, it has a section of its control dedicated to troubleshooting. This section includes the ability to view input signals to ensure they are received through the controller input bus, and the ability to force output signals for any duration of time so that you have sufficient time to test them. You can use a combination of troubleshooting the hard-wired part of the system and the controller part of the system.

You can troubleshoot the hard-wired part of the system by testing for 120 volt AC to ensure supply power is present. The display for the controller should be illuminated if power is present. If you think the proximity sensors are not picking up and sending the proper signal, you can stop the system when a container is in front of its proximity sensor and see if a signal is present at terminals −24 and input. The LED indicator on the proximity sensor should also be illuminated at this time. You can also advance the line until the gear tooth is located in front of the timing proximity sensor and make the same type of test to see if that proximity sensor is operating correctly.

From the troubleshooting menu, you can see on the display if each sensor is sending a signal and if the signal is making its way through the controller input bus. The output signals can be tested by manually sending a continual signal. Since this signal is sent to each output continually during the test, you can use a voltmeter to determine if the signal is reaching the cylinders and if the cylinders are activating. When you are testing the cylinders, you must remember to test the air supply to ensure sufficient air is present to cause the cylinders to activate correctly.

Some of the problems that may occur include the proximity sensors becoming loose and misaligned, a faulty proximity switch that loads down the power supply so that both proximity switches do not detect properly, mechanical misalignment in the carousel, and sensitivity adjustments set incorrectly in the proximity switches. Some minor timing adjustments may also be needed when higher speeds are used in the filler.

QUESTIONS

Note! You may need to review material in other chapters of this text to answer some parts of the questions. The case study is designed to ensure that you understand all of the technology involved in each application.

1. Explain the operation of a proximity switch. Include information about the types of proximity switches that can detect aluminum beverage cans and sensors that can detect plastic bottles.
2. Explain the operation of a shift register. Include the function of the CLOCK and DATA signals.
3. Explain why the length of the shift register may need to change as more stations are added to the carousel.
4. Explain why it is important for sensors to detect the absence of a container at the filler/ closer system.
5. List other applications that may need this type of high-speed detection technology.
6. Explain why a dedicated microprocessor is used as the controller rather than a PLC.

ALUMINUM ROLLING MILL

13.2.1 Overview of the System

Aluminum sheet material that is 0.105 in. thick is used to manufacture beverage cans. When the aluminum starts out, it is mined from bauxite and formed into ingots that are approximately 120 in. wide, 21.5 in. thick, and 16 ft long. It is reduced to 0.105 in. in two processes. In the first operation, a four-stand mill is used to reduced the thickness to 4.5 ins. During this operation, the ingot is moved back and forth under rollers in 17 passes where the screwdown system increases pressure on the rollers to reduce the thickness of the ingot. The rollers are driven by multiple 7000hp DC motors that use solid-state speed controls and direction controls.

The second process is a continuous rolling operation that uses five synchronous AC motors that are 5000 hp each. These motors are controlled by ac variable-frequency drives to control the speed and torque to ensure the aluminum is drawn in consistent thickness. When the aluminum exits this process, it is 0.105 in. thick and it is stored as large rolls so that it can be transported easily and used as aluminum can stock at container manufacturing factories.

This case study will explain the operation of the large variety of electronic sensors and controls used in the rolling process. The first part of this case study will explain the complete process and all of the equipment that is used, and the second part will provide in-depth discussion of the electronics used in each part of the system. Fig. 13-6 shows a diagram of the entire two-process mill. From this diagram you can see the ingots are cast and then passed to the scalper and heating furnace. After the ingot leaves the heating furnace, it moves to the first rolling mill where it is moved back and forth until it is reduced to 4.5 in. thick. When the stock comes out of this mill, it is rolled into large coils so that it can be handled more easily for the second part of the process.

During the second part of the process, the large rolls are fed into the double payoff reel station (1). The aluminum stock must move through the mill continuously as though it is one long sheet. This means that the individual rolls must be butt-welded together to act as one long sheet. The butt-weld equipment is located in the middle section of the finish mill (3) and the butt-welding process requires approximately 8 seconds and the line requires approximately 100 seconds to decelerate, stop the coil, butt-weld the ends, and accelerate back to speed. The large-strip accumulator is used just after the butt-weld section to store extra aluminum so that the finish end of the mill does not notice the feed end has been stopped to butt-weld another roll.

The accumulator uses tensioning controls, dancer rollers, and guide rollers to store eight loops of material up to 195 ft high to ensure that the finish end of the process has sufficient stock that can be supplied at a predetermined feed rate. When the accumulator is fully loaded it has approximately 4 minutes of stock that can be used while the next roll is butt-welded. The accumulator rollers are continually raised when a new roll is fed into the mill, and it is lowered to feed stock to the

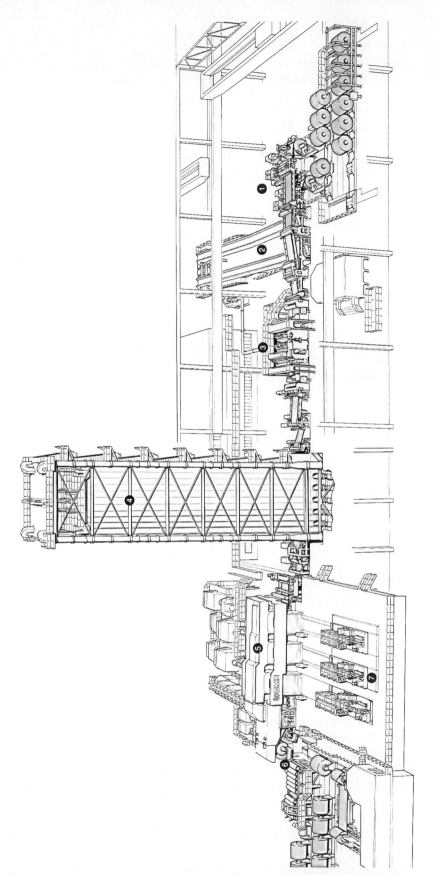

FIGURE 13-6 The diagram of two sections of the continuous tandem cold-rolling plant for aluminum strip container stock. (Courtesy of SMS Engineering.)

647

finish end of the mill. This means the accumulator is continually raised and lowered to meet the needs of the finish end of the line and the feed end of the line.

In the finish rolling section 5000hp AC synchronous motors control the speed of the aluminum. This section of the system uses a three-stand mill (5). Stands 1 and 2 are four-high mills, and stand 3 is a six-high mill. The screwdown system uses hydraulic servo valves to control the accurate positioning of the rollers.

When the aluminum is at its finished thickness, it exits the mill to the roll-up area. At this point it is rolled onto shipping rolls at the carousel reel and sleeve delivery (6).

13.2.2 Controls for the Cold-Roll Mill

The controls for the cold-roll mill include four 32-bit mainframe computers that store data for each ingot and send control variables for all of the motor drive speeds and the positions for the hydraulic servo valves for the screwdown controls. The mainframe computers also provide a host for multiple programmable logic controllers (PLCs) that are networked with man-machine interfaces (MMIs) to provide color graphic stations where operators and maintenance technicians can view the system. The MMIs also have computerized voice annunciation for alarm conditions. A fiberoptic conductor is used for the data highway network that the PLCs and motor drives are linked to.

The MMIs and PLCs provide the means to troubleshoot the system and determine the source of faults. The rolls maintain backup drives for each drive, so the mainframe computer must monitor the status of each drive and engage the backup drive if a problem occurs with any of the main drives.

The mainframe determines feeds and speeds from data that come from the analysis of the original ingot and roll of the thicker material that comes from the first process. When the roll is mounted in the finish mill, all of the computer data are sent to each drive and PLC as a feedforward control mode. The speed of the finish rolled stock is too fast for sensors to determine conditions and to feed back data to the computers so feedforward control must be used. The mainframe computers use sensor data to detect any variants as a roll is processed. These data are used for archival purposes so that changes to the control algorithm can be made if conditions warrant.

The PLCs use speed sensors, pressure sensors, position sensors, and torque sensors to keep track of the roll as it is reduced to the finish thickness. Microprocessor controlled systems control the hydraulic servo system to ensure the screwdown section reacts quickly enough. The PLC is typically too slow for high-speed position changes of the screwdown control of the finish mill rollers.

13.2.3 Typical Operation of the Mill

An overhead crane moves the rolls of aluminum from the first process to the finish mill. These rolls are placed on pallet conveyors, which move the rolls to coil cars. The coil cars are used to mount the rolls in place at the beginning of the finish mill. Two shears are used to trim the head and tail ends so that they will match up to the existing stock during the butt-weld operation. The scrap from the trim process is removed automatically and it is recycled. Two payoff reels are used so that one is in operation and one is being loaded at all times. This allows the sheet material to be fed into the process continually. Fig. 13-7 shows a diagram of the payoff reel station. In this diagram you can see that the lower reel is the active reel, and the

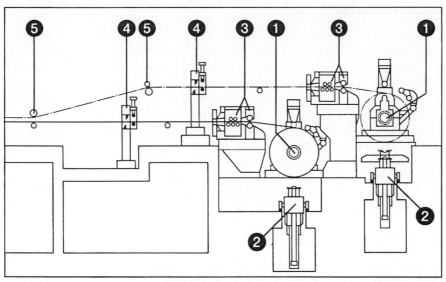

Double payoff reel station with automated scrap removal. Cut-off strip pieces fall onto the scrap roller tables from where they are automatically removed.

① Reel. ② Coil car. ③ Pinch-roll leveler. ④ Crosscut shears with scrap removal. ⑤ Pinch rolls.

FIGURE 13-7 A diagram that shows the upper and lower reels of the double payoff reel station at the start of the line. (Courtesy of SMS Engineering.)

upper reel is being prepared so that it will automatically feed the system when the first reel runs out.

The electronic controls of this part of the system are controlled through the PLC. The feed rollers are jogged to allow the start of the aluminum roll to feed uniformly into the pinch roller sections. After the front end of the roll is threaded into the rollers, sensors activate the automatic trimming operation where a hydraulic-operated shear trims the end and scrap falls to the scrap roller tables. The PLC controls the scrap removal process that includes trimming the scrap ends to smaller size so that they are easier to manage. The trimmed end is jogged to the point where the top and bottom reels meet. When one runs out, the other is automatically fed into the system where it moves directly to the butt-weld area so that it can be permanently welded to the end of the roll that is being processed. This causes the finish mill to see the supply of aluminum coming off of the rolls as one long piece rather than many short rolls.

13.2.4 Butt-Weld Section

When the front end of the new roll is ready to be welded to the tail end of the previous roll, one of the mainframe computers sends data through the network to the PLC that controls the butt-weld process. These data include the percentage of alloys in each of the rolls and the amount of current and pressure that will be needed during the welding process. The PLC controls all of the operations of the butt-weld process. Sensors connected to the PLC collect data from the welding process and these data are sent back to the mainframe computer to be used to make adjustments in future welds.

13.2.5 Accumulator Section

A diagram of the accumulator is shown in Fig. 13-8. The main part of the accumulator is the loop tower. The loop tower consists of eight deflector rollers (2) connected to a loop car (1). The loop car is raised and lowered with a rope winch as the system needs dictate. Three guide rollers are used inside the accumulator and two are used at the exit to ensure the roll stays in position as it moves through the accumulator.

Sets of bridle rollers (3) and dancer rollers (4) are used to maintain correct tension on the roll as it moves through the accumulator. These rollers are strategically placed at the entrance and exit of the accumulator and they are tensioned by fast-acting hydraulic servo valves that are controlled by dedicated microprocessor controllers. The mainframe computer and several PLCs combine to maintain control of the roll during this part of the process.

13.2.6 The Three-Stand Tandem Mill

Fig. 13-9 shows a diagram of the six-high mill that is used as mill 2 in the three-stand tandem mill. Mill 1 and mill 3 are four-high mills. In the diagram you can see why this mill is called a six-high mill, since it has six separate rollers. The aluminum sheet is threaded between the six rollers. The hydraulic screwdown mechanism is shown at the top of the stack of rollers (1). When the aluminum moves through this part of the process, pressure is applied to the screwdown mechanism and it is transferred to each of the rollers. The pressure on the rollers controls the

FIGURE 13-8 A diagram of the accumulator section of the mill. (Courtesy of SMS Engineering.)

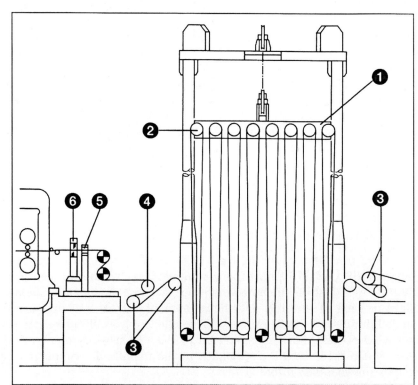

① Loop car with rope winch. ② Deflector rollers. ❸ Guide rollers. ③ Set of bridle rollers. ④ Dancer roll. ⑤ Measuring device. ⑥ Crosscut shear.

Stand 3, CVC 6-HS.

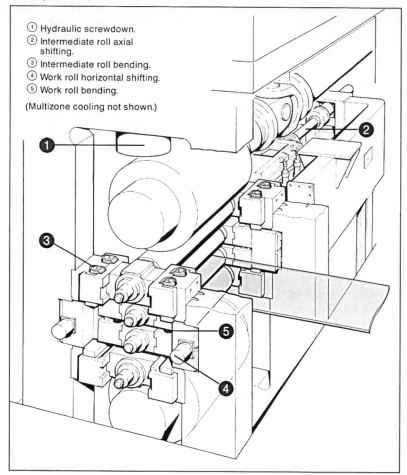

① Hydraulic screwdown.
② Intermediate roll axial shifting.
③ Intermediate roll bending.
④ Work roll horizontal shifting.
⑤ Work roll bending.

(Multizone cooling not shown.)

FIGURE 13-9 Diagram of a six-high mill. (Courtesy of SMS Engineering.)

thickness of the aluminum as it passes. A dedicated microprocessor is used to control the hydraulic positioning of each roller and the pressure that the screwdown mechanism puts on each of the rollers.

Since the roll of aluminum will heat up slightly as it passes through each set of rollers, the heat must be removed to maintain the uniformity of the aluminum roll. The heat is removed as the stock moves through a cooling zone immediately after the roller sections.

13.2.7 Speed Control for the Rollers

Speed control for the rollers is maintained by the mainframe computer by downloading a preset program into each drive. This means that the speeds and feeds for each roll drive are determined by a complex math calculation that gets original data from the analysis of the ingot. All the changes in the AC-frequency drives are determined before the aluminum enters the rollers. The changes are made automatically according to the program rather than by using sensors to determine whether or not changes should be made. The reason the speeds and feeds are determined by a computer calculation rather than feedback sensors is because the speed of the

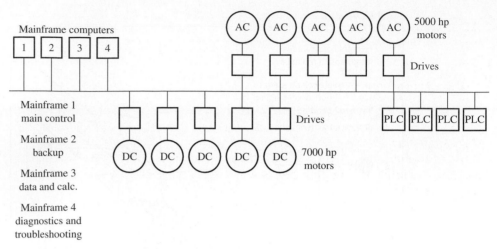

FIGURE 13-10 Diagram of computers, PLCs, and drives.

aluminum as it moves through the rollers is faster than the sensors could detect problems and feed back the signals to any computer or dedicated electronics.

After thousands of runs, the mathematical calculations in the mainframe computer are able to optimize changes of feeds and speeds to create the best results of the finished product.

13.2.8 Mainframe Computers, PLCs, and Motor Drives

Fig. 13-10 shows a diagram of the four mainframe computers and the PLCs and drives for the first and second sections of the process. The four mainframe computers are used to store data for each ingot and roll and to calculate speeds and feeds for each drive in the system. You can see that the first section of the system has eight 7000hp DC drives and the finish section of the system has five 5000hp AC drives. Multiple PLCs are used to control the scrap conveyors, butt-weld section, and other controls around the system.

Q U E S T I O N S

> **Note! You may need to review material in other chapters of this text to answer some parts of the questions. The case study is designed to ensure that you understand all of the technology involved in each application.**

1. Explain the operation of an AC variable-frequency drive.
2. Explain the operation of a DC variable-speed drive.
3. Explain why the speed of the aluminum material must be controlled as it moves through the mill.
4. Explain why it is not possible to use sensors in a feedback application for the rolling mill.
5. List other applications that may need this type of motor speed control technology.
6. Explain why the accumulator tower is required in the aluminum rolling mill.

MICROPROCESSOR CONTROLLED PLASTIC RESIN DRYER

13.3.1 Overview of the System

This application is of particular interest to industrial electronic students because it uses a large variety of electronic instrumentation and controls and it is controlled by a dedicated microprocessor. Another reason this application is useful to understand is because of the widespread use of plastic injection molding and blow molding processes where students will be working when they are on the job.

When plastic resin is used in plastic injection molding machines, its moisture content must be strictly controlled to prevent excess moisture from being released into the plastic part that is being formed. The moisture content of the raw material is controlled by a microprocessor controlled plastic resin dryer. Fig. 13-11 shows a diagram and a picture of a typical dryer. In the diagram you can see the dryer consists of a large hopper that holds 1–2 hours worth of raw product, and two identical moisture removal towers. The towers are connected with air hoses and valves so that one of the towers is the active tower for a period of time where it picks up moisture from the raw product, and the other tower is in *regeneration* to remove the moisture from the system that was picked up from the raw material. When a tower is active it means it is able to remove moisture, and when a tower is in the regeneration process it means that the moisture that it picked up from the raw material is being removed from the system. The regenerated tower will be ready to be used as the active tower when the active tower becomes moisture laden.

In the active tower dry air is passed over the raw material so that any moisture in the raw material can move to the dry air. You should remember from physics class that moisture will always move from air that is moisture laden to air that is dry. The movement of moisture will continue until both towers have the same amount of moisture in them. The amount of moisture in the air is called *relative humidity*, and the amount of moisture that air can hold will increase as the air is heated. When air is cooled, it will reach a temperature below which the moisture in the air begins to change from vapor form back to liquid form. The temperature where this occurs is called the *dew point*. The dew-point temperature will change with the relative humidity of the air. When the relative humidity of the air is high, the dew-point temperature will be relatively high (approximately 80°F), and when the relative humidity becomes lower the dew-point temperature may be as low as 50°F. This means that the temperature of the air in the tower that is being regenerated may need to be cooled below 50°F as the air in the tower loses its humidity so that the remaining moisture can be removed.

When the air in the active tower becomes moisture laden, air valves between the two towers switch so that the tower that was regenerating is now the active tower, and the tower that was active and became moisture laden is now set to regenerate.

The towers have special material installed in them that is called *desiccant*. Desiccant is able to absorb large amounts of moisture quickly from air, and release the moisture when the drier air is passed over it. The desiccant acts like a sponge

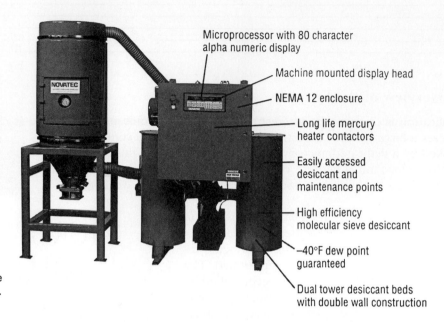

Microprocessor with 80 character
alpha numeric display

Machine mounted display head

NEMA 12 enclosure

Long life mercury
heater contactors

Easily accessed
desiccant and
maintenance points

High efficiency
molecular sieve desiccant

−40°F dew point
guaranteed

Dual tower desiccant beds
with double wall construction

FIGURE 13-11 Picture of the plastic resin dryer. (Courtesy of Novatec.)

in that it picks up the excess moisture from the air that is passed over the raw material when the tower is used as the active tower, and it gives up moisture quickly when the tower is in regeneration.

The dryer uses several concepts such as heating the air to ensure it picks up the maximum amount of moisture in the active tower, and cooling the air to ensure it gives up the maximum amount of moisture in the tower when it is regenerating. Other concepts include testing for ΔP (delta P) to determine when filters in the system are clogged, and testing temperatures and relative humidity to determine dew-point temperature.

During the regeneration process, the moisture-laden tower has dry air passed over it to pick up excess moisture. This air is then passed through a condenser section where the temperature of the air is dropped to a point below the dew point for the air, which will cause all moisture molecules in the air that are in vapor form to condense to moisture droplets so that they can collect against the condensing plates and run off as water droplets. This causes all of the moisture in the tower that is being regenerated to be removed from the desiccant beds so that the tower is ready to be used as the active tower again.

13.3.2 Basic Parts of the System

The picture in Fig. 13-11 of the resin dryer system shows its basic parts. From this picture you can see the two towers and the air hoses and valves that interconnect them. You can also see the main hopper where the raw material (plastic resin pellets) is stored. The dedicated microprocessor controller is mounted in the control cabinet with other electrical controls such as relays. The solenoid valves that switch the towers from active condition to regeneration are mounted in the hoses. The system also has a fan and electric heater to control the temperature of the air.

13.3.3 Control Parameters for the Dryer

The control parameters for the dryer include parameters for the tower that is active and parameters for the tower that is being regenerated. Fig. 13-12 shows a list of these

PARAMETERS

- Process Set-Point
- Process Temperature
- Bed temperatures
  Inlet & Outlet − Left Bed
  Inlet & Outlet − Right Bed
- Dew Point Sensor Temperature
- Process Dew Point
- Regen Heater & Motor Amps − each leg
- Process Heater & Motor Amps − each leg
- Regeneration & Process Filter Switch State
- Regeneration & Process Flow Switch State
- Regeneration & Process High Temp. Alarm
- Electrical Phase
- Regeneration Time Remaining
- L.H. and R.H. Regeneration State
- Process Heater Duty Cycle
- Auxiliary Input Status
- Monitor-number of resets
- Real Time Clock
- Highest & Lowest Process Temp. − current day
- Average Process Temp. − Previous Hour
- Energy Usage (KWH) − Current Day
- Energy Usage (KWH) − Previous Hour & Previous Day
- Time Right & Left Bed were on Process
- % L.H. & R.H. Bed Regeneration Time Used
- Head/Plate Communication Status
- Second Setpoint Status
- Cooling Coil Filter Status

FIGURE 13-12 List of control parameters for the resin dryer. (Courtesy of Novatec.)

parameters. The parameters in the active tower include the process temperature, bed temperature, amps for the bed heater, relative humidity, and filter status. The parameters for the tower when it is regenerating is the bed temperature, dew-point temperature, relative humidity, and regeneration timer. These parameters are all programmable through the microprocessor keypad and can be viewed on the microprocessor display.

13.3.4 Normal Operation of the Dryer

Fig. 13-13 shows a diagram of the dryer equipment that has specific areas numbered. These numbers will help you understand the operation of the system. The dryer is a closed-loop system where all air flow is strictly controlled for the purpose of removing moisture from unwanted areas (in the raw material) to the regeneration area where it is purged from the system. At point (1) the raw plastic resin is stored in a hopper where it will be fed from the bottom directly into the plastic injection molding machine. When new raw material is added to the hopper, it is dropped in through the top of the hopper lid by an air conveyor. The raw material is moisture laden and this moisture must be removed.

Dry air from the fresh regenerated tower passes through the resin material in the hopper at the bottom and exits at the top of the hopper through a flexible hose (2). Since the air is dry, and the raw resin has moisture in it, the moisture will move to the dry air naturally as the air moves past the resin. Next the air passes through a very fine filter (3) where all fine dust-type material is removed. Then it passes through a four-way valve (4) that directs the moisture-laden air to the tower that is acting as the active tower.

NOVATEC'S CLOSED LOOP DRYING SYSTEM:
PROVEN BEST YEAR AFTER YEAR

In the NOVATEC "closed loop" drying system moisture laden air exits from the hopper (1) passes through a flexible hose (2) and enters the two-stage filter (3) where virtually all fines carried from the hopper are trapped. The clean air then passes through a 4-way valve which diverts it to the on-line tower (4). The air then passes downward through a thick layer of desiccant (5). Unlike competitive units with thin "donut" desiccant beds and embedded heaters, the solid bed design of Novatec Dryers insures maximum air contact time, resulting in guaranteed minimum dew points of −40°F. The clean, dry air then passes through another 4-way valve (6) which directs it to the process air heater (7) which heats the air to the temperature selected for the material being processed (10). Another flexible hose (8) delivers the clean, dry air to the hopper (1) where the air diffuser (9), insures uniform drying without channeling.

Simultaneously, the desiccant in the second tower (11) is being regenerated. Ambient air is drawn through the reactivation filter (12) and heated by the reactivation heater (13). The heated air then passes through the lower valve (6) which directs it into the off-line tower and through the desiccant (14). The heated air purges the trapped moisture from the desiccant, exits through the upper valve (4) and is released into the atmosphere.

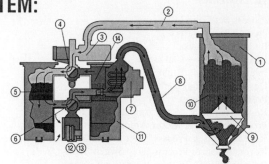

When the regeneration cycle is complete, the regeneration heater and blower are de-energized and the desiccant is allowed to cool statically before the next adsorption cycle begins. Unlike competitive units where the regeneration blower runs continuously using moist ambient air to cool the desiccant. Novatec's static cooling concept prevents any moisture build-up and completely eliminates dew point "bumps" during changeover.

Static cooling also results in long cycle times, a minimum of 4 hours per tower, which greatly reduces power consumption during regeneration, reduces maintenance and component wear to minimum levels, and significantly extends desiccant life.

FIGURE 13-13   Diagram of the internal parts of the resin dryer. (Courtesy of Novatec.)

When the moisture-laden air reaches the tower, it passes down over a thick layer of desiccant bed (5). The desiccant bed is maintained at approximately 40°F to ensure a temperature well below the dew point of the driest air. The low dew-point temperature and the time the air is in contact with the desiccant bed ensures that the maximum amount of moisture is removed from the air.

After the air passes through the tower, it moves to another four-way valve (6) where it is directed to a heater section (7). The heater raises the temperature of the air to approximately 300°F so that the air can absorb the maximum amount of moisture when it passes over the raw resin. The air hose (8) directs air from the heater section back to the hopper. Diffusers (9) in the hopper ensure that the air is evenly dispersed over the raw plastic resin in the hopper so that the maximum amount of moisture can be picked up by the air. At this point the air exits the hopper and continues the cycle until the desiccant in the tower is saturated.

While one tower is acting as the active tower to pick up moisture, the second tower is being regenerated so that moisture is removed. During the regeneration process, the second tower (11) receives ambient air from outside the system through a filter (12). This air is drawn across a reactivation heater (13) where its temperature is raised before it is directed across the moisture-laden desiccant in the tower that is being regenerated. As this air moves past the moisture-laden desiccant, the moisture from the desiccant moves naturally to the drier air. This air is moved past a condensing section where moisture is collected and then the air is purged to the atmosphere.

It is important to understand that a large fan moves the air through the entire system and four-way valves switch to create the paths for air through the active tower and the tower that is being regenerated. When the desiccant in the active tower becomes moisture laden, the four-way valves will switch so that the active tower becomes the regenerated tower and the previous regenerated tower becomes the active tower.

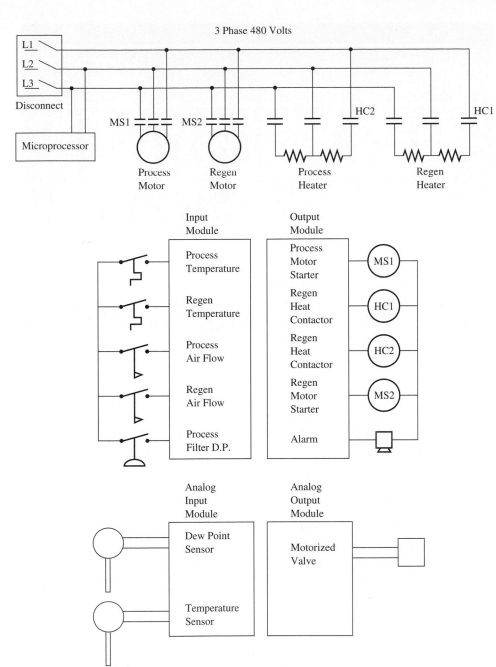

FIGURE 13-14 Connection diagrams for inputs and outputs to the plastic resin dryer.

13.3.5 Control Loops for the Dryer

The dedicated microprocessor controller in the dryer controls several closed-loop systems and several alarm points. One of the closed-loop systems is the temperature loop for process air. This loop uses thermocouples as the process variable (PV) and electric heating elements as the output device. A second loop is used to control the dew point in the active tower. The maximum temperature, air flow, and clogged filter conditions are monitored as alarm conditions. The microprocessor also controls the time that each tower is active and when the tower is regenerated.

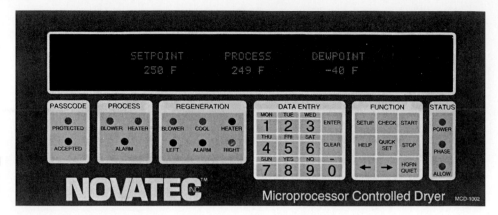

FIGURE 13-15 Keypad and display for the plastic resin dryer. (Courtesy of Novatec.)

13.3.6 Wiring Diagram for the Dryer

It is important to understand how the sensors and controls are wired to the system to be used as inputs and outputs for the dedicated microprocessor. This will also help you understand how the system is controlled during troubleshooting. Part of the system acts like a PLC where relay ladder logic is used to ensure that switches such as the start fan switch are directed through the microprocessor to control the fan motor starter coil. Other parts of the system act as a closed-loop controller where only the PV and the output must be identified. The microprocessor controls the remainder of the system including the setpoint, gain, reset, and rate of change.

Fig. 13-14 shows the wiring diagram for this system. This diagram will give you a better understanding of how the input sensors and outputs are connected to the controller. You will also get a better idea of which sensors are digital (on/off) in nature and which are analog. You can refer to the previous chapters to review the operation of any of the sensors or output devices.

Fig. 13-15 shows the keypad and display for the microprocessor controller for the resin dryer. You can see that this type of keypad and display are very similar to many others that you will find on industrial electronic controls. The keypad and display allow you to view and set all of the parmameters for the system.

The microprocessor allows the technician to see the status of each input as a 1 or a 0. If a switch is energized, the controller will display a 1. The controller also displays the values of all analog sensors so that they can be compared to the actual voltage the sensor sends. The controller can also turn on any output continually for troubleshooting. When troubleshooting this system, you can test all of the hard-wired inputs and outputs using traditional troubleshooting techniques, and you can troubleshoot any inputs and outputs connected to the processor using PLC troubleshooting techniques or microprocessor troubleshooting techniques.

QUESTIONS

Note! You may need to review material in other chapters of this text to answer some parts of the questions. The case study is designed to ensure that you understand all of the technology involved in each application.

1. Explain the operation of the resin dryer.
2. Identify the sensors that you would find on the resin dryer.
3. Explain why the temperature and dew point are controlled in the resin dryer.
4. Explain why two towers are used in the dryer.
5. List other applications that may need this type of temperature control technology.
6. Explain why closed-loop process control is used in this system for temperature control.

4

FINISHING AND BALANCING BRAKE ROTORS

13.4.1 Overview

The rotors for front disk brakes for automobiles must be finished to 0.0005 in. on the surface where the brake pads come in contact with the rotor and each rotor must be spin balanced to ensure that the rotor is perfectly balanced by weight after it is finished. The finishing process includes several CNC machines to provide the finish, and a several sensors to gauge the finish and to test the spin balance.

Fig. 13-16 shows the equipment and sensors for this application. In this diagram you can see the rough castings enter the first CNC machine that provides a rough cut. The second machine provides the finish cut on the bottom side of the rotor, and the third machine provides the finish cut on the top side of the rotor and the finish on the bearing race. The parts are moved from station to station with a conveyor and each machine has an automatic loading robot arm. The arm removes the finished part from the machine and locates the next part to be cut and loads it into the CNC machine's chuck. The last machine in this sequence provides the spin test and identifies the point on the rotor where material must be removed, indicating how much must be removed to make the rotor perfectly balanced. Each of the finishing machines provide gauging to ensure the surface that is finished is within specifications.

13.4.2 Electronics Used

The finishing and spin-balancing application uses three CNC machines that each have a two-axis robot loader. Each finishing machine uses a GE/FANUC controller that integrates a CNC controller with a PLC controller. This unique controller allows the CNC controller to operate the cutting and finishing parts of the equipment and the PLC to control all of the on/off and analog inputs and outputs through ladder logic to control conveyor speed, the robot loader arm, and other periphery equipment around the machine, such as the chuck.

Each CNC machine has two spindle heads so that it can cut two parts at the same time. Each spindle head has an X and a Z servomotor that provides two-axis motion. Since the part is round and it is rotated during the cutting cycle, a third cutting axis is not required. The spindle motor and servomotors for each CNC machine use an AC motor amplifier that is controlled by a CNC controller. Each servomotor has a tachometer for sensing speed and an encoder for sensing position. The spindle motor has only a tachometer since it only needs to sense rpm.

Each of the robot arms uses a servomotor that is controlled by a servo amplifier. The operation of the robot loader arm is controlled by a PLC, which determines the machine cycle and enables the robot to pick up the finished part and load a new part. The spin balancer is controlled by an independent PLC.

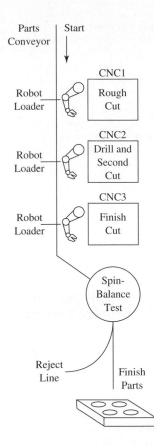

FIGURE 13-16 Diagram of the rotor-finishing and spin-balancing equipment.

13.4.3 Sensors for Balancing

When each rotor is finished, it is removed from the chuck in the CNC machine by the robot loader arm and placed on the conveyor. The conveyor moves the part to a spin-balancing test station. When the spin-balancing test begins, the rotor is rotated at high speed (500 rpm) for several seconds. If the rotor is not balanced correctly, it will provide pressure at the point where the imbalance exists. A load cell that contains strain gauges is used to measure the amount of pressure the imbalanced condition causes. An LVDT (linear variable differential transformer) is used to determine the exact location on the rotor where the imbalance occurs. The linear measurement is converted to a rotary measurement of 0° to 360°.

The values from the load cell and the LVDT are fed to a calculation in the PLC that determines the amount of material that must be removed from the rotor and the location on the rotor from 0° to 360° where the metal must be removed. After the metal is removed, the rotor is spin-balanced again. If the rotor is determined to be in balance, it is sent to the shipping area. If it is still out of balance, the rotor is considered to be a reject and it is removed from the line.

When the material is ready to be removed to balance the rotor, a servomotor rotates the rotor to the exact position between 0° to 360°. This servomotor has an encoder to aid in the precision positioning. The encoder and servo amplifier create a closed-loop positioning system to ensure the rotor is moved to exactly the precise position when excess material is machined off to balance the rotor.

A grinding tool is used to remove material from the rotor to cause it to be in balance. The amount of time and pressure applied to the grinding tools are

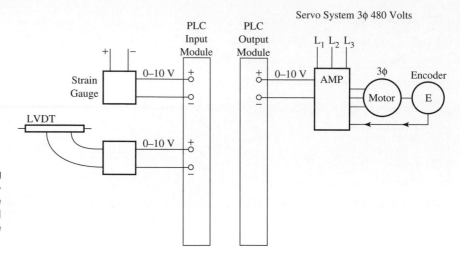

FIGURE 13-17 Wiring diagram of input sensors connected to the PLC input module and the servo system for the positioning loop.

determined from the PLC calculation so that sufficient material is removed to balance the rotor. Fig. 13-17 shows a diagram of the input sensors and the servo system used for the balancing system.

QUESTIONS

> **Note! You may need to review material in other chapters of this text to answer some parts of the questions. The case study is designed to ensure that you understand all of the technology involved in each application.**

1. Explain the operation of a load cell and the strain gauges inside them.
2. Explain the operation of an LVDT.
3. Explain why the positioning circuit for material removal for the balancing system uses a servo positioning motor.
4. Explain why the servo positioning system needs an encoder.
5. List other applications that may need this type of positioning control and sensing technology.
6. Explain the advantage of incorporating a CNC controller with a PLC controller to finish rotors and control loading and unloading robots.

SOLUTION TO JOB ASSIGNMENT

(You may need to complete Chapter 16 on Data Communications before you can complete the information about interfacing the equipment with other systems.)

Your report for the solution to the job for the beverage-filling application should identify the following electronic systems that are used: high-speed sensors to detect missing beverage bottles, dedicated electronic controller to control the no bottle/no fill function, motor drives to synchronize the line speed with the no bottle/no fill function, and synchronizing signals and shift register to synchronize the no-fill signal to the line speed. Your reports should include that this technology is well established; since it is a self-contained system with its own controller, the calibration and maintenance will be fairly easy.

Your report for the aluminum rolling mill application should identify the following electronic systems that are used: large AC and DC motor drives, on-line butt-welding technology for splicing the ends of rolls, multiple PLCs used in networks with motor drives, and large network integration to mainframe computers for downloading data for each run and for diagnostics and troubleshooting. Your report should include that this technology consists of several systems such as motor drives, PLCs, and computers that will make it rather sophisticated. The mainframe computers have diagnostic and troubleshooting capabilities so this system should be fairly easy to troubleshoot. This system should be capable of being interfaced with other factory systems since it has both PLCs and mainframe computers.

Your report for the plastic resin dryer for the plastic molding application should identify the following electronic systems: a dedicated microprocessor with a keypad and display to enter program parameters that control the process control functions for the dryer, and numerous temperature and air flow sensors for the process control system. Since the technology is a closed-loop system that is self-contained with its own microprocessor controller, it will be easy to calibrate, troubleshoot, and repair. This system should also be fairly easy to interface with other networks since it has a microprocessor.

Your report for the brake finishing and balancing application should identify the following electronic systems: CNC mills to accomplish the machining; PLCs to control the measuring and balancing system; and various load cells, torque, and position sensors to indicate whether the rotor is in balance or not. Since this system is a combination of different technologies such as CNC mills and PLCs, there might be some problems with calibration and troubleshooting. However, these problems could be worked out because of the capabilities of the controllers in these systems. These systems should be easy to integrate with other networks in the present company because of the CNC controllers and PLC controllers.

14

ROBOTS AND OTHER MOTION CONTROL SYSTEMS

OBJECTIVES

After reading this chapter, you will be able to:

1. Identify robots by their coordinate systems.
2. Identify robots by the control systems.
3. Identify the basic parts of a robot.
4. Explain the function of the teach pendant.
5. Use a diagram to explain the operation of a robot servo amplifier.

6. Connect input signals to robot input modules.
7. Connect output signals to robot output modules.
8. Explain the difference in how a robot program and a PLC program are executed.

664

You are working in a shop that has several robots used in material handling applications. It has been determined that robots could be used in several other applications in your shop such as machine loading, welding, or painting. Your assignment is to select the type of robots that you think would best accomplish one of these applications. You must also select the type of programming and determine the electronic signals that must be sent to each robot and received by each robot. You must also determine if a PLC is required to handle the signals between the robots and the machines used in the application. When you have collected all of the material, you must present your proposal to your shop quality improvement team. The proposal should include sketches of the types of robots you have selected, the list of signals that must be sent and received by each robot, and the type of programming each robot should have. Use one of the applications provided in Section 14.12.1, 14.12.2, or 14.12.3 as your project and review it to provide the details of the robot application.

(Courtesy of Motoman, Inc.)

14.1 OVERVIEW OF ROBOTS

The concept of how robots and other automated equipment should be used in industry has changed radically since the 1970s and 1980s when robots were first introduced into U.S. industry on a large scale. In the early 1980s most U.S. companies thought robots would be the answer to high labor cost and low production. Many companies had visions of replacing large numbers of workers with a robot that would be paid for in a few years and then work for free. What occurred nearly ended the "high-tech" automation revolution before it got started. The first robots that were installed did not match the application they were designed for, they broke down frequently, and they were difficult to program. This meant that three to five extra skilled workers (electronic technicians, electricians, machinists, programmers)

665

were needed to keep each robot running. Since robots broke down frequently, large amounts of downtime occurred and production schedules became unpredictable. Most robots were selected on the basis of return on investment (ROI) instead of what application they were to perform. This meant robots were being specified and purchased by the accounting department rather than by the people who understood the technology. In the middle of the 1980s nearly 100 companies were manufacturing and selling robots. At this time robots were sold like used cars and they were shipped to the industrial site with no programs installed, no end effectors, and no integration to other technology. This meant that the factory that purchased the robot needed to supply this technology or the new robot would sit in the corner.

In the late 1980s it became apparent that robots were actually an extension of present automation and that they must be integrated with existing human workers and existing automation to be successful. The number of companies manufacturing and selling robots began to combine until less than 25 of the best survived. Today the robot manufacturers work closely with technical systems houses to make sure their robots are programmed and integrated with their applications when they arrive on the factory floor. These systems are called *turnkey systems* because the factory personnel can simply turn the system on and run it after it has been installed and integrated and they do not need the high level of technology required to install, program, and integrate the robots.

Today robots are selected for the application they perform, the type of interface they provide to other systems, and the type of programming they use. If a robot can do a job, but it is difficult to change its program, it will not survive in a small company where an outside programmer is required each time a programming change is needed. If a robot can do the job, but it cannot be interfaced with existing machines or other automation, it will not be useful. Most robots today perform jobs that are unsafe for humans such as lifting heavy parts, spot welding, loading and unloading presses where moving parts can injure a person, welding or painting where fumes cause long-term health problems to humans, and jobs that require a high degree of speed or accuracy such as inserting parts in a printed circuit board, continuous welding, and painting. Fig. 14-1 shows examples of several types of modern robotic applications.

Today robots are thought of in the same terms as every other piece of equipment on the factory floor: Is it reliable, easy to integrate, easy to change its task as new products change, and easy to troubleshoot and repair to keep downtime to a minimum? If it is unreliable, it is no better than a conveyor with a worn belt that always breaks, or a hydraulic press that always leaks. The fact that many robots perform a more complex task is generally secondary when it is compared to other equipment in the process. The modern robot must be highly reliable and accurate, easy to program or change its program, and easy to interface to other machines it works with.

As an electronics technician, you will think of the robot in the same terms as any other electronically controlled machine on the factory floor. You will be expected to start it up, jog it, make simple program changes, test its input and output signals, and troubleshoot its electronic boards.

Robots can be categorized by the type of configuration of their arms (cylindrical, rectilinear, spherical, or jointed spherical) or they can be categorized by the type of drive (ac motors, dc motors, air cylinders, or hydraulics). They can also be categorized by the job they perform (palatizing, welding, spray painting, assembly, and so on) or by their size (small, medium, or large). Since some robots do not fit completely into one category, it becomes difficult to classify them, which brings a

FIGURE 14-1 Examples of robots used in welding, handling large parts, and pallatizing applications. (Courtesy of Fanuc Robotics North America, Inc. All rights reserved, and Courtesy of Motoman, Inc.)

certain amount of confusion. It is best to think of these types of classifications as general in nature and not be worried when a specific robot application does not fit exactly.

14.2 TYPES OF ROBOTS

Four basic types of robot configurations are in use today. Some robots are hybrids of one or more configurations. The configurations are called *cylindrical, rectilinear, spherical,* and *jointed spherical,* which is also called *articulated arm.* These robots are so named because their work envelope looks like the shape that describes them. For example, the work envelope of the cylindrical robot looks like a cylinder, and the work envelope of the rectilinear robot looks like a rectangle.

These robots can be powered by electric motors, electric actuators, hydraulic motors, hydraulic cylinders, or pneumatic cylinders. In some cases the electric motors rotate ball-screw mechanisms or rack and pinion systems to provide linear motion from the rotary motion of the motor. Sometimes linear stepper motors are also used to provide linear motion. In other applications motors are used with gears, belts, chains, and pulleys to provide a variety of linear and rotational motion.

Each axis a robot has is called a *degree of freedom.* For example, if a robot has three axes, it is said to have three degrees of freedom. The degrees of freedom include the axis that is provided by the wrist as well as the three axes provided by the main robot body. This means that a robot with three body axes (base, waist, and upper arm) and three wrist axes (pitch, roll, and yaw) will have six degrees of freedom.

One thing that you will notice about robots is that they have continually evolved since they were first introduced on a large scale in industry, and they continually evolve with changes that occur with technology. For example, the first electric robots were rather small and usually powered with DC motors since it was easier to control the speed and power of a DC motor. In the 1980s larger transistors that could handle over 1000 A and 500 volts were introduced, and AC variable-frequency amplifiers were designed. This allowed very large AC motors to be used on robots to replace the DC and hydraulic drives. You will see many of the changes that have taken place in the robot industry in the remainder of this chapter.

14.2.1 Cylindrical Robots

A typical *cylindrical robot* is shown in Fig. 14-2 and a diagram of this type of robot is shown in Fig. 14-3. From the diagram you can see that the motion of this robot is basically up and down at the main part of the body and circular at the base. The name *cylindrical* comes from the physical shape of the work envelope. The work envelope of this type of robot is in the shape of a cylinder.

The motion of the main arm is up and down. The robot can perform this motion by extending a cylinder that is built into the arm. In most cylindrical robots, the up-and-down motion is provided by a pneumatic cylinder, and the rotation is generally provided by a motor and gears. Any part of the robot that is moved by cylinders will generally move until it hits a stop. The location of the stop is determined by placing stop blocks or location pins. The cylinders are moved by pneumatic energy (air pressure) that is controlled by simple solenoid valves. The controller

determines the motion of the rotation by energizing the motor until the encoder determines the correct amount of movement has occurred.

Additional movement can be achieved by attaching a wrist to the end of the arm cylinder. In some robots the wrist is complex enough to provide one or more additional degrees of freedom. These are called *pitch* (up-and-down motion at the wrist), *roll* (rotational motion at the wrist), and *yaw* (side-to-side motion at the wrist). Wrists are available with one, two, or three of these motions depending on the expense of the robot and the application it is used for.

14.2.2 Rectilinear Robots

Another simple type of robot is called a *rectilinear robot* because all of its movements are along a straight line. Fig. 14-4 shows a picture of this type of robot, and Fig. 14-5 shows a diagram of this type of robot. In the diagram you can see that the motion of this robot is rather limited, but it is simple and inexpensive. The name *rectilinear* is used because the physical shape of the work envelope is a rectangle.

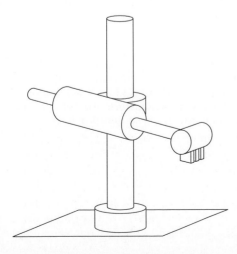

FIGURE 14-3 Motion of cylindrical robot. Notice the up-and-down motion and extend-and-retract motion of the arm are controlled by a pneumatic cylinder or ball screws. The rotation of the base of the robot is controlled by a motor and gear that make the arm swing in an arc.

FIGURE 14-4 A rectilinear robot that uses pneumatic cylinders to remove parts from a plastic injection molding machine. (Courtesy of Klockner Desma, KFD Sales and Service, Inc.)

This type of robot is used for fixed applications such as machine loading and unloading, palletizing, and other simple material handling functions. In each of these applications, the motion can be supplied with pneumatic cylinders, ball screws, or linear (stepper) motors. If pneumatic cylinders are used, they are generally allowed to move until they travel against a fixed stop. Intermediate points along the travel can be used as stop points when movable stop blocks are moved in and out of position with solenoid valves.

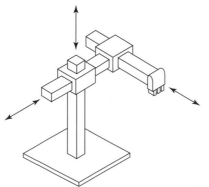

FIGURE 14-5 A rectilinear robot. Each of the axes on this type of robot can only move in linear motion. Ball screws, rack and pinion, or pneumatic cylinders are used as actuators. Arrows indicate the direction of motion each axis provides.

If a ball-screw mechanism or a linear stepper motor is used, a more precise positioning point can be achieved. In one simple application, this type of robot is used to unload clear plastic lenses for automobile headlights from a plastic press. If the press opened and the parts dropped directly on the conveyor while they were warm, they would become warped and not pass inspection. A rectilinear robot with a pneumatic cylinder for vertical motion and linear motion is used. A ball screw that is controlled by a stepper motor is used for the final travel to pick the part from the open mold. A vacuum-cup gripper is used to grasp the lenses as they are pulled from the mold and moved to a conveyor. Fig. 14-4 shows this type of robot. The total range of travel consists of eight steps starting with the robot arm waiting above the press. The steps include:

1. "Down" into press (after the mold is opened completely, the up/down cylinder is extended to lower the gripper to the right height to align with the part in the mold).

2. "Pick" forward (ball screw rotates clockwise and moves gripper toward part to "pick" it off the mold. When the gripper makes contact with the part, suction is turned on to hold the part).

3. "Pick" back (ball screw rotates counterclockwise and moves the part away from the mold so that it is clear to lift it out of the open clamps).

4. "Up" out of the press (up/down pneumatic cylinder is retracted to lift the part up and out of the press).

5. "Extend left" (extends linear motion air cylinder to move robot away from press over the conveyor).

6. "Down" on conveyor (after robot is moved over conveyor the up/down air cylinder extends and lowers the part to the conveyor where vacuum on the gripper is turned off and the part is released gently to the conveyor).

7. "Up" (after part is released, the up/down cylinder is retracted to lift the robot arm up).

8. "Retract right" (the linear motion cylinder is retracted to move the robot arm back to the starting position above the press).

Since all of these steps are sequential, the sequencer instruction in a PLC is often used as the controller. You should remember that the sequencer instruction was explained with several applications in the PLC chapter earlier in this text. If this type of robot stops, the technician can use the PLC to determine the step the sequencer is on and determine which output solenoid valves should be energized or de-energized.

14.2.3 Spherical Robots

A spherical robot is shown in the picture in Fig. 14-6. A diagram for this type of robot is shown in Fig. 14-7. From this picture and diagram you can see that the robot can move approximately 210° at its base, and its arm has up-and-down and in-and-out movements (extension and retraction). These movements allow the robot to easily reach every point in its envelope. You can see that the physical shape of the work envelope for this type of robot is spherical, which gives this classification its name.

This type of robot is now combined with multiple joints like a human (waist, upper arm, and wrist) to give it a greater range of motion so it can move its end effector into more locations.

FIGURE 14-6 Picture of spherical robot. (Courtesy of Fanuc Robotics North America, Inc. All rights reserved.)

14.2.4 Jointed Spherical (Articulated Arm) Robots

The *jointed spherical robot* is a more specialized spherical robot. It is also called an *articulated arm robot* by some manufacturers. One of the early articulated arm robots was called *selective compliance assembly robotic arm* (SCARA) robot. Fig. 14-8 shows a picture of this type of robot and Fig. 14-9 shows a diagram with an outline of the robot's work envelope. From the picture you can see that the waist axis is taller than one on a spherical robot, and the upper arm has a shoulder axis. These two changes allow the robot to reach higher in the work envelope, and lower to areas that are directly in front of itself. With the addition of a three-axis wrist, this type of robot is very useful in welding, painting, laser cutting, water jet cutting, gluing, and other applications that require a great deal of articulation (tight movements).

You should also notice that the jointed spherical robot has three large axes: The base, waist, and upper arm will provide the large or gross movements. And the three axes in the wrist (pitch, roll, and yaw) provide the fine motion. The combination of the two sets of axes allows the robot to travel large distances quickly, and then precisely drill a hole, or place a weld strategically. This has become the most widely used type of robot because it is so adaptable to a variety of applications. It is also easy to make changes in the application once the robot is in place. Fig. 14-10 shows an example of the three large axes (labeled *arm sweep, shoulder swivel,* and *elbow extension*) and the three fine axes (labeled *pitch, roll,* and *yaw*). The large axes represent the same movement that a human body makes with the waist,

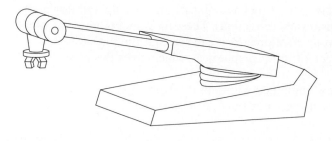

FIGURE 14-7 Diagram of spherical robot.

FIGURE 14-8 Picture of a jointed spherical robot. (Courtesy of Fanuc Robotics North America Inc. All rights reserved.)

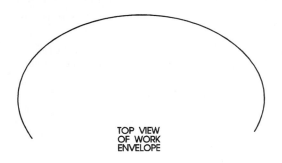

TOP VIEW
OF WORK
ENVELOPE

FIGURE 14-9 Diagram of a jointed spherical robot with a top and side views of its work envelope. (Courtesy of Fanuc Robotics North America Inc. All rights reserved. Labelling provided by author.)

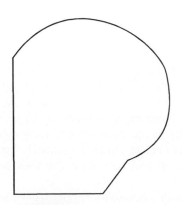

SIDE VIEW
OF WORK
ENVELOPE

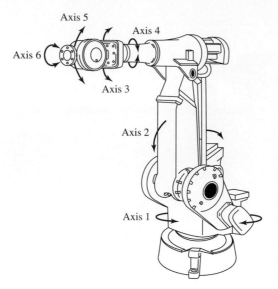

FIGURE 14-10 The large (gross) motion is controlled on the jointed spherical type robot by Axis 1 (base motion), Axis 2 (waist motion), and Axis 3 (upper arm motion). The fine motion is controlled by Axis 4, Axis 5, and Axis 6 which provide wrist roll, pitch, and yaw. (Courtesy of ABB Flexible Automation.)

upper body, shoulder, and elbow, and the three fine axes represent the same movement the human wrist makes.

The base allows the robot to rotate or swivel in approximately 300°. The waist allows the robot to bow or bend at the waist like a human. This provides the motion needed to reach points that are on the floor. The upper arm is similar to the human upper arm where the arm is attached to the shoulder. This axis allows the robot to reach upward and, combined with the waist, to reach low points in the work envelope.

The pitch, roll, and yaw movements of the robot wrist allow the robot to manipulate small tools or grippers into very small work spaces, providing movement similar to the human wrist.

14.3   TYPES OF ROBOT CONTROL

Robots can be classified by the type of control that is used to move them. The types of control that are commonly used are point to point, continuous path, *XYZ* coordinates, and positive stop. The point-to-point system records a series of locations during the programming function and the robot moves to these points when the robot is in run mode. The continuous-path control is used where the robot must follow an exact path such as in welding or spray painting. Each location in the path is recorded when the robot is in teach mode, and then the robot will follow each path during playback or run mode. The *XYZ* control uses mathematical coordinates to determine each point in the program. All points in the program are identified by their *X, Y,* and *Z* coordinates and these points can be entered manually into the program or by moving the robot to the location and recording the location through a teach pendant.

The positive-stop control is used primarily with pneumatic robots where an air cylinder is extended or retracted until it reaches a stop. This type of robot control is also called *bang-bang* because of the noise the robot makes when it hits a stop. The robot motion is determined by a sequencer that energizes each cylinder at the appropriate time. The positions for this robot program are determined by the placement of each stop.

14.3.1 Point-to-Point Robot Control

Most robots today operate on a system called point to point control. This type of robot control is achieved by moving the robot to a specific location (point) during programming and recording the coordinates of the point into memory by pressing a button on the teach pendant. During the programming phase of the project, all of the points are recorded in the order that the robot must move to them. When the robot runs the program, it moves sequentially from point to point. When the robot reaches a point, it can energize or de-energize any output signals to energize end effectors or send output signals that are used for interfacing to other equipment in the cell such as pneumatic cylinders that are used to move parts into location.

The most important part about the point-to-point program is that the robot can move one or all of its axes to move from one point to the next. The robot does not care if the travel between one point and the next is a straight path or if the motion has a slight arc in it. All that is important is that the robot stops when it reaches the next point. In most cases, the function the robot provides occurs after it reaches a point. For example, if the robot uses suction to pick up and place a part, it will turn on its suction when it reaches the point where it is ready to pick up the part, and it will turn off its suction when it reaches the point where it will drop off the part.

14.3.2 Continuous-Path Control

The continuous-path control is used when the action the robot must provide occurs at all times between points, such as spray painting, continuous cutting, continuous welding, or continuous gluing. Since this type of robot must follow a precise path when it is spray painting, each location in the path the robot takes to move from point to point is recorded during the programming phase of the project and replayed when the robot is in the run phase. Fig. 14-11 shows an example of a continuous-path robot that is used for spray painting.

The continuous-path robot also can be used in a gluing process where it must place a bead of glue around a complex contour like the clear lens of a headlight assembly for an automobile. In this application the glue is placed on the clear lens, and then the lens is placed against the remainder of the headlight assembly (the shell) where it is allowed to dry. The types of headlights commonly used on automobiles today have very complex contours to fit the design of the automobile to aid in aerodynamics. If you look at a variety of headlamps on cars today you can see that the path around the lens may be irregular with many curves. During the programming phase of the project, the robot is moved around the path with its gluing head in the correct position just above the surface. The technician who programs the robot puts the robot in teach mode, which disables the servomotors and allows the robot to be manually pulled through the path. This action would be similar to the process if the technician has a glue gun in hand and moved around the path placing glue in the correct position. Even though the servos are de-energized during this process, the encoders of each axis are still energized and they record the exact path as the robot is pulled through its program. During this teaching process the robot controller is recording every single point in this path as well as the speed at all times as the technician pulls the robot around the contours. The

FIGURE 14-11 Example of a continuous-path robot. (Courtesy of ABB Flexible Automation.)

robot controller also keeps track of when the glue gun is turned on to dispense glue and when it is turned off at any points while the robot is moving around the path. When the program is in the replay mode, the robot will follow the exact path at the exact speed the technician used while the robot was being taught.

This type of robot is fairly easy to program because no special programming language is needed to get the robot to repeat the exact path it was taught. The main drawback of this type of controller is that this type of programming requires large amounts of memory to record the exact path the robot was taught as well as the speed during each part of the program.

14.3.3 *XYZ* Control

XYZ control uses coordinates instead of points to identify each position the robot moves to during the time it runs its programs. The *XYZ* control is very similar to the point-to-point control except the entire work envelope is identified by its mathematical coordinates. Fig. 14-12 shows an example of a robot envelope defined as *XYZ* coordinates. The diagram in this figure is a top view of the robot work envelope. In this example the *X*-axis is the horizontal axis, and the *Y*-axis is the vertical axis. But it is important to remember that since this is a top view, the *Y*-axis is actually the front-to-back axis and it is sometimes called the *depth axis*. The

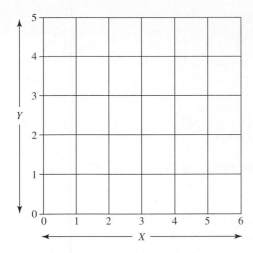

FIGURE 14-12   Example of a robot work envelope identified in *XYZ* coordinates.

axis that shows height is called the *Z*-axis. Since this robot has a pneumatic cylinder for the up/down motion, the *Z*-axis is not listed in the diagram. When you are programming you would simply indicate that the *Z* location is either up or down.

The size of each square in the grid is 1 centimeter (cm) and the middle of the grid is considered point *X*0, *Y*0. When the robot is programmed, points *X, Y,* and *Z* are indicated for each step. Since the entire work envelope is identified within the *X, Y,* and *Z* coordinates, the robot program can be written at a different location than right at the robot. The program could also be developed from a computer-aided design (CAD), computer-aided manufacturing (CAM) system. This means that a computer can help identify the points the robot must move to in a program, and by determining the *X, Y,* and *Z* coordinates of each position, they can be entered in the program in the sequence they occur.

When the robot executes the program, it will move in a straight line or arc as specified as the robot moves from point to point. This type of program is useful in ensuring the robot moves along a straight-line path to move from one point to another. This type of program is also useful when the robot must move in arc or circles because the radius or circumference can be specified mathematically. One advantage of this type of control is that the technician can determine where the robot should be at any time during the program by identifying the point in *XYZ* coordinates. This type of program is also used in computer numerical control (CNC) equipment.

Fig. 14-13 shows an example of this type of program where you can see that each step is numbered so the technician can identify them. Each step consists of a location identified in *XYZ* coordinates, and any codes to turn on (S80,1) or turn off any output signals (S81,1). The speed the robot should move between each step is identified as an *F code,* where *F* indicates a *feed rate.*

14.3.4 Positive-Stop Control

Positive-stop control is used in many robots that utilize pneumatic cylinders. This type of control energizes a solenoid valve to extend or retract the cylinder at the appropriate time. When the extend solenoid valve is energized, air is let into the pneumatic cylinder and the shaft inside the cylinder begins to extend and it moves until it hits a stop at the end of its travel. When the solenoid for retraction is energized, air is let into the other side of the pneumatic cylinder and the shaft will

N0	G01	F250		
	X156.56	Y285.46	Z450.40	
	S00			
N1	G01 F100			
	X159.50	Y369.40	Z990.38	
	S80,1	S60,20	S81,1	
N2	G01 F250			
	X974.38	Y405.90	Z436.43	
	S00			
N3	G01 F300			
	X948.37	Y059.48	Z928.08	
	S99			

FIGURE 14-13 Example of an *XYZ* program for a robot.

begin to retract until it hits the stop at the end of its travel. In some cases the stroke of the cylinder is short, and in other cases the stroke of the cylinder is long. When a short-stroke cylinder is used, the robot makes noises each time the cylinder hits a positive stop. This noise is referred to as a *bang* and this type of robot is sometimes called a *bang-bang* robot.

In some positive-stop robots, more than one stop can be used by placing the stops outside the robot. In these applications a rodless cylinder is used (see Fig. 14-14). The rodless cylinder does not have a rod that must extend from the cylinder.

FIGURE 14-14 **(a)** Examples of rod-less type pneumatic cylinders. **(b)** A cut-away diagram of a rodless cylinder. 1. Cylinder tube, 2. Head cover left, 3. Belt clamp, 4. Head cover WR, 5. Tube gasket, 6. Cushion ring, 7. Cushion seal, 8. Piston packing, 9. Scraper, 10. Belt separator, 11. Coupler, 12. Slide table, 13. Piston yoke, 14. Spring pin, 15. End cover, 16. Wear ring, 17. Piston, 18. Dust seal band, 19. Head cover right (Courtesy of SMC Pneumatics, Inc.)

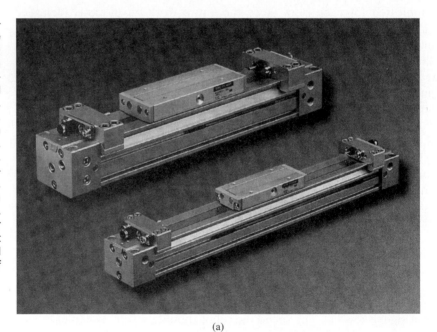

(a)

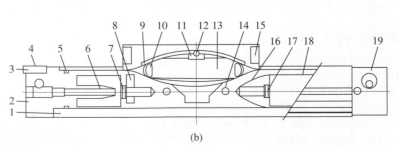

(b)

Instead it has a piston inside the cylinder tube that moves from end to end when air is used to move it. The rodless cylinder also has a coupler that is mounted on a rail on the outside of the rod. The coupler has a strong magnet that is attracted to the piston inside the cylinder. When the piston on the inside of the cylinder moves, the strong magnetic field also moves the coupler on the outside of the cylinder.

A rod-type pneumatic cylinder that is 8 ft long must have 16 ft of space to move in since the 8 ft rod must be able to be retracted into the cylinder and extended 8 ft out of the cylinder. The main advantage of the rodless cylinder is that the same 8 ft cylinder only needs 8 ft to operate in, since there is no rod to extend out of the cylinder. When the rod-type cylinder is fully extended, it may tend to droop because it is not supported. The coupler in the rodless cylinder is supported at all points along its travel. The programmable logic controller (PLC) is useful for this type of control. It can energize and de-energize the solenoids that route air to the correct part of the pneumatic cylinder in the exact order of sequence that is needed to move the robot into position.

14.4 TYPES OF ROBOT PROGRAMS

Typical point-to-point programs will contain a step number, positional data, operational codes, and a feed rate (speed) for each step. The positional data will list the number of pulses on the encoder that each motor (each axis) will display when the robot reaches the position. The pulses are considered raw positional data since they do not relate to a specific point. The robot controller records the pulse data from the encoders when the robot is programmed and it simply moves the robot motors until the encoders match for that position.

The *XYZ* program is similar to the point-to-point program except the positional data are listed as *XYZ* positions rather than the number of pulses each motor or axis will move. It is important to remember that the *XYZ* position is the place in the work envelope where the end of the robot wrist will be for that position. The important point to remember about an *XYZ* program is that you could measure the cell to determine the coordinates of any other points and enter them in the program and the robot would move to those points. This makes the *XYZ* program useful in programs that are developed from CAD drawings (CADCAM) or by off-line simulation software. CADCAM is computer-aided drafting (design) and computer-aided manufacturing.

The continuous-path program samples the position of the robot continually as it is moved through its path and records these data points in memory. The velocity of each motor is also recorded for each point that is recorded. When the robot is in the run mode, it repeats these points and motor velocities so that the robot repeats the program exactly the way that it was recorded. Anytime an input or output signal is received or sent during the time the robot is moving, it is recorded so that the robot will repeat this action when it is in the run mode. For example, the robot will remember where it is supposed to turn on a spray nozzle if it is spray painting, and repeat it at exactly the same point in the program each time the program is replayed.

The fixed-stop robot will generally use a program that allows you to put the axis moves in the sequence they are executed. Sometimes this type of program is actually a PLC program, or a sequencer of some type.

N0	G00	F5	
	T12356	W23546	U20940
	A39048	B-0904	
	S00		
N1	G01 F100		
	T13950	W30940	U29038
	A30040	B00345	
	S80,1	S60,20	S81,1
N2	G00 F7		
	T29438	W40590	U43643
	A39409	B93845	
	S00		
N3	G00 F8		
	T94837	W05948	U-29288
	A39045	B93083	
	S99		

FIGURE 14-15 Example of a point-to-point program.

You will begin to recognize the type of path control the robot is using and the type of program you would expect to see in the robot.

14.4.1 Point-to-Point and *XYZ* Programs

Point-to-point programs and *XYZ* programs will look very similar to each other. Fig. 14-15 shows an example of a typical point-to-point program. In this example you can see that each program step is numbered. Unlike a program written in BASIC, the line numbers are only for humans to use to identify the steps. The robot controller executes the steps in the order that they occur. From this example you can see that each also provides positional data either as an *XYZ* location or the pulse count of the encoder for each axis for a given location. Other information includes the speed, which will be called a feed rate or velocity, and the codes to energize or de-energize any outputs.

14.4.2 Continuous-Path Programs

Continuous-path programs are not used as often as they were during the 1980s because the newer robot programs can duplicate the robot motion during a continuous operation and use less memory. The main feature of the continuous-path program is that it recorded the robot position continually by recording the position frequently enough to cause the robot motion to look exactly like it was recorded. When you look at this type of program, you notice a large number of positional statements. Since the program records every robot position approximately 100 times a second, the program could contain thousands of positional statements. For this reason it is not necessary for humans to look at individual robot positions. If a move was not where it should be, the technician would erase the move for a section, and record that path again by moving the robot along the new path. Since the robot stored thousands of points for each program, these types of robots needed very large memory storage capability, which made them more expensive than the other types of robots. It should be noted that in the 1990s the point-to-point robot became more precise so that it can be programmed to execute many applications that previously had to be performed by a continuous-path robot. For this reason, continuous-path robots are seldom used in new applications.

Time-Driven SQO:

Step	Bit Addresses (Data Entry: ON = 1, OFF = 0)					Dwell Time (PR Value)
	A	B	C	D	E	
0	OFF	OFF	OFF	OFF	OFF	5 seconds
1	OFF	ON	OFF	ON	OFF	2 seconds
2	ON	ON	ON	ON	OFF	6 seconds
3	ON	OFF	OFF	ON	ON	10 seconds

FIGURE 14-16 Example of a positive-stop robot program.

14.4.3 Positive-Stop Programs and PLC Programs

Fig. 14-16 shows an example of a positive-stop program and Fig. 14-17 shows the statements of the robot's movement. You can see this program is from a sequencer instruction in a PLC that looks like a matrix. This program is for a three-axis robot that picks up parts from a machine and places them on a conveyor. From the example you can see that the robot has three output solenoids to control the three pneumatic cylinders. Each solenoid will extend a cylinder when it is energized and retract the cylinder when it is de-energized. The steps for the robot are listed at the side of the matrix and the outputs are listed across the top of the matrix.

FIGURE 14-17 A three-axis pneumatic robot that is used to unload parts from a machine. The parts are loaded into the machine automatically from a magazine located overhead. (Courtesy of Klockner Desma, KFD Sales and Service, Inc.)

14.4.4 BASIC Programs and C Programs

Very early robots in the 1970s and 1980s used BASIC programming language. The programs usually incorporated robot motion statements such as POSITION MOVE, PICK, GRIP, DROP, UP, DOWN, OUTPUT ON, OUTPUT OFF, and TEST INPUT to make the robot operational. Since these early robot applications were rather simple, the BASIC program was an acceptable way to operate a robot. It was also a logical program to use because thousands of engineers and technicians already understood BASIC programming. As robot applications became much more complex and faster speeds were needed, the BASIC program became ineffective.

Today the programs will look more like a C or FORTRAN program for industrial robots, and like PASCAL or FORTH programs for research robots or robots that use complex mathematical interfaces to cameras and other complex sensors. Each robot manufacturer started with the type of program such as C and added the input and output statements and other statements required to make the robot function in a modern industrial application.

14.5 COMPUTER NUMERICAL CONTROL (CNC) MACHINES

Computer numerical control (CNC) machines have been in use in industrial applications for nearly 30 years. These machines use lathes, mills, or turning centers and control them with computer control similar to robot axis control. Fig. 14-18 shows a CNC mill. The mill has a bed that travels right and left (the X-axis) and toward and away from the operator (the Y-axis). A spindle motor that holds the cutting tools is mounted above the table and it moves up and down to control the depth of cut, which is the Z-axis. These three motions allow the mill to cut parts in three dimensions (XYZ). This means that the CNC mill will operate very similarly to an XYZ robot.

Since the CNC mill has three axes, it will use three-axis amplifiers and three servomotors. The axis amplifiers and servomotors are similar to those found in robots. The CNC mill may also have variable-speed control on its spindle motor. The speed of the spindle motor can be controlled through gears as well as variable-speed control to control the cutting speed of the tools.

The CNC machine program will be similar to the XYZ program except that the CNC program usually has many more service codes than the robot. For example, a CNC mill has service codes to turn on the spindle, to turn on the coolant, to open and close the chuck, and other codes that control the program. Fig. 14-19 shows an example of a CNC program.

Some CNC equipment such as the Fanuc GE controllers incorporate motion control, velocity control, and positioning control in one part of the controller, and a programmable logic controller (PLC) in a second section of the controller. This arrangement allows data from the motion controller to be passed to the PLC to energize outputs such as clamp-close solenoids and conveyors or parts feeding systems and read input signals from limit switches, proximity switches, and photoelectric switches to indicate a part is ready to be loaded into the mill or a part is ready to be removed from the mill.

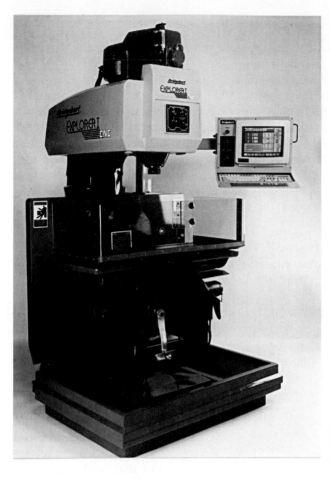

FIGURE 14-18 A CNC mill. (Courtesy of Bridgeport Machines Inc.)

N1;THIS PROGRAM IS FOR USE ON THE BRIDGEPORT MILL
N2G1X0.7525Y0.512Z0.5F80.
N3G1Z0.1F60.
N4G1Y0.2553
N5G1Z0.5F80.
N6G1X0.9836Y0.4864
N7G1Z0.1F60.
N8G1X1.035Y0.512
N10G1X1.1377
N12G1X1.1891Y0.4864
N14G1Y0.435
N16G1X0.9836Y0.3066
N18G1Y0.2553
N20G1X1.1891
N21G1Z0.5F80.
N22G1X1.3174
N23G1Z0.1F60.
N24G1X1.5229Y0.512
N25G1Z0.5F80.
N26G1X1.7026Y0.3836
N27G1Z0.1F60.
N231G1Z0.5F80.
N234G0M30
;END

FIGURE 14-19 A typical CNC program.

14.6 BASIC PARTS OF A ROBOT SYSTEM

Robots have basic parts that are common to all robots, which include the power supply; the main central processing unit (CPU); the teach pendant and operator panel; the axis control module with its amplifiers, motors, and feedback sensors; the input and output modules that are connected to the axis control module; the body parts such as base, waist, and arms; and the end effectors attached to the end of the robot arm. Fig. 14-20 shows how these basic parts relate to each other through the back plane of the racks. This section will explain the function of each of these basic parts.

14.6.1 Power Supply

The power supply for a robot is actually several different power supplies. Fig. 14-21 shows examples of the power distribution and transformers for a typical robot. The largest transformer and power supply provide voltage and current for all of the motors and voltage for all other power requirements in the robot. The main transformer for the robot is a multitapped three-phase transformer. From the diagram in the figure you can see that the transformer is tapped so it can receive incoming three-phase voltages of 575, 550, 500, 480, 465, 415/240, or 380/220 AC volts. This allows the robot to be connected to any one of these voltage sources. A second optional user-supplied transformer can also be used to provide voltage for a 120 volt AC utility receptacle, and any other 120 volt requirements. You should also notice that the three-phase voltage is first routed through a main circuit breaker

FIGURE 14-20 A diagram of how the basic parts of a robot are interfaced through the back plane of the rack that houses all of the modules. (Courtesy of Fanuc Robotics North America, Inc. All rights reserved. Labelling provided by author.)

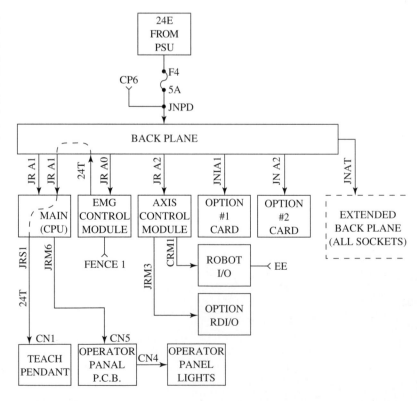

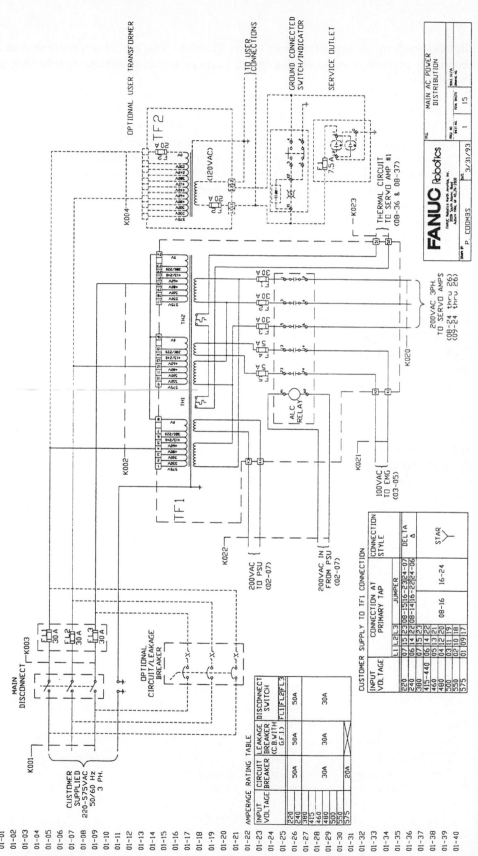

FIGURE 14-21 Example of all of the power supplies for a typical robot. (Courtesy of Fanuc Robotics North America, Inc. All rights reserved. Labelling provided by author.)

that is shown in the top left corner of the diagram. The multitap transformer is a means to provide voltage to the robot from nearly any voltage supply that is found worldwide.

The output voltages for the secondary side of the multitap transformer are sent as 230 volt AC three phase to the AC servo amplifiers, and a 230 volt AC single-phase circuit is sent to the power supply unit (PSU) module. A two-wire 110 volt AC circuit is also supplied to the emergency stop board (ENG).

Two circuits are sent to the transformer for safety and control. The first circuit is a thermal protection circuit that feeds voltage through a thermal overload element mounted in the transformer. This circuit is part of the diagnostic system to detect an overheated transformer. A second circuit is sent to the coil of a relay called the *power control relay*. This relay is shown directly below the main transformer. The signal to the coil comes from the PSU board to indicate the amount of voltage is correct and it is safe to send the secondary voltage to the robot systems. When the coil receives its signal, it closes five sets of contacts, enabling power for all parts of the robot.

The robot may also have several power supplies that are independently mounted throughout the robot cabinet, or they may be part of a single PSU module. Fig. 14-22 shows a typical PSU module that contains several different power supplies. From the diagram you can see that the PSU module provides voltages of +15 volts and −15 volts as a differential power supply that provides voltage for any analog circuits or op amp circuits, a +5 volt DC power supply that provides voltage for logic circuits, and a +24 volt DC power supply that provides voltage for a variety of internal robot boards and external signals. Additional voltages such as a 12 volt DC or 24 volt DC power supply is usually added by the user for any external signals such as solenoid valves or other interface signals to a PLC or other robots. Typically these power supplies are switching power supplies that were discussed in previous chapters. You can refer to these chapters if you need to review their internal circuitry.

FIGURE 14-22 Power supply module and the typical voltages it supplies. A voltage monitor and control board is used to monitor input and output voltages for the power supply. (Courtesy of Fanuc Robotics North America, Inc. All rights reserved. Labelling provided by author.)

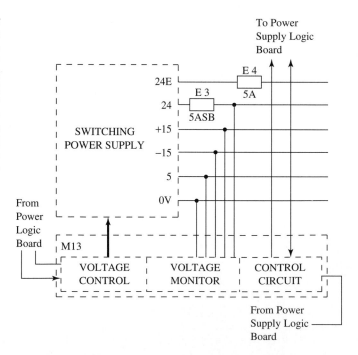

You should also notice that there are several control circuits connected to the PSU module. In the diagram in this figure you can see that a *voltage monitor and control board* is mounted directly under the power supply. The voltage monitor and control board consists of a voltage control circuit, a voltage monitor circuit, and a control circuit. The voltage control circuit receives its control signal from the PSU logic board. The PSU logic board constantly monitors the incoming power to the power supply and determines if any faults exist. When a fault occurs, this voltage control circuit will receive a fault signal. The voltage control circuit also monitors the output voltage from the PSU module to determine that all of the voltages remain within specifications. If low or high voltage is detected, a fault is generated and the control circuit part of the voltage monitor and control board opens the circuit to protect the robot from these conditions.

14.6.2 Controllers and Teach Pendants

The *controller* is the name for the cabinet that contains the microprocessor (CPU) for the robot and all of the input and output boards. The *teach pendant* is connected to the controller main CPU board to provide control of the robot for jogging and programming. Fig. 14-23 shows a typical controller cabinet and Fig. 14-24 shows a typical teach pendant. From the first figure you can see that the teach pendant is connected to the controller with a long cable. The long cable allows the technician to be close to the robot end effector when the robot is being programmed. The teach pendant allows the technician to jog the robot or to make changes in the robot program.

From the diagram of the teach pendant in Fig. 14-24, you can see that the

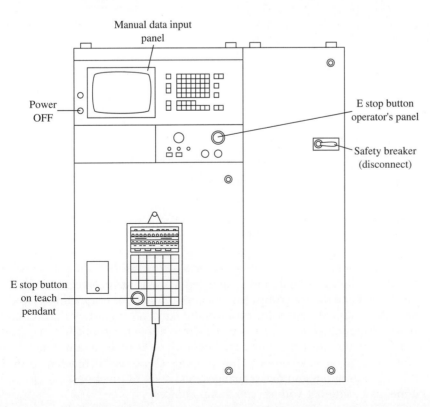

FIGURE 14-23 Picture of a typical robot controller. (Courtesy of Fanuc Robotics North America, Inc. All rights reserved. Labelling provided by author.)

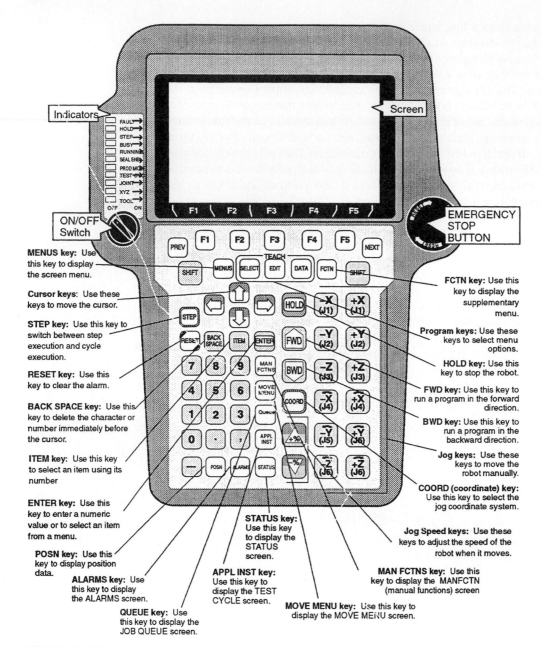

FIGURE 14-24 Picture of a typical robot teach pendant. (Courtesy of Fanuc Robotics North America, Inc. All rights reserved. Labelling provided by author.)

teach pendant has an LCD screen to display the information on the robot program, robot position, and input and output signal data the technician needs. The keys on the teach pendant are grouped according to function. For example, a set of number keys provides data entry for position data and other entries that require numbers. A set of motion control keys is provided to allow the robot to be jogged or moved during set up and programming functions. Sets of program control keys, cursor keys, and menu keys provide functions to allow the technician to move through the robot program and make changes, and a set of function keys (F1–F5) provides a means to interact with screen menu choices.

The teach pendant also has a set of 11 light-emitting diodes (LEDs) on the left side of the screen to indicate alarm conditions. The LEDs will illuminate when problems occur. An emergency stop button is also provided to allow the technician to stop the robot at any time.

Fig. 14-25 shows a detailed cut-away view of a typical robot cabinet. From this diagram of the controller you can see that the components in the cabinet are mounted according to their function. For example, three of the servo amplifiers are mounted in the top left part of the cabinet, and all of the input and output (I/O) modules are mounted in the top right part of the cabinet. The servo modules and I/O modules are rack-mounted modules so that each module can be removed and replaced quickly. All field wiring connections are made with a front-mounted wiring arm that quickly snaps away from the front of each module. This means that when a module is changed, the technician pulls the wiring arm forward so that the module can be removed. After the new module is replaced, the wiring arm is

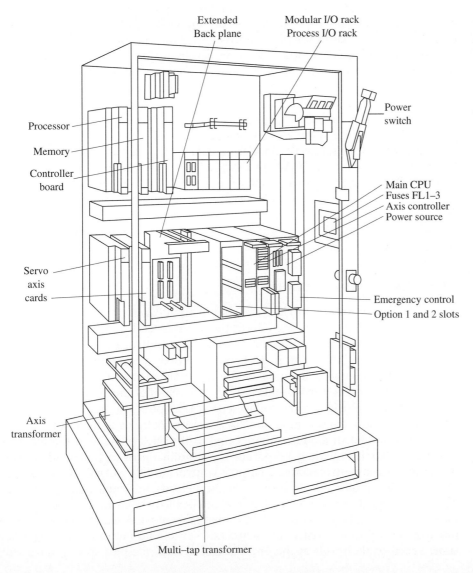

FIGURE 14-25 A cutaway view of a robot cabinet. (Courtesy of Fanuc Robotics North America, Inc. All rights reserved. Labelling provided by author.)

snapped back into place onto the front of the module. With this system the technician does not have to remove and replace any individual wires with a screwdriver.

The middle section of the cabinet contains the main central processing unit (CPU) module, the power supply unit (PSU) module, and the axis control board. These modules are all mounted in their own rack, which makes it easy to remove and replace each if one is suspected of being faulty. This rack has an extended back plane for additional option modules. The right side of the middle section of the cabinet contains additional servo amplifier modules. The exact number of servo amplifier modules will depend on the number of axes the robot has.

The bottom section of the cabinet has the main multitap transformer and the auxiliary transformer. As you know, the multitap transformer supplies stepped-down voltage for all of the servo amplifiers and the PSU. Voltage is also provided to cooling fans and auxiliary lighting for the cabinet. The three-phase AC voltage supply wires are fed into the robot through the holes in the bottom of the cabinet. These wires are connected to the bus terminals at the bottom of the robot where they are distributed throughout the remainder of the robot and transformers. A disconnect switch is provided on the top right corner of the cabinet. This switch allows all power to the robot to be shut off and locked out while the robot cabinet door is closed. The fuses for the robot panel are mounted in the middle section on the right-hand side where the technician can easily see if they are tripped. Each fuse has a trip indicator to indicate when the fuse has blown.

The controller on some robots has a front panel and a teach pendant through which the operator can enter data and program changes into the robot. In other robots, cabinets do not have a front panel, so the teach pendant is the only interface to the robot. In these systems the teach pendant is mounted in the front panel of the robot when it is not used as a remote device. This allows the teach pendant to function as the front panel. When the teach pendant is removed from the front panel, it allows the operator or technician to get close to the work envelope of the robot to get a close-up view of points where the robot must operate its end effector. For example, if a robot is used to spot-weld the frame of an automobile, it is important for the technician to be able to have the teach pendant nearby when the points are entered into the program.

14.6.3 Axis Control Module

The axis control module provides the circuits to integrate multiple axis so that the robot can move more than one axis at the same time. The axis control module is important because the velocity and torque commands of each axis amplifier and motor must be carefully controlled when the robot moves multiple axes at the same time. Fig. 14-26 shows a diagram of the axis control module with six amplifiers connected to it. A motor is connected to each amplifier and the feedback sensors for position and velocity may be connected to the amplifier or they may be connected directly to the axis control module depending on the manufacturer of the robot.

The axis control module is connected to the back plane with the main CPU. When the robot is executing a program, the coordinates of each new robot position is sent to the axis control module. The axis control module will calculate the number of pulses each motor should move and the velocity at which each motor should operate to provide smooth motion at the end of the robot arm. For example, if the base of the robot is to rotate 10°, the waist is to move 3 in., the upper arm is to move 12 in., and the wrist is to rotate 120°, all of the motors cannot run at the same speed or the result at the end effector is that one part of the move will be

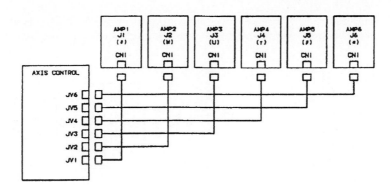

FIGURE 14-26 Axis control module with six motor amplifiers connected. (Courtesy of Fanuc Robotics North America, Inc. All rights reserved. Labelling provided by author.)

completed before the other moves are completed, which would result in jerky motion. The axis control module synchronizes all of the axes so the movement at the end of the robot arm is smooth.

It is also important to understand that some axes must move to accommodate the motion of other axes. For example, when the upper arm is to move up or down 2 in. and the wrist is to keep the same orientation, the wrist pitch and yaw motors must also be moved slightly to keep the wrist at the same orientation. The axis control module has a math coprocessor that uses complicated formulas (algorithms) to compute all of the motions and velocities that are involved in every move.

14.6.4 Amplifiers, Motors, and Feedback Sensors

The amplifier, motor, and feedback sensors can be proportional or set to operate in the on/off mode. If they operate in the on/off mode, the amplifier and motor will be energized to full speed and remain at full speed until they reach the programmed point. Then they are de-energized and stop at that position. The sensor only needs to indicate position because the speed (velocity) is not monitored since the amplifier and motor are energized to full speed. Only a few low-cost robots utilize this method of control.

14.6.4.1 DC Amplifiers ■ If the amplifier and motor operate as a proportional control, the motor will have an acceleration ramp and a deceleration ramp for each point the robot moves to. The proportional control uses both velocity sensors and positional sensors. Fig. 14-27 shows an example of this type of DC amplifier and motor diagram. From this diagram you can see that the amplifier for the DC motor is similar to the push-pull amplifiers and DC servo amplifiers shown in Chapter 11. From this diagram you can see that the AC supply voltage for the amplifier circuit is provided by an individual transformer secondary winding. Each motor on the robot will have its own transformer to supply AC voltage. You should notice that the transformer is a center-tapped transformer so the AC voltage can be rectified to full-wave DC with only two diodes. The DC voltage is provided to the transistor section of the amplifier as a bus connection. The transistor section uses two transistors as a push-pull amplifier. The control signal for the transistors is optically isolated so the amplifier transistors do not feed back excessive voltages to the axis control module. The axis control module will control the bias on the base of each transistor to provide sufficient voltage and current to allow the DC motor to accelerate or decelerate smoothly under all load conditions.

DV-M

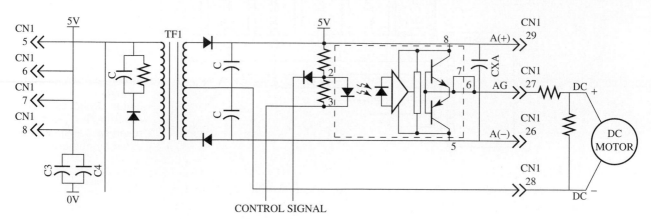

FIGURE 14-27 Electrical diagram of a DC servo amplifier. Notice that this type of push-pull amplifier uses a center-tap transformer to eliminate an additional pair of diodes in the rectifier section and a pair of transistors in the amplifier section. (Courtesy of Fanuc Robotics North America, Inc. All rights reserved. Labelling provided by author.)

You should notice that the positive terminal of the DC motor is connected to the transistors in the amplifier, and the negative terminal is connected to the center tap of the transformer. This allows the voltage to the positive lead of the motor to receive positive voltage from the top transistor to rotate clockwise, or negative voltage from the bottom transistor to rotate counterclockwise. The center-tap transistor allows this type of DC amplifier to operate with two fewer diodes and two fewer transistors, which cuts down on the number of components that can become faulty.

14.6.4.2 AC Amplifiers ■ Fig. 14-28 shows an example of a typical amplifier for an AC robot motor. The motor for this circuit is a three-phase AC motor. It has a DC generator that acts as a tachometer, which is directly connected to the shaft of the AC drive motor. An encoder is also connected to the AC motor shaft to provide positional data.

From the diagram in Fig. 14-28 you can see that the AC amplifier looks similar to the diagram for the variable-frequency drive that was explained in Chapter 12. The amplifier is powered with three-phase voltage on the left side of the diagram. The three-phase voltage wires are connected to a three-phase circuit breaker to provide overcurrent protection for the motor. It is important to note at this time that the circuit breaker provides physical protection, and the logic and fault circuits in the robot take samples of the current and voltage that the motor uses and interrupts all voltage to the motor if the current becomes too large and if the voltage is too high or too low. The value of the setpoints for the voltages and currents can be programmed through the teach pendant so that the levels can be adjusted to each application for the robot.

The voltage from the circuit breaker is sent to the three-phase bridge rectifier circuit. The bridge rectifier changes the three-phase AC voltage to DC voltage. A set of capacitors and an inductor are provided with the rectifier to filter the pulsing DC to pure DC. The DC voltage is provided to the output transistors as a bus. You should remember from the information about the AC variable-frequency drives that the diodes in the three-phase rectifier are mounted in molded plastic and are

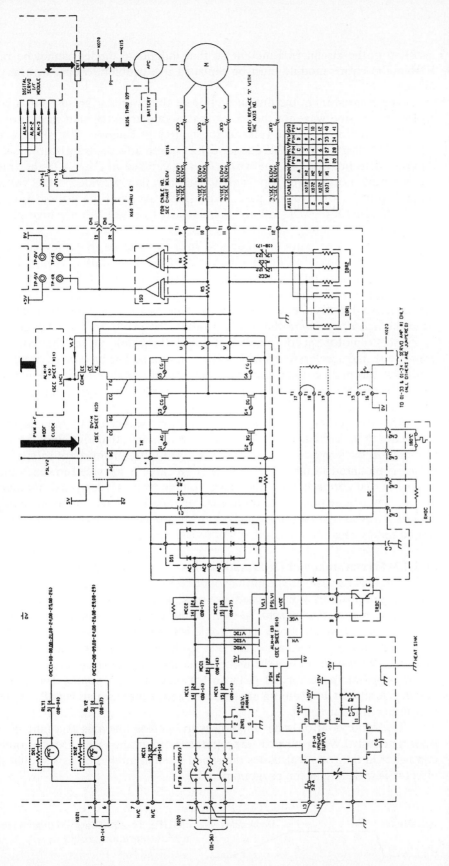

FIGURE 14-28 Electrical diagram of an AC amplifier for a three-phase AC drive motor. (Courtesy of Fanuc Robotics North America, Inc. All rights reserved. Labelling provided by author.)

693

installed in the circuit as a module so that individual diodes cannot be replaced. Rather the entire module must be replaced if a problem occurs with any of the diodes.

The transistors in the output section of the amplifier are used in a circuit to create an AC sine wave for the motor that is similar to the sine wave provided by an AC alternator. You can refer to the figures in Chapter 12 that show the AC waveforms for the variable-frequency drives to see how the transistors are switched on and off to provide the sine wave. As you can see in Fig. 14-28 the transistors are mounted in pairs. The top transistor of each pair provides + DC voltage that is used to make the positive part of the sine wave, and the bottom transistor of each pair provides − DC voltage that is used to make the negative part of the sine wave.

The transistor pairs are mounted in a module so that they can be easily removed and replaced when they become faulty. The transistors in the amplifier output section may be insulated gate bipolar transistors (IGBTs) or they may be darlington pairs. The base of each transistor is controlled by the base drive circuit that is shown directly above the transistors. The base drive circuit receives its signals from the *axis drive board* that converts all of the positional data signals from the processor to motor speeds, number of pulses (how long the motor should remain energized), and acceleration and deceleration ramps. The base drive circuit receives feedback information about motor current requirements and adjusts the base signal to each transistor to ensure that each motor receives the proper amount of torque to move its load smoothly.

The AC motor is shown in this diagram connected to the output of the transistor drive circuit. It is important to remember that the amplifier circuit is mounted in the servo axis board that is located in the main robot cabinet, and the motor is mounted in the robot axis. A large cable is provided for each motor on the robot that connects the three-phase voltage from the servo axis module to the motor. The connection at the robot uses threaded military-type connectors called *cannon plugs* so that the cable can be quickly and easily removed and replaced any time a motor must be removed.

14.6.4.3 Acceleration and Deceleration of AC and DC Servomotors in Robots ■

Since the axis control circuit in the robot controller has the ability to control the velocity and position of each axis the motors move, a velocity profile is used with each move signal that is sent to each robot motor to provide acceleration and deceleration during the move. This profile is shown in Fig. 14-29. You can see that the acceleration ramp occurs for the first 10–15% of the move and the deceleration occurs for the last 10–15% of the move. After the axis has accelerated, it will operate at the program speed for the distance between the two points. The acceleration and deceleration allow smooth movement for the robot as it moves all axes between the points in its program.

The robot controller has the power to operate one axis at a time or all axes simultaneously. When all axes are operated simultaneously, the microprocessor controller in the robot calculates the speed of each individual axis so that the end effector is traveling at the programmed speed.

Acceleration   Deceleration

FIGURE 14-29 Diagram of an acceleration and deceleration profile.

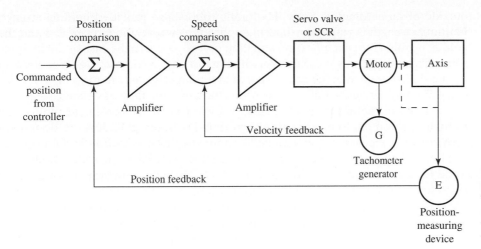

FIGURE 14-30 A diagram of the velocity feedback loop and the positional feedback loop for a robot axis controller.

14.6.5 Positional Feedback Loop and Velocity Feedback Loop

The robot controller needs to be able to control the speed (velocity) and the position of each axis. The robot axis controller uses a velocity feedback loop and a positional feedback loop (see Fig. 14-30). The command signal comes from the robot program in the form of a new position and feed rate (speed) for each axis. The controller will accelerate the motor and provide the designated control signal to ensure the axis is moving at the correct speed so that all of the robot's axes are synchronized. When the encoder indicates the robot is nearing the correct position for the move, the controller will begin the deceleration process and stop the motor at just the right position.

14.7 OTHER TYPES OF ROBOT ACTUATORS

Robots can use other types of actuators such as hydraulic and pneumatic actuators. In the previous section ac and dc servomotors were explained. In this section hydraulic and pneumatic actuators are explained and a short section that compares the advantages and disadvantages of the electric, hydraulic, and pneumatic actuators is provided.

14.7.1 Hydraulic Actuators

The hydraulic robot was popular in the 1970s and 1980s because of its ability to move very large payloads at a time when the size of electric motors for robotic servo systems was limited. For example, the hydraulic robot was used to pick up engine blocks that weighed over 800 lb. from an assembly line and place them on a pallet. These types of robots were also used extensively for spot-welding applications where the spot-welder weighed over 50 lb. The hydraulic robot could move the spot-welder quickly to nearly any position required to make a weld inside and outside of an automotive body. Hydraulic robots also provided holding strength. Holding strength is very important in applications where a load is moved and then held in position for an application and then moved again. In the case of the hydraulic robot, in its day it was the only type of actuator that could move a large load and then hold it extended in the correct position until it was ready to move again. Today electric motor driven robots can provide the same function less expensively.

Hydraulic actuators are available as cylinders to provide linear motion, and as hydraulic motors to provide rotary motion. Hydraulic cylinders are designed to provide large mechanical advantage. This means that the hydraulic fluid pressure can be applied against the large area of a cylinder, and it can provide a multiplying force of over ten times the force of the fluid. This allows hydraulic robots to lift large loads. The only drawback to this advantage is when the larger cylinders are used, they take extra time to fill with fluid, which causes the robot to be slower.

Hydraulic motors allow for the same advantage to be provided for rotary motion. The proportional hydraulic servo valve that you studied in Chapter 11 is used to provide fast response and precise positioning. The servo valve also provides positive stop for fluid flow, which allows the robot to hold its location without sagging when large loads are moved. This is one advantage that may make the hydraulic robot more suitable than an electric or pneumatic robot.

The main problem with the hydraulic robot is that the control system of servo valves and actuators is highly complex and requires special training and tools for repair. These types of systems also tend to leak more often, which causes additional problems of extended downtime and excessive maintenance costs.

14.7.2 Pneumatic Actuators

The most common type of pneumatic actuator is the pneumatic cylinder. The pneumatic cylinder operates in conjunction with a pneumatic solenoid to allow air to make a cylinder to extend or retract. The range of motion for the pneumatic cylinder is generally limited to full extension or full retraction of the cylinder, as described earlier in the bang-bang robot.

14.7.2.1 Stop Blocks for Pneumatic Robots ■ The main drawback to using pneumatic cylinders as actuators is that they do not provide a simple way to stop somewhere between the two extreme positions of fully extended and fully retracted. To correct this problem, robot designers found that *stop blocks* could be added to the pneumatic cylinder assembly to provide intermediate stops.

Fig. 14-31 shows a picture of a robot that uses stop blocks. The up/down axis for these robots uses pneumatic cylinders, which allows the cost of this type of robot to be less expensive. Generally when the pneumatic cylinder is used as the actuator, the axis has two points or stops: one at the top of the pneumatic cylinder stroke, and one at the bottom. Stop blocks that are 1 in. or 2 in. in thickness are added to the cylinder several inches above the bottom of the stroke. When the stop block is activated into position by extending another small pneumatic cylinder, the robot will stop at a point that is 2 in. above the bottom of its stroke, which is also 2 in. above the work surface. This position could be used if the pick-up point for the robot is higher than the drop-off point. If the stop block is retracted, the pneumatic cylinder is allowed to drop all the way to the bottom of its stroke.

Fig. 14-32a shows an example of the cylinder in the partially down position resting on the stop block to pick up a part from a raised work surface. Fig. 14-32b shows the suction cup resting in the fully down position on the lower surface. Notice that the stop block is retracted to enable the pneumatic cylinder to be lowered to

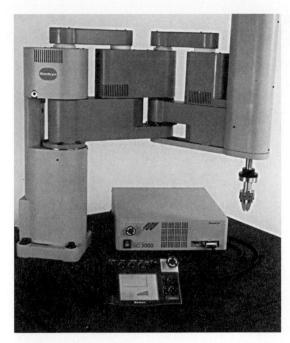

FIGURE 14-31 A robot that has a pneumatic cylinder for the up and down axis. The up and down axis uses stop blocks for multiple positions. The remainder of the axis motion is supplied by servomotors. (Courtesy of Sankyo Robotics.)

the fully down position. When the stop block is retracted, the suction cup can reach all the way to the bottom of the stroke. In this figure the stop blocks are shown at the bottom of the stroke. Some robots use stop blocks at the bottom of the pneumatic cylinder's stroke. The robot pictured in Fig. 14-31 has its stop block located at the top of its pneumatic cylinder.

14.7.3 Comparison of Robot Actuators

During the 1970s and 1980s more dc motors than ac motors were used on robots because ac variable-frequency drives had not been perfected for larger loads. The dc motors used on robots were mainly permanent magnet motors that required maintenance on their brushes and commutators, and they also were limited to smaller-size payloads. Today brushless servomotors and permanent magnet DC motors are popular on smaller robots that can attain high-speed travel. This is useful in applications where the robot must make multiple movements at high speed, such as inserting components in a printed circuit board. The table in Fig.

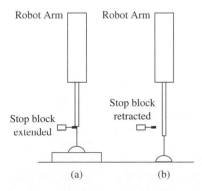

FIGURE 14-32 **(a)** Pneumatic cylinder that uses a stop block to stop suction cup 2 in. above the work surface to pick up a part. **(b)** The stop block is retracted so the pneumatic cylinder is extended all the way to the bottom of its stroke where it touches the work surface.

Actuator Type	Advantages	Disadvantages
Hydraulic	1. Able to handle large loads. 2. Able to hold large loads without sagging. 3. Safe in intrinsic environments like painting.	1. Leaks oil. 2. Requires excessive maintenance. 3. Expensive to purchase and operate.
Electric	1. Speed. 2. High degree of accuracy, resolution and repeatability. 3. Lower cost to purchase and operate. 4. Clean and quiet. 5. Modern ac and dc motors able to lift large loads.	1. May need mechanical brakes to help hold load. 2. Dc motors require brush maintenance.
Pneumatic	1. Low cost to purchase and operate. 2. Can be operated from PLC which is cheaper. 3. Clean (see exception #2). 4. Fast.	1. Not accurate, uses fixed stops for positioning. 2. Large cylinders require lubrication and the lubrication oil may leak on finished parts in the in work cell. 3. Little holding power. 4. Can't move large loads.

FIGURE 14-33   A table that compares electric, hydraulic, and pneumatic robots.

14-33 shows a comparison of the advantages and disadvantages of electric, hydraulic, and pneumatic robot actuators.

14.8   INPUT AND OUTPUT SIGNALS FOR ROBOTS

Each robot used in industry has the ability to send and receive a variety of signals. The signals can be broken into four categories: output signals to the robot arm, output signals to the work cell, input signals from the robot arm, and input signals from the robot work cell.

The signals to and from the robot work cell may include signals to and from a programmable logic controller (PLC), other robots, or from sensors and pilot devices or signals sent to output devices such as solenoids and relays. The input and output circuits for each of the robot input and output signals are routed through the cables connected to the input and output modules mounted in the robot control panel. Fig. 14-34 shows how these individual signals are routed to and from several robots that are part of a welding work cell. The welding work cell has several robots that are sequenced through a programmable controller. The first robot in the cell

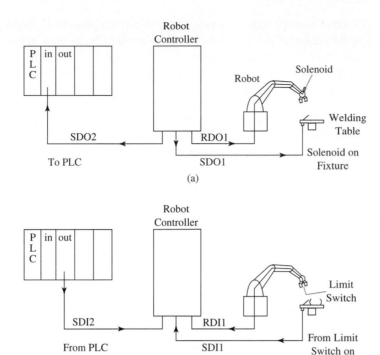

FIGURE 14-34 **(a)** Output signals sent from the robot to a solenoid on its arm, to a solenoid in the work cell, and to the PLC. **(b)** Input signals sent from a limit switch on the robot arm, from the fixture in the work cell, and from the PLC.

loads parts from a conveyor onto an indexing table so the parts will be located correctly for the second robot to begin welding on them.

From the diagram in Fig. 14-34a you can see that output 1 (RDO1) is sent from the robot through its arm to energize and de-energize the solenoid to turn on air for the vacuum gripper. Since this signal is an on/off type signal, it is referred to as a digital signal. Since it comes from the robot's arm, it is identified as a *robot digital output* (RDO). Output 2 is sent from the first robot's cabinet to the fixture on the indexing table to energize the clamp solenoid for the fixture. This signal is also an on/off signal, so it is a digital signal, but since it comes from the robot cabinet rather than the robot's arm, it will be called a *service digital output* (SDO).

At first to a new technician it does not become apparent why these two signals are identified differently since they are both outputs. But on closer examination you will see that the problem arises from where the signals are sent. Since the first signal is sent through the robot arm to the solenoid that is mounted on the end of its arm, the cable to deliver this signal must be routed through all of the moving parts of the robot's arm to enable the cable to move with the robot as it moves from point to point. The destination of the second output signal is on the indexing table that is located in the work cell. Since the indexing table does not move with the robot arm, its signal must come from a separate cable that runs from the robot cabinet where the output module is located to the service area of the work cell where the indexing table is located. This means that when one of the output signals is bad, you must first determine if its signal comes from the cable in the robot arm or from the cable that is routed to the work cell service area.

The third output signal is sent from the first robot to the PLC that acts as the cell controller where it provides sequential information to all the robots in the cell. This signal is sent through the output cable that is sent directly to the PLC through a second service cable, so it will also be called an SDO. You should be aware at this time that the signal between the robot and the PLC could be transmitted on a local area network (LAN). The operation of local area networks is explained in Chapter 16. As you can see now, the robot can have multiple service cables to send signals directly to other robots, the PLC, or parts of the work cell service area. Typically each cable is routed to its own set of input and output modules. For example, the first cable for the robot arm signals may be connected to output module 1. The cable for the work cell service area is connected to output module 2, and the cable to the PLC is connected to output module 3.

In Fig. 14-34b you can see the input signals to the first robot. Input 1 is a limit switch located in the vacuum gripper that is actuated when a part is picked up by the robot. This signal is an on/off signal and it is called a digital input signal. Since the limit switch is mounted in the gripper at the end of the arm, the wires that carry this signal must be routed through the robot arm to the robot cabinet. Since the cable is routed through the robot's arm, this signal will be called a *robot digital input* (RDI).

The second input signal comes from a limit switch mounted to the fixture on the indexing table. This limit switch indicates when the fixture is open or closed. Since it is an on/off signal it is also called a digital signal, but in this case the switch is located in the work cell service area so it is called a *service digital input* (SDI). The third input signal comes from the PLC to tell the robot that the second robot in the cell has completed its welding program and the part is ready to be picked up and moved to the next station in the cell. Since the parts-handling robot and welding robot both use the same area to perform their jobs, it is vitally important that the PLC is used to direct traffic so that only one robot can enter the work area at a time.

14.9 INPUT CIRCUITS FOR ROBOT SIGNALS

Fig. 14-35 shows examples of four types of circuits that are used in input modules to receive input signals. The diagram in Fig. 14-35a shows a circuit for a 24 volt DC source-current input. A switch is used only to indicate the side of the circuit where the input signal is coming from. This type of circuit expects the input terminal 1 to see +24 volts DC. You can see that the +24 volt DC signal will cause the LED status indicator and the LED in the optocoupler to illuminate. This signal can come from a switch in the work cell area, another robot, or from a PLC. When the LED in the optocoupler illuminates, it produces sufficient light to bias the phototransistor, which will allow current to flow through its collector–emitter circuit into the robot. The optocoupler provides isolation between the external input signal and the robot's internal circuits.

The diagram in Fig. 14-35b shows the circuit for a sink-current input. In this circuit, the input signal to terminal 1 must be −24 volts DC. You should notice that the LED used as the status indicator and the LED in the optocoupler are connected in reverse direction of the ones in the source-current circuit. The optocoupler also provides isolation in this circuit. It is important to understand that when you must use the output signal from another robot as the input signal of your robot, you may not be able to change the type of DC signal from current source to current sink. You will have to select the proper input module for your robot so

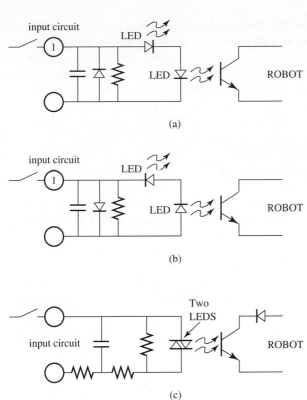

(a)

(b)

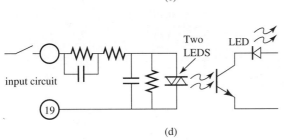

(c)

(d)

FIGURE 14-35 **(a)** Current-source circuit for 24 volt DC input to a robot. **(b)** Current-sink circuit for 24 volt DC input to a robot. **(c)** Input circuit for 24 volt DC current-source or current-sink signal. **(d)** Input circuit for 120 volt AC signal for a robot. (Courtesy of Fanuc Robotics North America, Inc. All rights reserved. Labelling provided by author.)

that the signal will interface properly. This problem also occurs in some smaller PLCs that do not have a choice of output modules, and you must live with the signal it sends the robot, selecting the proper module (current sink or current source) so that the signal will interface properly.

Fig. 14-35c shows a diagram of an input circuit that can receive either a current-source or a current-sink signal. This type of input module would be used if you have a variety of both current-source and current-sink signals that must be connected to the same robot and you do not want to purchase both a current-source input module and a current-sink input module. This circuit uses an optocoupler that has two LEDs connected back to back in parallel. This arrangement allows one of the LEDs to illuminate if the signal is current sink, and the other will illuminate if the signal is current source.

The diagram in Fig. 14-35d shows an input circuit for 120 volt AC input signals. This circuit uses several resistors to drop the voltage as it enters the circuit, and it also uses an optocoupler that has the same arrangement of its LEDs as the previous circuit. Since the LEDs are connected in inverse parallel to each other, they will illuminate anytime an AC signal is sent through it. The optocoupler provides circuit isolation between the AC input and the robot's internal circuitry. The input voltage of this circuit is 120 volts AC, which is generally used in all input (pilot) devices such as limit switches, push-button switches, and other control switches. Some factories prefer the 120 volt signal because this higher-voltage signal will work even when the contacts of the switches begin to become dirty or pitted.

14.10 OUTPUT CIRCUITS FOR ROBOT SIGNALS

Fig. 14-36 shows two typical circuits found in robots for output signals. Fig. 14-36a shows the circuit for a current-sink circuit. You should notice in this diagram that a 24 volt DC power source is needed for the circuit. The electronic circuit in the robot basically switches this power source on or off for the output device. The + terminal of the power supply is connected to terminal 2, and the − terminal of the power supply is connected to the common terminal identified as terminal 20. The load (such as a relay coil) should be connected between the center terminal and the common terminal.

The circuit consists of an optoisolation circuit that uses an LED to provide the light to activate the circuit. When the robot sends an output signal from its bus,

FIGURE 14-36 (a) A sink-current output module. **(b)** A source-current output module. **(c)** An output module for 120 volts AC. (Courtesy of Fanuc Robotics North America, Inc. All rights reserved. Labelling provided by author.)

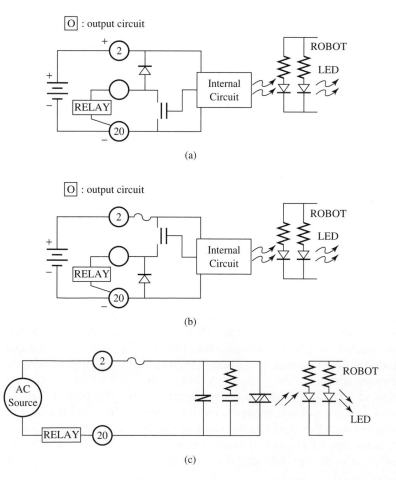

it will flow through the LED, which will cause it to illuminate and the light will activate the remainder of the circuit. When the circuit is activated, a solid-state device will allow current to flow to the output terminal of the circuit and through the load. A second LED is provided on the robot side of the circuit to illuminate as a *status indicator* to show that the signal is energized.

Fig. 14-36b shows the diagram of a current-source output circuit for a robot. The main difference between the current-source and the current-sink circuits is the way the current is switched to allow the polarity to be correct with the output device. If the device that receives the output signal is a relay coil, the polarity of the signal will not matter. But if the output signal is sent to another robot that expects a current-source input signal, you must use the current-source output module to ensure that the signal will interface correctly. It is important to understand that the wide variety of solid-state devices that is interfaced with robots and PLCs will require that the signal types be matched correctly, or the signals will not be transmitted correctly.

Fig. 14-36c shows the diagram of an output circuit for a robot that can send 120 volt AC signals. From this diagram you can see that the circuit uses a triac in the output section to accommodate the AC signal. The power source for this circuit is 120 volts AC and the output device (relay coil) is connected in series with terminal 4 and the voltage source. When the robot sends an output signal to the LED, it will illuminate and cause the triac to go into conduction, allowing AC current to flow to the output device. You can see that a variety of input and output modules is available for robots to ensure that they can have their signals interfaced to virtually any other electronic device or circuit in any industrial application.

14.11 INPUT AND OUTPUT MODULES

One problem that occurs in robotic applications is that an input circuit or output circuit will have a fault and need to be replaced while the robot is in production. Older robots used IC chips on an input/output (I/O) board for the interface for all of the I/O signals. When an input went bad, the entire I/O board needed to be replaced. This was complex and expensive because one board had all of the inputs and outputs for the entire robot. It was also difficult to troubleshoot some of the ICs because they did not have indicators to tell which one was faulty.

Newer robots have integrated I/O modules that are similar to those used in PLCs. The advantage of the I/O modules is that they have a status indicator for each circuit to indicate when they are operating correctly, and they usually limit the number of circuits to four or eight per module. The modules are mounted in racks like PLC modules so that the microprocessor in the robot can access all of the signals through the back plane of the rack. Since the I/O circuits are mounted on a module, the field wiring can be mounted to a field wiring arm that plugs into the module. The field wiring arm allows the technician to disconnect all wiring to each module by unplugging the field wiring arm on the faulty module and plugging it into the new module when it is placed in the I/O rack. This means that a technician can change a module in less than 5 minutes on a modern robot, which limits the problem of downtime. It is also easy to stock one or more of each module that the robot uses, since they are much less expensive than the earlier I/O boards. Fig. 14-37 shows the location of the I/O rack in a robot. The I/O rack is located near

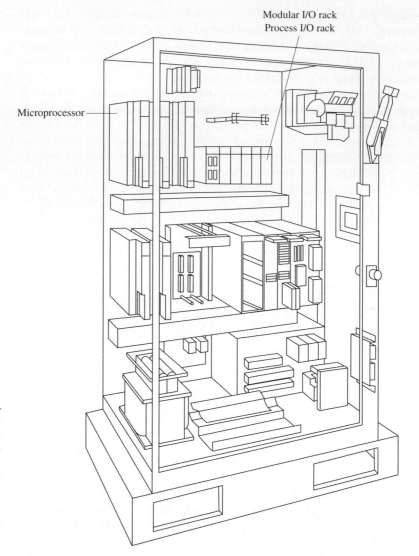

Modular I/O rack
Process I/O rack

Microprocessor

FIGURE 14-37 Location of I/O modules in the robot cabinet. The I/O modules are mounted near the processor for the robot. (Courtesy of Fanuc Robotics North America, Inc. All rights reserved. Labelling provided by author.)

the top of the cabinet just to the right of the servo amplifiers. This diagram is similar to the one shown in Fig. 14-25 except the I/O rack and the processor for the robot are the only parts of the diagram that are identified.

14.12 ROBOTS USED IN TYPICAL WORK CELLS AND MANUFACTURING CELLS

In modern factories, multiple robots can be connected together to perform a common function such as welding, painting, material handling, and machine loading or assembly. These functions can be modified slightly and a change of program in the robot allows the flexible manufacturing cell to change its job slightly. The robots are connected through a network or to PLCs that are used to pass information between them. The information includes data to indicate when one robot is in an exclusive space that is shared with another robot. It is important that both robots do not enter the same space at the same time, so the robots will lock out other robots from traveling in the space if it is already there. The next sections will present several typical applications where multiple robots work together to complete large jobs.

14.12.1 Automotive Welding Line

Robots have been used in spot-welding applications for many years. The major advantage of using the robot in this type of application is that the robot can move the 150 lb. spot-welders with relative ease. This job is very strenuous for humans and the fumes from the welding are a problem. Fig. 14-38 shows a typical spot-welding cell where eight robots are used to spot-weld automotive bodies as they pass by the robots on a continuous production line.

    The robots in this work cell have extended reach so that they can reach both inside and outside the body to make the welds. Since the automotive bodies are on a continuous production line that does not slow or stop while the robots are welding, the robots must have software that helps them track the body as it passes through the work cell.

    When the automotive bodies enter the work cell, the first robot on the left side and the first robot on the right side begin to track the frame and body and locate their first welds. Since the robot begins to track the frame and body as they move along, the robot will keep pace with the body and continue to apply welds to the exact location. When the body has moved far enough into the work cell, the second robot on the left side and the second robot on the right side begin to track

FIGURE 14-38 Application of a spot-welding work cell. (Courtesy of Motoman Inc.)

the body and finish making welds. After the body has traveled passed the first two sets of robots, the remainder of the robots continue to make welds until all of the welds have been made. The robot program has been timed so that it can easily complete all of its welds while a robot is in reach of the body.

In some welding cells each robot is programmed to make all of the welds for the robot on either side of it. This is necessary in the event that one of the robots has a failure. The line can be slowed down slightly while one robot completes welds for the robot that is next to it on that side. Most robots of this type have sufficient memory to store several different programs including programs for different types of body shapes. This allows the robot to use sensors to identify which body is arriving at the work cell and change to the program that matches the body. This function permits the factory to make more than one type of body style without major problems during the changeover.

14.12.2 Loading and Unloading Bonding Machines

Robots are used in machine loading and unloading applications where it may be dangerous for personnel to work, or in places where the parts that are loaded or unloaded are very heavy or bulky. Fig. 14-39 shows an example of a small robot work cell where two robots work together to bond two fiberglass automotive hood sections together. Several automobiles and vans now use fiberglass hoods instead of metal ones. One of the robots takes care of the material handling and press loading while the other robot applies glue to the sections prior to when they are bonded together. After the first robot applies glue, the second robot places the two hood sections together in a large press where 30,000 volts of RF energy are used to cure the glue and bond the two pieces of fiberglass together.

FIGURE 14-39 A robot work cell that includes a material handling robot and a gluing robot. The material handling robot loads parts into a bonding press.

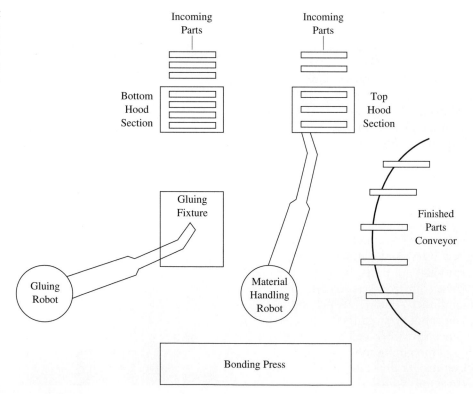

The diagram in this figure shows the location of the material handling robot and the bonding press and the gluing robot. The diagram shows the work cell from above, so it looks like the floor plan of the cell. When parts enter this work cell at the top of the diagram, the top half of the hood arrives at the top right side of the cell and the bottom half arrives at the top left side. The material handling robot has a two-sided vacuum gripper that picks up the bottom hood section with the bottom side of the gripper and picks up the top hood section with the top side of the vacuum gripper. When the cell is first started up, the material handling robot starts the cycle by picking up the bottom hood section and dropping it off at the gluing fixture. After the material handling robot drops off the bottom hood section, it would normally move to the press to unload cured and bonded hoods. Since this is the first cycle, it moves to pick up the top hood section and it will wait for the gluing process to be completed. In the second cycle and all other cycles, the material handling robot will have plenty of time to remove finished parts from the press while the gluing robot is applying the glue because the bonding cycle is rather lengthy.

When the gluing robot finishes its cycle, it signals the first robot to pick up the bottom hood section. When the material handling robot picks up the bottom hood section, it moves to the bonding press that is open and waiting for the parts to be placed in its fixture. The material handling robot places the top hood section into the top half of the bonding press fixture and signals to the press to activate its vacuum gripper to hold the top hood section in place. When the press vacuum gripper is activated, the bonding press signals the material handling robot that it is safe to release the top hood section, and the robot moves downward to place the bottom hood section into the bottom half of the bonding press fixture. Again the material handling robot signals the bonding press that the part is in position and the press should activate its vacuum gripper. When the press receives this signal, it activates its vacuum gripper and the robot releases the bottom hood section and moves out of the press. When the material handling robot is clear of the press, it signals the press to close and start its cycle. When the press closes completely, it applies pressure to the perimeter of the two hood sections and also provides 30,000 volts of 30 MHz RF signal. The high-frequency energy provides sufficient heat to bond the fiberglass hood sections and cure the glue so that the two hood sections become fused into one piece. The cycle time for the bonding process is approximately 30 seconds.

During the second robotic cycle and all succeeding robotic cycles, the material handling robot will return to pick up a bottom hood section while the press is in its bonding cycle. The material handling robot will place the hood in the gluing fixture and while it is waiting for the gluing robot to complete the gluing process, the material handling robot moves to the bonding press and waits for it to open. When the bonding press opens, the material handling robot removes the first completed part and places it on the finished part conveyor. After the material handling robot places the finished part on the conveyor, it goes back to pick up the bottom hood section from the gluing fixture and then it moves to pick up a top hood section. When it has both parts, it moves to the press and loads it.

When the press is loaded, the material handling robot moves to pick up another bottom hood section and places it in the gluing fixture. The material handling robot stays in this sequence for the remainder of its cycles.

Since this work cell requires complex timing between the material handling robot and the gluing robot, and the material handling robot and the bonding press, a cell controller must be used to control the timing of all the signals between these systems.

FIGURE 14-40 Typical spray-painting application. Several sets of robots are used together to provide a painting cell. (Courtesy of Fanuc Robotics North America, Inc. All rights reserved. Labelling provided by author.)

14.12.3 Robotic Painting Cells

Robots have been used in painting cells for a number of years. They can be set up to apply paint on automotive parts or bodies as they pass by on a continuous production line. The robots in this application must have a tracking software option so that they can locate the part that they are painting and follow it precisely as the part moves continually along the production line. Fig. 14-40 shows a typical robotic painting cell that uses four robots to paint automobiles as they pass through the cell on a continuous production line.

When an automobile body moves into the cell area, the first robot on each side of the painting line detects the body and begins tracking it as it moves through the cell. When the body reaches the correct location, the first robots on each side begin to apply paint. When the body moves far enough into the cell, the second robots on each side begin to track the body and begin to apply paint when the body reaches the correct location. When the body moves far enough through the cell, it will move out of reach for the first set of robots, and they will stop tracking and painting and return to their starting location while they wait for the next body.

The newest painting systems allow the robot to change paint colors quickly so that every other automotive body can be a different color. This technology also allows sensors on the body to indicate to the robots what the color should be.

14.13 REVIEW OF THIS CHAPTER

As a technician you will be expected to work on robots and their circuits as part of your job in a modern industrial facility. You may need to review the first part of this chapter to identify the type of robot that you are working with, or you may need to identify the type of actuators such as electric motors, hydraulics, or pneumatics that the robot uses. You may also need to review the section on ac and dc servo systems and their amplifiers, which are similar to the ac and dc motor amplifiers you learned about in Chapter 11. After you have been on the job for a while, you will begin to notice that much of the sensor technology and motor drive technology used in other applications of industrial electronics is also used with robotic applications. This will make you feel more comfortable when you have to work on different types of systems. At first you may be intimidated by the different types of robot program languages, but you will quickly find that all you need to be able to do as an electronics technician is to start up the robot, jog it around, and make minor changes to the program. As you gain experience, you will be requested to help design interface systems so robots can be integrated with other traditional applications in the factory.

SOLUTION TO JOB ASSIGNMENT

Your solution to the job should include a reference to one of the applications in Section 14.12. You can use information from Section 14.2 to select the type of robot that you will need to accomplish this job. You should include the advantages the type of robot you selected provides over other types of robots that you did not select. From the example in Section 14.9 you should have an idea of the input and output signals available in the robot. This section also provides an example of the interfacing function the PLC provides between the robot and the work cell, or between multiple robots. If your application includes more than one robot, a PLC should be included to help sequence the interface between the robots. The reason you may choose to use the PLC to interface two or more robots in a work cell is that the PLC program is a continuous-scan program, and it can receive or send a signal at any time. In comparison, the robot program is a step program, and the robot is only ready to receive or send a signal when it is on the correct step. This means that if two robots try to interface, it is possible that one will have to wait on the other to send or receive a signal. If the PLC is used, it can receive or send a signal at anytime to get around synchronizing problems between robots. The PLC is also used as an interface because it can be connected to a wide variety of color screens to provide operator information.

Your solution should include signals for your applications between the various parts of the work cell and to other robots if they are used. For example, typical signals between the cell and the robot and between the robot control cabinet and parts on the robot would include: part in place, open fixture, close fixture, turn on gripper, start weld, stop weld, and so on. Signals between two robots may include: robot in exclusive work space, robot out of exclusive work space, request to start job cycle, and job cycle complete.

1. Identify four types of robots by their coordinate systems.
2. Explain an advantage and a disadvantage of a continuous-path robot.
3. Explain the advantage of using *XYZ* control for a robot program.
4. Why are computer numerical control (CNC) systems discussed with robot systems?
5. Discuss the advantages and disadvantages between electric, hydraulic, and pneumatic robots.

T R U E   O R   F A L S E

1. _____ Stop blocks are used with pneumatic cylinders to provide an intermediate stop when a cylinder is extended.
2. _____ Robots need a special electronic circuit called an input module to accept a current-source signal, a current-sink signal, or a 120 volt AC signal.
3. _____ The input signal to the robot is isolated by stop blocks so that they do not damage the robot bus.
4. _____ Robot output signals provide isolation between the robot and the output signal user with optoelectronic devices.
5. _____ Input and output circuits in modern robots are mounted in modules to make them easier to remove and replace when they are suspected of being faulty.

M U L T I P L E   C H O I C E

1. Robots may use _____ as actuators.
   a. dc motors
   b. ac motors
   c. hydraulic actuators
   d. pneumatic actuators
   e. all the above
   f. only a and b
2. A teach pendant allows a technician to _____.
   a. jog the robot.
   b. make program changes.
   c. determine if dc motor brushes are bad.
   d. all the above.
   e. only a and b.
3. Stop blocks are used with a pneumatic arm on a robot to _____.
   a. protect output devices from transient voltages.
   b. provide intermediate stops for a pneumatic cylinder.
   c. provide an emergency stop function for pneumatic robots.
4. The PLC provides an ideal interface between multiple robots in a work cell because _____.
   a. the PLC program is a continuous-scan program and the robot program is sequential.
   b. the PLC has input and output modules and most robots do not.
   c. the PLC can handle both ac and dc signals and most robots cannot.
   d. all the above.

5. The robot cabinet provides a location to mount _____.
   a. amplifiers for the actuators.
   b. input and output modules for input and output signals.
   c. power supplies for amplifiers, motors, and I/O signals.
   d. all the above.

PROBLEMS

1. Sketch a typical DC amplifier for a dc robot motor and explain how all of the parts of the amplifier operate.
2. Sketch a typical AC amplifier for an ac robot motor and explain how all of the parts of the amplifier operate.
3. Use the diagram of the robot cabinet in Fig. 14-18 and list all the main parts included in the cabinet and explain their functions.
4. Use the robot in Fig. 14-9 and identify its basic parts and explain the function of each.
5. Sketch the following circuits for output signals from robots: current source, current sink, and 120 volt AC signal.
6. Sketch the following circuits for input signals to the robot and explain the current flow: a current-source circuit, a current-sink circuit, and a circuit for a 120 volt AC signal.
7. Use the diagram in Fig. 14-30 and explain how the position feedback loop is used with the velocity feedback loop to move the robot axis at the correct speed to the correct location.

REFERENCES

1. Malcolm, Douglas R., *Robots: An Introduction,* 2nd ed. Albany, NY: Delmar Publishing Inc., 1988.
2. Rehg, James, *Introduction to Robotics.* Englewood Cliffs, NJ: Prentice Hall Inc., 1985.

15

MOTOR CONTROL DEVICES AND CIRCUITS

After reading this chapter, you will be able to:

1. Identify pilot devices in a control circuit and their symbols in electrical diagrams.
2. Identify the load circuit and control circuit.
3. Explain the operation of a push-button switch, limit switch, flow switch, level switch, and pressure switch as they are used as pilot devices.
4. Identify a two-wire control circuit.

5. Identify a three-wire control circuit.
6. Explain the operation of a drum switch as it is used to reverse the operation of an AC or DC motor.
7. Explain the operation of a single-element and a dual-element fuse.
8. Identify the proper enclosure for specific industrial applications.

You are assigned to make changes to the control circuit of a main conveyor system that removes finished parts from a small assembly line. The present assembly line has one small conveyor that feeds parts to a main conveyor, and the main conveyor removes the parts from the assembly area. A second assembly line is being added and its conveyor will also feed parts to the main conveyor. You will need to add controls to make the two lines sequence so that they will not overload the main conveyor. The present control circuit is a two-wire control circuit with a selector switch that is turned on when the small conveyor needs to run (the main conveyor runs all of the time).

Your job assignment is to convert the control circuit for the small conveyor into a three-wire control circuit that has two separate sets of start and stop push buttons, and add a three-wire control circuit for the new conveyor that also has a set of start and stop push buttons at each end. The control circuits for the two small conveyors should also be sequenced so that the original conveyor is always started before the new conveyor can be started. You will also need to select the correct type of fuse to allow the 15 A motor on the new conveyor to start safely. You also need to select an enclosure for the motor starter that is rated for water-tight, dust-tight conditions for indoor applications. You will need to provide the electrical diagram of your proposed changes and a parts list for the fuses and enclosure.

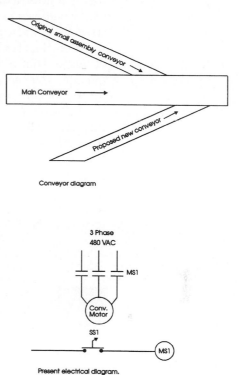

Diagram of two assembly conveyors and the main conveyors, and electrical diagram for job assignment 15.

15.1 INTRODUCTION

In this chapter we explain basic motor control circuits commonly in use on the factory floor. It is important to understand that many electronic circuits are used to supplement or replace standard motor control circuits that have been in operation for the last 30 years. These motor control circuits have been named and are easily identified so that when you hear about them on the job or when you see them on the job you will easily recognize them. You must also understand that in the past few years, companies have changed their concept of an electronics technician, and they now expect to hire each person to troubleshoot and repair both the motor control systems as well as the electronic systems. Traditionally troubleshooting the motor control systems has been handled by electricians, and the electronics technician would work on the electronic circuits. In the past ten years, companies have streamlined their personnel so that they will either employ an electrician that can work on the electronics or they will hire an electronics technician that can work on motor control circuits. For this reason it is vitally important that you learn as much as you can about standard motor control circuits, so that you can compete for the best-paying jobs.

The circuits in this chapter are identified so that you will be able to review them when you are on the job. This will be extremely useful when you find machinery that incorporates four or five of these basic circuits with additional complex circuits. In these circuits you will be able to gain knowledge that you can transfer to more difficult circuits for troubleshooting and repair. These circuits will include pilot devices and motor starters as well as different types of motors. If you do not understand the function or operation of relays and motor starters, you may need to review the chapters where they were introduced.

15.2 PILOT DEVICES

Pilot devices are a group of components that include push-button switches, limit switches, and other switches commonly found in motor control circuits. These switches are called pilot devices because they are rated for control circuit voltage and current. This means that these switches will normally have 115 volts AC and less than 1 A current flowing through their contacts. The control circuits that use pilot duty switches are used to energize or de-energize the coil of a relay, motor starter, solenoid, or an indicator lamp. All of the loads in these circuits use low current. Again you must understand that the L1-N voltage in your part of the country may be any value of voltage between 110–125 V AC. In this chapter all of these voltages will be referred to as 115 V AC.

Fig. 15-1 shows a set of multiple push-button switches and selector switches that are pilot devices used to control a large industrial machine. The push buttons are the most commonly used switches to start and stop a machine, and selector switches are used to select the operation of the machine for such functions as manual or automatic operation.

Fig. 15-2 shows an exploded view of a typical push-button switch installation. From this diagram you can see that the switch is actually an assembly of a number of parts including the push-button actuator that the operator presses on to activate the switch, a number of seals to keep moisture and other contaminants out of the switch assembly, and the contact block and contacts where the wires are actually connected. When the push-button activator is depressed, the contacts in the bottom of the switch will transition. If the switch has normally open contacts, they will go closed, and if the switch has normally closed contacts, they will go open. Some switches will have one set of contacts, while other switches have multiple sets of contacts. The important point to recognize is that all wires are connected to the contact block and the other parts of the switch can be changed without removing

FIGURE 15-1 A typical set of push-button switches and selector switches used to operate a machine. (Courtesy of Rockwell Automation's Allen Bradley Business.)

any wiring. This allows switches to be changed in the field with a minimum of down-time for rewiring.

15.2.1 Limit Switches and Other Pilot Switches

A variety of pilot switches is available for use in control circuits. One of the most popular is called a limit switch (see Fig. 15-3). These switches are mounted physically in the machine so that motion of the machine or the parts around the switch will cause the switch to activate. Fig. 15-3a shows a yoke limit switch. The yoke has a roller on each section of its activator arm. This arrangement is used where machine motion is in two directions. When the machine moves in one direction, it strikes the right-hand roller and causes it to switch to the right. This action causes the right roller to be pressed downward, and the left roller snaps to the up position where it is in position to detect machine motion when it moves back to the left. The yoke arm on switch will continue to detect motion to the right, and then to the left. This type of switch is useful for surface grinders where the part being

FIGURE 15-2 An exploded view of a typical switch. Notice that the switch activation push button is shown above the panel where it is accessible to the operator, and the remainder of the switch, including rubber seals, and the switch contacts are shown below the panel. (Courtesy of Rockwell Automation's Allen Bradley Business.)

FIGURE 15-3 **(a)** Fork lever roller yoke limit switch. **(b)** Roller arm limit switch. **(c)** Top roller limit switch. **(d)** Wobble lever actuated cat whisker limit switch. **(e)** Side roller limit switch. (Courtesy of Honeywell Micro Switch Division.)

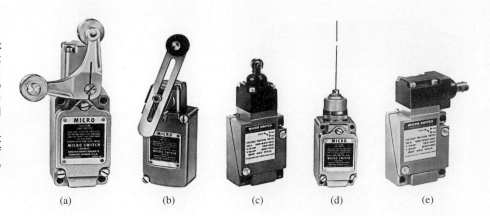

(a)   (b)   (c)   (d)   (e)

finished is moved back and forth under the grinding wheel. Each time the part moves to the end of its stroke, the switch is activated to the other direction.

Fig. 15-3b shows a limit switch with a single roller arm. This type of switch detects motion in only one direction. When the machine moves past the roller arm, it will move the arm to the right so that the switch is activated. Fig. 15-3c shows a top roller limit switch. This type of switch has a small roller mounted on a plunger arm. When the machine moves past the roller, it causes the plunger to depress to activate the switch. When the machine moves past the switch, a spring causes the plunger to move back upward into place to be reactivated. This type of switch can detect motion in either direction. A variation of this type of switch uses a blunt end on the plunger, which means the switch can only detect motion that is directed down on the plunger. This type of switch is generally mounted at the very end of machine travel, and when the machine reaches the end of its stroke, it depresses the plunger.

Fig. 15-3d shows a wobble lever actuated limit switch that is sometimes called a *cat whisker* limit switch. This type of limit switch can detect motion in any direction. This means that any movement of the cat whisker can cause the switch to activate. This type of switch is also much more sensitive than other types of limit switches. Fig. 15-3e shows a side roller limit switch. This type of limit switch is mounted in such a way as to detect end of travel. When the machine reaches the end of its travel, it will depress the plunger and cause the switch to activate.

15.2.2 Symbols for Limit Switches and Other Motor Control Devices

The limit switch can be connected as a normally open (NO) or a normally closed (NC) switch. It can also be used on a machine in such a way that the machine travel will hold the switch in the NO or NC position before the machine starts its travel. Fig. 15-4 shows the four ways you will find the electrical symbol for a limit switch. The top left switch symbol is for an *NO limit switch*. In this symbol notice that the switch contact arm is shown below the terminal on the right side of the switch, and that the switch is shown as an NO switch. The *NO held closed limit switch* is shown in the bottom left diagram. You should notice this is similar to the NO switch in that the switch contact arm is shown below the switch's output terminal on the right side. The major difference with this switch is that it is shown with NC contacts because the machine motion will keep this switch in its NC position.

It is important to remember when purchasing limit switches that the switch

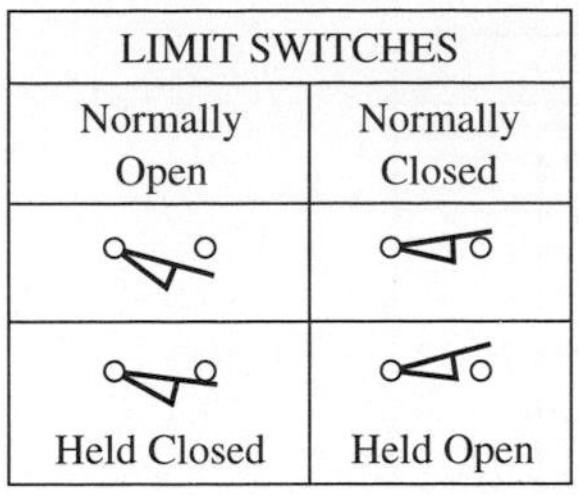

LIMIT SWITCHES	
Normally Open	Normally Closed
⊶°	⊶⊸
⊶⊸	⊶⊸
Held Closed	Held Open

FIGURE 15-4 The electrical symbol for NO limit switch, NO held closed limit switch, NC limit switch, and NC held open limit switch.

is only available as an NO or as an NC switch. The held open and held closed conditions occur when the switch is mounted into place on machinery and the location of the switch causes the machine to hold the switch in the activated position.

The switch in the top right section of this figure is an *NC limit switch*. This symbol is different in that the switch contact is shown on top of the switch's output terminal. When the machine motion activates the switch, it moves the contact arm upward to its open position. When this switch is wired in the field, you should select the NC contacts.

The symbol for the *NC held open limit switch* is shown in the bottom left corner of this figure. This symbol shows the switch contact arm in the open position and it is shown above the output terminal. When you wire this switch in the field, you would use the NC set of contacts, and the position of the machine when it is at rest would keep the switch held in the NO position.

15.2.3 Electrical Symbols for Other Pilot Devices

Other pilot devices such as pressure switches, flow switches, float switches, temperature-activated switches, and time-delay relays and lamps are also used in motor control circuits (see Fig. 15-5). These symbols are standard for pilot devices used in motor control circuits. When you encounter a typical motor control diagram on a blueprint, you can use the symbols in this figure to identify each type of switch. You should notice that the symbol for the part that causes the switch to activate is shown at the bottom of each symbol. For example, the symbol for a liquid-level switch is a circle, which represents the ball float typically found on this type of switch and the symbol for the pressure switch looks like the pressure dome. The symbol for each switch should look similar to the actuator for that switch. You should also notice that the contact arm for each switch will be shown in the NO or NC position.

When you examine each symbol carefully, you should notice whether its contacts are NO or NC. You should also understand that the switch arm can be drawn on top of or below the terminals. Each of these conditions allows four combinations of switches to be identified. It is important to understand these subtle differences. For example, Fig. 15-6 shows a pressure switch drawn with its contact arm in each of the four ways described previously, so each switch is different from the other. Fig. 15-6a shows the switch arm NC above the contact terminal and the switch is activated by pressure. This indicates that the switch is an NC switch, and an increase in pressure will cause the switch to open. It may help to remember the switch symbol by the type of application the switch would be used for. For example, this type of switch would be used as a high-pressure safety switch. When the pressure gets too large in a system, the switch opens and shuts off a compressor that is

STANDARD ELEMENTARY DIAGRAM SYMBOLS

The diagram symbols shown below have been adopted by the Square D Company and conform where applicable to standards established by the National Electrical Manufacturers Association (NEMA).

SWITCHES

DISCONNECT	CIRCUIT INTERRUPTER	CIRCUIT BREAKER W/THERMAL O.L.	CIRCUIT BREAKER W/MAGNETIC O.L.	CIRCUIT BREAKER W/THERMAL AND MAGNETIC O.L.	LIMIT SWITCHES		FOOT SWITCHES	
					NORMALLY OPEN	NORMALLY CLOSED	N.O.	N.C.
					HELD CLOSED	HELD OPEN		

PRESSURE & VACUUM SWITCHES		LIQUID LEVEL SWITCH		TEMPERATURE ACTUATED SWITCH		FLOW SWITCH (AIR, WATER ETC.)	
N.O.	N.C.	N.O.	N.C.	N.O.	N.C.	N.O.	N.C.

SELECTOR

SPEED (PLUGGING)	ANTI-PLUG	2 POSITION	3 POSITION	2 POS. SEL. PUSH BUTTON

2 POSITION: J K — I-CONTACT CLOSED

3 POSITION: J K L — I-CONTACT CLOSED

2 POS. SEL. PUSH BUTTON:

CONTACTS	SELECTOR POSITION			
	A		B	
	BUTTON		BUTTON	
	FREE	DEPRES'D	FREE	DEPRES'D
1 - 2	I			I
3 - 4		I	I	I

I-CONTACT CLOSED

PUSH BUTTONS / PILOT LIGHTS

MOMENTARY CONTACT				MAINTAINED CONTACT		ILLUMINATED	INDICATE COLOR BY LETTER	
SINGLE CIRCUIT		DOUBLE CIRCUIT	MUSHROOM HEAD	WOBBLE STICK	TWO SINGLE CKT.	ONE DOUBLE CKT.	NON PUSH-TO-TEST	PUSH-TO-TEST
N.O.	N.C.	N.O. & N.C.						

CONTACTS / COILS / OVERLOAD RELAYS / INDUCTORS

INSTANT OPERATING				TIMED CONTACTS - CONTACT ACTION RETARDED AFTER COIL IS:				COILS		OVERLOAD RELAYS		INDUCTORS
WITH BLOWOUT		WITHOUT BLOWOUT		ENERGIZED		DE-ENERGIZED		SHUNT	SERIES	THERMAL	MAGNETIC	IRON CORE
N.O.	N.C.	N.O.	N.C.	N.O.T.C.	N.C.T.O.	N.O.T.O.	N.C.T.C.					AIR CORE

TRANSFORMERS / AC MOTORS / DC MOTORS

AUTO	IRON CORE	AIR CORE	CURRENT	DUAL VOLTAGE	SINGLE PHASE	3 PHASE SQUIRREL CAGE	2 PHASE 4 WIRE	WOUND ROTOR	ARMATURE	SHUNT FIELD	SERIES FIELD	COMM. OR COMPENS. FIELD
										(SHOW 4 LOOPS)	(SHOW 3 LOOPS)	(SHOW 2 LOOPS)

SUPPLEMENTARY CONTACT SYMBOLS

SPST, N.O.		SPST N.C.		SPDT		TERMS
SINGLE BREAK	DOUBLE BREAK	SINGLE BREAK	DOUBLE BREAK	SINGLE BREAK	DOUBLE BREAK	SPST- SINGLE POLE SINGLE THROW
						SPDT- SINGLE POLE DOUBLE THROW
DPST, 2 N.O.		DPST, 2 N.C.		DPDT		DPST- DOUBLE POLE SINGLE THROW
SINGLE BREAK	DOUBLE BREAK	SINGLE BREAK	DOUBLE BREAK	SINGLE BREAK	DOUBLE BREAK	DPDT- DOUBLE POLE DOUBLE THROW
						N.O. - NORMALLY OPEN
						N.C. - NORMALLY CLOSED

FIGURE 15-5 Electrical symbols for pilot devices and other motor control devices. (Courtesy of SQUARE D COMPANY/GROUPE SCHNEIDER. "Square D Company/Groupe Schneider assumes no liability for accuracy of information".)

adding pressure to the system. The pressure switch in Fig. 15-6b shows an NO switch with the switch arm above the contact terminal. This type of switch will activate a circuit when pressure falls. An application for this type of switch is to be used to turn on a compressor when air pressure drops below a minimum, so the system will sustain the pressure. Since the switch arm is drawn above both of these switches, they may also be called *high-pressure switches*.

Fig. 15-6c shows an NC pressure switch with the switch arm below the contact terminal. This symbol indicates that when pressure decreases, the switch contacts will fall to the open position. This type of switch can be used as a low-pressure safety switch. For example, if a system is to maintain a minimum oil pressure, the switch will open if the oil pressure drops below the setpoint. Fig. 15-6d shows an NO pressure switch with the switch arm below the contact terminal. When pressure increases, it will cause the switch arm to go up and close. This type of switch can be used to start a motor when the pressure gets too high. Since the switch arm is shown below the contact on both of these switches, they may be referred to as low-pressure switches.

It is important to remember that the pressure switches are only sold with NO or NC contacts or the combination of one set of NO and one set of NC contacts. Again, like the limit switches, the conditions the switches are used in when they are mounted on machinery may refer to the function of the switch in the name the machine manufacturer gives the switch. For example, the switch may be called the compressor high-pressure cutout safety, or it may be called the low oil pressure protection switch. In both of these cases the switch is still a pressure switch and a set of NO or NC contacts is used to control circuit current when pressure increases or decreases.

You should now understand that whichever way the contact arm is drawn, a decrease in the physical parameter the switch is measuring will cause the switch arm to drop, and an increase will cause the arm to raise. The next section will show how all of the pilot devices are used together to make common motor control circuits.

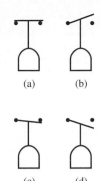

FIGURE 15-6 (a) NC pressure switch. (This switch may be referred to as an NC high-pressure switch.) **(b)** NO pressure switch. (This switch may also be referred to as an NO high-pressure switch.) **(c)** NC pressure switch. (This switch may be referred to as an NC low-pressure switch.) **(d)** NO pressure switch. (This switch may be referred to as an NO low-pressure switch.)

15.3 CONTROL CIRCUITS AND LOAD CIRCUITS

In motor control circuits, the components are broken into two groups: the *control components*, which are the switches and coils that control the load, and the *load components* such as motors that cause the system to perform some type of action. The part of the circuit that has the controls is called the *control circuit*, and the part of the circuit that has the motor in it is called the *load circuit*. Typically the switches and wires in the control circuit use lower voltages than the load and smaller currents than the load. In fact in most circuits, the load will be a single-phase or three-phase motor.

Fig. 15-7 shows a ladder diagram of a control circuit and a load circuit. You should notice in this figure that the control circuit is shown on the left, and it has a float switch connected to the relay coil. An NC set of overload (OL) contacts is shown connected in series with the motor starter coil. The load circuit is shown on the right side of the figure and you can see that it consists of a three-phase motor starter and a three-phase motor. It is important to note that the relay coil is technically a load, but it is considered part of the control circuit because of the function it performs.

The same control and load circuits are shown as a wiring diagram in Fig. 15-8. When this circuit is shown as a wiring diagram, you will notice that some of the control circuit components such as the motor starter coil are drawn where they are physically located on the motor starter. Since the motor starter contacts are part of the load circuit, it is hard to determine in a wiring diagram what components are part of the load circuit and what components are part of the control circuit. In some diagrams the load circuit is usually drawn with darker and wider lines than the wires of the control circuit to indicate the wires are larger. The remaining part of this chapter will explain the different types of control circuit configurations and load circuit configurations.

15.3.1 Two-Wire Control Circuits

The two-wire control circuit is commonly used in applications where the operation of a system is automatic and basically two wires are used to provide voltage to the load. This may include such applications as sump pumps, tank pumps, electric heating, and air compressors. In these systems you typically close a disconnect switch or circuit breaker to energize the circuit, and the actual energizing of the motor in the system is controlled by the operation of the pilot device.

The circuit is called a two-wire control because only two wires are needed to energize the motor starter coil. The circuit controls that are located prior to the coil will provide the operational and safety features, and the overload contacts that are located after the coil are used to protect the circuit against overcurrents.

Fig. 15-7 shows a ladder diagram of the two-wire control circuit for a pumping station and Fig. 15-8 shows the control circuit as a wiring diagram. You will notice that it is difficult to see the sequence of operation in the wiring diagram, so you generally use the ladder diagram to determine the sequence of operation. The wiring diagram shows the location of all wires and components.

Voltage to the pumping station is provided through a fused disconnect. After the disconnect is closed, the float switch in the diagram is in complete control

FIGURE 15-7 A ladder diagram of a control circuit and a load circuit.

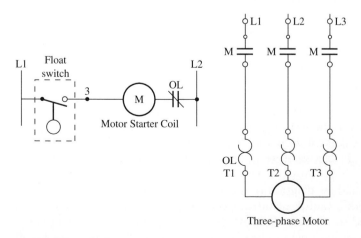

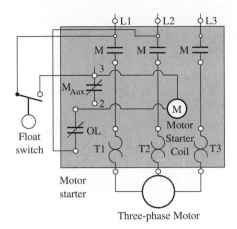

FIGURE 15-8 An example of a wiring diagram that shows a control circuit and a three-phase motor in the load circuit. Notice the motor starter is shaded in the wiring diagram so that you can locate it and all of its terminals.

of the motor starter. Notice that the float control is identified as a float switch. When the level of the sump rises past the setpoint on the control, its contacts will energize and the motor starter coil will become energized. The coil of the motor starter will energize and pull in the motor starter contacts, which will energize the motor. As the motor operates, it will pump the liquid to reduce the level in the sump. When the level of the sump is lowered, the level switch will sense the change in level and switch its contacts off, which will de-energize the motor starter coil. When the motor starter coil is de-energized, the motor starter contacts are opened, and the motor is turned off.

This on/off sequence will continue automatically until the disconnect is switched off or unless the motor starter overloads are tripped. If the motor draws too much current when it is pumping, the heaters will trip the overload contacts and open the control circuit, which will interrupt current flow to the motor starter coil. Since the overloads must be reset manually, the circuit will not become energized automatically when the overload is cleared. In this case, someone must physically come to the motor starter and press the reset button. At this time, the system should be thoroughly inspected for the cause of the overload. It would also be important to use a clamp-on ammeter to determine the full-load amperage (FLA) of the motor and cross-reference this to the size of the heaters in the motor starter to see that they match. The FLA should be checked against the motor's data plate. You should also take into account the service factor listed on the data plate when you are trying to determine if the pump motor is operating correctly.

Fig. 15-9 shows a more complex two-wire control circuit that is used to control an air compressor. You should notice that the wiring diagram for the control and load circuit is shown in Fig. 15-9a and the ladder diagram of the control circuit is shown in Fig. 15-9b. This circuit is still a two-wire control circuit and it is used to turn the air compressor on automatically when the pressure drops below 30 psi and to turn off the compressor when the pressure reaches 90 psi. Pressure switch A in this circuit controls the operation of the control at these pressures. Its high and low pressures are adjustable so that the system could be energized and de-energized at other pressures if the need arises. A hand switch is provided in this circuit that allows the system to be pumped to a predetermined pressure. The hand switch is intended to be used when the auto switch is not functioning properly or if you need to test the system.

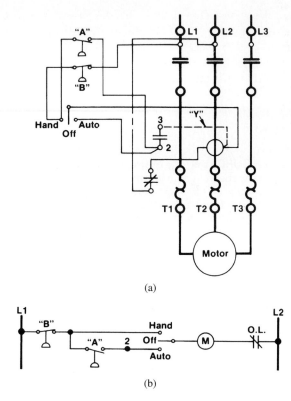

(a)

(b)

FIGURE 15-9 **(a)** A wiring diagram of a two-wire control circuit for a compressor motor. The compressor motor is controlled by a motor starter that is energized by one of two pressure switches. **(b)** A ladder diagram of the same circuit shown in the wiring diagram. In this circuit you can see a selector switch provides "hand" or "auto" control. Pressure switch A is in the circuit when the selector switch is in auto, and pressure switch B is in the circuit when the selector switch is in manual. (Courtesy of Rockwell Automation's Allen Bradley Business.)

Pressure switch B in this diagram acts as a safety for this circuit. This pressure switch is set at 120 psi and is not adjustable. It is in the circuit to prevent the tank pressure from rising too high. This could occur if the operational pressure switch became faulty and would not open when the pressure reached 90 psi. It could also be used to protect the system against overpressure when the switch is in the hand position. Generally, the safety switch is meant to protect the system against component failure or control failure.

If a pressure control would fail, the compressor would continue pumping air into the tank, which would allow its pressure to rise to an unsafe level. Since the operational switch is still closed, the air pressure in the tank could be increased to a level where the pump would cause the motor to stall if the safety switch were not in the circuit. This could build up pressure to several hundred pounds, which could cause lines and fittings in the air system to explode.

The nonadjustable pressure switch would act as a backup to the operational switch and trip off anytime the pressure reaches 120 psi. This switch is also different in that it is interlocked when it trips, so that it must be reset manually by having someone press the reset button. If you find the reset button activated on the safety switch, you must test the system thoroughly to determine why the operational switch did not control the circuit.

In normal operation, the motor starter would cycle the air compressor on and off to keep the pressure in the tank between the high and low setpoint that is set on the operational control. The motor starter overloads could also trip this circuit and require manual reset. This could occur if the motor was incurring overcurrent problems. The overcurrent could occur due to bad bearings or if someone physically depresses the switch. When this push button is released, the NO contacts return to their NO state.

The NO auxiliary contacts from the motor starter will close when the motor starter coil is energized. When they close they will provide an alternate path around the push-button contacts for current to get to the coil. These contacts are called seal or seal-in contacts when they are used in this manner. Sometimes the seal contacts are said to have memory, since they will maintain the last state of the push buttons in the circuit.

15.3.2 The Three-Wire Control Circuit

The *three-wire control circuit* is the most widely used motor control circuit. This circuit is similar to the two-wire circuit except it has an extra set of contacts that is connected in parallel around one of the original pilot switches to seal it in. The extra set of parallel contacts provides the third wire, which also gives this circuit its name. You should fully understand this circuit and learn to recognize it when it is shown as a ladder diagram and as a wiring diagram. In this way you will be able to understand the operation of the circuit wherever it appears.

One variation of the three-wire circuit is shown in Fig. 15-10. In this figure, stop and start push buttons are used as the pilot devices for control. Fig. 15-10a shows the wiring diagram for the three-wire circuit, and Fig. 15-10b shows the ladder diagram for the three-wire control circuit. In the ladder diagram you can see that the stop push button is wired NC. When the start push button is depressed, voltage will flow through its contacts to energize the motor starter coil. When the coil becomes energized, it pulls its main contacts closed to cause the motor to become energized, and it also pulls its NO auxiliary contacts that are connected

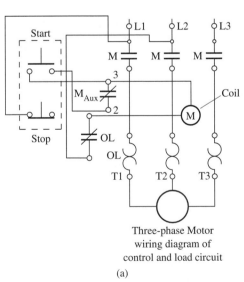

Three-phase Motor
wiring diagram of
control and load circuit

(a)

Ladder diagram of
control circuit

(b)

FIGURE 15-10 **(a)** A typical three-wire control system. This control circuit gets its name because of the auxiliary contacts that are connected in parallel with the start button. The auxiliary contacts seal in the circuit to keep the coil energized after the start push button is released. **(b)** Ladder diagram of a three wire control circuit. (Courtesy of Rockwell Automation's Allen Bradley Business.)

across the start push button to their closed position. When the auxiliary contacts close, they provide an alternate route for voltage to travel around the start push-button contacts, which will return to their NO position when the switch is no longer depressed. The auxiliary contacts are called the *seal-in* circuit or in some cases the *memory circuit*. If the auxiliary contacts are not used, you would need to keep your finger on the start button to keep the circuit energized.

You should notice in the three-wire circuit that the stop button is used to unseal the circuit and de-energize the motor starter coil. Anytime the stop button is depressed, the coil will become de-energized and the seal-in circuit will drop out. This means that the start push button must be depressed again to energize the circuit.

It is also important to understand in the three-wire circuit that the stop push button traditionally is always the first switch in the circuit. Most people assume this has something to do with safety. In reality, it has to do with economics. When you look at a motor starter, you will notice that the auxiliary contacts are physically located near the coil. If the start button is connected in the circuit next to the coil, the wire that is connected to the auxiliary contact on the right side (terminal 3) can actually be a short jumper wire that is connected directly to the left side of the coil. If the start button is not located next to the coil, the wires that connect the auxiliary contacts in parallel with the start push button will need to be as long as the distance between the start push button and the motor starter. This may be a distance of several hundred feet. This jumper wire is shown in the wiring diagram of the three-wire control circuit.

15.3.3 A Three-Wire Start/Stop Circuit with Multiple Start/Stop Push Buttons

Additional start and stop push buttons may be needed on a system that is very large. For example, if the system is installed over a large area, such as for a long conveyor system that may be over several hundred feet long, it would be inconvenient and unsafe to have one start and stop button at only one end of the system. To make the system more safe and to make it more convienient to start and stop the system, push-button switches can be installed every 40 feet. Each additional start button should be connected in parallel with the first start button, and all additional stop buttons should be connected in series with the original stop button. Fig. 15-11 shows examples of the additional start and stop push buttons in a three-wire circuit.

15.3.4 Three-Wire Control Circuit with Indicator Lamp

An indicator lamp can be added to a three-wire circuit to show when the coil in the circuit is energized or de-energized. The indicator lamp can be green to show when the system is energized and red to show when the circuit is de-energized. The lamp is usually mounted where personnel can easily see it at a distance. Sometimes the indicators are used where one operator must watch four or five large machines. After a machine has been set up, the operator will move on to the next machine. Since the installation is very large and spread out over a distance, the operator can watch for the indicator lamps to see if the machine is still in operation.

The indicator lamps can also be used by maintenance personnel when a machine has more than one motor starter. In this type of application each motor starter has an indicator lamp to indicate its coil is energized. This will help the

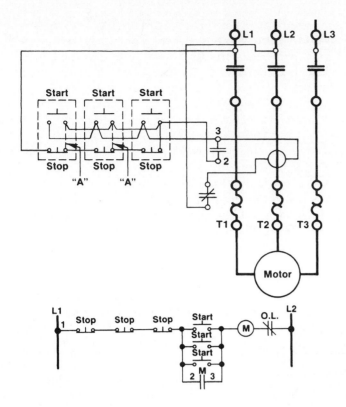

FIGURE 15-11 The wiring diagram for a three-wire control circuit with additional start and stop push buttons added to the circuit is shown at the top of this figure, and the ladder diagram of just the control circuit is shown at the bottom. (Courtesy of Rockwell Automation's Allen Bradley Business.)

maintenance personnel to begin testing for faults in the correct part of the circuit when the system has stopped or is not operating correctly. This is especially useful if the motor starter is mounted in a NEMA (National Electrical Manufacturers Association) enclosure where the technician cannot see through the enclosure door to verify if the starter is energized or de-energized.

Fig. 15-12 shows a wiring diagram and a ladder diagram of a circuit with an indicator lamp connected in the control circuit. The lamp is called a pilot light and it is connected in parallel with the seal-in contacts on the motor starter. When the motor starter closes, the lamp will be energized to indicate that the motor starter contacts are closed. Anytime the lamp is de-energized, the operator and maintenance personnel know that the motor starter is not energized. A press-to-test lamp could also be used in this circuit. The press-to-test lamp allows the operator and maintenance personnel to put their fingers on the lamp and depress the lens at any time to test it to see if it is operational. When the lamp lens is pressed, it will cause a special set of contacts in the base of the lamp holder to provide voltage instantly to the lamp and illuminate it. If the bulb is burned out, the lamp will not illuminate and the maintenance personnel or the operator can change the bulb. If the indicator is energized most of the time, such as in a continuous operation, the press-to-test lamp may not be necessary since the indicator will be energized most of the time.

Indicator lamps are available for 120 volt, 240 volt, 480 volt, and 600 volt AC, which provides them for any control circuit voltage. A wide variety of colored lenses is also available to indicate other conditions with the machine. These indicators can be connected across different individual motor starters in the machine to provide other information, such as hydraulic pump running, heaters energized, conveyor in operation, and other conditions that are vital to the machine. Fig. 15-13 shows an

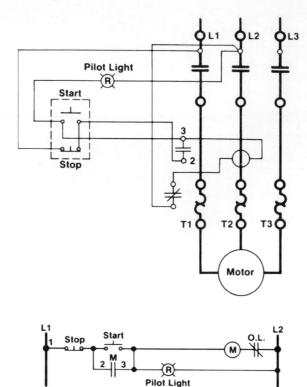

FIGURE 15-12 A wiring diagram and a ladder diagram of a three-wire control circuit with an indicator lamp added to show when the motor starter coil is energized. (Courtesy of Rockwell Automation's Allen Bradley Business.)

exploded view of a typical indicator lamp. In this diagram you can see that the lens is replaceable, and the part of the lamp that has wires connected to it can easily be removed from the socket part of the lamp in case the socket needs to be replaced.

15.3.5 Reverse Motor Starters

In previous chapters you have found that DC and AC motors can be reversed. These chapters provided the terminal connections for each type of DC motor, AC single-phase motor, and AC three-phase motor. The circuit shown in Fig. 15-14 shows a forward and reverse motor starter for an AC three-phase motor. When you must install or troubleshoot this circuit, you must be sure to provide interlocks so that the motor cannot be energized in both the forward and reverse directions at the same time.

From this diagram you can see that the control circuit has a forward and reverse push buttons. The forward and reverse push buttons both have NO and NC sets of contacts. In the ladder diagram each button switch shows a dashed line to indicate that both sets of contacts are activated by the same button. The forward and reverse push buttons are better defined in the wiring diagram, which shows that each switch has an open set and a closed set of contacts. The stop button is wired in series with the open contacts of both of these switches, so that the motor can be stopped when it is operating in either the forward or the reverse direction.

The operation of the push buttons is best understood through the use of a ladder diagram. The ladder diagram is used to show the sequence of the control circuit, while the wiring diagram shows the operation of the load circuit, which includes the motor and heaters for the overloads. The overload contacts are con-

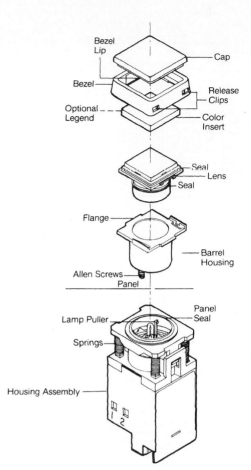

Bezel
Lip

Cap

Bezel

Release
Clips

Optional
Legend

Color
Insert

Seal

Lens

Seal

Flange

Barrel
Housing

Allen Screws

Panel

Panel
Seal

Lamp Puller

Springs

Housing Assembly

1   2

FIGURE 15-13 An exploded view of a typical indicator lamp. Notice the lens is replaceable so different color lenses can be used. The field wiring is connected to a terminal section that allows the lamp to be changed when it is damaged without removing and replacing the wiring. (Courtesy of Rockwell Automation's Allen Bradley Business.)

nected in the control circuit, where they will de-energize both the forward and reverse circuits if the motor is pulling too much current.

From the wiring diagram you can also see that two separate motor starters are used in the circuit. The forward motor starter is shown on the right side of the reverse motor starter. Each starter has its own coil and auxiliary contacts that are used as interlocks. The location of the auxiliary contacts is shown in the wiring diagram, but their operation is difficult to determine there. When you see the auxiliary contacts in the ladder diagram, their function can be more clearly understood.

This comparison of the ladder diagram and the wiring diagram should help you understand that you need both diagrams to work on the equipment. The ladder diagram will be useful in determining what should be tested, and the wiring diagram is useful in showing where the contacts you want to test are located.

You should also notice that the control circuit is powered from a control transformer that is connected across L1 and L2. The secondary side of the transformer is fused to protect the transformer from a short circuit that may occur in either coil.

15.3.6 Reversing Motors with a Drum Switch

The drum switch is a manual switch that allows you to manually reverse the direction in which a motor is turning. As you know, the switch contacts are open and closed

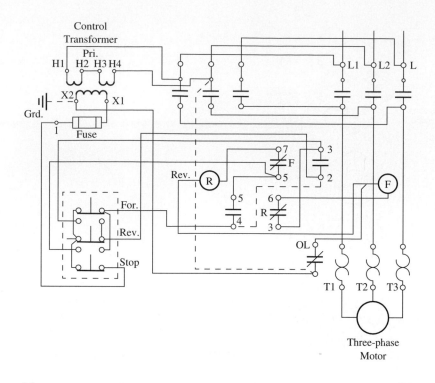

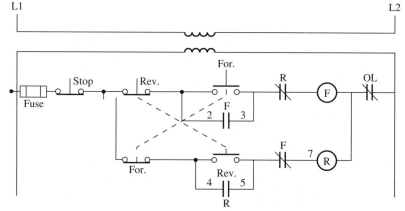

FIGURE 15-14 A wiring diagram and ladder diagram of a forward and reverse motor starter. Notice that you can clearly see the interlock system in the ladder diagram. (Courtesy of Rockwell Automation's Allen Bradley Business.)

manually by moving the drum switch from the off position to the forward or reverse position. Fig. 15-15 shows a picture of the drum switch and Fig. 15-16 shows a diagram of the drum switch contacts. Fig. 15-16a shows the drum switch contacts when the switch is in the reverse position, Fig. 15-16b shows the contacts when the switch is in the off position, and Fig. 15-16c shows the contacts when the switch is in the forward position. When the switch is in the *reverse position*, you should notice that terminal 1 is connected to terminal 2, terminal 3 is connected to terminal 4, and terminal 5 is connected to terminal 6. In the *off position*, all contacts are isolated from all other contacts. In the *forward position*, terminal 1 is connected to terminal 3, terminal 2 is connected to terminal 4, and terminal 5 is connected to terminal 6.

After you understand the operation of the drum switch in its three positions and the methods of reversing each type of motor, these concepts can be combined

FIGURE 15-15 A drum switch. Notice the handle requires the operator to manually change the position of the switch from forward to reverse or to the off position. (Courtesy of Eaton Corporation Cutler-Hammer Products.)

to develop manual reversing circuits for any motor in the factory as long as its full-load and locked-rotor amperage (FLA and LRA) do not exceed the rating of the drum switch. Fig. 15-17a shows an AC single-phase motor connected to a drum switch. Notice that the start winding must be reversed for the motor to run in the reverse direction, so the start winding is connected to terminals 3 and 2.

Fig. 15-17b shows a three-phase motor connected to a drum switch. You should remember that with this switch, any two lines to the motor can be swapped so the motor will change direction of rotation. Fig. 15-17c shows a DC motor connected to a drum switch. You should notice with the DC motor that the direction of current flow through the field is reversed to make the motor run in the opposite direction. You may need to study the connections inside the drum switch when it is in the forward or reverse position to fully understand how the drum switch is used with each of these motors to reverse the direction of their rotation.

These diagrams are especially useful for installation and troubleshooting of these circuits. The drum switch can be tested by itself or as part of the reversing circuit. The motors can also be disconnected from the drum switch and operated in the forward and reverse directions for testing or troubleshooting if you suspect the switch or motor of malfunctioning.

Handle end		
Reverse	Off	Forward
1 o——o 2	1 o o 2	1 o o 2
3 o——o 4	3 o o 4	3 o o 4
5 o——o 6	5 o o 6	5 o——o 6

FIGURE 15-16 **(a)** Contacts of a drum switch when it is switched to the reverse position. **(b)** Contacts of a drum switch when it is switched to the off position. **(c)** Contacts of a drum switch when it is switched to the forward position. (Courtesy of Eaton Corporation. Cutler-Hammer Products.)

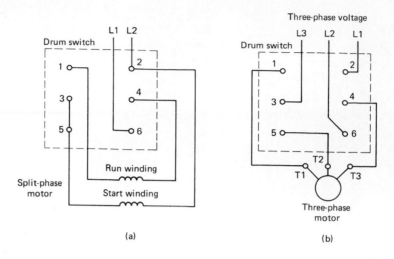

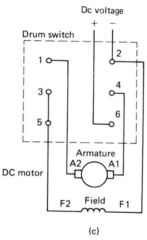

FIGURE 15-17 (a) A single-phase AC motor connected to a drum switch. **(b)** A three-phase AC motor connected to a drum switch. **(c)** A DC motor connected to a drum switch. (Courtesy of Eaton Corporation, Cutler Hammer Products.)

15.3.7 Jogging Control Circuits

In some applications, such as motion control, machine tooling, and material handling, you must be able to turn the motor on for a few seconds to move the load slightly in the forward or reverse direction. This type of motor control is called *jogging*. The jogging circuit utilizes a reverse motor starter to allow the motor to be moved slightly when the forward or reverse push button is depressed.

Another requirement of the jogging circuit is that the motor starters do not seal in when the push buttons are depressed to energize the motor when it is in the jog mode, yet operate as a normal motor starter when the motor controls are switched to the run mode. A ladder diagram of a jogging circuit is provided in Fig. 15-18. This diagram of the load circuit and control circuit is shown as an electrical wiring diagram with the location of each component. The control circuit is shown again as a ladder diagram so you can see the sequence of operation for the forward and reverse motor starters with the jog function.

The wiring diagram in Fig. 15-19 gives you a good idea of the way the jog/run switch operates. This switch is shown to the left of the motor starter in the

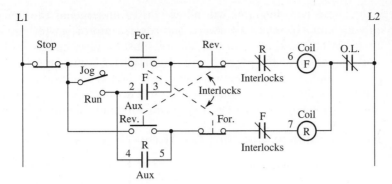

FIGURE 15-18 A ladder diagram of a forward and reverse jogging circuit. Notice the interlock between the forward push buttons and the reverse push buttons so that you cannot energize the forward and reverse motor starters at the same time. (Courtesy of Rockwell Automation's Allen Bradley Business.)

diagram. You can see that it is part of the start/stop station. The jog/run button is a selector switch that is mounted above the forward/reverse/stop buttons.

When the switch is in the jog mode, the selector switch is in the open position. From the ladder diagram, you can see that the jog switch is in series with both of the seal-in circuits, which prevents them from sealing in the forward or reverse push buttons when they are depressed. This means that the motor will operate in the forward direction for as long as the forward push button is depressed. As soon as the push button is released, the motor starter will become de-energized. This jog switch also allows the motor to be jogged from one direction directly to the other direction without having to use the stop button.

The motor is protected by the overloads that are connected in series with the forward and reverse motor starter coils. If the overload trips, the overload contacts in the control circuit will open and neither coil can be energized until it is reset.

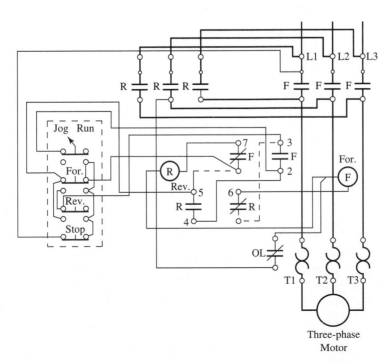

FIGURE 15-19 A wiring diagram of a forward and reverse jogging circuit. Notice that this diagram shows the location of the jog push button and the forward motor starter and reverse motor starter. (Courtesy of Rockwell Automation's Allen Bradley Business.)

These two diagrams will allow you to understand the operation of the jog circuit. You can make a forward and reverse motor starter circuit into a jogging circuit by adding the jog switch, but you must be sure that the motor and the motor starters are rated for jogging duty. Remember: Some motor starters and motors cannot take the heat that will build up when the motor is started and stopped continually during the jogging operation. The motor and the motor starters will be rated for jogging or plugging if they can withstand the extra current and heat.

15.3.8 Sequence Controls for Motor Starters

Sequence control allows a motor starter to be utilized as part of a complex motor control circuit that uses one set of conditions to determine the operation of another circuit. Fig. 15-20 shows an example of this type of circuit. The circuits in this figure are presented in wiring diagram and ladder diagram form. You will really begin to see the importance of the ladder diagram as it shows the sequence of operation, which would be difficult to determine from the wiring diagram. The wiring diagram is still important since it shows the field wiring connections and the locations of all terminals that will need to be used during troubleshooting.

The operation of the circuit in Fig. 15-20 shows two conveyors that are controlled by two separate motor starters. Conveyor 1 must be operating prior to conveyor 2 being started. This is required because conveyor 2 feeds material onto

FIGURE 15-20 Motor starters are sequenced so that motor starter 1 must be on before motor starter 2 is started. (Courtesy of Rockwell Automation's Allen Bradley Business.)

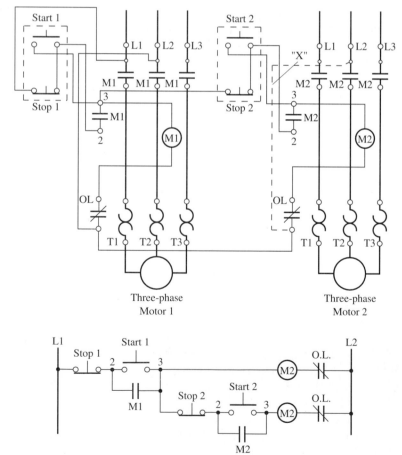

conveyor 1 and material would back up on conveyor 2 if conveyor 1 was not operating and carrying it away. The ladder diagram shows a typical start/stop circuit with an auxiliary contact being used as a seal around the start button. When the first start button is depressed, M1 will be energized, which will start the first conveyor in operation. M1 auxiliary contacts will seal the start button and provide the circuit power to the second start/stop circuit.

Since the second circuit has power at all times after M1 is energized, its start and stop buttons can be operated at anytime to turn the second conveyor on and off as often as required without bothering the first conveyor motor starter. Remember: This circuit requires conveyor 1 to be operating prior to conveyor 2, since conveyor 2 feeds material onto conveyor 1.

The circuit also protects the sequence if conveyor 1 is stopped for any reason. When it is stopped, the M1 motor starter becomes de-energized, and the M1 auxiliary contacts return to their open condition, which also de-energizes power to the second conveyor's start/stop circuit.

You should also notice that the power for this control circuit comes from the L1 and L2 of the first motor starter. This means that if supply voltage for the first conveyor motor is lost for any reason, such as a blown fuse or opened disconnect, the power to the control circuit is also lost and both motor starters will be de-energized, which will stop both conveyors. If the first motor draws too much current and trips its overloads, it will cause an open in the motor starter's coil circuit, which will cause the auxiliary contacts of the first motor starter to open and de-energize both motor starters.

If you need additional confirmation that the belt on the first conveyor is actually moving, a motion switch can be installed on the conveyor, and its contacts would be connected in series between the first start button and M1 coil. This would cause the first motor starter coil to become de-energized.

15.4 OTHER TYPES OF PILOT DEVICES

A wide variety of other pilot devices is commonly used in motor control circuits that are interfaced to industrial electronic circuits. These switches include pressure, level, and temperature switches as well as other types of switches. Fig. 15-21 shows pictures of three types of level switches and Fig. 15-22 shows example diagrams of each of these switches. The level switches are sometimes called float switches.

The first switch in Fig. 15-22a is used with a ball float that is attached to a long rod. The rod has adjustable cams attached at the bottom and the top of the rod. When the level in the sump raises, the float will lift the rod, and the cam that is attached to the lower part of the rod will move high enough to trip the switch handle to the up position, which closes the level switch contacts. When the pump runs, it will pump the sump level down and allow the float to drop. When the float drops low enough, the cam that is attached to the top of the rod will cause the switch handle to move to the down position, which will turn the pump off.

The switch in Fig. 15-22b uses a sealed float to activate the level switch. In this application, the switch is mounted in the tank at the precise level. When the level of the liquid in the tank increases above this level, the float will lift and cause the switch contacts to close and turn on the pump. When the pump runs, the level of the tank is decreased, which allows the float to drop and turn the switch off. It

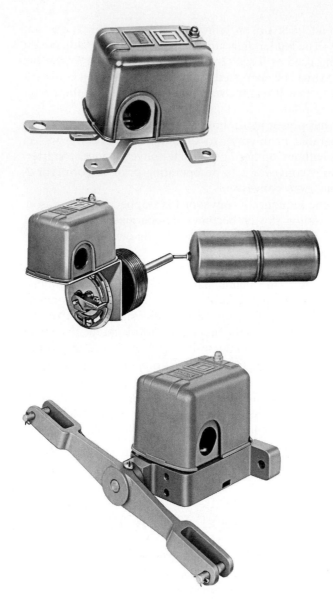

FIGURE 15-21 Three types of level switches. Applications and electrical diagrams of these switches are shown in the next figure. (Courtesy of SQUARE D COMPANY/ GROUPE SCHNEIDER. "Square D Company/Groupe Schneider assumes no liability for accuracy of information.")

is important to understand that the length of the switch arm will determine the amount of travel the float must make to turn the switch on and off. This distance is called the *control band* or *dead band*.

The float switch in Fig. 15-22c shows a pivot-arm level switch. An arm or a wheel can be attached to the pivot to cause it to activate. An arm is shown in the picture of this switch, and a wheel is shown in the diagram. A cable is threaded around the wheel or arm. A float is attached to one end of the cable and a weight is attached to the other. When the level of liquid in the tank changes, the cable will move up or down and cause the switch to turn on or off.

It is important to understand that since these switches are pilot switches, they can only safely switch 10–15 A. This means that they can be used to control small motors, or they can control larger motors by switching power on and off to a motor starter coil.

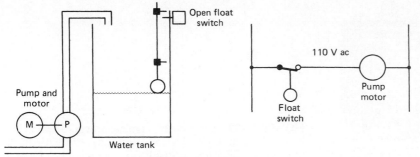

(a) Open float switch application and electrical diagram

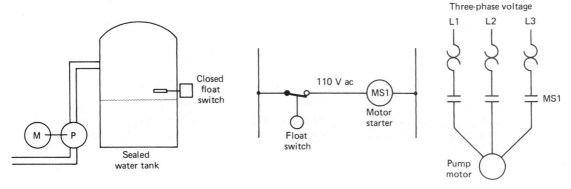

(b) Closed float switch application and electrical diagram

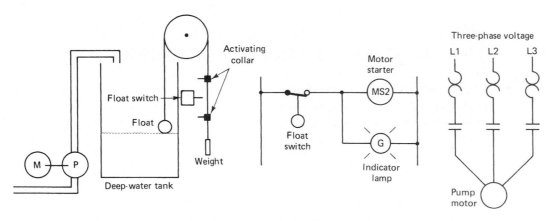

(c) Rod and chain float switch application and electrical diagram

FIGURE 15-22 **(a)** A float and rod are used with the level switch to turn it on and off. **(b)** A closed flow switch is mounted at the specific level that is controlled. **(c)** A float is attached to one end of a cable and a weight is attached to the other end. When the float moves up or down, the switch is activated. (Courtesy of SQUARE D COMPANY/GROUPE SCHNEIDER. "Square D Company/Groupe Schneider assumes no liability for accuracy of information.")

Another type of pilot switch is a pressure switch, which is used for sensing air pressure or water pressure and turning a set of contacts on or off. Fig. 15-23 shows a typical pressure switch. The pressure switch may have a single point where it is activated, or it may provide a span between the point where it turns on and off. If the switch has a span adjustment and a low-pressure adjustment, you should set the low-pressure adjustment where you want the switch to turn on, and adjust

FIGURE 15-23 Example of a typical pressure switch that is used as a pilot device. (Courtesy of SQUARE D COMPANY/GROUPE SCHNEIDER. "Square D Company/ Groupe Schneider assumes no liability for accuracy of information.")

the span to determine where you want the switch to turn off. For example, if you want the switch to control the pressure in a reservoir of an air compressor, you would set the low pressure to turn on when the pressure reaches 20 psi, and then set the span to 40 psi. The span of 40 psi would cause the switch to turn off when the air pressure reaches 40 psi more than the low-level setting. This means that the switch would turn on the compressor when the air pressure dropped to 20 psi, and it would turn the compressor off when the pressure reached 60 psi. Even though the switch has two points that turn it on and off, it will still only have one set of single-pole or double-pole contacts that is activated.

15.5 FUSES AND CIRCUIT BREAKERS

The motors, switch gear, transformers, and power distribution conductors must be protected against short-circuit current. The motors and other loads must also be protected against slow overcurrents or sustained overload that will allow them to overheat and become damaged. *Short-circuit current* is defined as any current that exceeds the normal full-load current of a circuit by ten times. When short-circuit current occurs unprotected, any conductor or switch gear that is involved will be severely damaged by excessive magnetic forces as well as by extremely high heat levels that will melt most metal objects. Fuses, circuit breakers, and magnetic overloads can provide protection for motors and other loads against slow overcurrents, and fuses and circuit breakers can provide short-circuit protection.

15.5.1 Slow Overcurrents

The slow overcurrent develops from overloading devices or from malfunctions in motors and other loads. For example, if a motor-driven conveyor is overloaded, the motor will be required to draw extra current to try and move the load. If the motor is not protected, it will draw the extra current and begin to overheat. If the transformer that supplies voltage to the motor is also fully loaded, it will become overloaded when the motor draws the extra current. If the overload condition continues for 10 to 20 minutes, enough heat will be built up in the motor and transformer to cause the insulation in both of these devices to break down and deteriorate.

The same problem will occur if the bearings in a motor become dry and begin to wear. After the bearing has operated without any lubrication, it will begin to heat up and seize on the motor shaft, which will in turn cause the motor to draw excessive current. This condition will cause the motor and transformer to overheat to a point where they are completely damaged because the overcurrent will continue as long as the motor is running.

15.5.2 Protecting Against Slow Overcurrent

In all these examples the problem is caused by the increase in normal operating current to the point where damage can occur. Circuit breakers can sense either the heat or magnetic forces as they increase beyond the maximum safe level, and fuses and overloads can sense increased heat that the overload creates. Separation of these devices will cause a set of contacts or a conducting element to open anytime the current increases above the safe level. This presents a problem with some loads, such as motors that have a very large inrush current when they start.

Fig. 15-24 shows a graph of the inrush current caused by a 208 volt, 5hp motor starting. You can see that the motor will draw 16.5 A at full load. This motor will draw up to 99 A when it starts, which is six times the amount of full-load current. This presents a problem in protecting against overcurrents because a circuit breaker or fuse that is sized to protect the motor during full-load current (16.5 A) would trip when the motor is started, and if they are sized to allow the motor to start (99 A), they will not provide adequate protection when the motor is running at full-load current.

Several solutions to this problem are available. One of them is a motor starter with heaters and overloads, and another is inverse-time circuit breakers. Each of these devices provides several minutes of time delay before they trip and take the motor off line. The theory of their operation involves allowing small overcurrents to exist for up to 4 or 5 minutes, and allowing larger overcurrents to exist for less than 10 seconds. These times are based on the amount of time a specific overload condition can exist before a motor begins to sustain damage. The devices must be sized properly to provide adequate protection, and may be adjusted slightly once they are installed. The only problem that remains with the inverse-time circuit breaker and overloads is that they cannot sense a short circuit and open the circuit fast enough to provide interruption capacity.

15.5.3 Short-Circuit Currents

When a conductor from one potential comes in contact with the system ground or a conductor from another potential, a short circuit can occur. The short circuit

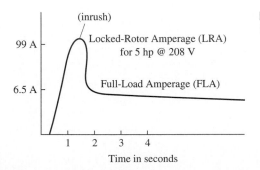

FIGURE 15-24 A graph of inrush current for a 208 volt, 5hp motor.

provides little or no impedance to the power source, so that the current can rise to 100 times the full-load current levels in one or two cycles. This means that currents of 50,000 A or larger are possible if the current is allowed to build.

Fig. 15-25 shows a graph of the current developed by a short circuit. Notice that the current continues to increase with each cycle of the ac voltage. As this current increases, it will build up powerful magnetic forces and tremendous amounts of heat energy that will cause the metal conductors and terminals to melt and explode.

A circuit breaker is an electromechanical device that requires approximately one-half of an AC cycle (positive or negative half of a sine wave) to sense the short circuit and another half-cycle to trip its mechanical contacts. The short-circuit current may still reach 40,000–45,000 A during this period of time before the circuit breaker can open its contacts fully. In some cases, the heat from this short-circuit

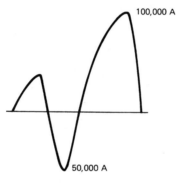

100,000 A

50,000 A

(a) Short-circuit current out of control

FIGURE 15-25 **(a)** Example graph of short-circuit current. **(b)** Example of graph of current when a circuit breaker is used to protect against short-circuit current. **(c)** Example graph of current when a fuse is used to protect against short-circuit current.

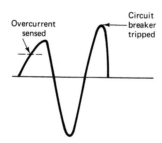

Overcurrent sensed

Circuit breaker tripped

(b) Short-circuit current interrupted during 1 1/2 cycles by a properly sized circuit breaker

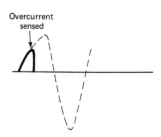

Overcurrent sensed

(c) Short-circuit current interrupted during first 1/2 cycle by a properly sized fuse

current is so intense that it will actually weld the contacts of the circuit breaker together so that they cannot open even though the circuit breaker's trip mechanism has activated. When this occurs, the short-circuit current continues to increase and eventually causes severe damage.

In comparison a fuse can sense the overcurrent as it begins to build and its fusable link will melt before the current increases to a dangerous level. A graph of the protection that a fuse can provide during a short circuit is also shown in Fig. 15-25. From this graph you can see that the fuse element will open as soon as the current reaches the overcurrent level. This means that the short-circuit fault will be sensed and opened in less than a half-cycle. The only problem with the single-element fuse is that it must be sized up to six times the full-load current level to allow the motor to start. This leaves two alternatives when providing short-circuit and overload current: fuses in combination with motor starters or circuit breakers, or the use of a dual-element fuse.

15.5.4 Single-Element Fuses

Single-element fuses basically provide overcurrent protection at one level. Fig. 15-26 shows several cut-away pictures of single-element fuses. From the Fig. 15-26a you can see that the single-element fuse is made of a single conducting element with several neck-down sections. The neck-down sections provide a point where

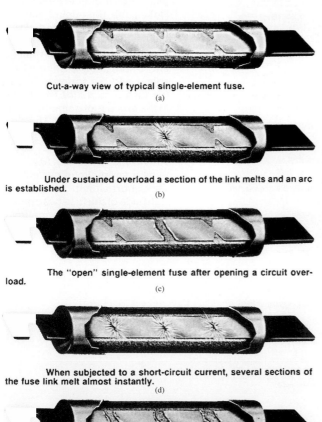

Cut-a-way view of typical single-element fuse.

(a)

Under sustained overload a section of the link melts and an arc is established.

(b)

The "open" single-element fuse after opening a circuit overload.

(c)

When subjected to a short-circuit current, several sections of the fuse link melt almost instantly.

(d)

The "open" single-element fuse after opening a shorted circuit.

(e)

FIGURE 15-26 **(a)** Cutaway picture of single-element fuse before it has developed an open. **(b)** Cutaway picture of a single-element fuse that shows an arc established at the middle neck-down section when the current is too large. **(c)** The middle neck-down section has completely opened after the current exceeded the maximum level. **(d)** Arcs are established at each neck-down section of the fuse when severe overcurrent is experienced. **(e)** All of the neck-down sections develop opens when the severe overcurrent continues. (Courtesy of Bussmann.)

the short-circuit current will be concentrated and cause the metal to melt. Since the element is made of thin metal, the larger the short-circuit current becomes, the more quickly the element will melt.

Fig. 15-26b shows an arc that is established at the center neck-down section when the fuse experiences a current that exceeds the current rating for the fuse. Fig. 15-26c shows how the arc continues to grow until the neck-down section is melted to cause an open.

Fig. 15-26d shows an arc established at each of the neck-down sections when the current is exceedingly large. When the large current is experienced, it may be large enough to cause the fuse to melt several or all of its neck-down sections at the same time. Fig. 15-26e shows a fuse with all of its neck-down sections open.

15.5.5 Dual-Element Fuses

The dual-element fuse provides the same short-circuit current protection as the single-element fuse along with time-delay protection against slow overcurrents. Fig. 15-27 shows cut-away pictures of a dual-element fuse. From these pictures you can see that the dual-element fuse provides a short-circuit element and an overload

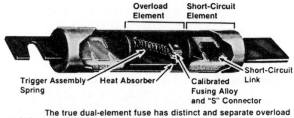

The true dual-element fuse has distinct and separate overload and short-circuit elements.
(a)

Under sustained overload conditions, the trigger spring fractures the calibrated fusing alloy and releases the "connector."
(b)

The "open" Dual Element fuse after opening under an overload condition.
(c)

Like the single element fuse, a short-circuit current causes the restricted portions of the short-circuit elements to melt and arcing to burn back the resulting gaps until the arcs are suppressed by the arc quenching material and increased arc resistance.
(d)

The "open" Dual Element fuse after opening under a short circuit condition.
(e)

FIGURE 15-27 **(a)** A cut-away picture of a dual-element fuse. Notice the time-delay element consists of a spring-loaded trigger mechanism that is held in place by solder. When the current is sufficient, it will melt the solder and the spring will open the fuse section. **(b)** The fuse is in the process of opening during a slow overcurrent. **(c)** The slow overcurrent section of the fuse has opened. **(d)** The short-circuit section of the fuse sustains a short-circuit current. **(e)** The short-circuit link on the right side of the fuse has opened after sustaining a short-circuit current. (Courtesy of Bussmann.)

element. Short-circuit elements (neck-down sections) are located on each end of the fuse, while the overcurrent element is located between them. The buildup will not damage the motor, and large overloads will be cleared quickly, since they will cause heat buildup that will permanently damage the motor. This design allows the motor to draw high inrush current, up to 500% of the full-load current for 10 seconds, which is adequate time to allow the motor to start. It will also allow the motor to develop extra horsepower to meet an increased load demand for several minutes, since the extra current that is drawn will not cause the fuse to open.

Fig. 15-27a shows a normal dual-element fuse and you can see the neck-down sections on each end and the time-delay element in the middle of the fuse. Fig. 15-27b shows the excess heat of a slow overcurrent as it begins to loosen the spring-loaded time-delay element. Fig. 15-27c shows the spring-loaded time-delay element has caused an open in the fuse.

Fig. 15-27d shows the dual-element fuse has experienced a short-circuit current and an arc is established in the neck-down section. Fig. 15-27e shows the neck-down section has opened and interrupted current flowing throught the fuse.

15.5.6 Interruption Capacity

The ability of the fuse to clear short-circuit currents safely is called its *interruption capacity*. The interruption capacity is listed as the maximum number of amperes that the fuse can safely clear. The interruption capacity of modern current-limiting fuses may be as large as 200,000 A. Fuses can also be used in the power distribution system for the expressed purpose of providing interruption capacity for protecting the system equipment and switch gear against large short-circuit currents.

Both the single-element and the dual-element fuse provide this safety feature by encasing the short-circuit element in silica sand in both types of fuses. When the short-circuit current is applied to the short-circuit element, tremendous amounts of heat are built up while the element is melting. If this heat is allowed to build up, gases can be released when the metal is melted and cause the casing of the fuse to rupture. When silica is placed around the short-circuit element, it will absorb the extra heat and use it to melt the sand into a semiliquid state. Since the silica is forced to change state, it will absorb more heat than the reaction can produce, which results in the excessive energy being controlled without damaging the fuse or the hardware and enclosures that are used to mount the fuse.

15.5.7 Types of Fuses

Several types of fuses are available for various applications in motor control systems. The two main types of fuses are cartridge and plug fuses. Fig. 15-28 shows several cartridge and plug fuses in different sizes. Some of these fuses provide blades to make connections in the fuse holders, and other fuses provide connections at the neck of the fuse. The plug fuse comes in several sizes, embossed on the current rating of the fuse. A rejection feature is provided on cartridge, blade, and plug fuses. Examples of the rejection features are also shown in this figure. The rejection feature prevents a fuse with a larger current rating from being substituted for the fuse of a smaller size that was originally specified.

This is accomplished by providing a matching rejection feature in the fuse holder that is mounted in the fuse box with the feature that is provided for the fuse size. This means that each size of fuse has a different type of rejection feature, or it is located at a different point on the fuse body for each amperage rating.

2.1 Buss Power Distribution Fuses

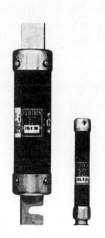

HI-CAP®
(Time-Delay)
KRP-C (600V)
601 to 6000A
200,000AIC
Current Limiting
UL Class L; CSA-HRC-L

The all-purpose silver linked fuse for both overload and short-circuit protection of high capacity systems (mains and large feeders). Time-delay (minimum of four seconds at five times amp rating for close sizing. Unlike fast acting fuses, pass harmless surge currents of motors, transformers, etc., without overfusing and any sacrifice of short-circuit current limitation (component protection). The combination use of ¹/₁₀ to 600 ampere LOW-PEAK dual-element time-delay fuses and 601 to 6000A KRP-C HI-CAP fuses is recommended as a total system specification. Easily selectively coordinated for blackout protection. Size of upstream fuse need only be twice that of downstream HI-CAP or LOW-PEAK fuses (2:1 ratio). HI-CAP fuses can reduce bus bracing; protect circuit breakers with low interrupting rating as well as provide excellent overall protection of circuits and loads.

LIMITRON®
(Fast-Acting)
KTU (600V)
601 to 6000A
Current Limiting
200,000AIC
UL Class L; CSA-HRC-L

Silver-linked fuse. Single-element units with no time-delay. Very fast-acting with a high degree of current limitation; provide excellent component protection. Particularly suited for protection of circuit breakers with lower interrupting ratings, and non-inductive loads such as lighting and heating circuits. Can be used for short-circuit protection only in circuits with the inrush currents. Must be oversized to prevent opening by the temporary harmless overloads with some sacrifice of current limitation. In motor circuits, must be sized at approximately 300% of motor full-load current and thus will not provide the overload protection of HI-CAP KRP-C fuses.

LIMITRON®
(Time-Delay)
KLU (600V)
601 to 4000A
Current Limiting
200,000AIC
UL Class L; CSA-HRC-L

10 seconds delay (minimum) at 500% of amp rating. Not as current limiting as KRP-C or KTU fuses.

LOW-PEAK®
(Dual-Element, Time-Delay)
LPS-RK (600V)
LPN-RK (250V)
¹/₁₀ to 600A
200,000AIC
Current Limiting
UL Class RK1; CAS HRC-I ("D")

High performance, all-purpose fuses. Provide the very high degree of short-circuit limitation of LIMITRON fuses plus the overload protection of FUSETRON fuses in all types of circuits and loads. Can be closely sized to full-load motor currents for reliable motor overload protection as well as backup protection. Close sizing permits the use of smaller and more economical switches (and fuses); better selective coordination against blackouts; and a greater degree of current limitation (component protection), LOW-PEAK fuses are rejection type but fit non-rejection type fuseholders. Thus, can be used to replace Class H, K1, K5, RK5 or other RK1 fuses.

FUSETRON®
(Dual-Element, Time-Delay)
FRS-R (600V)
FRN-R (250V)
¹/₁₀ to 600A
200,000AIC
Current Limiting
UL Class RK5; CSA-HRC-I ("D")

Time-delay affords the same excellent overload protection of LOW-PEAK fuses of motors and other type loads and circuits having temporary inrush currents such as caused by transformers and solenoids. (In such circuits, LIMITRON fuses can only provide short-circuit protection). FUSETRON fuses are not as fast-acting as LOW-PEAK fuses and therefore cannot give as high a degree of component short-circuit protection. Like LOW-PEAK fuses, FUSETRON fuses permit the use of smaller size and less costly switches. FUSETRON fuses fit rejection type fuseholders and can also be installed in holders for Class H fuses. They can physically and electrically replace Class H, K5, and other Class RK5 fuses.

LIMITRON®
(Fast-Acting)
KTS-R (600V)
KTN-R (250V)
¹/₁₀ to 600A
200,000AIC
Current Limiting
UL Class RK1; CSA HRC-I

Single-element, fast-acting fuses with no time-delay. The same basic performance of the 601-6000A KTU fast-acting LIMITRON fuses. Provides a high degree of short-circuit current limitation (component protection). Particularly suited for circuits and loads with no heavy surge currents of motors, transformers, etc. LIMITRON fuses are commonly used to protect circuit breakers with lower interrupting ratings. If used in circuits with surge currents (motors, etc.) must be oversized to prevent opening and thus only provide short-circuit protection. Incorporate Class R rejection feature. Can be inserted in non-rejection type fuseholders. Thus, can physically and electrically replace fast acting Class H, K1, K5, RK5, and other RK1 fuses.

ONE-
(General Purpose)
NOS (600V)
NON (250V)

¹/₁₀ to 600A
10,000AIC
Non Current
Limiting
UL Class H
(K-5 in Sizes 1-60A)

With an inter 10,000 amp not considering, Class H are used in available short-Single-element fuses do not delay. The 1-a 50,000 AIC, Class K-5.

FIGURE 15-28 Types of cartridge and plug fuses available for use in industrial circuits. (Courtesy of Bussmann.)

TIME

rupting rating of and generally current limit-ONE-TIME fuses circuits with low circuit currents. ONE-TIME incorporate time-60A ratings have and are U.L.

SUPERLAG
(General Purpose)
RES (600V)
REN (250V)
1 to 600A
10,000AIC
Non-Current Limiting
UL Class H

Time-delay is excellent for a class H fuse; affords slower response to temporary overloads. After opening, SUPERLAG fuse links can be replaced and the fuse reused.

Plug Fuses
125V 10,000 AIC

FUSTAT Type S fuses have a size limiting feature which prevents "overfusing." Dual element construction provides the time delay necessary for motor running protection. Sizes from ¼ thru 30 amps.

FUSETRON Type T fuses are similar to Type S fuses except for the Edison (light bulb type) base.

Type W fuses are non-time delay, used with non-inductive loads.

HI-CAP®
(Time-Delay)
JHC (600V) UL Class J Dim.
CSA HRC-I
HI-CAP JHC fuses are similar to Class J LIMITRON fuses except they have the advantage of time-delay permitting them to pass temporary overloads without oversizing. Offer both backup motor overload and short circuit protection. JHC fuses are listed by CSA and therefore comply with NEC Section 90-6.

LIMITRON®
(Quick-Acting)
JKS (600V)
1 to 600A
200,000AIC
Current Limiting
UL Class J; CSA HRC-I
JKS LIMITRON fuses are basically the same as RK1 LIMITRON fuses except somewhat smaller in physical size (but considerably larger than Buss T-TRON fast-acting fuses). JKS fuses are single-element units with no time-delay and are thus best applied in circuits free of the temporary overloads of motor and transformer surges. The smaller dimensions of Class J fuses prevent their replacement with conventional fuses.

T-TRON®
(Fast-Acting)
JJS (600V) 1-800A
JJN (300V) 1-1200A
200,000AIC
Current Limiting
UL Class T; CSA HRC-I
The space-savers. Counterpart of the KTN-R/KTS-R LIMITRON fuses but only one-third the size; thus, particularly suited for critically restricted space. A single-element fuse; extremely fast-acting. Provide a high degree of current limitation on short-circuits for excellent component protection. Must be over-sized in circuits with inrush currents common to motors, transformer, and other inductive components (will give only short-circuit protection). Commonly applied in electric heat circuits, load center, disconnect switches, meter, stacks, etc. The small size of T-TRON fuses permits them to be installed in panelboards and control centers for system upgrading when existing circuit breakers cannot safely interrupt larger available short-circuit currents.

LIMITRON®
(Fast-Acting)
KTK-R (600V)
1/10 to 30A
200,000AIC
Current Limiting
UL Class CC; CSA HRC-I
U.L. listed for branch circuit protection. A very small, high performance, fast-acting, single-element fuse for protection of branch circuits, motor control circuits, lighting ballasts, control transformers, street lighting fixtures. . . . A diameter of only ¹³⁄₃₂" and a length of 1½" give cost and space savings. A grooved ferrule permits mounting in "rejection" type fuseholders as well as standard non-rejection type holders.

CC-TRON™
(Time-Delay)
FNQ-R (600V)
¼ to 7½ A
200,000 AIC
Current Limiting
UL Class CC
Ideal for control transformer protection. Meets requirements of NEC 430-72 (b) & (c) and UL 508. It's miniature design and branch circuit rating make it ideal for motor branch circuit and short circuit protection required by NEC 430-52.

Type SC®
300V
1-60A
100,000AIC
Current Limiting UL Class G
A high performance general-purpose branch circuit fuse for lighting appliance, and motor branch circuits of 300 volts (or less) to ground. Fuse diameter is ¹³⁄₃₂"; lengths vary with ampere rating from 1⁵⁄₁₆ to 2¼" (serves as rejection feature and thus prevents dangerous oversizing).

Medium Voltage Fuses
R-Rated Fuses for Motor Circuits
    2400V
    4800V
    7200V
E-Rated Fuses for Potential Transformers
    2475V
    5500V
    8300V
    15,500V
E-Rated Fuses for Transformers & Feeder Protection
    2750V
    5500V
    8300V
    15,500V
Medium & High Voltage Links
    K, T, H, N Types

Buss Cable Limiters
UH Series
(250V)   For Copper
100 KAIC   or
K Series   Aluminum
(600V)   Cable
200 KAIC
Protect low voltage distribution networks and all types of service entrance cables. Totally self contained. UH Series is often used for residential applications.

K Series is used by utilities in downtown networks and for the protection of conductors between utility transformer & customer switchgear.

FIGURE 15-28 (Continued)

15.6 ENCLOSURES

Enclosures are required to house disconnects, motor starters, and other motor controls. The enclosures are available in a variety of designs that will protect the devices inside from all types of environmental conditions. Fig. 15-29 shows a list of typical enclosure types as classified by the National Electrical Manufacturers Association (NEMA), and Fig. 15-30 shows examples of each type of enclosure.

Type 1 enclosures are designed for general-purpose applications in indoor locations. This means that the enclosure should not be exposed to extreme conditions, such as excessive moisture. Type 2 enclosure is rated for drip-proof conditions that exist indoors. This means that some moisture may come in contact with the enclosure, but it is not approved where equipment must be washed down or steam cleaned daily.

Type 3 enclosures are designed for dust-tight, rain-tight, and sleet-tight conditions that exist outdoors, and type 3R rain-proof, sleet-resistant, and type 3S enclosures are designed for dust-tight, raintight, and sleet-proof conditions. These enclosures are intended for protection against falling rain, sleet, or dust. They are not

NEMA enclosures for starters	
Type	Enclosure
1	General purpose—indoor
2	Drip-proof—indoor
3	Dust-tight, raintight, sleet-tight—outdoor
3R	Rainproof, sleet resistant—outdoor
3S	Dust-tight, raintight, sleetproof—outdoor
4	Watertight, dust-tight, sleet resistant—indoor
4X	Watertight, dust-tight, corrosion resistant—indoor/outdoor
5	Dust-tight—indoor
6	Submersible, watertight, dust-tight, sleet resistant—indoor/outdoor
7	Class 1, group A, B, C, or D hazardous locations, air-break—indoor
8	Class 1, group A, B, C, or D hazardous locations, oil-immersed—indoor
9	Class II, group E, F, or G hazardous locations, air-break—indoor
10	Bureau of Mines
11	Corrosion-resistant and drip-proof, oil immersed—indoor
12	Industrial use, dust-tight, and driptight—indoor
13	Oiltight and dust-tight—indoor

FIGURE 15-29 List of enclosure types provided by the National Electric Manufacturers Association (NEMA). (Courtesy of Rockwell Automation's Allen Bradley Businesss.)

NEMA Type 4X
Non-Metallic,
Corrosion-Resistant
Fiberglass Reinforced
Polyester

Type 4X enclosures are intended for indoor or outdoor use primarily to provide a degree of protection against corrosion, windblown dust and rain, splashing water, and hose-directed water. They are designed to meet the hosedown, dust, external icing ■, and corrosion-resistance design tests. They are not intended to provide protection against conditions such as internal condensation or internal icing. Enclosure is fiberglass reinforced polyester with a synthetic rubber gasket between cover and base. Ideal for such industries as chemical plants and paper mills.

NEMA Type 6P

Type 6P enclosures are intended for indoor or outdoor use primarily to provide a degree of protection against the entry of water during prolonged submersion at a limited depth. They are designed to meet air pressure, external icing ■, and corrosion-resistance design tests. They are not intended to provide protection against conditions such as internal condensation or internal icing.

NEMA Type 7
For Hazardous
Gas Locations
Bolted Enclosure

Type 7 enclosures are for indoor use in locations classified as Class I, Groups C or D, as defined in the National Electrical Code. Type 7 enclosures are designed to be capable of withstanding the pressures resulting from an internal explosion of specified gases, and contain such an explosion sufficiently that an explosive gas-air mixture existing in the atmosphere surrounding the enclosure will not be ignited. Enclosed heat generating devices are designed not to cause external surfaces to reach temperatures capable of igniting explosive gas-air mixtures in the surrounding atmosphere. Enclosures are designed to meet explosion, hydrostatic, and temperature design tests. Finish is a special corrosion-resistant, gray enamel.

NEMA Type 9
For Hazardous
Dust Locations

Type 9 enclosures are intended for indoor use in locations classified as Class II, Groups E, F or G, as defined in the National Electrical Code. Type 9 enclosures are designed to be capable of preventing the entrance of dust. Enclosed heat generating devices are designed not to cause external surfaces to reach temperatures capable of igniting or discoloring dust on the enclosure or igniting dust-air mixtures in the surrounding atmosphere. Enclosures are designed to meet dust penetration and temperature design tests, and aging of gaskets. The outside finish is a special corrosion-resistant gray enamel.

NEMA Type 12

Type 12 enclosures are intended for indoor use primarily to provide a degree of protection against dust, falling dirt, and dripping noncorrosive liquids. They are designed to meet drip ■, dust, and rust-resistance tests. They are not intended to provide protection against conditions such as internal condensation.

NEMA Type 13

Type 13 enclosures are intended for indoor use primarily to provide a degree of protection against dust, spraying of water, oil, and noncorrosive coolant. They are designed to meet oil exclusion and rust-resistance design tests. They are not intended to provide protection against conditions such as internal condensation.

NEMA Type 1
Surface Mounting

Type 1 enclosures are intended for indoor use primarily to provide a degree of protection against contact with the enclosed equipment in locations where unusual service conditions do not exist. The enclosures are designed to meet the rod entry and rust-resistance design test. Enclosure is sheet steel treated to resist corrosion.

NEMA Type 1
Flush Mounting

Flush mounted enclosures for installation in machine frames and plaster wall. These enclosures are for similar applications and are designed to meet the same tests as NEMA Type 1 surface mounting.

NEMA Type 3

Type 3 enclosures are intended for outdoor use primarily to provide a degree of protection against windblown dust, rain, sleet, and external ice formation. They are designed to meet rain ■, external icing ■, dust, and rust-resistance design tests. They are not intended to provide protection against conditions such as internal condensation or internal icing.

NEMA Type 3R

Type 3R enclosures are intended for outdoor use primarily to provide a degree of protection against falling rain, sleet, and external ice formation. They are designed to meet rod entry, rain ■, external icing ■, and rust-resistance design tests. They are not intended to provide protection against conditions such as dust, internal condensation, or internal icing.

NEMA Type 4

Type 4 enclosures are intended for indoor or outdoor use primarily to provide a degree of protection against windblown dust and rain, splashing water, and hose-directed water. They are designed to meet hosedown, dust, external icing ■, and rust-resistance design tests. They are not intended to provide protection against conditions such as internal condensation or internal icing. Enclosures are made of heavy gauge stainless steel, cast aluminum or heavy gauge sheet steel, depending on the type of unit and size. Cover has a synthetic rubber gasket.

NEMA Type 3R, 7 & 9
Unilock Enclosure
For Hazardous
Locations

This enclosure is cast from "copper-free" (less than 0.1%) aluminum and the entire enclosure (including interior and flange areas) is bronze chromated. The exterior surfaces are also primed with a special epoxy primer and finished with an aliphatic urethane paint for extra corrosion resistance. The V-Band permits easy removal of the cover for inspection and for making field modifications. This enclosure meets the same tests as separate NEMA Type 3R, and NEMA Type 7 and 9 enclosures. For NEMA Type 3R application, it is necessary that a drain be added.

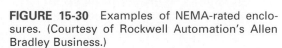

FIGURE 15-30 Examples of NEMA-rated enclosures. (Courtesy of Rockwell Automation's Allen Bradley Business.)

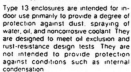

intended to prevent condensation from forming on the inside of the enclosure or on internal components in the enclosure.

If the application is located where windblown water or sleet may be encountered, a type 4 enclosure should be used. The type 4 and 4X enclosures are designed to protect against windblown dust, rain, or water from direct-hose-down conditions. This enclosure is specified by NEMA for water-tight, dust-tight, sleet-resistant indoor and outdoor applications. The 4X enclosure also provides protection in environments where corrosion resistance is required. The covers of these enclosures have protective gaskets that provide a barrier against these conditions. All hardware for conduits must match the same requirements to maintain the integrity of the design throughout the installation.

Type 5 enclosures are designed for dust-tight applications located indoors. This type of enclosure is designed to protect the switch and contacts from a buildup of dust that may prevent proper operation of the switches and starters that are enclosed.

Type 6 enclosures are designed to provide submersible, water-tight, dust-tight, sleet-resistant indoor and outdoor applications. The submersible water-tight specification allows the switch to be fully submersed in water to a limited depth and maintain water-tight integrity. This type of enclosure also depends on gaskets to maintain the water-tight condition. As with other enclosures, conduits and connectors must meet the same NEMA standard so that the entire installation maintains the stated protection.

Type 7 enclosures provide class I protection for group A, B, C, or D hazardous locations, with air-break protection for indoor locations. These enclosures are designed to prevent dangerous gases from penetrating the enclosure and coming into contact with the open parts of switches or other controls. This provides protection in explosive atmospheres. Class I locations are hazardous because flammable gases or vapors may be present in sufficient quantities to cause explosions. Type 8 enclosures provide class 1, group A, B, C, or D hazardous locations, oil immersed for indoor applications. This type of enclosure is very similar to the type 7 enclosure except that it will provide protection against oil immersion instead of air break. This allows the enclosure to be used where oil and machine coolants are used extensively.

Type 9 enclosures provide protection against class II, group E, F, or G hazardous locations and air break for indoor applications. Class II locations are hazardous because of the presence of combustible dust in quantities sufficient to explode or ignite.

Type 10 enclosures are rated for all applications within mines. This type of enclosure must be able to protect switch gear against explosive conditions. Type 11 enclosures provide corrosion resistance and drip-proof, oil-immersion protection for indoor applications. This type of enclosure is used where the vapors and fumes may be corrosive to switch gear or other motor controls that are mounted inside. The exterior of the enclosure is also resistant to corrosion from these fumes. This type of enclosure is also used for applications where machining operations are performed.

FIGURE 15-31 American Wire Gauge (AWG) wires sizing table provided by the National Electrical Code (NEC). (National Electrical Code®, and NEC®, are registered trademarks of the National Fire Protection Association, Quincy, MA 02269.)

Table 310-17. Allowable Ampacities of Single Insulated Conductors, Rated 0 through 2000 Volts, in Free Air Based on Ambient Air Temperature of 30°C (86°F)

Size	Temperature Rating of Conductor. See Table 310-13.						Size
	60°C (140°F)	75°C (167°F)	90°C (194°F)	60°C (140°F)	75°C (167°F)	90°C (194°F)	
AWG kcmil	TYPES TW†, UF†	TYPES FEPW†, RH†, RHW†, THHW†, THW†, THWN†, XHHW† ZW†	TYPES TBS, SA, SIS, FEP†, FEPB†, MI, RHH†, RHW-2, THHN†, THHW†, THW-2†, THWN-2†, USE-2, XHH, XHHW†, XHHW-2, ZW-2	TYPES TW†, UF†	TYPES RH†, RHW†, THHW†, THW†, THWN†, XHHW†	TYPES TBS, SA, SIS, THHN†, THHW†, THW-2, THWN-2, RHH†, RHW-2, USE-2, XHH, XHHW†, XHHW-2, ZW-2	AWG kcmil
	COPPER			ALUMINUM OR COPPER-CLAD ALUMINUM			
18	. . . .	. . . .	18	. . . .	. . . .	. . . .	. . . .
16	. . . .	. . . .	24	. . . .	. . . .	. . . .	. . . .
14	25†	30†	35†	. . . .	. . . .	. . . .	. . . .
12	30†	35†	40†	25†	30†	35†	12
10	40†	50†	55†	35†	40†	40†	10
8	60	70	80	45	55	60	8
6	80	95	105	60	75	80	6
4	105	125	140	80	100	110	4
3	120	145	165	95	115	130	3
2	140	170	190	110	135	150	2
1	165	195	220	130	155	175	1
1/0	195	230	260	150	180	205	1/0
2/0	225	265	300	175	210	235	2/0
3/0	260	310	350	200	240	275	3/0
4/0	300	360	405	235	280	315	4/0
250	340	405	455	265	315	355	250
300	375	445	505	290	350	395	300
350	420	505	570	330	395	445	350
400	455	545	615	355	425	480	400
500	515	620	700	405	485	545	500
600	575	690	780	455	540	615	600
700	630	755	855	500	595	675	700
750	655	785	885	515	620	700	750
800	680	815	920	535	645	725	800
900	730	870	985	580	700	785	900
1000	780	935	1055	625	750	845	1000
1250	890	1065	1200	710	855	960	1250
1500	980	1175	1325	795	950	1075	1500
1750	1070	1280	1445	875	1050	1185	1750
2000	1155	1385	1560	960	1150	1335	2000

CORRECTION FACTORS

Ambient Temp. °C	For ambient temperatures other than 30°C (86°F), multiply the allowable ampacities shown above by the appropriate factor shown below.						Ambient Temp. °F
21-25	1.08	1.05	1.04	1.08	1.05	1.04	70-77
26-30	1.00	1.00	1.00	1.00	1.00	1.00	78-86
31-35	.91	.94	.96	.91	.94	.96	87-95
36-40	.82	.88	.91	.82	.88	.91	96-104
41-45	.71	.82	.87	.71	.82	.87	105-113
46-50	.58	.75	.82	.58	.75	.82	114-122
51-55	.41	.67	.76	.41	.67	.76	123-131
56-60	. . . .	.58	.71	. . . .	.58	.71	132-140
61-70	. . . .	.33	.58	. . . .	.33	.58	141-158
71-80	. . . .	. . . .	.41	. . . .	. . . .	.41	159-176

†Unless otherwise specifically permitted elsewhere in this *Code*, the overcurrent protection for conductor types marked with an obelisk (†) shall not exceed 15 amperes for No. 14, 20 amperes for No. 12, and 30 amperes for No. 10 copper; or 15 amperes for No. 12 and 25 amperes for No. 10 aluminum and copper-clad aluminum.

747

Type 12 enclosures are intended for indoor applications to provide protection against dust, falling dirt, and dripping noncorrosive materials. They are not intended for use against direct spraying of these materials. The last classification of enclosures is type 13.

15.7 CONDUCTORS

One part of the power distribution system involved at all points in the system is the conductor. Conductors can be made of aluminum or copper wire. Aluminum is generally used for long-distance distribution because of its lighter weight. Once the power is inside the plant, copper conductors are generally used for distribution. Copper conductors will be solid in bus bar and busway applications, and they will be stranded conductors for all other applications.

The conductors are sized by the amount of amperage they can carry. A table of typical conductor sizes and their ampacities is provided in Fig. 15-31. The standard used to determine the size of each conductor is called the American Wire Gauge (AWG). These tables have been established by the National Electrical Code (NEC), and they are used to select the proper size conductor to carry the load.

Conductors are also classified by the type of insulation that is used as a cover. The type of cover also determines the voltage rating of the wire. Typical voltage ratings for conductors used in motor control applications include 300, 600, and 1000 V. The voltage and current ratings of the conductor should not be exceeded under any circumstances.

The type of covering is also listed in Fig. 15-31. Abbreviations for each of these coverings are listed in the NEC tables and help to determine the type of wire that should be selected for each application. The outer covering of the wire serves several purposes, including protecting the conductors from coming in contact with metal in the cabinet or other conductors. The cover also provides a location to stamp all specification data regarding the wire, including the voltage rating, AWG size, temperature specification, and type of covering.

15.8 LOCKOUT TAG-OUT

When you are working on a machine in a factory you will need to turn off its electrical power at the disconnect. You will also need to place a padlock on the disconnect to ensure no one will turn the power on until you are ready. The padlock will have a tag attached to it that has your name and picture on it and the nature of the work you are doing. When co-workers find the padlock and tag they will know who has the power turned off. This procedure is called lockout tag-out.

SOLUTION TO JOB ASSIGNMENT

The three-wire control circuit that you need to use is shown in Fig. 15-10 and in Section 15.3.2. Since your application needs additional start and stop switches connected to the circuit, you should refer to Fig. 15-11 and Section 15.3.3.

Since the two conveyors need to be sequenced so one will always be started before the other one can be turned on, you will need to refer to Fig. 15-20 and read Section 15.3.8 that explains sequence control.

Information regarding fuses is located in section 15.5, and information about enclosures is in Section 15.6. You can use the table in Fig. 15-29 to select the enclosure for the motor starters in your application. Use section 15.7 to provide information about conductor sizes and ratings, and Fig. 15-31 provides a table to select the correct conductor size.

QUESTIONS

1. Identify five types of limit switches and explain where each would be used.
2. Use the diagrams in Fig. 15-17 and explain how the drum switch can reverse the direction of rotation for a single-phase motor, a three-phase motor, and a DC motor.
3. Use the diagram in Fig. 15-18 to explain how the jog circuit must modify the traditional three-wire control circuit so it will not latch in.
4. Explain the operation of each of the three level-type pilot devices shown in Fig. 15-22.
5. Use Fig. 15-29 to select the correct enclosure for the following applications: dust-tight indoors, rainproof and sleet-resistant outdoors, general-purpose indoors, oil-tight and dust-tight indoors.
6. Explain the two types of interlocks that are used in the forward and reversing circuit shown in Fig. 15-18 that prevent the forward and reverse motor starter coils from being energized at the same time.
7. Explain how motor controls must interface with industrial electronic controls to make modern automated machinery run.
8. A dual-element fuse has a slow-blow section and a short-circuit section. Explain how each section works.
9. Explain what type of problem you would look for if you were troubleshooting the start/stop circuit in Fig. 15-10 if the motor starter coil would only remain energized as long as the start push button is depressed.
10. Explain how you would test the level switches shown in Fig. 15-22 if you suspected they were not closing to energize the pump.

TRUE OR FALSE

1. _____ A short-circuit current will occur if an L1 wire comes into contact with an L2 wire.
2. _____ When a slow overcurrent occurs the neck-down section of a dual-element fuse will open.

3. _____ A jogging circuit is used to interlock a motor starter during a sequence move.

4. _____ A three-wire control circuit uses NO auxiliary contacts of the motor starter to seal in the start push button.

5. _____ In the diagram in Fig. 15-8, the level switch will close when the level that is being sensed increases.

6. _____ A drum switch can be used to reverse the direction of a single-phase AC motor, a three-phase AC motor, or a DC motor.

7. _____ The auxiliary contacts found between terminals 2 and 3 on a motor starter are used to seal in the start push button in a three-wire control circuit.

8. _____ NEMA stands for National Electrical Manufacturers Association.

9. _____ When a switch like the one in Fig. 15-2 is found to be faulty on a piece of equipment, you must remove the wires from the contact block so they can be connected on the new switch you are installing.

10. _____ When a drum switch is used to reverse the direction of rotation of a single-phase AC motor, both the run windings and the start windings are reversed.

MULTIPLE CHOICE

1. Interruption capacity is _____.
   a. the ability of a fuse to withstand larger currents and clear them safely.
   b. the ability of a fuse to open during a slow overcurrent.
   c. the ability of a fuse to detect the difference between short-circuit current and slow overcurrent.

2. In the diagram in Fig. 15-20, motor starter M1 will always be running before motor starter M2 can become energized because _____.
   a. M1 coil requires a larger voltage than M2 coil.
   b. M1 coil is in series with M2 coil.
   c. M1 contacts are in series with M2 coil.

3. When two extra stop push buttons are added to a start/stop circuit, the stop buttons must be wired _____.
   a. in series with the other stop push button.
   b. in parallel with the other stop push button.
   c. in parallel with the start push button.

4. In Fig. 15-9 which pressure switch will protect the circuit when it is in the auto position?
   a. only pressure switch B.
   b. only pressure switch A.
   c. both pressure switches A and B.

5. When two extra start push buttons are added to a start/stop circuit, the start push buttons must be wired _____.
   a. in series with the other stop push button.
   b. in parallel with the other stop push button.
   c. in parallel with the start push button.

6. A _____ provides a sequential diagram of a circuit that is easy to troubleshoot.
   a. ladder diagram
   b. wiring diagram
   c. pictorial diagram

7. A 12-gauge wire with THWN insulation can safely carry _____ amps.
   a. 30
   b. 25
   c. 20

8. A 10-gauge wire is _____ than a 6-gauge wire.

    a. smaller

    b. larger

    c. can't tell unless you know the type of insulation on each wire.

9. A type _____ enclosure is water-tight, dust-tight, and corrosion resistant on indoor and outdoor applications.

    a. 3

    b. 4

    c. 4X

10. The NC contacts on a motor starter that are identified as OL are used to _____.

    a. open the control circuit when the motor draws too much current.

    b. open the load circuit when the motor draws too much current.

    c. open the control circuit when the motor starter coil draws too much current.

PROBLEMS

1. Draw the symbols for an NO limit switch, an NO held closed limit switch, an NC limit switch, and an NC held open limit switch.

2. Draw the symbols for NO and NC pressure switches that are used for a high-pressure application, and NO and NC pressure switches that are used for a low-pressure application.

3. Draw a two-wire control circuit of a limit switch and a motor as a ladder diagram.

4. Use the table in Fig. 15-31 to select a wire that will carry 30 A and has THHN insulation that can withstand 90°C.

5. Draw a two-wire control circuit of a limit switch and a motor as a wiring diagram.

6. Draw a ladder diagram of a three-wire control circuit that has a stop push button, a start push button, and a motor starter.

7. Draw a wiring diagram of a three-wire control circuit that has a stop push button, a start push button, and a motor starter.

8. Draw the diagrams of a drum switch when it is in the reverse position, the off position, and the forward position.

9. Draw a sketch of a single-element fuse and explain how it clears excess current.

10. Draw a sketch of a dual-element fuse and explain how it clears short-circuit current and how it senses slow overcurrent.

REFERENCES

1. Kissell, Thomas E., *Modern Industrial Electrical Motor Controls*. Englewood Cliffs, NJ: Prentice Hall Inc., 1990.

2. Maloney, Timothy J., *Industrial Solid-State Electronics*, 2nd ed. Englewood Cliffs, NJ: Prentice Hall Inc., 1986.

16

DATA COMMUNICATIONS FOR INDUSTRIAL ELECTRONICS

After reading this chapter, you will be able to:

1. Explain the seven-layer OSI model for network architecture and give an example of how each layer is used in factory data communications.
2. Explain how the ASCII code and the EBCDIC data formats are similar and how they are different.

3. List several example protocols such as MAP, TOP TCP/IP, and SNA and explain where they are used.
4. Sketch the RS232 interface and explain what each signal is used for.
5. Identify five types of network topologies and explain how each works.

Your immediate supervisor has read articles about connecting PLCs to networks to help provide maintenance supervision control of all of the machines on the factory floor and you are requested to see what is required to connect three PLCs in a small network for test purposes. The next day you are asked to attend a meeting on plantwide communications where you are asked to research the possibility of connecting the PLC network to the plantwide network so that production control and shipping and receiving can determine the status of production.

Your task is to select several examples of possible networks that you could use with the PLCs and to determine if they have the capability of connecting into the plantwide network. You are also asked to submit a list of data in each PLC that would be useful for the production planning and shipping and receiving departments. Be sure to include a sketch of the types of networks you have in mind and include any information to help the planning committee better understand data communications for the plant floor production equipment. You will also need to get a copy of the RS232 interface to show how many wires must be included in each interface between the PLCs and their network interface.

16.1 OVERVIEW OF DATA COMMUNICATIONS USED IN INDUSTRY

In the past few years data communications in industrial applications has expanded a hundredfold. Originally the data communications occurred only in the front office that sent daily or weekly reports to some remote headquarters. Today communications between programmable logic controllers and other electronic control systems on the factory floor is commonplace. Since programmable logic controllers (PLCs) are used on most machines for control, they also can provide secondary services such as sending and receiving information about the machine's status and operation. The information that is sent to the PLC may include information about maintenance conditions as well as production information. For example, it is commonplace to link each machine to a local area network (LAN) with a terminal located in the maintenance office and another terminal in production offices. The production personnel may send requests to make a specific number of parts, and then receive end-of-shift reports from each PLC as to how many good and bad parts that machine has produced. The terminal in the maintenance office receives information about machine faults and downtime. Other information about periodic maintenance information is also sent to the maintenance terminal.

When a machine has a fault and goes down, the maintenance office is notified automatically by the machine controller of the exact fault so that the proper personnel can be dispatched to work on the machine. The maintenance dispatcher can also determine priorities when more than one machine has a fault. Fig. 16-1 shows the diagram of a typical communications system used in a factory. In this diagram you can see that each machine has a PLC that is connected in a local area network (LAN). The LAN connects to one terminal that is also connected to a data network that reports information to production planning through an Ethernet link. The production planning office receives the data on a larger mainframe computer that is also connected to other computers at remote sites at other locations. This allows a company with various plant sites to connect all of them so that they can determine how many parts are made at any time and the status of orders from the time raw material is purchased to the point the finished parts are shipped.

At one time it was envisioned that all data about cycle times and other production related data would be moved from each PLC to the factory mainframe for storage. But it soon became apparent that the mainframe was completely bogged down with worthless information. Today the PLC analyzes its own data and sends only meaningful production reports that indicate the amount of time the machine was in production, the amount of good and bad parts, and the amount of energy usage.

This chapter will explain how data are sent and received from PLC to PLC and from other computers and networks in the factory. The easiest way to understand data communications as it applies to industrial applications is to follow the

FIGURE 16-1 A local area network of PLCs and a wide area network of Ethernet from factory to factory.

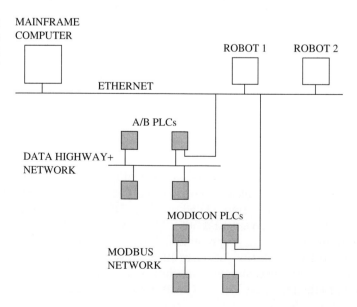

ISO (International Organization for Standards) model. The model the organization provided is known as the Open Systems Interconnect (OSI) reference model. You will see that communications for other business systems, such as bank teller machines and other automated transactions, use similar systems. However, the systems used in industrial machines have become more unique because of the wide variety of different types of machines and controls. You will also find that as the need for communications to spread farther and farther increases, systems have become more complex, and interfaces have multiplied so that incompatible systems can become compatible. This chapter will help you understand what is expected of you as a technician when the industrial electronic control system includes a data communications system.

You can see that it is important at this time to fully define what is meant by the term *data* in data communications. In industrial applications data may be as simple as signals that indicate a machine is running or stopped. Data may also include the number of parts a machine is requested to make or the number of parts the machine actually made, and the number of those parts that are good and bad. Data may also include the average cycle time for the machine, the ratio of uptime to downtime, or the ratio of production time to maintenance time. Data may also include a complete new PLC program for a machine that gets loaded from a remote source such as a maintenance or engineering office that is located in a separate part of the factory.

Other types of data may include an electronic form of a maintenance manual that is stored on a CD-ROM in a computer rather than as a hard-copy book located on a shelf. Data may also be a production request from factory headquarters to tell a machine to make a specific number of parts, or a final report for quality control on a shipment of finished parts.

In some cases the data will remain in the area of the machine for the operator, maintenance personnel, and quality personnel to use to solve problems on each machine as they arise. These data will also give a strong indication of how the machine is operating.

Data communications may also cross barriers between factory automation data and data communications for personnel such as E-mail systems and LANs between departments. Personnel in various departments may need to work together to install new production equipment, and they must share times and dates for time lines used in progress reports, and budget information as the project is brought on line. Each of these functions may use various parts of the total factory data communications system.

You will also see that some of the terminology and protocols for data communications originated when early telegraph and telephone systems were first invented in the late 1800s. Since these were established early and were enhanced over the years, they have been incorporated into modern data communications.

16.2 NETWORK ARCHITECTURES

In the early days of computers and PLCs, each electronics equipment manufacturer saw the need for getting data into and out of their machine controllers. The earliest attempts usually included a method that could only work on their brand of controller.

7	APPLICATION
6	PRESENTATION
5	SESSION
4	TRANSPORT
3	NETWORK
2	DATA LINK
1	PHYSICAL

FIGURE 16-2   The seven-layer OSI model.

Later when factories added several different brand names of controllers for their robots, CNC machines, and other PLCs, it became apparent that a common architecture must be used so that it would be easier to connect a wide variety of different brand names of equipment into a single network on the factory floor.

All electronic data communications equipment today use one or more layers of the seven-layer OSI model. The OSI model consists of seven layers and each application may use one or all of the layers. Fig. 16-2 shows a diagram of these layers and the remainder of this section will explain how the layers work together. If this is your first introduction to networks and data communications, you can make it easier to understand the ISO model if you think of the lowest levels as the simplest levels, and they will be used in almost every application. The upper four levels are more complex and are only used in more sophisticated systems. This means that you may work with only the first three or four layers most of the time. The layers also provide a means to visualize what each of the parts of the data communications system does.

16.2.1 The Physical Layer

The simplest layer is the physical layer and it has the job of moving the electrical signal from one point to another. This layer includes the wire or cable through which data are sent. This layer also includes the electrical on/off pulses that move through the wire. This level does not provide any error detection, error control, or sequencing of data. This layer is only concerned with the physical medium of transporting the signal from one point to another. The specifications for this layer will include wire specifications and cable termination specifications. As you can see, this layer is part of every data communications system but it is not much use by itself so it must be incorporated with the other layers of this model.

It is important to understand that as a technician you will become more familiar with the lower three layers of this model since you will be installing the physical medium (wires, cable, and fiberoptic cable) and you will troubleshoot and service hardware attached to them. An example of this layer is the blue, two-wire shielded cable that is used for network cables on most PLCs.

16.2.2 Data Link Layer

The data link layer combines with the physical layer to make a more reliable system in that it adds error detection and error recovery methods. An example of this layer would be parity check on a system where a scheme is devised to detect any lost bits of a transmission. The rules for this layer are usually called *protocols*. Many of these protocols existed from the beginning of telegraph and telephone communications in the late 1800s. Since they worked well over the years, they have become the basis for modern systems.

This layer will include a means to start and stop the data flow, a way to package each part of the data (size of the word), a means of sequencing when data are sent and when they are received, a means to acknowledge that data have been received, a means of providing timing that includes when the data are being sent and how long to wait if data have not arrived.

When you are first learning about data communications, it is easy to become confused about this layer because some modern systems automatically combine this layer with the layer above or below it and the separation between layers is not very clear. The main point for you to understand at this time is that it is not important for these layers to be separate, and that it may be advantageous to combine several layers. It also becomes confusing when some systems have incorporated standards from Europe with American standards because of advantages found in several different protocols. You will find that the protocols have evolved and changed to correct problems in data transmission and error detection as a technology has changed.

An example of the data link layer is when equipment incorporates parity check when it sends data from one unit to another. For example, when a controller has a microprocessor with a program stored in it, the data link layer is used when the program is downloaded to a disk or tape and the system uses an even or odd parity check. When parity check is used, the total number of high bits are added up in each packet that is sent and if even parity is used, the total number of high bits must equal an even number. If the number of high bits is an odd number, one extra high bit is added as a parity bit to make the total even. If the parity check is set for odd parity, the total of high bits will be an odd number, or one extra bit is added to make the total odd.

16.2.3 The Network Layer

The primary function of the network layer is to provide a delivery mechanism for the transport layer. As the name implies, this layer is used when the unit that is connected to the data communications system is part of a multiple-unit network rather than one single unit connected to only one other single unit. For example, five PLCs are connected in a LAN and production data such as parts counts are sent from each PLC to a main computer that keeps track of the total production of all five machines.

The network layer also provides a means to separate the upper layers from knowing anything about the transport layer and the physical layer. The network layer uses a combination of hardware and software to provide protocols like the x.21, x.25, and the x.75. These protocols provide means to encapsulate information into packets so that they can be sent over the network. At this point in the model, the technology is designed for all different types of networks. This layer can also be used when one network must interface with another network. For example, the PLCs on the factory floor can be connected to each other. Then the network layer software can provide a means for that network to be interfaced with the office network so that factory floor data (production times and counts) can be sent to the main office network where the information can be included in production reports and cost-accounting reports.

In simple terms this layer provides a means to encapsulate information in data packets so that one computer understands the information being sent to it, and the information packet may not be of any use for other computers in the network. For example, if an Allen-Bradley PLC5 is connected to the same network as an Allen-Bradley SLC500, a program for the PLC5 can be sent down the network and only the PLC5 will recognize the program. All other PLCs on the network will not recognize the program and they will not bother the packets.

16.2.4 The Transport Layer

The transport layer provides connections from one network to another. The transport layer also provides a means of keeping the upper layers from knowing what the lower layers are doing. This is the layer that begins to be different in various types of networks. The first three layers tend to be very specific and standards have been designed that are built into each network. This makes it possible to use a variety of different networks because they are fundamentally similar in the first three layers. In the transport layer and the upper layers, the systems tend to become very complex and may have a variety of differences.

For example, the transport layer for a simple network service may have to be complex if the lower layers are providing a simple interconnect. Since the interconnect is simple, the transport layer must take care of the complex interface to the upper layers if the service is also part of a larger network. On the other hand, if the network service is complex on the lower layers, the transport layer can be rather simple because many of the things that are important between the users are taken care of in the hardware area that is controlled in the lower layers.

An example of the transport layer would be a robot cell that is connected to the same wire as the PLC local area network. In this application the robot programs send and receive data that are only recognized by other robots of the same brand name. The program information can share transport with data from the PLC network on the same conductors. Since the data are packaged differently, the PLC data and the robot data will not get mixed up. In this case the hardware portion of the layer can be similar, such as voltage levels and parity check, and the software for the robot and PLC will be more complex to keep the data from each system from bothering data in the other system.

In more technical terms, the transport layer provides the ability to multiplex information on the network. It will also determine the delay between transmissions, and it can also set priorities for expedited services of specific users. Protocols such as ISO 8073 and CCITT X.24 are used. These protocols include methods of handling residual errors in transmission and recovery.

ISO 8073 defines three levels of service provided in the transport layer. The first level is called the *Type A system,* which is a network where the physical layer, data link layer, and the network layer are the most sophisticated and provide the most reliable service. This type of service allows the transport layer to be very simple because the lower layers provide error detection and correction. The second level is the *Type B system,* which is a system where the lower layers are somewhat less sophisticated and less reliable than Type A, since some residual error rate is acceptable but signal errors are corrected. The transport layer for this type of service must be more sophisticated to overcome these deficiencies. The *Type C* service is where the lower layers are the least reliable and least sophisticated. This means the transport layer for this service must be more complex and very sophisticated to compensate for the deficiencies of the lower layers.

CCITT X.224 uses five classes (Class 0 through Class 4) to describe the same types of services. In this protocol, Class 0 is the simplest transport layer because the hardware layers are more sophisticated, and Class 4 is the most sophisticated transport protocol because the hardware for this type of service is very simple.

16.2.5 The Session Layer

The session layer establishes sessions between users, maintains the sessions between users, and terminates the sessions between users. The major function of this layer also includes connections of two separate open systems. For example, a production supervisor or a maintenance supervisor on the factory floor can use a terminal from the factory floor network to obtain information from the office network to determine shopping dates of finished goods (see Fig. 16-3). ISO 8326 and CCITT.X212 are standards for the session layer. When one unit on one network requests data from a unit on a second network, the session layer directs the transport layer to make the proper connections and regulate the flow of data to smoothly interface the connections and data flow between the two networks.

16.2.6 The Presentation Layer

The presentation layer transforms data information so that one network can pass information to or receive information from another network. The protocols for this layer are contained in ISO 8823 and CCITT X.226. This layer works closely with the session layer to make the connection between systems work smoothly. This layer allows a variety of terminal displays to be used to access data across two different networks. The presentation layer makes the data format in the terminal acceptable to the mainframe computer on the other network. In some cases this layer provides the majority of data conversion so that equipment on different

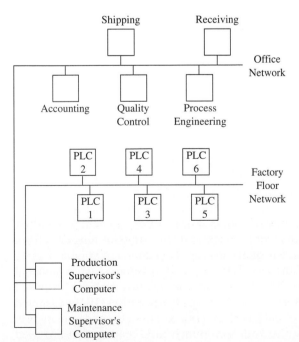

FIGURE 16-3 An example of the terminals used by the production supervisor and the maintenance supervisor on the factory floor to access data in the factory floor network and then connect to the office network with the same terminal. The session layer keeps data separate between the two networks so that the supervisors' terminal can be used on both networks with no modification.

networks will be compatible with each other. You can see that this layer is vital if you want to have the networks open to the largest number of different users.

16.2.7 The Application Layer

The application layer provides the functions of making all of the other layers compatible when users must interconnect between multiple networks. For example, if any of the lower layers does not provide a means of making a video terminal fully compatible with every application on a mainframe computer, the application layer can provide the final conversions and interfaces so that the terminal is fully compatible. One example of the application layer is that it can make a terminal that was installed on a factory floor network in the late 1980s compatible with the office network that was installed as an electronic mail system in the 1990s. Even though the terminal was designed and manufactured before the electronic mail system was installed, a means must be available to make all of the terminals in the older network compatible with all of the newest software on the newest networks. The protocols in the application layer provide this function.

16.2.8 Review of the ISO Model

At first glance the layers of the ISO model look very complex. As a technician you must understand that much of what happens in these layers is virtually invisible to you. For example, if a terminal is designed to work on an Ethernet network, you do not need to worry how it works technically. All you must be aware of is what type of interface boards or software are required to make the systems compatible.

If software indicates it is usable on a specific network such as a Novell network, all you need to do is make sure the computer with this software has the interface boards to connect it to the Novell network. And be sure any other computers that you want to communicate with are also connected to the Novell network and have this same software loaded. At this point the software can make the connections between the two computers almost invisible to the user. If problems arise, the software will have diagnostic functions loaded in it, or you can purchase network troubleshooting software to check the diagnostics for the Novell network or the Etheret network. You may only need to refer to the seven-layer ISO model when you are planning to interface more equipment to an existing network, or if you are planning to add another network. The next sections will explain more about the details of each layer that you may use.

16.3 DATA FORMATS

Data formats are important when information is sent from one system to another. As you now know, the presentation layer is concerned with transforming data from one format to another. The two main formats are ASCII (American Standard Code for Information Interchange) and the EBCDIC (Extended Binary Coded Decimal Interchange Code). The American National Standards Institute (ANSI) proposed the ASCII code, and International Business Machines Corporation (IBM) designed the EBCDIC code. The ASCII code and EBCDIC code are used to send information to printers and to send information to and from punch-tape readers that are used in older CNC (computer numerical control) machines, such as mills and lathes used on the factory floor in machining applications.

16.3.1 The ASCII Code

ASCII is a way to represent every letter of the alphabet (capital letters and lowercase letters) and any symbols such as @#$%& in a seven-bit binary format.

7	6	5	4	3	2	1
1	0	0	0	0	1	0

There are 128 letters, numbers, and characters in the ASCII code. Fig. 16-4 shows the American Standard Code for Information Interchange (ASCII) code. As you can see, each letter and character are shown with a decimal number value that is converted to binary and hexadecimal. The way you use this matrix is to find the letter and then find its coded number. The coded number is the hexadecimal value for the letter. From that point you can also find the binary value for the code number. If you wanted to find the coded value for any letter or character, you would locate the column number as the first digit in the code, and the row number as the second digit. For example, the capital letter *B* is found in column 4, row 2, so its coded number is 42. This means its hexadecimal value is 42 and when you convert this value to binary, the value is 1000010.

Bit 7			0	0	0	0	1	1	1	1
Bit 6			0	0	1	1	0	0	1	1
Bit 5			0	1	0	1	0	1	0	1
Bit / **Row Number**		Column Number	0	1	2	3	4	5	6	7
4 3 2 1										
0 0 0 0		0	NUL	DLE	SP	0	@	P	`	p
0 0 0 1		1	SOH	DC1	!	1	A	Q	a	q
0 0 1 0		2	STX	DC2	"	2	B	R	b	r
0 0 1 1		3	ETX	DC3	#	3	C	S	c	s
0 1 0 0		4	EOT	DC4	$	4	D	T	d	t
0 1 0 1		5	ENQ	NAK	%	5	E	U	e	u
0 1 1 0		6	ACK	SYN	&	6	F	V	f	v
0 1 1 1		7	BEL	ETB	'	7	G	W	g	w
1 0 0 0		8	BS	CAN	(	8	H	X	h	x
1 0 0 1		9	HT	EM	)	9	I	Y	i	y
1 0 1 0		A	LF	SUB	*	:	J	Z	j	z
1 0 1 1		B	VT	ESC	+	;	K	[	k	{
1 1 0 0		C	FF	FS	,	<	L	\	l	|
1 1 0 1		D	CR	GS	-	=	M	]	m	}
1 1 1 0		E	SO	RS	.	>	N	^	n	~
1 1 1 1		F	SI	US	/	?	O	–	o	DEL

First — hexadecimal digit

Second hexadecimal digit

FIGURE 16-4 The American Standard Code for Information Interchange (ASCII). You can find the ASCII code for any number by using the column number as the first digit and the row number as the second digit. This number is also the hexadecimal equivalent value for the letter or character.

You should notice that the ASCII code also contains control characters such as STX (start of text), SOH (start of header), and EOT (end of transmission) to help operate devices such as printers and punch readers to recognize the start and end of text and data.

16.3.2 The EBCDIC Code

IBM developed another system to represent letters and numbers called the Extended Binary Coded Decimal Interchange Code. This system uses eight binary bits to represent 256 different letters, numbers, and characters. Since the ASCII system uses seven bits, it is somewhat limited in that it can only represent 128 different characters. Fig. 16-5 shows the EBCDIC system in a chart so that you can determine the hexadecimal and binary value for each character. You can use this chart like the ASCII chart in that you locate the character and find the column number as the first digit of the code and the row number as the second digit. For example, the hexadecimal value for the capital letter B in the EBCDIC format is 43, and its binary equivalent is: 01000011.

8	7	6	5	4	3	2	1
0	1	0	0	0	0	1	1

FIGURE 16-5 The Extended Binary Coded Decimal Interchange Code (EBCDIC). The column number makes up the first digit of the code number and the row number makes up the second digit for the code number.

Bit 1		0	0	0	0	0	0	0	0	1	1	1	1	1	1	1	1
Bit 2		0	0	0	0	1	1	1	1	0	0	0	0	1	1	1	1
Bit 3		0	0	1	1	0	0	1	1	0	0	1	1	0	0	1	1
Bit 4		0	1	0	1	0	1	0	1	0	1	0	1	0	1	0	1
Bit 5 6 7 8	Column Number / Row Number	0	1	2	3	4	5	6	7	8	9	A	B	C	D	E	F
0 0 0 0	0	NUL	DLE			SP	&	-									0
0 0 0 1	1	SOH	SBA					/		a	j			A	J		1
0 0 1 0	2	STX	EUA		SYN					b	k	s		B	K	S	2
0 0 1 1	3	ETX	IC							c	l	t		C	L	T	3
0 1 0 0	4									d	m	u		D	M	U	4
0 1 0 1	5	PT	NL							e	n	v		E	N	V	5
0 1 1 0	6			ETB						f	o	w		F	O	W	6
0 1 1 1	7			ESC	EOT					g	p	x		G	P	X	7
1 0 0 0	8									h	q	y		H	Q	Y	8
1 0 0 1	9		EM							i	r	z		I	R	Z	9
1 0 1 0	A					¢	!	;.	:								
1 0 1 1	B					.	$	,	#								
1 1 0 0	C		DUP		RA	<	*	%	@								
1 1 0 1	D		SF	ENQ	NAK	(	)	–	'								
1 1 1 0	E		FM			+	;	>	=								
1 1 1 1	F		ITB		SUB	\		-									

First hexadecimal digit

Second hexadecimal digit

16.4 PROTOCOLS AND STANDARDS

You can see in the OSI seven-layer model that several protocols and standards are used. These protocols and standards have been determined by various companies and groups. Sometimes these standards like the EBCDIC code have become de facto standards that the majority of users have decided to use. In other cases the standards were never accepted on a wider basis so larger committees were formed to establish new standards. Equipment manufacturers, software writers, and end users were put on these committees to provide input and promote their points of view for the new standards. So far standards by ISO (International Standards Organization) and CCITT (Consultative Committee on International Telegraphy and Telephony) have been discussed. You will find that other interested groups like the Department of Defense (DoD) for the U.S. government, equipment manufacturers, software writers, and end users have also formed groups to establish standards. Generally these groups will establish several committees to provide a variety of standards and protocols. It is not important at this time that you understand all of the protocols and standards. When you get on the job, you will begin to work on specific systems, and at that time you will take a short course about the protocols and standards used in the system you are working on. Or you will receive a technical manual that provides the information that you need to install or troubleshoot the system.

16.4.1 Ethernet and IEEE 802 Standards

One such group of vendors got together in the early 1980s to design a set of standards for LANs. This group consisted of members from DEC (Digital Equipment Corporation), Intel Corporation, and Xerox Corporation. They designed standards for a LAN called Ethernet. During the same year the IEEE organization established a committee called the IEEE 802 Committee to establish similar standards for LANs. Even though Ethernet is used by a wide variety of users, the IEEE 802 standards for LANs have also been incorporated to a vast number of products. The standards for Ethernet and the standards that this committee provided apply to the data link layer and the physical layer of the seven-layer OSI model. You will find products today that use all or parts of both of these sets of standards. You will also find that from time to time these committees reconvene to update the standards to include changes in technology.

16.4.2 MAP Protocol

In the late 1980s General Motors Corporation found that their large automotive assembly facilities had a wide variety of robots, programmable controllers, CNC machining equipment, and other automated equipment that could not communicate with each other because of the lack of standards. During this time each equipment manufacturer was only concerned about getting data transferred between pieces of their own equipment. Their reasoning was market driven in that they calculated that if a company like GM purchased one piece of their brand of equipment, they would be forced to purchase only that brand of equipment. For example, if GM wanted to exchange data between a mill and a robot or a mill and a PLC, GM had to purchase the mill and PLC from the same vendor, such as Allen-Bradley or General Electric. This was the driving force for GM selecting General Electric

(GE) to be the prime vendor of automation equipment in the Saturn facility in Spring Hill, Tennessee. At the time the Saturn project was designed, GE had formed a partnership with Fanuc, a large Japanese manufacturer of robots and CNC machining controllers. GE combined its programmable controllers with the Fanuc controllers to make one large family of robots—PLC and CNC machining controllers that could pass data back and forth between each other because they all used similar standards and protocols.

In other GM assembly plants a large variety of Allen-Bradley PLCs and Fanuc robots were well established to control assembly applications that could not be performed by other manufacturers' equipment. This meant that if a robot was manufactured by Fanuc, and an assembly machine was controlled by an Allen-Bradley PLC, the two controllers could not pass data directly to each other. When GM determined this condition unacceptable, it established a set of standards and protocols called the Manufacturing Automation Protocol (MAP). It provides a means for all different types and brand names of equipment to talk to each other over LANs. Since this standard could be used in all areas of factory automation in the factories of other companies like Ford Motor and Chrysler, the MAP protocol has become widely accepted. The type of information that is passed through the MAP protocol is the number of parts to be made, the amount of cycle time, and other similar information.

The only problem that occurred with MAP is that technology expanded so quickly in the 1980s and 1990s that several versions such as MAP 2.0, MAP 2.1, and MAP 3.0 were quickly introduced in succession. Equipment vendors became confused about which standards to meet since the standards continually changed. Overall, the MAP protocol has provided the basis for standards for LANs used in factory automation today.

16.4.3 The Technical Office Protocol (TOP)

About the same time the automotive manufacturers were determining the standards for MAP, the Boeing aircraft company determined standards for connecting the vast number of computers used in the design and manufacturing of aircraft. Since this activity is more design oriented, the standards and protocols that resulted are focused on computers used in offices, such as computer-aided design and drafting (CADD), computer-aided manufacturing (CAM), and other office-related functions like shipping, receiving, and cost accounting. This protocol became known as TOP and its standards and protocols have become somewhat of a de facto standard for office LANs.

Since many industries have now combined traditional manufacturing with automated cells to create computer-integrated manufacturing (CIM), you likely will be asked to work on equipment that complies with TOP that may also be connected in a network with equipment that complies with MAP or other protocols.

16.4.4 Transmission Control Protocol and Internet Protocol (TCP/IP)

Another common standard and protocol that you will encounter on the equipment used on the factory floor is called TCP/IP (Transmission Control Protocol/Internet Protocol). This protocol is written for layer 4 and layer 3 of the OSI model. The transmission protocol (TCP) is specifically written for the transport layer of the model, and the internet protocol is written specifically for the network layer of the model. Actually the internet protocol resides on the upper portion of the network layer, and sets the standards for headers that are passed between nodes on the same network or between nodes on different networks.

Together the TCP/IP allows equipment from different manufacturers to send and receive data across different networks. This set of protocols was originally designed by the Department of Defense (DoD). It saw the need in the early 1980s to provide a means to allow computers from colleges and universities that were doing research for the DoD and other government agencies to connect with a wide variety of equipment found in government agencies. Government contractors needed to share information and data and they were not able to because of the incompatibility of their computers and networks. The DoD funded a network called Advanced Research Projects Network (ARPANET) that uses TCP/IP to make it possible for all types of computers and networks to access information from other networks that are operated by government agencies, contractors, and companies.

16.4.5 Systems Network Architecture (SNA)

IBM defined a group of standards and protocols for a network architecture that covered a complete line of products that it sold. SNA standards and protocols cover every layer of the ISO model except the physical layer. In fact the layers of SNA were developed prior to standardization by the OSI seven-layer model. As a technician you may run into equipment that uses the SNA protocols because it has become the de facto standard by virtue of the fact that IBM was the largest equipment vendor for a number of years when networks and data communications were being established.

16.4.6 Manufacturing Message Specification (MMS)

The *Manufacturing Message Specification* (MMS) is a seven-layer system based on the OSI seven-layer model. This system was designed to allow communications between dissimilar devices on the same network. The MMS fills the gap for factory automation where a wide variety of dissimilar devices is being asked to share data. For example, it is likely that a gauging and inspection system may be able to send data to any name brand of machining and manufacturing equipment. The manufacturing or machining equipment may have any brand of programmable controller or CNC controller, and it must be capable of receiving the inspection data to determine offsets if the machining tools are getting dull. The controller for the manufacturing center may need to send data between robots for material handling and to quality control computers to archive data that must be available to qualify parts to the vendor. Since all of this equipment can have different brands and generations of technology, they would not normally be able to pass data between themselves. The MMS makes this type of data communications possible.

16.4.7 European Standards: The Fieldbus

At the same time network standards and protocols were established in the United States, comparable standards and protocols were being established in Europe and Asia. The European standards use a system called Fieldbus that is similar to the OSI seven-layer model in function. The Fieldbus uses five layers to accomplish the same standards. They do this by combining the physical layer and data link layer into a set of standards called DIN V 19245 T1. DIN (Deutsches Institut for Normung) is a German standards institute. The Fieldbus also combines the session layer, the presentation layer, and the bottom part of the application layer into one layer called AP (automation protocol) layer. The last part of Section 16.4 will compare all of the American and European standards and protocols.

16.4.8 Profibus

Profibus is a network and data communications standard that is designed by Siemens of Europe that includes protocols that are established in the Fieldbus standards. In recent years Siemens Industrial Automation Group has purchased programmable controller technology from Texas Instruments and has combined this equipment with its European controllers. Since Siemens has been established in data communications and telecommunications in Europe, it had the depth of knowledge to make the necessary equipment and software required to connect its equipment with existing systems in the United States. You will find that a number of industrial electronic devices have been manufactured by European or Asian manufacturers and may conform to the standards of the original country as well as American automation standards.

16.4.9 Comparing Standards and Protocols

Fig. 16-6 shows a table of the standards and protocols that have been presented in this section. You can see how these standards and protocols resemble the original seven-layer OSI model. You can also begin to see where the different layers of each protocol may use or adopt existing standards and protocols rather than establish new ones. The objective of this section is to provide a way to identify how different standards and protocols relate to each other.

FIGURE 16-6 A comparison of popular architectures and protocols.

OSI		SNA		MAP		AP		TCP/IP		DNA		Fieldbus	
7	Application	7	████ Function Management Data Services	7	MMS ISO ACSE	7	MMS/STF	7	TFTP FTP SMTP TELNET	7	USER Network Application	7	MMS/STF
6	Presentation	6		6	ISO Presentation	6	SINEC AP	6		6		6	SINEC AP
5	Session	5	Data Flow Control	5	ISO Session	5		5		5	Network Services Protocol	5	
4	Transport	4	Transmission Control	4	ISO Transport	4	ISO Transport	4	TCP	4		4	L2 Transport
3	Network	3	Path Control	3	ISO Internet	3	ISO IEEE INACTIVE	3	IP	3	Transport	3	Inactive
2	Data Link	2	Data Link Control	2	ISO IEEE 802.2	2	ISO IEEE 802.2	2	Ethernet Version 2.0	2	Data Link	2	DIN V 19245 T.1
1	Physical	1	████	1	ISO IEEE 802.4	1	ISO IEEE 802.3 IEEE 802.4/7	1	ISO IEEE 802.4	1	Physical	1	

16.5 PARTS AND MEDIA OF THE PHYSICAL LAYER

As a technician you will spend most of your time working in the lower three layers of the OSI model. You will be installing and maintaining the cable, connections, and other parts of the physical layer. This section will provide more detail about this layer.

16.5.1 Transmission Media

Originally when networks and data communications were established as part of factory automation, wire was the only medium that was used to link equipment together. Fig. 16-7a shows a typical transmission cable for data communications. This is called coaxial cable and it has a single solid copper wire to carry information and a mesh cover to support the wire. A metal shield is also provided to keep any outside interference from getting to the solid conductor. It is important to understand that the metal shield should only be grounded at one end to prevent ground currents from circulating in the shield.

Another type of simple cable is the twisted pair cable (see Fig. 16-7b). The twisted pair may have an outside covering that is blue, and it is generally called *blue hose*. The twisted pair consists of two conductors that are twisted around each other, which tend to cancel any inductive signals that may try to get into the cable. The twisted pair also has a covering of metal foil that is located just under the plastic covering that functions as a shield. This shield is grounded at one end.

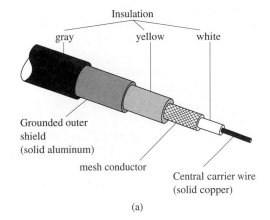

(a)

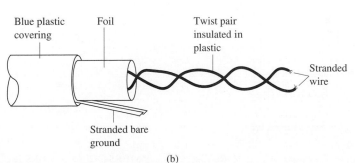

(b)

FIGURE 16-7 **(a)** Coaxial cable with its copper conductor and shield. **(b)** A twisted pair of wire used as physical medium. Notice the twisted pair has a metal foil covering. (Courtesy of Belden Wire & Cable Company.)

Today the physical medium, which is also called the *backbone* for the network, can also be fiberoptic cable. A wide variety of connecting devices is available to provide an interface between RS232 systems and RS422 system solid wire and fiberoptic conductors. (Refer to Chapter 7 to review fiberoptic cables.) As a technician you will be asked to make splices in this medium as well as maintain terminal boxes and termination points.

One advantage of these connectors is that a fiberoptic system can be installed on the factory floor after a wire-and-cable network has been operating for several years. This is especially useful if signal interference or noise on the factory floor cannot be eliminated. When the signal is traveling through the fiberoptic cable, it is impervious to induced voltages because the signal is traveling in the form of light.

16.5.2 Physical Cable Connections: The RS232 Interface

The RS232 interface was specifically designed as an interface between computers and modems. The modem provided access to the wire in the telephone system. The main reason for gaining access to telephone cables is that they are existing and they presently are connected to over 90% of residences and businesses in the United States. This meant that the means of transmitting data through the physical layer did not have to be installed so that a computer in Detroit could talk to another computer in Cleveland, since telephone lines already exist between these two cities and these lines are presently being maintained. The advent of the microprocessor chip and its integration into modems and other interface equipment have provided a number of enhancements to the original RS232 standard.

Before you begin to study the RS232 interface, you must understand that it is an extension of the original telegraph standard, so many of the signal descriptions and the system conditions were named prior to the introduction of computers and modems. The older systems also tended to be much less complicated; that is, one system was only capable of sending data and the other system was only capable of receiving data. Fig. 16-8 shows a typical system that uses an RS232 system. This system will be kept rather simple in that it contains only a personal computer at one end and a mainframe computer at the other end. The cable between each computer and its modem requires only twisted-pair conductors and it is shown as a single line in the diagram.

In this example you should notice that the personal computer is identified as a DTE (Data Terminal Equipment). The modem that it is connected into is designated as DCE (Data Circuit Equipment). At the other end of the system, the mainframe computer is designated as DTE and the modem that it is connected

FIGURE 16-8 An example of a personal computer connected by modem to a mainframe computer.

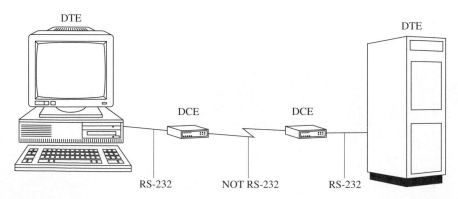

into is designated as DCE. The interface between each of the DTEs and DCEs is based on the RS232 standard. The cable between the two modems is standard telephone system cable.

16.5.2.1 The Mechanical Part of RS232 ■

The RS232 connections can be established with a 25-pin connector called a DB25. The connector can be a male connector with pins, or a female connector with jacks. Fig. 16-9 shows the arrangement of the 25 pins in the connector. You should notice that pins 1–13 are in the top row of the connector and pins 14–25 are in the bottom row. The connector is tapered slightly at the ends near the bottom row to make the physical outline of the connector such that it can be plugged into mating equipment in only one way. This is sometimes referred to as a physical *key*. The physical layout of the pins in the 25-pin connector

FIGURE 16-9   DB25 connector for RS232 interface.

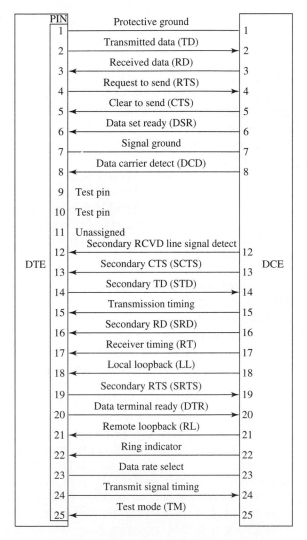

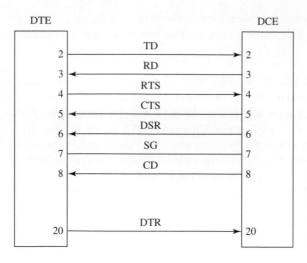

FIGURE 16-10 The most commonly used signals between DTE and DCE for the RS232 interface. The arrows show the direction in which the signal moves.

is identified in the Electronic Industries Association (EIA) standards and it is referred to as a DB25 connector. (*Note!* The DB25 pin connector is also used in RS422 interface and the RS366 interface and for parallel printers on computers.) It is also important to note that if the connector is identified as a DB25F it is a female connector, and if it is a DB25M it is a male connector.

To fully understand the RS232 interface, you will need to see which signals are sent from the DTE to the DCE and which signals are sent from the DCE to the DTE. Fig. 16-10 shows the names of each of the signals in the RS232 interface. This figure also identifies the direction of the signal. In this table you can see the signals are broken into four categories: ground, data, control, and timing. From this diagram you can see that terminal 2 is the transmitted data, and terminal 3 is the received data.

16.5.2.2 RS232 Connections Between DTE and DCE ■ The signals that are sent between the DTE and the DCE can generally be described by their function as part of a pair of signals. For example, if the DTE is sending data to the DCE, it must go out of the DTE on pin 2 transmitted data (TD), and it will be received in the DCE at pin 2 received data (RD). If the DCE sends a signal back to the DTE, the DTE will receive the signal on pin 3 received data (RD).

The control signals are also matched in pairs. For example, the request to send (RTS) must be acknowledged by the clear to send (CTS) signal, so the RTS of the DTE is sent to the CTS of the DCE, and vice versa.

The data set ready (DSR) signal is combined with the data terminal ready (DTR). This means that a DSR ready signal is sent from the DCE when it is ready to receive a signal and it waits on a DTR signal from the DTE. It is very imporant to understand that the timing of these signals is controlled automatically by software and as a technician all you need to do is execute the correct commands in the software to start the process. For example, if you want to transfer (save) a PLC program or a BASIC program from a microprocessor to a portable computer, the computer and the microprocessor must both have software that is specifically developed to transfer data in the RS232 format. The software is embedded in the PLC or microprocessor so you do not need to do anything for that end. The software for the portable computer must be loaded so that you can use it. When you are ready to start the transfer process, the portable computer sends the signal to the microprocessor. The microprocessor responds and starts sending data to the portable computer. The two systems transmit signals to control the stream of data so that the computer does not lose any of it. When the transfer is complete, the two systems shut down their respective ends of the RS232.

16.5.3 The Nine-Pin Connector for the RS232 Interface

Since not all of the signals in the RS232 are used in every application, it is possible to use a cable connector with only nine pins. In many industrial applications for data communications, such as saving a PLC program on a personal computer, the transmission is all in one direction, and the signal timing can be controlled from the software. The interface only needs pins 2 and 3, and signal ground. This means that all of the other pins can either be connected to each other or be grounded. Fig. 16-11 shows a diagram for a nine-pin connector for a DTE and a DCE. This diagram also shows the direction of the signals and which signals are connected to each other as jumpers in one end of each cable and which signals are grounded.

16.5.3.1 Limitations of the RS232 Interface ■ The RS232 interface is generally limited to 20 kbits/second and the distance is usually limited to 50 ft. This makes the RS232 interface utilize modems to extend the distance, which also makes the interface useless in *real-time* applications over longer signals from a PLC to indicate a machine has made exactly 100 parts. The one-hundredth part may have been made at 9:15 a.m., and by the time the modem sends the signal it will be several seconds later, or up to several minutes later. In the case of a production report, the real-time issue is not critical, but if the signal is used to try to sychronize two robots that are 100 ft apart at two ends of a work cell, the timing will not be accurate enough to provide true synchronization. Another problem would arise if the RS232 interface is used between a color graphics screen and the two robots. Since the robots are over 50 ft apart and the color graphics computer is another 50 ft, modems would need to be used. This would cause problems with sending and receiving signals in real time.

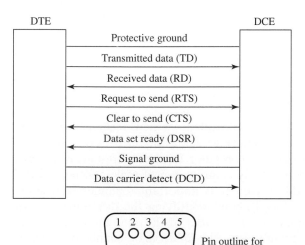

FIGURE 16-11 A diagram of a nine-pin DTE and DCE cable that uses the RS232 format. The nine-pin connector is called a DB9 connector.

16.5.4 The RS422 Interface

As you can see the RS232 has severe limitations of distance and speed, such as applications where a color graphics systems must send and receive data to a PLC or robot directly or over a small network. If the transmission rate is too slow, the screens cannot show real-time data. The term *real-time data* means that the color screen program has the ability to show a switch opening or closing at the same time that it physically occurs. In early color graphics systems, the color screen was interfaced with the PLC through an RS232 interface that was rather slow. This caused a short delay of 1–2 seconds between the time when a switch actually closed and when the signal was displayed on the color screen.

In newer systems it is likely that the color graphics system will monitor signals from several different PLCs and robots from a small network and produce the changes on the screen at speeds that look like real time. Since some of the robots and PLCs may be several hundred feet from each other and the color screen, the RS422 interface allows the signals to transmit 10 Mbits/second over longer distances up to 4000 ft. Fig. 16-12 shows an example of the pin assignment for the RS422 signals on a DB25 pin connector. The diagram also shows the RS422 pin assignments for a DB9 pin connector.

It is important to understand that the RS422 interface does not use any of the pins in the DB25 connector that the RS232 interface uses. For this reason it is possible for a vendor to send out an RS232 signal or an RS422 signal from the same DB25 port.

16.5.5 Other Common Connectors Used in Industrial Data Communications

At times you will find other types of male or female connectors used in vendor equipment that have 15 or 37 pins. The 37-pin connectors are typically used for the RS449 interface and the 15-pin connectors are used for RS232 and RS422 interfaces. It may be of interest to know that Allen-Bradley uses the 37-pin connector for its PLC programming software that resides on personal computers. The 37-pin female connector is mounted on an interface card that is located in one of the computer slots. A matching 37-pin male connector is connected on the cable that connects the personal computer to the PLC5 programmable controller. The interesting detail of this connector is that only three of the 37 pins in the DB37 pin connector are used. This means that for all practical purposes, the connector could have been a DB9 connector, except the protocol suggests a 37-pin format be used.

16.5.6 De Facto Standards for DB25 Connectors for Modern Personal Computers

One of the most daunting tasks for a new student or new technician is to be requested to complete the simple task of plugging a cable into the back of a computer for a printer or other device. When the computer is turned around, up to five connectors are found and they may be any combination of male and female 25-, 15-, and 9-pin connectors. For this reason it is important to understand some de facto standards that have developed about identifying the different types of connectors. The DB25 female connector in the rear of the computer is generally designated as the 25-pin parallel printer port. The DB25 male connector is designated as the *Com Port terminal*. It can be designated on most computers as Com Port 1 or Com Port 2. The DB9 male connector is also a Com Port connector and

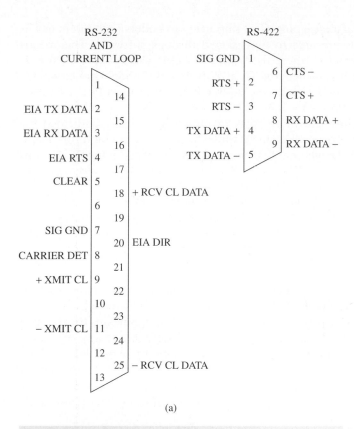

(a)

(b)

FIGURE 16-12 (a) The RS422 interface shown with pin connection diagrams for a DB25 pin connector and a DB9 pin connector. **(b)** An RS232/422 interface board for an IBM compatible computer. (© Copyright 1995 Omega Engineering, Inc. All rights reserved. Reproduced with the Permission of Omega Engineering, Inc., Stamford, CT 06907.)

it is usually designated as a mouse port. (It is important to understand that in newer computers the Com Ports are able to be changed through software. This means that you may find one computer where the 9-pin connector is designated Com Port 1 and the 25-pin connector is designated Com Port 2, and the identical computer nearby may have its software designate the 9-pin connector as Com Port 2 and the 25-pin connector as Com Port 1.)

The female connector with three rows of pins is specifically dedicated for video. This is the port that you will connect the cable to from the monitor. These connectors are specifically designed so that they cannot be mistaken for RS232 connections.

16.6 TYPICAL NETWORK TOPOLOGIES

Local area networks (LANs) can be classified as *baseband* or *broad-band* networks. A baseband network is usually limited to equipment that is connected over short distances. Since this type of network system uses the entire bandwidth for the data signal at any one time, only one signal can be on the network at any one time. The advantage of this type of network is that it can support many devices. The disadvantage of the system is that no two devices can be active on the network at the same instant. This means that a device signs on the network and delivers its data or request for data in a short burst and then clears off the network and waits for a reply.

A broad-band network, on the other hand, depends on multiplexing techniques to split the network into multiple channels. This allows signals to be on the network at the same time, which allows simultaneous services to multiple users. Typical broad-band networks may be capable of carrying a different signal on each channel. The broad-band network can also use amplifiers to boost weak signals to allow them to travel longer distances.

16.6.1 Types of Network Topologies

Fig. 16-13 shows four types of *network topologies* that are commonly used in industrial data communications systems. Network topology means how the computers are actually connected to each other. This is similar to connecting electrical components in series or parallel or as a compound series-parallel circuit. You will see that there are advantages and disadvantages for each type of topology. As a technician you will not be responsible for designing these systems. Instead you will order the parts and make the connections according to a diagram that is specified by an equipment manufacturer for the specific type of network topology its equipment uses. The following information is provided so that you will have an idea of how the computers are connected for each type of network.

Fig. 16-13a shows a bus network. This type of network is identified by one long spine with user devices connected along the length of the network. Fig. 16-13b shows a branching tree network. The tree network is identified by the long spine and the branches that flow from the spine. Each branch can also have branches.

Fig. 16-13c is called a star network. The star network is identified by a central node on the network with all other branches of the network emanating from the central node. The computers at the ends of each branch can fail without harming

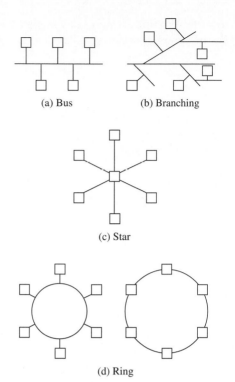

(a) Bus   (b) Branching

(c) Star

(d) Ring

FIGURE 16-13 **(a)** A bus network. **(b)** A branching tree network. **(c)** A star network. **(d)** Ring networks.

data that are passed to other branches. If the computer that acts as the central node fails, the entire network will become disabled.

Fig. 16-13d shows two types of ring networks. The ring network is identified by a main spine that is closed in the shape of a ring. In both variations, data information flows in only one direction around the ring. The first variation of this type of network shows the main network spine connected in a circle and each node is a tap off of the main spine, much like the bus network. The second variation of the ring network shows each tap as an integral part of the ring. In this variation, each computer is a vital link to pass data to any of the other computers in the network. This type of ring network is not used very often because the failure of any one computer will stop data from being passed beyond that failed computer. In the first variation, each computer is only tapped to the spine, so if a failure occurs, only that computer is affected.

Each example of these types of network topologies allows a problem to occur at any one node without affecting the operation of all of the other nodes.

16.7 TYPICAL FACTORY AUTOMATION NETWORKS

As a technician you will install and repair several types of data communications networks that are used to connect PLCs, robots, and other industrial electronic devices. Typical networks may be vendor specific or they may be more universal in nature such as the Ethernet. Fig. 16-14 shows an example of these types of

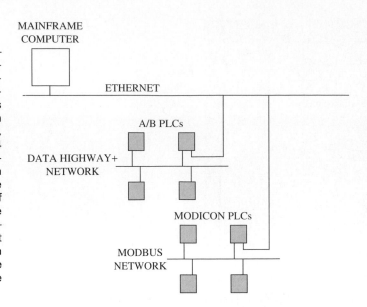

FIGURE 16-14 A typical network in an automated factory. Four Allen-Bradley PLC5 programmable controllers are connected to a Data Highway + network, and three Modicon 984 programmable controllers are connected to a MODbus network. One of the PLC5 and one of the 984 programmable controllers have an Ethernet card in them so that each PLC network can be connected to the main network for the factory.

networks. In this example you can see that all of the Allen-Bradley PLCs are connected to the Allen-Bradley proprietary network called the *Data Highway +*. The Data Highway + is a bus network that allows the personal computer with the programming software to act like a terminal node on the network. This allows the programming terminal to get on the network and connect to any PLC on the network. In this application other equipment has Modicon PLCs controlling them and these PLCs are connected to a *MODBus +* network, since the Modicon PLCs are not able to connect to the Allen-Bradley Data Highway + network. Since it is important to get data between the machines that have Allen-Bradley controllers and the machines that have Modicon controllers, a third network called Ethernet is installed that is more universal. It will allow connections between different brands of equipment. One of the Allen-Bradley controllers in the network will have an Ethernet card installed in it so it can act as a gateway for the Data Highway + network. One of the Modicon controllers will also have an Ethernet card installed so it can act as a gateway. For example, if a machine that is controlled by an Allen-Bradley PLC has completed a final step on a part and the part is ready to move the next machine that is controlled by a Modicon PLC, the Allen-Bradley PLC must send this signal to the Allen-Bradley that has the Ethernet card in it. This PLC sends the signal over the Ethernet to the Modicon PLC that has an Ethernet card, and then this PLC sends the signal over the MODbus + network to the Modicon PLC that controls the machine.

Fig. 16-15 shows an example of an Ethernet card that can be installed in an Allen-Bradley PLC or in a Modicon PLC. The Ethernet runs throughout the factory floor and it also has a drop in the main offices, in production planning, in maintenance supervision, and in the shipping and receiving area. This allows all of these areas of the factory to check on the status of the machines and determine the number of parts that are being made.

16.7.1 Strengths of the Ethernet

The Ethernet is generally used for a factorywide network because it is inexpensive and it can handle large amounts of data at high speeds. The typical Ethernet can

FIGURE 16-15 A typical Ethernet card for a PLC. (Courtesy of Rockwell Automation's Allen Bradley Business.)

run at speeds of 10 MBaud. It can have over 100 nodes and maintain speeds of 20 mS with 1024 bytes per transmission.

When other pieces of automation equipment are installed, they can have Ethernet communications cards installed in them, so that they can also be tapped onto the Ethernet network. It is important at this time to understand that most vendor-specific networks like the Data Highway + or the MODbus are designed to take advantage of the processors in their own equipment, which means they will not work well with other equipment. The vendor-specific network is also much slower and its data capacity will be much smaller than an Ethernet. The Ethernet is specifically designed to handle large numbers of nodes and large amounts of data moving across it.

16.8 THE BROAD-BAND NETWORK

The broad-band network may reach beyond the physical building where the manufacturing occurs. In some corporations, the broad-band network spans several cities so that two or more manufacturing sites can be connected to the corporation mainframe computers. The broad-band network can operate at speeds of up to 10 Mbaud and it should be MAP compatible. The host for the broad-band network will be a larger computer system such as an IBM mainframe, HP mainframe, or DEC mainframe. It is important to understand that companies like the automobile manufacturers may have three or four factories making parts like headlights or taillights at one site, and they are assembled at another site 40 miles away. It is important that both the factory making the parts and the assembly plant both know how the other plant is progressing on the schedule for that day and for the next day. The broad-band network allows data to be passed between several different factory sites.

16.9 TROUBLESHOOTING NETWORKS

When a network has a failure, you will be asked to locate the problem and repair it quickly so that data can begin to flow again. In most cases, factory automation is based on a concept called *distributive control*. This concept ensures that the controller on each piece of equipment in the factory can still operate and produce parts or product when the node on the network that controls that machine fails. In the early days of factory data communications, large portions of the factory and large numbers of machine controllers were all dependent on one or two computers that controlled network functions. If one of these computers failed, all of the machines connected to that network would automatically stop, and production would come to an abrupt halt. You can see that this could cause lost production time of 8–20 hours with the failure of a single card in any of the computers. Designers soon found that it is more prudent to leave all control of the machine in the PLCs, and have the PLCs send or receive data with the network in such a way that if the network failed, the PLCs could keep the machines running. The worst-case scenario in this system would mean that you may not receive production data during the network failure, but since the PLC stores these data, they could be shipped to the necessary computers on the network at a later time when the network is up and running again. This type of system gives the technicians a little more time to test and repair network failures.

16.9.1 Test Equipment

When failures do occur, many of them can be located with software that runs in the computer with the network software. In the early days, the technician would need to have a full understanding of the number of nodes on the network and use voltmeters and oscilloscopes to trace each node and determine which node had failed and how much of the network was not operating. Later each node had a number of LEDs on the face of the node equipment in the network. This meant that the technician would need to physically visit each piece of equipment and check the LEDs to see which node had failed.

Today most networks have diagnostic software loaded on one or more of the node computers. The network also runs diagnostic routines on the network to monitor continually for any irregular activity or faults. When a failure is detected anywhere on the network, retry routines and reset routines in the software try to clear the problem automatically before a critical failure occurs. If the failure is due to a loose cable or a faulty interface card or component, the diagnostic software will determine the failure and post the exact point in the network where the failure has occurred. A failure notice is printed to all screens and to the network node that is used by the technicians to monitor the network. In some cases a simple reset or retry will cure the problem and a message is printed saying the fault has occurred and has been cleared or reset.

If the failure is critical, the technician is sent notice of the failure. At this point the technician can use the computer that is connected to the network specifically for troubleshooting and testing for problems. Nearly all networks have their own proprietary troubleshooting software, and other systems like the Ethernet have commercial software available to troubleshoot each part of the network. This software basically goes out and *polls* each node on the network and looks for a response. This means the diagnostic computer sends a packet to each node on the network and looks for the response. If one of the nodes is not responding, the diagnostic computer sends the packet several times again to ensure the problem is not a temporary condition. The technician can make a selection in the software program of the diagnostic computer to specify the number of retries.

When the diagnostic computer has determined the nature and location of the fault, it will identify on the screen which node the technician needs to physically visit. At this point the software will also indicate the nature of the problem and suggest additional tests or repair procedures. The technician must then visit the suspect node and verify the condition of all cables and interface hardware. When the source of the problem is detected, the technician can remove and replace the faulty parts and put the system back into service.

In some cases the problem is not easily found, and the technician must use a multichannel oscilloscope with storage capabilities. The storage scope allows the technician to sample the signal activity on a cable over a very short period of time and then store this activity so it can be displayed in a manner that allows the technician to look for signals that have been sent. The strength of the signal and the timing of the signal can be compared to other known good signals that are sampled on other channels of the scope.

Test and diagnostic equipment is also available that can produce a signal and inject it into the network cables so that the signal can be traced through the network. At each node of the network the signal can be analyzed to see how much its strength has degenerated.

In some applications you will be asked to locate a fault in a new cable installation between two devices that are using an RS232 interface. Since the type of signal, the timing, and the pin assignment of each signal is usually known in an RS232 interface, a diagnostic device called a *breakout box* can be used to see whether or not the signal is present. Fig. 16-16 shows a typical breakout box. The simplest breakout boxes provide a means to isolate each of the 25 pins in a DB25 cable connection, or provide jumpers between pins that should be connected together or that should be grounded. This allows the technician to test for each type of signal and determine if the cable connector is soldered correctly. One of the biggest problems with equipment that is connected to RS232 systems is that wires are connected to the cable ends in the field by the technician who is installing the interface and the equipment vendor does not specify which pins are used and which ones are jumpered. This means that the technician must use the breakout box to get the interface to operate, and then use the information gain in the prototype connection to make the permanent cable connectors.

The technician can use DB25 to DB9 or DB15 adapters to make the breakout box fit to any male or female DB25, DB15, or DB9 connectors so that other interfaces can be tested. It is important to understand that the breakout box is designed to indicate if a signal is present on a specific pin in the interface connector,

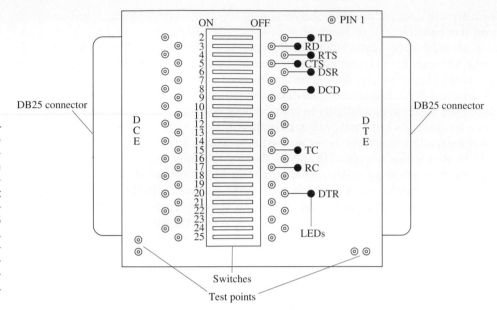

but the breakout box cannot indicate critical timing functions or the actual strength of a signal. A storage oscilloscope must be used if these conditions are suspected.

16.10 REVIEW OF THIS CHAPTER

You will find at times that you will need to work with network and data communications systems for PLCs and other microprocessor controlled equipment. When you are first introduced to the communications applications, you should try to gather information about the systems such as what type of network is being used, and what software is running. You should also try to determine what medium (wire or fiberoptic cable) is being used to transmit and receive data. You should also try to determine the number and type of equipment on each node and the distance between the nodes in the network. If the system is a specific brand name, you may need to know if the network is designed specifically for the equipment or if it is a generic network such as an Ethernet that allows a variety of brand-name equipment to be connected to it. In some cases a blueprint will be available that indicates how the system was originally installed.

After you have determined all of the details about the system, you may try to sketch the layout of the network to get an idea of the topology that is used. This will help you to determine the limitations the network has. Next you can determine if any diagnostic software is available on any of the nodes and use that computer to determine the status of the specific problem. Then you can gather any manuals or manufacturer's specifications about the network and the individual equipment connected to the network. In some cases in newer systems the entire technical manual is stored on a computer disk and it can be viewed directly on the computer screen, or it can be printed locally at the printer connected to your computer.

If the problem is too technical, you may not be able to solve it on the first try and you will need to call a *technical help line* that will allow you to talk directly to a technician at the company that designed the equipment. If they are not able to provide sufficient information to troubleshoot and repair the problem, you may need to request a field service technician to be sent to your factory to help you locate the problem. The field service technician has received many hours of training with the engineers who designed the equipment and the network and may have experienced a similar problem in a different factory.

It is very important at this point to understand that you cannot possibly understand all of the technology needed to troubleshoot every piece of equipment that you come in contact with today. Rather you must have a sufficient base of electronics knowledge so that a support technician can explain possible problems in technical terms over the telephone. Then you should be able to perform the tests that are suggested and review the results. This knowledge base will also be of use when you must work with a field service technician. You will come to understand that some of the best training is obtained by working with or observing the field service technician troubleshoot problems on your equipment.

If you determine that you need additional knowledge, you may need to attend short training courses provided by the equipment vendors. You will also need to collect technical data about the systems you have. This means that you will need to read articles and technical reference materials. Then store the information so you can refer to it when necessary. For example, if the problem you are troubleshooting includes a DB25 connector, you may need to refer to the detailed figure in this chapter that identifies each pin and what signal is transmitted on it.

As you gain more knowledge, you will become more involved in designing new data communications systems to provide data to others in your company who need to know about production and quality information for the products your company makes.

SOLUTION TO JOB ASSIGNMENT

Your solution should include one or more examples from Fig. 16-1, Fig. 16-3, Fig. 16-13, or Fig. 16-14, you could also use examples which include networks with PLCs and an Ethernet for the entire factory. You should explain that PLCs may use proprietary vendor networks for the PLC network. You could add an Ethernet card as shown in Fig. 16-15 in the PLC rack. Each PLC vendor makes an Ethernet card so that it is simple to add the card to the PLC rack.

The sketch of the network should also include the drops on the Ethernet for the other parts of the factory that need to be interfaced, such as production control and shipping and receiving.

The information that is needed for the RS232 interface is included in Figs. 16-9, 16-10 and 16-11. You can explain that the RS232 may need to be used to interface electronic control devices at the low end of the vendor networks.

QUESTIONS

1. Use the hexadecimal values for the ASCII code to spell your first and last name.
2. Explain what the TCP/IP protocol is.
3. Explain what the SNA protocol is.
4. Use the hexadecimal value for the EBCDIC data format to spell your first and last names.
5. Explain how the EBCDIC data format and the ASCII are fundamentally different.
6. How can a computer be used to troubleshoot a network?
7. Explain what the MAP protocol is.
8. Explain what the TOP protocol is.
9. Compare the limitations of the RS422 with the RS232 interface.
10. Explain why a broad-band network may be used with the Ethernet network.

TRUE OR FALSE

1. _____ The RS422 is used instead of the RS232 because it can be used over a longer distance.
2. _____ Data that are sent to and from a PLC may include production counts, production times, and the amount of energy used to produce the parts.
3. _____ DTE stands for data terminal equipment and DCE stands for data current equipment.
4. _____ The bus, tree, star, and ring are types of network topologies.
5. _____ An Ethernet is a type of network that allows different proprietary networks such as the Allen-Bradley Data Highway + and the Modicon MODbus + to be interfaced so that they can send data to each other.

MULTIPLE CHOICE

1. One limitation of the RS232 interface is distance. A(n) _____ is usually used with the RS232 interface to transmit these types of data over longer distances.
   a. Ethernet card
   b. modem
   c. DB25 male connector
2. The three lowest levels of the seven-layer OSI model are the _____.
   a. session layer, data link layer, and the network layer.
   b. physical layer, data link layer, and the network layer.
   c. applications layer, presentation layer, and the session layer.
3. A modem is a device that _____.
   a. only sends data over telephone wires.
   b. only receives data over telephone wires.
   c. sends and receives data over telephone wires.
4. The _____ of the seven-layer ISO model includes the wire and cables.
   a. physical layer
   b. data link layer
   c. network layer
5. In the transport layer the simplest and least reliable service is provided by a _____ service.
   a. Type A
   b. Type B
   c. Type C

PROBLEMS

1. Draw a sketch of the seven-layer OSI model for network architecture.
2. Sketch the RS232 protocol with the send data, receive data, RTS, CTS, DSR, SG, CD, and DTR and indicate in which direction each signal moves. Use the DTE as a point of reference.
3. Draw a sketch of a breakout box and explain how it is used.
4. Sketch a data communications system that has two different brands of PLCs networks, each having five PLCs. The PLC networks are connected to an Ethernet network.

REFERENCES

1. *Data Acquisition*. Omega Engineering Inc. P.O. Box 4047, Stamford, CT 06907-0047, 1994.
2. Helmers, Scott A., *Data Communications*. Englewood Cliffs, NJ: Prentice Hall, 1989.
3. *Siemens Industrial Networks Course Manual*, Siemens Industrial Automation, Johnson City, TN 37605, 1992.

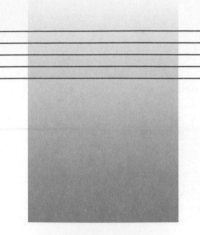

APPENDIX

Symbols, Terms and Definitions

NPN Bipolar Transistor	P-Channel Enhancement Mode MOS FET Dual Gate	Silicon Unilateral Switch (SUS)	Optoisolator with Photo-Darlington Output, No Base
PNP Bipolar Transistor	Silicon N-Type Unijunction Transistor (UJT)	Silicon Asymmetrical Switch (SAS)	Optoisolator with Photo-Darlington Output, and Base
NPN Darlington Transistor	Silicon P-Type Unijunction Transistor (UJT)	PIN Diode	Optoisolator Triac Driver
PNP Darlington Transistor	Programmable Unijunction Transistor (PUT)	Varactor	Optoisolator SCR Driver
N-Channel J FET	Rectifier or Diode	Varistor	AND Gate
P-Channel J FET	Full-Wave Bridge Rectifier	Metal Oxide Varistor (MOV)	OR Gate
N-Channel Depletion Mode MOS FET Single Gate	Zener Diode	Light Emitting Diode (LED)	NAND Gate
P-Channel Depletion Mode MOS FET Single Gate	SCR (Thyristor)	Photodiode	NOR Gate
N-Channel Enhancement Mode MOS FET Single Gate	Triac (Thyristor)	NPN Phototransistor, No Base Connection	Inverting AMP
P-Channel Enhancement Mode MOS FET Single Gate	Diac (Bilateral Trigger Diode)	NPN Phototransistor, with Base Connection	Non-Inverting AMP
N-Channel Depletion Mode MOS FET Dual Gate	Silicon Controlled Switch (SCS) (Thyristor)	Optoisolator with Photodiode Output	Exclusive OR Gate
P-Channel Depletion Mode MOS FET Dual Gate	Silicon Controlled Switch (Transistor) (SCS)	Optoisolator with Phototransistor Output, No Base Connection	
N-Channel Enhancement Mode MOS FET Dual Gate	Silicon Bilateral Switch (SBS)	Optoisolator with Phototransistor Output, and Base Connection	

(Courtesy of Philips Semiconductors)

Diodes, Rectifiers, Thyristors

C_t — Total Capacitance - The Total Small-Signal Capacitance Between The Diode Terminals.

d_I/d_t — Rate Of Change Of Current Versus Time.

d_V/d_t — Rate Of Change Of Voltage Versus Time.

I_F — Forward Junction Current - The Value Of DC Current That Flows Through A Semiconductor Diode Or Rectifier Diode In The Forward Direction.

I_{FRM} — Peak Forward Current Repetitive Peak - The Peak Value Of The Forward Current Including All Repetitive Transient Currents.

I_{FSM} — Forward Surge Peak Current - Maximum (Peak) Surge Forward Current Having A Specified Waveform And A Short Specified Time Interval.

$I_{GT\ Min}$ — Gate Trigger Current - Minimum Gate DC Current Required To Trigger The Device Under The Conditions Specified.

$I_{GO\ Max}$ — Peak Gate Turn-Off Current - Maximum Negative Gate Current Required To Switch Off.

I_H — Holding Current - Anode Current Necessary To Maintain On-State.

I_O — Average Rectifier DC Forward Current - The Value Of The Forward Current Averaged Over A 180° Conduction Angle At 60 Hz.

I_R — DC Reverse Current - Value Of DC Current That Flows Through The Diode In The Reverse Direction. (Leakage Current.)

I_{RM} — Maximum Reverse DC Current - The Respective Value Of Current That Flows Through The Junction In The Reverse Direction.

I_{trms} — Continuous On-State Current.

I_{tsm} — Surge (Non-Repetitive) Peak On-State Current - A Surge Current Of Short-Time Duration.

L_S — Series Inductance - The Inductance Between The Terminals On The Diode.

PRV — Peak Reverse Voltage - Maximum Repetitive Peak Reverse Blocking Voltage That May Be Applied To The Anode-Cathode Of The Device.

R_S — Series Resistance - The Total Small Signal Resistance Between The Diode Terminals.

T_A — Ambient Temperature - The Air Temperature Measured Below A Device, In An Environment Of Substantially Uniform Temperature, Cooled Only By Natural Air Convection And Not Materially Affected By Reflective And Radiant Surfaces.

T_C — Case Temperature - The Temperature Measured At A Specified Location On The Case Of A Device.

T_J — Semiconductor Junction Temperature.

T_Q — Turn Off Time.

t_{rr} — Reverse Recovery Time - The Time Required For The Current Or Voltage To Recover To A Specified Value After Instantaneous Switching From A Stated Forward Current Condition To A Stated Reverse Voltage Or Current Condition.

V_B — DC Breakdown Voltage - Value Of Voltage Measured At The Point Which Breakdown Occurs With The Diode Reverse Biased.

$V_{(BR)R}$ — Static Reverse Breakdown Voltage - The Value Of Negative Anode-To-Cathode Voltage At Which The Differential Resistance Breakdown Between The Anode And Cathode Terminals Changes From A High Value To A Substantially Lower Value.

V_{DRM} — Repetitive Peak Off-State Voltage - Maximum Instantaneous Value Of The Off-State Voltage That Occurs Across The Devices, Including All Repetitive Transient Voltages, But Excluding All Non-Repetitive Transient Voltages.

V_F — Forward Voltage - The Voltage Drop In A Semiconductor Diode Resulting From The Respective Forward Current.

V_{FM} — Maximum Forward Voltage - The Voltage Drop In A Semiconductor Diode Resulting From The Respective Forward Current.

V_{GFM} — Maximum Forward Gate Voltage - Maximum DC Forward Gate Voltage Permitted To Produce A Specified Forward Gate Current.

$V_{GO\ Max}$ — Peak Gate Turn-Off Voltage - Maximum Reverse Gate Voltage Required To Switch Off.

V_{GRM} — Maximum Reverse Gate Voltage - Maximum Peak Reverse Voltage Allowable Between The Gate Terminal And The Cathode Terminal When The Junction Between The Gate Region And The Adjacent Cathode Region Is Reverse Biased.

V_Z — Zener Regulator Reference Voltage - Value Of DC Voltage Across The Diode When It Is Biased To Operate In Its Breakdown Region.

$\Delta V_Z/\Delta T$ — Change In Zener Voltage To Change In Temperature.

Transistors

BV_{CBO} — Collector To Base Breakdown Voltage - Voltage Measured Between Collector And Base With Emitter Open.

BV_{CEO} — Collector To Emitter Breakdown Voltage - Voltage Measured Between Collector And Emitter With Base Open.

BV_{CER} — Collector To Emitter Breakdown Voltage - Voltage Measured Between Collector And Emitter When The Base Terminal Is Returned To The Emitter Terminal Through A Specified Resistance.

BV_{CES} — Collector To Emitter Breakdown Voltage - Voltage Measured Between Collector And Emitter With The Base Terminated Through A Short Circuit To The Emitter.

BV_{CEV} — Collector To Emitter Breakdown Voltage - Voltage Measured Between Collector And Emitter When A Specified Voltage (V) Is Applied Between The Base And Emitter.

Transistors (cont'd)

BV_{CEX} — Collector To Emitter Breakdown Voltage - Voltage Measured Between Collector And Emitter When The Base Is Terminated Through A Specified Load (X) To The Emitter.

BV_{DSS} — Drain To Source Breakdown Voltage - Voltage Measured Between The Drain And Source Terminals With The Gate Short-Circuited To The Source Terminal.

BV_{EBO} — Emitter To Base Breakdown Voltage - Reverse Voltage Measured Between Emitter And Base With The Collector Terminal Open.

BV_{GSS} — Gate To Source Breakdown Voltage - The Breakdown Voltage Between The Gate And Source Terminals With The Drain Terminal Short-Circuited To The Source Terminal.

C_{ISS} — Input Capacitance - The Capacitance Between The Terminals (Gate And Source) With The Drain Short-Circuited To The Source.

C_{RSS} — Reverse Transfer Capacitance - The Capacitance Between The Drain And Gate Terminals.

f_T — Gain Bandwidth Product - Frequency At Which Small-Signal Gain Becomes Unity.

gfs — Forward Transfer Conductance - Common Source Forward Transconductance.

G_{PE} — Power Gain Emitter Output.

h_{FE} — DC Current Gain - The Ratio Of Collector Current To Base Current At A Specified Collector-Emitter Voltage.

I_B — DC Base Current - Value OF DC Current Into The Base Terminal.

I_C — DC Collector Current - Value OF DC Current Into The Collector Terminal.

I_{DSS} — Zero Bias Drain Current - Amount Of Current Which Flows In The Drain When The Gate Is Connected To The Source.

N.F. — Noise Figure.

P_D — Average Power Dissipation.

P_{IN} — Signal Input Power To Device.

P_{OUT} — Signal Output Power.

r_{DSS} — Drain-Source On-State Resistance.

V_{CC} — DC Supply Voltage Applied To The Collector Terminal.

Special Purpose Devices

BV_{CER} — Breakdown Voltage Between Collector And Emitter With A Specified Resistor Between Base And Emitter.

BV_{GKF} — Gate To Cathode Forward Breakdown Voltage.

BV_{GKR} — Gate To Cathode Reverse Breakdown Voltage.

h_{FE} — DC Current Gain - Ratio Of DC Output Current To The DC Input Current.

$I_{BO+(Max)}$ — Maximum Forward Breakover Current.

$I_{BO-(Max)}$ — Maximum Reverse Breakover Current.

I_E — Value Of The DC Current Into The Emitter.

I_{EO} — Emitter Current With One Base Open.

I_G — DC Gate Current - The DC Current Flowing Through The Gate As A Result Of Applied Gate Voltage.

$I_{T\ PK}$ — Total Peak Current.

I_V — Valley Current - The Valley Current Is The Emitter Current At The Second Lowest Current Point.

η — Intrinsic Stand Off Ratio.

R_{BBO} — Base 1 To Base 2 Resistance With Open Emitter.

V_{AK} — Anode To Cathode Voltage - The Maximum Value Of Voltage Applied Between Anode And Cathode Without Failure.

$V_{(BO)+}$ — Forward Breakover Voltage.

$V_{(BO)-}$ — Reverse Breakover Voltage.

ΔV_F — Forward Breakback Voltage.

ΔV_R — Reverse Breakback Voltage.

V_{GT} — Gate Trigger Voltage - The Gate Voltage Required To Produce The Gate Trigger Current.

Opto Electronic Devices

I_D — Dark Current - The Current Which Flows In A Photodetector When There Is No Incident Radiation On The Detector.

I_{FT} — Input Trigger Current - Emitter Current Necessary To Trigger The Coupled Device.

I_L — Light Current - The Current That Flows Through A Photo Sensitive Device When It Is Exposed To Illumination.

P_t — Total Device Power Dissipation.

Response Time — The Time It Takes The Device To React To An Incoming Signal.

Rise Time (t_r) — The Time Duration During Which The Leading Edge Of A Pulse Is Increasing From 10% To 90% Of Its Maximum Amplitude.

V_{ISO} — DC Isolation Surge Voltage - The Dielectric Withstanding Voltage Between The Input And Output.

λ_P — Wavelength At Peak Emission - The Wavelength At Which The Power Output From A Light-Emitting Diode Is Maximum.

θ_{HI} — Half-Intensity Beam Angle - The Angle Within Which The Radiant Intensity Is Not Less Than Half Of The Maximum Intensity.

(Courtesy of Philips Semiconductors)

MOTOROLA
SEMICONDUCTOR ▬▬▬
TECHNICAL DATA

Designers Data Sheet

STUD MOUNTED
FAST RECOVERY POWER RECTIFIERS

. . . designed for special applications such as dc power supplies, inverters, converters, ultrasonic systems, choppers, low RF interference, sonar power supplies and free wheeling diodes. A complete line of fast recovery rectifiers having typical recovery time of 150 nanoseconds providing high efficiency at frequencies to 250 kHz.

**FAST RECOVERY
POWER RECTIFIERS
50-600 VOLTS
30 AMPERES**

**CASE 42A-01
DO-203AB
METAL**

Designer's Data for "Worst Case" Conditions

The Designers Data sheets permit the design of most circuits entirely from the information presented. Limit curves — representing boundaries on device characteristics -- are given to facilitate "worst case" design.

MECHANICAL CHARACTERISTICS
CASE: Welded, hermetically sealed
FINISH: All external surfaces corrosion resistant and readily solderable
POLARITY: Cathode to Case
WEIGHT: 17 Grams (Approximately)
MOUNTING TORQUE: 25 in-lbs max.

*MAXIMUM RATINGS

Rating	Symbol	1N3909	1N3910	1N3911	1N3912	1N3913	MR1396	Unit
Peak Repetitive Reverse Voltage Working Peak Reverse Voltage DC Blocking Voltage	V_{RRM} V_{RWM} V_R	50	100	200	300	400	600	Volts
Non-Repetitive Peak Reverse Voltage	V_{RSM}	75	150	250	350	450	650	Volts
RMS Reverse Voltage	$V_{R(RMS)}$	35	70	140	210	280	420	Volts
Average Rectified Forward Current (Single phase, resistive load, T_C = 100°C)	I_O	30						Amps
Non-Repetitive Peak Surge Current (surge applied at rated load conditions)	I_{FSM}	300						Amp
Operating Junction Temperature Range	T_J	-65 to +150						°C
Storage Temperature Range	T_{stg}	-65 to +175						°C

THERMAL CHARACTERISTICS

Characteristic	Symbol	Max	Unit
Thermal Resistance, Junction to Case	$R_{\theta JC}$	1.2	°C/W

*ELECTRICAL CHARACTERISTICS

Characteristic	Symbol	Min	Typ	Max	Unit
Instantaneous Forward Voltage (i_F = 93 Amp, T_J = 150°C)	v_F	–	1.2	1.5	Volts
Forward Voltage (I_F = 30 Amp, T_C = 25°C)	V_F	–	1.1	1.4	Volts
Reverse Current (rated dc voltage) T_C = 25°C	I_R	–	10	25	µA
T_C = 100°C		–	0.5	1.0	mA

*REVERSE RECOVERY CHARACTERISTICS

Characteristic	Symbol	Min	Typ	Max	Unit
Reverse Recovery Time (I_F = 1.0 Amp to V_R = 30 Vdc, Figure 16)	t_{rr}	–	150	200	ns
(I_{FM} = 36 Amp, di/dt = 25 A/µs, Figure 17)		–	200	400	
Reverse Recovery Current (I_F = 1.0 Amp to V_R = 30 Vdc, Figure 16)	$I_{RM(REC)}$	–	1.5	2.0	Amp

*Indicates JEDEC Registered Data for 1N3909 Series.

1N3909 thru 1N3913, MR1396

FIGURE 1 – FORWARD VOLTAGE

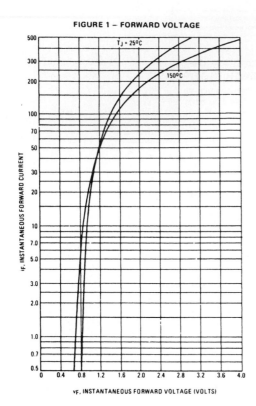

FIGURE 2 – MAXIMUM SURGE CAPABILITY

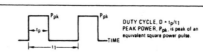

NOTE 1

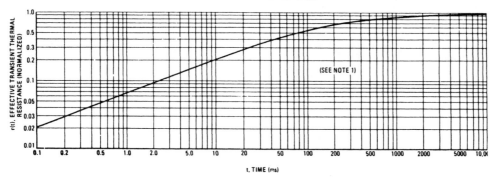

DUTY CYCLE, D = t_p/t_1
PEAK POWER, P_{pk}, is peak of an equivalent square power pulse.

To determine maximum junction temperature of the diode in a given situation, the following procedure is recommended:

The temperature of the case should be measured using a thermocouple placed on the case at the temperature reference point (see Note 3). The thermal mass connected to the case is normally large enough so that it will not significantly respond to heat surges generated in the diode as a result of pulsed operation once steady-state conditions are achieved. Using the measured value of T_C, the junction temperature may be determined by:

$$T_J = T_C + \Delta T_{JC}$$

where ΔT_{JC} is the increase in junction temperature above the case temperature. It may be determined by:

$$\Delta T_{JC} = P_{pk} \cdot R_{\theta JC} [D + (1 - D) \cdot r(t_1 + t_p) + r(t_p) - r(t_1)]$$

where
  $r(t)$ = normalized value of transient thermal resistance at time, t, from Figure 3, i.e.
  $r(t_1 + t_p)$ = normalized value of transient thermal resistance at time $t_1 + t_p$

FIGURE 3 – THERMAL RESPONSE

(SEE NOTE 1)

(Copyright of Motorola. Used by permission.)

General-Purpose Rectifiers

Motorola offers a wide variety of low-cost devices, packaged to meet diverse mounting requirements.

All devices are connected cathode-to-case or cathode-to-heatsink, where applicable. Reverse polarity may be available on some devices upon special request. Contact your Motorola representative for more information.

Table 7. General-Purpose Rectifiers

V_{RRM} (Volts)	I_O, Average Rectified Forward Current (Amperes)[1]					
	1	3	3	6	12	20
	59-03 (DO-41) Plastic Cathode = Polarity Band	60-01 Metal Style 1	267-03 Plastic Cathode = Polarity Band	194-04 Plastic Style 1	245A-02 (DO-203AA) Metal Style 2	
50	1N4001[3]	1N4719	1N5400	MR750	MR1120 1N1199,A,B	MR2000
100	1N4002[3]	1N4720	1N5401	MR751	MR1121 N1200,A,B	MR2001
200	1N4003[3]	1N4721	1N5402	MR752	MR1122 1N1202,A,B	MR2002
400	*1N4004*[3]	*1N4722*	*1N5404*	*MR754*	*MR1124 1N1204,A,B*	*MR2004*
600	1N4005[3]	1N4723	*1N5406*	MR756	MR1126 1N1206,A,B	MR2006
800	1N4006[3]	1N4724	—	MR758	MR1128	MR2008
1000	*1N4007*[3]	*1N4725*		*MR760*	*MR1130*	*MR2010*
I_{FSM} (Amps)	30	300	200	400	300[9]	400
T_A @ Rated I_O (°C)	75	75	$T_L = 105$	60	—	—
T_C @ Rated I_O (°C)	—	—	—	—	150	150
T_J (Max) °C	175	175	175	175	190	175

[1] I_O is total device output.
[3] Package Size: 0.120″ max diameter by 0.260″ max length.
[9] IFSM is for MR1120 series, 1N1199 = 100, -A = 240, -B = 250.

(Copyright of Motorola. Used by permission.)

Schottky Rectifiers

SWITCHMODE™ Schottky power rectifiers with the high speed and low forward voltage drop characteristic of Schottky's metal/silicon junctions are produced with ruggedness and temperature performance comparable to silicon-junction rectifiers. Ideal for use in low-voltage, high-frequency power supplies, and as very fast clamping diodes, these devices feature switching times less than 10 ns, and are offered in current ranges from 1 to 600 amperes, and reverse voltages to 200 volts.

In some current ranges, devices are available with junction temperature specifications of 125°C, 150°C, 175°C. Devices with higher T_J ratings can have significantly lower leakage currents, but higher forward-voltage specifications. These parameter tradeoffs should be considered when selecting devices for applications that can be satisfied by more than one device type number.

All devices are connected cathode-to-case or cathode-to-heatsink, where applicable. Reverse polarity may be available on some devices upon special request. Contact your Motorola representative for more information.

Table 4. Schottky Rectifiers

V_{RRM} (Volts)	I_O, Average Rectified Forward Current (Amperes)[1]							
	1		**3**				**5**	**6**
	59-04 Plastic Cathode = Polarity Band	403A-03 SMB Cathode = Notch	267-03 Plastic Cathode = Polarity Band		403-03 SMC Cathode = Notch	369A-13 DPAK Style 3	60-01 Metal Style 1	369A-13 DPAK Style 3
20	1N5817	MBRS120T3	1N5820	MBR320	MBRS320T3	MBRD320	1N5823	MBRD620CT
25								
30	1N5818	MBRS130LT3* MBRS130T3	1N5821	MBR330	MBRS330T3	MBRD330	1N5824	MBRD630CT
35						MBRD330L*		
40	1N5819	MBRS140T3	1N5822	MBR340	MBRS340T3	MBRD340	1N5825	MBRD640CT
45						MBRD340L		
50	MBR150			MBR350		MBRD350		MBRD650CT
60	MBR160			MBR360		MBRD360		MBRD660CT
70	MBR170			MBR370				
80	MBR180			MBR380				
90	MBR190			MBR390				
100	MBR1100	MBRS1100T3		MBR3100				
I_{FSM} (Amperes)	25	40	80	80	80	75	500	75
Max V_F @ $I_{FM} = I_O$	0.6[2] $T_L = 25°C$	0.6[2]/0.395* $T_C = 25°C$	0.525[2] $T_L = 25°C$	0.74[2] $T_L = 25°C$	0.525[2] $T_L = 25°C$	0.45/0.390* $T_C = 125°C$	0.38[2] $T_C = 25°C$	0.85 $T_C = 125°C$
T_J (Max) °C	125	125	125	150	125	150	125	150

[1]I_O is total device output current.
[2]Values are for 40 volt units, lower voltage parts exhibit lower V_F.

MOTOROLA
SEMICONDUCTOR ▰▰▰▰▰▰▰▰
TECHNICAL DATA

5 Watt Surmetic 40
Silicon Zener Diodes

. . . a complete series of 5 Watt Zener Diodes with tight limits and better operating characteristics that reflect the superior capabilities of silicon-oxide-passivated junctions. All this in an axial-lead, transfer-molded plastic package offering protection in all common environmental conditions.

Specification Features:
● Up to 180 Watt Surge Rating @ 8.3 ms
● Maximum Limits Guaranteed on Seven Electrical Parameters

Mechanical Characteristics:

CASE: Void-free, transfer-molded, thermosetting plastic
FINISH: All external surfaces are corrosion resistant and leads are readily solderable
POLARITY: Cathode indicated by color band. When operated in zener mode, cathode will be positive with respect to anode
MOUNTING POSITION: Any
WEIGHT: 0.7 gram (approx)

**1N5333B
thru
1N5388B**

**5 WATT
ZENER REGULATOR
DIODES
3.3–200 VOLTS**

**CASE 17-02
PLASTIC**

MAXIMUM RATINGS			
Rating	Symbol	Value	Unit
DC Power Dissipation @ $T_L = 75°C$	P_D	5	Watts
Lead Length = 3/8″			
Derate above 75°C		40	mW/°C
Operating and Storage Junction Temperature Range	T_J, T_{stg}	− 65 to +200	°C

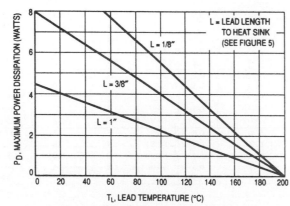

Figure 1. Power Temperature Derating Curve

1N5333B thru 1N5388B

ELECTRICAL CHARACTERISTICS ($T_A = 25°C$ unless otherwise noted, $V_F = 1.2$ Max @ $I_F = 1$ A for all types)

JEDEC Type No. (Note 1)	Nominal Zener Voltage V_Z @ I_{ZT} Volts (Note 2)	Test Current I_{ZT} mA	Max Zener Impedance		Max Reverse Leakage Current		Max Surge Current I_r, Amps (Note 3)	Max Voltage Regulation ΔV_Z, Volt (Note 4)	Maximum Regulator Current I_{ZM} mA (Note 5)
			Z_{ZT} @ I_{ZT} Ohms (Note 2)	Z_{ZK} @ I_{ZK} = 1 mA Ohms (Note 2)	I_R μA	@ V_R Volts			
⇒ 1N5333B	3.3	380	3	400	300	1	20	0.85	1440
1N5334B	3.6	350	2.5	500	150	1	18.7	0.8	1320
1N5335B	3.9	320	2	500	50	1	17.6	0.54	1220
1N5336B	4.3	290	2	500	10	1	16.4	0.49	1100
1N5337B	4.7	260	2	450	5	1	15.3	0.44	1010
⇒ 1N5338B	5.1	240	1.5	400	1	1	14.4	0.39	930
⇒ 1N5339B	5.6	220	1	400	1	2	13.4	0.25	865
1N5340B	6	200	1	300	1	3	12.7	0.19	790
1N5341B	6.2	200	1	200	1	3	12.4	0.1	765
⇒ 1N5342B	6.8	175	1	200	10	5.2	11.5	0.15	700
⇒ 1N5343B	7.5	175	1.5	200	10	5.7	10.7	0.15	630
⇒ 1N5344B	8.2	150	1.5	200	10	6.2	10	0.2	580
1N5345B	8.7	150	2	200	10	6.6	9.5	0.2	545
1N5346B	9.1	150	2	150	7.5	6.9	9.2	0.22	520
⇒ 1N5347B	10	125	2	125	5	7.6	8.6	0.22	475
1N5348B	11	125	2.5	125	5	8.4	8	0.25	430
⇒ 1N5349B	12	100	2.5	125	2	9.1	7.5	0.25	395
⇒ 1N5350B	13	100	2.5	100	1	9.9	7	0.25	365
1N5351B	14	100	2.5	75	1	10.6	6.7	0.25	340
⇒ 1N5352B	15	75	2.5	75	1	11.5	6.3	0.25	315
⇒ 1N5353B	16	75	2.5	75	1	12.2	6	0.3	295
1N5354B	17	70	2.5	75	0.5	12.9	5.8	0.35	280
⇒ 1N5355B	18	65	2.5	75	0.5	13.7	5.5	0.4	265
1N5356B	19	65	3	75	0.5	14.4	5.3	0.4	250
⇒ 1N5357B	20	65	3	75	0.5	15.2	5.1	0.4	237
1N5358B	22	50	3.5	75	0.5	16.7	4.7	0.45	216
⇒ 1N5359B	24	50	3.5	100	0.5	18.2	4.4	0.55	198
⇒ 1N5360B	25	50	4	110	0.5	19	4.3	0.55	190
⇒ 1N5361B	27	50	5	120	0.5	20.6	4.1	0.6	176
1N5362B	28	50	6	130	0.5	21.2	3.9	0.6	170
⇒ 1N5363B	30	40	8	140	0.5	22.8	3.7	0.6	158
⇒ 1N5364B	33	40	10	150	0.5	25.1	3.5	0.6	144
⇒ 1N5365B	36	30	11	160	0.5	27.4	3.3	0.65	132
⇒ 1N5366B	39	30	14	170	0.5	29.7	3.1	0.65	122
1N5367B	43	30	20	190	0.5	32.7	2.8	0.7	110
⇒ 1N5368B	47	25	25	210	0.5	35.8	2.7	0.8	100
1N5369B	51	25	27	230	0.5	38.8	2.5	0.9	93
1N5370B	56	20	35	280	0.5	42.6	2.3	1	86
1N5371B	60	20	40	350	0.5	42.5	2.2	1.2	79
⇒ 1N5372B	62	20	42	400	0.5	47.1	2.1	1.35	76
1N5373B	68	20	44	500	0.5	51.7	2	1.5	70
1N5374B	75	20	45	620	0.5	56	1.9	1.6	63
1N5375B	82	15	65	720	0.5	62.2	1.8	1.8	58
1N5376B	87	15	75	760	0.5	66	1.7	2	54.5
1N5377B	91	15	75	760	0.5	69.2	1.6	2.2	52.5
1N5378B	100	12	90	800	0.5	76	1.5	2.5	47.5
1N5379B	110	12	125	1000	0.5	83.6	1.4	2.5	43
1N5380B	120	10	170	1150	0.5	91.2	1.3	2.5	39.5
1N5381B	130	10	190	1250	0.5	98.8	1.2	2.5	36.6
1N5382B	140	8	230	1500	0.5	106	1.2	2.5	34

(continue

⇒ **Preferred part**

(Copyright of Motorola. Used by permission.)

MAXIMUM RATINGS

Rating	Symbol	Value	Unit
Collector-Emitter Voltage	V_{CEO}	300	Vdc
Collector-Base Voltage	V_{CBO}	300	Vdc
Emitter-Base Voltage	V_{EBO}	6.0	Vdc
Collector Current — Continuous	I_C	500	mAdc
Total Device Dissipation @ T_A = 25°C Derate above 26°C	P_D	1.0 8.0	Watt mW/°C
Total Device Dissipation @ T_C = 25°C Derate above 25°C	P_D	2.5 20	Watts mW/°C
Operating and Storage Junction Temperature Range	T_J, T_{stg}	−55 to +150	°C

THERMAL CHARACTERISTICS

Characteristic	Symbol	Max	Unit
Thermal Resistance, Junction to Ambient	$R_{\theta JA}$	125	°C/W
Thermal Resistance, Junction to Case	$R_{\theta JC}$	50	°C/W

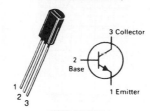

MPSW42★

CASE 29-05, STYLE 1
TO-92 (TO-226AE)

ONE WATT
HIGH VOLTAGE
TRANSISTOR

NPN SILICON

★This is a Motorola
designated preferred device.

ELECTRICAL CHARACTERISTICS (T_A = 25°C unless otherwise noted.)

Characteristic	Symbol	Min	Max	Unit
OFF CHARACTERISTICS				
Collector-Emitter Breakdown Voltage(1) (I_C = 1.0 mAdc, I_B − 0)	$V_{(BR)CEO}$	300	—	Vdc
Collector-Base Breakdown Voltage (I_C = 100 μAdc, I_E = 0)	$V_{(BR)CBO}$	300	—	Vdc
Emitter-Base Breakdown Voltage (I_E − 100 μAdc, I_C − 0)	$V_{(BR)EBO}$	6.0	—	Vdc
Collector Cutoff Current (V_{CB} − 200 Vdc, I_E − 0)	I_{CBO}	---	0.1	μAdc
Emitter Cutoff Current (V_{EB} = 6.0 Vdc, I_C = 0)	I_{EBO}	—	0.1	μAdc
ON CHARACTERISTICS				
DC Current Gain (I_C = 1.0 mAdc, V_{CE} = 10 Vdc) (I_C = 10 mAdc, V_{CE} = 10 Vdc) (I_C = 30 mAdc, V_{CE} = 10 Vdc)	h_{FE}	25 40 40	— — —	—
Collector-Emitter Saturation Voltage (I_C = 20 mAdc, I_B = 2.0 mAdc)	$V_{CE(sat)}$	—	0.5	Vdc
Base-Emitter Saturation Voltage (I_C = 20 mAdc, I_B = 2.0 mAdc)	$V_{BE(sat)}$	—	0.9	Vdc
SMALL-SIGNAL CHARACTERISTICS				
Current-Gain — Bandwidth Product (I_C − 10 mAdc, V_{CE} 20 Vdc, f 20 MHz)	f_T	50	—	MHz
Collector-Base Capacitance (V_{CB} = 20 Vdc, I_E = 0, f = 1.0 MHz)	C_{cb}	—	3.0	pF

(1) Pulse Test: Pulse Width ≤ 300 μs, Duty Cycle < 2.0%.

(Copyright of Motorola. Used by permission.)

MPSW42

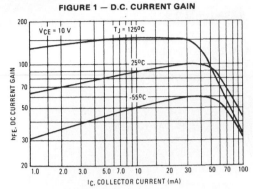

FIGURE 1 — D.C. CURRENT GAIN

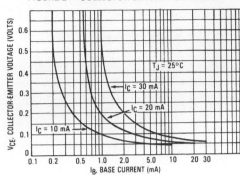

FIGURE 2 — COLLECTOR SATURATION REGION

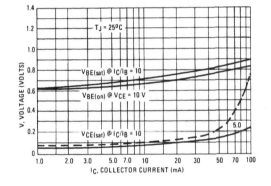

FIGURE 3 — ON VOLTAGES

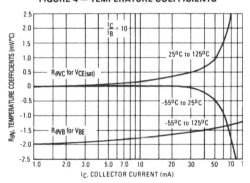

FIGURE 4 — TEMPERATURE COEFFICIENTS

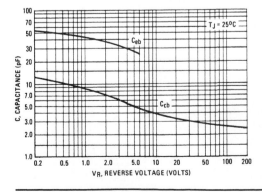

FIGURE 5 — CAPACITANCE

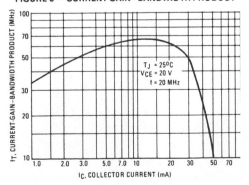

FIGURE 6 — CURRENT GAIN - BANDWIDTH PRODUCT

Motorola Small–Signal Transistors, FETs and Diodes Device Data

High-Beta Transistor Modules Selector Guide

Powerex through cooperative efforts with its business partner Mitsubishi Electric is proud to present continued innovation in Bipolar Transistor Technology with it's complete line of newly developed High-Beta Darlington Transistor Modules.

The development of the High-Beta Series was undertaken in response to the customer desires for:

☐ Simplified Base Drive

☐ Improved Switching Speed

☐ Increased Ruggedness

Powerex Engineered Solutions

☐ H_{fe} increase to 10X providing—90% base drive reduction

☐ $T_{(s)}$ (storage time) reduction of 10-20% providing—reduced switching power loss

☐ RBSOA and SCSOA (Reverse Bias and Short Circuit Safe Operating Areas) increase of up to 20%—providing increased ruggedness

(Courtesy of POWEREX, Inc.)

High-Beta Module Line Up

Type	Current Rating I_C (Amperes)	Voltage V_{CES} (Volts)
Single High-Beta Module S†		
KS824501HB	10	600
KS524502HB	20	600
KS524503HB	30	600
KSF22005	50	250
KS524505HB	50	600
KS624530HB	300	600
KS621K30HB	300	1000
KS621230HB	300	1200
KS624540	400	600
KS621K40HB	400	1000
KS621240HB	400	1200
KS624550	500	600
KS621K60HB	600	1000
KS621260HB	600	1200
KS621K80HB*	800	1000
KS621280HB*	800	1000
K621K1KHB	1000	1200
KS62121KHB	1000	1200
Dual High-Beta Module D†		
KD224503HB	30	600
KD224505HB	50	600
KD221K05HB	50	1000
KD221205HB	50	1200
KD224575HB	75	600
KD221K75HB	75	1000
KD221275HB	75	1200
KD324510HB	100	600
KD421K10HB	100	1000
KD421210HB	100	1200
KD324515HB	150	600
KD421K15HB	150	1000
KD421215HB	150	1200
KD424520HB	200	600
KD621K20HB	200	1000
KD621220HB*	200	1200
KD624530HB*	300	600
KD621K30HB	300	1000
KD621230HB	300	1200
KD621K40HB*	400	1000
KD621240HB*	400	1200
KD621K50HB*	500	1000
KD621250HB*	500	1200

Type	Current Rating I_C (Amperes)	Voltage V_{CES} (Volts)
6 in 1 High-Beta Module E†		
KEE24501HB	10	600
KED235A1HB	15	500
QM15TG 9B	15	500
KED245A1HB	15	600
KE721KA1HB	15	1000
KE7212A1HB	15	1200
KED23502HB	20	500
QM20TG 9B	20	450
KED24502HB	20	600
KEF24503HB	30	600
KE524503HB	30	600
KE721K03HB	30	1000
KE721203HB	30	1200
KET24505HB	50	600
KE524505HB	50	600
KE921K05HB	50	1000
KE921205HB	50	1200
KET24575HB	75	600
KE524575HB	75	600
KET24510HB	100	600
KE524510HB	100	600
KET24515HB	150	600
KE524515HB	150	600
High-Beta Module M†		
KMG24501HB	10	600
KMK245A1HB	15	600
KMK24502HB	20	600

* Under Development

Note:
600V = 3 Stage Darlington/1000 and 1200V = 4 Stage Darlington, KE5 = All Screw Terminals, KEF or KET = TAB Signal Terminal.

† Second Digit
    S = Single
    D = Dual
    E = 6 in 1
    M = Includes Diode Converter

The curves at the right provide a comparative analysis of some of the key High-Beta parameters as compared to the conventional bipolar devices:

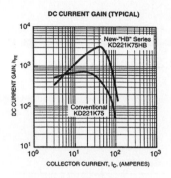

DC CURRENT GAIN (TYPICAL)

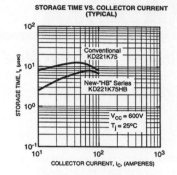

STORAGE TIME VS. COLLECTOR CURRENT (TYPICAL)

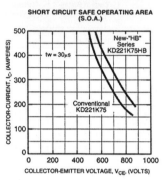

SHORT CIRCUIT SAFE OPERATING AREA (S.O.A.)

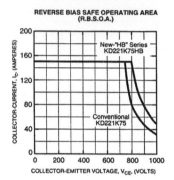

REVERSE BIAS SAFE OPERATING AREA (R.B.S.O.A.)

Application of Power Devices

— Current Product Range

- - - - - - - Future Development Plan

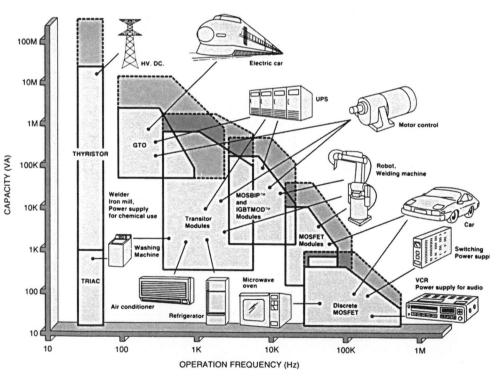

(Courtesy of POWEREX, Inc.)

Metal–Can Transistors

Metal–can packages are intended for use in industrial applications where harsh environmental conditions are encountered. These packages enhance reliability of the end products due to their resistance to varying humidity and extreme temperature ranges.

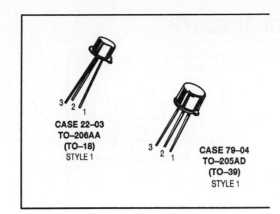

CASE 22–03
TO–206AA
(TO–18)
STYLE 1

CASE 79–04
TO–205AD
(TO–39)
STYLE 1

Table 26. Metal–Can General–Purpose Transistors

These transistors are designed for DC to VHF amplifier applications, general–purpose switching applications, and complementary circuitry. Devices are listed in decreasing order of $V_{(BR)CEO}$ within each package group.

Device Type	$V_{(BR)CEO}$ Volts Min	f_T @ I_C MHz Min	mA	I_C mA Max	h_{FE} @ I_C Min	Max	mA
Case 22–03 — TO–206AA (TO–18) — NPN							
2N3700	80	80	50	1000	50	—	500
BC107	45	150	10	200	110	450	2.0
BC107B	45	150	10	200	200	450	2.0
2N2222A	40	300	20	800	100	300	150
BC109C	25	150	10	200	420	800	2.0
Case 22–03 — TO–206AA (TO–18) — PNP							
2N2906A	60	200	50	600	40	120	150
2N2907A	60	200	50	600	100	300	150
2N3251A	60	300	10	200	100	300	10
BC177B	45	200	10	200	180	460	2.0
Case 79–04 — TO–205AD (TO–39) — NPN							
2N3019	80	100	50	1000	100	300	150
2N3020	80	80	50	1000	40	120	150
2N1893	80	50	50	500	40	120	150
2N2219A	40	300	20	800	100	300	150
Case 79–04 — TO–205AD (TO–39) — PNP							
2N4033	80	—	—	1000	25	—	1000
2N4036	65	60	50	1000	40	140	150
2N2904A	60	200	50	600	40	120	150
2N2905A	60	200	50	600	100	300	150
2N4032	60	—	—	1000	40	—	1000

(Copyright of Motorola. Used by permission.)

Thyristor Triggers

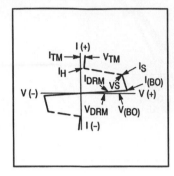

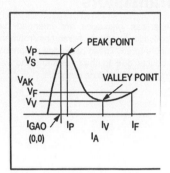

Table 5. SIDACs

High voltage trigger devices similar in operation to a Triac. Upon reaching the breakover voltage in either direction, the device switches to a low-voltage on-state.

Device Type	I_{TSM} Amps	V_{BO} Volts	
		Min	Max
Case 267-03/1			
MKP3V110	20	100	120
MKP3V120	20	110	130
MKP3V130	20	120	140
MKP3V240	20	220	250
MKP3V260	20	240	270
MKP3V270	20	250	280
Case 59-04/1			
MKP1V120	4	110	130
MKP1V130	4	120	140
MKP1V140	4	130	150
MKP1V240	4	220	250
MKP1V260	4	240	270
MKP1V270	4	250	280

Table 6. Programmable Unijunction Transistor — PUT

Similar to UJTs, except that I_V, I_P and intrinsic standoff voltage are programmable (adjustable) by means of external voltage divider. This stabilizes circuit performance for variations in device parameters. General operating frequency range is from 0.01 Hz to 10 kHz, making them suitable for long-duration timer circuits.

Device Type	I_P			I_V	
	$R_G =$ 10 kΩ	$R_G =$ 1 MΩ	I_{GAO} @ 40 V	$R_G =$ 10 kΩ	$R_G =$ 1 MΩ
	μA Max	μA Max	nA Max	μA Min	μA Max
Plastic TO-92 (Case 29-04/16)					
2N6027	5	2	10	70	50
2N6028	1	0.15	10	25	25

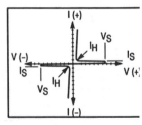

Table 7. Silicon Bidirectional Switch (SBS)

This versatile trigger device exhibits highly symmetrical bidirectional switching characteristics which can be modified by means of a gate lead. Requires a gate trigger current of only 250 μAdc for triggering.

Device Type	V_S Volts		I_S μA Max	I_H mA Max
	Min	Max		
Plastic TO-92/TO-226AA (Case 29-04/12)				
MBS4991	6	10	500	1.5
MBS4992	7.5	9	120	0.5
MBS4993	7.5	9	250	0.75

Silicon Controlled Rectifiers
Reverse Blocking Triode Thyristors

**C35
Series**

SCRs
**35 AMPERES RMS
50 thru 800 VOLTS**

. . . designed primarily for half-wave ac control applications, such as motor controls, heating controls and power supplies; or wherever half-wave silicon gate-controlled, solid-state devices are needed.
- Glass Passivated Junctions and Center Gate Fire for Greater Parameter Uniformity and Stability
- Blocking Voltage to 800 Volts

**CASE 263-04
STYLE 1**

MAXIMUM RATINGS (T_J = 25°C unless otherwise noted.)

Rating		Symbol	Value	Unit
Peak Repetitive Forward and Reverse Blocking Voltage, Note 1 (Gate Open, T_C = −65 to +125°C)		V_{DRM} or V_{RRM}		Volts
	C35F		50	
	C35A		100	
	C35B		200	
	C35D		400	
	C35M		600	
	C35N		800	
Peak Non-Repetitive Reverse Voltage (T_C = −65 to +125°C, t < 5 ms)		V_{RSM}		Volts
	C35F		75	
	C35A		150	
	C35B		300	
	C35D		500	
	C35M		720	
	C35N		960	
RMS On-State Current (All Conduction Angles)		$I_{T(RMS)}$	35	Amps
Peak Non-Repetitive Surge Current (One cycle, 60 Hz)		I_{TSM}	225	Amps
Circuit Fusing (t = 8.3 ms)		I^2t	210	A^2s
Peak Gate Power		P_{GM}	5	Watts
Average Gate Power		$P_{G(AV)}$	0.5	Watt
Peak Reverse Gate Voltage		V_{GRM}	5	Volts
Operating Junction Temperature Range		T_J	−65 to +125	°C
Storage Temperature Range		T_{stg}	−65 to +150	°C

Note 1. V_{DRM} and V_{RRM} for all types can be applied on a continuous basis. Ratings apply for zero or negative gate voltage; however, positive gate voltage shall not be applied concurrent with negative potential on the anode. Blocking voltages shall not be tested with a constant current source such that the voltage ratings of the devices are exceeded.

(Copyright of Motorola. Used by permission.)

C35 Series

THERMAL CHARACTERISTICS

Characteristic	Symbol	Max	Unit
Thermal Resistance, Junction to Case	$R_{\theta JC}$	1.7	°C/W

ELECTRICAL CHARACTERISTICS (T_C = 25°C unless otherwise noted.)

Characteristic		Symbol	Min	Typ	Max	Unit
*Peak Repetitive Forward Blocking Current		I_{DRM} or				mA
(V_{AK} = Rated V_{DRM} or V_{RRM}, T_C = +125°C,		I_{RRM}				
Gate Open)	C35F,A		—	—	13	
	C35B		—	—	12	
	C35D		—	—	6	
	C35M		—	—	5	
	C35N		—	—	4	μA
(V_D = Rated V_{DRM}, T_C = 25°C, Gate Open)	All Devices		—	—	10	
Average Reverse Blocking Current		$I_{RRM(AV)}$				mA
(V_R = Rated V_{RSM}, T_C = +125°C, Gate Open)	C35F,A		—	—	6.5	
	C35B		—	—	6	
	C35D		—	—	4	
	C35M		—	—	2.5	
	C35N		—	—	2	
(V_R = Rated V_{RRM}, T_C = 25°C, Gate Open)	All Devices		—	—	10	μA
Peak On-State Voltage		V_{TM}	—	—	2	Volts
(I_{TM} = 50.3 A peak, Pulse Width ≤ 1 ms, Duty Cycle ≤ 2%)						
Gate Trigger Current (Continuous dc)		I_{GT}				mA
(V_D = 12 Vdc, R_L = 50 Ω)			—	6	40	
(V_D = 12 Vdc, R_L = 50 Ω, T_C = −65°C)			—	—	80	
Gate Trigger Voltage (Continuous dc)		V_{GT}				Volts
(V_D = 12 Vdc, R_L = 50 Ω, T_C = −65°C to +125°C)			—	—	3	
(V_D = Rated V_{DRM}, R_L = 1000 Ω, T_C = 125°C)			0.25	—	—	
Holding Current		I_H	—	—	100	mA
(V_D = 24 Vdc, Gate Supply = 10 V, 20 Ω, 45 μs minimum pulse width, I_T = 0.5 A)						
Critical Rate of Rise of Forward Blocking Voltage		dv/dt				V/μs
(V_D = Rated V_{DRM}, T_C = +125°C)	C35F,M,N		10	—	—	
	C35A,B		20	—	—	
	C35D		25	—	—	

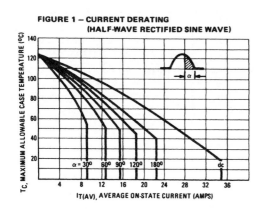

FIGURE 1 — CURRENT DERATING
(HALF-WAVE RECTIFIED SINE WAVE)

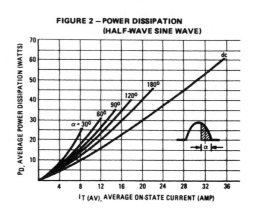

FIGURE 2 — POWER DISSIPATION
(HALF-WAVE SINE WAVE)

SCRs

Silicon Controlled Rectifiers

Table 1. SCRs — General Purpose Metal/Plastic Packages
0.8 to 55 Amperes RMS, 25 to 800 Volts

V_{DRM} V_{RRM} (Volts)	On-State (RMS) Current			
	0.8 AMP			1.5 AMPS
	T_C = 58°C		T_C = 80°C	T_C = 50°C
	CASE 29-04 TO-226AA (TO-92) STYLE 10		CASE 318E-04 SOT-223 STYLE 10	CASE 29-04 TO-226AA (TO-92) STYLE 10
	Sensitive Gate			
25	MCR102 2N5060	BRX44/BRY55-30[4]		
50	MCR103 2N5061	BRX45/BRY55-60[4]		*MCR22-2*
100	*MCR100-3* *2N5062*	BRX46/BRY55-100[4]		*MCR22-3*
200	*MCR100-4* *2N5064*	BRX47/BRY55-200[4]	*MCR08BT1*	*MCR22-4*
400	*MCR100-6*	BRX49/BRY55-400[4]	*MCR08DT1*	*MCR22-6*
500		BRY55-500[4]		
600	*MCR100-8*	BRY55-600[4]	*MCR08MT1*	*MCR22-8*

Maximum Electrical Characteristics

I_{TSM} (Amps) 60 Hz	10	15 150[3]	10	15 150[3]
I_{GT} (mA)	0.2			
V_{GT} (V)	0.8			
T_J Operating Range (°C)	−65 to+110	−40 to+125	−40 to+110	−40 to+125

[3]Exponential decay 2 µs wide at 5 time constants, f = 12 Hz.
[4]European Part Numbers. Package is Case 29 with Leadform 18. Case style is 3.

Table 1. SCRs — General Purpose Metal/Plastic Packages

V_{DRM} V_{RRM} (Volts)	On-State (RMS) Current				
	25 AMPS	**35 AMPS**			**40 AMPS**
	$T_C = 70°C$	$T_C = 65°C$			$T_C = 80°C$
	CASE 174-04 STYLE 1	CASE 263-04 STYLE 1	CASE 263-04 STYLE 1	CASE 311-02 STYLE 1 Isolated	CASE 221A-04 TO-220AB STYLE 3
50			MCR70-2A		
100	2N3870		MCR70-3A	2N6171	
200	2N3871	MCR3935-4A		2N6172	*MCR264-4*
400	2N3872	MCR3935-6A	MCR70-6A	2N6173	*MCR264-6*
600	2N3873	MCR3935-8A		2N6174	*MCR264-8*
800	MCR3835-10	MCR3935-10A			*MCR264-10*

Maximum Electrical Characteristics

I_{TSM} (Amps) 60 Hz	350	350 850(2)	350	400
I_{GT} (mA)	40	30	40	50
V_{GT} (V)	MCR3835/MCR3935/2N 1.5/1.5/1.6	1.5	1.6	1.5
T_J Operating Range (°C)	−40 to +125			

I_{TSM} is in Amps, I_{GT} in mA, V_{GT} in V.

(2)Peak capacitor discharge current for $t_w = 1$ ms. t_w is defined as five time constants of an exponentially decaying current pulse (crowbar applications).

Triacs
Silicon Bidirectional Triode Thyristors

. . . designed primarily for industrial and military applications for the control of ac loads in applications such as power supplies, heating controls, motor controls, welding equipment and power switching systems; or wherever full-wave, silicon gate controlled solid-state devices are needed.

- Glass Passivated Junctions and Center Gate Fire
- Press Fit — MAC6400
  - Stud — T6410
  - Isolated Stud — T6420
- Gate Triggering Guaranteed in All 4 Quadrants
- Isolated Stud for Ease of Mounting

MAC6400
T6410
T6420
Series

TRIACs
40 AMPERES RMS
200 thru 800 VOLTS

MAXIMUM RATINGS (T_J = 25°C unless otherwise noted.)

Rating	Symbol	Value	Unit
Peak Repetitive Off-State Voltage, Note 1 (T_J = −65 to +110°C, Gate Open) MAC6400B, T6410B, T6420B MAC6400D, T6410D, T6420D MAC6400M, T6410M, T6420M MAC6400N, T6410N, T6420N	V_{DRM}	 200 400 600 800	Volts
On-State Current RMS T_C (Pressfit) = 70°C (Conduction Angle = 360°) T_C (Stud) = 65°C	$I_{T(RMS)}$	40	Amps
Peak Surge Current (Non-Repetitive) (One Full Cycle, 60 Hz)	I_{TSM}	300	Amps
Circuit Fusing (t = 8.3 ms)	I^2t	375	A^2s
Peak Gate Power (Pulse Width = 10 μs)	P_{GM}	40	Watts
Average Gate Power	$P_{G(AV)}$	0.75	Watt
Peak Gate Current (Pulse Width = 1 μs)	I_{GTM}	12	Amps
Operating Temperature Range	T_C	−65 to +110	°C
Storage Temperature Range	T_{stg}	−65 to +150	°C
Stud Torque	—	30	in. lb.

THERMAL CHARACTERISTICS

Characteristic	Symbol	Max	Unit
Thermal Resistance, Junction to Case Pressfit Stud Isolated Stud	$R_{\theta JC}$	0.8 0.9 1	°C/W

Note 1. V_{DRM} for all types can be applied on a continuous basis. Blocking voltages shall not be tested with a constant current source such that the voltage ratings of the devices are exceeded.

CASE 263-04
STYLE 2
T6410
STUD

CASE 174-04
STYLE 3
MAC6400
PRESS FIT

CASE 311-02
STYLE 2
T6420
ISOLATED STUD

MAC6400 • T6410 • T6420 Series

ELECTRICAL CHARACTERISTICS (T$_C$ = 25°C unless otherwise noted.)

Characteristic	Symbol	Min	Typ	Max	Unit
Peak Blocking Current	I$_{DRM}$				
(V$_D$ = Rated V$_{DRM}$, Gate Open) T$_C$ = 25°C		—	—	10	μA
T$_C$ = 100°C		—	—	4	mA
Maximum On-State Voltage (Either Direction)	V$_{TM}$	—	1.4	1.6	Volts
(I$_T$ = 100 A Peak)					
Gate Trigger Current (Continuous dc), Note 1	I$_{GT}$				mA
(V$_D$ = 12 Vdc, R$_L$ = 30 Ohms)					
V$_{MT2}$(+), V$_G$(+)		—	15	50	
V$_{MT2}$(+), V$_G$(−)		—	30	80	
V$_{MT2}$(−), V$_G$(−)		—	20	50	
V$_{MT2}$(−), V$_G$(+)		—	40	80	
V$_{MT2}$(+), V$_G$(+), V$_{MT2}$(−), V$_G$(−), T$_C$ = −65°C		—	—	125	
V$_{MT2}$(+), V$_G$(−), V$_{MT2}$(−), V$_G$(+), T$_C$ = −65°C		—	—	240	
Gate Trigger Voltage (Continuous dc)	V$_{GT}$				Volts
(V$_D$ = 12 Vdc, R$_L$ = 30 Ohms) T$_C$ = 25°C		—	1.35	2.5	
T$_C$ = −65°C		—	—	3.4	
(V$_D$ = Rated V$_{DRM}$, R$_L$ = 125 Ohms, T$_C$ = 110°C)		0.2	—	—	
Holding Current (Either Direction)	I$_{HO}$				mA
(V$_D$ = 12 Vdc, Gate Open,					
Initiating Current = 500 mA) T$_C$ = 25°C		—	25	60	
T$_C$ = −65°C		—	—	100	
Gate Controlled Turn-On Time	t$_{gt}$	—	1.7	3	μs
(V$_D$ = Rated V$_{DRM}$, I$_T$ = 60 A, I$_{GT}$ = 200 mA, Rise Time = 0.1 μs)					
Critical Rate-of-Rise of Commutation Voltage, On-State Conditions	dv/dt(c)	—	5	—	V/μs
(di/dt = 20.4 A/ms, Gate Unenergized, V$_D$ = Rated V$_{DRM}$,					
I$_{T(RMS)}$ = 40 A) T$_C$ = 65°C					

Note 1. All voltage polarities referenced to main terminal 1.

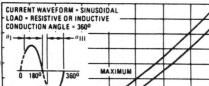

FIGURE 1 — ON-STATE POWER DISSIPATION

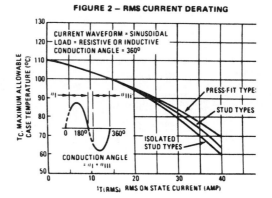

FIGURE 2 — RMS CURRENT DERATING

(Copyright of Motorola. Used by permission.)

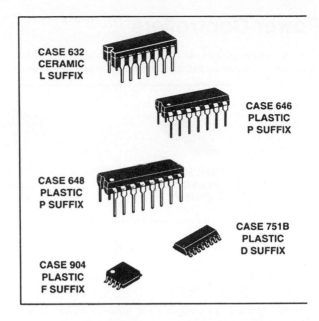

Miscellaneous Amplifiers

Motorola provides several Bipolar and CMOS special purpose amplifiers which fill specific needs. These devices range from low power CMOS programmable amplifiers and comparators to variable-gain bipolar power amplifiers.

MC3405/MC3505 Dual Operational Amplifier and Dual Voltage Comparator

This device contains two Differential Input Operational Amplifiers and two Comparators; each set capable of single supply operation. This operational amplifier-comparator circuit will find its applications as a general purpose product for automotive circuits and as an industrial "building block."

Table 5. Bipolar

Device	I_{IB} (µA) Max	V_{IO} (mV) Max	I_{IO} (nA) Max	A_{vol} (V/mV) Min	Response (µs) Typ	Supply Voltage Single	Supply Voltage Dual	Suffix/ Package
MC3405	0.5	10	50	20	1.3	3.0 to 36	± 1.5 to ± 18	L/632, P/646
MC3505	0.5	5.0	50	20	1.3	3.0 to 36	± 1.5 to ± 18	L/632

Table 6. CMOS
MC14573 Quad Programmable Operational Amplifier
MC14576B/MC14577B Dual Video Amplifiers
MC14575 Dual Programmable Operational Amplifier and Dual Programmable Comparator

Function	Quantity Per Package	Single Supply Voltage Range	Dual Supply Voltage Range	Frequency Range	Device Number	Suffix/ Package
Operational Amplifiers	4	3.0 to 15 V	± 1.5 to ± 7.5 V	DC to 1.0 MHz	MC14573	D/751B, P/648
Video Amplifiers	2	5 to 12 V[1]	± 2.5 to ± 6 V[2]	Up to 10 MHz	MC14576B MC14577B	P/626, F/904
Operational Amplifiers and Comparators	2 and 2	3 to 15 V	± 1.5 to ± 7.5 V	DC to 1.0 MHz	MC14575	D/751B, P/648

[1]5.0 to 10 V for surface mount package
[2]± 2.5 to ± 5 V for surface mount package

Power Controllers

An assortment of battery and AC line-operated control ICs for specific applications is shown. They are designed to enhance system performance and reduce complexity in a wide variety of control applications.

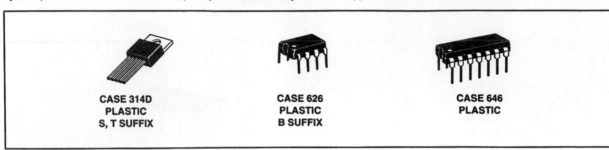

CASE 314D
PLASTIC
S, T SUFFIX

CASE 626
PLASTIC
B SUFFIX

CASE 646
PLASTIC

Zero Voltage Switches

CA3079/CA3059
T_A = −40° to +85°C, Case 646

These devices are designed for thyristor control in a variety of AC power switching applications for AC input voltages of 24 V, 120 V, 208/230 V, and 227 V @ 50/60 Hz.

- **Limiter-Power Supply** — Allows operation directly from an AC line.

- **Differential On/Off Sensing Amplifier** — Tests for condition of external sensors or input command signals. Proportional control capability or hysteresis may be implemented.

- **Zero-Crossing Detector** — Synchronizes the output pulses to the zero voltage point of the AC cycle. Eliminates RFI when used with resistive loads.

- **Triac Drive** — Supplies high-current pulses to the external power controlling thyristor.

- **Protection Circuit** (CA3059 only) — A built-in circuit may be actuated, if the sensor opens or shorts, to remove the drive circuit from the external triac.

- **Inhibit Capability** (CA3059 only) — Thyristor firing may be inhibited by the action of an internal diode gate.

- **High Power DC Comparator Operation** (CA3059 only) — Operation in this mode is accomplished by connecting Pin 7 to Pin 12 (thus overriding the action of the zero-crossing detector).

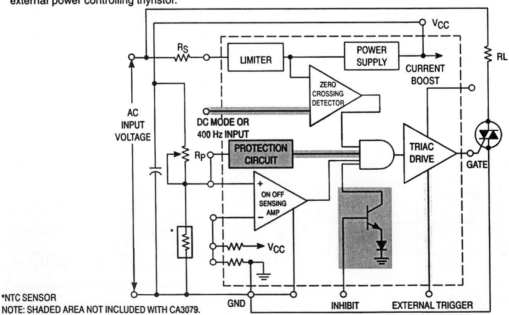

*NTC SENSOR
NOTE: SHADED AREA NOT INCLUDED WITH CA3079.

(Copyright of Motorola. Used by permission.)

Thermocouple Reference Tables

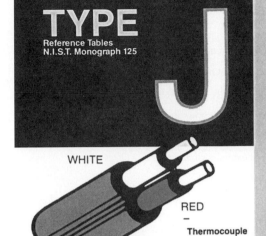

TYPE J

Reference Tables
N.I.S.T. Monograph 125

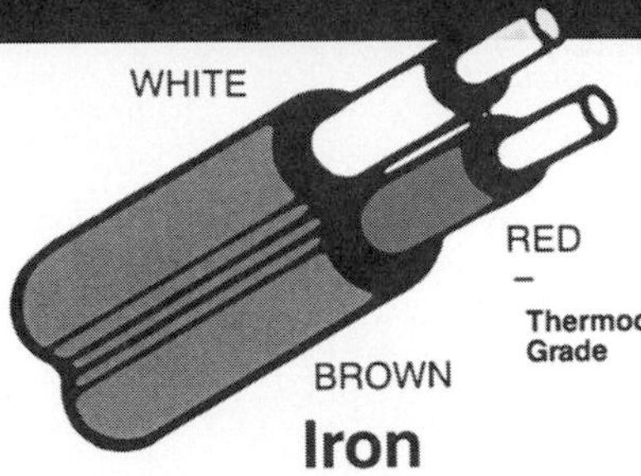

WHITE
RED
–
BROWN
Thermocouple Grade

Iron vs. Copper-Nickel

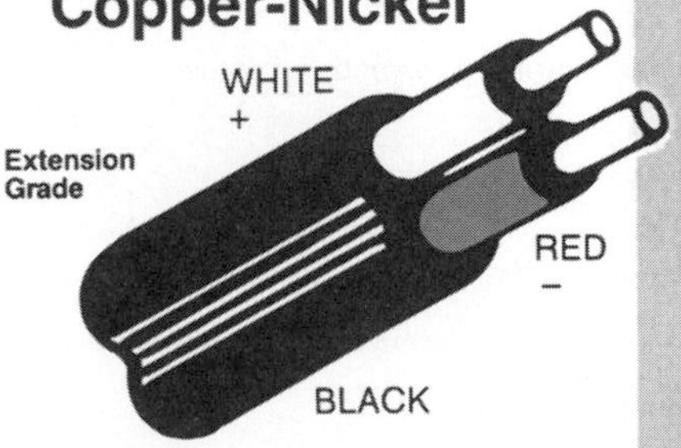

WHITE
+
Extension Grade
RED
–
BLACK

MAXIMUM TEMPERATURE RANGE

Thermocouple Grade
32 to 1382°F
0 to 750°C

Extension Grade
32 to 392°F
0 to 200°C

LIMITS OF ERROR
(whichever is Greater)
Standard: 2.2°C or 0.75%
Special: 1.1°C or 0.4%

COMMENTS, BARE WIRE ENVIRONMENT:
Reducing, Vacuum, Inert; Limited Use in Oxidizing at High Temperature; Not Recommended for Low Temperatures

TEMPERATURE IN DEGREES °C
REFERENCE JUNCTION AT 0°C

THERMOELECTRIC VOLTAGE IN ABSOLUTE MILLIVOLTS

DEG C	0	1	2	3	4	5	6	7	8	9	10	DEG C
-210	-8.096											-210
-200	-7.890	-7.912	-7.934	-7.955	-7.976	-7.996	-8.017	-8.037	-8.057	-8.076	-8.096	-200
-190	-7.659	-7.683	-7.707	-7.731	-7.755	-7.778	-7.801	-7.824	-7.846	-7.868	-7.890	-190
-180	-7.402	-7.429	-7.455	-7.482	-7.508	-7.533	-7.559	-7.584	-7.609	-7.634	-7.659	-180
-170	-7.122	-7.151	-7.180	-7.209	-7.237	-7.265	-7.293	-7.321	-7.348	-7.375	-7.402	-170
-160	-6.821	-6.852	-6.883	-6.914	-6.944	-6.974	-7.004	-7.034	-7.064	-7.093	-7.122	-160
-150	-6.499	-6.532	-6.565	-6.598	-6.630	-6.663	-6.695	-6.727	-6.758	-6.790	-6.821	-150
-140	-6.159	-6.194	-6.228	-6.263	-6.297	-6.331	-6.365	-6.399	-6.433	-6.466	-6.499	-140
-130	-5.801	-5.837	-5.874	-5.910	-5.946	-5.982	-6.018	-6.053	-6.089	-6.124	-6.159	-130
-120	-5.426	-5.464	-5.502	-5.540	-5.578	-5.615	-5.653	-5.690	-5.727	-5.764	-5.801	-120
-110	-5.036	-5.076	-5.115	-5.155	-5.194	-5.233	-5.272	-5.311	-5.349	-5.388	-5.426	-110
-100	-4.632	-4.673	-4.714	-4.755	-4.795	-4.836	-4.876	-4.916	-4.956	-4.996	-5.036	-100
-90	-4.215	-4.257	-4.299	-4.341	-4.383	-4.425	-4.467	-4.508	-4.550	-4.591	-4.632	-90
-80	-3.785	-3.829	-3.872	-3.915	-3.958	-4.001	-4.044	-4.087	-4.130	-4.172	-4.215	-80
-70	-3.344	-3.389	-3.433	-3.478	-3.522	-3.566	-3.610	-3.654	-3.698	-3.742	-3.785	-70
-60	-2.892	-2.938	-2.984	-3.029	-3.074	-3.120	-3.165	-3.210	-3.255	-3.299	-3.344	-60
-50	-2.431	-2.478	-2.524	-2.570	-2.617	-2.663	-2.709	-2.755	-2.801	-2.847	-2.892	-50
-40	-1.960	-2.008	-2.055	-2.102	-2.150	-2.197	-2.244	-2.291	-2.338	-2.384	-2.431	-40
-30	-1.481	-1.530	-1.578	-1.626	-1.674	-1.722	-1.770	-1.818	-1.865	-1.913	-1.960	-30
-20	-0.995	-1.044	-1.093	-1.141	-1.190	-1.239	-1.288	-1.336	-1.385	-1.433	-1.481	-20
-10	-0.501	-0.550	-0.600	-0.650	-0.699	-0.748	-0.798	-0.847	-0.896	-0.945	-0.995	-10
0	0.000	-0.050	-0.101	-0.151	-0.201	-0.251	-0.301	-0.351	-0.401	-0.451	-0.501	0
DEG C	0	1	2	3	4	5	6	7	8	9	10	DEG C
0	0.000	0.050	0.101	0.151	0.202	0.253	0.303	0.354	0.405	0.456	0.507	0
10	0.507	0.558	0.609	0.660	0.711	0.762	0.813	0.865	0.916	0.967	1.019	10
20	1.019	1.070	1.122	1.174	1.225	1.277	1.329	1.381	1.432	1.484	1.536	20
30	1.536	1.588	1.640	1.693	1.745	1.797	1.849	1.901	1.954	2.006	2.058	30
40	2.058	2.111	2.163	2.216	2.268	2.321	2.374	2.426	2.479	2.532	2.585	40
50	2.585	2.638	2.691	2.743	2.796	2.849	2.902	2.956	3.009	3.062	3.115	50
60	3.115	3.168	3.221	3.275	3.328	3.381	3.435	3.488	3.542	3.595	3.649	60
70	3.649	3.702	3.756	3.809	3.863	3.917	3.971	4.024	4.078	4.132	4.186	70
80	4.186	4.239	4.293	4.347	4.401	4.455	4.509	4.563	4.617	4.671	4.725	80
90	4.725	4.780	4.834	4.888	4.942	4.996	5.050	5.105	5.159	5.213	5.268	90
100	5.268	5.322	5.376	5.431	5.485	5.540	5.594	5.649	5.703	5.758	5.812	100
110	5.812	5.867	5.921	5.976	6.031	6.085	6.140	6.195	6.249	6.304	6.359	110
120	6.359	6.414	6.468	6.523	6.578	6.633	6.688	6.742	6.797	6.852	6.907	120
130	6.907	6.962	7.017	7.072	7.127	7.182	7.237	7.292	7.347	7.402	7.457	130
140	7.457	7.512	7.567	7.622	7.677	7.732	7.787	7.843	7.898	7.953	8.008	140
150	8.008	8.063	8.118	8.174	8.229	8.284	8.339	8.394	8.450	8.505	8.560	150
160	8.560	8.616	8.671	8.726	8.781	8.837	8.892	8.947	9.003	9.058	9.113	160
170	9.113	9.169	9.224	9.279	9.335	9.390	9.446	9.501	9.556	9.612	9.667	170
180	9.667	9.723	9.778	9.834	9.889	9.944	10.000	10.055	10.111	10.166	10.222	180
190	10.222	10.277	10.333	10.388	10.444	10.499	10.555	10.610	10.666	10.721	10.777	190
200	10.777	10.832	10.888	10.943	10.999	11.054	11.110	11.165	11.221	11.276	11.332	200
210	11.332	11.387	11.443	11.498	11.554	11.609	11.665	11.720	11.776	11.831	11.887	210
220	11.887	11.943	11.998	12.054	12.109	12.165	12.220	12.276	12.331	12.387	12.442	220
230	12.442	12.498	12.553	12.609	12.664	12.720	12.776	12.831	12.887	12.942	12.998	230
240	12.998	13.053	13.109	13.164	13.220	13.275	13.331	13.386	13.442	13.497	13.553	240
250	13.553	13.608	13.664	13.719	13.775	13.830	13.886	13.941	13.997	14.052	14.108	250
260	14.108	14.163	14.219	14.274	14.330	14.385	14.441	14.496	14.552	14.607	14.663	260
270	14.663	14.718	14.774	14.829	14.885	14.940	14.995	15.051	15.106	15.162	15.217	270
280	15.217	15.273	15.328	15.383	15.439	15.494	15.550	15.605	15.661	15.716	15.771	280
290	15.771	15.827	15.882	15.938	15.993	16.048	16.104	16.159	16.214	16.270	16.325	290
300	16.325	16.380	16.436	16.491	16.547	16.602	16.657	16.713	16.768	16.823	16.879	300
310	16.879	16.934	16.989	17.044	17.100	17.155	17.210	17.266	17.321	17.376	17.432	310
320	17.432	17.487	17.542	17.597	17.653	17.708	17.763	17.818	17.874	17.929	17.984	320
330	17.984	18.039	18.095	18.150	18.205	18.260	18.316	18.371	18.426	18.481	18.537	330
340	18.537	18.592	18.647	18.702	18.757	18.813	18.868	18.923	18.978	19.033	19.089	340
350	19.089	19.144	19.199	19.254	19.309	19.364	19.420	19.475	19.530	19.585	19.640	350
360	19.640	19.695	19.751	19.806	19.861	19.916	19.971	20.026	20.081	20.137	20.192	360
370	20.192	20.247	20.302	20.357	20.412	20.467	20.523	20.578	20.633	20.688	20.743	370
380	20.743	20.798	20.853	20.909	20.964	21.019	21.074	21.129	21.184	21.239	21.295	380
390	21.295	21.350	21.405	21.460	21.515	21.570	21.625	21.680	21.736	21.791	21.846	390
400	21.846	21.901	21.956	22.011	22.066	22.122	22.177	22.232	22.287	22.342	22.397	400
410	22.397	22.453	22.508	22.563	22.618	22.673	22.728	22.784	22.839	22.894	22.949	410
420	22.949	23.004	23.060	23.115	23.170	23.225	23.280	23.336	23.391	23.446	23.501	420
430	23.501	23.556	23.612	23.667	23.722	23.777	23.833	23.888	23.943	23.999	24.054	430
440	24.054	24.109	24.164	24.220	24.275	24.330	24.386	24.441	24.496	24.552	24.607	440
450	24.607	24.662	24.718	24.773	24.829	24.884	24.939	24.995	25.050	25.106	25.161	450
460	25.161	25.217	25.272	25.327	25.383	25.438	25.494	25.549	25.605	25.661	25.716	460
470	25.716	25.772	25.827	25.883	25.938	25.994	26.050	26.105	26.161	26.216	26.272	470
480	26.272	26.328	26.383	26.439	26.495	26.551	26.606	26.662	26.718	26.774	26.829	480
490	26.829	26.885	26.941	26.997	27.053	27.109	27.165	27.220	27.276	27.332	27.388	490
500	27.388	27.444	27.500	27.556	27.612	27.668	27.724	27.780	27.836	27.893	27.949	500
510	27.949	28.005	28.061	28.117	28.173	28.230	28.286	28.342	28.398	28.455	28.511	510
520	28.511	28.567	28.624	28.680	28.736	28.793	28.849	28.906	28.962	29.019	29.075	520
530	29.075	29.132	29.188	29.245	29.301	29.358	29.415	29.471	29.528	29.585	29.642	530
540	29.642	29.698	29.755	29.812	29.869	29.926	29.983	30.039	30.096	30.153	30.210	540
550	30.210	30.267	30.324	30.381	30.439	30.496	30.553	30.610	30.667	30.724	30.782	550
560	30.782	30.839	30.896	30.954	31.011	31.068	31.126	31.183	31.241	31.298	31.356	560
570	31.356	31.413	31.471	31.528	31.586	31.644	31.702	31.759	31.817	31.875	31.933	570
580	31.933	31.991	32.048	32.106	32.164	32.222	32.280	32.338	32.396	32.455	32.513	580
590	32.513	32.571	32.629	32.687	32.746	32.804	32.862	32.921	32.979	33.038	33.096	590
600	33.096	33.155	33.213	33.272	33.330	33.389	33.448	33.506	33.565	33.624	33.[illegible]3	600
610	33.683	33.742	33.800	33.859	33.918	33.977	34.036	34.095	34.155	34.214	34.273	610
620	34.273	34.332	34.391	34.451	34.510	34.569	34.629	34.688	34.748	34.807	34.867	620
630	34.867	34.926	34.986	35.046	35.105	35.165	35.225	35.285	35.344	35.404	35.464	630
640	35.464	35.524	35.584	35.644	35.704	35.764	35.825	35.885	35.945	36.005	36.066	640
650	36.066	36.126	36.186	36.247	36.307	36.368	36.428	36.489	36.549	36.610	36.671	650
660	36.671	36.732	36.792	36.853	36.914	36.975	37.036	37.097	37.158	37.219	37.280	660
670	37.280	37.341	37.402	37.463	37.525	37.586	37.647	37.709	37.770	37.831	37.893	670
680	37.893	37.954	38.016	38.078	38.139	38.201	38.262	38.324	38.386	38.448	38.510	680
690	38.510	38.572	38.633	38.695	38.757	38.819	38.882	38.944	39.006	39.068	39.130	690
700	39.130	39.192	39.255	39.317	39.379	39.442	39.504	39.567	39.629	39.692	39.754	700
710	39.754	39.817	39.880	39.942	40.005	40.068	40.131	40.193	40.256	40.319	40.382	710
720	40.382	40.445	40.508	40.571	40.634	40.697	40.760	40.823	40.886	40.950	41.013	720
730	41.013	41.076	41.139	41.203	41.266	41.329	41.393	41.456	41.520	41.583	41.647	730
740	41.647	41.710	41.774	41.837	41.901	41.965	42.028	42.092	42.156	42.219	42.283	740
750	42.283	42.347	42.411	42.475	42.538	42.602	42.666	42.730	42.794	42.858	42.922	750
760	42.922											760
DEG C	0	1	2	3	4	5	6	7	8	9	10	DEG C

Note: NIST (formerly NBS) is planning on revising the Thermocouple Reference Tables, and OMEGA will publish these new tables as soon as they are released.

ACRONYMS

ANSI American National Standards Institute

ASBS asymmetrical silicon bilateral switch

BCD binary coded decimal

BFET bipolar field effect transistor

CAD computer-aided design

CAM computer-aided manufacturing

Cds cadmium sulfide

CdSe cadmium selenide

CMOS complementary metal-oxide semiconductor

CMRR common-mode rejection ratio

CPU central processing unit

CSCR capacitor start, capacitor run motor

CSI inverter current-source input

CSIR capacitor start, induction run motor

CUJT complementary unijunction transistor

DIN Deutsche Industrie Normenausschuss

DIP dual inline package

DOS disk operating system

EBCDIC Extended Binary Coded Decimal Interchange

ECKO eddy current killed oscillator

EIA Electronics Industry Association

emf electromotive force

EMR electromagnetic radiation

EPROM erasable programmable read-only memory

FIFO first-in, first-out

FLA full-load amperage

GFI ground-fault interrupter

GTOs gate turn-off devices

HVT high-voltage transistor

IC integrated circuit

IEC International Electrotechnical Committee

IGBT insulated gate bipolar transistor

ISA Instrumentation Society of America

ISO International Organization for Standards

J-FET junction field effect transistor

LAN local area network

LAPUT light-activated programmable unijunction transistor

LAS light-activated switch

LASCR light-activated SCR

Laser light amplification stimulated emission of radiation

LCD liquid crystal display

LED light-emitting diode

LRA locked-rotor amperage

LVDT linear variable differential transformer

LVT low-voltage technology

MAP Manufacturing Automation Protocol

MLS modulated light source

MMI man-machine interface

MMS Manufacturing Message Specification

MOSFET metal-oxide semiconductor field effect transistor

MOV metal-oxide varistor

NC normally closed

NCTO normally closed, timed open

NEMA National Electrical Manufacturing Association

NO normally open

NOTC normally open, timed closed

OSI Open System Interconnect

PAL programmable array logic

PLA programmable logic array

PID proportional, integral, and derivative

PLC programmable logic controller

PLD programmable logic device

PSC permanent split-capacitor motor

PSI pounds per square inch

PUT   programmable unijunction transistor

PWM INVERTER   pulse-width modulation inverter

QAD   quarter amplitude decay

RDI   robot digital input signal

RDO   robot digital output signal

ROI   return on investment

RTD   resistive temperature detector

RTU   remote transmitter unit

SAS   silicon asymmetrical switch

SBS   silicon bilateral switch

SCARA   selective compliance assembly robot

SCR   silicon controlled rectifier

SCS   silicon controlled switch

SF   service factor

SNA   System Network Architecture

SSR   solid-state relay

TCP/IP   Transmission Control Protocol/Internet Protocol

TFD   thin film detector

TOP   Technical Office Protocol

TTL   transistor-transistor logic

UJT   unijunction transistor

UPS   uninterruptible power supply

VA   volt ampere

VVI   variable-voltage input

GLOSSARY

ac-coupled amplifier   An amplifier with a capacitive-coupled input to filter out the dc component of the input signal.

ac input circuit   A circuit on a PLC I/O module that converts ac signals from a pilot switch voltage level to backplane logic-level dc signals.

ac input module   An I/O module for a PLC or other controller that contains circuits that convert ac signals to backplane logic-level dc signals.

ac output circuit   A circuit on a PLC I/O module that converts backplane logic-level dc signals to field-level ac voltage.

ac output module   A PLC I/O module that contains circuits that convert backplane logic-level dc signals to field-level ac voltage.

across-the-line starter   A motor starter that allows full voltage to be applied to a motor when the contacts of the starter are closed.

actuator   A device that converts an electrical signal into motion.

address   A character string that uniquely identifies a memory location in a PLC, computer, or other programmable system.

alternating current   Current that changes from a positive level to a negative level periodically. Example: a 60 Hz alternating current change from positive to negative 60 times per second.

ambient temperature   The temperature of the medium (air, water, earth).

American wire gauge (AWG)   A standard used for identifying the size of electrical conductors. Gauge numbers have an inverse relationship to size; larger numbers have a smaller cross-sectional area.

analog circuit   A circuit in which the signal can vary continuously between specified limits from some minimum to some maximum.

analog input module   A PLC I/O module that contains circuits that convert analog dc input signals to digital values.

analog output module   A PLC I/O module that contains circuits that convert digital values to an analog dc signal.

analog-to-digital conversion (A/D conversion)   A circuit that converts analog values to digital (numerical) values.

angstrom   Unit of measure for the wavelength of light (10 Å = 1nm).

ANSI   American National Standards Institute. An organization in the United States that develops standards for industry.

armature   The moving part of a magnetic circuit, such as the rotating part of a motor or a generator or the movable iron part of a relay.

ASCII   American Standard Code for Information Interchange. It is a seven-bit code with an optional parity bit used to represent alphanumerics, punctuation marks, and control-code characters.

backplane   The back side of a chassis of a PLC rack that provides electrical interconnection between the module and the chassis.

battery backup   A battery or set of batteries that will provide power to the memory of PLCs and other processors when the main power is off.

BCD   Binary coded decimal. A numbering system used to express individual decimal digits (0–9) in four-bit binary notation.

binary   A base 2 numbering system (using only the digits 0–1).

bit   Binary digit. The smallest unit of information in the binary numbering system, represented by the digits 1 and 0.

braking   A method of stopping or reducing the time required to stop an ac or dc motor, which can be accomplished in several ways including: dynamic braking, regenerative braking, and friction braking.

brush   A carbon conductor that maintains an electrical connection between stationary and moving parts of a motor or generator or other type of rotating machine.

buffer   In software terms, a register or group of registers used for temporary storage of data.

bus   Single or multiple parallel paths for power or data signals to which several devices may be connected at the same time.

bus topology   A topology in which all stations are connected in parallel to a medium.

byte   A string of eight bits, operated as a unit.

CAD   Computer-aided design (also known as computer-aided drafting). A system developed to create drawings on a computer.

CD-ROM   Read-only memory on a compact disk.

central processing unit (CPU)   The microprocessor that is the main controller for a PLC system or other computer controlled system.

centrifugal switch   A switch that is connected in series with the start winding of a single-phase ac motor and is mounted in its end plate. The purpose of this switch is to open its contacts at approximately 90 percent of full rpm to disconnect the start winding of the motor from the source of voltage. The switch will close when the motor is de-energized and slows to a stop.

circuit breaker   An electrical device specifically designed to protect against overcurrent. The circuit breaker is similar to a fuse in that it protects, but it is different in that it can be reset.

closed-loop system   A control system that uses a feedback signal.

CMOS   Complementary metal-oxide semiconductor.

CNC   Computerized numerical control. A control system where a computer guides a machine like a mill or lathe to various mathematical points in a two- or three-dimensional plane.

cogging   A condition in which a motor does not rotate smoothly, but "steps" or "jerks" from one position to another during shaft revolution.

cold-junction compensation   An electronic circuit that is used with a thermocouple to simulate the value of a reference signal of 32°F or 0°C.

common-mode rejection   The ability of a differential analog input to cancel a common-mode signal expressed in dB.

common-mode voltage   A voltage that appears in common with both input terminals of a differential analog input with respect to ground.

commutator   A cylindrically shaped assembly that is fastened to the shaft of a motor or generator and is considered part of the armature assembly. It consists of segments of "bars" that are electrically connected to the two ends of one or more armature coils.

contactor   One or more sets of contacts that are opened or closed by a magnetic coil. The contactor is similar to a relay but normally larger. By definition, the contactor is rated for more than 15 A.

converter   A device or circuit for changing ac to dc. This is accomplished through a diode rectifier or thryristor rectifier circuit.

counter emf   The induced voltage in an ac or dc motor that opposes the applied voltage.

CPU Central processing unit (*see* central processing unit).

CRT terminal A display terminal containing a cathode ray tube.

CTS Clear-to-send. A signal that tells the transmitting device to start transmitting data.

current limiting An electronic method of limiting the maximum current available to the motor.

current transformer A transformer that is used primarily as a step-down transformer to indicate the amount of current flowing in a motor drive or other large-current application.

damping The reduction in amplitude of an oscillation.

dark operate Pertaining to a photoelectric control that energizes its output when the light intensity on the photodetector is at a low level.

data A general term for any type of information. In a more specific sense, the term *data* refers to information in a particular context such as the amount of parts a machine has made, the amount of time on a time-delay circuit, or other machine values.

dead band The values to which a system input can be changed without causing a corresponding change in system output.

delta connection A three-phase connection where windings are connected in series with the power applied. The windings are connected in the shape of a triangle or the Greek letter delta (Δ).

deviation Difference between the setpoint and process variable in a control system. Also called *error*.

di/dt The instantaneous rate of change in current over time.

diac A two-terminal electronic device that is specifically designed to fire in both the positive and negative directions when its applied voltage reaches a predetermined amount.

digital signal A signal that has only two distinct states: off and on.

digital-to-analog converison (D/A conversion) A digital value whose instantaneous magnitude is converted to its equivalent analog signal. Typical digital values range from 0–4095 or 0–999 in PLCs and the equivalent analog value ranges from 4–20 mA or 0–10 V dc.

DIN Deutsche Industrie Normenausschuss. A German agency that establishes European standards.

diode A two-terminal solid-state semiconductor that allow current flow in one direction.

DIP Dual inline package. A configuration in which printed circuit components are built with two parallel rows of pins so that they can be easily mounted and soldered to printed circuit boards.

direct current (dc) Current that flows in only one direction.

disable To inhibit logic from being activated.

discrete An electronic device that has an individually distinct identity as opposed to being part of an integrated circuit.

discrete circuit A circuit built from separate components opposed to being part of an integrated circuit.

discrete signal A signal that has only two distinct states: off and on.

disk drive The device that reads data from or writes data to a disk.

diskette (floppy disk) A thin, flexible disk coated with magnetic oxide and used to store data.

documentation A collection of printed pages for a PLC that may include the ladder diagram, program comments, data tables, and other diagrams to provide reference information for operation and troubleshooting.

DOS Disk operating system. A set of computer commands specifically designed to aid in setting up files and directories in the hard drive or floppy drive of a computer. The commands also are designed for copying or moving programs and files between memory devices.

download Transferring a program from a terminal to control unit such as a programmable controller or CNC controller.

DSR   Data-Set-Ready. A signal that indicates the modem is connected, powered up, and ready for data transmission.

DTE   Data-Terminal-Equipment. Equipment that is attached to a network to send or receive data or both.

DTR   Data-Terminal-Ready. A signal that indicates the transmission device (terminal) is connected, powered up, and ready to transmit.

duty cycle   The ratio of working time to total time for a motor or machine that operates intermittently.

dv/dt   1. The instantaneous rate of change in voltage over time. 2. A network consisting of a resistor and capacitor that can help protect a SCR from excessive *dv/dt,* which can result from line voltage spikes.

dwell   A programmed time delay.

dynamic braking   A method of braking a DC motor which involves disconnecting the motor windings from applied voltage and reconnecting them across a set of resistors.

eddy current   Current induced in components from the movement of magnetic fields.

EEPROM   Electrically erasable PROM. A type of PROM that can be erased and reprogrammed by electrical signals. It is also a nonvolatile type memory.

EIA   Electronics Industries Association. A U.S. agency that sets electrical/electronic standards.

emf   Electromotive force, which is another term for voltage or potential difference.

enable   To activate logic by the removal of a suppression signal.

enclosure   The metal box or housing in which equipment is mounted, available in designs for various environmental conditions.

encoder   Any feedback element that converts linear or rotary position to digital signals (pulses). Encoders are available as absolute and incremental devices.

EPROM   Erasable programmable read-only memory. A PROM can be erased with ultraviolet light, then reprogrammed with electrical signals.

error   The difference between the setpoint signal and the feedback signal (process variable). Error can be specified as SP-PV or as PV-SP. In a motion control system, error is specified as the difference between the command signal and the feedback signal.

Ethernet network   A local area network with a baseband communications rate of 10 M bits.

false   A signal that is assigned the digital value of 0.

fault   Any malfunction that interferes with normal system operation.

feedback   The signal or signals used to sense motion, machine control, or process control of the controlled variable and sent to the summing junction or controller to be compared to the setpoint or command signal.

FET   Field effect transistor.

field   The stationary electrical part of a dc motor.

field control   A method of controlling dc motor speed by varying the field current in the shunt field winding.

FIFO   (first-in, first-out) A method of ordering data items and storing them in the order in which they are received.

filter   A device that passes a signal or a range of signals and eliminates all others.

floating-point number format   A data storage format that includes the location of the decimal point by expressing the power of the base. (Example: 2357 would be shown as 2.357 E-3 and 0.0023 would be shown as 2.3 E3.)

flow meter   An instrument that measures and indicates the rate of flow of a liquid or gas.

frame size   The physical dimension and size of a motor, usually expressed as a NEMA design code number. This allows interchangeability between motors made by different manufacturers.

frequency   The number of periodic cycles per unit of time.

fuse   A device designed as a one-time protection against overcurrent or short-circuit current.

gain   The ratio of the magnitude of the output signal with respect to that of the input signal.

gauge factor   The ratio of the change in resistance to the change in length in a strain gauge.

GFI   Ground-fault interrupter. A protection device that senses small amounts of ground currents and opens the circuit. The GFI must be manually reset.

GTO   Gate turn-off or gate turn-on power semiconductor device.

Hall effect   A condition in a semiconductor where current flowing perpendicular to a magnetic field produces a (small-voltage) potential difference perpendicular to both the magnetic field and the current flow.

hard contacts   Any type of physical switching contacts as compared to soft programmed contacts in a PLC program.

HDLC   High-level data-link control. A communications protocol sanctioned by the International Standards Organization (ISO) that defines procedures for the data link and physical layers for the seven-layer ISO model for communications.

hexadecimal numbering system   A base 16 numbering system which uses the symbols 0, 1, 2, 3, 4, 5, 6, 7, 8, 9, A, B, C, D, E, F for numerals.

holding current   1. The amount of current required to keep a thyristor in its conductive state after its firing signal is removed. 2. The current used to hold relay contacts in their activated condition.

home position   A reference position for all absolute positioning movements. Usually defined by a home limit switch and encoder marker. Normally set at power up and retained as long as the control system is operational.

horsepower (hp)   A unit of power. 1 hp = 33,000 ft-lb/min = 746 W.

IC   Integrated circuit. A solid-state device that includes combinations of circuit elements (resistors, capacitors, transistors) that are fabricated on a single continuous substate.

IEC   International Electrotechnical Commission.

infrared   Invisible light radiation starting at a wavelength of 69 nm (6900 Å).

input device   A digital or analog device such as a limit switch, push-button switch, pressure sensor, or temperature sensor that supplies input data through an input circuit to a programmable controller.

instrumentation amplifier   A differential amplifier specifically designed for use in measuring circuits.

integer   Any positive or negative whole number including zero.

intelligent I/O modules   A PLC module that uses a microprocessor.

interface   A hardware circuit or software that allows the interaction of two or more separate systems.

interlock   A switch or other device that prevents activation of a piece of equipment when another piece of equipment is operating.

International Standards Organization (ISO)   An organization established to promote development of international standards.

intrinsic safety   A design technique applied to electrical equipment and wiring for hazardous locations. It is based on limiting electrical and thermal energy to a level below that required to ignite hazardous atmospheric mixtures.

intrinsic standoff ratio   The ratio of the triggering voltage to the applied voltage in a UJT.

inverter   1. A circuit designed to change dc voltage to ac voltage. 2. An ac adjustable-frequency drive. 3. A particular section of an ac drive where dc voltage is changed to ac voltage..

I/Os (inputs and outputs)   Input and output signals for a PLC or robot.

isolation transformer   A transformer that provides isolation from one circuit to another.

JIC   Joint Industrial Council. An organization of manufacturing concerns that establishes standards.

jog   A momentary state of being enabled. When a jog push button is depressed, a motor or machine is energized (enabled), and when the push button is released, the motor is de-energized.

ladder diagram   An industry standard for representing relay control logic. The name comes from the fact that the overall form of the diagram looks like a wooden ladder.

LAN   Local area network. A network limited to a local geographical area such as an office or a factory.

LASCR   Light-activated SCR. A SCR that can be activated into conduction when a sufficient level of light is received on its junction.

latching relay   A relay that maintains a given position by mechanical or electrical means until released mechanically or electrically.

leading-edge triggering   A programming technique of triggering some step by means of an off-to-on transition of an input signal.

LED   Light-emitting diode. A two-lead solid-state PN junction device that produces a small amount of light when forward biased.

light operated   Pertaining to a photoelectric control that energizes its output when the light intensity on the photodetector is at a high level.

limit switch   An electrical switch that is actuated by motion of a machine or equipment contacting the switch.

linear stepper motor   A stepper motor whose motion is in a straight line instead of rotary.

lockout   A safety procedure that ensures that the source of supply voltage for a machine is disconnected and locked in the off position so that personnel can safely work on the equipment.

logic   General term for relay circuits, digital circuits, and programmed instructions to perform required decision-making and computational functions.

loop   1. (closed-loop system) A control system that has an output device and a process variable signal supplied by a sensor that is used as feedback. 2. A sequence of instructions that executes repeatedly until a terminating condition is satisfied. 3. A closed-loop or open-loop system.

LRA locked-rotor amperage (current)   The amount of current a motor winding draws when power is initially applied and the rotor has not begun to rotate. This would be the same amount of current the winding would draw if the rotor stopped rotating.

LSB   Lease significant bit. The bit that represents the smallest value within a string of bits.

LVDT   Linear variable differential transformer. A linear position sensor that uses a transformer whose output is proportional to the position of its core.

M (mega)   1. A prefix used with units of measurement to designate a multiple of 1,000,000 (1 million). 2. An amount of memory equal to $2^{20} = 1,048,576$ bits, bytes, or words in denoting size of a block of data memory.

master control relay (MCR)   A hard-wired relay that controls voltage to a PLC system or other automated system so that voltage can be de-energized by any series-connected emergency stop switch. Whenever the master control relay is de-energized, its contacts open to remove the source of power from all I/O circuits, sensors, and actuators.

microprocessor   A central processing unit that is manufactured on a single integrated circuit (or on only a few integrated circuits) by utilizing large-scale integration technology.

microstepper   The process of dividing steps of a stepper motor into substeps.

mnemonic   A term (or abbreviation) that is simple and easy to remember used to represent a complex or lengthy set of information. For example, the mnemonic for the Timer On Delay instruction for a PLC is TON, the mnemonic for an Up Counter instruction is CTU, and the mnemonic for the subtraction instruction is SUB.

modem   Modulator/demodulator. Equipment that connects data terminal equipment to a communications line.

module   An interchangeable plug-in circuit within a larger (modular) assembly.

MOS   Metal-oxide semiconductor. A semiconductor device in which an electric field controls the conductance of a channel under a metal electrode called a gate.

MSB   Most significant bit. The bit representing the greatest value within a string of bits.

NEC   National Electrical Code. A set of regulations governing the construction and installation of electrical wiring and apparatus. The NEC is established by the National Fire Protection Association.

NEMA standards   National Electrical Manufacturers Association. This group provides standards for electrical equipment approved in the United States.

network   A series of stations (nodes) connected by some type of communication medium. The network may be made up of single or multiple links.

network layer   The third layer of the ISO open-systems interconnected reference model for data communications. It provides routing and relaying services associated with all of the layers of that station and is responsible for setting and resetting parameters and obtaining reports of error conditions.

normally closed contacts   A set of contacts on a relay or switch that is closed when the switch or relay is in the de-energized state.

normally open contacts   A set of contacts on a relay or switch that is open when the switch or relay is in the de-energized state.

octal numbering system   A base 8 numbering system which uses only the digits 0–7.

off   A term used to designate the 0 state of a bit; or the inoperative state of a device.

on   A term used to designate the 1 state of a bit; the operative state of a device; the state of a switch or circuit that is closed.

one-shot   A programming technique that sets a bit to the on state for only one program scan.

on line   Refers to equipment or devices that are in direct interaction communications. For example, connecting a programming computer to a PLC to make changes in the PLC program.

op amp   An operational amplifier. A high-gain stable linear dc amplifier that is designed to be used with external circuit elements.

optocoupler   A light-emitting diode and a light-detecting device sealed together in an integrated package. The input circuit and output circuit are electrically isolated since they are only connected with light.

output device   Devices such as a solenoid or motor starter that are connected to the output circuit of a programmable controller, robot CNC machine.

photoelectric control   An electronic device that recognizes changes in light intensity and converts these changes into a change in electrical output state.

pilot circuit   The portion of a control circuit that carries the controlling signal for a device which, in turn, controls the primary current.

PLC controller   A programmable logic controller such as an Allen-Bradley programmable controller.

plugging   Reversing either line voltage polarity or phase sequence on a motor so that the motor develops a countertorque that exerts a retarding force to brake the motor.

position loop   A feedback control loop that uses an encoder or resolver as a feedback device to indicate mechanical position.

position transducer   An electronic device (e.g., encoder or resolver) that measures incremental or absolute position and converts this measurement into a feedback signal.

positive feedback   A feedback signal from the output that is added to the input signal.

power   Work done per unit of time. Measured in horsepower or watts: 1 hp = 33,000 ft-lb/min = 746 W.

power factor   A measurement of the time phase difference between the voltage and current in an ac circuit. It is represented by the cosine of the angle of this phase difference. Power factor is the ratio of real power (in watts) to apparent power (in volt-amperes).

power supply   A device that converts available power to a form that a system can use; usually converts ac power to dc power.

pressure switch   A switch that is activated at a specific pressure.

printed circuit board   A board (card) made up of a nonconductive layer sandwiched by conductive layers that are etched to form circuit connections between connection points where components can be mounted.

processor   The decision-making and data storage sections of a programmable controller or computer.

programmable controller   A solid-state control system that continually scans its user program. The controller has a user-programmable memory for storage of instructions to implement specific functions such as I/O control, logic, timing, counting, report generation, communications, arithmetic, and data file manipulation. A controller consists of a central processor unit, input/output interface, and memory. A controller is designed as an industrial control system. The program is generally in the form of ladder logic.

program mode   On a programmable controller, a mode in which the ladder logic is not executed and all outputs are held off.

PROM   Programmable read-only memory. A type of ROM that requires an electrical operation to store data. In use, bits or words are read on demand but never changed.

proportional band   The range of values through which the output is proportional to the change of the input variable.

proportional, integral, derivative control (PID)   A method of control that uses proportional (gain), integral (reset), and derivative (rate) to provide control of the response rate of the output signal.

Proportional control causes an output signal to change as a direct ratio of the error signal variation.

Integral control causes an output signal to change as a function of the integral of the error signal over the time duration.

Derivative control causes an output signal to change as a function of the rate of change of the error signal.

protocol   A set of conventions governing the format and timing of data between communications devices.

proximity switch/sensor   A switch/sensor that is actuated when an actuating device is moved near it, without physical contact.

PWM   Pulse-width modulation. A technique used to eliminate or reduce unwanted harmonic frequency when inverting dc voltage to sine wave ac.

RAM   Random access memory. The type of memory in which each storage location is by X/Y coordinates, as in core or semiconductor memory.

rectifier   A device that conducts current in only one direction, thereby transforming alternating current to direct current.

register   A memory word or area for temporary storage of data used within arithematical, logical, or transferral functions.

resolution   The smallest distinguishable increment into which a quantity can be divided.

resolver   A transducer using magnetic coupling to measure absolute rotary position. The resolver creates a fixed sine wave and one or more variable sine waves that are compared to create a position within 360 rotary degrees. It requires an analog signal interface and special conditioning electronics.

ring topolgy   A network where signals are transmitted from one station and relayed through each subsequent station in the network.

rms   Root-mean-square. The effective value of an alternating current corresponding to the dc value that produces the same heating effect. The rms value is computed as the square root of the average of the squares of the instantaneous amplitude for one complete cycle. For a sine wave the rms value is 0.707 times the peak value.

ROM   Read-only memory. A type of memory with data content that cannot be changed in normal mode of operation. In use, bits and words are read on demand but not changed.

RS232-C   An EIA standard that specifies electrical, mechanical, and functional characteristics for serial binary communications circuits in a point-to-point link.

RS422   An EIA standard that specifies electrical characteristics of balanced voltage digital interface circuits in a point-to-point link.

RTD   Resistance temperature detector. A wire-wound resistance with a moderate positive coefficient of resistance.

RTS   Request to send. A request from the module to the modem to prepare to transmit. It typically turns on the data carrier.

SCADA   Supervisory control and data acquisition.

scanner   A photoelectric control that contains the light source and the detector in the same housing.

SCR   Silicon controlled rectifier. A solid-state unidirectional latching switch.

sensor   A device that detects or measures something and generates a corresponding electrical signal to an input circuit or controller.

serial   Pertaining to time-sequential transmission of data over a single conductor.

service factor (SF)   When used on a motor nameplate, a number that indicates how much above the nameplate rating a motor can be loaded without causing serious degradation (i.e., a motor with 1.15 SF can produce 15 percent greater torque than one with a service factor of 1.0).

setpoint   The value selected to be maintained by an automatic controller.

slip   The difference between rotating magnetic-field speed (synchronous speed) and the rotor speed of an ac induction motor. Usually expressed as a percentage of synchronous speed.

soft start (motor starting)   An electronic motor starting circuit where less than full voltage is applied to the motor during starting.

star configuration   In an arrangement of parallel (bus) connections a physical configuration such that each device is connected on the bus at the same junction of conductor segments.

star connection   The arrangement of phase windings in a poly-phase circuit in which one end of each phase winding is connected to a common junction. In a three-phase circuit, it is sometimes called a wye connection.

star topology   A network where all devices are connected to a central or master communication device that routes messages.

status indicator   LED or other type of indicator that is illuminated when an input circuit or output circuit is energized.

summing point (summing junction)   A mixing point where the setpoint and process variable feedback signals are compared.

surge   A transient wave of voltage, current, or power.

surge suppression   The process of absorbing and clipping voltage transients on an incoming ac line or control circuit. MOVs (metal-oxide varistors) and specially designed *R-C* networks are usually used to accomplish this.

synchronous   A type of serial transmission that maintains a constant time interval between successive events.

synchronous speed   The speed of an ac induction motor's rotating magnetic field. It is determined by the frequency applied to the stator and the number of magnetic poles present in each phase of the stator windings.

tachometer   A precision dc generator used to provide velocity feedback.

TCP/IP   Transmission Control Protocol/Internet Protocol. A transport-layer protocol and a network-layer protocol developed by the Department of Defense.

thumbwheel switch   A multiposition rotary switch with a sprocket that is stepped forward or backward by using a finger or thumb to rotate it. The face of the switch displays numbers 0-9.

toggle   To switch alternately beween two possible selections.

topology   The way a network is physically structured. Example: a ring, bus, or star configuration.

torque   A turning force applied to a shaft, tending to cause rotation. Torque is equal to the force applied, times the radius through which it acts.

transmitter (XMTR)   A device that sends data.

triac   A solid-state bidirectional latching switch that provides full-wave control of ac power. The terminals of a triac are Anode 1, Anode 2, and Gate.

true   A signal that is high or on and is assigned the value of 1.

truth table   A matrix that describes a logic function by listing all possible combinations of inputs, and by indicating the outputs for each combination.

TTL   Transistor-transistor logic. An integrated circuit with its inputs and outputs directly tied to transistors.

twin-axial cable   A transmission line made up of a twisted pair of insulated conductors centered inside and insolated from a conductive shield.

UART   Universal asynchronous receiver/transmitter. An interface device for serial/parallel conversion, buffering, and adding check bits.

UJT   Unijunction transistor.

UL   Underwriters Laboratories (an approval agency).

USART   Universal synchronous/asynchronous receiver/transmitter. A UART with the added capability for synchronous data communication.

variable-speed drive   A controller that changes the amount of field voltage for a dc motor or the frequency to an ac motor to cause a change in its speed.

velocity loop   A feedback control loop in which the controlled parameter is motor velocity. A tachometer is usually used as a feedback device.

watchdog timer   A timer that monitors a cyclical process and is cleared at the conclusion of each cycle. If the watchdog runs past its programmed time period, it will cause a fault.

zener diode   A diode that conducts current when it receives reverse voltage that is in excess of its zener voltage value.

INDEX

821